Biology of Disease

Fundamentals of Biomedical Science

Biology of Disease

Dr Andrew Blann

PhD FRCPath FRCP (Ed) FIBMS aFHEA FRStatsSoc CSci
Visiting Reader in Biomedical Science, Huddersfield University
Huddersfield, UK

OXFORD

UNIVERSITY PRESS

Great Clarendon Street, Oxford, OX2 6DP,
United Kingdom

Oxford University Press is a department of the University of Oxford.
It furthers the University's objective of excellence in research, scholarship,
and education by publishing worldwide. Oxford is a registered trade mark of
Oxford University Press in the UK and in certain other countries

Published in the United States of America by Oxford University Press
198 Madison Avenue, New York, NY 10016, United States of America

British Library Cataloguing in Publication Data
Data available

Library of Congress Control Number: 2023946622

ISBN 978-0-19-883423-6

Printed in the UK by
Bell & Bain Ltd., Glasgow

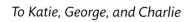

To Katie, George, and Charlie

Preface

The evolution of the practice of clinical medicine continues, with barriers between the various practitioners being periodically revised, remodelled, and removed. One consequence of this is that the knowledge and skills basis of these practitioners also needs to be revisited as new methods and techniques appear which must be addressed. No longer can one particular profession claim exclusive rights over a certain clinical procedure or investigation, or the knowledge base on which it relies.

The objective of this book is to provide a broad overview of the pathological process in the most common organic diseases that develop in those of us who seek the help of the healthcare profession—and of the scientific basis underlying those disease processes. The layout is driven by data from the UK's Office for National Statistics, which annually publishes details of the most common causes of death, these being cancer (Chapters 3–5), cardiovascular disease (Chapters 6–8), and respiratory disease (Chapter 13). These chapters are complemented by those on disorders appearing less frequently. A caveat with this method is that it fails to address those conditions that are not necessarily fatal but are nonetheless worthy of our attention. A further point is that very rare diseases are unlikely to be described in detail, if at all, mainly because they demand expert attention, and so are beyond the scope of this volume.

Each chapter will address a certain set of conditions to a common pattern that covers the aetiology, diagnosis, and management of these diseases. In this way the practitioner will be able to appreciate not only how to treat the disease, but also its cause, the latter being a crucial part of the formulation of this treatment. This approach has been taken so that the general text will be of value to undergraduate and postgraduate students who will be future practitioners in the healthcare industry, as well as to practitioners already working in the field today. Finally, it should be noted that all forms of diagnosis and management, especially pharmaceuticals and their dosages, are subject to change—a feature of modern medicine that is always with us.

An introduction to the Fundamentals of Biomedical Science series

Clinical scientists and biomedical scientists form the foundation of modern healthcare, from cancer screening to diagnosing HIV, from blood transfusion to food poisoning and infection control.

Without scientists, the diagnosis of disease, the evaluation of the effectiveness of treatment, and research into the causes and cures of disease would not be possible.

However, the path to becoming a scientist is a challenging one: trainees must not only assimilate knowledge from a range of disciplines but must understand—and demonstrate—how to apply this knowledge in a practical, hands-on environment. The UK's Institute of Biomedical Science (IBMS) is the leading professional body for clinical and biomedical scientists. One of its key roles is the provision of training material, guidelines, and, in collaboration with Oxford University Press, a comprehensive series of textbooks aimed at those taking IBMS examinations.

The *Fundamentals of Biomedical Science* series is written to reflect the challenges of biomedical science education and training today. It blends essential basic science with insights into laboratory practice to show how an understanding of the biology of disease is coupled with the analytical approaches that lead to diagnosis.

The series provides coverage of the full range of disciplines to which a clinical or a biomedical scientist may be exposed—from microbiology to cytopathology to transfusion science. Alongside volumes exploring specific biomedical themes and related laboratory diagnosis, an overarching *Biomedical Science Practice* volume provides a grounding in the general professional and experimental skills with which every laboratory scientist should be equipped.

Produced in collaboration with the Institute of Biomedical Science, the series

- understands the complex roles of scientists in the modern practice of medicine;

- understands the development needs of employers and the profession;

- places the theoretical aspects of biomedical science in their practical context.

Acknowledgements

It is my pleasure to thank all those who have shown me the utmost patience and support over my career, including Marc, Geoff, Lizzie, Phil, David (GI), Tommy, Paul, Brian, Gareth, and many others. But my principal thanks go to Charles and Greg, for their confidence in my potential, which I hope has been justified.

Contents

An introduction to the biology of disease

You are going to die. I absolutely guarantee it. However, naturally we all hope that this will not be for a good many years yet, but unless you meet with a fatal accident, you are not likely to just 'die'. Your death is likely to be preceded by a slow but steady deterioration in your health, quite likely accompanied by some disease. As the body ages, it simply can't do some of the things it was once able to—this is not the same as disease, and indeed we now talk of healthy aging. There are many, many different types of disease, and the aim of this chapter is to provide an appreciation of their biological basis. Whilst subsequent chapters in this book will address the fundamental biological aspects of different groups of human diseases, the present chapter will cover broad, underlying concepts, and will use examples to illustrate these.

Isaac Newton, possibly the first of the modern scientists, taught us that nothing comes from nothing—or to put it another way, everything is caused by something—and this is certainly the case with disease. The investigation and treatment of disease follows scientific principles, and accordingly the *Oxford English Dictionary* (*OED*) defines a scientist as 'a person who conducts scientific research or investigation'. Medicine is defined as 'the science or practice of the diagnosis, treatment, and prevention of disease' by the *OED*. These definitions include a large group of professionals, and it is globally recognized that those who wish to undertake these principles, and so become practitioners of medicine, are trained to do so to at (at least) undergraduate, and often also at postgraduate, levels of education. The complexity of the body and its diseases demands that these practitioners undertake specialist training to enable them to focus on specific aspects.

Since 'biomedical' may be defined as 'of or relating to biomedicine; of or relating to both biology and medicine', and 'clinical' as 'of or relating to the sickbed', then between them we have biomedical scientists and clinical scientists. However, there is considerable overlap in the training and practice of these two groups. Furthermore, other practitioners with scientific (biomedical) training include pharmacists (persons qualified to prepare and dispense drugs), physicians (persons legally qualified to practise medicine, especially those specializing in areas of treatment other than surgery), physiotherapists (persons who treat people using physiotherapy), and surgeons (medical practitioners who specialize in surgery). All these groups will study the same basic concepts of anatomy, physiology, and pathology, but each subsequently undertakes in-depth studies on their particular specialization. This model of education is employed not only in the UK, but globally.

The broad objective of this book is to provide these practitioners with a view of how organic disease arises, how it causes problems within the body, how the disease is recognized, and how it is treated.

Learning objectives

After studying this chapter, you should confidently be able to:

- appreciate the diversity of disease
- explain the meaning of pathophysiology
- describe the major causes of mortality and morbidity
- outline how aetiology and pathophysiology are linked to diagnosis
- summarize the major form of the treatment of disease

1.1 Defining health and disease

Health and disease can be difficult concepts to define, and this difficulty increases with the age of the individual. This is the matter of some debate and concern as many older people note temporary loss of memory that they (and perhaps others) erroneously interpret as early onset **Alzheimer's disease** (see section 16.4.2). Certainly, the older one becomes, the more likely it is that disease will appear, followed by death, although some come to death without any preceding disease. Some diseases have almost no physical consequences for the individual, whilst others bring considerable pain and discomfort, and adversely affect the quality of life—this is called morbidity. The presence of two or more independent disease processes—such as of the skin and the kidney—in the same individual is common and known as co-morbidity, and this naturally leads to difficulties in both diagnosis and treatment. Therefore, we can perhaps describe a spectrum or journey along which we have all embarked (Figure 1.1).

Health is often defined as the absence of disease and an individual may be in good health if there are no impediments to his or her functioning or survival. The World Health Organization (WHO) defines health as 'a state of complete physical, mental and social well-being and not merely the absence of disease or infirmity'. This definition is quite demanding, and it is likely that most of us would not be considered 'healthy' on the basis of these criteria. However, the WHO definition is useful since it acknowledges the importance of psychological and social wellbeing in the maintenance of health. Perhaps a more realistic definition considers that health is a condition or quality of the human organism which expresses adequate functioning under given genetic and environmental conditions. This definition implies that an individual may be considered healthy even if compromised in some way. An example here may be someone with **Down's syndrome** who might well be considered healthy under the latter definition but not under that of the WHO.

Implicit in many of these definitions of health is the concept that the efficient performance of bodily functions takes place in the face of a wide range of changing environmental conditions—physiology. Health in this context may be regarded as an expression of adaptability, and disease as a failure of this adaptability, such as to a lack of water or the appearance of a new and dangerous microbe. Disease can also be defined as a pattern of responses to some form of insult or injury resulting in disturbed function and/or structural alteration in certain organs or tissues. The study of disease is pathology, which we put together with physiology to make '**pathophysiology**'—how the disease influences how the body performs.

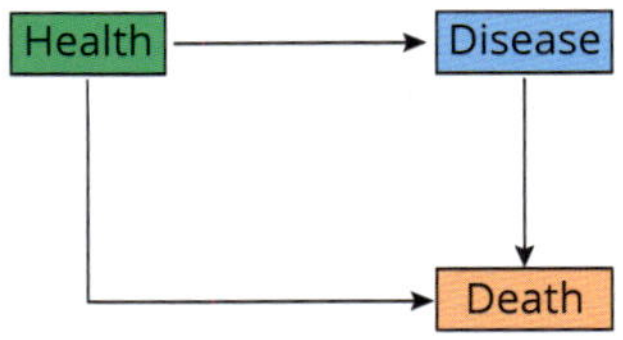

FIGURE 1.1
Health, disease, and death may be seen as a spectrum. Although disease may be treated and perhaps cured, with the individual reverting back to health, death is certainly not reversible.

Key points

Pathology is the study of disease. Pathophysiology is how disease impacts on the body.

SELF-CHECK 1.1

What is the difference between mortality and morbidity?

1.2 The classification of disease

1.2.1 The WHO classification of disease

The WHO's system for classifying disease is the International Statistical Classification of Diseases and Related Health Problems (the ICD); the latest version, ICD-11, is freely available online. It classifies all major human diseases and conditions, broadly by anatomy, and so provides a global framework for the study of disease (as summarized in Table 1.1).

TABLE 1.1 The ICD-11 classification of diseases

Section	Examples
1. Certain infectious or parasitic diseases	Cholera, typhoid and paratyphoid fever, botulism, salmonella, rotavirus, norovirus, cytomegalovirus, syphilis, tuberculosis, leprosy, HIV, influenza, malaria, sepsis
2. Neoplasms	Of the brain or CNS, haemopoietic or lymphoid tissues, breast, digestive organs, urinary tract, bronchus or lung, prostate
3. Diseases of the blood or blood-forming organs	Anaemias or other erythrocyte disorders, coagulation defects, diseases of the spleen
4. Diseases of the immune system	Primary and acquired immunodeficiency, autoimmune disease (lupus erythematosus, systemic sclerosis), allergic or hypersensitivity disorders, eosinophilia
5. Endocrine, nutritional, or metabolic diseases	Disorders of the thyroid, parathyroids, adrenals, and pituitary. Diabetes mellitus, undernutrition, overweight or obesity, inborn errors of metabolism
6. Mental, behavioral, or neurodevelopmental disorders	Schizophrenia, catatonia, mood disorders, bipolar disorder, feeding or eating disorders, autism, dementia e.g. due to Alzheimer's, Parkinson's disease
7. Sleep–wake disorders	Insomnia, hypersomnolence, sleep-related breathing disorders, delayed sleep–wake phase disorder
8. Diseases of the nervous system	Multiple sclerosis, epilepsy, cerebral palsy, cerebrovascular diseases (stroke), Parkinsonism, tremor, headache disorders, prion diseases, myasthenia gravis, motor neurone disease
9. Diseases of the visual system	Disorders of the eyeball, glaucoma, strabismus. Disorders of the conjunctiva, cornea, and lens
10. Diseases of the ear or mastoid process	Otitis externa, deformity of the pinna, otitis media, mastoiditis, otosclerosis, acquired and congenital hearing impairment
11. Diseases of the circulatory system	Hypertension, hypotension, acute and chronic ischaemic heart disease, coronary artery disease, pericarditis, heart valve disease, cardiomyopathy, arrhythmia, heart failure
12. Diseases of the respiratory system	Tonsillitis, pharyngitis, bronchitis, emphysema, asthma, cystic fibrosis, chronic obstructive pulmonary disease, respiratory failure, pneumonia (bacterial, viral, or fungal)
13. Diseases of the digestive system	Diseases of the oesophagus, stomach, duodenum, small intestines, appendix, liver, gall bladder, pancreas, peritoneum, large intestines. Inflammatory bowel disease
14. Diseases of the skin	Those diseases attributable to bacterial, viral, fungal, or parasitic infections, inflammatory dermatoses, benign proliferations, neoplasms, and cysts of the skin
15. Diseases of the musculoskeletal system or connective tissue	Osteoarthritis, reactive arthropathies, rheumatoid and psoriatic arthritis, polymyalgia rheumatica, gout, certain crystal arthropathies, axial and ankylosing spondylitis

(Continued)

TABLE 1.1 Continued

Section	Examples
16. Diseases of the genitourinary system	Endometriosis, menopausal disorders, recurrent pregnancy loss, diseases of the prostate, benign and inflammatory breast disease, hypertrophy of the breast, kidney failure
17. Conditions related to sexual health	Hypoactive arousal; orgasmic, erectile, and ejaculatory dysfunctions; sexual pain disorders; gender incongruence
18. Pregnancy, childbirth, or the puerperium	Abortion; ectopic or molar pregnancy; the oedema, proteinuria, hypertension, and hyperglycaemia of pregnancy; pre-eclampsia; obstetric haemorrhage; caesarean section
19. Certain conditions originating in the perinatal period	Injury to the foetus or neonate relating to the pregnancy, labour, or delivery. Birth injury, infection of the foetus or neonate, haemolytic disease of the foetus or newborn
20. Developmental anomalies	Structural abnormalities, e.g. cleft lip or palate, brachydactyly, polydactyly, macrocephaly. Chromosomal anomalies, e.g. Down's syndrome.
21. Symptoms, signs, or clinical findings not elsewhere classified	Those relating to blood and blood-forming elements; the immune system, endocrine, nutrition, or metabolic disease; the visual, respiratory, or digestive system
22. Injury, poisoning, or certain other consequences of external causes	Injuries to defined anatomical regions, e.g. the head, neck, elbow, or ankle. Burns, frostbite, harmful effects of substances, effects of a foreign body
23. External causes of morbidity or mortality	Unintentional causes, e.g. transport injury, fall, submersion or falling into water, assault. Intentional self-harm
24. Factors influencing health status or contact with health services	Contact with health services for purposes of examination or investigation. Donors of organs or tissues. Risk factors associated with infectious or certain other conditions
25. Codes for special purposes	International and national provisional assignment of new diseases of uncertain aetiology and emergency use, COVID-19, post-COVID-19 condition
26. Supplementary chapter: Traditional medicine conditions	Traditional medicine disorders linked to specific organs (e.g. liver, heart, spleen, lung, kidney, etc.) or multi-organ systems (e.g. skin and mucosa, musculoskeletal system)
V. Supplementary section for functioning assessment	Assessment of cognition, mobility, self-care, getting along and life activities, seeing and hearing functions
X. Extension codes	Medicaments such as anticoagulants, vitamins, insulin, levothyroxine, insulin, vaccines (e.g. to SARS-CoV-2 and influenza)

This table is edited for clarity.

For full details, see the ICD website: https://icd.who.int/browse11/l-m/en.

Although thorough, several sections of ICD-11 will not be considered in this book, partially because the classification system considers not only disease, but also related problems. For example, section 23 considers external causes of morbidity and mortality, with sections on transport accidents; accidental drowning; assault; and exposure to smoke, fire, and flames. Whilst these may well bring the individual to the healthcare system, these are not diseases in the sense that **cancer** and **rheumatoid arthritis** are diseases. Similarly, section 22 considers effects of a foreign body entering through a natural orifice; burns and corrosions of external body surface; and frostbite. Again, whilst they may often be seen in many accident and emergency units, these conditions will not be part of our discussion as there is little place for pathophysiology.

Table 1.1 plays an important part in this book because it provides a framework for subsequent chapters. However, not all sections and diseases will be given equal space, as, for example, in Western Europe and North America the presence of disease caused by parasitic worms is unlikely to be a frequent problem for the healthcare professional. Nevertheless, where appropriate, the defence of the body against these and related pathogens will indeed be discussed.

Key points

The classification of disease continues to be modified in view of advances in medical science and fresh perspectives. A good example of this is the COVID-19 pandemic that started in early 2020, and so is unnamed in ICD-10, but which appears in ICD-11. The causative virus and the consequences of infections are discussed in section 13.6.

1.2.2 Mortality

Not everybody who is ill can be treated. In some cases, the side effects of the treatments may be so severe, and the patient so frail, that the effect of the treatment (such as cytotoxic chemotherapy for leukaemia) may hasten the patient's death. But which diseases warrant treatment? A simple answer to this is those that are an immediate risk to life, followed by those likely to cause disease in the future, then those that bring considerable morbidity that is not life-threatening. It follows that abnormal signs, symptoms, and conditions that bring an increased risk of neither mortality nor morbidity are unlikely to attract the attention of the healthcare profession unless linked to other conditions. For example, it may be that the skin disease **vitiligo**, which itself has no major pathological consequences (and so is described as benign), can be considered by the patient to be so disfiguring that they become clinically depressed, and if so, a psychiatric opinion may be sought.

Mortality in the UK

An additional important aspect of the book will be causes of death as defined by the United Kingdom's Office of National Statistics (the ONS) (Table 1.2), which itself follows the system of the ICD. Close attention to this table also shows differences between the sexes. The proportion of men dying of cancer and those dying of circulatory disease both exceed that of women, by around 10% and 14% respectively, although the biggest differential is those dying of COVID-19 (15%). Conversely, the proportion of women dying of mental and behavioural disorders, of disease of the nervous system, and

TABLE 1.2 Leading causes of death in England and Wales, 2021

Cause	All persons	%	Males	% of males	Females	% of females
All causes	585,484	100.00	297,488	100.00	287,996	100.0
Cancer	148,109	25.3	78,969	26.5	69,140	24.0
Circulation	134,174	22.9	72,498	24.4	61,676	21.4
COVID-19	67,057	11.5	36,627	12.2	30,430	10.6
Respiratory	54,799	9.4	28,064	9.4	26,735	9.3
Mental and behavioural disorders	41,076	7.0	14,729	5.0	26,347	9.1
Nervous system	37,888	6.5	16,226	5.4	21,662	7.5
Digestive system	28,470	4.9	14,719	4.9	13,751	4.8
Endocrine, nutritional, and metabolic	9,815	1.7	5,003	1.7	4,812	1.7
Genitourinary	8,616	1.5	4,044	1.4	4,572	1.6

Figures do not sum to 100% as only selected categories are presented.

© Office for National Statistics. Source: https://www.ons.gov.uk/. Licensed under the Open Government Licence 3.0.

of disease of the genitourinary system exceeds that of men by 82%, 39%, and 14% respectively. Missing from this table (and so of less prominence in this book) are external causes of death that do not have a clear organic pathological basis. These include accidents and self-harm, linked to 16,180 and 4,976 deaths respectively in England and Wales in 2021, which sum to 3.6% of all deaths. Disease of the circulation, cancer, and COVID-19 are the leading causes of death in the USA and Europe (Eurostat 2023, Xu et al. 2022).

Although of immense value, this system has shortfalls. For example, suppose a man with haemophilia had such a profuse bleed that blood transfusion was necessary. Next, suppose that blood was contaminated with the human immunodeficiency virus (HIV), which led to the acquired immunodeficiency syndrome (AIDS), and death due to an overwhelming infection such as septicaemia or pneumocystosis. Without the haemophilia, there would have been no haemorrhage and no blood transfusion, and so no HIV infection or AIDS, thus no death from infection. Which was the final cause of death? In retrospect it depends entirely on how the death certificate was filled out, and the extent to which that person adhered to the WHO/ICD rules.

The ONS gives us other interesting data, such as the effect of age on mortality. Naturally, the frequency of deaths rises with decade of life, and in all but the final step (age 85 or more) male deaths exceed those of females (Figure 1.2). Note also the increased frequency of neonatal deaths.

Key points

In 2020 and 2021, the leading causes of death in both sexes in the UK were cancer, cardiovascular disease, COVID-19, and respiratory disease, together named on two-thirds of death certificates.

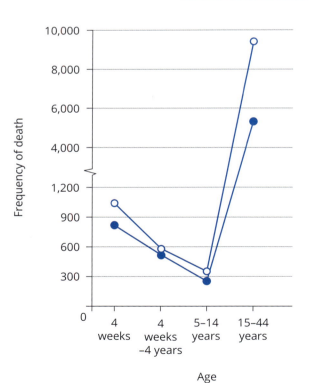
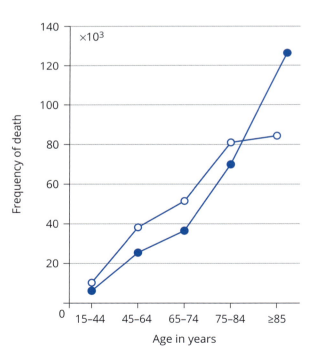

FIGURE 1.2
Frequency of death by subject's age. Note that the open circles represent males, and the closed circles represent females.

TABLE 1.3 Global changes in cause of death 1990–2017

1990	2007	2017
1. Neonatal disorders	1. Neonatal disorders	1. Ischaemic heart disease
2. Lower respiratory infections	2. Lower respiratory infections	2. Neonatal disorders
3. Diarrhoeal disease	3. Ischaemic heart disease	3. Stroke
4. Ischaemic heart disease	4. Diarrhoeal disease	4. Lower respiratory infections
5. Stroke	5. HIV/AIDS	5. Diarrhoeal disease
7. Tuberculosis	6. Stroke	8. HIV/AIDS
19. HIV/AIDS	10. Tuberculosis	11. Tuberculosis
28. Diabetes	17. Diabetes	15. Diabetes

GBD 2017 Causes of Death Collaborators. Global, regional, and national age-/sex-specific mortality for 282 causes of death in 195 countries and territories, 1980–2017: a systematic analysis for the Global Burden of Disease Study 2017. Lancet 2018: 392; 1736–88. © 2018 The Author(s). Published by Elsevier Ltd.

Global mortality

In 2018, *The Lancet* published data on the global burden of disease, reporting cause-specific mortality based on data collected in 2017, and which therefore provides a perspective for data from the UK in Table 1.2. Of the 55,945,700 global deaths, infectious diseases (tuberculosis, malaria, etc.) caused 12.6%, neoplasms 17.1%, cardiovascular disease 31.8%, chronic respiratory disease 7%, neurological disease 5.5%, digestive disease 4.2%, and injuries 8%. Other data describe the leading causes of deaths from 1990, 2007, and 2017 (selected data shown in Table 1.3). Over this period, deaths due to measles fell from 9th in the rankings, to 21st, to 39th, whilst lung cancer increased from 17th, to 14th, to 12th.

SELF-CHECK 1.2

What are the major differences between deaths in the UK and global deaths?

1.2.3 Morbidity

There is more to disease than death—there are many conditions (such as **osteoporosis**, **osteoarthritis**, and **deep-vein thrombosis**) that adversely influence quality of life, such as bringing pain, discomfort, and loss of mobility, but in themselves are rarely and directly fatal (Table 1.4). These are all examples of morbidity. Accordingly, diseases that do not typically lead directly to death, but instead to an increased burden of ill health—that is, morbidity—will be described.

Many textbooks place a disproportionate weight on certain diseases that have a low frequency in the population. For example, almost twice as many women die from bladder cancer as do from cancer of the cervix, but the latter often has more presence in the public's medical mind. Although more men die of breast cancer than die of testicular cancer, the former has almost no public recognition. A leading UK charity, the British Lung Foundation, estimates that **asthma**, which carries a great deal of morbidity, is present in five million people in the UK (perhaps 8% of the population), but in England and Wales in 2021 was named on only 1,146 of death certificates, a frequency of 0.022%. Similarly, although it has been estimated that three million people in the UK (5% of the population) have osteoporosis, it was named as causing only 443 deaths, a frequency of 0.015%.

This textbook aims, broadly, to allocate its text proportionally in relation to the frequency of disease (be it morbidity or mortality) in the population, so that the most common diseases will be emphasized. However, this approach will be amended as there is frequently an overlap of aetiological processes

TABLE 1.4 Examples of morbidities that are generally unrelated to mortality

Acne	Allergy	Anxiety
Asthma	Back pain	Breast pain
Chronic fatigue syndrome	Dizziness	Eczema
Fibroids	Headache	Hearing impairment
Osteoarthritis	Osteoporosis	Overweight
Post-natal depression	Psoriasis	Sleeping disorders
Urinary incontinence	Visual impairment	Vitiligo

This list is intended neither to be exhaustive nor to imply that persons with these morbidities are protected from mortality. Indeed, the reverse is often the case, in that morbidity (especially where multiple) is often a risk factor for mortality.

and actual disease. For example, smoking causes both lung cancer and heart disease (as well as many other pathologies). A further problem is that there is no centralized authoritative source for the population frequency of morbidity—the ONS is concerned only with mortality. This is compounded by the fact that many people have more than one morbidity—that is, co-morbidity—but usually only the major disease appear on the death certificate.

Unsurprisingly, the greater the extent of co-morbidity, the greater the likelihood of these factors compounding to increase the risk of mortality—for example, being overweight is a major risk factor for cancer, osteoarthritis, and diabetes. The picture is further complicated by the fact that overweight is an established risk factor for obesity, deep-vein thrombosis, and asthma, but is not referred to on death certificates. Osteoporosis in itself is an infrequent cause of mortality, but, following an accidental fall, it may cause fractures that bring about immobility which, in turn, may precipitate a fatal **pulmonary embolism**. It is therefore the latter that is the ultimate cause of death, not the precipitating factor of osteoporosis (or any other cause of a fall).

Paying close attention to Table 1.2, mental and behavioural disorders are named on 7.0% of death certificates, a figure that exceeds that of the digestive, endocrine, and genitourinary systems. These disorders (such as Alzheimer's disease and **Parkinson's disease**) often have a genetic background (see Chapter 16), but as yet there are no reliable routine laboratory tests that help the practitioner. However, complex techniques such as magnetic resonance imaging can help.

1.2.4 The causes of disease

Almost all the diseases and conditions described in Tables 1.1, 1.2, and 1.4 can be described in terms of a spectrum between genetic factors and environmental factors. An example of the former is **Huntington's disease**, caused by a mutation in *HTT*, located at 4p16.3, and coding for huntingtin, a protein of unknown function. The disease is always fatal, with a mean life expectancy of some 20 years from diagnosis. There are many environmental factors causing numerous diseases, such as pathogenic microorganisms and industrial and social toxins, exposure to which inevitably leads to disease. Examples of the former include hepatitis B and C viruses, of the latter alcohol and tobacco smoke: avoidance of these factors considerably reduces the risks of cancer of the liver and lung. Males carrying the *BRCA-1* gene are at extremely low risk of breast cancer, as oestrogen, which is only present in high levels in females, is a cofactor and prompts the oncogenesis. A proof of concept of this are drugs such as tamoxifen, which, as aromatase inhibitors, suppress oestrogen levels and so reduce risk of breast and other malignancies. Conversely, there is evidence for the hypothesis that, through middle age, women are relatively protected from cardiovascular disease by oestrogen, a benefit that expires with the menopause, which brings the rate of disease closer to that of men of a similar age.

Midway between the two are diseases where there are both genetic and environmental components, an example being deficiency of glucose-6-phosphate dehydrogenase. This is coded for by

G6PD at Xq28, a loss-of-function mutation in *G6PD*, and so the enzyme deficiency is relatively mild, but a severe **haemolytic anaemia** may follow an infection, a **diabetic ketoacidosis**, and ingestion of certain foods, notably fava beans. Additional examples will be presented in chapters to come.

1.3 Aetiology, presentation, diagnosis, and management

As far as is possible, each chapter will address each of these components in turn.

1.3.1 Aetiology

The exact process by which the particular illness comes about—the aetiology—is crucial, and one of the many triumphs of science is our ability to discover those factors causing disease. The process of discovery may well have taken decades and involved thousands of scientists working in diverse areas such as **epidemiology**, cell biology, and biochemistry, using methods such as population studies, animal models, tissue culture, and molecular genetics. Once the aetiology has been established, understand the pathophysiology will follow, and so the **diagnosis**. The aetiology and pathophysiology will generally point to management.

1.3.2 Presentation

Patients present themselves to their general practitioner or to an accident and emergency (A&E) unit of a local hospital.

Primary, secondary, and tertiary care

Much of the practice of modern medicine can be described at three levels: primary, secondary, and tertiary care. The former equates to the position where the individual presents to their healthcare professional, probably a medically qualified general practitioner (GP) and his/her team (such as nurses and healthcare assistants). The patient (as the individual may well become) may be diagnosed (often with additional help, such as the pathology lab and imaging by X-ray and ultrasound) and treated by the GP. However, should the problem with the patient be serious and/or problematical, he or she will most likely be referred to a consultant-lead team at a hospital, at which point secondary care takes over. A key factor is diagnosis, which is best made via a full understanding of the pathophysiology of the disease or syndrome the patient is presumed to be developing. Once the diagnosis has been made, according to the principles of medical science, then the disease can be managed by appropriate treatments. The practitioner at any level cannot proceed to address their patient's issues without a firm view of the problem (i.e. a preliminary diagnosis). Only when this has been confirmed can management (in the form of treatment) begin.

In an ideal world, the treatment will be curative, and this is clearly the goal. However, should the patient's problem be particularly difficult, it may be necessary to pass them to a specialist tertiary referral centre, possibly at another hospital or other healthcare centre, which is likely to be dedicated to their particular condition. These pathways are summarized in Figure 1.3.

1.3.3 The professionals

Dozens of specialist and generalist professionals comprise the workforce of a modern healthcare system. The modern and astute practitioner in secondary and tertiary care (be they a nurse, therapist, scientist, physician, pharmacist, or surgeon) must come to grips with the concept of the pathophysiology of human disease—that is, how the disease affects the working of the body—and then act on this information as their speciality requires. Some interact early in the patient's journey, others diagnose and monitor, and some treat. Once the disease process is understood, the patient will interact with a

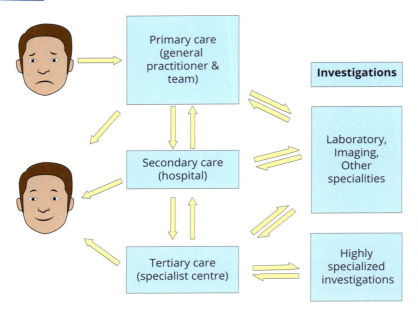

FIGURE 1.3

Pathways of healthcare. The patient first consults their general practitioner in primary care, and then hospital, and, perhaps, more specialist units, depending on the severity of their condition.

host of other professionals. A crucial early step when meeting a possible patient is for the professional to define an exact diagnosis.

1.3.4 Diagnosis

The information required to make a diagnosis comes from a number of sources: the patient, the laboratory, imaging, and elsewhere.

Signs and symptoms

The first place to start is with a medical history (if present—and likely to be more complicated with the aged). The patient themselves will come with a list of symptoms (i.e. what they feel is wrong with them), such as cough, tiredness, aches and pains, nausea, increased/decreased frequency of urination and defaecation, etc. The problem is that all these symptoms are vague, non-specific, and can alter over the course of the interview, or over time. If the patient gives a poor history, then the information is unlikely to be reliable. However, in many cases the patient provides a very clear list of symptoms that directly help the diagnosis.

Step two is a physical examination of the patient, looking for unambiguous physical signs, the implications of which the patient themselves may not be aware of. Once more, these signs can be vague and non-specific (such as a rash or muscle **atrophy**), or can be more precise, such as **jaundice** (which may indicate problems with the liver) or an irregular pulse (common in the **arrhythmia** atrial fibrillation). It is likely that their history, signs, and symptoms will direct the practitioner to a number of potential diagnoses. These can be confirmed or refuted by calling upon other branches of healthcare, such as **electrocardiography**, lung function, the laboratory, and imaging.

Initial investigations

In primary care, the patient's temperature will be taken, and he/she is likely to be able to provide a sample of urine, which may provide useful information (Table 1.5). Urine may also be assessed for cloudiness/clarity, and odour (e.g. ammonia, pear drops/acetone), and microscopy can also be useful, although the latter is rarely performed in primary care. The practitioner is also likely to measure blood pressure, and possibly listen for chest sounds with a stethoscope, and obtain samples of blood for routine screening tests.

TABLE 1.5 Urine analysis

Analyte	Pathology
Bilirubin	Liver disease/haemolysis
Ketones	Diabetes
Leukocyte	Bacterial infection
Nitrites	Bacterial infection
pH	Acidity
Protein	Renal damage
Specific gravity	Dehydration
Sugar	Hyperglycaemia/diabetes

Should the primary care practitioner feel that the patient's interests are best served in secondary care, then they will refer the patient. Upon the patient's arrival in secondary care (such as an outpatient clinic), much of the above is likely to be repeated.

The laboratory

In a hospital pathology laboratory, scientific staff perform analyses, techniques, and other methods for making and/or confirming a diagnosis. The crucial role of laboratory science in today's healthcare system is amply demonstrated by the importance placed on the pathology laboratory, where this science is practised. The scientists in the path lab are not generally concerned directly with patient care—this is the role of clinical practitioners who see patients on the ward and in the clinic. Instead, laboratory scientists provide their colleagues with precise and accurate information regarding the health status of their patients. However, this is not to say scientists never leave the laboratory—they may service point-of-care devices and micro-analysers used around their hospital (such as in intensive care or in accident and emergency), or participate in an oral anticoagulant clinic.

The laboratory offers testing in various areas of pathology, which include cell pathology (also known as histology), clinical chemistry (or biochemistry), cytopathology (or cytology), haematology, immunology, microbiology, and virology. Some large establishments may have a laboratory dedicated to molecular genetics, and there will certainly be reference laboratories within a network (Table 1.6). Some tests are quantitative, in that they provide an exact number (such as that of the haemoglobin level of 90 g/l supporting the diagnosis of anaemia), whilst others provide a yes/no answer (as with the presence of increased deposits of iron in the liver denoting haemochromatosis (Figure 1.4)).

The laboratory offers a host of general and specific investigations of great value to a number of possible diagnoses. Indeed, it has been estimated that the laboratory provides 75% of the information required to make a diagnosis and guide management. These will be described as each particular disease/condition arises in the chapters that follow.

Imaging

This umbrella term includes X-rays, **ultrasound** (including **echocardiography**), **angiography**, isotope scans, **magnetic resonance imaging (MRI)**, **positron emission tomography (PET)** scanning, and several other investigations. Imagers both control complex machines that provide insights into the internal workings of the body and interact directly with the patient—those who use radiation are radiographers. The history of X-rays spans over 120 years, and this technique continues to provide essential information on the skeleton and major internal organs such as the heart and lungs (Figure 1.5). It can indicate the presence of a tumour, pneumonia, or an unusually large heart (cardiomegaly).

By integrating X-rays taken from several different directions, a more informative picture can be constructed. This process of **computer-assisted tomography (CAT, or CT)** scanning provides more information, but at the cost of exposing the patient to higher doses of ionizing radiation, an established

TABLE 1.6 Tests performed in a pathology laboratory

Department	Test	Value
Cellular pathology	Histological examination of tissues	Presence of malignant tissues
Clinical chemistry	Urea and electrolytes	Renal function
Cytopathology	Fine-needle aspiration	Presence of malignant cells
Haematology	Full blood count	Presence/absence of an infection
Immunology	Presence of antibodies to one's own cells and tissues (autoantibodies)	Diagnosis of autoimmune disease such as rheumatoid arthritis
Molecular pathology	Mutation in a certain gene	Diagnosis of familial hypercholesterolaemia
Medical microbiology	Identification of a particular bacterium	Infection with the bacterium and possible treatment
Transfusion science	Detection of alloantibodies	Potential for successful blood transfusion
Virology	Presence of antibodies to a particular virus	Previous infection with that virus and presumed protection from future infection

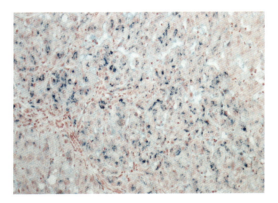

FIGURE 1.4

Histological demonstration of iron in a section of liver. High levels of iron, as defined by Perls' stain (colour blue) support the diagnosis of haemochromatosis.

From Orchard and Nation, *Histopathology*, second edition, Oxford University Press, 2018.

FIGURE 1.5

A chest X-ray.

Shutterstock.

cause of cancer. Nevertheless, this technique is now an indispensable part of the investigation of many diseases (Figure 1.6).

The advantage of ultrasound is that it does not use ionizing radiation and is considerably more flexible than X-rays/CT as the ultrasonographer can actively search for a particular problem. Figure 1.7 shows an abdominal ultrasound from a patient who complained of intermittent right-sided pain, and in whom the final diagnosis was cholelithiasis (gallstones). More complex imaging includes visualization of the lumen of blood vessels—the process of angiography, which is particularly valuable in investigating coronary artery disease.

(a)

(b)

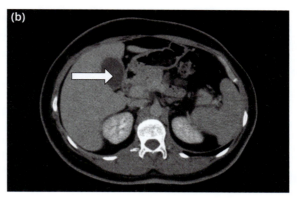

FIGURE 1.6

A patient about to undergo a CT scan (a) and their result (b). The scan in (b) highlights the bone in white but also shows an abnormality in the liver (indicated by the arrow).

(a) Getty Images, (b) iStock.

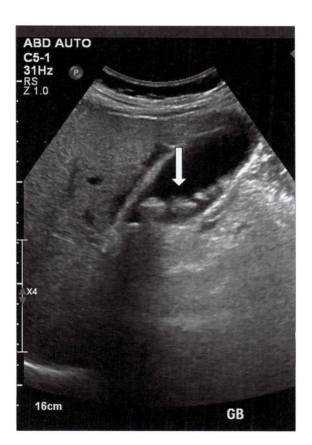

FIGURE 1.7

Ultrasound of the gall bladder. You can see several gallstones in the scan, one is indicated by the arrow.

Shutterstock.

Key points

Diagnosis relies on a number of factors, including signs, symptoms, the laboratory, and imaging.

What are the major disciplines within the pathology laboratory?

Other specialist services

A heart complaint will inevitably need an electrocardiogram (ECG). Taking blood from a vein in the arm (venepuncture) may cause discomfort but is not considered dangerous. However, several imaging procedures do bring a real (but many would say clinically insignificant) risk of malignancy due to ionizing radiation (as in X-rays and CT). The functioning of the heart (its chambers and valves) can be assessed by echocardiography (shortened to echo, using ultrasound). As with the path lab, each of these imaging investigations (and others) will be described for the particular disease or condition in the chapters to come. There are numerous other specialist services available to the practitioner, which include examination of lung function, audiology, and ophthalmology, each with its own highly technical procedures.

1.3.5 The value of an accurate diagnosis

It is an exceptional sign, symptom, or investigation that is perfect—lack of specificity is common. In examining these, we address a number of issues.

Sensitivity and specificity

A problem with signs and symptoms is that many lack a firm definition and can even alter over the course of the history-taking, or over time. Consider a symptom such as a recurrent cough, which could be due to a chest infection. Not everyone with a chest infection has a cough, and not everyone with a cough has a chest infection—so how reliable is this symptom? Many presenting with lung cancer also have a persistent cough. We can assess the value of a sign, symptom, or an investigation with four metrics:

- True positive: those people we are absolutely sure have the disease and who exhibit the particular sign or symptom or are positive for the investigation.

- False positive: those displaying the sign or symptom, or where the investigation is positive, but where we are assured that these people do not have the disease.

- True negative: those we are absolutely sure do not have the disease.

- False negative: those who do have the disease, but do not display the sign or symptom, or who test negative in the investigation.

We can extend this principle with two key metrics:

- Sensitivity is an indicator of how good a test, sign, or symptom is at identifying people who are ill/diseased.

- Specificity is a measure of a lack of the disease or health.

Now consider someone wearing glasses or contact lenses—probably 100% of people who do so have some problem with their eyes, but not everyone with a problem with their eyes wears glasses or contact lenses. People who wear contact lenses will almost always have problems with their eyes and so this is indicative of good sensitivity. Not wearing glasses or contact lenses would indicate that the person has good eyesight and so be a good measure of specificity. However, this approach will not always be 100% correct. People may wear contact lenses instead of glasses, and some (without any ophthalmological disease) to change eye colour.

Consider the data in Table 1.7, of a cohort of 301 people, of whom we are absolutely sure that 238 have an abnormal pathology (and so are patients), whilst 63 are free of the abnormality, and so are normal (being described as controls). Now consider that we have a new investigation, sign, or

TABLE 1.7 Sensitivity and specificity

New test	Pathology (as defined by the gold-standard method)		Total
	Abnormal (i.e. patients)	Normal (i.e. free of this pathology)	
Positive/present	213 (true positive)	23 (false positive)	236
Negative/absent	25 (false negative)	40 (true negative)	65
Total	238	63	301

symptom and want to see how it compares to the 'gold standard' of the known pathology. We find that the new test is positive in 236 of the cohort (who we would therefore consider to be patients) and is negative in 65 (so would describe these as normal). So the new test gives almost exactly the same result as the gold-standard pathology.

However, there are discrepancies—the new sign/symptom/test falsely classifies 23 normal subjects as patients and 25 patients as normal. Of the 238 patients we know have the disease, the new test correctly identified 213, a proportion of 213/238, which is 89.5%. This figure is the sensitivity of the new test. So, in 1,000 people who would have this disease the new test would incorrectly diagnose 105 patients as being healthy. Similarly, we know there are 63 people who do not have the pathology, but the test correctly identified only 40 of them, giving a proportion of 40/63, which translates to 63.5%. This figure is the specificity of the new test. Therefore, this new test would incorrectly identify 365 out of 1,000 healthy people as being diseased. We describe the group of 213 patients as the true positives, the 63 normal subjects as true negatives, the 25 incorrectly identified by the new test as being normal as false negatives, and the 23 incorrectly identified by the new test as patients (i.e. that have the disease) as false positives.

Positive and negative predictive value

From all these data we get a number of other metrics for the sign/symptom/investigation, such as the positive predictive value (PPV), the negative predictive value (NPV), and the positive likelihood ratio (+LR).

- The PPV, defined as true positives/(true positives + false positives), tells us the likelihood that the patient actually has the disease if the test is positive. In our example this would be 238/(238+23) = 238/261 = 0.91. Ideally, there would be no false positives (i.e. 238/238+0), so a result as close to 1 as possible is desirable. It is important to note that PPV is affected by both specificity and sensitivity; the higher these are the higher the PPV. But it is also highly affected by the prevalence of the disease being tested for. This is why when screening for diseases it is best to do so in populations where the disease is more common. So, for example, it makes more sense to screen for HIV in high-risk populations such as men who have sex with men and intravenous drug users.

- The NPV, defined as the true negatives/(true negatives + false negatives), tells us the likelihood that the patient does not have the disease if the result is negative. In our example this would be 63/(63+25) = 63/88 = 0.72. Similarly, with no false negatives, the result would be 1. Again, it is affected by disease prevalence.

- The +LR, defined as sensitivity/(1 – specificity), tells us how much more likely it is that the patient who tests positive actually has the disease, compared with the patient who tests negative. In our example this would be 0.895/(1–0.635) = 2.45, and we can say that this test result is found 2.45 times more commonly in the diseased group compared to the control, normal, healthy group. The higher the +LR, the better the test.

An important point about all these metrics is that there is no absolute and clear rule about what is acceptable and what is not. All tests and procedures must be judged on their own merits and in the light of other information. Nevertheless, the objective of all these figures is to help the practitioner to decide what weight of relevance is deserved by the new test. However, the data in the table are only applicable for the populations from which they were drawn—note that the number of patient cases exceeds the number of controls by 3.77 times, so that there is the possibility of error in that the number of controls is not as representative as the number of patients. Best practice is to have roughly equal numbers of each group of subjects.

1.3.6 Management

Once a diagnosis has been formed, and with the patient's agreement, treatment may begin, and as with diagnosis, there are several aspects. The first point considers whether the patient can treat themselves, such as via diet and exercise in the obese or overweight with **dyslipidaemia** and hypertension, reducing the use of alcohol, and stopping smoking. Should these be ineffective, other approaches will be necessary.

Chemotherapy

By far the most common form of treatment is with drugs, also described as pharmacotherapy or chemotherapy (the latter being therapy with a chemical, which is what drugs are, i.e. complex chemicals). In theory, any drug treatment is technically chemotherapy, although the latter is often thought of in terms of cancer. It follows that patients may be surprised (and possibly concerned) to learn that they are being prescribed chemotherapy, when in fact this may be something well known (and less concerning), such as aspirin and antibiotics. There are of course thousands of drugs (and certain other treatments) available to practitioners, and all are listed on the UK's British National Formulary (the BNF), available free online. However, no one drug is completely safe (generally viewed as lack of side effects) and effective in all patients, so the practitioner may have to trial an assortment of agents until a suitable and effective treatment is found. Among the major roles of the pharmacist are providing advice on the correct drug, and its dose, for the patient, and ensuring that the use of chemotherapy is safe and follows guidelines.

Surgery

All surgery carries a risk. This is obvious in terms of operations inside the abdomen and thorax, but even a simple procedure, such as removal of a skin tag, carries a risk of infection. Nonetheless, should there be no appropriate medical approach with drugs, or if such an approach fails, then surgery is inevitable. There are several surgical specializations, such as in tumour excision (oncology), hip or knee replacement (**orthopaedics**), and coronary artery bypass grafting (cardiac surgery). A surgeon may also be called on by those thought of as being medically healthy, for example to perform a **caesarean section** on otherwise healthy women should delivery fail to proceed, or in cases of trauma as may follow a road traffic accident. If the surgical procedure is particularly protracted and/or is painful, an anaesthetist will be called upon. Post-surgical care will inevitably employ the laboratory to assess renal and liver function, etc., check for infections, and to determine the scale of any blood loss.

Other management options

Physiotherapy can be used in its own right to correct certain conditions (and ideally prevent surgery) but is also valuable in a post-surgical setting to accelerate patient recovery and so discharge. The same principles may apply to occupational therapy, important in many aspects, such as in getting the patient back home in a safe condition and other approaches. Radiotherapy may also be used as a stand-alone treatment and may be used before chemotherapy (to shrink a tumour) and after (to minimize the viability of any remaining neoplastic tissue). All options will be considered for each patient and perhaps used in parallel or series as particular cases apply. Specialist and referral centres are likely to have their own investigations linked to their particular speciality.

What are the major techniques used for imaging?

Key points

The management of disease requires combinations of chemotherapy, surgery, physiotherapy, occupational therapy, radiotherapy, and other treatments.

The UK's National Institute for Health and Care Excellence (NICE) is the leading body for guidance on the diagnosis and treatment of a great many conditions. Clinical guidelines refer to, for example, cancer, chronic fatigue syndrome, wound management, and stroke. NICE also publishes technology appraisals of new drugs, and documents on health protection (e.g. drug misuse); lifestyle and wellbeing (sexual health); and on service delivery, organization, and staffing. Other countries have their own authorities, such as the Food and Drug Administration (FDA) in the USA.

Why is the aetiology of a disease so important?

1.3.7 The division of labour

The previous sections have described a path along which the patient and their disease will travel, often from primary to tertiary care (Figure 1.3) and which can be further described in a number of discrete stages of presentation, diagnosis, treatment, and management.

The practice of modern medicine has traditionally seen a strict categorization between different groups of staff with different roles, such as those with scientific, nursing, or medical qualifications. There is a degree of logic to this separation of skills, as scientific knowledge of precisely how a microorganism becomes pathogenic is not required in the medical decision about which antibiotic or antiviral to prescribe, and vice-versa. Having said this, the laboratory scientist will, for example, need some knowledge of pharmacy to appreciate the effects of chemotherapy on the liver and bone marrow. Thus the late stages of the patient journey in the clinic are impossible without an understanding of the initial stages, which will have been elucidated by scientists.

Furthermore, there is also often overlap between the latter stages of the pathway in that certain subgroups of scientists may be directly involved in treatment (e.g. pharmacists, radiographers) or indirectly involved (e.g. laboratory scientists) in monitoring the effects of treatment (Figure 1.8).

However, the reduction of the complexity of modern medicine to Figure 1.8 inevitably requires some short cuts: it is certainly not the case that medical staff need be unaware of the aetiology of the disease that burdens their patient, and likewise the signs and symptoms of a patient's illness can be important in enabling the laboratory scientist to correctly interpret results.

1.4 What this book hopes to achieve

The objective of this book is to provide a joined-up approach to the major aspects of human disease—how it arises, how it causes discomfort, how it is diagnosed, and how it is treated—to enable the various healthcare professionals to be better practitioners. The chapters that follow will cover the majority of human disease, and for each disease, aspects of aetiology and pathophysiology will be followed by diagnosis and so clinical aspects of treatment and general management. These will be discussed in the chapters that follow.

- Chapter 2 introduces basic concepts of disease.

- Chapters 3, 4, and 5 consider the general aetiology, diagnosis, and treatment of the different forms of cancer.

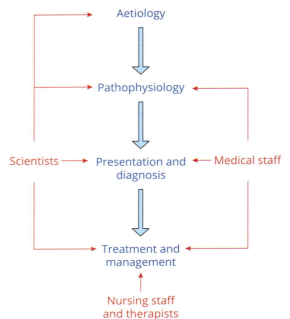

FIGURE 1.8
The pathway of modern medicine and major staff roles.

- Chapters 6, 7, and 8 discuss cardiovascular disease—its aetiology, risk factors, and clinical features.

- Chapters 9 and 10 focus on the aetiology and clinical features of inflammatory and infectious diseases, and immunological disease.

- Chapter 11 discusses diseases of the blood.

- Chapter 12 considers the aetiology, diagnosis, and management of malnutrition and other diseases of the digestive system.

- Chapters 13 and 14 describe the aetiology, diagnosis, and management of respiratory disease (including COVID-19) and disease of the genitourinary system.

- Chapter 15 examines the aetiology, diagnosis, and management of endocrine, nutritional, and metabolic disease.

- Chapter 16 looks at the aetiology, diagnosis, and management of chromosomal, genetic, and metabolic disease.

1.4.1 A caveat

The difficulty with the ONS system (and therefore of this book) for describing mortality is exemplified by the study of respiratory disease in Chapter 13. Leaving out lung cancer (covered in Chapter 4), **pleurisy** (Chapter 10), and pulmonary embolism (Chapter 11), what remains include conditions such as pneumonia (although many with COVID-19 are at risk of pneumonia, as explained in Chapter 13), **emphysema**, and chronic obstructive pulmonary disease. Asthma and bronchitis would also be part of this chapter, but they are also inflammatory diseases (Chapter 10). Many of the diseases of the digestive system, the focus of Chapter 12, are due to cancer (Chapter 4), to **autoimmune disease** (such as certain liver and intestinal conditions), and to infectious disease (as in viral hepatitis) (both Chapter 10). Major forms of disease that can be present in different organs are summarized in Table 1.8, which is intended to be neither extensive nor exclusive.

TABLE 1.8 Organs that are the subject of different disease processes

Organ	Disease processes
Liver	Cancer, autoimmune disease, metabolic disease, infectious disease, genetic disease
Blood	Cancer, autoimmune disease, infectious disease, genetic disease, nutritional disease
Lung	Cancer, thrombosis, infectious disease, genetic disease
Thyroid	Cancer, autoimmune disease, nutritional disease
Stomach	Cancer, autoimmune disease, infectious disease

Consequently, in order to avoid duplication, a particular disease process in a particular organ will be described only once. Wherever potential duplication may occur, it will be highlighted in the text. This may lead to some confusion: one may consider disease of the reproductive system to include the role of the pituitary. Alternatively, others may expect this organ to be discussed in the chapter on endocrine disease.

Chapter summary

- We inevitably move from health to certain types of disease, and then to death. Pathology is the study of disease, whilst pathophysiology considers how the disease process causes damage to the body.

- The International Classification of Diseases (ICD) is a system placing common diseases together according to anatomy and pathophysiology.

- In the UK, the Office of National Statistics (ONS) uses a very similar system, publishing data on deaths due to particular diseases. Whilst reporting the various causes of death is of course important, a considerable degree of ill health is not necessarily life-threatening, and is described as morbidity.

- Successful treatment of morbid and mortal conditions follows understanding of how disease develops, and so considers epidemiology, pathophysiology, and aetiology.

- Practitioners aim for a quick and accurate diagnosis (using signs, symptoms, the laboratory, imaging, etc.), then design a management scheme which is likely to involve features such as patient self-help, surgery, radiotherapy, and physiotherapy.

- Leading objective diagnostic methods in imaging include ultrasound, X-ray, computerized tomography, magnetic resonance imaging and nuclear medicine, whilst those of the laboratory include haematology, biochemistry, microbiology, immunology, histology, cell pathology, and molecular genetics.

Further reading

- GBD 2017 Causes of Death Collaborators. Global, regional, and national age-/sex-specific mortality for 282 causes of death in 195 countries and territories, 1980–2017: a systematic analysis for the Global Burden of Disease Study 2017. Lancet 2018: 392; 1736–88.

Useful websites

■ International Classification of Diseases (ICD): **https://icd.who.int/en**

■ The Office of National Statistics: **www.ons.gov.uk/**

■ British National Formulary: **https://www.bnf.org/products/bnf-online/**

■ National Institute for Health and Care Excellence: **www.nice.org.uk**

 # Discussion questions

1.1   Which of the most common causes of mortality and morbidity are avoidable?

1.2   What are the roles of the various professional groups in effecting and monitoring the different forms of treatment?

2

Basic concepts of disease

Chapter 1 introduced some of the concepts of health and disease; the classification of disease; mortality and morbidity; pathophysiology; and the presentation, diagnosis, and management of disease. This chapter will build on those aspects to provide a more comprehensive view of what disease is and how it develops. In doing so we must first understand what we mean by normal, and what we mean by healthy (which are very often not the same). An important tool in understanding disease is epidemiology, and we conclude the chapter by looking at how this branch of biomedical science informs our views of health and illness.

Learning objectives

After studying this chapter, you should confidently be able to:

- explain simple concepts of health and disease
- describe the major aspects of how disease is caused and develops
- appreciate the importance of leading statistical analyses
- understand key features of clinical medicine
- outline the importance of epidemiology

2.1 Who or what is normal?

People who do not have any disease are often described as 'normal' and are therefore presumed to be healthy. In a medical setting a person who does not complain about a particular condition (such as chest pain, a symptom of an underlying medical condition) is considered **asymptomatic**. However, this does not mean that this person is necessarily free of disease—only that their lifestyle is not negatively affected by any signs or symptoms of disease at the moment. It is also crucial to understand that normality does not necessarily equal health but merely indicates the frequency which with a condition occurs in a defined population. Some conditions or diseases—such as being overweight or obese—can in fact be so frequent in populations that they could be considered 'normal'. This shows that normality is not always desirable or healthy. A good example of this is high serum cholesterol, which is asymptomatic and widespread in the population, and therefore might be thought of as 'normal', but predicts and contributes to cardiovascular disease. A further aspect to consider is the life cycle: for example, the degree of strength and fitness in the young allows them to counter threats to health (such as excessive drinking of alcohol) that their parents and (more so) their grandparents cannot address without undesirable side-effects. Similarly, age brings a reduced capacity for physical exercise, an activity which the young can use to offset a high calorific intake.

Data can be classified as categorical or continuously variable. Examples of the former are categories such as male/female, alive/dead, and present/absent regarding a disease. Data of this nature are often reported as a percentage. Examples of the second type include height, weight, age, serum cholesterol, and the length of a pregnancy. The data cannot be categorized, and can be anywhere along a scale (such as pregnancy, generally 0–41 weeks). In presenting continuously variable data, we refer to the central point and the variation. One of the ways we can view the frequency of a continuously variable index is in the form of a graph, which we call a distribution. There are a number of different ways in which we can say that a certain feature is distributed within a population.

2.1.1 The normal distribution

Figure 2.1 shows the distribution of the level of serum cholesterol in a population of 1,315 people, which follows a symmetrical curve, like the cross-section of a small hill. The vertical (y) axis is the frequency (the number of people with a particular result), which is greatest in the middle of the figure (the central point); the horizontal (x) axis shows the scale of the effect in question, which in this case is serum cholesterol.

The figure consists of a number of 'towers', and the tallest of these represents the largest number of people with that result. The central, or middle, point of this set of data, which in this case is the average value (the mean), is 4.9 mmol/l. This particular shape is referred to by different names, including the **normal distribution**, a bell-shaped curve, and the Gaussian distribution. But whatever name is used, this distribution is important as it represents one way in which we can visualize how the values of a particular index are spread in a population.

The distribution is called 'normal' because the greater majority of sets of this type of data (e.g. height, weight, haemoglobin, number of books in UK homes) are distributed in this way. Its name does not make a statement about what is normal or abnormal about the data themselves.

The distribution in Figure 2.1 includes a small number of low values on the left side of the diagram (~2 mmol/l) and an equal number of high values on the right (~8 mmol/l). But while the data set has values in a range from around 2 to 8 mmol/l, most of its values are in the middle, between 4.2 and 5.6 mmol/l. The mathematical expression of this variation, or spread, is the **standard deviation (SD)**. The numbers in the box to the right-hand side of the graph of Figure 2.1 tell us the SD is (to two significant figures) 1 mmol/l (rounding down 1.0078 to 1.0). The mean plus or minus (+/−) two SDs should include 95% of the results, which is 2.9 to 6.9 mmol/l in this case. The remaining 5% of the values fall outside this range.

2.1.2 The non-normal distribution

There are several alternatives to the most common—or normal—pattern of distribution. Triglycerides, another type of fat (lipid) found in the blood that are also referred to as triacylglycerides, are a good example of an alternative pattern. In this distribution, the individual data points are not found in a symmetrical spread so that a roughly equal number of points is found on either side of the most frequent result. Instead, the data are skewed in a way that shows them shifted over to the left, meaning that the majority of data points are found concentrated to the left of the midpoint of the data set while the number of data points to the right is lower and distributed across a wider range. This distribution is called non-normal, but it does not make any statement on whether individual data points are actually normal or abnormal (Figure 2.2).

A typical data set of triglyceride results taken from a large population may show 1.7 mmol/l as the most frequent result. This value would then be described as the median, which is defined as the middle point when all individual data points are ranked in order from smallest to largest. The smallest value may, for instance, be 0.5 mmol/l, while the highest one could be 11.0 mmol/l. The median of 1.7 is then not in the mathematical middle between 0.5 and 11.0, in contrast to the mean result from the normally distributed cholesterol shown in Figure 2.1, where 4.9 is close to the middle of the full range of 2.2 to 8.0, that number being 5.1.

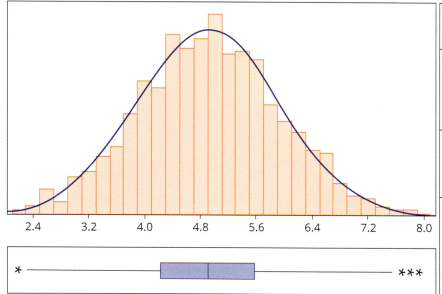

Summary for Total cholesterol (mmol/L)

Anderson–Darling normality test	
A-squared	0.25
P-value	0.758
Mean	4.9066
SD	1.0078
Variance	1.0158
Skewness	0.041250
Kurtosis	−0.168053
N	1315
Minimum	2.2000
1st quartile	4.2333
Median	4.9000
3rd quartile	5.5667
Maximum	8.0000
95% Confidence interval for mean	
4.8520	4.9611
95% Confidence interval for median	
4.8333	4.9667
95% Confidence interval for SD	
0.9707	1.0479

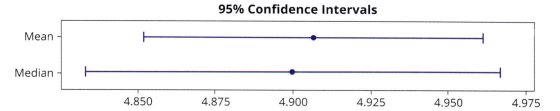

95% Confidence Intervals

FIGURE 2.1

The normal distribution. A key aspect of a normal distribution is that the highest histogram bar (around 5) is roughly in the centre of the graphic. The continuous blue line is the estimation of a normal distribution. This panel includes more complex statistical terms that do not concern us at present (but see sections 2.1.4 and 2.6.3).

Based on a graphic generated by Minitab™ software.

A second point regarding non-normal data is the variation, which is taken to be the **inter-quartile range**, or IQR, and which is determined in a different manner to the measure of variation in a normally distributed data set (the SD). The IQR is the value of the 25th and 75th individual data points when all are ranked in order (the median being the 50th value). The box on the right of Figure 2.2 tells us that the first quartile (i.e. the 25th point) is 1.1 and the third quartile (i.e. the 75th point) is 2.4. So the interquartile range is 1.1–2.4. At the bottom of Figures 2.1 to 2.3 is a box titled 95% confidence intervals, with blue lines that in Figure 2.1 fully overlap, whilst in Figure 2.2 they markedly do not. This is further evidence that Figure 2.1 has a normal distribution and that Figure 2.2 has a **non-normal distribution**. Full details of **confidence** are provided in section 2.6.3.

An alternative to the non-normal distribution in Figure 2.2 is when the data is skewed to the right, as in Figure 2.3, which shows the number of weeks of a pregnancy. Here, the 'tallest' histogram bars are on the right-hand side of the graph. As before, the box on the right gives us some statistical data, the key indices being the central point (the median, 38 weeks) and the variation (the IQR, 35–39).

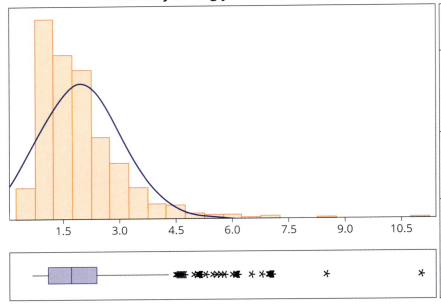

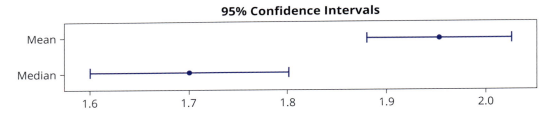

FIGURE 2.2
The non-normal distribution of triglyceride data. In contrast to the normal distribution (Figure 2.1), the highest histogram is not in the centre of the diagram but is skewed over to the left. Note that, unlike in Figure 2.1, the continuous blue line does not track the histogram bars that closely.

Based on a graphic generated by Minitab™ software.

2.1.3 Normal versus non-normal

The difference between these two distributions is not merely of academic interest but has two important consequences.

Statistical analysis

When seeking a numerical difference between two sets of data, the distribution must be assessed as it will determine the statistical test. Two sets of data, both with a normal distribution, will be analysed by **Student's t test**, but if one or both sets of data have a non-normal distribution, the **Mann–Whitney U test** is applied. Use of an inappropriate test may well lead to an incorrect result (more detail in section 2.6.3).

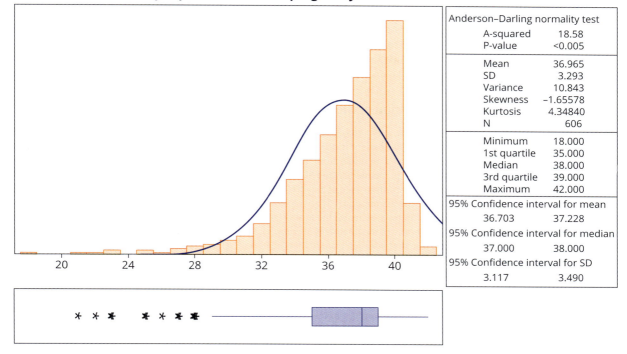

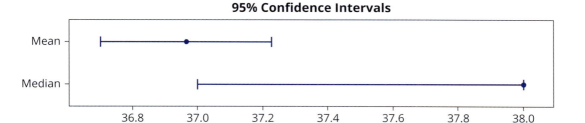

FIGURE 2.3

The non-normal distribution of pregnancy data. This data set data is clearly skewed to the right, indicating a non-normal distribution, and the continuous line does not track the histogram bars as precisely in the normal distribution in Figure 2.1.

Based on a graphic generated by Minitab™ software.

Distribution of a set of continuously variable data can been determined by a number of methods. The least scientific is simply to look at a graphic (as in Figures 2.1, 2.2, and 2.3 where the distribution is obvious), but formal methods include the size of the SD compared to the mean; the difference between the mean and the median (small in a normal distribution, large in a non-normal one); and the use of additional testing, such as the **Anderson–Darling test** and the Kolmogorov–Smirnov test. Results for these can be determined from the summary information provided by the software as shown in the box on the right side of each figure.

Study the boxes on the right side of Figures 2.1 and 2.2. Look carefully at the relative differences between their mean and SD, between the mean and the median, and at the results of the Anderson-Darling test. What conclusions about distribution can you draw? Now look at Figure 2.3 and draw a further conclusion.

Determination of the reference range

Scientists working on indices with a continuous variation need to know the extent to which a result from a particular patient is normal. This will have been decided by experienced practitioners and researchers based on results from hundreds or even thousands of essentially normal individuals, and we often look to the median and SD, or median and IQR, for guidance. However, if someone's result is above or below the range of mean +/- two SDs, this does not automatically mean that this result is abnormal—this assumption is frequently made but incorrect. It is true, though, that the likelihood that a result is abnormal increases the further it falls outside of the relevant range.

The objective of the treatment of raised cholesterol or any abnormal result is generally to reduce the result so that (ideally) it falls within the normal range. However, describing a result as abnormal can be unhelpful and possibly detrimental, often placing undue pressure on the patient and practitioner. In other situations, it may be very difficult to successfully treat an abnormal result, such that is does not get into the normal range, leading to a sense of failure. Accordingly, the preferred descriptor is *reference range*, or perhaps even the *target range*, instead of the *normal range*. But it is also important to consider exactly where the reference range came from—if it comes from apparently healthy hospital workers (as is common), this can cause bias as this group are healthier than the general population as any apparently healthy population will include some whose disease is, as yet, asymptomatic.

2.1.4 Categorical data

Nearly all people have either two X chromosomes, or one X and one Y chromosome, giving us the two mutually exclusive categories of sex (female and male, respectively). A small percentage of the population has only one X chromosome (XO, Turner syndrome), an extra X chromosome (XXY, Klinefelter syndrome), or an extra Y chromosome (XYY, Jacobs syndrome). This shows that statistics split into binary categories of sex that we typically rely on, such as the data on deaths from the UK's Office of National Statistics used throughout this book, are not without limitations. Similarly, we can classify people as those who have been given a diagnosis of a certain disease, and those who have not been given that diagnosis. Incidentally, this does not mean people without the diagnosis do not have the disease. We can represent data of this nature in a number of ways, such as in a pie chart (Figure 2.4), or in histograms (Figure 2.5). The former gives a very rough estimate—and in this

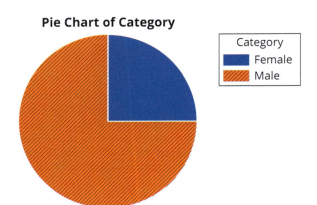

FIGURE 2.4

Pie chart of the frequency of males and females in a population.

Based on a graphic generated by Minitab™ software.

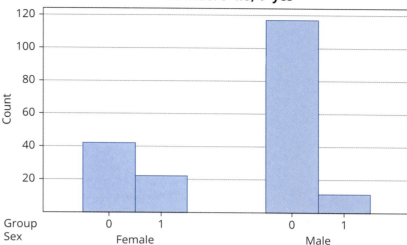

FIGURE 2.5

Histogram plot of the frequency of arthritis in a group of females and males.

Based on a graphic generated by Minitab™ software.

instance the frequencies are 75% male and 25% female—but does not tell us the exact number of people in this analysis.

The histogram plot of Figure 2.5 shows the frequency of females and males according to the category of arthritis, and also shows the exact count of people in each disease category. With access to the raw data, a formal statistical analysis can be performed, which shows a significant difference in the proportions: only around 8% of the 110 men have arthritis, whereas a third of the 60 women have this diagnosis. A statistical test (the **chi-square test**) will tell us the probability that this difference is significant (discussed in section 2.6.3).

Cross reference

Turner syndrome, Klinefelter syndrome, and Jacobs syndrome will be discussed in more detail in Chapter 16.

Cross reference

Full details of statistical analysis can be found in the book *Data Handling and Analysis*, which is listed in the Further Reading list at the end of this chapter.

Key points

Choice of the correct statistical test is crucial, as an incorrect choice very possibly leads to a false result.

2.1.5 Hypothesis testing

The statistical tests we have just looked at—the t test, the Mann–Whitney U test, and the chi-square test—are used to determine whether a difference between two sets of data is genuine. The formal way of doing this is hypothesis testing. We first set the null hypothesis, which is that any difference between two sets of data is not statistically significant. As an example we may state that our null hypothesis is

BOX 2.1 Software for statistical analysis

The practitioner must be aware of basic statistical principles and is likely to be tested on these using software, of which there are many forms. Some simple calculators are available freely online, but more complex packages exist, such as Minitab (the author's favourite for relatively simple analyses). Packages such as R and SPSS are the domain of the professional statistician and are unlikely to be used by healthcare professionals.

that men are not taller than women. We then pose an alternative hypothesis, which is that men are taller than women. We then collect data on the heights of a sufficiently large number of men and women, and perform the appropriate statistical test. Should the result of the test show that men are the same height as women, the data then support the null hypothesis and refute the alternative hypothesis. However, should our data show that men are taller than women, these data support the alternative hypothesis and refute the null hypothesis. At this point it is worth pointing out that statisticians rarely prove or disprove a particular hypothesis—they provide evidence that supports or fails to support the hypothesis.

2.1.6 Assay performance and quality control

Additional aspects of statistics are important to quantitative scientists, such as those in a haematology or biochemistry laboratory, where concepts such as reproducibility are crucial. Suppose in one laboratory, repeated analysis of the same sample gives results of 25, 24, 26, 23, 25, 25, and 26. The mean (SD) of this data is 24.9 (1.1), and dividing the SD by the mean gives us the coefficient of variation (CV) of 4.4%. In a different laboratory, the same sample and methods give results of 25, 27, 24, 28, 23, 25, and 22, and so the same mean (24.9) but a larger SD (2.1), and consequently a CV of 8.5%. Many would consider the methods of the second laboratory to be inferior to those of the first laboratory (and possibly unacceptable). Similar types of statistics are used to monitor laboratory performance over a defined time period, as required in quality control and quality assurance.

2.2 The development of disease

The body is, of course, fragile, but it has a remarkable ability to repair itself when traumatized, especially in the young. A feature of the elderly is that this resilience slowly falls, and so disease susceptibility increases with age. However, should this damage be too great, such as in road traffic accidents or traumatic violence, repair may be impossible so that death may follow. But damage can also come from other directions—some insidiously slow and others acute—causing malfunctions in the body that we interpret as disease. In some cases, disease is caused by an external factor or factors acting on an otherwise healthy body, an excellent example of this being a pathogenic microbe such as a bacterium or a virus. Failure to combat a pathogenic organism can be fatal, and there may be reasons why the body cannot do this.

Some disease can be apparent at, or soon after, birth, in which case it is said to be congenital. Infant syphilis is a good example of an acquired congenital disease. Other features present at birth or early in life may well be due to a genetic abnormality, of which cleft lip and **haemophilia** are good examples. Disease due to alterations in an otherwise healthy genome can appear at any age, and the adverse effects of acute radiation on genes were proven in atomic bomb survivors, who suffered a variety of different cancers. But the effect of radiation need not be acute, as a low level of exposure to ionizing radiation over years also leads to gene-based disease.

The development of acquired disease may be brief. For example, the effect of a venomous snake bite or a poison can become manifest within minutes. However, identifying the exact start of a disease process can be problematical, such as in rheumatoid arthritis, where the disease process may have been slowly developing for months, or even years, before the appearance of symptoms. Someone infected with hepatitis B virus undoubtedly acquired it at some precise point, but if the symptoms were mild, it may not have brought that person to our profession. However, over the years the virus would slowly cause liver damage, then **cirrhosis**, and ultimately hepatocellular carcinoma, and the consequences of any of these steps (haemorrhage, abdominal pain, jaundice) would certainly bring the patient to our attention.

2.2.1 How does disease arise?

A great many external agents and stimuli (pathogens: literally disease-causing) are implicated in the causation (aetiology) of human illness, although in many cases it is not possible to discover a single causative agent which is directly linked to disease when present. The degree to which a pathogen initiates disease is its *penetrance*: within a population, some will be susceptible (and suffer disease) whereas others will be relatively resistant, leading to a form of Darwinian natural selection within a given population.

For example, exposure to a pathogenic microorganism such as an influenza virus does not invariably result in disease in everybody, but in some it is fatal. Not everybody becomes ill to the same degree—a variety of other factors are also important for establishment of infection, including the status of the host immune system and the size of the infective dose (of viruses). Exposure to the virus is therefore considered the most important causal factor (perhaps described as necessary, obligatory, primary, or crucial) and the other factors (the state of the host's immune system, co-morbidities such as lung disease) may be defined as secondary or contributory causal factors. A further good example of a single pathogen causing disease is alcohol. Over the years, an excessive alcohol intake effectively poisons the liver, leading to jaundice, cirrhosis, and then irreversible liver failure (as do hepatitis B and C viruses), curable only by transplantation. However, alcohol is also associated with other disease, such as certain other cancers, especially of the intestinal tract and of the pancreas.

Other disease, such as **atherosclerosis**, inevitably arises from the interaction of several different factors. These are the established risk factors of diabetes, hypertension, smoking, and high levels of cholesterol in the blood (as discussed in Chapter 7). It is likely that a small degree of all four risk factors is as likely to cause disease as is the strong effect of only one of the risk factors by itself. In this respect we look to the 'multi-hit' or 'multi-stage' hypothesis of the aetiology of a disease, which is often evident in cancer, where mutations in certain genes may also be pathogenic (i.e. are **oncogenes**, discussed in Chapter 3). Certain genes have a relatively minor effect on the development of a particular disease, whilst others act with a high degree of penetrance, inevitably causing disease (Chapter 16).

There are many instances where the total risk of a disease compounds, depending on the exposure to different 'amounts' of each risk factor. Influenza and atherosclerosis are examples of diseases for which the causal factors involved are well established. For other diseases, identification of the causal factors has proved more difficult and the search for these factors continues to represent a significant challenge to medical science. Often the initial search for a causal factor begins by the examination of the patterns of disease within human populations. This process, epidemiology, is the subject of section 2.4 that follows. Epidemiological studies usually include analysis of the rate of mortality (death) and of morbidity (illness).

2.2.2 Disease in a population

As discussed in Chapter 1, many people beyond middle age carry asymptomatic disease and at least one morbidity, of which connective tissue disease is common and serves as an example of disease in a population.

Osteoporosis (loss of strength in the bones) and osteoarthritis (inflammation of, and so destruction of, generally large weight-bearing joints, such as of the hip and the knee) are relatively common diseases. In the UK it is estimated to cause around half a million broken bones annually, whilst fragility fractures present with rates of 3,840 and 9,860 per million person/years in men and women respectively, with rates in White women being 4.7 times those in Black women (Wilson-Barnes et al. 2022).

Although associated with considerable pain, loss of mobility and other morbidity, these diseases rarely in themselves cause death. However, the major cause of osteoarthritis is being overweight and obesity, major risk factors for diabetes. So it may be that someone with osteoarthritis will succumb to the heart disease of which diabetes is a major risk factor. Diseased joints can be replaced by surgeons, but no surgery is free of risk. Indeed, orthopaedic surgery and obesity are independent risk factors for deep-vein thrombosis, a condition that is rarely on a death certificate, but the closely related pulmonary embolism (a clot in the lung) is frequently fatal. This risk of thrombosis is so great that drugs (anticoagulants) must be given to the patient to prevent this from happening (Chapter 11).

Rheumatoid arthritis and **systemic lupus erythematosus (SLE)** are other inflammatory diseases of the skin, bone, and joints that cause a great deal of pain and discomfort (Chapter 10). However, grossly painful, swollen, and misaligned joints, or a skin rash, by themselves, are very rarely a direct cause of death. Both carry an increased risk of cardiovascular disease, whilst SLE is associated with renal failure, so that the latter conditions inevitably find their way onto the death certificate in preference to the arthritis or SLE. Influenza and asthma appear in the official ONS listings as diseases of the respiratory system. Each winter, millions of people suffer from colds and influenza. However, in England and Wales in 2021, influenza appeared on the death certificates of only 35 people (0.006% of all deaths), even though one source suggests that some 5 million people suffer this condition. These situations illustrate the gross discrepancies between the occurrence of disease in a population and the extent to which a particular disease contributes to mortality.

2.2.3 Causes of death

The ultimate cause of death is irreversible cardiac arrest, but there are very many factors that lead to this (cancer, inflammation, diabetes, etc.). Not all disease (i.e. morbidity) causes death, and not all deaths are caused by disease (e.g. some are caused by accidents). In 2021, 585,484 people in England and Wales died, and the causes of their deaths were reported and collated. The dominant causes of death are cancer, diseases of the circulation (the heart and blood vessels, together called the cardio-vascular system), and respiratory disease (Table 1.2). Together, these account for over two-thirds of all deaths. Furthermore, there are several instances where there are differences between the sexes, and by age. One of the greatest differences between the sexes is in death due to mental and behavioural causes, which is cited in death certificates of more than twice as many women as men.

SELF-CHECK 2.2

Which causes of disease can be avoided, and which cannot?

2.2.4 Types of aetiological factors

Aetiological factors may be broadly divided into *endogenous* factors (those which create a disturbance or imbalance from within) and *exogenous* factors (those which threaten existence from the outside).

Endogenous factors

Some endogenous factors are relatively easy to identify. Some, such as chromosome abnormalities (Turner, Down, Fragile X, and other syndromes) and specific gene mutations (**muscular dystrophy**, DiGeorge syndrome) can be detected in the neonate or infant, whilst others may appear later in life (such as Huntington's and Alzheimer's diseases). These conditions are described in Chapter 16. However, the principal endogenous disease is cancer, described in Chapters 3, 4, and 5, which can appear at any age, as can many connective-tissue diseases, described in Chapter 10.

Exogenous factors

These may be described as pathogens acting on an otherwise healthy body, although disease may also follow from the absence of a factor(s), as in vitamin deficiency and reduced-calorie malnutrition. In England and Wales, certain infectious and parasitic diseases (intestinal, tuberculosis, meningitis, viral hepatitis) together were linked to 5,368 deaths in 2017, that being 1% of all deaths. Notably, this proportion fell in the first COVID-19 year of 2020 to 0.71% of all deaths, rising slightly to 0.94% in 2021. However, globally, communicable infectious diseases caused 12.6% of all deaths in 2017—clearly a far greater issue (Table 2.1). There is often apparent overlap as some genetic abnormalities have been shown to be the result of exposure to mutagens in the environment. An example of this is ionizing radiation causing leukaemia, another is excess alcohol and certain viruses causing liver cancer. Some genetic disorders may therefore ultimately be attributed to exposure to exogenous factors.

Exogenous factors may not only be defined as poisons and microorganisms. The well-known unhealthy lifestyle, consisting of choices of excess alcohol, smoking, poor diet, and lack of exercise, is a cause of certain cancers, cardiovascular disease, and respiratory disease. Figure 2.6 summarizes exogenous and endogenous factors.

TABLE 2.1 Global deaths caused by infectious diseases in 2017

Infectious disease	Number of deaths	% of infectious disease
Lower respiratory tract infections	2,558,600	36.2
Enteric diarrhoeal diseases	1,569,600	22.2
Tuberculosis	1,183,700	16.8
HIV/AIDS	954,600	13.5
Malaria	619,800	8.8
Meningitis	288,000	4.1
Enteric typhoid and parathyroid	135,900	1.9
Acute hepatitis (71% B virus)	126,400	1.8
Syphilis	113,500	1.6
Measles	95,300	1.3
Encephalitis	92,400	1.3
Whooping cough	91,800	1.3

GBD 2017 Causes of Death Collaborators. Global, regional, and national age–sex-specific mortality for 282 causes of death in 195 countries and territories, 1980–2017: a systematic analysis for the Global Burden of Disease Study 2017. Lancet 2018: 392; 1736–88. © 2018 The Author(s). Published by Elsevier Ltd.

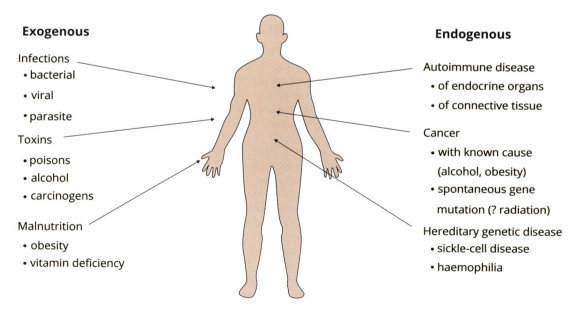

Exogenous

Infections
• bacterial
• viral
• parasite

Toxins
• poisons
• alcohol
• carcinogens

Malnutrition
• obesity
• vitamin deficiency

Endogenous

Autoimmune disease
• of endocrine organs
• of connective tissue

Cancer
• with known cause
  (alcohol, obesity)
• spontaneous gene
  mutation (? radiation)

Hereditary genetic disease
• sickle-cell disease
• haemophilia

FIGURE 2.6

Causation of disease. Diseases may be seen as being exogenous (caused by an external agent) or endogenous (developing within the body without the involvement of an outside factor).

We will examine the cause of disease in more detail in section 2.6, looking more precisely at how we can be more informed about how particular pathogens cause disease.

Key points

The aetiology of a particular disease can be complex and arise from several different mechanisms.

2.3 Clinical aspects of disease

Diseases can be classified in a number of ways, such as by aetiology, as we have already discussed, but another is how the person themselves comes forward to the healthcare professional with a problem they have. These are often classified on the basis of the outward signs and symptoms produced by disease (such as jaundice, a rash, tiredness and lethargy, bruising, or a pain in the chest). However, the problem with this approach is that many different diseases may cause the same signs and symptoms—that is, the signs are not specific to that disease. For instance, jaundice may be caused by alcoholic liver disease, viral hepatitis, or by cancer of the liver.

Others classify disease on anatomical or pathological criteria (such as heart disease), but most are classified on symptomatic criteria irrespective of the causative agents involved. Thus, a number of causal factors are implicated in the development of various types of carcinoma of the lung, of which a common presenting feature is a persistent cough, bringing up phlegm, or pain on deep breathing. Causes include cigarette smoke, asbestos, coal smoke, and other atmospheric pollutants, but patients with the symptoms of lung cancer are grouped together irrespective of the involvement of one or more of these agents. Conversely, disease caused by the bacterium *M. tuberculosis* may produce different clinical manifestations (in the bone, lung, and/or liver) in different individuals, yet all are classified as forms of tuberculosis.

2.3.1 Identifying disease in an individual

We have seen how diseases are classified but have not yet considered how a particular group of features is first designated as a disease state. As discussed in Chapter 1, healthcare professionals use signs (noted in a physical examination), symptoms (what the patient complains of), and a range of laboratory and clinical tests (such as an X-ray or a blood test) to determine whether a patient has a given disease. In some cases, it may be necessary to take a small sample of tissue (a biopsy) and then examine this tissue in the histology laboratory. This will be useful in determining whether a growth is benign or malignant.

The taking of a thorough clinical history for the patient will determine whether there has been exposure to any potential aetiological agents, and this is generally done in clinic. The existence of certain predisposing conditions, including a family history, may make the development of a disease more likely. Examples of these risk factors include certain genetic disorders, lifestyle, psychological and personality profile, age, and environmental factors such as climate and pollution. Furthermore, the presence of one disease (e.g. diabetes) may predispose a patient to the development of further disease (in this case, problems with the eye).

One of the early steps in identifying a disease is to establish a range of diagnostic possibilities from which the eventual diagnosis will be selected. The final diagnosis may sometimes be established shortly after clinical presentation, in other cases perhaps only after extensive use of laboratory and clinical tests, or occasionally may never be identified during the lifetime of the patient. It must be remembered that disease is a dynamic process and the indicators of disease may vary as the disease progresses. Furthermore, in some patients, particularly those who are elderly, different diseases may coexist, thus confusing the diagnostic processes.

2.3.2 Screening

If a particular test, process, or examination has a very strong association with a particular disease, it may be worthwhile checking out an entire population for this feature. The strength of this approach, which is called screening, is that it will identify people who, unknown to them, are at risk of developing the particular disease at some point in the future. Accordingly, treatment can begin as soon as is practicable, which may reap dividends years in the future with the avoidance of serious disease, and the related costs. Examples of this approach are screening women for lumps in the breast which, if present, may be cancer; use of the PSA blood test to screen men for prostate cancer; and screening for abnormalities in faeces to test for colorectal cancer. If screening indicates a likelihood of disease, other investigations may follow, and if present, the early surgical removal of the tumour may be advantageous, and/or chemotherapy and/or radiotherapy commenced.

However, the key factor is the number of cases of breast, prostate, or colorectal cancer discovered compared to the number of men and women screened. This process can be made more efficient by screening only those susceptible high-risk groups, such as those with a strong family history. Thus a young man whose grandfathers both suffered with premature heart disease (such as a heart attack at age 55) may be screened for high serum cholesterol, which can have a genetic basis and so part-explain his grandfathers' disease. If present, high cholesterol can be reduced by a drug of the statin class, also likely to reduce the risk of future cardiovascular disease in the young man. Likewise, a young woman whose grandmothers, mother, and aunts all suffered clots in their veins (a deep-vein thrombosis, DVT), possibly before or after childbirth, may be carrying the factor V Leiden (FVL) gene, and so opportunistic screening may be an advantage. If the gene is present, this young woman may benefit from anti-coagulation in her own reproductive history and may be advised to use non-hormonal contraceptives. However, as discussed earlier with reference to pathogens, penetrance also applies to genes.

Gene mutations that are fully penetrant include the one that causes haemophilia in males. One cannot have 'some' of this mutation; it is either 100% present or 100% absent because the gene is present on the X chromosome. However, the genotype (the presence of the gene) may not necessarily have a great effect on the phenotype (the whole body). Haemophilia is an example of a gene that has 100% penetrance into the phenotype, but it is important to recognize that this does not always cause haemorrhage. Other examples of genes with a high degree of penetrance include those which cause Huntington's chorea, **cystic fibrosis**, and muscular dystrophy. An example of a gene that has a weak or low penetrance is Factor V Leiden (FVL), present in its carrier status in ~5% of the UK population, a very high figure indeed. The fact that it does not always cause a DVT is because the effect of the gene is weak. It follows that for the FVL to contribute to a DVT, other factors are required, such as age. The obese who carry this gene are at greater risk of a DVT than someone who is obese but is free of the gene. These are classic examples of the potential interaction between genes and the environment, and will be revisited in chapters to come.

2.3.3 Treatment of disease

We have already looked briefly at treatment (drugs, surgery, physiotherapy, etc.), the objective of which is to cure the disease, in Chapter 1. However, the disease may only be partially cured, in which case the patient is said to be in remission, and the disease may return (as will the patient), in which case they have relapsed. This may be because the treatment has been terminated prematurely, perhaps because the patient's symptoms have abated. A common example of this is treatment of a chest infection with antibiotics—should the patient feel better, they may fail to complete the course of antibiotics, assuming they have been cured. However, this may allow a small residual number of the bacteria to re-establish the infection after a second period of incubation. Another example of the presumption of a curative treatment is surgical excision of a cancer—merely removing a tumour does not guarantee cure, as micro-metastases may already be present elsewhere in the body, and these could make their presence felt perhaps years later.

If the disease is serious and life-threatening, such as cancer, then the longer-term success of the treatment may be quantified as 'quality-of-life years' that have been saved. This, of course, requires some estimation of how long the patient is expected to live, and is quantified as prognosis. However, not everybody can or will be treated—some may decline treatment. In the elderly, the toxic effects of some forms of chemotherapy can actually cause more disease (renal and liver failure) and may even be fatal: a younger patient would be able to withstand this toxicity.

Some diseases (such as atherosclerosis) are not amenable to cure, but the progression of the disease and its symptoms may be controlled or suppressed by the administration of drugs or by surgery. Some patients with advanced and incurable disease may receive palliative treatment only with the aim of relieving their symptoms for the remainder of their life.

2.3.4 Mechanisms of disease

Elucidation of the aetiology is the major focus of biomedical research, since understanding of the biology of disease will contribute to an improvement in healthcare, in terms of both the prevention of disease and its treatment. In numerous cases this has proved to be correct. For example, the recognition that type 1 diabetes mellitus results from damage to insulin-secreting cells in the pancreas led to the development of insulin-replacement therapy. Similarly, without the observations by histopathologists of deposits of cholesterol in the walls of arteries of those dying of a myocardial infarction, and the elucidation by biochemists of the metabolic pathway for the synthesis of cholesterol, we would not have the cholesterol-lowering statin class of drug which has prevented tens of thousands of heart attacks and strokes. However, an understanding of the biology of many important diseases (such as certain types of cancer) is only beginning to emerge, and for the most part this has not yet contributed greatly to a decline in mortality rates.

Key points

Disease can be classified in the clinic by signs and symptoms. Confirming disease calls for investigations such as blood tests, imaging with X-rays, or histological examination of a biopsy of tissue. Screening for hidden disease can be very effective. Once confirmed, treatment is often initiated, but not everybody can be treated.

SELF-CHECK 2.3

What do we mean by 'penetrance' of a microbial pathogen and of a mutated gene?

2.4 Causation of disease

Having examined various aspects of the aetiology of disease, we now embark on a formal study of the causation of disease, and in particular the use of **correlation**, a much-misused word. It refers to the extent to which two indices, such as height and weight, or perhaps systolic blood pressure and diastolic blood pressure, are interrelated. It tells us nothing about which index actually causes which. In the first example, does increasing height cause increasing weight? Or does increasing weight cause increasing height? The former seems more likely, as the taller someone grows then the heavier they become. But in the blood pressure example, there is no established causal relationship—it merely happens that the systolic measurement correlates with the diastolic measurement by virtue of physiology (and, where it applies, by pathology).

One of the most bizarre correlations is the very strong relationship between the increasing divorce rate and the domestic consumption of bananas in the UK during the 1950s and 1960s. Initially, few would consider this relationship to be causal. However, it could be that there is a mysterious chemical in bananas that causes people to become dissatisfied with their spouse, and so seek to divorce them. Conversely, newly divorced people may express their new status by buying more bananas. Fortunately, there are now formal rules to define whether index X is causally related to index Y. These include the requirement that X always precedes Y in time, and that changing X always changes Y.

A prime example of these requirements is the risk factors for atherosclerosis and heart attack or stroke. The correct statistical descriptor of a potential link between two continuously variable indices (such as systolic and diastolic blood pressure, height and weight, and bananas and divorce) is correlation, as will be explained in section 2.4.3.

Thus care is required as in many instances the finding of a relationship between a proposed mechanism and disease may be by association, not by cause. We must rigorously test our assumptions with well-designed studies and the appropriate statistics. In addressing causality we refer to two key aspects of pathology: those of Koch and those of Bradford Hill.

2.4.1 Koch's postulates

In 1884 the great pathologist and microbiologist Robert Koch, with another eminent biologist, Friedrich Loeffler, described a set of characteristics that must apply to an organism (here referring to a microorganism such as a bacterium) that is said to cause a disease. To paraphrase, these were:

1. The organism must be present in all cases of the disease, and absent from those free of the disease.

2. A diseased host (in a clinical setting, a patient; in an experimental setting, an animal that has been deliberately infected) must provide a sample of the organism, which must then be grown in tissue culture.

3. The cultured organism must then cause disease when passed into a healthy individual (generally an animal, but human cases can apply).

4. The organism must then be isolated from this second infected animal/person and be identical to the organism in step 1.

These steps were ground-breaking in seeking to determine the scientific cause of an infection, and many are still relevant today and apply in a broader setting, such as with viral pathogens. However, we now recognize faults in the postulates, such as in step 1 where several infectious organisms are known to be carried in those who do not suffer the disease—the asymptomatic carriers (as in cholera, hepatitis B virus, and *Salmonella typhi* (carried by 'typhoid Mary')). A further problem is the postulate that the presumed agent must always cause the disease when introduced into a healthy organism. It is not possible to suffer from food poisoning unless certain microbes are present (*necessary cause*), but the presence of these microbes alone may not always be a *sufficient cause* for the food poisoning—a view that reflects sensitivity and specificity (section 1.3.3). The clinical expression of disease may depend on other characteristics in the infected or exposed person, such as genetic susceptibility, a compromised immune system, over-crowded living conditions, or malnutrition.

2.4.2 Bradford Hill criteria

An updated version of **Koch's postulates** are the nine criteria published by the English medical statistician Sir Austin Bradford Hill in 1965. These may be considered if one is to be assured of a cause (such as, in Koch's mind, a pathogenic bacterium) and its effect (such as infection). However, some criteria have failed the test of time, and some duplicate each other. Nevertheless, several remain robust.

1. **Time**. The subject must have been free of the proposed cause (which may be difficult to prove) before the exposure to that cause. It can also be difficult to define exactly when exposure to the pathogen began. The effects of the cause must then follow some time later, such as cigarette smoking leading perhaps years later to myocardial infarction or lung cancer.

2. **The strength of the association**. Clearly, the stronger the link, the more likely it is that the cause is genuine. There are numerous metrics to define these links (**relative risk**, **odds ratio**, etc.), as we shall examine in sections to come. Bradford Hill refers to the data that mortality of chimney sweeps from scrotal cancer was some 200 times greater than that of workers who were not specifically exposed to tar or mineral oils, a fold increase many would describe as unarguable (see section 3.8, where Box 3.1 applies).

3. **Consistency**. A crucial component of all scientific work is that its validity is confirmed independently by other groups. When new discoveries are announced, it may take years before they are repeated and so accepted by the community at large. However, should a study fail to confirm the original findings, this does not necessarily invalidate them, as it is rare for two populations to be identical and so give identical results.

4. **Specificity**. This criterion is less firmly held nowadays, as it is clear that many pathogens cause different clinical manifestations of disease (such as atherosclerosis causing myocardial infarction as well as stroke), and conversely, a disease may be caused by several different agents (such as the effects of different hepatitis viruses on the liver).

5. **Dose-response relationship**, or **biological gradient**. The greater the physical mass or number of pathogenic particles (toxins, microorganisms), the greater the effect on the disease process. It is established that the higher the blood pressure, the greater the risk of stroke, and that an increased use of tobacco leads to an increased frequency of lung cancer. The reverse also applies: if the cause is removed or relaxed, there should be a linked benefit (as where, after months or perhaps years of smoking, quitting reduces the risk of myocardial infarction).

6. **Plausibility**. The relationship between tobacco smoking and lung cancer is plausible because of the data on carcinogens in cigarettes. As regards atherosclerosis, there are ample data from various animal models to confirm the hypothesis that the number of risk factors is linked to myocardial infarction. However, there are several instances where laboratory data have effectively rejected a previously well-supported hypothesis, so there must be caution.

7. **Is there coherence?** Here, we look for a general agreement across different types of study: cohort, case-control, intervention, outcome/survival analysis, animal models, tissue culture, etc. However, Bradford Hill himself advised caution in this respect, noting (at the time) that arsenic causes skin cancer in humans but not in certain animals.

8. **Experiment**. This is perhaps a little vague, as experimentation of numerous types has already been covered in other points. Bradford Hill suggests intervening to reduce pathogens (e.g. reducing dust in workshops, changing lubricating oils).

9. **Analogy**. This regards considering the possibility that a very minor change in the pathogen should result in a broadly similar response in the organism. This might work for chemical carcinogens, where the effects of different atoms in different molecular positions may be tested.

These two lists (and especially the **Bradford Hill criteria**) are not absolute rules to be obeyed, but are simply guidelines that we may wish to adopt. However, the widest scientific agreement is likely to be awarded to a study that can address (or at least consider) as many of these points as possible.

Key points

Koch and Bradford Hill each devised criteria to help us in examining the cause of a disease.

2.4.3 Correlation

This word has a precise statistical meaning, and must not be misused. In the three examples previously described, (systolic and diastolic blood pressure, height and weight, and bananas and divorce), as one index rises, so does the other. The strength of this relationship can be mathematically measured as the correlation coefficient (r). This number varies between 0 and 1, where 0 is no relationship, and 1 is a perfect relationship. However, in practice, no two indices ever correlate with a coefficient of 1, but may correlate when r = 0.72, as in Figure 2.7, which is a scatterplot of the correlation between factors A and B. The dots are the individual data points, the line is the *line of best fit*, which is computed as that which minimizes differences between pairs of points. Although the relationship between A and B seems to be obvious, how can we be sure? We use the expression *probability* to quantify the likelihood

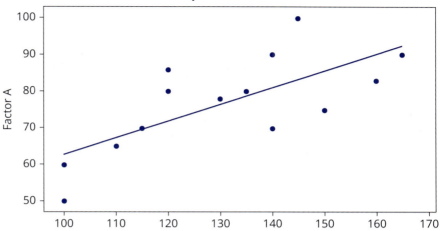

FIGURE 2.7

Scatterplot of factor A and factor B, with the line of best fit.

Based on a graphic generated by Minitab™ software.

that a link, difference, or relationship is genuine, and the cut-off point is that should the probability be > 95%, we are assured the data are robust. The probability is abbreviated to p, and the 95% to a decimal of 0.05. Thus we (generally) accept $p < 0.05$ as a benchmark for a genuine result, so that data returning $p = 0.055$ are not statistically significant. The smaller the p value, the more assurance we have that the data are acceptable.

The statistical software that generated Figure 2.7 also gave a probability (p) value, that being $p = 0.004$, which is to say we are 99.6% sure the relationship is genuine.

Now consider Figure 2.8, a scatterplot of factors C and D, where $r = 0.84$, indicating a close relationship between the two factors with a probability of $p = 0.038$, meaning we are 96.2% confident of a robust link. This difference in p values follows from the power of the two studies, where there are 14 pairs of A and B, but only 6 pairs of C and D. Therefore, we might consider the possibility that the study of the relationship between C and D is flawed (and perhaps should not even have been attempted) because the number of analyses is too small. This small sample size has given us a risk of a **false positive** error. Another way of looking at this is to say that the study is underpowered.

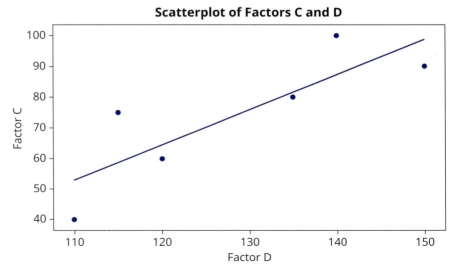

FIGURE 2.8

Scatterplot of factor C and factor D.

Based on a graphic generated by Minitab™ software.

In a third scatterplot (Figure 2.9), the relationship between factor E and factor F is actually less strong than that between factors A and B or even C and D, and where r = 0.67. The relationship between the data points and the line of best fit also seems poor. However, the presence of 200 data points provides much more power and drives the probability of a false positive result down to p < 0.001 (i.e. very unlikely), and so there is a confidence of 99.9% that the link between E and F is genuine. But common sense tells us that the A/B link *seems* stronger than the E/F link, so we must not always robotically believe all statistical output.

The links between these three sets of data are all positive—as one increases, so does the other. However, it may be that as one index rises, the other one falls, as illustrated in Figure 2.10.

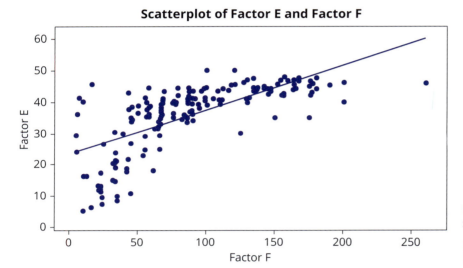

FIGURE 2.9
Scatterplot of factor E and factor F.

Based on a graphic generated by Minitab™ software.

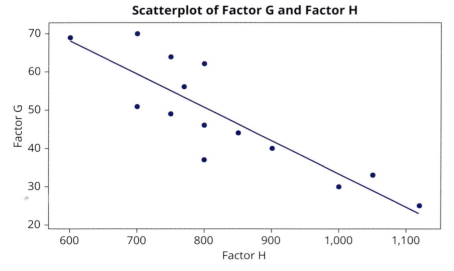

FIGURE 2.10
Scatterplot of factor G and factor H.

Based on a graphic generated by Minitab™ software.

In this Case, we describe the relationship as inverse, and place a minus sign before the correlation coefficient, which, in this case, is r = −0.86. This data may be the distance (factor H) in metres that was run by a group of athletes of different ages (factor G) in a set time period. Clearly (and intuitively) the younger the athlete, the further they can run, although there are cases where an older athlete outruns a younger athlete. We would predict that, generally, age has a causative impact on running, not vice-versa. One's ability to run does not define one's age.

Key points

The likelihood that a link, relationship, or difference is genuine is given by its probability, denoted by p. If we find p < 0.05, the data are robust, and the smaller the p value (e.g. p = 0.003), the greater the probability of a true result.

SELF-CHECK 2.4

What do you understand of (a) a correlation coefficient of r = 0.56 and p = 0.065, and (b) a correlation coefficient of r = −0.65 and p = 0.012?

2.5 Epidemiology: General concepts

Epidemiology (the scientific study of diseases and how they are found, spread, and controlled in groups of people) is perhaps the oldest branch of pathology, and clearly derives from epidemics (of disease) that would periodically sweep through ancient civilizations, and are described in old texts such as the Bible. Epidemics contrast with disease that is endemic—meaning disease is always present in a population, generally at a relatively low level. Examples of endemic diseases in the UK include viral infections such as chickenpox and measles, the latter of which will continue to pose a problem until it is eliminated by vaccination (as smallpox was in 1977). If the disease is extensive, across several continents and possibly the world, it may be described as a pandemic—examples being the cholera pandemic of 1816–26, the influenza outbreak of 1918–19, HIV/AIDS (c.1970 to the present), and the COVID-19 pandemic that began in 2020 (as discussed in Chapter 13).

2.5.1 A historical perspective

Epidemics of the Middle Ages include the Black Death (in Europe c.1345–1355) and the Great Plague of London (c.1665–6). They were of great value to early scientists, who formed the view that something must be causing the disease to spread, leading to the germ theory of disease, which was subsequently refined (by scientists such as Semmelweis, Pasteur, Koch, and Lister) into the biomedical and clinical science we recognize today.

We owe a considerable debt to John Graunt (1620–74) who meticulously collected and published reports on births and deaths in numerous London parishes. Of particular interest is that of December 1665, at the height of the plague of 1665–6 (caused by the bacterium *Yersinia pestis*) (see Table 2.2).

Understandably, the most frequent cause of death is plague (over 70% of deaths), whilst it may be that *ague and fever* (5.4%) is also part of plague. *Consumption and tissick* killed 4.9%, whilst *teeth and worms* were responsible for 2.7% of deaths. Although many causes are recognizable (e.g. leprosy, rickets) or translatable today (e.g. *bloody flux* is probably haemorrhagic diarrhea, *meagrom* could be interpreted as migraine), the pathophysiology of others, such as *Kings Evill* and *frightened*, seems obscure. Although these 'Bills of Mortality' had been published for decades, it was the scale and precision of the data collected during the plague that set the foundations for subsequent schemes, such as the establishment, in the Victorian era, of the General Registrar Office system for the collection of data on births, marriages, and deaths in the UK. In 1970 the Office became part of the newly created

TABLE 2.2 Bill of Mortality for December 1665

Cause	Number	Cause	Number	Cause	Number
Abortive and stillborn	617	Aged	1,545	Ague and fever	5,257
Bedrid	10	Bleeding	16	Bloody flux, scowring & flux	185
Burnt and scalded	8	Cancer, gangrene and fistula	56	Cold and cough	68
Collick and wind	134	Consumption and tissick	4,808	Distracted	5
Dropsie and timpany	1,478	Drowned	50	Executed	21
Flox and Small Pox	655	French pox	86	Frightened	23
Found dead in streets, etc.	20	Gout and sciatica	27	Grief	46
Gripping in the guts	1,288	Hang'd and made away themselves	7	Jaundics	110
Kill'd by several accidents	46	Kings Evill	86	Leprosy	2
Lethargy	14	Livergrown	20	Meagrom and headach	12
Measles	7	Murdered and shot	9	Overlaid and starved	45
Palsie	30	Plague	68,596	Poysoned	1
Rickets	557	Rising of the lights	397	Rupture	34
Scurvy	105	Shingles and swine pox	2	Sores, ulcers, broken and brusied limbs	82
Spleen	14	Spotted feaver and purples	1,929	Stopping of the stomach	332
Stone and strangury	98	Teeth and worms	2,614	Vomiting	71

Office of Population Censuses and Surveys (OPCS), with the Registrar General in overall charge, which in 1996 developed into the present Office of National Statistics (ONS). The World Health Organization (WHO) has developed a more formal method—the ICD—enabling deaths from all over the world to be compared (discussed in section 1.2.1 of Chapter 1, where Table 1.1 applies).

Whilst Graunt's work should not be downplayed, Dr John Snow (1813–58; Figure 2.11) is widely credited as being the founder of modern epidemiology and (therefore) of public health, as he formed a hypothesis, and then acted. Observing an outbreak of cholera in the Soho district of London in 1854, he concluded that the source was contaminated water. His petitioning of the local council to remove the handle of the water pump may have been instrumental in reducing the epidemic, although this may also have been partly due to the exodus of the nearby population, who would therefore have saved themselves. Snow also collected data on the daily progress of the epidemic, and the lag period between the development of the disease and the death of the patient (Figure 2.12). Later work showed the causative agent to be the bacterium *Vibrio cholerae*.

Epidemiology is a powerful tool for determining the aetiology of disease, which therefore provides clues to ways of enhancing healthcare. Accordingly, it has value in the wider sense of public health in that it can detect disease and patterns of developing diseases that may not be visible on a small scale, such as in local populations. The sections that follow will include examples of how epidemiological studies have contributed to our understanding of pathology, and so how the general health of a particular nation can be improved. But before doing considering these, we must acquaint ourselves not only with the specialized vocabulary of epidemiology, but also with how key terms are calculated.

FIGURE 2.11
Dr John Snow (1813–58), founder
of modern epidemiology.

Courtesy of the US National Library of
Medicine.

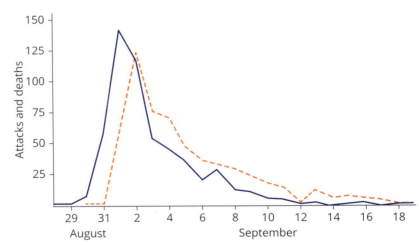

FIGURE 2.12
Cholera deaths. Snow's data on the daily
rates of attacks (blue solid line) and deaths
(orange dashed line) due to cholera. Note
the rapid development on the left but the
slow right-sided washout, typical of such
an epidemic.

SELF-CHECK 2.5

What other data could Snow have collected that would have been helpful in further analysis of his
results and conclusions?

2.5.2 The mathematics of epidemiology (part 1)

In addition to his work in Soho, Snow also collected cholera death data in areas of London supplied
with water from different suppliers. The Southwark and Vauxhall Company supplied local river water
considered by Snow to be impure and contaminated with sewage, whereas the water supplied by the

Lambeth Company was sourced far upstream and free of London sewage. The rate of cholera deaths in areas supplied by the Southwark and Vauxhall Company was 71.4/10,000 houses, compared to 5.4/10,000 houses in areas supplied by the Lambeth Company—the former had an incidence rate 13.2 times that of the latter.

Let us suppose that a researcher forms a hypothesis that the regular smoking of a least ten cigarettes a day is important in liver disease. Over a period of 6 months, she questions 122 attenders at a liver clinic, of whom 22 admit cigarette smoking, a rate therefore of 22/122 = 18.8%, which seems large. But this is meaningless unless compared to another group with whom to control the data from the liver patients, such as attenders at an ophthalmology clinic. In posing the same question she finds that 14 of 87 patients (17.2%) admit to being smokers. In this example, the liver patients are cases, and the ophthalmology patients are the controls. Formal statistical analysis of this data finds no significant difference. However, the sample sizes are small, and we cannot tell if the two groups are matched for potential confounders such as sex, age, and other disease. However, if we know that the rate of smoking in the general population who were not attending a hospital clinic was 30%, we could speculate about the rate of smoking in hospital attenders and non-attenders. This comparison of three groups of patients is very simple, and can hardly be described as epidemiology, but it serves to illustrate a number of concepts.

2.5.3 Incidence and prevalence

These are measures of the illness in the population and are usually expressed as either incidence rates or prevalence rates. The incidence of a disease is the number of new cases occurring over a specified time period, which may be days, weeks, months, or years, and the method for determining this factor is given by equation 1. Incidence is a measure of the risk of contracting or developing a disease since it concerns only new cases, not those who already have the disease. It follows that the precise health status of each individual must be rigorously defined when data collection begins, and those with existing disease excluded. In many cases, one would need to qualify this data by describing the time period involved, e.g. 45 per 10,000 per month.

Equation 2.1

$$\text{Incidence rate} = \frac{\text{Number of subjects developing the disease in a given time period}}{\text{Total population at risk over the same time period}} \times 10,000$$

In contrast, the prevalence rate includes everybody with the disease at a set time point, divided by the total number of people in the sample pool at that time (Equation 2.2). Snow's water/cholera data are an example of prevalence, as they simply counted the number of people with cholera and divided it by the number of houses.

Equation 2.2

$$\text{Prevalence rate} = \frac{\text{Number of cases}}{\text{Population at risk}} \times 10,000$$

There is a close, and often confusing, relationship between incidence and prevalence, and the extent to which the disease being studied is endemic. The latter may well give a prevalence rate of 90 cases per 10,000 of the population. This could be regarded as the baseline rate, and is crucial in public health, because it gives a starting point. Thus if the incidence rate is prospectively assessed, and found to be 125 cases per 10,000 per month (i.e. a 30% increase), it could be argued that this represents a fresh epidemic, which must therefore have a cause. It also follows that these studies must be large enough to be able to exclude those with existing disease and still provide robust data. A large sample size is also likely to overcome bias and gives greater confidence that any result is genuine.

2.5.4 Standard mortality rates

Disease is of course to be avoided, but most of us would prefer to avoid death. Accordingly, death rates are important, and in many cases 'death' can be substituted for 'disease', which can be seen as a prevalence rate without the time period. There were 585,484 deaths in England and Wales in 2021, and assuming a population of 60 million gives us an annual death rate of 97.6 per 10,000. The crude death rate brings us immediately to a very important metric—the standardized mortality rate (SMR) (Equation 2.3), generally taken to be per year.

Equation 2.3

$$\text{Standardized mortality rate} = \frac{\text{Number of deaths in the population}}{\text{Number in the population}}$$

The SMR is important as it allows us to track death rates over time and by other factors, such as decade of life and geography. Table 2.3 shows data from the ONS for 2015, with the total number of people, of males, and of females, in England and Wales, and the related SMR per 1,000 of the

TABLE 2.3 **The SMR by 5-year blocks in England and Wales in 2015**

Age group	Total population	Males	Females	SMR all subjects	SMR males	SMR females
1–4	2,750,423	1,409,084	1,341,340	0.14	0.16	0.13
5–9	3,346,252	1,712,653	1,633,599	0.08	0.09	0.07
10–14	3,003,377	1,537,159	1,466,218	0.09	0.10	0.08
15–19	3,235,761	1,662,164	1,573,597	0.24	0.31	0.16
20–4	3,634,161	1,851,251	1,782,910	0.33	0.47	0.19
25–9	3,758,172	1,886,352	1,871,821	0.45	0.62	0.28
30–4	3,707,645	1,843,946	1,863,699	0.63	0.82	0.45
35–9	3,451,597	1,717,430	1,734,168	1.01	1.27	0.75
40–4	3,638,122	1,801,171	1,836,951	1.38	1.72	1.05
45–9	3,918,536	1,931,347	1,987,190	2.16	2.68	1.66
50–4	3,863,188	1,905,596	1,957,592	3.30	3.95	2.67
55–9	3,343,049	1,650,611	1,692,438	5.17	6.32	4.05
60–4	2,963,155	1,450,912	1,512,242	8.11	9.88	6.42
65–9	3,058,665	1,485,037	1,573,628	11.95	14.53	9.52
70–4	2,306,033	1,096,990	1,209,043	22.49	27.44	17.99
75–9	1,829,526	839,298	990,227	35.17	42.33	29.10
80–4	1,339,965	574,481	765,484	63.93	76.85	54.23
85–9	820,468	310,302	510,166	119.46	141.82	105.85
90–4	596,890	225,744	371,146	128.33	125.42	130.10

SMR presented as per 1,000 of the population.

© Office for National Statistics. Source: https://www.ons.gov.uk/. Licensed under the Open Government Licence 3.0.

population. Unsurprisingly, the SMR rises consistently with age, and the SMR of males exceeds that of females, until age 90–4. Because of the special conditions associated with the first year of life, data collection is a little different, just as data on those aged 95 or more are merged, so no clear 5-year grouping can be determined.

The ONS data can also be presented over time. Data in Table 2.4, showing SMRs for the years 2014–17, indicate that in almost all cases, the SMR of females is less than that of their age-matched male counterparts, and that, broadly speaking, there is no clear pattern or trend in these data. There is, of course, a huge temptation to try to find something interesting in these data (which, without a firm hypothesis, should be resisted) such as the 50% reduction in the SMR for those aged 1–4 years from 2014 to 2017. But this must be countered by the fact that these groups are comprised of very small numbers, which can very easily lead to bias. In fact, the median difference in all SMRs over all age ranges is precisely zero, as eventually everybody dies.

TABLE 2.4 SMR in England and Wales 2014–17

Age	Males				Females				All			
	2014	2015	2016	2017	2014	2015	2016	2017	2014	2015	2016	2017
< 1	4.1	4.4	4.2	4.4	3.6	3.3	3.6	3.6	7.7	7.7	7.8	8.0
1–4	0.2	0.2	0.2	0.1	0.1	0.1	0.2	0.1	0.3	0.3	0.4	0.2
5–9	0.1	0.1	0.1	0.1	0.1	0.1	0.1	0.1	0.2	0.2	0.2	0.2
10–14	0.1	0.1	0.1	0.1	0.1	0.1	0.1	0.1	0.2	0.2	0.3	0.2
15–19	0.3	0.3	0.3	0.3	0.2	0.2	0.2	0.2	0.5	0.5	0.5	0.5
20–4	0.5	0.5	0.5	0.5	0.2	0.2	0.2	0.2	0.7	0.7	0.7	0.7
25–9	0.6	0.6	0.6	0.6	0.3	0.3	0.3	0.3	0.9	0.9	0.9	0.9
30–4	0.8	0.8	0.8	0.8	0.4	0.4	0.5	0.4	1.2	1.2	1.3	1.2
35–9	1.1	1.1	1.1	1.1	0.7	0.7	0.6	0.7	1.8	1.8	1.7	1.8
40–4	1.7	1.7	1.8	1.7	1.1	1.0	1.1	1.1	2.8	2.7	2.9	2.8
45–9	2.5	2.5	2.5	2.6	1.6	1.6	1.6	1.6	4.1	4.1	4.1	4.2
50–4	3.6	3.7	3.7	3.7	2.4	2.5	2.5	2.5	6.0	6.2	6.2	6.2
55–9	5.8	5.8	5.8	5.7	3.8	3.8	4.0	3.7	9.6	9.6	9.8	9.4
604	9.4	9.5	9.4	9.2	6.1	6.1	6.2	6.0	15.5	15.6	15.6	15.2
65–9	14.2	14.3	14.5	14.5	9.2	9.5	9.6	9.4	23.4	23.8	24.1	23.9
70–4	23.7	23.6	23.3	22.7	15.8	15.9	15.7	15.1	39.5	39.5	39.0	37.8
75–9	39.6	40.6	40.4	39.8	27.2	27.9	27.7	27.7	66.8	68.5	68.1	67.5
80–4	70.2	72.3	69.7	69.4	51.2	53.8	50.9	50.6	121.4	126.1	120.6	120.0
85–9	124.7	129.4	124.3	124.1	95.4	102.0	97.5	97.5	220.1	231.4	221.8	221.6
≥ 90	227.4	244.4	233.0	237.0	199.0	219.3	207.1	214.1	426.4	463.7	440.1	451.1

TABLE 2.5 Age distributions in selected countries

Age band (years)	Sweden	UK	Isle of Man	Brunei	Costa Rica
0–14	17.1	17.6	16.3	22.4	22.1
15–24	10.8	11.5	11.0	16.1	15.2
25–54	39.0	39.7	37.8	47.2	44.0
55–64	11.9	12.7	13.8	8.3	10.0
65 and older	20.6	18.5	21.1	5.9	8.8
Median age	41.1	40.6	44.6	38.7	32.6

Age band data are % of the population.

Data source: https://www.cia.gov/the-world-factbook/.

Key points

Major data points in epidemiology are mortality rates; a representation of the latter is the standardized mortality rate (SMR). Researchers need to be aware of possible problems with data collection, analysis, and interpretation, such as that correlation does not imply causation.

Even though raw data are robust, danger still lies in the interpretation. On the Isle of Man, in 2017, the death rate was 10.2/1,000 people, markedly greater than in the UK as a whole, with a rate of 9.34/1,000, a relative increase of 8.5%. This does not imply that the Isle of Man is a dangerous place to be, or that the people who live there carry more disease. A close examination of the demographics of the island shows that this rate can be almost completely accounted for by the more elderly population on the Isle of Man. This demographic difference may be accounted for by the influx of those retiring to the Isle of Man, and the likelihood of the young emigrating to find employment or for education.

In an international perspective, in 2015 the death rate in Brunei was 3.51/1,000 of the population, and in Costa Rica it was 4.88. These data may be compared to death rates in the UK and Sweden, which are both 9.3/1,000. Can it be that the two developed countries are more dangerous or unhealthy places to live? Of course not: the data are for the entire population, who are far older in the UK and Sweden than in Brunei and Costa Rica. Thus the difference in death rates in both these analyses can be explained almost entirely by demographics (Table 2.5). As the frequency of death increases with age, the developed countries will automatically have an apparently worse death rate.

SELF-CHECK 2.6

In Table 2.5, the percentage of people in each age band in each country fluctuates. Why is this the case?

2.6 Types of epidemiological studies

There are numerous types of epidemiological studies. Some find that they can be classified as being descriptive, aetiological, evaluative, clinical, or focused on health services. Others describe two types: cohort and case-control. However, a common view is that epidemiology studies observe, and record, but do not intervene. The latter is the preserve of the clinical trial, which asks fundamentally different questions, such as whether drug A is better than drug B. Epidemiology can certainly answer the same question, but such data will be less robust. Conversely, epidemiology can answer questions on a much larger number of issues than clinical trials can.

2.6.1 Cohort studies

A cohort is a single large group of individuals in whom observations are made, and conclusions are drawn, such as that of Snow and his observations on cholera and drinking water. If large enough, cohort studies can identify links between different factors, such as obesity and endometriosis, but such a study cannot define causation—does obesity cause endometriosis or does endometriosis cause obesity?

A more comprehensive cohort study proceeds over a defined time period and the same observations are made and differences analysed. Accordingly, the study may be described as 'follow-up', 'prospective', or 'longitudinal'. For example, to answer the cause/effect question of endometriosis and obesity, a sufficiently large group of women free of both conditions would need to be studied over a long period of time to see which of these conditions was the first to develop. Problems may arise in that some participants may choose to end their participation in the study, may die, or may move away and so be 'lost to follow-up'.

Based on the initial observations, it may be that a cohort can be subdivided so that additional questions may be answered. The authors of one of the earliest cohort studies (Doll & Bradford Hill, 1956) reported their investigation of the smoking habits of British doctors, focusing on lung cancer. Two years previously, they had sent a questionnaire to 59,600 men and women on the medical register. Fifty-three months later, of the 34,494 men who replied, there were 1,714 deaths. Analysis showed a clear dose–response effect of the amount smoked for lung cancer and coronary thrombosis. Taking the trend of grammes of tobacco smoked daily as 0 (i.e. non-smokers), 1–14 grams, 15–25 grams, and 25+ grams, then the SMR due to lung cancer per 1,000 men in these four groups is 0.07, 0.47, 0.86, and 1.66 respectively—a clear trend. For coronary thrombosis, the data are 4.22, 4.64, 4.60, and 5.99—not as clear a trend but still very significant. This study effectively ended any discussion of whether cigarette smoking was linked to lung cancer, and prompted additional studies on the links between cigarette smoking and myocardial infarction.

A further example is the Whitehall Study, reporting data on thousands of civil servants. One sub-study comprised 17,530 men, in whom, after ~7.5 years, there were 1,086 deaths (0.83% per year), and of these 462 (0.35% per year) were described as coronary heart disease (CHD). Analysis according to grade of occupation shows clear trends in CHD deaths across three age bands and across four occupation bands. Part of these trends may be accounted for by cigarette smoking, which is more than twice as common in the 'other' grades (60.9%) as in administrators (28.8%), and blood pressure (inversely associated with grade), and physical activity that is undertaken in leisure time (positively linked to grade). There was no clear relationship with body mass index, but (against expectation) plasma cholesterol increased with occupation grade. The raw data from Marmot et al. are available free online via Pub Med.

2.6.2 Case-control studies

Case–control studies are the most common type of clinical research in biomedical science. Generally, researchers ask more precise questions and test predefined hypotheses on a group of subjects in whom they are most interested (the cases). But a second group must also be studied, which provide an alternative feature of the study (the controls). In many studies the cases are a group of patients, and the controls are a group of healthy individuals, but others may compare patients who have disease A (such as rheumatoid arthritis) with other who have disease B (such as osteoarthritis). More expansive studies may compare two disease groups with each other and with a healthy control group. But regardless of design, an absolute requirement in case–control studies is that the groups are matched for certain factors likely to have a bearing on the results, such as age and sex, or that these are at least addressed. This principle is applicable in studies of, for example, lung cancer or bronchitis, but also applies widely in other conditions, such as matching for menopausal status and parity in studies of reproductive or breast disease. Failure to match for these confounders may lead to bias and so a false positive (finding a link where there should be none) or a false negative result (failing to find a link where one should be present). Cohort studies may also have a case–control aspect. In the study on the smoking habits of British doctors, for instance, the non-smokers are effectively the healthy control group against which the smokers are compared.

An example of a case–control study is that of Mousa and colleagues, who recruited 400 women in their first pregnancy from a single centre, 200 of whom (the cases) had non-alcoholic fatty liver disease (NAFLD), whilst the remainder were free of the disease (and so were the controls). They reported that the women with NAFLD were more likely to have gestational diabetes mellitus, pre-eclampsia, and hypertension, but an equivalent frequency of preterm delivery and foetal growth retardation compared to the healthy women. They also found that the cases had higher levels of plasma glucose, serum total cholesterol, and serum uric acid, but equivalent levels of haemoglobin and serum albumin (Table 2.6).

These results are observational, and so do not tell us of cause and effect, but they may indicate. NAFLD was determined, and blood pressure and the blood tests were taken during the first trimester of the pregnancy, the latter defining hypertension and gestational diabetes, whereas pre-eclampsia, pre-term delivery, emesis gravidarum, and foetal growth retardation were assessed later in the pregnancy, or at birth. Therefore there is some difficulty in determining cause and effect, as, for example, it cannot be stated that the women with NAFLD had raised uric acid before their pregnancy. Likewise, was raised cholesterol a contributor to NAFLD, or vice versa? The link between hypertension and pre-eclampsia is established, and is very likely to be causal.

SELF-CHECK 2.7

Perform a simple calculation on the data in Table 2.6 to estimate the fold increase in each of the four post-partum observations in the women with NAFLD. Which of these four metrics is greatest?

TABLE 2.6 Non-alcoholic fatty liver disease and pregnancy

	200 women with NAFLD	200 healthy women
Baseline observations		
Age (years)	26.3 (3.8)	25.9 (3.8)
Gestational diabetes (yes/no)	66/134	20/190
Plasma glucose (mmol/dl)	4.9 (0.5)	4/7 (0.4)
Albumin (g/l)	39 (12)	41 (10)
Total cholesterol (mmol/l)	6.3 (2.0)	4.8 (1.4)
Uric acid (mmol/dl)	0.37 (0.13)	0.28 (0.12)
Haemoglobin (g/l)	140 (41)	134 (29)
Post-partum observations		
Pre-eclampsia (yes/no)	50/150	24/176
Pre-term delivery (yes/no)	22/178	16/184
Foetal growth retardation (yes/no)	26/174	20/180
Emesis gravidarum (yes/no)	16/184	7/193

Data are mean with (standard deviation) or number of women.

Adapted from Mousa N, Abdel-Razik A, Shams M, et al. (2018) Impact of non-alcoholic fatty liver disease on pregnancy, British Journal of Biomedical Science 75: 4; 197–9, DOI: 10.1080/09674845.2018.1492205. Published by Frontiers Media SA.

2.6.3 The mathematics of epidemiology (part 2)

It is not enough merely to observe and report on difference. The data must be subject to rigorous statistical testing, and there are many such tests. However, a crucial point is that the correct test must be applied to the data. For example, section 2.1.1 dealt with continuously variable data that have a normal distribution (and so are presented as mean with standard deviation (SD)), whilst section 2.1.2 discussed data with a non-normal distribution (presented as median with interquartile range (IQR)). Different tests and hypotheses were also introduced.

The t test

This test relies on the difference between the means of two sets of data, their SD, and the sample sizes. The continuously variable data in Table 2.6 (age, glucose, albumin, cholesterol, uric acid, and haemoglobin), all of which have a normal distribution, are compared using a t test. Levels of cholesterol are some 30% higher in NAFLD (with means of 6.3 vs 4.8), and in uric acid (0.37 vs 0.28) at 32%. In both cases, the result of the t test is p = 0.001, which tells us that we can be 99.9% sure that both these differences are genuine.

The χ^2 (chi-square) test

The parallel test for data of a categorical nature is the χ^2 (chi-square) test, which compares the data that we have observed with those we would expect to find should there be no difference, and (again) takes the sample sizes into account. Once more we can use the data from Mousa et al in Table 2.6 to illustrate this test, and to help it is convenient to lay out the data in what is called a 2 × 2 table, as shown in Table 2.7. The rate of pre-eclampsia in the NAFLD group is 50/200, = 25%, but in the controls it is 24/200, = 12%, which points to a 2.1-fold increase due to pre-eclampsia. Application of the χ^2 test to this data gives us p = 0.001, very much smaller that the cut-off of p < 0.05, which tells us that we can be 99.9% confident that the difference is real.

But what about the data on emesis gravidarum (the clinical term for morning sickness)? Here, the difference of 2.3 is greater (16 vs 7), so does that mean this is *more* significant? The χ^2 test gives p = 0.053, telling us the difference is only just not-significant, in that we are only 94.7% sure the difference is genuine. The reason of this discrepancy is that the number of women is far smaller (16 vs 7) than in the pre-eclampsia comparison (50 vs 24 women). This brings us to the concept of confidence.

Confidence

This metric has two components: the degree of difference and the number of people or samples that have been analysed (the sample size), and the bigger the sample size, the greater the statistical power. If the difference is large, we don't need a large sample size to give us sufficient confidence that the difference is genuine. But if the difference is small, we need many more data points for the required degree of confidence. We often cite a 'confidence interval' (CI), that gives us an idea of how sure we

TABLE 2.7 Mousa et al.'s data in a 2 × 2 table

	NAFLD	Healthy controls
Pre-eclampsia	50	24
Normal pregnancy	150	176

NAFLD = non-alcoholic fatty liver disease. Data are number of women in each group.

Adapted from Mousa N, Abdel-Razik A, Shams M, et al. (2018) Impact of non-alcoholic fatty liver disease on pregnancy, British Journal of Biomedical Science, 75: 4; 197–9, DOI: 10.1080/09674845.2018.1492205. Published by Frontiers Media SA.

are of the validity of a set of data. We have already met this metric, albeit briefly, in sections 2.1.1 and 2.1.2 with their discussion of the 95% CI of the mean or median.

Suppose we measure the height of seven men, and find the average to be 1.73 m—how confident can we be that this height is accurate? Statistical confidence follows directly from the size of the study. With a sample size of only seven men, the confidence that the mean result of 1.73 m is a good reflection of the true height of these seven men is low—the true result may lie anywhere between 1.67 and 1.81 m. Statisticians have a consensus on how to present these data, and take a level of confidence of 95% as a standard, so this range (1.67–1.81) is the 95% CI. The more men in the study, the more accurate the mean will be, and so the smaller the 95% CI. So with 20 men, we have more confidence that our result is true, but the real result may still vary between 1.70 and 1.76 m. By extending the sample size to 100 men, we would be much more confident that 1.73 m is indeed a true estimate, but there is still always a range, which in this case may be 1.72 to 1.74 m.

An oft-cited metric is the mean $\pm$ SD, and the mean $\pm$ 2 SDs. In a continuously variable data set with a normal distribution, the mean $\pm$ 2 SDs includes 95% of all the data points. This is *not* the same as a 95% CI. The SD is an intrinsic quality of the data set that does not vary with the number of data points (that is, the sample size), whereas the 95% CI is strongly influenced by the number of data points.

Returning to Mousa et al.'s data on pre-eclampsia and NAFLD, analysis of the data in Table 2.7 tells us that there is a large and statistically significant difference among the frequencies in the four boxes. But what is the likelihood, for an individual woman with NAFLD, that she is at risk of pre-eclampsia? This can be calculated in terms of the relative risk when compared to women free of NAFLD.

Relative risk

Data is presented in a similar but extended form of that in Table 2.7, as shown in Table 2.8, where the participants are classified into one of four groups (hence 2 × 2). The relative risk (which may be described as the risk ratio) can be calculated as in Equation 2.4.

Equation 2.4

$$\text{Relative risk} = \frac{A/(A+C)}{B/(B+D)}$$

Substituting the data in Table 2.7 into Equation 2.4, we get (50/(50+150))/(24/(24+176)), which becomes (50/200)/(24/200) and so 0.25/0.12 and finally 2.08 (to three significant figures). Thus the relative risk of pre-eclampsia in NAFLD is slightly over twice that in women free of NAFLD. A problem with this analysis, which should be addressed in prospective studies, is that it cannot define absolute risk. Since (in this setting) the risk of pre-eclampsia in a healthy pregnancy is 0.12, and the risk in NAFLD is 0.25, it follows that the absolute risk increase (also described as the risk difference, or attributable risk) is 0.25 minus 0.12, that is, 0.13. The same principle applies to the decrease in the relative risk, with appropriate changes to the various components of the equations. Although relative risk is a popular and inherently simple concept, it has issues, and accordingly, some prefer to use the odds ratio.

TABLE 2.8 Form of data presentation in a case–control study

	With the disease, or exposed to a risk factor	Free of the disease, or not exposed to a risk factor	Total
Cases/patients (where the outcome is present)	A	B	A+B
Controls (where the outcome is present)	C	D	C+D
Total	A+C	B+D	A+B+C+D

The odds ratio

A question that arises is whether the number of cases and controls do indeed fairly represent a large sample of all such cases, such as the population as a whole. Another is the time scale of the observation, whether the study is prospective or retrospective. An alternative to relative risk is the odds ratio, which can address some of these issues. The odds ratio is the odds (or likelihood) of something happening in one population compared to the odds of it happening in a different (possibly a control) population. We still bring our observations into a 2 × 2 table as in Table 2.7, but the arithmetic is simpler, as in Equation 2.5. So for the pre-eclampsia data in NAFLD, the odds ratio becomes 50 (= A) × 176 (= D)/24 (= B) × 150 (= C) = 8,800/3,600 = 2.44, somewhat larger than the relative risk. This approach is particularly useful if the disease in question is rare.

Equation 2.5

$$\text{Odds ratio} = \frac{AD}{BC}$$

Another way of looking at these data is that, of 200 women with NAFLD, 50 had pre-eclampsia and 150 did not. So the odds of getting pre-eclampsia are 50/150, = 0.333. Similarly, of 200 healthy women, 24 had pre-eclampsia whereas 176 did not, giving the odds for getting pre-eclampsia in this group as 24/176 = 0.136. Dividing these two odds (0.333 by 0.136) to derive a ratio gives us 2.45, almost exactly the same as in Equation 2.5.

Although crude odds ratios may be determined on a hand-held calculator (as can the various risk metrics), how confident are we that the odds ratio of 2.08 is accurate and statistically significant? Calculations are more often performed on a computer with some statistical software, which is generally automatically programmed to give the 95% CI as well as the odds ratio (which abbreviates as OR), perhaps as OR 2.08 (95% CI 1.55–2.95). The fact that both parts of the 95% CI are > 1 tells us the significance is $p < 0.05$.

SELF-CHECK 2.8

Shortly after a lead-smelting foundry began to work on a new type of ore, 56 foundry workers out of 201 complained of symptoms of a cough and/or a painful chest. The owners contacted a public health consultant, who found that of 85 administrators and drivers, 13 made the same complaint. From this data, construct a 2 × 2 table, and use it to estimate (a) the risk of the symptoms in each group, (b) the relative risk of a cough of chest pain in a foundry worker, and (c) the linked odds ratio.

2.6.4 Survival analysis

Survival analysis focuses on the time taken for a particular event to occur. Although described in terms of survival in a life-threatening situation, this does not need be death: it may be any precise and exact health-related event. In cardiovascular disease, the event may be a heart attack or a stroke; in transplantation, it may be the rejection of an organ. The endpoint of a particular study may be the proportion of subjects having or surviving an event by a certain time, or perhaps the length of time that a patient can survive whilst being free of the event.

In clinical research, survival analysis may well be called upon to test the efficacy of a new drug or drug regime, or a surgical procedure, when compared with the existing drug regime/surgery (often called standard care). At the onset of a study, subjects (often patients) may be randomized to one of two or more different treatments. As the study progresses, the patients are regularly reviewed (perhaps every 3 months) to determine which of them has suffered an endpoint. At the conclusion of the study, the outcomes of patients on the different treatments are compared, and conclusions are drawn.

As one of the outcomes is an exact endpoint (it is either present or not) there are similarities with binary logistic analysis. One of these is that the calculation of the risk of an individual suffering an endpoint is allied to that of the odds ratio. In survival analysis, however, this risk is called a **hazard ratio**.

Clearly, if we describe something as a hazard, it must be worth avoiding, and the factor worth avoiding is often an adverse endpoint such as death. However, an endpoint may also be another aspect of healthcare, such as being discharged from hospital a given number of days after surgery. The key aspect about a hazard ratio is that the numbers of endpoints in two or more different groups are collected over a set period of time, such as 2 years. Although the rate at which hazardous endpoints develop is reasonably easy to determine (e.g. 54 events from 463 = 11.7%), this does not account for the time period over which the data are collected. So if the observation period is, say, 700 days, then the rate becomes 11.7/700 = 0.0167% per day, which converts to 6.1% per year. Compare this rate with that for another group (perhaps with a different disease or being treated in a different way) and you have the hazard ratio, which effectively compares the number of events with that expected from a null hypothesis (that there is no difference between the two rates).

However, this explanation is too simple, as the correct analysis of survival data is more complex and beyond the means of a hand-held calculator. It demands certain statistical software (such as that of Cox, hence Cox regression analysis) that takes account of the length of time for each endpoint to appear, and, if they are provided, other features such as the age of the subjects. The same software will also provide the 95% CI of the ratio.

A worked example

Let us consider the discovery of a new molecule in the blood, substance X. Preliminary case–control studies reported high serum levels in patients with cancer. But do these high levels predict outcome? A null hypothesis states that levels of this molecule will not differ according to the long-term outcome of patients with this cancer. An alternative hypothesis states that high levels are found in patients who succumb earlier to an adverse event, such as death or the requirement for surgery. Another way of putting this is that the hypothesis tests the proposition that increased levels are a predictor of poor outcome.

Following ethics committee approval, substance X is measured in 300 patients with the particular cancer, and their disease is monitored monthly over the next 2–3 years. This number of patients (300) is the sample size, and is determined by strict mathematics, depending on the hypothesized difference in levels between the two groups, and the variance of the data (generally, their standard deviation).

At the conclusion of the study, the patients are sorted into two groups depending on their initial substance X result. In this case the cut-off point of 0.24 ng/ml would be arrived at by sensitivity, specificity, and positive and negative predictive values (Chapter 1, section 1.3.3), so there will be 150 patients in each group. As each particular patient suffers an endpoint, the date of this event is plotted according to whether the level of substance X was above or below the cut-off point. The graph is called a Kaplan–Meier plot (Figure 2.13), also often referred to as a life table, where the two sets of survival curves are compared by complex mathematics called a log-rank test.

Interpretation

It appears that low levels of Substance X have some kind of protective role in survival. The rate of decline of the upper (black) line (substance X < 0.24) is markedly less acute than that of the lower (blue) line (substance X > 0.24). Note that the two lines are roughly straight: the rate of the development of endpoints is constant throughout the whole 800 days of the study. Overall, endpoints appear at a rate of about 1.75% per 200 days when substance X < 0.24 ng/ml, and at about 4.5% per 200 days when substance X > 0.24 ng/ml. The ratio of the rate at which these hazardous endpoints accrue by virtue of levels of substance X is therefore in the order of 2.6. However, by adjusting for the differences in the times to each endpoint, the final hazard ratio may well be a little different, such as 1.82.

Discussion

The probability of there being a meaningful overall difference between these lines is $p < 0.01$, but at what exact time point does this difference become apparent? By doing individual analyses at regular intervals, perhaps every 200 days, the slow increase in significance can be determined. This is relatively easy to do as long as the sample size is known. If there were 150 patients in each group at baseline, the difference in the accrual of endpoints could be plotted in a number of ways, such as by the χ^2 (chi-square) test (Table 2.9). This indicates that the difference between the two groups became

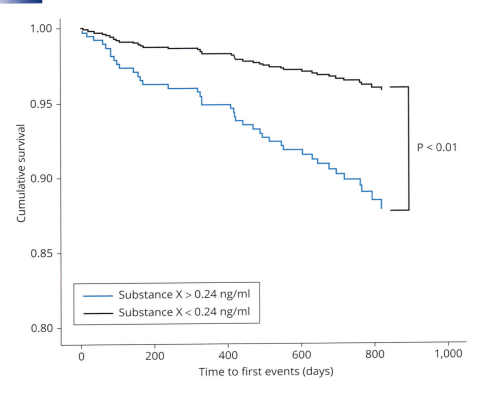

FIGURE 2.13
Kaplan–Meier plot of survival
according to levels of substance X.

From Blann, *Data Handling and
Analysis*, second edition, Oxford
University Press, 2018.

TABLE 2.9 χ^2 **data on survival outcome**

Day	Low levels of substance X		High levels of substance X		p value
	Event	No event	Event	No event	
200	2	148	4	146	0.409
400	3	147	7	143	0.198
600	5	145	12	138	0.080
800	7	143	18	132	0.022

Data are number of patients with or free of an event.

significant at, perhaps, somewhere around days 625–75. However, there are several other ways of seeking a difference between the groups, such as sensitivity, specificity, and odds ratio. One can also (and probably, inappropriately) probe these data for other metrics. If we crudely divide the number of events in those with high levels (4, 7, 12, 18) by the number in those with low levels (2, 3, 5, 7) at days 200, 400, 600, and 800 respectively, we get a fold increase of 2.0, 2.3, 2.4, and 2.6, a clear trend. If we then put these data into Equation 2.4, we get 1.34, 1.42, 1.45, and 1.50—another clear trend but with lower metrics. The same applies to Equation 2.5, which gives 2.0, 2.4, 2.5, and 2.8 respectively. This again underlines the importance of justifying and using the correct test for the type of data. Overall, these data may lead to a different type of treatment in those with high levels of substance X compared with those with a low level.

TABLE 2.10 Meta-analysis

Study	Sample size (cases/controls)	Odds ratio (95% CI)	p value
A	79/84	1.26 (0.93–1.80)	0.075
B	156/145	1.31 (0.89 –2.16)	0.081
C	226/210	1.22 (0.95–2.01)	0.062
A+B+C	461/439	1.25 (1.05–1.90)	0.043

Assumptions and caveats

The analysis makes many assumptions, the main one being that there are no major differences in standard clinical, demographic, and laboratory indices between those with a high or low level of substance X at baseline. These indices include age, sex, smoking, the stage of the disease, and many other factors that vary according to the nature of the disease. For example, in a study of cardiovascular disease, all groups would also need to be matched for factors such as hypertension, diabetes, and serum lipoproteins. Accordingly, there must be data which show that substance X does not vary with these factors.

A second assumption is that there are no changes in the treatment of the cancer (probably surgery, radiotherapy, and chemotherapy) in the two groups over the 800 days. In an ideal world, all those who started the study will remain contactable for the entire 800 days. Regrettably, some patients may be lost to follow-up or decline to take further part, and it is conceivable that this subgroup may be different from those who complete the study, leading to difficulties in interpretation.

2.6.5 Meta-analysis

Despite having a large number of people in a particular study, results often fail to reach the required $p < 0.05$ level of significance, and a value such as $p = 0.06$ suggests a false negative that may have been significant with a larger sample size. Unfortunately, clinical research may take years to complete, and researchers can't simply go out and recruit additional patients. One way around this is to merge data from individual independent studies (should they exist) into a **meta-analysis**. Suppose workers in three different countries all test the same hypothesis that a certain drug causes lung fibrosis, as defined by lung function tests, and that their design is case–control with patients taking the drug versus patients not taking the drug (Table 2.10).

In this example, studies A, B, and C all (only just) fail to find a significant effect of the drug, suggesting the drug is not linked to fibrosis. However, by pooling the three studies, with the additional power, a significant effect is found. There are serious shortcomings with this approach, the major one being the assumption that the three studies are exactly the same and recruited exactly the same type of patients, etc. Should these studies have been conducted in Lagos, Stockholm, and Singapore, critics may point out the racial disparity, although others may consider this as a strength.

Key points

Epidemiology is the study of disease in large populations and gives us information of a most broad nature. Although very useful, findings from epidemiology do not necessary apply to an individual.

SELF-CHECK 2.9

What are the differences between an odds ratio and a hazard ratio?

2.7 Case studies in epidemiology

We conclude our exploration of the basic concepts of disease with in-depth cases studies.

2.7.1 Cancer

Cancer deaths by age and sex

Tables 2.11, 2.12, and 2.13 show data on selected cancers in England and Wales in 2021. Overall, more men die of cancer than do women (Table 2.11), the difference being ~14%, and of those cancers presented, men seem to be markedly more prone to oesophageal cancer (~138% increase), whilst the increased frequencies in colorectal/anal and pancreatic cancer are 19% and 1.2% respectively.

Breaking these data down by particular cancer reveals age-specific differences (Table 2.12). Taking the incidence at years 35–44 as a reference, unsurprisingly the frequency of all cancers increases with age, although the age gradient for colorectal cancer is less acute than those of oesophageal and pancreatic cancer. Of the five sex-specific cancers (Table 2.12), that of the prostate kills three times as many men as ovarian cancer kills women. Furthermore, the increase in the rate of prostate cancer with age is very marked, whilst cancer of the testis is very much a disease of the middle-aged.

From the public health viewpoint, breast cancer is some three times as important as ovarian cancer, which in turn approaches being five times as deadly as cancer of the cervix.

Changes with time

Table 2.13 shows the frequency of three of the most common cancer causes of death from 2015 to 2021 (intervening years are omitted for clarity). Over the 6 years, the total number of deaths increased by 10.8%, but death due to all malignant cancers increased by -0.4%, implying an increase in deaths

TABLE 2.11 Cancer deaths in 2021

Cancer	Age band	All subjects		Men		Women	
All malignant	All ages	144,501		76,989		67,512	
neoplasms	35–44	2,002	1	849	1	1,153	1
	55–64	18,241	9.1	9,736	11.5	8,505	7.4
	75–84	47,452	23.7	25,950	30.6	21,502	18.6
Colorectal and anal	All ages	15,291		8,320		6,971	
	35–44	338	1	172	1	166	1
	55–64	1,917	5.7	1,153	6.7	764	4.6
	75–84	4,761	14.1	2,647	15.4	1,614	9.7
Oesophageal	All ages	6,870		4,836		2,034	
	35–44	56	1	42	1	14	1
	55–64	1,076	19.2	834	19.9	242	17.3
	75–84	2,239	40.0	1,552	36.9	687	49.1
Pancreatic	All ages	8,829		4,441		4,388	
	35–44	74	1	51	1	23	1
	55–64	1,292	17.4	730	14.3	563	24.4
	75–84	2,956	39.9	1,491	29.2	1,465	63.7

Data are numbers of deaths with fold increase in deaths in the older groups compared with the younger.
© Office for National Statistics. Source: https://www.ons.gov.uk/. Licensed under the Open Government Licence 3.0.

TABLE 2.12 Sex-specific cancer deaths in 2019

	Age band	Incidence	
Breast*	All ages	9,592	
	35–44	355	1
	55–64	1,556	4.4
	75–84	2,442	6.8
Cervix	All ages	753	
	35–44	101	1
	55–64	138	1.4
	75–84	106	1.05
Ovary	All ages	3,324	
	35–44	42	1
	55–64	555	13.2
	75–84	1,091	26.0
Prostate	All ages	10,359	
	35–44	2	1
	55–64	523	261.5
	75–84	3,742	1,871.0
Testis	All ages	65	
	35–44	10	1
	55–64	13	1.3
	75–84	11	1.1

* Women only, who comprise 99.3% of all cases. Data are numbers of deaths with fold increase in deaths in the older groups compared with the younger.

© Office for National Statistics. Source: https://www.ons.gov.uk/. Licensed under the Open Government Licence 3.0.

due to other causes. Over the same time period, prostate cancer deaths fell by around 2%, whilst breast cancer deaths in women fell by 5.9%. Trachea, bronchus, and lung cancer fell by 7.6%, but the number of deaths due to colorectal cancer increased by 6.1%.

Death is inevitable, so the increase in cancer deaths may be because people have avoided deaths by other routes, such as respiratory or cardiovascular disease. Indeed, deaths from circulatory and respiratory disease compared to total deaths in men and women fell in a direct linear manner from 2013–19. The total deaths due to cancer over this period fell by 1%, but those of circulatory deaths fell by 12.3% and of respiratory disease by 6.8%. A likely reason for both these reductions is improved understanding (and avoidance) of the risk factor, and better treatment once the disease has become manifest. Analysis of this nature in 2020 and 2021 is compounded by the effects of the COVID-19 pandemic: in these two years total deaths, cancer deaths, and circulatory deaths all increased, but respiratory deaths fell, as discussed in Chapter 13. Some examples of international perspectives of epidemiology, demonstrating changes in global deaths, are presented in section 1.2.2.

SELF-CHECK 2.10

Comment on the changes in Table 2.13 with respect to aetiology and public health.

TABLE 2.13 Cancer deaths from 2015–21

	2015	2017	Change From 2015	2019	Change from 2017	2021	Change from 2019
All persons							
Deaths	528,507	532,130	+0.7%	529,553	−0.5%	585,484	+10.6
All cancer	147,472	149,354	+1.3%	150,550	+0.8%	148,109	−1.6
Trachea, bronchus, and lung	30,466	30,079	−1.2%	29,418	−2.2%	28,147	−4.3%
Colorectal	14,413	14,873	+3.2%	15,278	+2.7%	15,291	<0.1%
Men							
Prostate	10,579	10,755	+1.7%	10,867	+1.1%	10,359	−4.7%
Women							
Breast	10,191	10,147	−0.4%	10,058	−0.9%	9,592	−4.6%

Data are numbers of deaths with percentage change for the previous data set.

© Office for National Statistics. Source: https://www.ons.gov.uk/. Licensed under the Open Government Licence 3.0.

Key points

Cancer rates vary greatly between the sexes and with age, over decades, and with geography (the latter possibly linked to local risk factors). Since death is inevitable, a fall in rates due to one disease may facilitate a rise due to another disease.

2.7.2 Framingham

In 1948, US researchers at the University of Boston established a healthcare project in an entire community—the small town of Framingham in Massachusetts, New England. It started by simply asking a random sample of 5,209 men and women (out of a population of some 28,000) who had not yet developed overt symptoms of cardiovascular disease or suffered a heart attack or stroke some basic questions about their health, with a focus on the heart.

Over the years more questions were added, which also became more complicated, other investigations were added, and blood was taken for assorted blood tests. Eventually, the weight of information gathered justified analysis, which produced evidence unequivocally confirming the general theory of atherosclerosis that had been developing over the previous 150 years. However, it also produced new perspectives on this disease. Framingham is now rightly recognized as perhaps the best-known epidemiological study of a single population.

The study has been running for such a long time that the children of the original cohort are now adults and so represent a unique opportunity to see how the health of parents impacts upon that of their children. The Framingham Offspring Study, initiated in 1971, now has over 5,000 participants. This process is continuing with the third-generation study, which will ask whether or not grandparents' health has any impact on that of their grandchildren. With advances in technology over 50 years, such as DNA profiling, the whole Framingham project has considerable and unique power. Key results from the Framingham Heart Study include:

The 1960s

- Cigarette smoking, increased cholesterol, high blood pressure, obesity, and abnormalities on an electrocardiogram are all found to increase the risk of heart disease.

- Physical activity reduces the risk of heart disease.

The 1970s

- High blood pressure increases the risk of stroke.

- Atrial fibrillation increases the risk of stroke five-fold.

- Menopause increases the risk of heart disease.

- Psychosocial factors affect heart disease.

The 1980s

- High levels of HDL cholesterol protect against heart disease.

The 1990s

- Left ventricular hypertrophy increases the risk of stroke.

- Hypertension leads to heart failure.

- A simple algorithm for predicting coronary disease involving risk factor categories is developed to predict coronary heart disease risk in patients without overt disease of this nature.

- The lifetime risk at age 40 years of developing heart disease is one in two for men and one in three for women.

The 2000s

- High-normal blood pressure is found to increase risk of cardiovascular disease.

- The blood test serum aldosterone predicts future risk of hypertension in non-hypertensive individuals.

- The lifetime risk of becoming overweight is > 70%, that for obesity ~ 50%.

- Social networking appears to be relevant in obesity, but also in helping to stop smoking.

The 2010s

- Obstructive sleep apnoea (snoring) increases the risk of stroke.

- Having a first-degree relative with atrial fibrillation increases the risk for this disorder.

- The occurrence of stroke by age 65 years in a parent increases the stroke risk in offspring three-fold.

The Framingham study also found numerous fascinating health issues not related to the heart, such as that genes may play a role in Alzheimer's disease, that parental dementia may lead to poor memory in middle-aged adults, and genes that link puberty timing and body fat in women. Researchers associated with the Framingham Heart Study have contributed to the discovery of hundreds of new genes underlying the major risk factors of heart disease.

Framingham will quite possibly be best known for generating a number of risk factor equations, generally based on age, diabetes, smoking, treated and untreated systolic blood pressure, total cholesterol, HDL cholesterol, with BMI replacing lipids in a simpler model. This model has spawned numerous risk calculators, such as QRISK, recommended by certain publications of the National Institute for Health and Care Excellence (NICE) (see Chapter 8). Equations of this type are freely available online.

Although Framingham has in itself generated a host of powerful findings, it is not perfect in that its conclusions may not necessarily be applied broadly. For example, the socio-economic, genetic, and ethnic homogeneity of its subjects means that the conclusions may be applicable only to relatively affluent communities that are predominantly of a White European background. This is especially true of its risk factor formulae, which need to be independently validated if used in other groups.

> **Key points**
>
> Epidemiology can be used to pinpoint a problem in a community, suggest a solution, and monitor the effect of the intervention.

2.7.3 Diabetes and heart disease in thalassaemia

Thalassaemia is a genetic disease of red blood cells that in its most severe form produces such a profound anaemia that regular blood transfusions are called for. This leads to the build-up of excess iron in various organs, such as the heart and pituitary, causing additional disease (as discussed in Chapter 11). Researchers tested the hypothesis that diabetes is associated with a higher risk of heart complications in patients with thalassaemia major. A sub-hypothesis was that the extent of damage was related to the degree of myocardial iron overload. The hypothesis was tested in a cohort of 957 patients with thalassaemia major, gathered from 68 different thalassaemia centres. Of these patients, 86 had an additional diagnosis of diabetes, allowing a case–control study, selected results of which are shown in Table 2.14.

According to these data, thalassaemic patients with diabetes are more likely to have had their spleens removed, to have underactive gonads, and to be positive for hepatitis C virus RNA (potentially derived from blood transfusions). One could therefore say that these are all risk factors for the presence of diabetes, but it cannot be said that they predict the development of diabetes. Now consider the frequencies for splenectomy: 81.4% in diabetes and 54.6% in non-diabetes, which is highly significant (p = 0.001). In fact, the relative rate of splenectomy is 49% higher in diabetes when compared to non-diabetes. The paradox of a smaller relative difference of splenectomy (49%) being significant where a larger relative difference in hypertension (72%) is not significant can be explained in terms of the sample size and power—there are far fewer subjects with hypertension than who have had their spleens removed.

The researchers subsequently evaluated the impact of diabetes on heart disease, and the extent to which it was related to the burden of iron within the heart, which was assessed as myocardial iron overload (Table 2.15).

TABLE 2.14 Comparison of selected demographic and clinical data in thalassaemia patients with and without diabetes

	Patients with diabetes (n = 86)	Patients free of diabetes (n = 709)	Relative difference	p value
Sex (male/female) (%)	40.7	49.4	1.21	0.129
Splenectomy (%)	81.4	54.6	1.49	0.001
Hypertension (%)	3.8	2.2	1.73	0.416
Hypogonadism (%)	41.9	19.5	2.14	< 0.001
Hepatitis C virus positive (%)	54.4	40.0	1.36	0.01
Myocardial iron overload (%)	66.3	53.6	1.24	0.026

Data are mean (SD) or numbers of patients (%).

Pepe A, Meloni A, Rossi G, et al. (2013), Cardiac complications and diabetes in thalassaemia major: a large historical multicentre study. Br J Haematol, 163: 520–7. https://doi.org/10.1111/bjh.12557.

TABLE 2.15 Analysis of diabetes in the cardiac disease of thalassaemic patients

	Adjusted for non-myocardial iron overload		Adjusted for non-myocardial iron overload, age, and other endocrine disease	
	Odds ratio (95% CI)	p value	Odds ratio (95% CI)	p value
Cardiac complications	4.23 (2.65–6.76)	<0.001	2.84 (1.71–4.69)	<0.001
Arrhythmias	4.09 (2.16–7.74)	<0.001	2.21 (1.12–4.37)	0.023
Heart failure	3.14 (1.87–5.26)	<0.001	2.33 (1.33–4.06)	0.003
Myocardial fibrosis	2.12 (1.24–3.63)	0.006	1.91 (1.11–3.29)	0.021
Right ventricular dysfunction	1.82 (1.01–3.30)	0.048	1.33 (0.71– 2.49)	0.366

Pepe A, Meloni A, Rossi G, et al (2013), Cardiac complications and diabetes in thalassaemia major: a large historical multicentre study. Br J Haematol, 163: 520–7. https://doi.org/10.1111/bjh.12557.

The data in the middle columns indicate that all five types of heart disease are independently related to diabetes once adjusted for non-myocardial iron overload. The most powerful findings are that diabetes is most strongly related to cardiac complications and to arrhythmias. Both of these have an odds ratio of over 4, which can be translated as both carrying over a four-fold risk (over 400%). The probability of the presence of heart failure and myocardial fibrosis is less (over three-fold (actually 314%) and over two-fold (actually 212%) respectively), whilst the presence of diabetes increases the risk of having right ventricular dysfunction by 'only' about 82%.

However, many of these different types of heart disease are likely to share the same pathophysiology, and may also be related to age and other endocrine disease. The right side of Table 2.15 shows analyses that have been adjusted for these confounders (also called covariates). Although cardiac complications and arrhythmias are still related to diabetes, this risk has fallen to between two- and three-fold. The risk for the other types of heart disease has also fallen, and in the case of right ventricular dysfunction this risk has become insignificant. This finding implies that some of the risk of the different types of heart disease was linked to the covariates, and in the case of right ventricular dysfunction this link is so strong that it can no longer be considered as an independent link with diabetes.

2.7.4 Haematological malignancy and racial/ethnic groups

The frequency of blood cancer, principally of white blood cells (leukaemia, lymphoma, and myeloma) varies markedly throughout the world, possibly as a result of racial, ethnic, and geographical factors. The latter can be controlled for by studying these diseases in a single location, therefore leaving only racial (genetic) and ethnic factors. Researchers tested the hypothesis that there are differences in the frequencies of the various white blood cell cancers, by probing a database of cancer registrations in England collected over the years 2001 to 2007. The self-defined ethnic and racial groups were White, Indian, Pakistani, Bangladeshi, Black African, Black Caribbean, and Chinese. Taking National Census data, the frequencies of these seven groups in England were taken to be 93%, 2.2%, 1.5%, 0.6%, 1%, 1.2%, and 0.5% respectively. The rate of cancers in White people was taken as a reference point to which the other groups were compared.

In broad terms, the analysis for cancer in these groups is relatively simple. For example, assuming no effect of race between the White people and the Chinese people, if there are 930 cancers in the former, then there should be five (0.5%) in the latter. It follows that finding ten cancers in the group of

Chinese people implies a doubling of the rate ratio, which may be statistically and pathogenically significant in that being of Chinese origin may (in this setting alone) be a risk factor for cancer. Conversely, finding only one cancer implies that being of Chinese origin is protective against cancer.

However, as in other analyses, we must consider confidence, and because of multiple testing, the researchers opted for a more demanding CI than usual—that of 99% instead of 95%. If present, this brings a probability that any difference is meaningful at $p < 0.01$, as opposed to the standard $p < 0.05$. Similarly, because of the potential confounding covariates, the raw data were adjusted for age, sex, and income, and presented in terms of 100,000 people. The results of the analysis of cancer of mature B-lymphocytes are shown in Table 2.16.

In interpreting this data, we consider whether the rate ratios are different and if the CIs overlap with 1. For example, the 99% CIs of the rate ratio for these cancers in Indian people are 0.70 (99% CI 0.63–0.78) in men and 0.75 (99% CI 0.64–0.89) in women and are significant as these CIs do not include 1 (unlike, for example, 0.48–1.16).

These data suggest that Indian ethnicity protects against cancer as compared with White ethnicity. Pakistani ethnicity seems to carry no such advantage, whereas only Bangladeshi men had a reduced rate of cancer compared with White men. Being of Black African origin carries an increased risk in both sexes (by 26% in men and 65% in women), whilst being of Chinese origin carries a reduced risk in both cases (by 43% in men and 34% in women). In pooling the three South Asian ethnicities and the two Black ethnicities, relative risks of the cancers differed markedly (higher in the group of Black people, $p < 0.001$).

In many cases, the reasons for these differences are unclear, but some may be related to under-reporting of their disease by certain groups. Alternatively, differences in rates may be due to other disease such as infection with HIV, religious observations (and so lifestyle), increased rates of smoking, and the consumption (or not) of alcohol, high-meat, and high-fat diets by different groups. Some of

TABLE 2.16 Frequency of mature B-lymphocyte cancers according to race/ethnicity

Race/ethnicity	Men			Women		
	Number of cases	Incidence of cancer*	Rate ratio (99% CI)	Number of cases	Incidence of cancer*	Rate ratio (99% CI)
White**	32,330	12.5	1.00 (0.98–1.02)	25,511	8.1	1.00 (0.98–1.02)
Indian	334	7.0	0.70 (0.63–0.78)	254	5.9	0.75 (0.64–0.89)
Pakistani	262	12.5	1.09 (0.93–1.27)	136	7.1	0.96 (0.77–1.20)
Bangladeshi	49	5.8	0.58 (0.40–0.84)	34	6.3	0.74 (0.48–1.16)
Black African	179	15.5	1.26 (1.04–1.53)	155	11.6	1.65 (1.34–2.03)
Black Caribbean	325	11.9	1.01 (0.87–1.16)	263	8.9	1.13 (0.97–1.33)
Chinese	56	7.1	0.57 (0.40–0.80)	49	5.0	0.66 (0.45–0.95)

* Adjusted to 100,000 subjects. CI = confidence interval. ** White people, as the largest population in this study, are taken to be the reference group to which other groups are compared.

Shirley MH, Sayeed S, Barnes I, et al. (2013), Incidence of haematological malignancies by ethnic group in England, 2001–7. Br J Haematol 163: 465–77. https://doi.org/10.1111/bjh.12562.

these may account for the differences between the Indian, Pakistani, and Bangladeshi people, who are effectively the same racial group. The impressive sample size (almost 60,000 people) provides considerable power, and the authors can be very sure of the conclusions.

SELF-CHECK 2.11

In Table 2.16, with reference to the White population (because it is the largest group in the study), determine which men and women in the other ethnic groups are at highest and lowest risk of cancer.

Key points

Intensive study of a small population can be very productive, but extrapolation of small studies to wider populations may be inadvisable.

2.7.5 Lessons from epidemiology

The healthcare system exists to ease pain and disease where it can, but cannot do everything. Epidemiology has taught us much. In the Victorian era, Snow showed the link between polluted water and cholera. More recently, epidemiology contributed to establishing the fact that for men, avoiding the risk factors for atherosclerosis in middle and late middle age will reduce the risk of disease in decades to come, whilst women need to have their breasts checked regularly. For both sexes, avoiding obesity, not smoking, taking exercise, and moderating alcohol intake all bring considerable benefit, but sooner or later one of the major disease processes will strike.

A further aspect of epidemiological and statistical analysis is to determine how much of an effect can be accounted for by all the known variables. For example, risk factor equations are imperfect as there are still deaths in those having a seemingly good profile. This indicates that there are many unknown factors yet to be discovered. Indeed, while we can now look back on our predecessors, the healthcare professionals defining disease during the London plague of the 17th century (Table 2.2) and struggling to find some science behind the deaths of their populations, so our successors will find those unknowns that at present elude us.

The purpose of epidemiology is not so much as simply to record who died of what, but to discover patterns of death (and disease) so that preventative measures can be started that will improve the general health of a nation.

 ## Chapter summary

- Measuring rates of disease and health in populations allows comparisons between them to be made and trends to be detected.

- Data collected are helpful in decision-making on the distribution of health and medical services, as well as suggesting further ideas for investigation in the aetiology of disease and the promotion of health.

- Descriptive, observational, and experimental studies can be used to test hypotheses derived from such data accumulation.

- Statistically significant associations between risk and health condition may indicate a causal relation, but even after allowance is made for possible bias and confounding effects, a degree of uncertainty may remain in attributing actual cause to a particular agent.

- Epidemiology provides information about populations and frequently confirms views derived from smaller studies. Accordingly, this information is often used to determine health policy.

Further reading

- Blann, A. D. *Data handling and Analysis*. Second edition, Oxford University Press, 2018.

- Bradford Hill, A. The environment and disease: association or causation? Proc Royal Soc Med, 1965: 58; 295–300.

- Doll, R. The lessons of life: Keynote address to the nutrition and cancer conference. Cancer Res, 1992: 52 (Suppl.); 2024–29s.

- Doll, R. and Bradford Hill, A. Lung cancer and other causes of death in relation to smoking. Br Med J, 1956: ii; 1072–81.

- Forouzanfar, M. H., Foreman, K. J., Delossantos, A. M., et al. Breast and cervical cancer in 187 countries between 1980 and 2010: a systematic analysis. Lancet, 2011: 378; 1461–84.

- Inglis, T. J. Principia aetiologica: taking causality beyond Koch's postulates. J Med Microbiol, 2007: 56; 1419–22.

- Jefferson, T. and Demichelli, V. Experimental and non-experimental study designs. J Epidemiol Commun Health, 1999: 53; 51–4.

- Marmot, M. G., Rose, G., Shipley, M., and Hamilton, P. J. S. Employment grade and coronary heart disease in British civil servants. J Epidemiol Commun Health, 1978: 32; 244–9.

- Mousa, N, Abdel-Razik, A, Shams, M, et al. Impact of non-alcohol fatty liver disease on pregnancy. Br J Biomed Sci, 2018: 75; 197–9.

- Rothman, K. J. *Epidemiology: An Introduction*. Second edition, Oxford University Press, 2012.

- Rothwell, P. M., Wilson, M., Elwin, C. E., et al. Long-term effect of aspirin on colorectal cancer incidence and mortality: 20-year follow-up of five randomised trials. Lancet, 2010: 376; 1741–50.

- Saracci, R. *Epidemiology: A Very Short Introduction*. Oxford University Press, 2010.

- Sheridan, S., Pignone, M., and Mulrow, C. Framingham-based tools to calculate the global risk of coronary heart disease: a systematic review of tools for clinicians. J Gen Intern Med, 2003: 18; 1039–52.

Useful websites

- Framingham Heart Study: **www.framinghamheartstudy.org**
- Office for National Statistics: **www.ons.gov.uk**

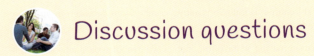

Discussion questions

2.1 How do you explain the changes in the rates of cancer deaths from 2015–21 as described in Table 2.13?

2.2 How does knowledge of the aetiology of disease help diagnosis and management?

3

The aetiology of cancer

Cancer is the leading cause of death in England and Wales, and some 25% of people worldwide suffer this disease, with 900 new cases diagnosed daily in the UK. The UK's Office for National Statistics refers to this group of diseases as '**neoplasms**' (= new growth), reporting 148,109 deaths in 2021 (25.3% of all deaths), with 144,501 (97.6%) being malignant, the remainder being *in situ* and benign neoplasms, and those of uncertain or unknown behaviour. But what these data do not tell us is the degree of morbidity, which will be considerable. In the early stages there may be depression, and in the later stages physical and psychological pain, both of which can be marked. Even if the patient succumbs to another disease, cancer is a huge burden for the patient, the healthcare industry, and society. The purpose of this chapter is to explain how normal tissue develops into a cancer.

Learning objectives

After studying this chapter, you should confidently be able to:

- explain the regulation of cell signalling and why it is important in cancer

- understand the fine details of the control of the cell cycle, and the consequences of loss of control

- describe the roles of chromosomal and gene malformations in the development of cancer

- outline the process of apoptosis

- explain the processes of carcinogenesis and how it drives malignant transformation

- summarize the concepts of clonality, angiogenesis, and metastasis

3.1 Basic concepts in cancer

Cancer is an umbrella term for numerous diseases, often with widely differing presentations, signs and symptoms, aetiology, pathophysiology, morbidity, and mortality. The study of cancer is oncology (from the Greek *onkos*—tumour, volume, or mass), but is also taken to include diagnosis, treatment, and prevention. The common feature of cancer is inappropriate growth, although there are many instances where an inappropriate growth occurs in other diseases—but in those cases it is limited and hardly dangerous. Some define a neoplasm (or tumour, from the Latin for swelling) as an abnormal growth of tissue, a definition that excludes a consideration of the degree of danger, whilst others define neoplasm as a growth with malignant potential and accept the view that some tumours are benign. It could be argued that a cyst (a fluid-filled ball that, in a cancer setting, is inevitably benign) is a tumour, as is a lipoma (an abnormal growth of fatty tissue), or a uterine fibroid (a mass of knotted fibrous tissue), but none are potentially fatal. It is now accepted that a working definition of cancer is an abnormal growth that spreads from its origin in a single cell, tissue, or organ, moving to other sites (the process of **metastasis**) and that ultimately causes serious disease.

But whatever the type of cancer, a key component is the conversion of a normal cell or cells into an abnormal cell or group of cells with uncontrolled and inappropriate growth, the process of transformation. A leading question facing oncologists is why a normal cell should be transformed, and so become abnormal. Here we are guided by Isaac Newton in that all change is caused by something, so a key feature of the early stages of cancer is to determine its cause. Certain key concepts in cancer are as follows:

- Normal cells respond to signals from other cells in a certain way: the biochemistry of these signalling pathways is often aberrant in cancer cells.

- One of the primary reasons for this is that the genes for the molecules in pathways are mutated in such a way as to promote the inappropriate growth of the cell.

- If a normal cell is damaged or otherwise malformed, it undergoes self-destruction by the process of **apoptosis**. Cancer cells have avoided this process, and so continue to flourish.

- The abnormal growth of the cancer cells forms a tumour. Parts of the tumour can escape and become lodged in other tissues. This process of metastasis leads to secondary tumours, and so the spread of the cancer.

In antiquity, cancer was described (and so named) because, like the crab, it too has a central body (the cellular origin of the cancer, that being the primary tumour) with radiating legs (the metastases, sending tumour cells out to other tissues to form secondary tumours). However, some cancers do not metastasize, but remain in their original site, becoming increasingly large so that their mere presence causes other, possible fatal, disease, for instance by pressing on other organs or nerves, or obstructing blood flow. We will now embark on a discussion of these main concepts.

Key points

Although definitions may differ, we will define a cancer as an abnormal growth (a tumour) that becomes malignant, with deleterious consequences for the body.

3.2 Physiological cell signalling

Before embarking on a discussion of those processes that lead to cancer, the key features of the normal function of the cell must be reviewed. Only then can we appreciate the complex nature of the development of this disease. This is not a textbook on cell biology, and so a basic understanding of the structure (nucleus, cytoplasm, organelles, etc.), function (contraction, storage, protein synthesis, etc.), reproduction (DNA, mitosis, meiosis, etc.), and metabolism (ATP, Krebs cycle, etc.) of the cell is expected.

Many cells perform their functions in a form of steady-state dynamics, but there are times when there is a need to respond to external changes, such as an acute inflammation or after a sugar-rich meal. This may include the need to reproduce, as part of growth or as part of repair following damage. The signals received by the cell are therefore likely to modify cellular metabolism, function and movement, and gene expression and development. The changes in the function of the cell follow a strictly regulated pathway with a series of well-defined steps.

3.2.1 Step 1: Receptor/ligand binding

Signals (in effect, instructions) from the external environment are received at the surface of the cell by glycoprotein receptors embedded in the cell membrane. Almost all **receptors** are specific for the chemical nature of the particular signal, technically described as a **ligand** (the primary messenger).

The external factor may be soluble molecules such as hormones, the external components of other cells, or components of a semi-solid extracellular matrix such as collagen or actin. Receptors are complex molecules which span the cell membrane: the extracellular (outside) section binds the ligands, whilst the intracellular (inside) component is often linked to a series of other molecules, such as enzymes and phospholipids. Many soluble ligands are known to be growth factors (such as erythropoietin, vascular endothelial growth factor (VEGF), and epidermal growth factor (EGF)), chemokines (interferon, interleukins), and hormones (insulin, oestrogens, growth hormone). However, there are numerous examples of non-soluble ligands that are part of the cell membrane of other cells that provide activation signals, such as those of the Notch family.

3.2.2 Step 2: Second messengers

Once the primary messenger ligand has bound its receptor, the latter changes its physical conformation such that the intracellular sections of the molecule respond to and/or act upon other molecules within the cytoplasm and/or that are fixed to the internal face of the cell membrane. In turn, this results in a cascade of biochemical reactions, often resulting in changes in the function of the cell (e.g. protein synthesis) or in cell division. The molecules that pass the information from the cell membrane to the nucleus are called **second messengers**. The entire process is called **signal transduction**, and one process can involve several independent second messenger pathways. As these pathways, of which there are hundreds, perhaps thousands, can be very complex, we will focus only on those that are most well-established and relevant to carcinogenesis.

G proteins linked to cAMP

One major class of receptors is characterized by links with G proteins, a large and diverse family of molecules, so named because they hydrolyse guanine triphosphate to guanine diphosphate. Ligands for G protein receptors include epinephrine and glucagon. The functions of one group of G proteins are initiated when ligand–receptor binding activates the membrane enzyme adenylyl cyclase. Signal transduction follows with the formation (from ATP) of the second messenger cyclic adenosine monophosphate (cAMP), which in turn activates the enzyme protein kinase A. However, some G protein-coupled receptors inhibit adenylyl cyclase. Almost all second messenger enzyme are kinases—they catalyse the addition of a phosphate group (inevitably from ATP) to their substrate. A further G protein is Ras, a small enzyme attached to the cell membrane, and an early part of a chain of cytoplasmic second messengers that downstream include molecules such as B-Raf and the mitogen-activated protein (MAP) kinases, ERK (extracellular signal-regulated kinases), and MEK (an acronym derived from MAPK/ERK kinase). We will hear more of these molecules in sections to come.

G proteins linked to lipids

Activation of a second type of G protein results in phospholipase C converting its substrate, the lipid phosphatidyl inositol 4,5-bisphosphate, into two important second messengers—inositol triphosphate (IP$_3$) and diacylglycerol—which diffuse freely into the cytoplasm. Increased levels of IP$_3$ have a number of consequences, one of which is to increase levels of another second messenger, the calcium cation (Ca^{2+}), by release from intracellular stores, from the endoplasmic reticulum via a gated calcium channel, and by import via calcium channel activity at the cell membrane. As many enzymes are calcium-dependent, increased Ca^{2+} results in the activation of numerous pathways in various cells, some being by activation of protein kinase C, which (once activated by diacylglycerol) can then phosphorylate other substrates. Examples of the consequences of this rise in Ca^{2+} include the promotion of glycogenolysis in hepatocytes, and the acetylcholine-induced synthesis of nitric oxide by endothelial cells.

SMADs

These signal transductors molecules (of which there are at least eight) are named as they bear similarities to molecules involved in determining a small body size in the nematode C. *elegans*, and in *Drosophila*. SMAD represents an acronym of 'small' (from small body size) and 'Mothers Against

Decapentaplegic'. Ligands for this pathway include transforming growth factor-β (TGF-β) and bone morphogenetic protein (BMP), which binds to the TGF-β receptor. Once the ligand has bound its receptor, the latter is activated, and this process then allows a serine/threonine kinase to phosphorylate a SMAD molecule. This passes to the nucleus and, with more molecules, binds to the regulator region of a particular gene, over 500 of which are linked to this pathway. SMADs are relevant for cancer as many tumours have DNA changes that result in inactivation in the TGF-β receptor or in one of the many SMAD molecules, leading to abnormal cell growth. One of these mechanisms is an irregularity in the control of the cell cycle (see section 3.3.4).

Phosphorylation

The process of adding a phosphate group to a protein (**phosphorylation**) has already been mentioned, and as it is a common feature of signal transduction, some brief discussion is warranted. Mediated by a kinase, phosphorylation confers on its substrate a number of features, a crucial one being changes that result in it uncoupling from a second molecule to which it is bound, which is then free to perform other functions. Examples include the phosphorylation of tumour suppressor protein Rb in the Rb/E2F complex, so that E2F (a **transcription factor**) is released, and of a molecule named mouse double minute 2 (mdm2) in the mdm2/p53 complex, so that p53 (another **tumour suppressor**) is released. The importance of these two pathways is explained in section 3.4.1, and phosphorylation is more fully described in Figure 3.14. Kinases may be free in the cytoplasm or bound to the intracytoplasmic domain of receptors.

STAT

Another family of intracellular messengers are the signal transductor and activator of transcription (STAT) family, with at least seven members. The best example of a ligand–receptor combination in this pathway is that of erythropoietin and its receptor. The intracytoplasmic domain of the receptors binds the enzyme Janus kinase (abbreviated to JAK), of which at least three types are recognized. JAK is interesting because, once activated by binding its ligand, it self-phosphorylates, which then allows the phosphorylation of a tyrosine residue on one of the seven types of STAT molecule. The phosphorylated STAT molecules dimerize and move to the nucleus where they activate transcription. As we shall see in Chapter 11, this pathway is altered in certain haematological malignancies.

Tyrosine kinase

This important enzyme is found bound to the intracytoplasmic section of a group of 50–60 receptors (Receptor tyrosine kinase, RTK), these being the first step in the regulation of a host of processes, such as cell proliferation and metabolism. The insulin receptor is a member of this family, and therefore has a role in glucose metabolism. Other ligands for members of the RTK family include epidermal growth factor (EGF), fibroblast growth factor (FGF), and VEGF, many of which have their own receptors. In common with erythropoietin, its receptor, and JAK (as in the previous paragraph), occupancy of an RTK by a ligand permits the phosphorylation, and so activation, of tyrosine kinase. Signal transduction pathways linked to this enzyme include the MAP kinase, IP_3, and diacylglycerol second messengers. One type of RTK, the human epidermal growth factor receptor-2 (HER2), is overexpressed in around 25% of cases of breast cancer (see section 4.6.3).

NFκB

Although nuclear factor κB (NFκB) was discovered and so named because of its role in regulating the synthesis of the κ-light chain of immunoglobulins in B lymphocytes, it is important in several signal-transduction pathways. NFκB is formed of two proteins, p50 and p65, of which there are several isoforms, all coded for by genes on different chromosomes. It is constitutively expressed and found in the cytoplasm, bound loosely to an inhibitor, unsurprisingly named inhibitor of κB (IκB, of which there are again several isoforms), that renders it inactive. Once the system is activated, typically by a cytokine binding its receptor at the cell membrane, the key enzyme IκB kinase phosphorylates IκB,

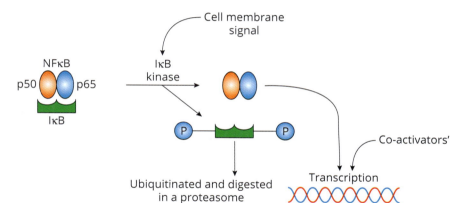

FIGURE 3.1
Activation of NFκB. Once freed from its inhibitor, NFκB migrates to the nucleus to participate in transcription.

which results in its being uncoupled from NFκB, and ubiquitinated (addition of the protein ubiquitin, which marks it for degradation). Once freed from the restraining effects of its inhibitor, NFκB migrates through a nuclear pore and, with coactivators, initiates transcription by binding to response elements of the DNA (Figure 3.1).

SELF-CHECK 3.1

In simple terms, name and briefly describe the key steps by which a cell responds to its environment.

3.2.3 Step 3: Effects on the nucleus

Certain second messengers migrate across the nuclear membrane into the nucleus. One such messenger, protein kinase A, activates (by phosphorylation) a further messenger, the cAMP response element-binding protein (CREB). The latter binds to a section of the DNA that regulates the transcription of several genes, such as those of certain hormones. The end-effector of the Ras–BRAF–MEK pathway, EKR, moves to the nucleus where it phosphorylates other transcription factors. Phosphorylated STAT is a further example of a second messenger actively transported to the nucleus where it binds to and activates certain sections of DNA. The JAK–STAT system part-regulates immune responses and is involved in the control of blood cell production in the bone marrow. As we will see, errors within the JAK–STAT pathway may lead to clinical problems, with increased numbers of red blood cells, white blood cells, and platelets, some of which may be malignant.

3.2.4 Regulation of signal transduction

These processes are not without control: there are many inhibitors and regulators of second messengers and enzymes. At the cell surface, hormones epinephrine, glucagon, and ACTH all bind to a receptor and stimulate the G protein complex, and so the activation of adenylyl cyclase and increased cAMP. This process is countered by the effects of inhibitory hormones prostaglandin E_1 and adenosine, which, following binding to their receptor, result in the inhibition of adenylyl cyclase and so failure to generate cAMP. The activity of certain SMAD molecules is inhibited by regulators called Ski and SnoN, which are also involved in many other physiological and pathological processes. Intracellular calcium levels are partly regulated by the protein calmodulin. Just as failure of the brakes to inhibit the speed of a car can lead to a road traffic accident, failure of a regulator to inhibit cell growth can lead to cancer.

3.2.5 Section conclusion

The section above is of course grossly simplified, and much (for the sake of brevity and clarity) is omitted. For example, NF-κB is linked to over 150 genes involved in infection and immunity alone and can be activated by the occupancy of the tumour necrosis factor alpha (TNF-α) receptor, the

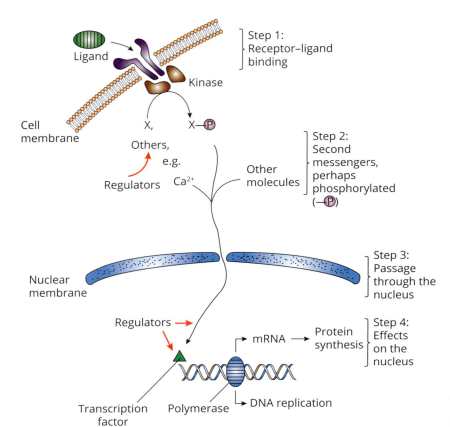

FIGURE 3.2

Steps involved in signal transduction. Information is received at the cell surface and is passed, via second messengers, to the nucleus where transcription or replication is initiated.

interleukin-1 receptor, and toll-like receptors. Furthermore, these pathways are far from specific, and to a degree may be described as incestuous. For example, the MAPK–ERK pathway can also activate members of a family of transcription factors called Myc (or c-Myc), with possible roles in carcinogenesis discussed in section 3.9.5. NFκB can be activated by several second messengers and can have various effects on the nucleus. Nevertheless, the broad processes described above are key to understanding normal cell functioning, and so to understanding the significance of errors, the consequences of which include cancer. Indeed, treatment of certain cancers is aimed at ligands, receptors, and second messengers. Finally, many second messengers amplify their downstream substrates, leading to a cascade of instructions. Figure 3.2 summarizes and simplifies certain steps in signal transduction, whilst Table 3.1 summarizes receptors, ligands, second messengers, and transcription factors.

Now we have covered some basics of the molecular aspects of cell signalling, we are able to move to study the cell as a whole.

Key points

Regulation of the function of the cell can be described in three stages: (i) a ligand binding its cell surface receptor, (ii) activation of second messengers within the cytoplasm, and (iii) passage of these messengers to the nucleus where they effect changes such as alteration in protein synthesis and in cell division.

TABLE 3.1 Selected receptors, ligands, and their second messengers

Ligands	Receptors	Enzymes	Second messengers/ transcription factors	Examples of linked cancers
Epinephrine, calcitonin, glucagon	G protein-coupled receptors	Adenyl cyclase, phospholipase C	cAMP, Ras, MAPK, CREB, IP3, diacylglycerol	Pituitary, thyroid, prostate
TGF-β, BMP, activins	TFG-β receptor	Serine/threonine kinase	SMADs, Ras, MAPK, ERK	Pancreas, head and neck
Epo	Epo receptor	JAK	STATs	Erythro-leukaemia
TGF-α, VEGF, EGF	Receptor tyrosine kinase	Tyrosine kinase	Ras, MAPK, IP3, diacylglycerol	Breast, stomach, lung
TNF-α, interleukins	TLRs, interleukin receptors	I-κB kinase	NF-κB	Pancreatic, cervical, colon

This table is illustrative and does not imply that certain pathways are specifically or exclusively linked in biochemistry or cancer. Epo = erythropoietin. TLR = toll-like receptor.

3.3 Mitosis and the cell cycle

Although perhaps the most well-known part of the cell cycle, mitosis is in fact only a part of the process, the major section being interphase.

3.3.1 Cell replication

Whilst in development *in utero*, the embryo needs to grow, with cells as yet of unclear function differentiating into defined cells, tissues, and organs in the foetus, and then the neonate, child, and adult. It does this by replicating all its cells by mitosis. This process continues in all cells until some point after birth, when some cells, notably those of the central nervous system (brain and spinal cord), will have completed their allotted reproductive life. Conversely, millions of blood cells need to be replaced each day, so that rapid and wholesale replication is vital (Chapter 11). However, apart from these polar extremes of replication, almost all other cells continue to reproduce themselves throughout life, but this process slows down over the decades at different rates in different organs. At some point (generally, years 10–13 or thereabouts), other cells in specialized organs (the gonads) undergo a different type of replication, leading to the generation of gametes (sperm and ova), by meiosis. The process also continues for decades, but in females this stops at around the age of 50 to 55 at the menopause.

Apart from reproduction, there are several reasons why a cell needs to replicate: in the adult, one of the most common is to replace cells that are redundant, damaged or have been destroyed, or perhaps shed as part of a physiological process (such as the renewal of skin cells, and of the endometrial cells of the uterus after menstruation). In almost all cases, cells reproduce when instructed to do so by soluble signals (such as growth factors) that bind cell surface receptors or by contact with other cells via specialized structures embedded in the cell membrane (as outlined in section 3.2). In some cases, cells of a particular form and function arise from a small number of self-perpetuating stem cells (as in the case of blood cells). In other cases, cells are constantly generated from a layer of germinal cells (such as epithelial cells of the skin, situated on a basement membrane in the dermis).

All cell processes are controlled by the nucleus, an organelle composed of chromosomes that in turn are composed of DNA (60%), RNA (5%), and proteins (35%). Once a cell has received its instruction from second messengers to reproduce, a highly regulated series of steps involving the replication of all material inside the cells, including organelles and chromosomes, begins—the process of mitosis, which can be viewed in terms of the cell cycle. The allied process of meiosis follows many of the same steps but has its own final pathways.

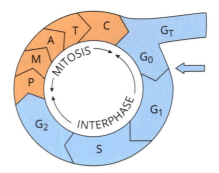

FIGURE 3.3

The cell cycle. The cell cycle consists of a series of stages: interphase (with stages G_0, G_1, S, G_2, and G_T) and mitosis (itself consisting of prophase (P), metaphase (M), anaphase (A), telophase (T), and cytokinesis (C)). The arrow on the right indicates the point where the cell is induced into a cycle by signal transduction pathways.

The cell cycle consists of two main stages: interphase and the M phase. The first stage, interphase, may in addition be broken down into additional stages, which in sequence are Gap 0 (G_0), Gap 1 (G_1), Synthesis (S), and Gap 2 (G_2). Similarly, the M (for mitosis) stage can be broken down into other steps, these being prophase, metaphase, anaphase, telophase, and cytokinesis (Figure 3.3).

3.3.2 Interphase

G_0 is a resting, or quiescent phase, where the cell may be seen as being between mitoses: cells continue with their own particular function (such as an endocrine cell synthesizing and then releasing a hormone). Interphase can last for a considerable time, and possibly (such as in neurones and other non-regenerating cells) for the remainder of a cell's lifespan. If so, some describe this stage as G_T, where T implies terminal, to emphasize that the cell is at the end of its mitotic life. It could be argued that terminally differentiated cells of the nervous system are in G_T. What this means in practice is that once the cell is dead, it cannot be replaced, which explains why diseases such as stroke are not only debilitating but are also irreversible in that the tissues or processes controlled by that part of the brain cannot return to normal function.

At some point in G_0, the cell receives its instructions to reproduce, as described in section 3.2 and indicated in Figure 3.2. Once this signal has been received, stage G_1 begins, and although the timescale for this varies from cell to cell, it is said to be the longest phase, often lasting up to 9 or 10 hours. This process is characterized by a marked increase in transcription and translation as the cell increases its rate of protein synthesis and the generation of new organelles and increases its size to about twice that in its G_0 phase in preparation for mitosis. All these metabolic processes demand a great deal of energy (as ATP) and raw building blocks, required by the nucleus in generating new infrastructure in advance of chromosome replication. This period sees the accumulation of high concentrations of nucleotides and allied molecules needed for the replication of its chromosomes, and centrosomes are replicated. Moving from stage to stage requires specific signals, and if these appropriate signals are not received (as will be discussed in section 3.3.4), the cell can return to G_0.

Once sufficient building blocks (nucleotides, construction enzymes, etc.) have been generated, the cell enters the S phase, in which it duplicates its DNA by semi-conservative Meselson–Stahl replication, a process lasting 5–9 hours (Figure 3.4).

Normally, the process of DNA replication is robust and accurate, but should there be errors, these can be recognized by regulators, and the cell cycle is put on hold, at which point errors may be corrected. Alternatively, if the errors cannot be corrected, the cell may be marked for elimination (called apoptosis, which is discussed in section 3.4). Once chromosomes have been faithfully duplicated as two sister chromatids, protein and organelle synthesis continues, and the cell enters G_2 (for maybe 4 hours) and makes final preparations for actual cell division.

Cross reference

Dedicated textbooks on cell biology, such as Cooper's *The Cell: A Molecular Approach* (eighth international edition) from which this figure was taken, provide further detail on DNA replication.

SELF-CHECK 3.2

In one short sentence for each, summarize the four stages of interphase.

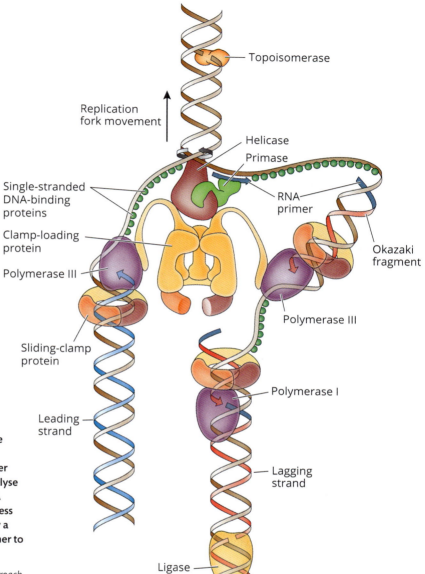

FIGURE 3.4

DNA replication. At the top, topoisomerase is involved in unwinding the DNA strand, helicase in separating the two strands. Other enzymes and non-enzymatic proteins catalyse the formation of a duplicate strand of DNA from each of the original strands. The process is completed at the bottom of the figure by a ligase, which joins the single strands together to form a double strand.

Adapted from Cooper, *The Cell: A Molecular Approach*, eighth edition, Oxford University Press, 2019.

3.3.3 The M (mitosis) stage

This stage, lasting perhaps an hour, consists of five substages, which are summarized in Figure 3.5:

- Prophase is characterized by a breakdown in the nuclear membrane, and the condensed chromatin unravels into discrete chromosomes, a process which can be recognized by light microscopy. Two centrosomes each move towards polar extremes of the cell, and spindles begin to form from recycled microtubules.

- During metaphase, chromosome migration proceeds, and at its conclusion, all sister chromatids are aligned in the centre of the cell at the metaphase plate.

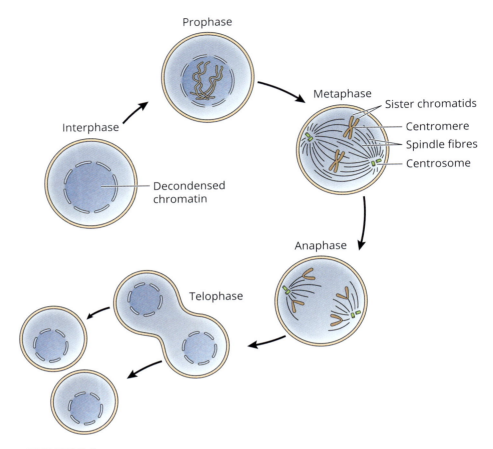

FIGURE 3.5

The four stages of mitosis. Once the nuclear membrane dissolves at interphase, prophase (the first stage) sees the centromere of each sister chromatid linked to each of the centrioles. In metaphase, the sister chromatids line up in the centre of the cell in the metaphase plate. In anaphase, sister chromatids separate into individual chromosomes, and one each is drawn to opposing poles of the cell. In telophase, the chromosomes collect together into an embryo nucleus, and a new cell membrane forms on the old metaphase plate as a precursor to cytokinesis.

Adapted from Cooper, *The Cell: A Molecular Approach*, eighth edition, Oxford University Press, 2019.

- In anaphase, the spindle fibres contract so that one of each of the sister chromatids is pulled away from the metaphase plate in opposite directions, and towards the centromeres, leading to the development of two clusters of chromosomes.

- During telophase, the two clusters (each with one of the replicated pairs of chromosomes) form a more condensed ball, and the spindle fibres disintegrate. A new nuclear membrane forms around each cluster, creating a nucleus.

- Cytokinesis is characterized by the development of a ring of actin and myosin around the waist of the cell that contracts to cleave the cell to form two independent and identical daughter cells. This is followed by the resumption of G_0.

Each time a cell goes through mitosis, sections of the ends of the chromosomes (the telomeres) lose some of their DNA, so that the entire chromosome becomes shorter. These terminal sections of multiple tandem hexanucleotide repeats of DNA are perhaps 11 kilobases long at birth, falling to around

4 kilobases in the elderly. However, these can be replaced by the enzyme telomerase. The Hayflick limit refers to the number of times that a normal cell can go through mitosis *in vitro*, that number in adult cells being 40–50 (more in embryonic and foetal cells), at which point cell division stops. The two processes are related, in that the Hayflick limit is linked to the length of the telomere. However, overactivity of telomerase in rebuilding the telomeres can extend a cell's lifetime, a phenomenon common in many cancer cells as a way of avoiding senescence.

Key points

The cell cycle consists of a series of stages characterized by cell growth, chromosome replication and separation, and finally the formation of two daughter cells.

3.3.4 Control of the cell cycle

The progress of the cell through the stages described in the previous sections is not haphazard but is tightly controlled by a series of regulators. Should these regulators fail to be satisfied that the procedure is working correctly, the cell will not progress to the next stage and is likely to be destroyed. There are three checkpoints to ensure that the cell is ready to enter the stage that follows, and that there is no damage to the DNA. If there is damage, the purpose of these checkpoints is to put the cell cycle on hold so that repair enzymes can correct the problem (if possible).

Checkpoint 1, towards the G_1/S transition, is the position in G_1 at which the cell is committed to the cell cycle and so enters the S phase. It can only be passed once sufficient organelles and nucleotide building blocks have been generated, the latter being required for DNA replication.

Checkpoint 2, at G_2, checks cell size, and all freshly generated DNA for damage, and if none is found the cell proceeds to the M stage.

Checkpoint 3, in metaphase, ensures all chromosomes are aligned correctly at the metaphase plate, roughly in the centre of the enlarged cell, and are connected to their microtubule spindles, allowing the cycle to proceed to anaphase.

The key gatekeepers at each stage are cyclins.

Cyclins and CDKs

There are several regulators of the cell cycle, but the most important are cyclins (levels of which rise and fall at different stages of the cell cycle) and cyclin-dependent kinases (CDKs, levels of which remain fairly constant throughout the cycle). Notably, many members of the CDK family (of which there are at least 13) show homology to the second messenger mitogen-activated protein kinase (MAPK: described in section 3.2.2), implying common evolutionary function.

The cyclins comprise a family of proteins, each individual cyclin being a specific cofactor in the activation of an otherwise dormant CDK. Once activated, the cyclin-CDK complex phosphorylates its substrate on a serine or threonine molecule. This leads to the movement of the cell to the next stage of the cell cycle, involving other cyclins and CDKs (Table 3.2). The M stage regulation occurs in the middle of metaphase and, if successful, permits passage to anaphase. Accordingly, the cyclin B/CDK1 may be described as a maturation-promoting factor, mitosis-promoting factor, or anaphase-promoting complex.

Cell cycle regulators

In addition to positive regulators (promoting transcription), there are also negative or inhibitory regulators (preventing transcription) that act on the cyclin/CDK complexes (Table 3.2). Two families of small proteins are of interest: those of the INK4 (<u>IN</u>hibitors of CD<u>K4</u>) family and those of the CIP/KIP (<u>C</u>DK <u>I</u>nteracting <u>P</u>rotein, and <u>K</u>inase <u>I</u>nhibitory <u>P</u>rotein) family. The former (p15, p16, p18, p19—the

TABLE 3.2 Regulators of the cell cycle

Cycle stage	Cyclin	CDK	Inhibitors
Mid G_1	Cyclin D	CDK 4, CDK6	p15, p16, p18, p19
Late G_1/S	Cyclin E	CDK2	p21, p27, p57
S	Cyclin A	CDK 2	p21, p27, p57
G_2	Cyclin A	CDK1	p21, p27, p57
M	Cyclin B	CDK 1	p21, p27, p57

numbers refer to the approximate molecular weight) all have the capacity to inhibit CDK4 and CDK6, and also suppress other transcription factors such as cMyc and the ubiquitous NFκB. Inhibitors in the CIP/KIP family (p21, p27, and p57) may, between them, inhibit all of the cyclin/CDK complexes. Many of these molecules are active in other pathways, such as the participation of p57 in neuronal development and erythropoiesis. Since all these molecules are produced by the transcription and translation of their genes, it follows that control of these genes is crucial in regulating the cell cycle. It is a convention that a gene/protein pair is named with the former in italics. For example, *p53* is the gene that codes for the protein p53.

Implications for cancer: Tumour suppressors

Roles for these molecules and pathways in cancer arise from the observation that alterations or **deletion** of the section of chromosome 9p21 where the coding genes *INK4b* and *INK4a* are located are linked to numerous malignancies, such as familial predisposition to melanoma. Should there be damage to the DNA, protein p53 (a 53 kDa product of *TP53* at 17p13.1) can (once activated by kinase phosphorylation) stop the replication cycle at the G_1/S checkpoint by inhibiting p21, ostensibly to allow for the DNA to be repaired. However, should this process fail, and the cell cycle continue, it may allow precancerous cells to escape removal. p53 has many other functions, probably linked to various isoforms, such as initiating the repair of damaged DNA. We shall hear much more of this very important molecule, and those with similar functions, in sections to come.

This brings us to a key point in cancer genetics: the tumour suppressor. Loss of the regulation of the p16 protein (and many, many others) permits the development of a tumour, such that in physiology it suppresses tumourigenesis, and so is seen as a tumour suppressor. It follows that a gene coding a tumour suppressor molecule (e.g. p16) is therefore a tumour suppressor gene (in this case, *CDKN2A*, coded at 9p21.3). It also follows that other molecules acting on p16 are likely to be tumour suppressors, as indeed is the case for *p53* and its p53 product. As explained in more depth in sections 3.9.2 and 3.9.3, pRb and *Rb1* are also tumour suppressors.

The importance of the previous section on regulation is therefore in its malfunction. Failure of the regulatory system (perhaps by loss of a tumour suppressor gene or protein) may lead to inappropriate cell growth, which, in turn, may lead to cancer. Indeed, anti-cancer drugs that target CDKs have been developed. By inhibiting the cyclin D/CDK4/6 complex with a pharmaceutical (as do p21 and p16 naturally), Rb will not be phosphorylated, and so will fail to release its E2F partner. Alternatively, failure of any part of the cell cycle (such as errors in DNA replication and chromosome allocation) has serious consequences for the viability of the cell and is likely to lead to it being eliminated. The leading process of the elimination of damaged cells is called apoptosis.

Key points

The cell cycle is controlled by regulators such as cyclins and cyclin-dependent kinases. Checkpoints at key steps ensure that the cell is stable and can proceed to the next stage.

What is the relationship between cyclins and the CDKs, and what is their function?

3.4 Apoptosis

As discussed in Chapter 1, you are going to die, perhaps by accident, violence, disease, or simply old age. Cells also die, and some do so by accident (a trauma), by being poisoned by a toxin, by being attacked or destroyed by a phagocyte, or by apoptosis (sometimes described as programmed cell death, or cell suicide). Cells that have failed to correct damage to their DNA by normal repair mechanisms (details of which are to follow) will also self-destruct using this process. Apoptosis is separate from cell senescence, where a normal cell stops replication with shut-down of the cell cycle, perhaps to allow repair to proceed. This, of course, can be entirely physiological and reversible, and may be viewed as cell cycle stage G_0. Apoptosis, in contrast, is a series of regulated steps concluding with the activation of caspase enzymes that effectively digest the cell from the inside.

Apoptosis is not cancer-specific: there are a number of situations in which too many cells are produced, and those that are unwanted need to be eliminated. An example of this is the production of lymphocytes that fail to bind to an antigen, and so are surplus to requirements, and those that react to 'self' antigens (as discussed in Chapter 9). Cells may also be induced into apoptosis if infected with a virus (such as HIV) or an intracellular parasite (such as mycobacteria, chlamydia, and rickettsia). Cells that avoid or escape from the normal processes that lead to their elimination often become immortal, which many have achieved *in vitro*. Perhaps the best-known immortal cell line is HeLa cells, which originate from Henrietta Lacks, an African American woman with cervical cancer (from which she died in 1951) whose cells were used for research without her consent.

How does apoptosis proceed? Once more, we refer to Newton in that everything is caused by something, and there are numerous molecular changes that cause a cell to become apoptotic. Just as we will die if denied oxygen and glucose, so the cell will die if it fails to receive these nutrients, but also if it fails to receive crucial 'stay alive' signals from growth factors. Two major routes to apoptosis are recognized: the intrinsic and extrinsic pathways, although they have a common (execution) terminal pathway.

3.4.1 The intrinsic pathway

Almost inevitably, cells will come under stress from a variety of non-specific sources, such as increased temperature, lack of nutrients, toxins, viral infections, hypoxia, and radiation. Some will successfully manage this stress, whilst others will not. The mitochondrion is particularly susceptible to adverse changes to the cell. Should these changes be sufficiently adverse, the membranes around this organelle increase in permeability, and certain molecules pass into the cytoplasm. These are detected and may trigger apoptosis.

Section 3.3.4 on the control of the cell cycle describes p53, a cyclin/CDK regulator at the G_1/S checkpoint. Under resting conditions, p53 is bound by its inhibitor, mdm2, a small (56kDa) molecule coded for by *MDM2* at 12q15, described in section 3.2.2. However, in certain circumstances, such as DNA damage, mdm2 is phosphorylated by a kinase such as ATM (ataxia telangiectasia mutated), coded for by *ATM* at 11q22.3 (further detail in section 3.5.6). Once phosphorylated, mdm2 uncouples from p53, which is therefore liberated (and stabilized) to act in a variety of ways, one being the arrest of the cell cycle (fully explained in section 3.9.3). Should this damage be irreparable, p53 mediates certain other changes that involve molecules such as the Bcl-2 family (among them Bcl-xs, Bad, and Bax). Members of the Bcl-2 family are present on the outer membrane of the mitochondrion, the consequences of their activation being the loss of cytochrome C from within the organelle. Once in the cytoplasm, the cytochrome C binds to another molecule called apoptotic protease activating factor (Apaf-1) to form an apoptosome, which activates members of the caspase group of enzymes, principally pro-caspase-9. However, other members of the Bcl-2 family, such as Bcl-XL (which represses cell death activity), are anti-apoptotic, and so are likely to regulate this process.

3.4.2 The extrinsic pathway

This pathway is characterized by the binding of either of two receptors at the cell surface by their ligands. The Fas receptor is also charmingly known as the death receptor, as binding of the receptor with its ligand, Fas, leads to the activation of the death-inducing signalling complex. Apoptosis may also be triggered by the interaction between TNF-α and its receptor. The latter has strong homology to the Fas receptor, so both are part of the same family of receptors. Once TNF-α has docked into its receptor, second messenger-like intermediates such as the TNF-α receptor-associated death domain and FAS-associated death domain (FADD) are activated. Thus, in both cases, a series of interactions follow receptor/ligand binding, not dissimilar to the signal transduction pathways described in section 3.2.2. FADD converts one of the (pro)caspases to its active form, the principal one being pro-caspase 8.

3.4.3 The final common pathway

Thus, both pathways result in the activation of a different pro-caspase. Caspase 8 and caspase 9 may be regarded as initiators, as they activate 'executioner' caspases 3, 6, and 7. The active form of these caspases leads a cascade of many other enzymes that effectively digest the cell from within, leading to characteristic changes that can be followed by light microscopy and other techniques. These changes include cell shrinkage and rounding, condensation of the chromatin, and dissolution of the nuclear membrane. The DNA is cleaved into short fragments, and cytoplasmic 'blebs' of cell membrane extrusions are formed, which are pinched off and appear as apoptotic bodies, to be readily detected and removed by phagocytes (neutrophils, monocytes, and macrophages). Eventually, the entire cell fragments, and most is recycled. These three pathways are summarized in Figure 3.6.

3.4.4 Regulation of apoptosis

Naturally, such a crucial process cannot be allowed to proceed without regulation. As mentioned, negative (preventative) regulators include Mcl-1 and Bcl-2, and positive (promoting) regulators include FADD and NF-κB. Apoptosis may also be regulated by miRNAs (to be discussed in section 3.8.4). Certain of these miRNAs act to suppress Mcl-1 and Bcl-2: therefore, loss of these miRNAs due to aberrant or absent p53, or deletion of 13q14 (the locus of the tumour suppressor Rb), leads to anti-apoptosis and so to the possible survival of cells (perhaps proto-tumour cells) that would otherwise be marked for elimination. A more detailed examination of Rb follows in section 3.9.2.

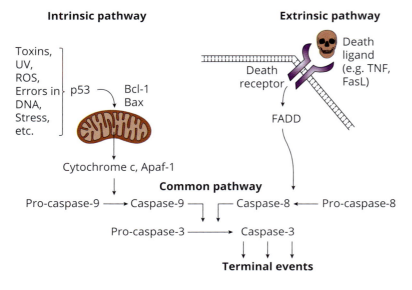

FIGURE 3.6

The three apoptosis pathways. The two pathways to apoptosis operate via the mitochondrion and cell surface receptors respectively. The former focuses on nuclear p53 and the mitochondrion, and the latter on 'death' signals. Both pathways promote the activation of caspases and so, in the common pathway, the initiation of the autodigestion of apoptosis.

3.4.5 Other forms of cell death

We recognize several other types of cell death.

- Necroptosis may be viewed in terms of acute inflammatory damage to a cell, perhaps when it is under attack from a pathogen. Like apoptosis, it is highly regulated, and responds to 'death' signals that result from binding of TNF-α to its receptor (i.e. TNFR1). However, in necroptosis the caspase system is not activated, but an alternative signalling system, via a receptor-interacting protein kinase (RIPK), activates a protein called 'mixed lineage kinase domain-like pseudokinase' (coded by *MLKL* at 16q23.1), which damages the cell membrane leading to loss of cytoplasm and organelles and so lysis.

- Eryptosis is the self-destruction of damaged erythrocytes (red blood cells). The mechanism involves an increase in cellular calcium that in turn leads to loss of potassium and so deterioration of the integrity of the cell membrane. This results in characteristic membrane blebbing (due to a breakdown in the cell's cytoskeleton), a reduction in cell volume, and the appearance of phosphatidylserine at the cell surface. These changes mark the cell for elimination by phagocytes such as neutrophils and macrophages.

- Cell death in parthanatos is also independent of caspase, being mediated by poly-ADP-ribose polymerase, an enzyme that generates poly-ADP-ribose. This induces the passage of apoptosis-inducing factor (AIF, an oxidoreductase coded by *AIFM1*, on the X chromosome) from the mitochondrion to the nucleus where it promotes DNA fragmentation. Triggers of parthanatos include oxidative stress.

- Ferroptosis causes cell death by the iron-dependent formation and accumulation of **reactive oxygen species (ROS)** such as hydroperoxides, part-facilitated by glutathione depletion. p53 (described in section 3.9.3) may be a regulator of ferroptosis.

- One of the responses of neutrophils to microbial pathogens is to generate and extrude a meshwork of decondensed chromatin (from its own fragmented nucleus), antimicrobial proteins, and cytokines, together described as a neutrophil extracellular trap (NET). Should these changes lead to the death of the cell, the process may be termed NETosis.

- Pyroptosis is a process by which a macrophage self-destructs when faced with an overwhelming intracellular infection, perhaps by *Salmonella* or *Listeria* species. Sensor proteins activate caspase-1 and caspase-11, which themselves generate a gasdermin pore that integrates into the cell membrane, leading to swelling and rupture.

- It could be argued that necrosis is a form of cell death that cannot be classified into any of the above. In contrast with apoptosis, where the cell sheds membrane microparticles and shrinks, the necrotic cell swells and bursts, liberating its potentially toxic components. These may have further consequences, such as in inflammation.

Key points

Apoptosis is a process by which a damaged cell can be removed. Initiating pathways (some of which rely on receptor–ligand binding) culminate in the activation of caspases that digest the cell from within.

SELF-CHECK 3.4

Under what circumstances is apoptosis necessary?

3.4.6 Autophagy

Autophagy is the cell's response to two processes: the removal of damaged organelles and superfluous proteins, and the need to downregulate metabolism in response to starvation or pathogenic processes such as hypoxia. The lysosome is the primary organelle in the process, and once the cell has stabilized, it can resume its physiological function. However, should autophagy be intense or prolonged, the cell cannot survive and passes into apoptosis or necrosis. The key mediator is Beclin-1, aberrations in which may be linked to stomach, breast, colon, and other cancers.

3.5 Chromosomes

A key aspect of cancer is the chromosome and the genes within it. As regards chromosomes, we can describe changes as quantitative (in the number of chromosomes and/or large sections of particular chromosomes) and qualitative (in a small number of nucleotides; see section 3.6). Quantitative changes may be described in major groups as follows, the first two being the most common.

3.5.1 Translocation

Chromosome strands may come apart but can be rejoined by repair enzymes. However, if incorrectly repaired, these breaks can lead to a particular type of error called a **translocation**, which occurs when a relatively large section of DNA, often at one of the two ends, is transferred from one chromosome to another. These changes can be explained in terms of letters of the alphabet, where each letter represents a section of a chromosome and its genes. For example, a break-point between the 'W' and 'X' nucleotides from a sequence 'UVWXYZ' on one chromosome will provide two loose ends, that is, UVW– and –XYZ. Similarly, a break-point in another sequence 'ABCDEF' on a different chromosome will produce two other loose ends, such as ABC- and -DEF. Misaligned repair enzymes may then rejoin ABC- with -XYZ, and UVW- with -DEF, giving two new sequences of ABCXYZ and UVWDEF.

This process may therefore bring together two sections of DNA which separately perform perfectly normal functions, but that together form a completely new functional sequence (a gene), such as if XXXDANGEXXX is present on one chromosome, and OOOROUSOOO is present on another. A reciprocal translocation between the two chromosomes may create the neogene sequence XXXDANGEROUSOOO on one and XXXOOO on the other, the former with an implied problem. Note that in this example, unlike the ABCXYZ example, the mutated chromosomes are of unequal length. Figure 3.7 illustrates this process.

One of the principal examples of the pathology of a translocation is the formation of the fusion of *BCR* (normally coding the protein 'breakpoint cluster region') and *ABL* (originally described in the

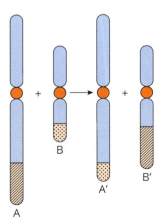

FIGURE 3.7

Translocation of two sections of DNA. This figure shows an example of a reciprocal translocation. The diagonally shaded section of the terminal part of chromosome A is swapped with the dotted section at the terminal end of chromosome B. This creates two new chromosomes, A' with the translocation of the dotted section of chromosome B, and chromosome B', with the translocation of the diagonally shaded section of chromosome A.

FIGURE 3.8
Inversion of a section of DNA. The left image shows a paracentric inversion within a single chromosomal arm, and the right shows a pericentric inversion that involves both arms.

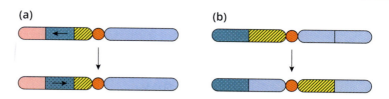

Abelson murine leukaemia virus, and normally coding a tyrosine kinase) to form *BCR-ABL*, which causes a leukaemia, and is therefore an **oncogene**, that is, a gene that causes cancer, of which there are many examples, as we shall see in sections to come.

3.5.2 Deletion

A deletion may bring together two normal DNA sequences that are far apart on the chromosome to create a new abnormal DNA sequence. For example, the section XXX may be deleted out of the sequence OOODANGEXXXEROUSOOO, which would then read as OOODANGEROUSOOO. Deletions may also nullify vitally important sequences, such as OOOCRUCIALOOO becoming simply the non-functioning OOOCRUOOO. In some cases the extent of the deletion can be mapped with precision, such as that in 22q11.21 of base pairs 18,894,835–20,311,763. Should a deletion result in an inhibitor regulatory molecule being inactivated, then what it should have controlled will continue to function, perhaps inappropriately. As we shall see, this is particularly important in genes that control the cell cycle, some of which are described as tumour suppressor genes. Loss of these genes therefore permits malignancies to develop. Deletions of <5 million base pairs are microdeletions.

3.5.3 Inversion

An **inversion** results in parts of the DNA sequence running in the opposite direction to how they should run in, such as the correct OOODANREGOUSOOO forming an incorrect OOODANGEROUSOOO sequence. This only happens where there are two break-points within a certain chromosome. There are two types of inversions: those that occur only within one arm of a chromosome are described as paracentric, but if an inversion includes the centromere, it is pericentric, and is due to breaks in both arms (Figure 3.8).

3.5.4 Duplication

A **duplication**, or amplification, is present when a section of DNA is repeated, often alongside the primary sequence, such as XXXABCDEXXX becoming XXXABCDEABCDEXXX. Repeats may not necessarily be contiguous—that is, a sequence like XXXABCDEXXXABCDEXXX is also possible. Transcription of both genes will generate twice as much product (be it mRNA or protein) as normal, which may disrupt the physiology of the cell, leading to disease.

Table 3.3 shows examples of nucleotide mutations and what can result from these.

3.5.5 Nomenclature and mapping

The chromosome abnormalities we have just discussed can be named within a system including the first letter or letters of the particular abnormality (t, del, inv). The precise site of an abnormality can be named and so mapped by the chromosome(s) and arm(s) involved. So t(9;22)(q34;q11) tells us of section q34 of chromosome 9 being translocated (t) to chromosome 22, and that section 11 of chromosome 22 is translocated to chromosome 9. This results in the formation of a malignant gene on chromosome 22 (the Philadelphia chromosome), causing a form of leukaemia (i.e. it is an oncogene),

TABLE 3.3 Examples of nucleotide mutations and their consequences

Form of mutation	Nucleotide sequence	Amino acid sequence	Note
Wild-type	AAA CAU UUA	Lys-His-Leu	Normal sequence and tripeptide
Deletion	AAA-UUA	Lys- Leu	A trinucleotide deleted
Duplication	AAA CAU CAU UUA	Lys-His-His-Leu	Insertion/amplification
Inversion	AAA UAC UUA	Lys-Tyr-Leu	Reversal of a trinucleotide sequence
Functional SNP	AAA UUC UUA	Lys-Phe-Leu	Results in a different peptide product
Nonsense SNP	AAA CAU UGA	Lys-His-STOP	Leads to a truncation, appears as a deletion
Silent SNP	AAG CAU UGA	Lys-His-Leu	Lys can be code by different triplets so no change in peptide

SNP = single nucleotide polymorphism. The nucleotide codons are from the mRNA.

as explained in sections 3.9.1 and 11.4.2. An example of a deletion is del(2)(p16.3-21), a 4.8 Mb section that encodes at least 27 known genes, and leads to the development of Lynch syndrome, a form of colorectal cancer. The shorthand inv(16)(q13;q22) tells us of paracentric inversion of the section bounded by q13 and q22 within chromosome 16. This inversion is linked to a good outcome in acute myeloid leukaemia.

Sections of chromosomes can be classified by their affinity for dyes, leading to light and dark bands, as explained in section 16.1. These bands are numbered sequentially from the centromere to the telomere. Figure 3.9 summarizes these mapping criteria showing the location of *p53* at 17p13.1.

3.5.6 Damage repair

Damage to nuclear and mitochondrial DNA does not automatically lead to errors and so disease. Indeed, it has been estimated that there are in the region of 10,000–15,000 instances of DNA damage in an adult cell each day (and this estimate may be low). Whilst this is insignificant in view of the three billion base pairs in our DNA, consequences can be fatal if the damage is to a crucial gene. We have already noted several agents that cause various forms of damage, such as radiation, industrial chemicals, and viruses.

Fortunately, there are repair systems that recognize damage, such as helix distortions and single- and double-strand breaks in the DNA (as in the four types of chromosomal breaks described above), and there are many enzymes that correct this damage. Examples of the latter include the ATM,

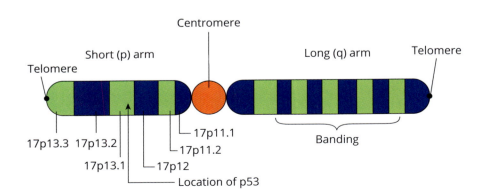

FIGURE 3.9
Location of *p53* in chromosome 17. The banding is for illustrative purposes only and does not imply accurate positioning and size.

SIRT6, ADP-ribose polymerase 1 (PARP-1), ATR (ataxia-telangiectasia and Rad-related) kinases, and nucleotide excision repair enzymes. These enzymes activate the p53/cyclin/CDK regulators to arrest the cell cycle so that repair can go ahead (section 3.3.4). Errors in the DNA sequences coding for p53, and so errors in the protein itself, may lead to failure to activate the repair mechanism for correcting damaged DNA. Thus, failure of damage-repair enzymes to perform their function could lead to inappropriate and potentially dangerous changes to certain genes. Conversely, overactivity of PARP-1 is linked to a wide range of cancers, whilst del(11)(q22-23) points to a deletion in the q22-q23 region of chromosome 11, an abnormality found in around a third of patients with chronic lymphocytic leukaemia. *ATM* maps to this location, its product regulating part of the cell cycle to prevent processing of damaged DNA and induce apoptosis if the repair fails. Deletion of q22-23 therefore removes the genes for the enzymes that provide this safety net, leading to the malignancy.

As is described in Chapter 16, chromosomal malformation of this nature is also important in a host of non-neoplastic diseases.

SELF-CHECK 3.5

Use t(9;22)(q34;q11) to explain the meaning of oncogene.

Key points

The major chromosomal lesions are translocations, inversions, deletions, and duplications. Translocations can form neogenes that may cause malignancy, whilst deletions in sections that contain tumour suppressor genes may also be the basis of a neoplasm.

3.6 Genes

The second part of the understanding of the genetics of cancer is in qualitative changes, most of which occur in one particular nucleotide of the gene. However, there are numerous examples of changes to a small number of nucleotides, some being micro-deletions.

The genome (the sum of all nucleic acids in the nucleus and mitochondrion) is estimated to contain some 20,000–21,000 protein-coding genes, around 22,000 non-protein coding genes, and some 15,000 pseudogenes. The latter are non-functioning, probably because of inconsequential change in the DNA. The greatest number (2,000) of protein-coding genes are on chromosome 1 (the largest autosome), the least on chromosome 21, whilst the sex chromosomes X and Y have around 840 protein-coding genes and 60 protein-coding genes respectively. The density of genes (i.e. number of genes per unit length of chromosome) is not equal: chromosome 9 is relatively gene-dense, whilst chromosomes 13, 18, and 21 are relatively gene-poor. This is reflected in the G-banding—dark bands generally contain more adenine and thymine nucleotides and are relatively gene-poor. The mitochondrial chromosome carries some 37 genes, of which 13 code for protein.

3.6.1 Nucleotides, the genetic code, and the central dogma

The carriage of information by genes is made possible because DNA is composed of a series of four different nucleotides (abbreviated by A, T, G, and C), the order of which forms a particular gene (such as ATATACGAT). Nuclear DNA is transcribed into a chain of messenger RNA (mRNA, in whose sequence T is replaced by U), and at the ribosome, the mRNA sequence is translated, and the protein is built.

The code for this translation lies in the sequence of a codon of three nucleotides on the mRNA. Transfer RNA molecules (tRNA) carry an amino acid at one end and a set of three nucleotides (an anti-codon) at the other. The tRNA expressing the anticodon AUA (ATA in the DNA) is linked to the amino acid isoleucine, UAC codes for tyrosine, and GAU codes for aspartic acid. Each tRNA anticodon seeks its complimentary mRNA codon. Thus, the nuclear gene ATATACGAT transcribes to an mRNA chain of AUAUACGAU that in turn builds the tri-peptide isoleucine-tyrosine-aspartic acid.

Not all the DNA in a chromosome codes for a gene, and within a gene, not all the DNA ultimately codes for a protein. Indeed, it has been estimated that as much as 98% of the genome consists of non-coding DNA (erroneously once called 'junk DNA'), leaving 2% of DNA coding for around 21,000 protein-coding genes. Some non-coding sections of DNA are described as tandem repeats, which may also be known as minisatellites, whilst others code not for proteins but for types of RNA required for protein translation. Many genes consist of sections of DNA (exons) that code for protein, inter-spersed with non-coding sections (introns). Each gene has a 'switch-on' region (a promoter) to which a transcription factor binds to initiate replication or transcription. Many promoters are characterized by a series of thymidine and adenine bases—the TATA box—to which the enzyme generating mRNA (RNA polymerase) can bind.

Just as not all DNA codes for mRNA, not all mRNA is translated into protein. There are so-called untranslated sections at each end of the chain, and there are often sections of RNA in the middle of the message that are non-coding. These non-coding sections are excised (or spliced out) by RNAase enzymes so that only a true and fully functioning message is translated at the ribosome. The proteins coded for by the DNA can have several functions, such as the immunoglobulins, digestive enzymes, hormones, and albumin that are destined for export. Others, such as the protein component of ribo-somes, and second messenger enzymes, remain within the cell. Some proteins become constituent parts of the cell membrane, forming receptors for chemical messages from other cells, or as channels or gates regulating the passage of ions, water, glucose, and other chemicals in and out of the cell.

3.6.2 Variation in genes

Many of the genes (often very important ones) that code for various proteins are invariant between members of our species, whilst other genes coding for different proteins are slightly different, leading to variety. In some cases, these different variants seem to have no functional consequences, whilst in others a high degree of variation is crucial. Perhaps the best example of this is in the genes for parts of the immunoglobulin molecules that recognize foreign material (such as bacteria and viruses), where a high degree of variability is desirable. In other cases, there are variants of genes that code for the proteins that fold to form enzymes, and as a result these have different enzyme activity. A good example of this are those genes that code for the metabolic enzyme cytochrome P450, one form of which has a much-reduced catalytic activity. This variation is described as polymorphism.

From a Darwinian perspective, polymorphisms are desirable as they are means by which a popula-tion of individuals will vary in their ability to respond to new and possibly challenging environmental situations, and some polymorphisms will cope well. Polymorphisms in certain genes result in different eye colours, but these seem to have no physiological or Darwinian effect. However, equally, certain polymorphisms can result in functional changes to proteins that are undesirable (such as complete loss of enzyme activity, or changes to growth patterns), and these can be lethal.

3.6.3 Mutations

Another way of looking at polymorphisms is to view them as mutations. Consider a normal gene (e.g. that above, with the nucleotide sequence ATATAGGAT) that codes for part of the active site of a perfectly well functioning enzyme that we may describe as Xase, coded for by *Xase*. Should there be a change in the series of nucleotides that make up this gene to, shall we say, ATATAGGAG, this may well lead to a change in the structure of the active site of the enzyme, which we will call Xase^m. Because this change has occurred at a single position in the sequence, we call it a point mutation, or perhaps a **single nucleotide polymorphism (SNP)**. We would then say that Xase^m is a mutated form of Xase, although others may say that *Xase* and *Xase^m* are simply alleles. Without wishing to get too bogged

TABLE 3.4 Mutations in different forms of cancer

Cancer	Number of mutations reported
Retinoblastoma	1–90
Medulloblastoma	7–200
Testicular cancer	100–1,000
Leukaemia	100–1,100
Colon cancer	850–5,000
Pancreatic cancer	2,000–11,000
Melanoma	5,000–80,000
Lung cancer	11,000–200,000

down in semantics, the point is that if the altered gene $Xase^m$ coding for $Xase^m$ results in a disease, then the altered gene is a mutation.

Therefore, mutations in DNA cause cancer. However, there is enormous variation in the type of mutation, ranging from a change in one single nucleotide to changes in large sections of chromosomes (translocations etc., as depicted in Figure 3.7). Furthermore, there is great variation in the number of mutations found in the tissues of a particular cancer (Table 3.4), although not all of these mutations are carcinogenic—the table merely gives an indication of the susceptibility of the normal tissue to mutation. There is also variation in a particular mutation in different cancers—*RAS* mutations are found in 60% of pancreatic cancers, 50% of colon cancers, and 20% of lung cancers, but in only 1% of renal cancers, whilst mutated *BRCA1* is present in 3–8% of cases of breast cancer, but in 18% of ovarian cancers.

SELF-CHECK 3.6

How many genes are in the human genome?

Key points

The sequence of nucleotides in the DNA defines genes for protein products. Variations in these sequences (mutations) lead to abnormal genes that may code for abnormal proteins, which may in turn cause diseases such as cancer and sickle-cell anaemia.

3.7 Summary of basic cell function

The purpose of sections 3.2–3.6 is to provide a brief synopsis of the ways in which the normal cell functions. Regulation of the cell begins with signals received at the cell surface by receptors (perhaps for growth factors or hormones), which are passed by second messengers to the nucleus, where effector functions proceed. Other signals may induce the replication/reproduction of the cell, which passes through several stages, controlled by cyclins and related enzymes.

When a cell and/or its DNA are irreparably damaged, it is eliminated by the process of apoptosis. Errors in chromosome numbers or the sequence of DNA lead to mutations in genes which, in turn, may lead to disease. Cancer can develop from any combination of errors in the way in which the cell functions and replicates. Chromosomal and gene errors are a frequent finding in many diseases,

TABLE 3.5   Examples of chromosome and gene abnormalities and their associated diseases

Abnormality	Lesion: disease
Translocation	t(9;22)(q34;q11) and chronic myeloid leukaemia
	t(2;3)(q13;p25) and follicular thyroid cancer
Deletion	del(2)(p16.3-21) and Lynch syndrome
	del(7)(q11.23) and William syndrome
Inversion	IVS6 in Xq26.2-26.3 and Lesch–Nyhan syndrome
	ALK intron 8/19 breakpoints and lung adenocarcinoma
Duplication	*cMYC* in numerous cancers
	Sections of *FLT3* at 13q12.2 and acute myeloid leukaemia
SNP	R273C in *p53* at 17p13.1 and prostate cancer
	E6V in *HBB* at 11p15.4 and sickle-cell anaemia

and so these principles will surface many times in almost all chapters of the current volume and are the focus of Chapter 16. Examples of disease linked to chromosome and gene defects are shown in Table 3.5.

Now that we are more informed about the physiology of the cell, we are in the position to be able to examine in more detail how cancer arises, the process of carcinogenesis.

3.8 Carcinogenesis

Carcinogenesis is the process by which cancer develops. Should there be an identifiable external cause (such as an industrial chemical or exposure to high-energy radioactivity), that agent is a carcinogen. Indeed, we read in section 2.4.2 of the huge historical risk of scrotal cancer in chimney sweeps, and now have the mechanism to explain this fact, that being the carcinogens in soot (see Box 3.1).

Many use mutagen and mutagenesis interchangeably with carcinogen and carcinogenesis, whilst others would argue that there are subtle differences. However, not all cancers have an immediately identifiable physical carcinogen. Some mutations, such as those in *BRCA1* (from BReast CAncer type 1, a tumour suppressor gene mapped to 17q21.31) result in a loss of function in its product (BRCA, part of a DNA-repair complex) and so cause a greatly increased risk of breast and ovarian cancer. Furthermore, these mutated genes are heritable, which implies that they are stable, and so could, in theory, have developed thousands of years ago. One possible reason why these deleterious genes are still maintained in the genome is that they are late acting, and do not cause cancer until after the woman's reproductive life is past its peak, thus bypassing neo-Darwinian natural selection. Many carcinogens will act on a large number of cells, and whilst many resist the transformation, generally only one will succumb and transform into a cancer cell. This first cell is often described as a cancer stem cell which gives rise to a clone and so a tumour (discussed in section 3.9).

A further important point is the implication that only one carcinogenic event is sufficient. Whilst this is certainly true in most cancers, there are examples of tumours (such as a tumour of the eye—a **retinoblastoma**) where two or more carcinogenic events are required—this is called the 'two-hit' **Knudson hypothesis**. A further refinement of this hypothesis is the point that although one 'standard' carcinogen is needed, the second stimulus need not necessarily be a carcinogen—it may also be some other pseudo-physiological factor, such as increased levels of a blood cofactor, or body temperature. We shall return to this hypothesis in subsequent sections of the chapter. Carcinogens can be classified as endogenous or exogenous.

BOX 3.1 *Chimney-sweep cancer: A chronology*

1775: Pott notes chimney sweeps are at increased risk of scrotal cancer.

1890: Spencer proposes that sweat running down the torso causes soot to accumulate in the scrotal rugae.

1894: von Volkmann describes scrotal cancer in those exposed to paraffin and tar.

1921: Kennaway shows that coal has a cyclic hydrocarbon that is carcinogenic.

1922: Passay exposes mice to an extract of soot: they develop skin tumours.

1930s: Kennaway, Cook, and Hieger demonstrate that polycyclic aromatic hydrocarbons (PAHs) such as benzene, benzapyrene, and anthracene are carcinogenic in mice, causing skin tumours.

1950s–1960s: Numerous workers produce animal, *in vitro*, and biochemical evidence supporting the hypothesis that PAHs have the potential to cause human cancer.

1975: McCann and colleagues demonstrate that PAHs are mutagenic to DNA.

1982: Eisenstadt and colleagues report that PAHs cause alterations in nucleotide chemistry, thereby linking them with potential DNA mutation and carcinogenesis.

1991: Puisieux and colleagues provide data supporting the hypothesis that benzapyrene induces mutation in *p53* leading to the loss of its tumour suppressor functions.

3.8.1 Endogenous carcinogens

These are carcinogens that arise from within the body. One of the ways in which the body defends itself from microbiological attack is by the effects of reactive oxygen species (ROS) such as the hydroxide radical and superoxide generated by phagocytes. Unfortunately, in addition to causing damage to microbes (and, ideally, their elimination), they can also damage healthy tissue, and also DNA, causing the breaks and mutations that can lead to malignant transformation. High levels of ROS within the cytoplasm (that cannot be addressed by the cell's natural antioxidant defence pathways) can be generated by hyperoxia and ultraviolet light, both mechanisms that can trigger apoptosis.

Excess levels of hormones can also cause cancer, such as in the case of oestrogen and breast cancer, and erythropoietin in a range of cancers, probably as they act as growth factors and thus essentially overstimulate the cell. As part of the natural distribution, there are inevitably some individuals whose DNA is 'naturally' fragile and prone to excessive error breaks that seem to have no cause and bring a risk of oncogenesis. This would be compounded by failure of DNA repair enzymes, which are under genetic control themselves, to correct the problem.

3.8.2 Exogenous carcinogens

These arise outside the body. Probably the most effective and well-known chemical carcinogens (certainly from a public health perspective) are found within tobacco smoke and alcohol. To these can be added hundreds of industrial chemicals (such as benzene, dioxin, and asbestos) and heavy metals (cadmium, nickel, arsenic, and chromium) known to cause cancer. Although there were suspicions of the deleterious effects of radioactive elements and ionizing radiation in the first part of the 20th century, their carcinogenic effects were unequivocally demonstrated following the events in Japan in

August 1945. It follows that certain isotopes of radioactive elements (radium, uranium, plutonium, etc.) are carcinogens. 'Biological' exogenous carcinogens (microbial pathogens and their products) are discussed in section 3.8.3.

Key points

Carcinogens are endogenous or exogenous agents that directly or indirectly cause damage to DNA and so initiate carcinogenesis. Examples include noxious chemicals in tobacco smoke, radioactivity, and certain viruses.

3.8.3 Mechanisms

Whatever the causative agent, a common theme is that the carcinogen, directly or indirectly, disrupts the normal functioning of the cell. This may occur by altering the second messengers involved in signal transduction, and/or at the level of the DNA in the nucleus. A normal gene that has been altered in such a way that it causes cancer is an oncogene and, if present, the related product would be an oncoprotein. There are several direct and indirect mechanisms by which carcinogens cause cancer.

High-energy ionizing radiation (UV light, X-rays, emissions from radioactive elements, especially those emitting high-energy alpha and beta particles) can directly damage DNA by breaking down bonds between and within nucleotides. The radiation may also act indirectly by promoting ROS that themselves can cause nucleotide bonds to break and so possibly lead to incorrect reform/repair.

Industrial carcinogens

We have already noted the carcinogenic potential of heavy metals and polycyclic aromatic hydrocarbons such as benzene (Box 3.1) and the products of the combustion of tobacco. All have the capacity to interfere with second messengers (such as p53), and penetrate the nucleus and so directly damage the DNA, such as by intercalating and alkylation of nucleotide bases, or (in the case of transition metals) providing the template for bond rearrangement.

Aflatoxin

This potent natural carcinogen is derived from certain moulds which may grow on a wide variety of foodstuffs. Taken in sufficient quantities, it may cause acute liver failure, but more commonly, repeated low doses are linked to liver cirrhosis and then carcinoma. The suggested mechanism lies with mutations in p53 (as with so many other cancers). Nitrites taken in the diet may be converted in the liver to carcinogenic nitrosamines.

Hormones and growth factors

Oestrogen, insulin, and erythropoietin are hormones that prompt their target cell into action, and some cells rely on a constant signalling to remain functional (that is, as 'stay alive' or growth factors). Indeed, there are physiological and pathological consequences of falling levels of these hormones. Similarly, there are consequences of high levels, which effectively overstimulate the cell. Although oestrogen can also promote damaging ROS *in vitro* that could act as a direct mutagen on DNA, a more likely mechanism is overstimulation of the target breast cell by binding to the oestrogen receptor. In support of this is the fact that a key aspect of the treatment of breast cancer (as explained in Chapter 4) is whether the tumour expresses a high number of oestrogen receptors. Furthermore, long-term use of synthetic oestrogens in contraception or in hormone replacement therapy is linked to the development of cancer of reproductive tissues (breast, ovary, endometrium). The use of the hormone erythropoietin to treat a certain type of anaemia also resulted in an increased frequency of cancers.

An explanation for this is that the tumours expressed receptors for erythropoietin, and so regard the hormone as a growth factor.

Receptors

Section 3.2.1 focused on the interaction between a ligand and its receptors, and the immediately preceding paragraph describes how increased levels of the ligand can be pathological. The same is true for increased expression of receptors, which will therefore increase binding of ligands and so also overstimulate the cell, perhaps leading to cancer. This is a well-characterized feature of oncology, an example being the overexpression of HER2, the receptor for epidermal growth factor, which can guide treatment (Figure 3.10).

Microorganisms

The bacterium *Helicobacter pylori* is believed to cause gastric cancer by enhanced production of ROS, although an alternative aetiology is a hyperactive inflammatory response. There is also evidence that *Salmonella typhi*, *Streptococcus bovis*, and *Chlamydia pneumonia* are linked to malignancies. Certain parasitic worms are also carcinogenic; the most well-known link is that between certain *Schistosoma* species and bladder cancer.

Several **oncoviruses** have been described (Table 3.6). As a retrovirus, hepatitis C virus inserts sections of its nucleic acid directly into the host genome, perhaps near to a **proto-oncogene** that is then transformed to a full oncogene. Other viruses disrupt second messengers, and many cause disease not directly and immediately related to cancer (such as jaundice and cirrhosis in hepatitis viruses), although over decades of infection they may indeed transform into cancer. Infection with the human papilloma virus (HPV, linked to 5% of all cancers) produces pre-cancerous growths (warts) and many forms of genital, mouth, and anal cancers. One possible aetiology is that HPV produces proteins E6 and E7 that (in common with other oncoviruses) respectively inactivate p53 and pRb, with the consequence that they fail to prevent damaged cells from continuing the cell cycle and transforming.

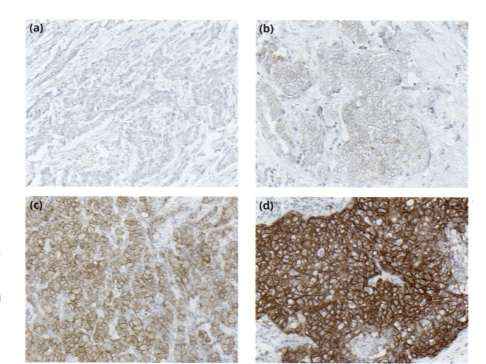

FIGURE 3.10

Expression of increased levels of HER2. Panels (a) to (d) demonstrate increasing expression (negative, 1+, 2+, and 3+) of HER2 in breast cancer.

From Orchard & Nation, *Histopathology*, second edition, Oxford University Press, 2018.

TABLE 3.6 Oncoviruses

Virus	Cancer
Epstein–Barr virus	Burkitt's lymphoma, nasopharyngeal carcinoma
Hepatitis B virus, hepatitis C virus	Hepatocellular carcinoma
Herpes simplex virus	Kaposi sarcoma, multiple myeloma
HTLV-1	Adult T cell leukaemia/lymphoma
Papilloma virus	Cervical, skin, and laryngeal cancer

However, the fact that not all oncoviruses always cause a cancer in their hosts implies that other mechanisms may also be important, fulfilling Knudson's two-hit hypothesis. An example of this is that the virus may be held in check by the immune system, only to become a functional oncovirus should it escape from immunosurveillance during a period of immunosuppression.

Between them, the infectious agents described in Table 3.6 are estimated to be linked to 15% of cancers. However, none of them should be carcinogenic if the existing DNA repair mechanisms are intact and functional (covered in section 3.3.4). Failure to repair damaged DNA with the correct sequence (possibly because of defects in the repair mechanism, or failure of apoptosis) may create an oncogene. The errors of translocation, deletion, and inversion (possibly caused by radiation) can all lead to the development of an oncogene. A subtler mutation is a proto-oncogene—a normal gene that is easily mutated into an oncogene. There are numerous examples of these, for instance the genes that code for second messengers such as Myc, Ras, and p53.

SELF-CHECK 3.7

Give an example of each of the major types of carcinogen.

3.8.4 Epigenetics and non-coding RNA

The genome is more than simply the sum of its nucleotide sequences, and errors in genes need not only be caused by changes in the sequence of bases in the DNA. **Epigenetics** refers to the alterations of structural aspects of genes and histone proteins that influence transcription, and DNA repair and replication. The DNA sequence itself is not changed. Epigenetics has attracted interest as defects are potentially reversible (perhaps by pharmacological intervention), unlike genetic defects. Addition or removal of methyl, acetyl, phosphate, and ubiquinone to the components of chromatin have all been described, whilst agents that induce epigenetic changes include hypoxia, inflammation, hormones, and aromatic hydrocarbons such as benzene.

Although only some 2% of the human genome codes for protein, it has been suggested that 70% to 90% of the mammalian genome is transcribed, leaving much RNA unaccounted for. Although some of these 'missing' transcripts are transfer and ribosomal RNA, much of the remainder are now being characterized, and may be grouped according to size. Four major forms of epigenetics are recognized.

DNA methylation

The most common modification of DNA is methylation of a cytosine nucleotide (where adjacent to a guanine nucleotide) by a DNA methyltransferase (DNMT, of which there are several isoforms) to form 5-methyl-cytosine. As the cytosine and guanine bases are linked by a phosphate group, the target of the transferase is denoted as CpG, and within the human genome some 70% of CpG sites are methylated. This is important because CpG sites (or 'islands') are often found near the start of transcription sites known as promoter sequences. Since these promoter regions (40% of which are linked to a CpG island) are part of the system for gene transcription, then too much (hypermethylation) or too little (hypomethylation) of this process can lead to changes in the expression of the relevant gene, most changes being to induce its silences. Errors in methylation can also lead to the removal of a section of 'healthy' DNA, and so an alternatively spliced transcript. This therefore is a form of gene deletion. A further problem is that once methylated, the methylcytosine can shed its amine group, and transition to a thymine reside, with further consequences for the local DNA sequence.

The most well-characterized epigenetic hypermethylation changes result in the silencing of tumour suppressor genes such as *BRCA1* (resulting in an increased risk of breast and ovarian cancer), *MLH1* (MutL homolog 1, linked to colon cancer), and the von Hippel–Landau gene, the importance of which is discussed in section 3.10.2. Blood cancer such as leukaemia may be linked to loss of function mutations in methyltransferase gene *DNMT3A*, (coding for the enzyme DNA (cytosine-5)-methyltransferase-3a) which seems to increase proliferation and inhibit the differentiation of immature blast cells.

Histones

These proteins are the major non-DNA component of chromatin and ensure efficient packaging of the DNA (without which the sum length of DNA in a cell would be 2 metres), but also determine which sections of DNA are available to transcription factors or other regulators. Thus, changes to histones may lead not only to disordered DNA folding and nucleosome instability, but also to exposure of the nucleotide sequences to changes that may lead to errors of transcription and translation.

Histones are a family of molecules with dozens of isoforms that can be further grouped into H1, H2A, H2B, H3, and H4, coded for by gene clusters at 6p22.2, 1q42, and 1q42.13. The base unit is cylinder/sphere octamer of two each of H2A, H2B, H3, and H4 subunits around which the DNA winds. The nucleosome is completed by an H1 linker molecule that ties off the DNA strands, and a 10–60 base-pair section of DNA linking two nucleosomes (Figure 3.11).

Each subunit has a protruding tail that can be subject to modifications that include phosphorylation, ubiquitination, and hyper- and hypo- methylation and acetylation, changes mediated by enzymes. For example, histone-lysine N-methyltransferase 2D catalyses a methyl group on to a lysine residue on H4. Acetylation confers a positive charge and makes for looser packing of nucleosomes, which in turn makes the DNA more accessible for transcription.

These enzymes are frequently overactive in prostate, breast, skin, lung, colorectal, and liver cancers, resulting in increased cell growth and the promotion of malignant transformation. Stress conditions may result in the upregulation of enzymes to remove the methyl group (i.e. histone lysine demethylase), a change also implicated in carcinogenesis. Therefore, the balance between these methyl transferases and demethylases is an important regulator of cell-cycle gene expression and chromatin structure.

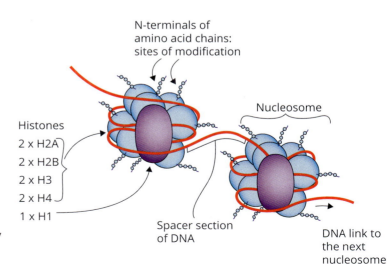

FIGURE 3.11
DNA looping around nucleosomes: Exposed lysine and arginine residue of the histone tails may be modified, leading to change in the unravelling processes prior to transcription.

In an acetylation, the enzyme histone acetyltransferase (HAT) catalyses the transfer of an acetyl group from acetyl-coenzyme A to a lysine residue on a histone, whereas histone deacetylase (HDAC) removes an acetyl group. As with the lysine methyl transferases, the balance between these lysine acetyl transferases is also important in the regulation of the cell cycle. Mutations in genes (such as *EP300*) encoding these acetyltransferases have been identified in leukaemia, colorectal cancer, and breast cancer. The consequences of all these histone modifications include altering the transcription activity of a gene, repression of genes, and marking sites of DNA damage. Unsurprisingly, these histone-modifying transferase enzymes are now the object of a class of inhibitors being trialled in sarcoma, myelodysplastic syndromes, myeloma, various forms of follicular and diffuse B-cell and T-cell lymphoma, and breast cancer. In some clinical trials these inhibitors are being used alone, and in others they are adjuvants to standard chemotherapy such as dexamethasone and tamoxifen.

MicroRNAs

microRNAs (miRNAs) are small (19–25 nucleotide) regulatory sections of RNA coded for by nuclear DNA and transcribed by RNA polymerase II. Each of them has a unique name consisting of 'miR' followed by a hyphen and a number. About 2,500 miRNAs are present in the genome, a figure that compares with 20,000–25,000 protein-coding genes, and their function is usually to silence the mRNA they complement, either by RNA cleavage or by suppressing translation. However, there are also instances in which a miRNA increases the expression of its target mRNA, and some miRNAs have DNA as their target (an example of the latter is miR-373). The initial miRNA transcript is of a hairpin structure of some 70 nucleotides that is subsequently edited by enzymes Drosha (within the nucleus) and Dicer (in the cytoplasm) into a short, straight chain. Once edited, miRNAs combine with an argonaute to form the RNA-induced silencing complex, a macromolecule, when the target mRNA is degraded (Figure 3.12).

Genes that code for miRNAs can either be found by themselves or in **polycistron** clusters, such as the miR-17-92 cluster at 13q31-q3, which consists of miRs 17, 18a, 19a, 19b, 20a, and 92a. Each miRNA targets several mRNAs, and it has been suggested that more than a third of our genes are miRNA targets. A large proportion of miRNAs are associated with the CpG islands of histone methylation described above, implying a form of epigenetic regulation, errors in which may be related to carcinogenesis. For example, in T cell lymphoma there may be methylation of the DNA that codes for miR-203, a change not seen in normal T lymphocytes, whilst an early event in some cases of breast cancer is the epigenetic silencing of several miRNAs.

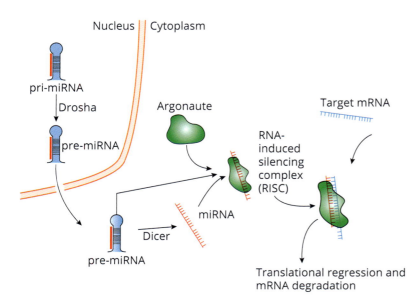

FIGURE 3.12

Formation and function of miRNAs. Pri-miRNA are processed in the nucleus, pre-miRNAs in the cytoplasm, by Drosha and Dicer enzymes respectively. The mature miRNA forms the RNA-induced silencing complex (RISC) with an argonaute protein, although Dicer may generate miRNAs as part of the RISC. The target miRNA complexes with its complementary miRNA within the RISC and is degraded.

These small molecules have been implicated in numerous cancers, such as subtypes of lymphoma and leukaemia that exhibit a deregulated miR expression pattern (miR-34a, miR-29, and miR-17-5p) in patients with deleted and/or mutated *p53*. These miRNAs target genes with roles in various pathologies, and cell-cycle (*E2F1*, *p21*, *Cyclin D1*) and hypoxia-related genes (*HIF-1α*). There are several instances of the downregulation of miRNAs in a number of cancers, and others where there are high levels of miRNAs (Table 3.7).

One of the first miRNAs in cancer to be reported was in chronic lymphocytic leukaemia (CLL). miR15 and miR16 are both coded at 13q14 and are either deleted or downregulated in more than two-thirds of cases of CLL. Their expression has been found to inversely correlate with the expression of the protein Bcl-2, which has roles in apoptosis (as discussed in section 3.4.1). The role of miR-650, which is coded for by a gene located near the immunoglobulin kappa light chain (IgLλ) in CLL provides a further case model. The relationship between the two may be that the IgLλ promoter is an enhancer element for miR-650. The link with transformation is that miR-650 reduces the expression of CKD1, which plays a key role in the control of the cell cycle checkpoint at G_1/S. CLL patients with higher expression of miR-650 showed improved survival rates and a longer time to first treatment, which suggests that it has clinical relevance and a role in management. The oncogenic potential of miR-650 extends to colorectal, breast, and gastric cancer and many other malignancies.

There are now hundreds of reports of other miRNAs in various cancers and with different effects on the cell. The fact that many miRNAs are specific for certain cells and organs, and that levels are abnormal in their particular malignancy, gives rise to a value in diagnosis. For example, miRNA let-7 is downregulated in lung cancer, so that knowledge of levels in a biopsy sample from a presumed malignancy may be helpful in diagnosis. Similarly, that ratio between miR-196a and miR-217 may help differentiate pancreatic carcinoma from pancreatitis. Downregulation of miR-600 is linked to a poor prognosis in breast cancer, identifying patients in need of greater care. Advances in the laboratory now permit isolation and quantification of miRNA by polymerase chain reaction (PCR) from standard histology paraffin-wax embedded tissues that are up to a decade old.

In some cases, the sensitivity and specificity of miRNA measurable in plasma exceeds that of routine cancer markers, leading to the possibility that they may direct diagnosis and the assessment of management. For example, not only is serum miR-155 diagnostic for CLL, but higher levels are also associated with poor response to treatment. Serum miR-150 and miR-342 levels are raised in acute myeloid leukaemia but return to normal levels in patients in complete remission. Several miRNAs can be detected in urine and have value in diagnosing bladder cancer. Furthermore, direct targeting of certain miRNAs (e.g. miR-122 in hepatitis C virus infection, hoping to reduce the risk of cirrhosis and hepatocellular carcinoma) is a potential novel treatment.

miRNAs have roles in other forms of physiological gene regulation (such as those of miR-23a, miR-221, and miR-222 in erythropoiesis) and in other pathology (such as miR-451 in haemoglobinopathy;

TABLE 3.7 miRNA changes in selected cancers

Cancer	miRNA Downregulated/ underexpressed	miRNA Upregulated/ overexpressed
Lung	Let-7, 29, 143, 145, 155	21, 26a, 205
Breast	Let-7, 125a, 126a, 143, 145, 155, 181, 200	21, 26a
Prostate	Let-7, 15a, 16-1, 101, 125a, 125b, 181, 205	21, 106b-93-25 cluster
Colon	Let-7, 29, 155	21, 26a, 106b-93-25 cluster
Acute myeloid leukaemia	221, 222	Let-7, 125a, 125b
Myeloma	15a, 16-1	26a, 106b-93-25 cluster

Note: This table is intended to be neither comprehensive nor exhaustive.

miR-155, miR-22, and miR-133 in heart failure; miR-145 in lung cancer; and miR-223 in chronic kidney disease). miRNAs are also of direct value in clinical care. In glioblastoma, an inhibitor of miRNA-21 in combination with targeted anti-EGFR treatment improves outcome compared to single-agent therapy, whilst the combined miR-99a/Let-7C/miR-125b signature may identify which patients with metastatic colorectal cancer are good candidates for treatment targeting EGFR. Table 3.7 presents changes in certain miRNAs in selected cancers.

Small non-coding RNAs comprise some 40% of all non-coding RNAs (the remainder being lncRNAs—see below), and of these miRNAs are the most common, themselves making up 40%. Other small ncRNAs include small nuclear RNAs (20%), small nucleolar RNAs (10%), and ribosomal RNA (5%). The remainder consists of small-interfering RNAs, piwi-interacting RNAs, transfer RNAs, circular RNAs, and miscellaneous poorly characterized forms.

Long non-coding RNAs

This group consists of long (200 to thousands of nucleotides) non-coding RNA (lncRNAs) molecules, many of which may have a role in carcinogenesis as they are found in certain cancers, similar to miRNAs. However, the number of lncRNAs in the genome (50,000–60,000) is much higher than that of miRNAs and protein-coding genes. But although there are many more lncRNA species than miRNAs, there is currently no clear numbering system for them comparable to that for miRNAs; instead, they are named by abbreviations that reflect their origin. The lncRNAs are a highly heterologous group with considerable variability of structure and are very likely to have multiple functions (Figure 3.13).

The observation that NFκB has binding sites in non-coding as well as coding sections of the genome has led to the observation that it may regulate lncRNAs, and it is now known that p53 is another regulator of certain other lncRNAs. Other inducers of lncRNAs include high temperature, hydrogen peroxide, high glucose, hypoxia, hyper-osmotic stress, and heavy metals, many of which also influence miRNAs. The >200 nucleotide size of lncRNAs gives a three-dimensional structure that permits several interactions unavailable to miRNAs. These include interacting with proteins, DNA, in peptide coding, and (possibly) action as a catalytic RNA. lncRNAs may also regulate other epigenetic processes, such as miRNAs.

An example of this is increased lncRNA EWSATI in nasopharyngeal carcinoma that in turn increases levels of miR-326/330-5p clusters that target cyclin gene D1. Furthermore, lncRNAs Xist (a major

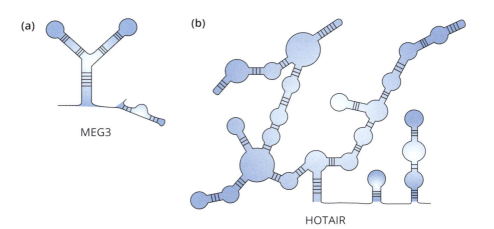

(a)

MEG3

(b)

HOTAIR

FIGURE 3.13
Secondary structures of various lncRNAs. (a) The two sections of MWG3 are separated by ~500 nucleotides. (b) The very large size of HOTAIR brings with it the possibility of multiple functions. These secondary structures are unlikely to reflect tertiary structures.

effector of X chromosome inactivation) and ANRIL (abbreviated from antisense non-coding RNA in the INK4 locus) have effects in histone modification (and many other functions such as gene silencing and heterochromatin formation).

LncRNAs are altered in several cancers and are likely to be involved in carcinogenesis. In breast cancer, upregulation of lncRNA MALAT1 regulates cancer cell migration and cell cycle progression, particularly in tumours that are oestrogen receptor and HER positive tumours. lncRNAs H19 and DSCAM-AS1 are also highly expressed in oestrogen receptor positive breast tumours and cell lines. Numerous abnormal lncRNAs have been described in bladder cancer: H19, MALAT1, SNHG16, TUG1, and UCA1 are upregulated, whilst BANCR and MEG3 are downregulated. Some, such as TUG1, are linked to poor overall survival, pointing to value in diagnosis and management. Mechanistic links for others are being elucidated. For example, lncRNA UCA1 (urothelial carcinoma associated 1, arising from a bladder cancer cell line) acts via WNT signalling to increase second messengers ERK1/2, MAPK and Pi3-k/AKT kinase (see section 3.2.2), leading to increased transcription factor P300 and so promotion of cell cycle progression. UCA1 is also altered in acute myeloid leukaemia, breast, and colorectal cancer. In non-small cell lung cancer it targets miR-193a-3p, thereby promoting tumour progression, whilst in colorectal cancer it has the same effect by targeting miR-204-5p.

Like miRNAs, lncRNAs also appear to have a role in physiology, for example, in interacting with key erythroid transcription factors MYC, GATA, TAL1, and KLF1. One lncRNA species, lncRNA-EC7, is an enhancer of genes coding for the major erythrocyte membrane protein Band 3. However, unlike miRNA, there is evidence that some lncRNAs are translated, although the stability and function of their protein products is unclear. There is also evidence of the involvement of lncRNAs in physiological regulation, such as the promotion of erythroid progenitor survival, erythroblast enucleation, and activation of α-globin gene expression, and in various other diseases, such as Parkinson's disease, Prader–Willi syndrome, Huntington's disease, Alzheimer's disease (all described in Chapter 16), and cardiovascular disease.

miRNA/lncRNA interactions

There are several examples of situations in which both species of non-coding RNAs compete for the same mRNA. In colorectal cancer, lncRNA NEAT1 competes with miR-34a for the mRNA from *SIRT1* coding for a deacetylating enzyme, whilst in osteosarcoma, lncRNA ODRUL competes with miR-3182 for the mRNA from *MMP2*, coding for type IV collagenase. In breast cancer, miR-148 activity represses the expression of lncRNA HOTAIR, whilst lncRNA H19 can be a precursor for miR-675 (Figure 3.12), but also acts as a sponge to miR-130b-3p.

Key points

Epigenetic changes include methylation of DNA, methylation and acetylation of histones, and the activities of miRNAs and lncRNAs are additional causes of abnormal gene activity that can lead to carcinogenesis.

SELF-CHECK 3.8

How do carcinogens cause cancer?

3.9 Genetic drivers of malignant transformation

We have established, in previous sections, that there are numerous ways in which the healthy genome can be transformed, and so how malignancy can develop. We will now examine these in detail in a number of examples.

3.9.1 *Bcr-Abl* and leukaemia

ABL-1 (or perhaps *Abl-1*, the name derived from its location in the Ableson virus), found at 9q34.1, codes for ABL-1, a nuclear tyrosine kinase that may have a role in regulating the cell cycle. ABL-1 also interacts with many other genes, including *BRAC1*, and those coding for second messengers such as *PAK-1*, and *Rb-1* (see section 3.9.2). The function of *BCR* (or *Bcr*, an abbreviation of breakpoint cluster region), found at 22q11.2, is unclear. However, like *ABL*, *BCR* interacts with numerous other genes, including those for second messengers *Grb2* and *PIK₃CG*.

A reciprocal translocation between chromosomes 22 and 9 brings together *BCR* and *ABL-1* to form a new gene, *BCR-ABL*. As the break-points are 9q34.1 and 22q11.2, the translocation is defined as t(9;22)(q34;q11) and is referred to as the Philadelphia chromosome, that being a remodelled and slightly shorter version of chromosome 22. The reciprocal translocation is a slightly longer version of chromosome 9. The new *BCR-ABL* gene codes for a fusion protein BCR-ABL, a novel form of the tyrosine kinase coded for by *ABL-1*, and which (due to the proximity of *BCL*) is permanently active, and accordingly, *ABL-1* may be described as a proto-oncogene (Figure 3.14).

As the tyrosine kinase is a major second messenger, the relevant downstream growth signals are phosphorylated and so permanently switched on, leading to uncontrolled cell division. Should this mutation occur in a bone marrow stem cell, it promotes the production of immature (and therefore non-functioning) blast cells of the myeloid lineage, which most commonly manifests as chronic myeloid leukaemia (section 11.4.3 of Chapter 11).

3.9.2 *Rb1* and retinoblastoma

Retinoblastoma has two forms: a non-familial disease that affects only one eye, and a rare familial form with tumours in one or both eyes that manifests in childhood. The latter is caused by mutations in either of two genes: *Rb1* (or perhaps *RB1*, coded at 13q14.1-q14.2) and *MYCN* (at 2p24.3, coding N-myc proto-oncogene protein). *Rb1* codes for a second messenger protein Rb1 (or simply RB) involved in G_1/S cell cycle regulation (see section 3.3). In the nucleus, Rb1 is complexed to a transcription activating factor of the E2F family, which has its own binding partner (DP), needed for the cell cycle to progress from G_1 to the S phase. When phosphorylated by the kinase portion of cyclin D-Cdk4/6, pRb uncouples from E2F/DP, which is then able to direct the cell into the S phase.

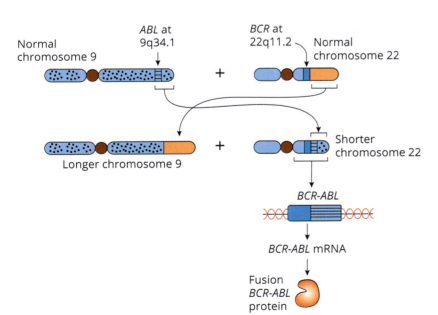

FIGURE 3.14

Formation of *BCR-ABL*. Part of chromosome 22 is transferred to chromosome 9, and vice-versa. This brings together *BCR* that was on chromosome 22, and *ABL* that was on chromosome 9 to form a new gene, *BCR-ABL*.

E2F also activates members of subsequent steps of the cell cycle, these being cyclin A and cyclin E (see Figure 3.14).

Rb1 has two alleles: mutations leading to an inactive form of Rb (such as its inability to bind to E2F/DP) therefore allow the latter to promote unregulated cell cycling. The results of this include the ophthalmological tumour that names it—that is, retinoblastoma—but also an increased risk of other cancer later in life. Indeed, those with the familial form of the disease suffer a six-fold increased rate of other cancers compared to the nonfamilial form. The exact role of the *MYCN* mutations is unclear. The molecular genetics of *Rb1* was one of the first to be elucidated, and since mutation errors in this gene led to tumour development, in its normal wild-type form it was described as a tumour-suppressor gene, and its product, pRb, as a tumour-suppressor protein. It is now known to interact with more than 200 proteins, including histone transferases described in the previous paragraphs, and DNA repair factors such as BRCA1.

3.9.3 *p53* and numerous cancers

We have already met this molecule in a number of preceding sections (e.g. those on the cell cycle, section 3.3, and apoptosis, section 3.4). *p53* (or perhaps *TP53*), located at 17p13.1, codes for p53, a molecule with many important second messenger functions. Once phosphorylated by a kinase such as MAPK, p53 can activate DNA repair mechanisms, arrest cell cycle progression at G_1/S, and initiate apoptosis. Indeed, p53 is said to regulate over 500 genes, such as in metabolism, apoptosis, and senescence. Consequently, errors in p53 physiology have many implications, such as permitting damaged cells to continue their cell cycle.

p53 is mutated in over 50% of human cancers. For example, the hereditary Li–Fraumeni syndrome is linked to an increased risk of sarcoma, leukaemia, and breast and adrenal cancer, most of which are caused by missense mutations in *p53* and so non-functioning p53. Without the regulatory presence of p53, oncogenesis can proceed. However, *p53* can be mutated by a number of factors, with the same general outcome of an increased risk of cancer, principally of the colon, breast, and lung. This may be because p53 is reported to regulate genes coding for pro-apoptotic molecules such as BAK and BAX.

Figure 3.15 summarizes several aspects of sections 3.8, 3.9.2, and 3.9.3, focusing on the relationship between Rb and p53, and their roles in the cyclin/CDK regulation in the cell cycle progression from G_1 to S phases.

FIGURE 3.15

The relationship between Rb and p53. Cell stress, damage, etc. (perhaps by hypoxia, UV light, inflammation) releases p53 from its inhibitor mdm2. p53 is now free to promote p21, one of the CIP/KIP inhibitors of the cyclin D-Cdk4/6 complex, which would otherwise catalyse the phosphorylation of Rb. Once phosphorylated, pRb disconnects from transcription factor E2F, which then transcribes genes whose products promote the passage of the cell into S phase. Should p53 fail to activate p21, or should p21 itself be non-functional, then the cyclin/CDK complex is free to promote E2F release from Rb, and possible uncontrolled and inappropriate cell cycle progression, which may be carcinogenic. Thus, both p53 and Rb are tumour suppressors.

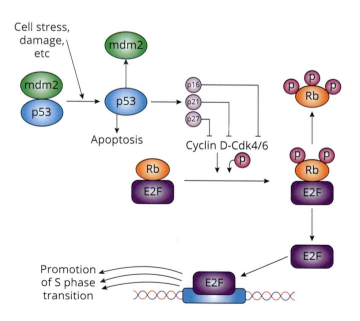

3.9.4 *Bcl-2* and follicular lymphoma

As indicated in section 3.4, apoptosis is tightly regulated by a number of factors. One such molecule, Bcl-2 (so named as it was first described in a B-cell lymphoma and coded for by *BCL-2* at 18q21) has anti-apoptotic activity. The apparatus for generating immunoglobulin heavy (IgH) chains at 14:q32 is highly active in lymphocytes involved in antibody production within lymph nodes. Some of this production will be redundant, and so the production of these unwanted lymphocytes needs to be removed by apoptosis. In a similar manner to Bcr-Abl, a translocation t(14;18)(q32;q21) can place the promoter region of IgH genes next to *Bcl-2*. This generates a new *Bcl-IgH* (onco)gene whose product, high levels of Bcl-2, effectively prevents these redundant lymphocytes from going into apoptosis, and so permits the transformation of malfunctioning lymphocytes into a lymphoma.

Although this translocation is present in 70–95% of cases of follicular lymphoma, the fact that not every case of t(14:18) leads to follicular lymphoma promotes the view of the multiple-hit model of carcinogenesis. Should t(14:18) be the primary hit, secondary hits could be provided by other genetic lesions, such as those of tumour suppression (e.g. *p53* and *Myc*), histone modification (KMTD2), and apoptosis regulation (*FAS/FASR*). Notably, high levels of Bcl-2 protein are also present in lung, breast, prostate, colorectal, and bladder cancers.

3.9.5 *c-Myc* and Burkitt lymphoma

c-Myc (or sometimes *Myc*, at 8q24.21) is part of a family (*Myc*, *MycN*, and *MycL*) of genes that code for transcription factors activating many different sections of DNA, recruiting other molecules involved in chromatin remodelling and the initiation of RNA polymerase mediated transcription. When activated by the EGF/MAPK/ERK pathway, *Myc* codes for a transcription factor (Myc) that has been linked to processes such as histone acetylation, cyclin upregulation, and, ultimately, cell replication. Myc may also interact with certain cyclin/CDK complexes and the CDK inhibitor p27 at several stages in the cell cycle. Amplification of *c-Myc*, and so high levels of Myc, are found in ovarian, colon, breast and other cancers.

As with Bcl-2 and follicular lymphoma, a t(8;14)(q24;q32) translocation can place *c-Myc* next to the *IgH* promoter locus at 14q32 to create a hybrid neogene of *Myc/IgH* that results in the overactivity of *c-Myc* and so increased generation of Myc. This translocation is present in around 85% of cases of Burkitt lymphoma, the remaining 15% being linked to other translocations, such as those involving genes for immunoglobin light chains kappa (t(2;8)(p12;q24)) and lambda (t(8;22) (q24;q11)).

Notably, this association is particularly strong in areas of equatorial Africa where both the Epstein–Barr virus (EBV) and malaria are endemic. If this link is causal, it provides the support for the multi-hit model of carcinogenesis, as both these pathogens, alone and in combination, could induce a state of immunosuppression in the patient and so permit the 'escape' of a potentially malignant lymphocyte that is the basis of a **lymphoma**. However, cases do occur outside Africa and in the absence of EBV.

Key points

There are numerous examples of links between tumour suppressors, oncogenes, their oncoprotein products, and malignant transformation.

SELF-CHECK 3.9

Why do we describe *p53* and *Rb* as tumour suppressors?

3.9.6 Summary of cell transformation

The mutations that imply genetic instability that drives mutations we have just described (and many others) confer on the cell a number of features that enable the successful transformation of a malignant cell into a tumour. Hanahan & Weinberg (2011) summarized these as follows:

1. The basis of all neoplasia is their ability to rapidly divide, and so sustain proliferative capacity. This can proceed via several pathways, such as the cancer cells secreting growth factors in a form of auto-stimulation or inducing nearby normal cells to secrete these factors. Tumour cells may also upregulate growth factors receptors at the cell surface, making them hyper-responsive.

2. The loss of control over the cell cycle by tumour suppressors such as Rb and p53 is an important factor. Normally, these molecules part-regulate the cyclin–CDK pathways and the repair of damage. Without them, potential carcinogenic events may proceed.

3. Another key aspect is the avoidance of cell death by apoptosis, perhaps by failing to respond to the Fas pathway, or changes in the balance of pro- and anti-apoptotic signals from Bcl-2 family members. Pathway complexity is underlined by the participation of p53 in apoptosis, whilst cancer cells can use autophagy to protect themselves from toxic process that would kill a normal cell.

4. Telomeres at the ends of chromosomes protect the DNA from erosive damage. The length of these nucleotide sections slowly shortens each time a cell replicates, and their full erosion leads to senescence and apoptosis. Cancer cells can replenish their telomeres, thus avoiding this problem, which can be interpreted as enabling replicative immortality.

5. An early laboratory finding was that certain transformed cells often failed to adhere to a solid substratum (loss of anchorage dependence), which we can now recognize as part of the process of the ability to be mobile, and to invade other tissues and send out metastases, potentially linked to the loss of adhesion molecule E-cadherin. Loss of function of this and similar adhesion molecules has been observed in several types of neoplasia.

6. One of the mechanisms by which the hyperactive tumour cell ensures sufficient energy (often in anaerobic conditions) is by reprogramming its metabolic pathways, such as by increasing its GLUT receptors to import more glucose, and in upregulating its glycolytic pathways.

7. The long-noted observation of cells of the innate (macrophage) and adaptive (lymphocyte) arms of the immune system within and around tumours was presumed to be an anti-tumour response. Paradoxically, there are now data to suggest that inflammatory cells contribute to the success of the tumour, perhaps by secreting growth factors and supporting the dissolution of the extracellular matrix. There is also evidence that local fibroblasts may be co-opted to support the tumour. These, and other features, have led to the view of a pro-tumour microenvironment.

Each of these processes is likely to have arisen following different mutations, and the scheme described here does not imply that they develop in any set order, although some may be important.

3.10 Clonality, tumour development, and metastasis

The previous sections have focused on the process by which a normal cell becomes a cancer cell (possibly a cancer stem cell). In the final section of this chapter we will look at the growth of tumours, and how they spread from their origin to other sites—the process of metastasis. In almost all cases, cancer transformation occurs in one single cell, whose rate of growth exceeds that of its normal neighbours to form a single nodule (a clone) of identical abnormal cells, and as it grows it will form a tumour. Some describe this single transformed cell as a cancer stem cell, whereas others use the term to describe a cell that seeds a metastatic colony. However, as the nodule of tumour cells grows, it faces several logistical issues.

3.10.1 Metabolism

As one of the definitions of a cancer cell is its increased growth rate, it follows that it must have changes to its metabolism to effect these changes, described by some as a reprogramming. Alterations in both **substrate-level phosphorylation** and **oxidative phosphorylation** pathways are known, whilst cancer cells are also more efficient at obtaining energy and materials for building blocks from other sources, notably amino acids. Cancer cells therefore exhibit greater metabolic heterogeneity, flexibility, and adaptability than normal cells, a phenomenon also expressed by cells undergoing metastatic transformation.

Glycolysis

A common characteristic of cancer cells is the increased rate of glycolysis, even in the presence of abundant oxygen, producing lactate and so other intermediates for the synthesis of macromolecules from lipid, amino acid, and nucleic acid components. Examples of this include increased hexokinase and phosphofructokinase activities. Several parts of these metabolic pathways are promoted by second messengers such as MYC (activating the transcription of genes for almost all glycolytic enzymes) and KRAS (promoting the import of glucose by upregulating cell membrane GLUT1). Increased glycolysis also produces pyruvate (with increased pyruvate kinase activity), the substrate for the acetyl group carried into the Krebs cycle by coenzyme A (i.e. acetyl-CoA).

The Krebs cycle

Alterations in enzymes of the Krebs cycle (aconitase, succinate dehydrogenase, fumarate hydratase, etc.) have been implicated in neoplastic transformation in a variety of different models of carcinogenesis. An example of this is gain-in-function mutations in genes coding for isocitrate hydrogenases that accelerate ATP production but which also increase histone and DNA methylation. Furthermore, hypoxia promotes isocitrate dehydrogenase-dependent carboxylation of alpha-ketoglutarate to citrate, so supporting cell growth and viability. In addition to its role in promoting glycolysis, MYC also enhances that uptake of glutamine, an additional energy source, and its transfer into the mitochondrion where it feeds into the Krebs cycle.

3.10.2 Angiogenesis

The problems for a growing tumour arise when its size (perhaps 1–2 mm^3) outstrips its ability to be provided with enough oxygen and nutrients, the former diffusing in from the local blood supply. The same problem arises with a difficulty in removing the waste products of its metabolism. Failure of the young tumour to solve these problems will ultimately lead to its necrosis and death. Lack of oxygen in this setting is hypoxia, and tissues that are hypoxic may also be described as ischaemic (pertinent in cardiovascular disease). In order to facilitate an improved blood supply (and so counter the hypoxia), the most successful tumours engineer the development of their own capillary network to deliver their metabolic requirements and remove waste products. This process, **angiogenesis**, is found in many different physiological states, such as in the embryo and foetus, in the reconstruction of the endometrium after menstruation, and in wound healing, where new blood vessels are required. Closely related terms are vasculogenesis (which some reserve for embryonic blood vessel formation) and neovascularization.

Hypoxia as a driver

Hypoxia stimulates the production of transcription factors, the principal being hypoxia-inducible factor (HIF), of which there are at least six isotypes. One such, HIF-2α, regulates the production of erythropoietin by cells of the renal cortex and elsewhere. A gain of function mutation in its gene, *EPAS1* at 2p21, resulting in high levels of the protein, will lead to increased production of erythropoietin and so a high red blood cell count (erythrocytosis). Another, HIF-1α (coded for by

HIF1A on 14q23.2) induces the production of growth factors to stimulate new blood vessel formation. These include VEGF, angiopoietins 1 and 2, fibroblast growth factor, and PDGF. The former, of
which there are six isotypes, is the leading growth factor in angiogenesis, and acts on its target cells
(inevitably endothelial cells) via one of three specific receptors. Other molecules also have a role,
such as regulating remodelling matrix metalloproteinases, which degrade supporting tissue around
the blood vessel wall to allow endothelial cells to form new capillaries. Alongside endothelial cells,
pericytes are very likely to have a role, and fibroblasts are required for the formation of a basement
membrane and a supporting extracellular matrix to enable capillary formation that will develop into
arterioles and venules.

The most successful tumours secrete copious growth factors and other molecules to promote the
development of their own microcirculation. Indeed, these can be cancer markers, as high levels of
growth factors in the plasma must arise somewhere, and this may be an as-yet undiscovered tumour.
However, increased VEGF in pregnant women is clearly physiological (and very desirable) and circulating VEGF levels are highest in the newborn and fall as the infant develops.

Pathogenic angiogenesis

HIF-1α is overexpressed in almost all large, solid tumours (colon, breast, prostate) and consequently has
clinical implications, as does increased VEGF, perhaps the primary physiological and pathological growth
factor. Indeed, VEGF is the target for a number of drugs, perhaps the best-known being bevacizumab,
first licensed for use in colorectal cancer alongside conventional chemotherapy, a use later extended to
lung, breast, and other cancers. Angiogenesis may also be a response to certain other non-cancer pathological conditions, such as in endometriosis, proliferative retinopathy, and cardiac ischemia. In macular
degeneration, excess VEGF drives the proliferation of retinal capillaries, causing loss of vision: anti-VEGF
mAbs have been successful in this condition. There may also be a place for miRNAs, as certain of these
small RNA molecules have been linked to the promotion and inhibition of angiogenesis.

The von Hippel–Lindau gene (*VHL*) at 3p25.3 codes a protein (pVHL) that is important as part of an
enzyme complex (E3 ubiquitin ligase (EC 6.3.2.19)) that mediates the attachment of ubiquitin to certain
proteins, such as hydroxylated HIF-1α. Once ubiquitinated, such proteins are directed towards proteasomes for degradation. A mutation in *VHL* leads to an abnormal ubiquitin ligase that fails to ubiquinate
HIF-1α, which therefore remains active. Increased levels of HIF-1α can lead to increased levels of erythropoietin and other growth factors that have HIF-responsive elements, such as VEGF and PDFG, and so
the increased angiogenesis required for the growth of the tumour. Increased growth factor activity leads
to VHL disease, characterized by an increased risk of high blood pressure, visual disturbances, and a
variety of conditions such as stroke, myocardial infarction, haemangioblastoma, retinal hypervascularity,
and renal cell carcinoma. Accordingly, pVHL may (like pRb and p53) be described as a tumour suppressor and the mutated *VHL* an oncogene. However, growth factor secretion may also be driven by aberrant
gene activity, such as by *Ras* and *Myc*, and as discussed, growth factors may also arise from normal cells of
the peri-tumour microenvironment. Figure 3.16 illustrates the process of angiogenesis.

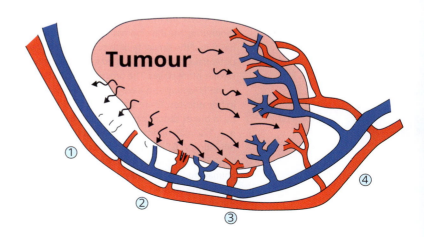

FIGURE 3.16
Stages in tumour-driven angiogenesis.
Prompted by hypoxia, at the left-hand side
(1) angiogenic growth factors (represented by
arrows) diffuse towards nearby blood vessels.
Endothelial cells, pericytes, and fibroblasts in
turn respond and build capillaries up a growth
factor gradient towards the tumour (2). This
proceeds with the development of capillaries
into arterioles and venules (3), which develop
further and mature into the tumour itself (4).

3.10.3 Metastasis

Metastasis is a key pathophysiological feature of cancer—some would say *the* key factor that distinguishes cancer from non-malignant tumours—and is clinically critical as up to 50% of patients have detectable metastases at diagnosis. Not only does the primary tumour grow *in situ*, but it also sends out fragments to colonize other tissues, and so set up a secondary tumour elsewhere. For those solid tumours of an epithelial nature (which comprise over 75% of all cancers), the key step is the development of a mesenchymal state (hence epithelial to mesenchymal transition: EMT), where metastatic cells acquire invasive and migratory potential. An EMT is also present physiologically in embryology, and in wound healing, fibrosis, and repair. An early theory of metastases was 'seed and soil', whereby the tumour sent out many seeds, but only those that fell on agreeable soil (in other tissues) would take root and succeed. Some commentators take a broad view of the cancer stem cell theory, extending it from those cells in the earliest stages of transformation to metastatic cells that form new colonies.

A further theory is of dormancy, which describes a lag phase between the arrival of the seed cells and the growth of a tumour. An extension of this view suggests that the local microenvironment at first resists the malignant cell but is eventually turned into an ally. This has some plausibility, as there is evidence that the microenvironment of the primary tumour supports its growth (see point 7 in section 3.9.6), and thus the 'soil' where the secondary tumour is developing (consisting of macrophages, endothelial cells, fibroblasts, and adipocytes) should also be supportive. Although not all tumours become metastatic, it is almost inevitable. An exception to this is carcinoid, a form of neuroendocrine tumour, generally of the intestinal tract, that often behaves in an indolent manner, and may be benign although, rarely, metastases may develop (discussed in section 5.13.2).

Mechanisms

The biology of metastasis is complex, and there are many plausible mechanisms, most of which rely on the proposition of different gene mutations driving particular changes to the cell biology of the tumour. Several steps are recognized:

- Loss of the cell's natural basal–apical polarity, due to changes in the cytoskeleton

- Secretion of enzymes (such as matrix metalloproteinases, elastases, collagenases, etc.) to digest the extra-cellular matrix, allowing the cell to escape the primary tumour

- Loss of control of adhesion molecules such as E-cadherin, potentially under the control of core transcription factors of the Zeb, Goosecoid, Twist, Slug, and Snail families

- Increased activity of growth factors (such as TGF-β and fibroblast growth factor) and other signalling pathways (Notch, MAPK).

These factors enable the seeding malignant tumour cells to escape into the lymphatics and the blood, some as single cells, others as clusters, and the number of circulating tumour cells (CTCs) has been linked to the number of metastatic sites. CTCs may be quantified by a machine that is a cross between a fluorescence flow cytometer and a haematology analyser, and their detection may be helpful in diagnosis and monitoring the development of the disease and the effects of treatment. More than one commentator has noted the similarity between Darwinian natural selection and the evolution of cells with metastatic potential. Others have pointed to Knudson's two-hit hypothesis, but in the case of metastases this would involve more stages. The process of metastatic development is summarized in Figure 3.17.

miRNAs may also have a place, as *in vitro* data points to a role for miRNA-10 in regulating the EMT in hepatocytes. There are many data from breast cancer: miR-31 promotes the tumour-seeding ability in breast cancer cells; EMT may be linked to miR-200, miR-205, and miR-22; whilst miR-105 and miR-122 may be important in metastasis. In this respect, mR-29b suppresses metastatic activity by inhibiting VEGF, PDGT, and other regulators.

Much of our knowledge of gene interactions in basic cancer cell biology arises from *in vitro* and animal studies that may not translate to a clinical setting. For example, in a mouse breast cancer model, the combination of inhibitors of ATR and nuclear kinase Wee1 lead to tumour shrinkage and the suppression of metastasis. This primary research may lead to the development of clinically effective therapeutics.

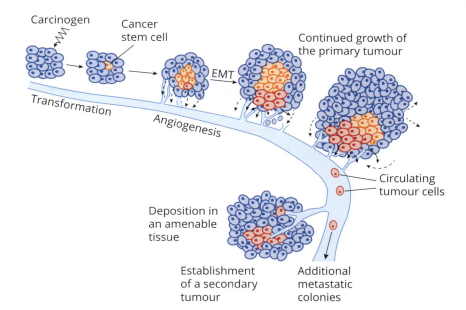

FIGURE 3.17

Sequential steps in the development of metastatic disease. Dashed arrows: angiogenesis. Solid arrows, secretion of factors promoting metastatic transformation.

Clinical perspective

The site(s) of these metastases varies with the type of tumour and its original site (the 'primary') and can be unpredictable. However, broad concepts do apply. For example, breast cancer often metastasizes to the lungs, liver, brain, and bones; ovarian and colorectal cancer to the liver; prostate cancer to the lung, bone, and brain; melanoma to lung, bone, and brain; bladder cancer to the brain and the lung; with pancreatic and oesophageal cancer often metastasizing to the liver. The exact nature of the metastasizing tissue may be determined histologically (Figure 3.18).

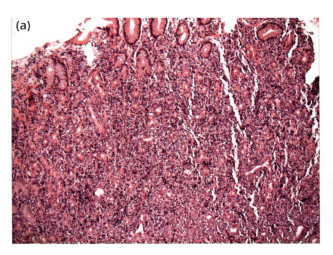
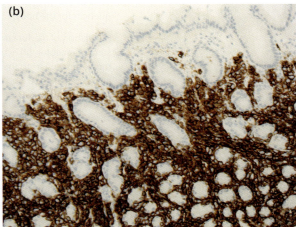

FIGURE 3.18

Lymphoma metastases in the stomach. A patient with non-Hodgkin's lymphoma complained of stomach pain. An endoscopy found gastric ulcers, that were then biopsied and stained. (a) shows a standard haematoxylin and eosin showing invasion of normal gastric tissues with cells of an unknown origin. (b) shows that immunocytochemistry of the same tissue with an antibody to a B lymphocyte marker (brown colour) confirmed the invading cells as lymphoma metastases.

From Orchard & Nation, *Histopathology*, second edition, Oxford University Press, 2018.

In some cases, the secondary growth may cause the signs and symptoms that bring the patients to our profession, with an unknown primary cancer. However, tumour cells in a secondary growth generally retain the characteristics of the primary growth. As in Figure 3.18, should surgeons successfully remove the primary tumour, the secondary growth may still prove an issue, perhaps years later. This is important because the laboratory can demonstrate if a tumour surgically removed from the lung is, in fact, a lung cancer. If the cells in that tissue resemble, for example, breast tissue, this will provide a clue as to the location of the primary.

SELF-CHECK 3.10

Explain why slowly increasing levels of serum VEGF in a patient with cancer are a concern.

Key points

Progression of cancer development to metastases is a complex, orchestrated process involving detachment of a metastatic cell or cells from the primary tumour, escape into the circulation or lymphatics, and establishment in an amenable distant tissue where it flourishes.

Chapter summary

- Cancer is the result of the malignant transformation of a normal cell, with deleterious consequences for the body.

- Regulation of the function of the cell can be described in three stages: (i) a ligand binding its cell surface receptor, (ii) activation of second messengers within the cytoplasm, and (iii) passage of these messengers to the nucleus where they effect changes such as alteration in protein synthesis and in cell division.

- The cell cycle is controlled by regulators such as cyclins and cyclin-dependent kinases. Checkpoints at key steps ensure that the cell is stable and can proceed to the next stage.

- Apoptosis is a process by which a damaged cell effectively commits suicide. Initiating pathways (some of which rely on receptor–ligand binding) culminate in the activation of caspases that digest the cell from within. There are other forms of programmed cell death, such as necroptosis.

- Factors that cause cancer (carcinogens) may be classified as endogenous (e.g. ROS) or exogenous (e.g. industrial chemicals). The prefix onco- refers to a factor that promotes carcinogenesis, e.g. oncovirus, oncogene.

- Epigenetics refers to non-DNA changes in chromosomes, e.g. to individual nucleotides, and to histone proteins. These may be mediated by small and long non-coding RNAs.

- As the tumour grows, it forms its own vascular supply—the process of angiogenesis—and later spreads to other organs or tissues, the process of metastasis.

Further reading

- Felsher DW. Role of MYCN in retinoblastoma. Lancet Oncol. 2013: 14; 270–1.

- Hamilton E, Infante JR. Targeting CDK4/6 in patients with cancer. Cancer Treatment Revs 2016: 45; 129–38.

- Hu G, Niu F, Humbug BA, et al. Molecular mechanisms of long non-coding RNAs and their role in disease pathogenesis. Oncotarget 2018: 9; 18648–63.

- Hanley DJ, Esteller M, Berdasco M. Interplay between long non-coding RNAs and epigenetic machinery: Emerging targets in cancer? Phil. Trans. R. Soc. B 2018: 373; 20170074.

- Hanahan D, Weinberg RA. Hallmarks of cancer: The next generation. Cell. 2011: 144; 646–74.

- Jackson M, Marks L, May GWH, Wilson JB. The genetic basis of disease. Essays in Biochem. 2018: 62; 643–723.

- Liu Q, Zhang H, Jiang X, et al. Factors involved in cancer metastases: A better understanding to 'seed and soil' hypothesis. Molecular Cancer 2017: 16; 176.

- Yokoyama T, Takehara K, Sugimoto N, et al. Lynch syndrome-associated endometrial carcinoma with MLH1 germline mutation and MLH1 promoter hypermethylation: A case report and literature review. BMC Cancer 2018: 18; 576. doi: 10.1186/s12885-018-4489-0.

- Robinson EK, Covarrubias S, Carpenter S. The how and where of lncRNA function: An innate immune perspective. Biochem. Biophys. Acta, Gene Regul. Mech. 2020: 1864; 194419.

- Uroda T, Anastasakou E, Rossi A, et al. Conserved pseudoknots on lncRNA MEG3 are essential for stimulation of the p53 pathway. Molecular Cell 2019: 75; 982–95.

- Rumgay H, Murphy N, Ferrari P, et al. Alcohol and Cancer: Epidemiology and Biological Mechanisms. Nutrients. 2021: 13; 3173. doi: 10.3390/nu13093173.

Discussion questions

3.1 What techniques are needed to investigate chromosomal and genetic abnormalities?

3.2 How can metastatic disease be prevented and treated?

4

Clinical oncology 1: Basic concepts and common cancers

Malignant neoplasm (for all intents and purposes, cancer) was the principal cause of 144,501 deaths in the England and Wales during 2020, that being 24.7% of all deaths. The World Health Organization (WHO) global data estimate that it caused almost 10 million deaths in 2020, the most common cancer being that of the lung, with 1.8 million deaths. However, these figures do not include all cases of cancer as not everybody diagnosed with cancer dies from it. Indeed, the UK's Office for National Statistics (ONS) also provides a cancer registry, showing that some 300,000 people are diagnosed with this disease each year. Furthermore, there is a great deal of morbidity present before the terminal stages of the disease, which may be decades after diagnosis. Therefore, the burden of cancer certainly falls on at least a third of the population, to say nothing of its effect on their colleagues, friends, and family, perhaps bringing the total to half of the population.

Cancer can be present in almost any organ, and accordingly, as there will be much to discuss, the topic will be spread over two chapters. The purpose of this first chapter is to introduce, in initial sections, features common to all cancers. Sections that follow will explain the clinical aspects of the biological basis of four of the major cancers (lung, colorectal, prostate, and breast)—how they present, are diagnosed, and managed. However, blood cancer is described in Chapter 11, alongside other blood disease.

Learning objectives

After studying this chapter, you should confidently be able to:

- describe basic concepts in the prevention, epidemiology, and classification of cancer
- explain leading methods in the diagnosis and treatment of cancers
- understand key features of lung, colorectal, prostate, and breast cancer

4.1 Basic concepts in clinical oncology

4.1.1 Prevention is better than cure

It has always been evident from individual diseases and general observations that prevention is better than cure, but in 2011 Parkin et al. provided hard evidence for this, reporting that the exposure to less than optimal levels of 14 factors was responsible for 43% of UK cancers in 2010 (45% in men, 40% in women), that being around 134,000 cases. Of these, tobacco smoking is by far the most important, being linked to 60,000 cases (19% of all new cancers), so that optimum exposure level is nil, as it is for alcohol, exogenous sex hormones, infections, radiation, and occupational exposures. Overweight and obesity are the second most powerful risk factors, being linked to 5.5% of cancers. A body mass index (BMI) <25 kg/m^2 is preferred for minimizing the risk of colorectal cancer, whereas <21 kg/m^2 gives best protection from breast cancer. Similarly, breastfeeding for at least 6 months, and ≥30 minutes of physical exercise five times a week, are linked to a reduction in the risk of cancer. Other factors linked to an increased risk include a diet deficient in fruit and vegetables, intake of dietary salt >6g per day, and <23g of dietary fibre per day. With its global view, the WHO recommends vaccinating against the hepatitis B virus and the human papilloma virus for liver cancer and cervical cancer, as these cause ~25% of deaths in low- and middle-income countries.

4.1.2 Initial observations in cancer

Despite the variety in the aetiology and pathology of different cancers, there are many semi-specific clinical findings common to certain malignancies and which support a diagnosis. Leaving aside clear organ-specific signs and symptoms (such as haematuria (blood in the urine) in renal or bladder cancer, jaundice in liver cancer), many patients present with tiredness and lethargy, possibly traced to a **normocytic anaemia**. A leading cause of unexplained venous thromboembolism is cancer—thus a deep-vein thrombosis or pulmonary embolism may be the first indicator of a neoplasm. Should there be metastatic disease to the bones, then pain is likely, with stress fractures as the tumour weakens the internal structure. But overall, many of these signs and symptoms have poor sensitivity and specificity (section 1.3.5).

Once the disease is established in secondary care with a firm diagnosis, a consultant will generally take responsibility for management, supported by a multi-disciplinary team (MDT). This group comprises surgeons, physicians, pathologists, nurses (in particular, a clinical nurse specialist), scientists, radiologists, and other healthcare professionals. Each of them will have their own particular and specialist part to play in ensuring overall best practice.

4.1.3 Epidemiology, incidence, and mortality

Data on cancer come in many forms, the most objective being those provided by the WHO and the UK Government's ONS with its figures on mortality and a cancer registry—the latter recording those who have been newly diagnosed. An additional aspect is data on survival, taken to be from the date of diagnosis. However, this information is often gathered over different time periods, and so many data sets may not be directly comparable. Other data come from other UK governmental bodies such as Public Health England, and from charities such as the British Lung Foundation and Cancer Research UK that, although unofficial, nonetheless have their place.

Not all cancers are invariably fatal—a subset are described as *in situ*. These are at their earliest stages, having not yet spread from their initial source in an organ or other tissues, and so are usually curable. The best-known example of this is the ductal **carcinoma** *in situ* (DCIS) of breast cancer, often detected by mammographic X-ray screening. However, benign breast disease, including DCIS, still brings a four- to five-fold increase in the risk of developing breast cancer, which has a major bearing on cancer statistics. Cancers also vary considerably in their potential to cause death. Table 4.1 shows ONS data from England and Wales, with the number of deaths in 2021 and the number of new cases in 2019.

Dividing deaths by new cases generates a crude survival index, showing that ~44% of new cases will ultimately succumb to their disease, and so 56% of people survive their disease. However,

TABLE 4.1   Mortality of malignant cancer

Cause	Deaths	%	New cases	%	Survival index
Malignant neoplasms	144,501	100	327,174	100	0.44
Bronchus and lung	28,147	19.5	41,897	12.8	0.67
Colorectal	15,291	10.6	39,458	12.1	0.39
Blood*	11,634	8.1	28,942	8.8	0.40
Prostate	10,359	7.2	49,744	15.2	0.21
Breast**	9,664	6.7	50,338	15.4	0.19
Pancreas	8,829	6.1	9,705	3.0	0.91
Oesophagus	6,870	4.7	8,182	2.5	0.84
Liver	5,375	3.7	6,015	1.8	0.89
Bladder	4,947	3.4	9,389	2.9	0.53
Brain	3,864	2.7	4,803	1.5	0.80
Kidney	3,575	2.5	10,720	3.3	0.33
Stomach	3,325	2.3	5,876	1.8	0.57
Ovary	3,324	2.3	6,899	2.1	0.48
Oral cavity and throat***	2,953	2.0	8,851	2.7	0.33
Uterus	2,266	1.6	8,579	2.6	0.26
Melanoma	2,138	1.5	15,989	4.9	0.13
Mesothelioma	1,939	1.3	2,374	0.7	0.82

* Includes myeloma, lymphoma, leukaemia, and others. ** Women only. *** Includes the lip, mouth, pharynx, and larynx. Figures for deaths do not sum to 100% as only selected categories are presented. Survival index = deaths/new cases, so that a high index implies poor prognosis.

© Office for National Statistics. Source: https://www.ons.gov.uk/. Licensed under the Open Government Licence 3.0.

the point that must be made is that the data on deaths are from 2021, whereas those from the new cases are from 2019—a considerably disparity. The first five cancers comprise over a half of all cancer deaths.

Survival is better if the tumour is localized and detected early; the 10-year survival of testicular cancer in the UK is 98%, whereas for pancreatic cancer it is 1%, and notably the latter has the most adverse survival index. Conversely, 86.6% of those diagnosed with a melanoma survive. These UK data may be compared to the global data from the WHO's Global Cancer Observatory (Table 4.2). The prime difference is that stomach and liver cancer are more frequent globally, the latter probably due to hepatitis viruses. Globally, cancer of the cervix is a much greater problem, being the cause of 604,127 new cases (3.1% of all new cases) and 341,831 deaths (3.3%), giving it a survival index of 0.56, which compares to 0.29 in England and Wales. Conversely, global pancreatic cancer causes 4.7% of deaths (466,003 cases)—so it is less of an issue than in the UK, where it causes proportionately 30% more deaths.

TABLE 4.2 Global data on cancer, 2020

Cause	Deaths	%	New cases	%	Survival index
All cancers	99.5	100.0	192.9	100.0	0.52
Lung	18.0	18.0	22.1	11.4	0.81
Colorectal	9.3	9.4	19.3	10.0	0.48
Liver	8.3	8.3	9.0	4.7	0.92
Stomach	7.7	7.7	10.9	5.6	0.71
Breast	6.8	6.9	22.1	11.7	0.30
Oesophagus	5.4	5.5	6.0	3.1	0.90
Prostate	3.7	3.8	14.1	7.3	0.26

Raw data $\times 10^5$.

The Global Cancer Observatory—All Rights Reserved, December, 2020.

A major disadvantage with this approach focusing on death and newly diagnosed cancer is that it fails to fully address the morbidity of cancer. The morbidity may be described in terms of quality of life (QoL), where some cancers, often those that are metastatic, are sufficiently aggressive to bring a poor QoL, such as by causing pain, whereas others are relatively 'tolerable', with a better QoL.

Sex

Overall, male malignant cancer deaths (76,989) exceed those of females (67,512) by 14% in 2021 (Table 4.3, for England and Wales). After adjusting for this sex difference, the leading cancer, lung/bronchus, is broadly equivalent between the sexes, but almost all other cancers kill more males: mesothelioma four times as many, and oesophageal, bladder, and lip/mouth/pharynx cancer twice as many.

However, pancreatic cancer kills proportionately more females, even after adjustment. Although prostate cancer kills more males than breast cancer kills females, proportionately more females die of breast cancer. Deaths from prostate and breast cancers exceed those of colorectal cancer in their respective sexes by 24.5% and 37.6% respectively.

SELF-CHECK 4.1

What are the most common cancers?

Age

Unsurprisingly, death and the incidence of cancers rise with age, possibly because of the increased rate of genomic mutations (Chapter 3), although there is a small but significant peak in infant and childhood cancer at ages 5 to 9, mostly of the brain and blood. Table 4.4 shows how the frequency of common cancers increases at each age band.

Note the increased relative rate of breast cancer in younger women compared with prostate cancer in men. However, also note that in some cases, numbers fall in those aged over 80, implying that the disease does not strike this population—an incorrect assumption brought about by a smaller number of people in these age bands. To get round this, data can be plotted as the number of deaths relative to the number of people in each age band, the standardized mortality ratio (SMR) per 100,000 of that population, as shown in Figure 4.1.

TABLE 4.3 Leading cancer causes of death in males and females in England and Wales, 2021

Cancer	Males	Females	% difference	% of males	% of females	% difference
Bronchus and lung	14,784	13,363	10.6	19.2	19.8	3.1
Prostate	10,359	–	–	13.4	–	–
Breast	72	9,592	133.2	<0.1	14.2	142.0
Colorectal	8,320	6,971	19.3	10.8	10.3	4.9
Blood*	6,651	4,983	33.5	9.5	7.4	28.3
Pancreas	4,441	4,388	1.2	5.8	6.5	12.1
Oesophagus	4,836	2,034	137.8	6.3	3.0	110.0
Liver**	3,239	2,136	51.6	4.2	3.2	31.2
Bladder	3,383	1,564	116.3	4.4	2.3	91.3
Brain	2,280	1,584	43.9	3.0	2.3	30.4
Kidney	2,261	1,314	72.1	2.9	1.9	52.6
Stomach	2,123	1,202	76.6	2.8	1.8	55.6
Ovary	–	3,324	–	–	4.9	–
Lip, mouth, pharynx	2,081	872	138.6	2.7	1.3	107.7
Mesothelioma	1,592	347	358.8	2.1	0.5	320.0
Uterus	–	2,265	–	–	3.4	–
Melanoma	1,247	891	40.0	1.6	1.3	23.1

* Lymphoid, haemopoietic, and related tissues (principally leukaemia and lymphoma). ** Includes intrahepatic bile ducts.

Figures do not sum to 100% as only selected categories are presented. © Office for National Statistics. Source: https://www.ons.gov.uk/. Licensed under the Open Government Licence 3.0.

Figure 4.1 shows the age-standardized rates of bronchial, tracheal, and lung cancer increase in parallel in each sex, but more so in men. It also indicates that breast cancer in women develops earlier than does prostate cancer in men, but that the latter causes far more deaths in the older men than does the former in older women.

Caveats

Any data set is only as good as the method for collecting that data. The ONS data sets include the ICD classification 'malignant neoplasms of ill-defined, secondary, and unspecified sites', which killed 9,009 people (4,253 men and 4,756 women). This group is therefore the sixth largest (6.2% of cancer deaths), exceeding those of the pancreas, oesophagus, liver, etc., and many are likely to reflect metastatic disease in which the primary tumour cannot be located.

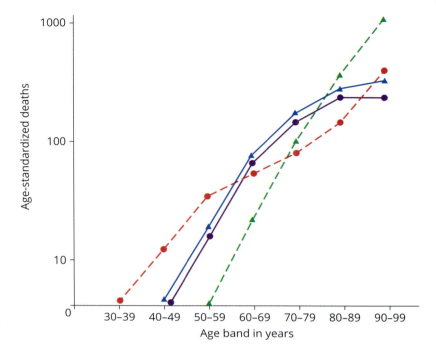

FIGURE 4.1
Relationship between age and deaths from certain cancers. Solid lines: bronchial, tracheal, and lung cancer; dashed lines: breast and prostate cancer. Closed circles: females; triangles: males.

TABLE 4.4 Age-stratified deaths of selected cancers

Age group (years)	All malignant cancers		Lung, bronchus, and trachea		Breast	Prostate
	Males	Females	Males	Females	Females	Males
30–34	189	254	7	8	68	0
40–44	532	726	49	42	218	2
50–54	2,117	2,293	377	313	595	44
60–64	5,898	3,887	1,270	1,108	777	356
70–74	12,100	9,675	2,731	2,546	1,080	1,273
80–84	12,722	10,872	2,397	2,183	1,254	2,097
90+	6,578	7,064	751	864	1,132	1,914

Data are deaths in England and Wales in 2021.

© Office for National Statistics. Source: https://www.ons.gov.uk/. Licensed under the Open Government Licence 3.0.

Key points

The incidence and deaths from different cancers vary with age, sex, and geography, and there is a marked variation in the likelihood of a cure.

4.1.4 Classification of cancer

The vocabulary of oncology can be difficult to grasp, but there is structure. For example, *myo-* refers to muscle (cardiomyocyte: heart muscle cell), the suffix *-oma* to a mass, generally a tumour. Hence leiomyoma is a tumour of smooth muscle cells, whilst a rhabdomyoma is a tumour of striated muscle cells. An osteosarcoma is a cancer of the bone, a haemangioma is a mass of red blood cells in a knot of blood vessels (angio = blood vessels), a carcinoma is a tumour based on epithelial cells, hence lung carcinoma. An adenoma is a tumour of a gland (and therefore secreting a soluble product such as mucus), and should that gland be of epithelial tissue, it is an **adenocarcinoma**. The meaning of seemingly random collections of letters defining certain disease will be explained (where possible) as they arise. For example, leukaemia is so named (Greek for white is leukos, blood is haima) because of the high number of white blood cells in the blood.

4.2 Common clinical aspects of cancer

Whilst specific cancers will be diagnosed in a certain manner, and each has its own dedicated forms of treatment, there are features in common that will apply to many malignancies, regardless of their tissue basis or aetiology. In the GP's surgery or outpatient clinic, a preliminary diagnosis rests on signs, symptoms, some simple investigations, and medical history, which can later be confirmed or refuted with investigations such as imaging and in the laboratory. Few signs and symptoms are unequivocal; most are semi-specific, and provide supporting evidence of disease in a particular organ. Alternative and differential diagnoses must always be considered in the early stages of an investigation. Guideline NG12 from the National Institute for Health and Care Excellence (NICE) offers guidelines on the recognition of suspected cancer and its referral to specialist carers.

Common features of management include surgery, chemotherapy, and radiotherapy, often in combination. However, in certain circumstances, radiotherapy or chemotherapy may precede surgery, the objective being to weaken the tumour before it is excised.

4.2.1 Imaging

As introduced in section 1.3.4, there are several forms of medical imaging: radiography includes X-ray (including fluoroscopy), computer/computed axial (or assisted) tomography (CAT, or CT scanning), and positron-emission tomography (PET scanning), all of which rely on ionizing radiation and so bring an inherent risk. Others are ultrasound and magnetic resonance imaging (MRI). Nuclear medicine refers to the use of radioactive substances placed inside the body, such as in PET scanning and cardiac perfusion, which have applications in several neoplastic and non-neoplastic conditions. A practical application of imaging for tumours is to minimize, and hopefully eradicate, futile surgery.

PET scanning relies on the release and detection of positrons from a radioactive source, the most common being fluorine-18. Cells and tissues with a high metabolic activity absorb and use glucose more so than normal tissue. This means that, when conjugated to a glucose analogue such as fluoro-deoxyglucose, the radiolabel will concentrate in highly active cells, such as the heart and tumours. It may also detect lymph node involvement in lung and other cancers. Other radiolabels are used for specific purposes, such as carbon-11 for tumours of the adrenal cortex. Different imaging platforms can be combined to provide improved detail. For example, combined PET/CT scanning is superior to other formats in the detection of **metastases**. Figure 4.2 shows a PET scan of the upper body.

Key points

Imaging consists of a range of versatile techniques that can provide essential information of the pathology of the skeleton and internal organs.

SELF-CHECK 4.2

What is ionizing radiation, and why is it linked to cancer?

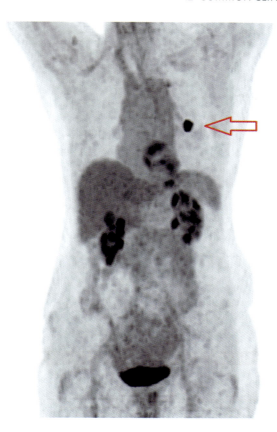

FIGURE 4.2
PET scan. The dark areas indicate where the isotope is present, as in the kidneys and bladder, as is expected as it is excreted. However, the dense circular area in the left lung (arrowed) is not expected and is likely to be a tumour.

Radiologie München.

4.2.2 Cancer staging

Determining the extent or severity of the disease—its stage—is a crucial part of clinical decision-making and relies on a number of factors. Imaging is particularly useful in determining the stage of the disease in solid-tissue malignancies (less so in blood cancers, or those of the central nervous system). Pre-intervention staging is important because it will direct the different types of treatment; staging of the tumour once it has been removed is relatively easy and may also help direct further treatment.

By consensus of several national and international oncology groups, the letters TNM describe the size and invasiveness of the tumour (T), the involvement of local lymph nodes (N), and the presence or absence of distant metastases (M). Each part of the triad has further delineations:

- T0 = no evidence of tumour, Tis = tumour *in situ*, whilst T1, T2, T3, and T4 are progressively larger or more infiltrating tumours, and TX is used when the primary tumour cannot be assessed.

- N0 = no lymph nodes involved, N1, N2, and N3 denote progressively more, and more distant, lymph node involvement.

- M0 is no distant metastases, M1 reports the presence of distant metastases.

Thus T2N2M1 describes a situation where the tumour has a modest size, where there are a small number of distant lymph nodes involved, and where there are distant metastases. This may give a poor prognosis, depending on the nature of the cancer. Further classifications include elevated serum markers, vascularity, and grade of the tumour (low or high). Not all imaging tools can provide definitive information on all three aspects, so combinations of modalities are often used, providing more precise information. Should staging be done on clinical features alone, the prefix c may be used (e.g. cT2N0M0), whereas staging defined with knowledge of the cell pathology (obtained by surgery) may carry the prefix p (e.g. pT2N1M0).

(a)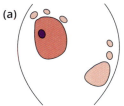

(b)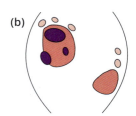

(c)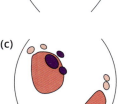

(d)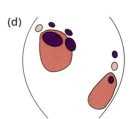

FIGURE 4.3

The four stages of cancer in a fictitious organ. In stage (a) the tumour is relatively small and is only within the organ, whereas stage (b) is characterized by a second tumour in the same organ, or the first tumour having breached the outside of the organ. Stage (c) is defined by the involvement of nearby tissues, almost always lymph nodes, and in stage (d) the disease has spread to distant lymph nodes or to a distant organ.

Staging is relatively specific for particular organs. For example, a key point in lung cancer is whether the disease has spread to a second lobe, or if there are metastases on the abdominal side of the diaphragm. Figure 4.3 represents four stages in a 'typical' organ.

SELF-CHECK 4.3

What are the differences between cT1N1M0 and pT3N0M1?

4.2.3 The laboratory

Section 1.3.4 and Table 1.6 introduced certain aspects of the laboratory as essential tools in most diseases, more so in cancer. Laboratory information is valuable because of its very high degree of objectivity, accuracy, and reproducibility. However, the practitioner is always responsible for the interpretation of the data for their particular patient. Of particular value are relatively organ-specific blood markers, and non-specific and general indicators.

Cancer markers

Certain molecules are more highly expressed and/or are secreted by malignant cells compared to normal cells, an example being 'cluster of differentiation' (CD) molecules. Although a relatively simple classification system, individual CD molecules can have widely varying functions, and many are interrelated. Some are enzymes, others are adhesion molecules, and some have signalling function. Originally developed to differentiate leukaemia and lymphoma cells from normal white blood cells, the system now extends to non-leukocytes as well. These include red blood cells and platelets, and non-blood cells such as epithelial and endothelial cells, fibroblasts, smooth muscle cells, and cancer cells such as melanoma and neuroblastoma. Several other molecules relating to cancer can be determined in the plasma by immunoassay or other methods (Table 4.5).

However, each method/marker combination has variable sensitivity and specificity as there are numerous instances where increased markers are present in benign diseases. An example of this is prostate-specific antigen, with increased levels in prostate cancer, **prostatitis**, and **benign prostatic hyperplasia**. Some markers have no biological function, whereas others participate directly in pathophysiology. Very high levels of insulin secreted by a pancreatic **insulinoma** can cause hypoglycaemia, and very high gamma-globulins in a **myeloma** cause increased blood viscosity leading to renal failure and stroke. These markers are commonly determined in haematology, biochemistry, or immunology laboratories.

TABLE 4.5   Selected serum cancer markers

Marker	Cancers
Alpha-feto protein (AFP)	Liver, cancer of reproductive organs
CA 15.3	Breast
CA 19-9	Pancreatic, colorectal, and other intestinal malignancies
CA 125	Ovary, lung, breast, and intestinal
Carcinoembryonic antigen (CEA)	Colorectal, breast, ovary, uterus, pancreas, stomach, liver, oesophagus
Gamma-globulins	Blood cancers, but principally myeloma
Prostate-specific antigen (PSA)	Prostate

Note the semi-specific nature of certain markers.

The histology laboratory will analyse biopsy and excised tissues for the presence of malignant cells using routine dyes such as haematoxylin and eosin and special methods such as immunocytochemistry, indirect immunofluorescence, and *in situ* hybridization (Figure 4.4). Tissue markers overexpressed by malignant cells include adhesion molecules and cytokeratins (Table 4.6). Each marker will be discussed in more detail in the section on its associated cancer.

Routine blood markers

A full blood count (FBC), urea and electrolytes (U&Es), liver function tests (LFTs), and calcium (at least) will be ordered for every investigation of cancer, and regularly thereafter. An FBC brings information on red blood cells (hence anaemia), white blood cells (infection and inflammation), and platelets (potential for haemorrhage) (Chapter 11). These indices are also useful surrogates for the cytotoxic potential of certain forms of chemotherapy, as the bone marrow is particularly sensitive to many of these drugs, resulting in suppressed **haemopoiesis**. LFTs inform on the integrity of the liver, and of these, bilirubin is useful and is tracked by the clinical sign of jaundice. U&Es check renal function, whilst calcium is particularly helpful in bone disease. Other routine markers have a role in specific

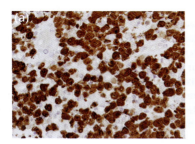
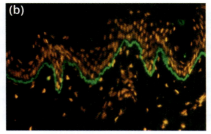
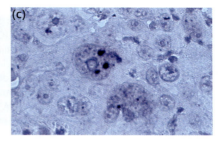

FIGURE 4.4

Special techniques in histology. (a) immunochemistry of strong CD246 (ALK-1) expression in a lymphoma, (b) indirect immunofluorescence of a fluorescein-labelled anti-IgG showing autoantibodies to basement membrane in pemphigoid, (c) presence of intra-nuclear human papilloma virus using an oligonucleotide probe.

From Orchard & Nation *Histopathology*, second edition, Oxford University Press, 2018.

TABLE 4.6 Selected cancer tissue biomarkers

Biomarker	Application
CD3, CD5, CD21, CD45, CD128	Lymphoma and lymphoid invasion into other tissues
Cytokeratins 1, 5, 10, 14, 18	Pancreas, bile duct, salivary gland, cancers of the bladder, naso-pharynx, thymus, and some mesothelioma and breast cancers
CA125	Adenocarcinoma of the colon, breast, lung, uterus, ovary
Calretinin	Mesothelioma and some lung tumours
CEA	Ductal carcinoma of the breast, lung, and colorectal carcinomas
Epithelial growth factor receptor	Simple and squamous epithelium and therefore a variety of carcinomas
Oestrogen and proges-terone receptors	Certain breast cancers
Ki-67	To assess proliferation rate
Melan A, S100	Malignant melanoma
Von Willebrand factor	Marks endothelial cells, so used to assess vascularity and angiogenesis

cancers, such as FBC in leukaemia. On the whole, routine markers lack sufficient sensitivity and specificity to be helpful in particular cancers, but may help in assessing metastatic disease, and more so in the effects of chemo- and radiotherapy (e.g. raised LFTs and liver toxicity).

Many cancers induce a general low-grade acute-phase response, so that markers such as CRP may be increased. This does not necessarily imply the presence of an infection, although it may do if the patient's immune system has been compromised by chemotherapy and/or radiotherapy. Combinations of serum markers may also be useful, such as that of CA 125, prealbumin, apolipoprotein A-1, β-2 microglobulin, and transferrin in predicting the risk of ovarian cancer in the face of a pelvis mass.

Molecular pathology

As discussed in Chapter 3, the genetic basis of cancer is established, so that single-nucleotide poly-morphisms (SNPs) and other mutations can be helpful and even definitive. Indeed, chronic myeloid leukaemia (CML) was once defined by leukocyte morphology on a blood film, whereas now it is defined entirely by the BCR/ABL mutation. The biology of this mutation is discussed fully in Chapter 3, section 3.9.1. Therefore, a blood film that resembles CML but lacks BCR/ABL cannot be CML, and an alternative diagnosis is required. Numerous mutations are linked to certain cancers, but sensitivity and specificity must be addressed as expression overlaps for different malignancies. For example, methyla-tion of DNA is linked to biomarker MLH1 (coded for by *MLH1*) in endometrial and ovarian cancer, whilst the latter is also linked to *BRCA1*, itself also linked to breast cancer. Considerable advances in methodology such as next generation sequencing and microarray analysis have led to the develop-ment of multi-target panels that analyse several genes at the same time.

In contrast to the long-established pathology tests outlined above, performed in all hospitals, molecular genetics are generally performed in a small number of specialist genomic laboratory hubs, each located in a centre of high population. These units also offer testing for non-cancer conditions such as cystic fibrosis and muscular dystrophy, and for chromosome conditions such as Klinefelter's syndrome, as discussed in Chapter 16.

As discussed in section 3.8.4, microRNAs (miRNAs) can be regulator genes (particularly, for this chapter, tumour-suppressor genes), or act as oncogenes, and so may also be biomarkers. Abnormal plasma or serum levels are present in certain malignancies, so that panels of circulating miRNAs could be used as diagnostic and prognostic indicators. For instance, the presence of bladder cancer can be predicted by the expression of miR-126 and miR-152 in urine, and miR-21 levels have been linked to relapse-free survival in diffuse B-cell lymphoma. There is indication that those miRNAs with specific roles in **carcinogenesis** may become the object of a new class of therapeutics.

4.2.4 Surgery

In many cases, surgery is the preferred option for a number of reasons, primarily that it may be curative if the tumour is in an early stage. However, it is not always possible: the tumour may be inaccessible or too diffuse to ensure complete excision, or the patient may not be fit enough for the rigours of surgery. In this respect the surgeon will call on the laboratory to determine (for example) if the patient has a sufficiently high haemoglobin, and a sufficiently low creatinine. Fortunately, emerging advances in surgical technique, such as keyhole surgery, have brought many otherwise difficult cases to the operating theatre. Keyhole surgery is often preferred to 'open' surgery as it is generally quicker (so less time under general anaesthetic) and safer (less chance of infection), with reduced need for post-surgical anticoagulation and loss of blood.

> **SELF-CHECK 4.4**
>
> Why might the laboratory be called upon following a relatively simple surgery?

4.2.5 Chemotherapy

Treatment with chemicals can be broadly divided into two categories: those with systemic effects on all body cells and those whose effects are targeted. Many are oral, but others need to be infused intravenously, intramuscularly, as a subcutaneous depot, or (very rarely) into the cerebrospinal fluid or **peritoneal cavity**. Historically, drugs have targeted DNA, with the objective of reducing cell proliferation, but more recently advances have produced drugs that act on other metabolic processes within the cell (second messengers, microtubule formation) and on receptors expressed on the outer surface of the cell membrane. Whilst most drugs are directly cytotoxic, some act by inducing the cell to move into apoptosis. Many drugs are given in blocks, or cycles, perhaps of several weeks' duration, interspersed with weeks or months of drug-free recovery.

Drugs with effects on all cells

Observations in the last century that the unfortunate exposure of otherwise healthy people to mustard gas was accompanied by a fall in their white blood cell count led to the development of drugs such as cyclophosphamide, chlorambucil, and melphalan. These exert their effect by attaching an alkyl group to DNA, which renders mitosis impossible and so leads to the death of the cell, probably by apoptosis.

Denying folic acid to cells results in failure to generate the components of DNA and so the arrest of the cell cycle, leading to the development of folate antagonists such as methotrexate. Other drugs, such as 6-mercaptopurine followed, as did entirely new classes of drugs, such as the vinca alkaloids (e.g. vincristine) and taxanes (paclitaxel) that interfere with the microtubules that draw chromosomes to the centrioles in the mitotic anaphase. Many drugs interfere with DNA synthesis and replication, such as enzyme inhibitors 5-fluorouracil (5-FU: of thymidylate synthase) and etoposide (of **topoisomerase**), and drugs based on platinum (e.g. cisplatin and carboplatin). Several anti-metabolites are based on 5-FU, including pro-drugs tegafur and capecitabine. These must be given with folinic acid or uracil to rescue the particular metabolic pathway where they operate and may be given in a combined preparation e.g. irinotecan = 5-FU + folinic acid.

The mode of action of early forms of cancer chemotherapy was to destroy malignant cells comprising the tumour, hence the term cytotoxic chemotherapy. The principle was that as the tumour was growing more rapidly than normal tissues, drugs targeting those cells that were actively and rapidly growing should be successful in destroying the malignant cells. Many of these drugs are described as

TABLE 4.7   Common side effects of chemotherapy

Short-term	Nausea, vomiting, fever, bone marrow suppression, fluid retention, alopecia, skin rash, cystitis, infertility, venous thromboembolism, cellulitis, cardiotoxicity, neurotoxicity
Long-term	Pulmonary fibrosis, heart and renal failure, cardiomyopathy, secondary cancers (lymphoma, bladder cancer)

Note: some of these side effects are also present in radiotherapy.

anti-metabolites, as, being similar to molecules already within the cell (such as purines and pyrimidines), they effectively interfere with the cell's growth cycle, principally mitosis. Whilst many of these drugs are very successful in reducing tumour load, there are serious undesired side effects (Table 4.7).

Unfortunately, drugs targeting cells with high mitotic activity (ideally just the tumour cells) are indiscriminate in that they also target normal, healthy cells that also have high mitotic activity. An example of this is the bone marrow, charged with generating millions of cells daily. Therefore, a major side effect of certain classes of cytotoxic chemotherapy is reduced red blood cells (leading to anaemia), white blood cells (leading to infections), and platelets (leading to haemorrhage). Consequently, the number of these cells must be monitored regularly with a full blood count. Should cell numbers fall too low, the chemotherapy may need to be reduced to allow blood counts to improve.

Targeted drugs

Despite the side effects outlined in Table 4.7, chemotherapy is not simply sophisticated poison: newer drugs target specific cancers and their metabolism. Dozens of monoclonal antibodies (mAbs) are widely used in cancer and in many other diseases, such as COVID-19 (Box 4.1). However, although mAbs are specific for their antigen (which would be located on the external surface of the target cell, such as CD20 on malignant B lymphocytes), that same antigen is very often found at a low level on normal cells (in this case, healthy B lymphocytes), so that there may still be undesired (side) effects.

Approximately 70% of breast cancer cells overexpress the oestrogen receptor, and many may also express the progesterone receptor, and accordingly drugs aimed at these and related molecules are effective therapeutics. Undoubtedly the best-known drug in this class is tamoxifen, which targets the oestrogen receptor. Similarly, endocrine antiandrogenic therapy can be directed towards prostate cancer, some targeting the hypothalamus–pituitary–gonad axis.

BOX 4.1   Monoclonal antibodies

A normal response to an infectious agent includes the generation of a spectrum of antibodies, each with varying specificities (hence polyclonal), that have an important role in the destruction of the pathogen. This can be exploited by using antibodies from those who had contracted a particular infection to treat others with that infection, an example being 'convalescent' serum, as used to treat an infection with the Ebola virus. However, the very strength of a polyclonal antiserum comprising many different antibodies is also a disadvantage in that it lacks specificity for a precisely defined antigen, a weakness in certain circumstances where exquisite specificity is required.

This problem was resolved in 1975 with the development of monoclonal antibodies (mAbs) by Kohler and Milstein, for which they were awarded a Nobel Prize. mAbs have revolutionized several areas of biomedical science, an example being immunoassay for small molecules such as protein hormones. A further area is in treatment of a variety of diseases, as will be described numerous times in this book. Over 80 mAbs have been licensed for use in transplantation rejection, thrombosis, cancer, rheumatoid arthritis, bone loss, hypercholesterolaemia, bacterial and viral infections, migraine prevention, osteoporosis, multiple sclerosis, macular degeneration, sickle-cell disease, and asthma.

Further developments in targeted treatment are agents specific for molecules involved in signal transduction and those acting on the nucleus. A translocation forms a neogene, *BCR-ABL*, in CML, which translates a new form of a tyrosine kinase (see section 3.2.2) in excessive amounts, and whose structure and function are very close to those of natural tyrosine kinase. Various forms of this enzyme can be inhibited by a number of drugs. However, as in many other instances, the drug is indiscriminate as to the cell it enters, so it has the potential to suppress the healthy (and desirable) function of the enzyme, and so lead to undesirable side effects. This may be tolerated by the patient if the drug is effective against the leukaemia. Other targeted drugs include those directed towards histone deacetylase and the **proteasome**. By their very nature, targeted therapies have different toxicities. mAbs may induce an allergic response whereas other drugs cause cardiac pathology and secondary cancers. Table 4.8 summarizes major classes of chemotherapeutics. Our colleagues in industry have established certain suffixes and other clues to inform us of the nature of a given drug. The suffix -mab fixes the drug as a monoclonal antibody and that of -nib as an inhibitor of a second messenger, whilst vincristine, vindesine, and vinblastine are derivatives of the vinca alkaloid *Catharanthus roseus*.

4.2.6 Radiotherapy

X-rays are valuable tools for studying bone and, to a lesser extent, soft tissues, as discussed in section 4.2.1. Unfortunately, the high-energy wavelengths of X-rays mean that they can ionize certain atoms and molecules (hence ionizing radiation), and this can cause the generation of toxic chemicals such as reactive oxygen species (ROS). As discussed in sections 3.5 and 3.6, these can cause further damage, especially to DNA, and so increase the potential for cancer, although in practice this risk is very small for most patients.

Modification of the diagnostic X-ray generator produces a device (a linear accelerator) that can emit a controlled beam of X-rays and focus this beam on certain tissues. The objective of this is to exploit the damaging effects of the rays, and so destroy those cells onto which they are directed, hence radiotherapy. This is certainly effective, but unfortunately the damaging rays must pass through normal, healthy cells and tissues which are also damaged. One way of getting around this problem is to 'fire' beams of radiation from specific directions that converge on the site of interest. The ultimate version of this is where the generator moves in an arc around the body, with the tumour at the centre of the virtual circle. Nearby normal tissues will be shielded, as much as is possible, but damage

TABLE 4.8 Common chemotherapeutic agents

Class	Drugs
Alkylating agents	Cyclophosphamide, melphalan, chlorambucil
Antimetabolites	Methotrexate, 5-fluorouracil, gemcitabine, capecitabine
Anti-microtubules	Vincristine, vinblastine, paclitaxel, docetaxel
Anti-topoisomerases	Etoposide, doxorubicin
Antibiotics	Daunorubicin, bleomycin, epirubicin
Platinums	Cisplatin, carboplatinum, oxaliplatin
Monoclonal antibodies	Trastuzumab, rituximab, obinutuzumab, bevacizumab
Endocrine therapy	Tamoxifen, goserelin, fulvestrant, anastrozole
Direct second messenger inhibitors	Imatinib, dasatinib, everolimus, sorafenib
CDK4/6 inhibitors	Palbociclib, ribociclib, abemaciclib

to the skin above the tumour with erythema, desquamation, and ulceration is almost inevitable. Should the radiotherapy be near the vertebrae, pelvis, and (especially) the sternum, there may be suppression of haemopoiesis, whilst fatigue is a common and poorly understood side effect. Thus, those cancers most likely to be treated with radiotherapy are solid tumours whose location can be precisely defined. Radiotherapy is often given as a series of exposures, perhaps three or four daily, each lasting a few minutes, and possibly over the course of a few weeks.

An alternative to the guided radiotherapy technique is to place small nuggets of radiochemicals within the body—**brachytherapy**. This has the advantage of being able to place the source of the radiation, which will emit beta particles and/or gamma rays, close to the tumour, where it remains. Accordingly, great attention must be paid to the half-life and emitting energy of the radioactive source. Breast cancer and prostate cancer are particularly suited to this tool.

4.2.7 Immunotherapy

The concept of treating the tumour as a foreign body has long been attractive, targeting proposed tumour-specific antigen(s). Although cancer often evokes a non-specific acute phase response, with raised levels of white blood cells and pro-inflammatory cytokines, this need not necessarily be immunological in the classical sense of defence from viruses and bacteria. The concept of immunological attack by the host's immune system has followed the observation of many white blood cells within tumours. Similar to chemotherapy (section 4.2.5), immunotherapy may be general or specific, the former comprising broad strategies to activate leukocytes with cytokines such as interleukins and γ-interferon, although these may also directly target the tumour. A non-specific anti-cancer therapy uses a vaccine against Bacillus Calmette–Guerin (BCG), most commonly used against tuberculosis.

Specific immunotherapy relies on the presence of a cancer antigen on the surface of the tumour. In this strategy, some of the patient's tumour cells are removed (by standard surgery or biopsy), separated from normal tissues, and then presented to a micro-collection of the patient's own immune cells. The principle is that a population of cytotoxic leukocytes will be formed, which can then be expanded (with the use of growth factors and interleukins), harvested, and returned to the patient, with the infused activated and primed leukocytes seeking out and destroying the tumour.

A more recent development is chimaeric antigen receptor-modified T cell therapy (CAR-T), where a CAR-encoding DNA cassette is inserted into a patient's T cells, expanded, and returned. The engineered T cells are programmed to attack and destroy the tumour cell bearing the particular antigen. This technique has provided partial or complete remission in >50% of chemotherapy-resistant leukaemias, and in a number of solid tumours. The antigens in question may be CD molecules, adhesion molecules, or growth factor receptors. An extension of this principle is the engineering of a patient's own NK cells, i.e. CAR-NK.

4.2.8 Combinations of treatments

As discussed, each of the three major forms of treatment can be used in series, but rarely concurrently. However, side effects are still important, and in some respects may be predicted. For example, gastrointestinal toxicity of combined chemo- and radiotherapy in cervical cancer is linked to an SNP in a gene coding for glutathione transferase, an enzyme that contributes to antioxidant defence. The specificity of mAbs can be exploited in that they can be linked to radiochemicals (e.g. yttrium-90) or cytotoxic drugs (e.g. an anti-microtubule agent) to deliver the active drug directly to the malignant cell.

Key points

Chemotherapy (treatment with drugs) is an effective, yet often deleterious, therapeutic option: some drugs are non-specific, acting on all body cells, whilst others are designed to precisely target the tumour. The practitioner may be constantly aware of the risk:benefit ratio in this, and indeed, in all forms of treatment.

Having described the major general aspects of the detection and management of cancer, we now move to examine each of the major malignancies in turn, and in order of the frequency of the deaths they cause.

4.3 Lung cancer

The leading cause of cancer death globally (18.1% of all cancer deaths) and in England and Wales (19.0%), lung cancer naturally attracts considerable attention from government bodies, cancer charities and professional associations (e.g. British Thoracic Society), many of whom have published guidelines. Those from NICE include NG122, on the diagnosis and management of lung cancer, and many documents (TA184, TA190, TA227, and others) on chemotherapeutics. Much of the text that follows derives from these (and other) such documents.

4.3.1 Anatomy and aetiology

The ONS and many other data-collection systems often fail to separate lung cancer from bronchial or bronchiolar cancer, and sometimes also from tracheal cancer. However, the UK's leading cancer charity, Cancer Research UK, estimates that some 4.5% of lung and bronchial cancer is due to bronchial cancer alone, but it is unclear whether this can be extrapolated to cancer deaths. In 2021, there were 11 deaths from tracheal cancer in England and Wales. Accordingly, when lung cancer is referred to, it may include a small proportion of bronchial and tracheal cancer. The number of deaths linked to malignant neoplasm of the trachea, bronchus, and lung has been falling year on year from 2016 to 2021, possibly due to an increased recognition of tobacco smoking as the leading addressable risk factor for this type of cancer.

The bronchus is a simple structural tube of connective tissue lined with **columnar epithelial cells** enabling the passage of gases between the lungs and the mouth/nose. Rings of cartilaginous tissues extend down from the trachea to maintain patency, and both left and right bronchi split down to bronchioles that feed lung lobules. Conversely, the alveoli of the parenchyma of the lung, where gas exchange occurs, are formed of **squamous epithelia** that interact intimately with the endothelial cells of the pulmonary capillary tree. This is pertinent as squamous cell epithelia are markedly more susceptible to carcinogens than are columnar epithelia.

Environmental risk factors for lung cancer include tobacco smoke (linked to 9% of all deaths, 15–20% of all cancers, and >50% of lung cancers), ionizing radiation (4.7%), and industrial toxins. Lack of fruit and vegetables is linked to 8.8% of cancers, and certain occupations are linked to 13.2%. It follows that perhaps half of lung cancers are linked to genetic mutations not easily explained by risk factors. There are relatively few well-established gene mutations with clear roles in carcinogenesis in this organ, although there are numerous examples of oncogenes acting alone or in combination, and some of these have a role in choice of treatments. As outlined in Chapter 3, pathology at the level of the cell includes epigenetic changes to histones and mutations in genes for second messengers such as *p53* (mutated in 46% of cases), *KRAS* (32%), and genes for epidermal growth factor receptors (27%), and their pathways. Some of these have impact into treatment choices in this and other cancers.

SELF-CHECK 4.5

Give examples of non-specific and targeted chemotherapies for lung cancer.

4.3.2 Presentation

The primary semi-specific presenting feature of lung cancer is a persistent cough, often with wheezing and shortness of breath (dyspnoea) and unexplained weight loss. These common symptoms point to several differential diagnoses, the most pressing being a chest infection (exclusion of which can be determined with a course of antibiotics) and unexplained fevers. Other differential diagnoses include **mesothelioma**, chronic obstructive pulmonary disease, and pulmonary fibrosis. Asthma may be excluded by testing for potential **allergens** (sections 10.2.1 and 13.5.2). More serious indicators include coughing up blood (haemoptysis), possibly with mucus/sputum, although this may not arise from lung tissue.

Perhaps the principal physical sign is nail clubbing, where the ends of the fingers and/or toes become enlarged, although (in common with many, many other signs and symptoms) there are sensitivity and specificity issues, in that not everyone with lung cancer has finger clubbing and not all cases of finger clubbing are related to lung cancer. Differential diagnoses in this respect include certain forms of heart disease, particularly those where there is cyanosis (bluish tinge to the skin, but most often evident at the lips).

As the tumour grows, it may impact on other structures within the chest, such as blood vessels and lymphatics, the former causing cardiac symptoms (such as chest pain) and worsening the breathing-related symptoms, and possibly enlarged lymph nodes around the neck and collarbone area (supra-clavicular lymphadenopathy). A routine blood test may reveal an anaemia, accounting for fatigue, tiredness and lethargy, a mildly elevated erythrocyte sedimentation rate, and thrombocytosis. Should a sufficient number and quality of these signs and symptoms be present in those over 40 (especially with unexplained haemoptysis), an urgent (within 2 weeks) X-ray is required.

4.3.3 Diagnosis

Initial investigations

An initial X-ray may identify an abnormality often referred to as a shadow, which may be a tumour (Figure 4.5). In advanced cases, X-rays may help locate metastases in other parts of the skeleton. An additional tool is a lung function test, another a bronchoscope, the latter to determine physical changes to the left and right bronchial tubes. However, as a chest X-ray has a high **false-negative** rate, more complex investigations are inevitably needed, and these include CAT/CT scanning and MRI, to include contrast-enhanced CT of the chest, liver, adrenals, and lower neck. This is particularly important in detecting possible metastases. Combined PET/CT will be required to assess any potential **nodal** diseases. Screening of high-risk subjects (age $\leq$55, $\leq$30 pack-years smoking, quit smoking <15 years) with low dose CT is more likely to diagnose lung cancer at an early stage, and this approach is associated with a 20% reduction in mortality.

The laboratory

Despite the value of imaging, the ultimate diagnosis is by **endobronchial** ultrasound-guided tissue biopsy of the tumour, which is not always possible as the tumour may be inaccessible. Laboratory examination of a tissue biopsy of a presumed tumour will determine the cellular basis of the pathology based on the size of the malignant cells—hence the distinction between small-cell lung carcinoma (15% of cases) and non-small-cell lung carcinoma (NSCLC—85%), which can be further classified as squamous cell carcinoma, adenocarcinoma, or large-cell carcinoma. From the gathered tissue,

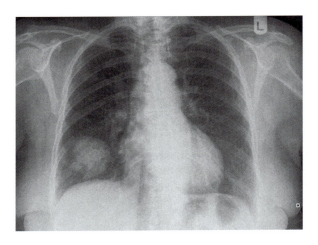

FIGURE 4.5

Lung cancer. There is a clear mass in the right lung.

Khalid T, Phelan B, & Yousif A. Omental Metastasis from ALK-positive Lung Cancer—A Case Report. Egypt J Intern Med 32: 8 (2020); https://doi.org/10.1186/s43162-020-00010-3.

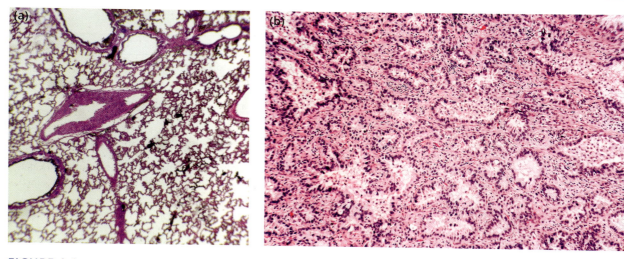

FIGURE 4.6

Lung tissue. (a) A section of normal lung tissue. Note the large air spaces (alveoli). (b) A section of lung tissue heavily infiltrated with a non-small cell carcinoma, with almost complete loss of air spaces.

(a) Shutterstock, (b) iStock.

methods in molecular genetics can also determine mutations in a number of key genes, knowledge of which will be pertinent to certain treatments. Figure 4.6 shows a section of normal lung tissues, and a section of tissue from a lung cancer.

Routine blood tests are not especially helpful, but molecular genetics show promise. A panel of genes (*CDKN1A*, *E2F1*, *ERCC1*, *ERCC4*, *ERCC5*, *GPX1*, *GSTP1*, *KEAP1*, *RB1*, *TP53*, *TP63*, and *XRCC1*, several described in Chapter 3) in bronchial epithelial cells are able to predict lung cancer with sensitivity and specificity that exceeds 90%. Other panels can test for *EGFR*, *ALK*, *BRAF*, *KRAS*, and *MET*, and for *ROS1*, *RET*, *EML4-ALK*, *NTRK1*, *NTRK2*, *NTRK3*, and *MET*. These panels can help diagnosis and management.

Using a panel of miRNAs including miR-205, miR-210, and miR-708, squamous cell carcinoma may be distinguished. Similarly, miR-21, miR-31, and miR-201 may be useful in differentiating lung cancer from other pulmonary disease, whilst tissue expression of miR-935 is reduced in NSCLC and is linked to survival.

4.3.4 Management

Should the patient be a smoker, they must be advised to stop, possibly with the support of nicotine replacement. As with most cancers, surgery is a leading option (if practicable) and it may be curative. An entire lobe of the particular lung may be removed (a lobectomy), but more complex operations remove additional tissue and/or lymph nodes, or perhaps, in exceptional cases, the entire lung. However, before that decision is made, further investigations are inevitable, as knowledge of the TNM stage is required to determine an appropriate surgical approach. Should any enlarged intrathoracic nodes have been discovered on imaging (such as PET/CT), these will need to be biopsied as results of this procedure are likely to influence treatment as well. Similarly, the pathology of any potential distant metastases will need to be determined.

Once a biopsy of the malignant tissue has been obtained, the laboratory will characterize the tumour as being of a small-cell or non-small-cell phenotype, and determine whether it is squamous or not and whether there is a mutation in one of certain genes likely to have contributed to carcinogenesis. Many cancers are linked to mutations in the epidermal growth factor receptor–tyrosine kinase pathway (EGFR–TK)—mutations often present in NSCLC. Accordingly, patients whose locally advanced metastatic cancer bears this mutation (around 17% of NSCLCs in England and Wales) are more likely to benefit from an EGFR–TK inhibitor. A further gene of interest is *ALK* (anaplastic lymphoma kinase), a mutation which, when merged with another gene, generates a fused molecule linked to 3–5% of cases of NSCLC. The expression of a mutation in the gene for programmed death ligand-1

TABLE 4.9 Potential chemotherapy for lung cancer

Pathology	Possible initial agents
Advanced non-squamous NSCLC with *EGFR–TK* mutations	TK inhibitors afatinib, osimertinib, gefitinib, erlotinib
Advanced NSCLC with *ALK* rearrangement,	ALK inhibitors crizotinib, ceritinib, alectinib, brigatinid
PD-L1 expression	Pembrolizumab, atezolizumab, nivolumab
Squamous NSCLC tumours that express PD-L1 ≥50%	Pembrolizumab, then gemcitabine (a nucleoside analogue) or vinorelbine (inhibits tubulin)
Squamous NSCLC tumours that express PD-L1 <50%	Gemcitabine or vinorelbine, and cisplatin or carboplatin (which cross-link DNA).
Small-cell lung cancer	A cisplatin-based regime

(PDL-1), increased levels of which may allow the tumour to evade immune surveillance, also directs treatment.

Chemotherapy

NICE has pointed to numerous options for chemotherapy, depending on histology and molecular genetics (Table 4.9, see NICE TAs for further details), such as pemetrexed (a folate antimetabolite), docetaxel (a microtubule inhibitor), or topotecan (a topoisomerase inhibitor). In some patients, four to six cycles of radiotherapy twice daily after the second round of chemotherapy may be called for. **Dexamethasone** is the agent of choice for symptomatic brain metastases, whilst bone metastases may be treated with **bisphosphonates**, radiotherapy, or denosumab (a mAb that targets the receptors linked to NFκB and TNF). This regime may be used with bone metastases from other primary tumours such as those of the breast and prostate.

Surgery

Once the decision to operate has been made (generally on firm objective data), the question that remains is whether the patient is sufficiently fit (strong enough) for the rigours of surgery. The 'thoracoscore' provides indication of the predicted death rate for surgical patients and those in intensive care units, considering age, sex, an assessment of physical status, **dyspnoea**, priority of surgery, procedure class, diagnosis group, and co-morbidities (smoking, history of cancer, chronic obstructive pulmonary disease, hypertension, cardiovascular disease, diabetes, obesity, and alcoholism). NICE recommends the use of this score, or similar, to estimate risk of death in NSCLC patients, although fitness for surgery can be assessed in a number of ways, such as lung function by **spirometry**, and cardiovascular function by a shuttle-walking test.

For those cases of NSCLC in whom surgery is declined or is contraindicated, radical radiotherapy, possibly with chemotherapy, should be offered. Precise protocols will vary, but radiotherapy is generally fractionated, perhaps into 20 sessions over 4 weeks or 30–33 sessions over 6–6/½ weeks. Similarly, patients with good performance status should be offered postoperative chemotherapy, such as one based on cisplatin. Conversely, in some cases chemoradiotherapy can be scheduled 3–5 weeks before surgery.

Key points

Lung cancer is the most common malignancy, and generally carries a poor prognosis as it is often diagnosed late, and accordingly can be difficult to treat successfully.

CASE STUDY 4.1 Lung cancer

Early in February, a 60-year-old market-gardener presents to his GP with a 4-week history of a troublesome dry cough, and denies a recent cold, flu, or any fevers. He had smoked around twenty cigarettes a day from his teens until around age 50, then cut down slowly, and has now not smoked for 5 years. There is no outstanding previous medical history, although the man seems reluctant to offer information: his father died of a stroke aged 75, and his 85-year-old mother is living in sheltered accommodation. He drinks ~12 units of alcohol a week. The practice nurse performs a lateral flow test for COVID-19 (which is negative), takes his temperature, measures his blood pressure, and performs a dipstick analysis of his urine: all are unremarkable. The GP prescribes a 10-day course of a common antibiotic, and as he has not seen the man for a number of years, orders a routine chest X-ray and routine blood tests, and invites him to return in 2 weeks. Upon doing so, the man admits some weight loss, and progressive weakness and tiredness after a day's work, which he ascribed to increasing age, but the cough persists and sometimes he brings up yellow/green sputum. The blood results are all within the reference range, but the red blood cell count and the haemoglobin result are both at the lower end of normal. The GP telephones the local hospital to upgrade the X-ray request to urgent and invites the man to return a week after the X-ray.

The patient (as he has now become) returns to his GP and receives the news of an abnormality in the X-ray suggestive of lung cancer, although the GP is quick to point out that there are alternative diagnoses. A rapid referral to a consultant chest physician at the local hospital is arranged, and upon meeting her, the patient is scheduled for an additional X-ray, CT, MRI, and lung function tests. A bronchoscope is arranged and, guided by the imaging results, finds an abnormality deep in the bronchial tree, and a biopsy is taken and sent to the laboratory. Once all the tests have been completed, an MDT is convened to discuss the case, make a diagnosis, and formulate options referring to NICE NG122.

Imaging reports a well-delineated mass 2 cm × 3 cm × 3 cm in the middle lobe of the right lung (thus stage cT1 or T2), with the probable involvement of three local lymph nodes, possible **lymphadenopathy** in the clavicular region (stage N2b or N3), and the possibility of excessive pleural fluids. The left thorax is normal. Together, these suggest at least a TNM stage IIB, possibly stage IIIa cancer, hence cT2bN1M0. Routine histology reports an NSCLC, and molecular genetics reports rearrangement in ALK at 2p23.2 that codes for a tyrosine kinase of the same name, i.e. ALK. The MDT recommends surgery to the primary tumour, for which the patient feels he is not ready, but he will consent to chemotherapy and radiotherapy. In accordance with NICE TA500, this is a course of ceritinib (an inhibitor of ALK), starting on 150 mg a day, rising every 2 weeks to 750 mg. Clinical trials have shown that this drug more than doubles progression-free survival compared with standard chemotherapy, and is superior to another ALK inhibitor, crizotinib (Soria et al. 2017 and Li et al. 2019). The radiotherapy is 55 Gy in 20 fractions over a 4-week period.

The side effects of hair loss, some diarrhoea, and general gastrointestinal discomfort are tolerated by the patient, and liver function tests rise a little, and haemoglobin and the white blood cell count both fall by about 10%, all established but undesirable side effects of this form of chemotherapy. After 10 weeks, a CT/MRI finds no change in the size of the primary tumour, but indicates that one of the lymph nodes has enlarged. This prompts an ^{18}F-fluorodeoxyglucose PET scan combined with CT, which is equivocal, reporting the possibility of early disease in some bones. Accordingly, with the stage now at least cT2bN1M1, a change of therapy is in order.

The patient is scheduled for a 4-week course of radiotherapy directed at the tumour and those bones where metastases are suspected, followed by the combination of nintedanib (another tyrosine kinase inhibitor described in NICE TA347) 200 mg orally twice a day on days 2 to 21 of a 3-week cycle, plus docetaxel (75 mg/m^2 intravenously) on day 1 of the cycle. This regime is followed for two cycles, but the patient's white blood cell count starts to fall, so the nintedanib is reduced to 150 mg twice a day, and the docetaxel to 60 mg/m^2 for another two cycles. Two weeks after the final cycle, imaging reveals that the tumour had reduced by 50% and that all previously affected lymph nodes are smaller.

This good news gives the patient the confidence for surgery with curative intent, which he requests. His thoracoscore is good, shuttle walk test and spirometry acceptable. The surgery (a lobectomy) goes well, but he needs two units of packed red blood cells as post-operative haemoglobin was suboptimal. After 2 days in intensive care, he moves to a surgical ward for a further 10 days and with the help of physiotherapists, haematologists, and pharmacists is discharged on oral anticoagulants and analgesics. Once recovered, he continues the cycles of adjuvant chemotherapy, following which he is maintained on a lower dose of these drugs. In counselling with his Macmillan nurse, he is still haunted by the fear of metastatic disease, and that the lymph nodes may become malignant, and so is prepared for additional cycles of high-dose chemotherapy and/ or radiotherapy, should these be necessary. Five years later, he is well.

SELF-CHECK 4.6

What factors influence the choice of therapy that may be offered to someone with lung cancer?

4.4 Colorectal cancer

Colorectal cancer brings the second-highest risk of death both globally (9.4%) and in England and Wales (10.3%). There is a marked global geographical variation, the incidence being highest in Australia and New Zealand (367/million), Europe (304/million), Eastern Asia (265/million), and North America (262/million), and lowest in Africa (78/million) and South-Central Asia (49/million) (Mattiuzzi et al. 2019). As with lung cancer, there are several NICE documents pertinent to this section. These include NG151 'Colorectal cancer', and technology assessments TA61, TA100, and TA176 on chemotherapy. Unlike lung cancer, the number of deaths from colorectal cancer in England and Wales has slowly been rising, increasing by 4.2% from 2016 to 2021, whereas deaths from all malignant neoplasms fell by 0.9%.

4.4.1 Anatomy and aetiology

The large intestines, approximately 1.4–1.6 metres long (depending on body size and sex), comprise several sections. A short caecum transforms to the longest section, the colon, further classified as ascending, transverse, and descending. The terminal sigmoid colon transforms into the rectum. The function of the colon is to extract water from remnants of the digestive **chyme** and compact it to form faeces. The colon secretes mucus to aid the passage of the developing faecal matter. The latter is stored in the rectum (perhaps 15–20 cm long) until expelled via the anus (a collection of ring-like sphincter muscles) by a collection of voluntary and involuntary nervous system messages acting on striated muscles of the pelvic floor.

Histology is of a multi-layered muscular tube with a luminal lining of columnar epithelial tissue arranged in invaginations called crypts (the mucosa). The middle layer of circular, spiral, and longitudinal muscle is enclosed by a serosal layer with blood vessels, lymphatics, and nerves that is in contact with the peritoneal cavity. It is the intestinal epithelial cells that become malignant and, should this be a mucus-secreting goblet cell, a large mass of mucus may be secreted along with the faeces, which can be symptomatic. The dominant form of cancer is therefore a carcinoma and, should it involve goblet cells, it is an adenocarcinoma.

Risk factors include old age, male sex, overweight and obesity (linked to 13% of cancers), excessive use of alcohol (11.6%), smoking (8.1%), insufficient dietary fibre (12.2%), and a diet rich in red meat (21.1%). Certain intestinal diseases, such as **Crohn's disease**, **ulcerative colitis**, and allied inflammatory disease, are often linked with, and may predict, colorectal cancer.

A key feature is the **intestinal polyp**, very commonly found in the middle-aged bowel (hence polyposis), some of which oversecrete mucus, and if so, can be described as benign adenomas, and there is compelling evidence in many cases that they are the first stage of the disease. Whilst there are numerous examples of a single mutation directly causing a disease, there are also instances in which a number of different mutations are required in a specific sequence in order for carcinogenesis to proceed—an extension of Knudson's multi-hit hypothesis. Many harvested polyps have lost a tumour suppressor gene (APC: adenomatous polyposis coli, at 5q22.2) but remain benign. However, activation of K-ras (a mutated oncogene form of the physiological G-protein second messenger Ras on 12p12.1 (section 3.2.2) which is present in 20% of colon cancers and codes for an abnormal GTPase), followed by loss of a tumour-suppressor gene (imaginatively named deleted in colorectal cancer, hence DCC on 18q21.2), are linked to the development of an adenoma. Further changes, including loss of tumour-suppressor gene p53 (on 17p13.1), can convert this relatively benign adenoma into a full-blown carcinoma.

In this respect, familial adenomatous polyposis, characterized by large numbers of luminal polyps, is an autosomal dominant condition, with the risk of polyps transforming into early colorectal cancer. It is linked to the loss of function in APC described above. A second autosomal dominant form, non-polyposis Lynch syndrome, is caused by mutations in DNA mismatch repair genes, most commonly MLH-1 (at 3p22.2) and MSH-2 (at 2p21). Those taking aspirin for conditions such as cardiovascular disease have a reduced risk of colorectal cancer, and so people with Lynch syndrome are advised to take 600 mg daily. Nevertheless, most cases are non-hereditary (i.e. are sporadic), with the disease developing only after a series of mutations. This process hypothesizes sequential changes to *APC*, followed by *KRAS*, then *TP53*, and finally in *DCC*. Other cases of colorectal cancer are linked to *MUTYH* at 1p34.1, coding a DNA base excision repair enzyme, and to the tumour-suppressor *PTEN* at 10q23.31, coding a phosphatase.

4.4.2 Presentation

Semi-specific presenting features include abdominal pain, constipation, diarrhoea, and mucus and blood in the faeces (which may be red-streaked), the latter more obvious if the tumour is close to the anus. However, if there is bleeding in the upper intestines (perhaps due to gastric or duodenal disease as well as the proximal colon), faeces may be black. As with many cancers, non-specific symptoms include nausea, vomiting, weight loss, and tiredness, the latter possibly resulting from a normocytic anaemia. Those with a number of symptoms should be referred urgently to secondary care (within 2 weeks) if they:

- are aged ≥40 and with unexplained weight loss and abdominal pain, or

- are aged ≥50 with unexplained rectal bleeding, or

- are aged ≥60 with iron-deficient anaemia or changes in the bowel habit, or

- have a positive faecal occult blood test.

4.4.3 Diagnosis

The process of diagnosis follows a number of steps, starting with testing for faecal occult blood (FOB), and then moving to blood tests, and imaging of the abdomen and the lumen of the bowel.

Faecal occult blood

The presumption underlying the FOB test is that the pathology (of which there are several different forms) causes loss of blood into the lumen of the intestines, which then mixes with the general digestive process and is expelled with the faeces. A simple chemical method relies on free haem catalysing the action of hydrogen peroxide on substrates based on guaiacin, which turns a blue colour (Figure 4.7). A negative FOB does not rule out colorectal cancer, but a positive result will lead to the next stage of investigation.

Other FOB tests are based on the detection of haemoglobin by antibodies (hence **immunocytochemical** tests: see NICE DG56). Faecal calprotectin is positive in inflammatory bowel disease and is used to exclude this differential diagnosis.

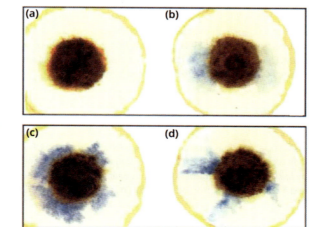

FIGURE 4.7

The Faecal occult blood test. Preparation A is negative; the blue colour around B, C, and D denotes positive results.

Schapiro M, Allison JE. 'The Role of Fecal Occult Blood Testing in Screening for Colorectal Cancer. Colorectal Cancer: An Update for Diagnosis and Prevention, Series #3.' (2007).

BOX 4.2 Screening for colorectal cancer

There is considerable scope for screening for colorectal cancer in the general population with FOB testing, but more pertinently in those at increased risk. The latter include those aged over 60–74, and those with a family history. In a follow-up study, screening every 2 years brought (at least) a 15% reduction in colorectal cancer deaths, and in another, cancer was present in 12%, and an adenoma in 50% of those with an abnormal result. However, a 20-year follow-up study found no difference between the incidence of invasive cancer in those screened compared with that in a control group not screened.

The laboratory

In common with all other cancers, the laboratory provides a full blood count, urea and electrolytes, and liver function tests at a minimum. Semi-specific tests include serum **carcino-embryonic antigen (CEA)**, CA-19.9, cytokeratin-1, and mucin-1: all are increased in colorectal cancer compared to benign lesions, and several also predict TNM tumour stage, lymph node invasion, and metastases. However, as with almost all investigations, issues of sensitivity and specificity arise as increased levels may also be present in several other cancers and some inflammatory diseases. Indeed, raised CA19-9 also marks pancreatic cancer, whilst raised CEA may also be present in breast, lung, pancreatic, and some thyroid cancers. Nevertheless, a combination of these four semi-specific markers into a mathematically derived score helps the diagnosis and possibly the TNM stage.

In molecular genetics, two multi-target panels can help with diagnosis and management. One panel assesses *BRAF*, *MLH1*, *MSH2*, *MSH6*, *PMS2* (all cited by NICE DG27 in respect of Lynch syndrome), *KRAS*, *NRAS*, *POLD1*, and *POLE*; whilst another can report on *NTRK1*, *NTRK2*, and *NTRK3*. A further important gene is *EGRF*, as particular variants have roles in choice of chemotherapy. Clinical research indicates that serum levels of miR-409, miR-7, and miR-93 may be useful in colorectal cancer, and so one day may enter routine practice.

Visual inspection

The lumen of the bowel may be viewed by a colonoscopy or sigmoidoscopy (generally the terminal third). These procedures introduce a flexible tube with a camera (and possibly a biopsy grab) into the anus, which is then inserted into the rectum and colon in order to look for potential lesions. Should such a lesion be encountered, it may be biopsied and tissues sent to the laboratory to determine the extent of a malignancy, although if the lesion is a large and unambiguous tumour, a biopsy may be academic. However, processing a biopsy in the laboratory can determine the cellular nature of the tissue and mutations in genes of interest. As yet, and in contrast to lung cancer, genetics do not have a major influence on management, with the exception of *KRAS*. The inspection may reveal variable numbers of polyps, and if large ones are encountered, these may also be biopsied or removed by a snare. Figure 4.8 shows two colonoscopy results: one normal and one with an abnormality, the precise nature of which can only be determined by a biopsy.

Imaging

Although colonoscopy and sigmoidoscopy can make the diagnosis, they are unable to define the TNM stage, which may be achieved by contrast-enhanced CT of the chest, abdomen, and pelvis.

In rectal cancer, an anticipated resection margin, tumour, and lymph node staging can be detected by MRI. Endorectal ultrasound is an option if MRI shows that the disease is amenable to local excision, and a barium enema may be called for in those with co-morbidities. Figure 4.9 shows an MRI image of a rectal tumour.

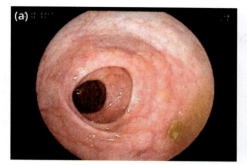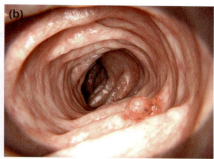

FIGURE 4.8
Colonoscopy. (a) A normal colon. (b) A colon with an abnormality (right).

Shutterstock.

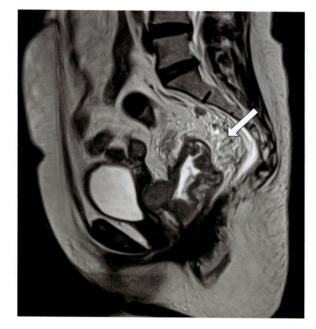

FIGURE 4.9
MRI of a rectal tumour, which is indicated by the arrow.

Getty Images.

4.4.4 Management

In common will all major diseases, a multidisciplinary team (MDT) will convene to survey the results of the investigations, consider alternative diagnoses, and so come to a working plan. Then, considering the physical state of the patient, a combination of the three major treatment options will be discussed, dependent on stage, and put to the patient, and (after consent) possibly also to their family and/or carers.

Once the MDT offers surgery to the patient, and they accept, this may proceed promptly, or may be preceded by some chemotherapy and/or radiotherapy. The excised tumour (Figure 4.10) will be processed for standard haematoxylin and eosin staining (Figure 4.11), and from this the cancer can be staged.

The 'historical' system of Dukes was developed in the 1930s and considers the invasiveness of the tumour once it has been removed, with four stages of progressively more advanced disease. In stage A, the disease is present only in the inner (luminal) layers whereas in stage B the tumour extends into the muscular layers. In stage C there is local lymph node involvement, and in stage D there is metastatic disease in other organs. As such, the Dukes system laid the way for the development of

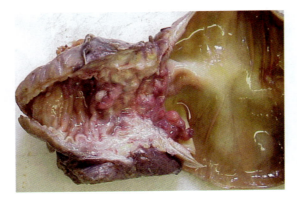

FIGURE 4.10

An excised bowel tumour. The tumour itself is on the left (with whiteish tissues, normal tissue on the right).

From Orchard & Nation, *Histopathology*, second edition, Oxford University Press, 2018.

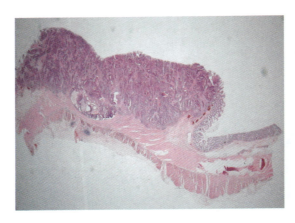

FIGURE 4.11

Histological processing of a bowel tumour. Haematoxylin and eosin staining of a section of the tumour (left) and normal tissue (right), showing a moderately differentiated adenocarcinoma invading up to the muscularis propria.

From Orchard & Nation, *Histopathology*, second edition, Oxford University Press, 2018.

other scores, using sophisticated imaging, evolving into the TNM system, and still retains some value today. Application of the TNM system to colorectal cancer and its 5-year survival may be described with the progressive expansion of the tumour through the layers of the bowel and beyond (Table 4.10, Figure 4.12).

Managing local disease

If the disease is low-risk and operable (e.g. stage cT1–T2, cN0, M0), pre-surgical radiotherapy or chemotherapy is generally unnecessary. Removal of the tumour can take an endoscopic and/or surgical approach, the former being transanal and so minimally invasive (i.e. the abdomen is not opened), the latter an open resection. Further aspects include the possible need for a stoma, potential for lymph node removal (which will inform further on staging), possible need to convert to open surgery, length of hospital stay (1–2 days for endoscopic, 5–7 for open surgery), external scarring, and complications (bleeding, anal incontinence, perforation, etc.).

Should the disease be more extensive (e.g. cT1–T2, cN1–N2, M0 or cT3–T4, any cN, M0), then pre-operative radiotherapy or chemoradiotherapy should be offered. Options for chemotherapy include capecitabine (1,000–1,250 mg/m^2 orally od (od = once daily)) and intravenous oxaliplatin (an alkylating agent: 85 mg/m^2 every 2 weeks) for 3 months; a combination of oxaliplatin, 5-FU, and folinic acid (500 mg/m^2) for three to 6 months; or a single agent such as capecitabine for 6 months. In some cases, a **stent** may be placed for bowel obstruction in palliative cases, or in cases of curative intent where the obstruction is acute. Patients with M0 disease who have had potentially curative surgical treatment should be followed up for the first 3 years with serum CEA and CTs of the chest, abdomen, and pelvis.

TABLE 4.10 The TNM system in colorectal cancer

Tumour stage	Nodal stage	Metastases stage	Approximate Dukes stage	Pathology	Approximate 5-year survival
T1	N0	M0	A	Tumour is present only in the mucosa	>90%
T2	N0	M0	B1	Tumour is also present in the muscularis propria	70–85%
T3	N0	M0	B2	Tumour is also present in the serosal layer	55–65%
T4	N0	M0	B2	Tumour extends outside the normal boundary of the bowel	55–65%
Any T	N1	M0	C1	1–4 lymph nodes involved	45–55%
Any T	N2	M0	C2	>4 nodes involved	20–30%
Any T	Any N	M1	D	Metastases	<10%

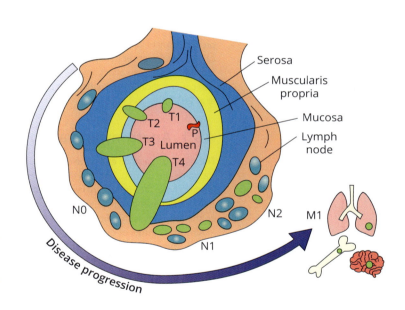

FIGURE 4.12
The TNM system as it may be applied to colorectal cancer. Key features in the development of metastatic colorectal cancer: T = tumour; N = (lymph) node; M = metastases, such as (M1) to the lung, bone, and brain; P = polyp.

Managing metastatic disease

All those with metastatic disease (e.g. liver, lung, peritoneum) should be tested for *RAS* and *BRAF V600* mutations. High-quality CT imaging is essential, such as of the liver, as this is an important metastatic target, although the chest, abdomen, pelvis, and brain may also be imaged. This imaging may indicate PET/CT scanning of the whole body, and isolated and well-delineated metastases may be amenable to resection. Chemotherapy options include those for moderate disease, but also tegafur (another pro-drug of 5-FU), often used with uracil. Should some of these regimes be poorly tolerated (i.e. cause major side effects), raltitrexed (an inhibitor of thymidylate synthase) is an option. Many are used combined with mAbs bevacizumab (targeting VEGF), cetuximab, and panitumumab (both targeting EGFR). Notably, these monoclonal antibodies can only be used in patients with normal (i.e. non-mutated, wild-type) *KRAS*, as they are ineffective in those with a mutation in this gene.

Ongoing care and support

Patients will be offered a full range of psychological and medical support. They may expect a change in bowel habits, which may be profound. If so, dietary management, laxatives, and anti-bulking, and anti-diarrhoeal and anti-spasmodic agents will be offered.

Key points

Colorectal cancer is the second most common cancer and cause of cancer death. It may be screened for by testing for faecal occult blood, and early disease may be cured by surgery.

SELF-CHECK 4.7

Briefly describe the stages of colorectal cancer.

4.5 Prostate cancer

A key document in this section is NICE NG131, but there are also several technology appraisals of potential chemotherapeutics, and interventional procedure guidelines on ultrasound, brachytherapy, electroporation, cryoablation, and cryotherapy. In England and Wales, over 10,000 men die annually from prostate cancer, a figure that has changed little since 2014, which makes it the second most frequent cause of male cancer deaths (after lung/bronchus, which leads to death in over 14,000 men each year). Prostate cancer can affect anyone with a prostate, including transgender women who have had gender-confirmation surgery, where the prostate is usually retained and may still become malignant (Sharif et al. 2017).

4.5.1 Anatomy and aetiology

The prostate is a small **exocrine** organ the size of a large walnut that lies at the base of the bladder, surrounding the urethra. The two ejaculatory ducts from the testes join the urethra within the prostate. It is also linked to the seminal vesicles, and together they provide an alkaline fluid to promote sperm survival within the female reproductive tract. The tissue basis of the prostate is of pseudostratified epithelium, although other types of epithelial may also be present, supported by fibrous tissue.

  Almost all forms of disease of this organ include cancer, inflammation (prostatitis), and abnormal growth (benign prostatic hyperplasia, BPH), the latter two therefore being differential diagnoses. Prostatitis may be diagnosed by bacteria and/or white blood cells (bacturia and leucuria respectively), and there may be a fever and an acute phase response, BPH by imaging and by the exclusion of other possible diagnoses. There are few specific risk factors, and non-specific factors include obesity, smoking, and increasing age, but also being of African descent. Changes in genes such as *TMPRSS2*, *BRCA1*, *RAD51*, and *FANCL*.

4.5.2 Presentation

Unsurprisingly, certainly in the early stages, the signs and symptoms of prostate cancer are all genito-urinary. There may be changes in urination—of volume, or frequency, and at night—and some may be painful. Rarely, there may be blood in the urine (haematuria, also present in bladder and renal cancer) or ejaculate. Other symptoms include erectile and ejaculation dysfunction. As these are also symptoms of other prostate disease, more information is needed to exclude alternative diagnoses. The GP may undertake some formal testing, such as for the prostate-specific antigen (PSA, as described in section 4.2.3), blood test, and digital rectal examination (DRE), where the practitioner estimates the size of the prostate with their fingers. If positive, an urgent referral to secondary care will follow.

4.5.3 Diagnosis

If not done in primary care, PSA (a 26 kDa peptidase coded for by *KLK3* at 19q13.33) will be measured and DRE performed, although testing for other markers such as spermine and spermidine may be useful. Should these tests be abnormal, imaging is called for with multiparametric MRI, and possibly trans-rectal ultrasound. However, 11–28% of men whose MRI suggests the growth is low-risk actually do have clinically significant cancer. Should the MRI suggest a moderate or high-risk growth, the patients will be offered a trans-rectal biopsy, which may be guided by ultrasound.

Biopsy

This procedure, under local anaesthetic, may take up to a dozen or so small pieces of the prostate, which will go to the laboratory where histology will provide a Gleason score, of which there are two parts. The tissue is scored on a scale of 1 to 5 for the degree to which it resembles normal prostate tissue (score 1), or malignant tissue not recognizable as normal (score 5). The second part records the highest grade of the non-dominant cell pattern, also on a scale of 1–5. So a low grade result would be reported as Gleason 2+3=5, that being low risk, whereas a Gleason 4+4=8 is more likely to lead to surgery.

The procedure is not without risk: 44% report pain and in 15% this will last 2 weeks, 20% will develop a fever, 66% will have haematuria, 37% blood in their faeces, and 90% blood in their first post-biopsy ejaculate, although all these will be temporary. More serious is the 1% risk of sepsis, as the rectum (through which the biopsy needle passes) is rich in bacteria, so that antibiotic cover is necessary. However, as with almost all procedures, there is the risk of false negative in that clinically active cancers may be missed. Indeed, 18–23% of men with a low-risk MRI image get a biopsy result of a clinically insignificant prostate cancer. The MRI will also contribute to staging by providing opinions about local lymph nodes. If the PSA is very high, and there is bone pain suggestive of metastases, a PET isotope bone scan may be offered. Figure 4.13 shows a section of the prostate with normal tissue and a malignant section.

Should the patient be deemed low-risk, serial PSA tests are undertaken, so that an increase in PSA over time (the PSA velocity) can be determined in primary care every three to 6 months. Should the velocity exceed >40 ng/ml/year, biopsy should be offered, and if positive, surgery.

4.5.4 Management

Prostate tumours exhibit remarkable variation in their growth rates—some progress very slowly whilst others are aggressive. Several post-mortem studies have reported apparently asymptomatic prostate cancer in men who have died of other disease. Balanced against the argument for surgery are the risks of this procedure, of which two need to be addressed.

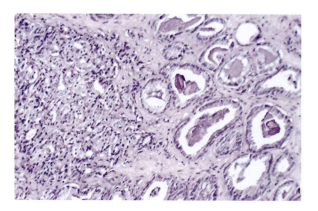

FIGURE 4.13

The prostate gland. Malignant tissue on the left, normal tissue on the right.

From the website of the National Cancer Institute (https://www.cancer.gov).
Photographer: Otis Brawley.

Erectile dysfunction

An erection requires increased arterial blood flow into the **corpus cavernosum**, which is governed by nerves that run from the spine and alongside the prostate. Surgery to the latter may therefore lead to damage or loss of these nerves, and so erectile dysfunction, leading the surgeon to take more care with 'nerve-sparing' prostatectomy. Even if the surgery leaves the nerves undamaged, there will be a period of impotence that may last weeks or years, or which may become permanent. First-line treatment is oral phosphodiesterase inhibitors (such as sildenafil) which bypass the nervous stimulation of the relevant arteries to act directly on the endothelium, which induces relaxation of the arterial smooth muscle cells and so increased blood flow into the penis. Should this fail, pellets infused with prostacyclin can be passed down the urethra (wherein the drug diffuses into the corpus cavernosum to increase arterial blood flow), or prostacyclin can be injected directly into the penis, having the same mechanism. The last resort is surgery, with the placement of a semi-rigid rod, or of balloons that can be filled with saline to produce an erection-like state. Use of a vacuum pump may also help restoration of a normal erection regardless of pharmacotherapy. A prostatectomy is also effectively a vasectomy, so patients will be offered sperm banking.

Incontinence

Removal of the prostate may also remove some of the sphincter muscles that regulate urine flow out of the bladder, requiring a urinary catheter to be inserted through the penis into the bladder, to allow the **anastomosis** of the base of the bladder and the urethra to grow together. As with loss of erection, a degree of incontinence is inevitable, but in many men this will resolve with time and pelvic floor muscle exercise, although in a minority full continence will never return, requiring the lifetime use of incontinence pads. Men with intractable stress incontinence should be offered surgery for an artificial urinary sphincter.

Options

The MDT will advise the patient of the following options:

- Watchful waiting describes a strategy where the emphasis is not so much on curing the cancer as attempting a degree of control. It is aimed at patients with localized cancer who do not ever wish to have curative treatment. It therefore excludes radiotherapy and surgery but may include chemotherapy.

- Active surveillance is a curative strategy in patients with localized or intermediate-risk localized prostate cancer, where only those tumours showing signs of progressing are acted upon with radical treatment. In the first year, PSA is tested every 3–4 months, DRE at 12 months. Thereafter, PSA every 6 months, DRE annually.

- Radical prostatectomy, offered to all patients with localized high-risk cancer.

- Radical radiotherapy, offered to all patients with localized high-risk cancer.

In both these latter cases, patients with intermediate-risk cancer may be offered active surveillance if they do not want to have radical treatment. As informed patient choice is paramount, some may decline treatment altogether, or chose to move between options.

Surgery

As with other cancer, knowledge of the stage is crucial as it drives decision-making by the MDT and the patient. Often because of the knowledge of an active tumour of unknown activity in their prostate, or of a clinically insignificant cancer (which may, of course, transform), many men opt for surgery, despite the consequences previously described. Should the excised tumour be particularly aggressive, and should an MRI have suggested lymph node involvement, these may also be removed. The preferred surgical technique is keyhole, which consists of five incisions in the abdomen, one being very close to the umbilicus. A more profound surgical treatment offered to all men with metastatic disease, as an alternative to hormonal therapy, is bilateral orchidectomy (castration).

Radiotherapy

This option is offered to men with high-risk prostate cancer, but patients should also be advised of a small increase in the risk of colorectal cancer and of other bowel problems. The latter include radiation-induced enteropathy, and if suspected should be investigated with flexible sigmoidoscopy to exclude inflammatory bowel disease or malignancy and ascertain the nature of the radiation injury. Pelvic radiotherapy could be considered in patients with locally advanced cancer who have a >15% risk of pelvis lymph node involvement and who are to receive hormonal therapy and radical radiotherapy. An alternative or adjunct to external beam radiotherapy is brachytherapy, but in high-risk patients this should not be the only strategy.

Chemotherapy

This can be viewed as in two parts: anti-androgen (hence hormonal therapy) and standard cytotoxics, although they may both be part of the same regime. The former relies on the presumption that the tumour is part-driven by testosterone, and that this can be suppressed with anti-androgens such as bicalutamide (casodex, 150 mg od). Anti-androgens may be offered for 6 months before or after radiotherapy and may be considered for up to 3 years in high-risk localized cancer. For those on intermittent hormonal therapy, PSA should be measured every 3 months, and if there is symptomatic progression, or the PSA is ≥10 ng/ml, the **androgen** deprivation therapy should be restarted. The common side effect of hot flushes may be treated with medroxyprogesterone (20 mg daily for 10 weeks), or cyproterone acetate (50 mg twice daily for 4 weeks). Patients will also need to be made aware of the deterioration in sexual function whilst on this therapy, treatment for which is as for prostatectomy.

A further risk factor of hormone therapy is osteoporosis, and if this arises, bisphosphonates (which induce bone-remodelling osteoclasts to go into apoptosis) or denosumab (a monoclonal antibody RANKL (receptor activator of NF–kB ligand) inhibitor that suppresses osteoclast development: 60 mg intravenously once every 6 months) may be offered. Those offered the anti-androgen bicalutamide for over 6 months should have prophylactic radiotherapy to breast tissues to reduce the risk of gynaecomastia. If this fails, the patient may be prescribed tamoxifen.

The primary cytotoxic chemotherapeutic is docetaxel (a micro-tubule inhibitor), and in newly diagnosed metastatic disease it is given in six 3-weekly intravenous cycles of 75 mg/m^2, potentially with prednisolone. This regime can also be offered to patients free of significant co-morbidities and non-metastatic disease who are about to start long-term anti-androgens and have high-risk disease (e.g. stage T3/T4 or Gleason 8–10 or PSA >40 mg/ml). Use of docetaxel extends the timing of disease progression from 5 years to 6 years, but at a cost of side effects that include febrile neutropenia (in 15%), feeling unusually weak or tired (not necessarily linked to anaemia), or with gastrointestinal symptoms (both 8%), respiratory symptoms (5%), and problems with the nervous system (4%).

Castration-resistant metastatic disease should be treated first with docetaxel (possibly with prednisolone), and then with abiraterone (an anti-androgenic, 1 g od), and those with hormone-relapsed cancer may be offered a corticosteroid such as dexamethasone (0.5 mg od) as a third-line agent after androgen-deprivation and anti-androgen therapies. Other drugs in this area include enzalutamide (an antagonist of the nuclear androgen receptor). If bone metastases are present, bisphosphonates and zoledronic acid may be offered.

Combination therapy

Patients with intermediate and high-risk localized cancer may be offered androgen-deprivation therapy alongside radiotherapy.

Relapse

The standard monitoring aftercare is a PSA, increasing levels of which after radical therapy are described as a biochemical relapse, and is seen in 25% of patients. However, this does not necessarily call for a change in the treatment plan, but an isotope bone scan may be offered if symptoms or PSA trends are suggestive of metastases. The PSA doubling time should be estimated from at least three

CASE STUDY 4.2 *Prostate cancer*

In his late 50s, an apparently healthy man is diagnosed with a mild heart condition requiring 75 mg aspirin and 40 mg atorvastatin daily. Upon reaching age 60, he requests a PSA test whilst having a blood test to check if the atorvastatin is effective. Two weeks later, he is surprised to learn levels are high at 5.95 ng/ml (reference range <4.0). Two weeks after that, a repeat blood test reports levels of 5.85 ng/ml. After a further 2 weeks, he visits a consultant urologist, and on reflection recalls some erectile problems. The urologist performs a DRE, with unequivocal results—that is, a swollen prostate—and orders an MRI. This reports a diffuse loss of normal signal in the peripheral zone bilaterally, and at the mid-level towards the apex there is a more discreet focus which is suspicious. The consultant recommends a trans-rectal biopsy. Fourteen small pieces of prostate are sent to the laboratory, which makes the diagnosis of prostatic adenocarcinoma (Figure 4.14).

The sum of the biopsy results provides a Gleason score, which brings together the degree of involvement of the tumour within the prostate, and in this case the score is 3+4 = 7 (the maximum being 10), which implies a moderately to poorly differentiated tumour. This information is fed into the NHS 'Predict Prostate' scoring system for patients without metastatic disease in whom conservative management and radical treatments are options. With his profile, he has an 86% chance of surviving 10 years, and a 69% chance of surviving 15 years with conservative treatment. With radical treatment, these figures are 90% and 77% respectively. However, these benefits come at a cost: that of erectile dysfunction at 3 years being 27% with conservative treatment, 56% with nerve-sparing prostatectomy, and 39% with radiotherapy. The figures for incontinence (wearing one or more pads in the last 4 weeks) being <1%, 20%, and 3% respectively, whilst bowel issues (bloody stools about half the time or more frequently) are <2%, <2%, and 7% respectively. He opts for surgery.

A month later, he is admitted to hospital, where bloods are taken and he is swabbed for MRSA. The following morning, he undergoes nerve-sparing keyhole prostatectomy, which takes around 90 minutes. After a further 90 minutes in post-op recovery, he is returned to the ward with a venous catheter for the self-regulated delivery of morphine for analgesia. As expected, recovery is uneventful and he is anti-coagulated with a low molecular weight heparin, and post-operation haemoglobin has hardly deviated. The drain is removed after 48 hours, and he is discharged after 5 days. The urethral catheter is removed in outpatients after a further 2 weeks.

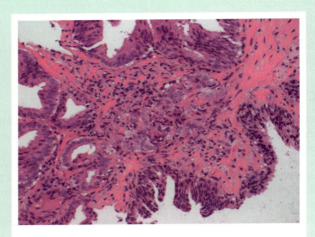

FIGURE 4.14

Prostate cancer. Haematoxylin- and eosin-stained prostate biopsy material with changes consistent with a malignancy.

From Orchard & Nation, *Histopathology*, second edition, Oxford University Press, 2018.

Upon review with the surgeon a few weeks later, he learns of the histology report of his prostate, which confirms the biopsy, and in which was found a relatively well-defined neoplasm occupying perhaps 20% of the organ. Fortunately, the tumour was completely enclosed within the prostate (hence T1 or T2), and as there was no evidence of lymph node involvement, no further treatment is called for. However, less good news is that the final pre-operative PSA level was 8.81 ng/ml, meaning that levels had increased by 2.9 ng/ml (50%) in 6 months (equivalent to 5.8 ng/ml/year), this being indicative of a rapidly growing tumour and so a poor prognosis. He is given further appointments for blood tests for PSA, and for incontinence and erectile dysfunction clinics. The former offers pelvic floor exercises, the latter initially phosphodiesterase inhibitors (e.g. Viagra, which proves to be ineffective) and intra-cavernosal injections of prostacyclin (Viridal, which are effective).

Two years later he is reliant on intra-cavernosal injections for an effective erection (natural erections are slowly returning but of insufficient quality for satisfactory sexual intercourse), is very mildly incontinent (dry 98% of the time), and PSA levels are undetectable. He will continue to have his PSA monitored every 6 months for life, as rising levels indicate production from a second source, most likely metastases.

tests over 6 months, and radical radiotherapy to the prostatic bed may be offered. Hormonal treatment is not recommended in biochemical relapse unless there is symptomatic local disease progression, any proven metastases or a PSA doubling time of less than 3 months.

A key tool is investigation of cells bearing the prostate-specific membrane antigen (PSMA, with biological function as a glutamate carboxypeptidase), coded for by *FOLH1* at 11p11.12. Once serum levels of PSA have reached a cut-off point (typically 0.2 ng/ml), complex PET/CT imaging with isotopes linked to PSMA is used to detect foci of activity which inevitably indicate a metastasis. A developing mode is to use next-generation sequencing for specific gene lesions, such as microsatellite instability-high or mismatch-repair deficient tumours. If present, the mAb pembrolizumab, targeting the programmed cell death protein 1 receptor (also used in many other cancers), may be an option. A further potential therapy is the beta-emitter ^{177}lutetium or the alpha-emitter ^{225}actinium that, when linked to a ligand that itself can bind to PSMA, can deliver radiotherapy specifically to the malignant cell.

Key points

Prostate cancer has one well-established risk factor—being of African descent – although age is also important.

SELF-CHECK 4.8

List any five major medical and surgical aspects of prostate cancer.

4.6 Breast cancer

Although almost all literature refers to breast cancer in women, much of what follows is also relevant to the small number of men who die from the disease. NICE documents NG101 on early and locally advanced breast cancer, CG81 on the diagnosis and management of advanced breast cancer, and CG164 on familial breast cancer are relevant. In England and Wales, the number of deaths from breast cancer has varied at over 10,000 from 2013–20, but fell by 6.8% to 9,592 women from 2020 to 2021 (19.3% of all female cancer deaths). It is the second most frequent cause of female cancer deaths, after bronchus/lung, which killed 13,360 women (13.9% of all female cancer deaths). In 2020, breast cancer killed 684,996 women globally, that being 6.9% of all cancer deaths and 15.5% of all female cancer deaths.

4.6.1 Anatomy and aetiology

Almost all breast tissue consists of adipose tissue, glandular tissue, supporting fibrous connective tissue (Cooper's ligaments), and ducts leading from the milk-producing glandular tissue to the nipple. **Lactocytes** constitute an alveolus, or lobule, of which there are 8–15 in a group, arranged in a spoke-like formation with a duct at the centre, which merge and lead to a lactiferous sinus at the base of the nipples. Montgomery's glands are modified sebaceous glands that support the nipple in breastfeeding and provide cues to the infant. Sensory nerves lead from the nipple and surrounding areola, whilst lymphatics lead to **axillary** and para-sternal lymph nodes.

Like prostate disease, conditions pertaining to the breast include cancer, inflammation (mastitis), and benign hyperplasia, but unlike the prostate, many people of both sexes complain of breast pain (mastodynia), the aetiology of which is unclear. Breast cancer itself can be viewed as being *in situ* or being malignant. *In situ* disease is benign (failing to cause death), although it may transform to a malignancy; most breast cancer (around 87%) is malignant. Arising from epithelial glandular tissues, the underlying malignancy is an adenocarcinoma, which is slightly different to the benign form arising in epithelial tissue of the ducts carrying milk from the lactiferous alveoli to the nipple—hence ductal carcinoma *in situ*, or DCIS. However, it is recognized as a pre-cancerous state, which in 25% of cases will transform to malignant breast cancer.

The leading aetiology is female sex hormones, and therefore also the presence of ovaries, as oophorectomy (removal of the ovaries) for ovarian cancer also markedly reduces the risk of breast cancer. Conversely, post-menopausal hormone-replacement therapy increases the risk, accounting for 3.2% of cases. Amendable risk factors include lack of cardiovascular exercise (responsible for 3.4% of cases), alcohol (6.4%), and overweight and obesity (8.7%, where a BMI of <21 kg/m^2 provides most protection). Not having breastfed is linked to 3.1% of cancers, as is lack of childbearing. Together these lifestyle and environmental factors are linked to 26.8% of breast cancer in women. The role of molecular genetics in breast cancer is described in section 4.6.3.

4.6.2 Presentation

Breast cancer can become manifest in a number of signs and symptoms—lumps, bumps, changes to the skin and subcutaneous tissue, liquids (blood, milk, tissue fluid) expressed from the nipple—all features at or near the surface. Unfortunately, disease may also arise deep in the breast, so that it may be undetectable until at an advanced stage, such as when spreading to axillary (armpit) lymph nodes, causing lymphadenopathy. Lesions may be of discrete masses, and so relatively easy to detect by examination, but also as two-dimensional sheets or strands of neoplastic tissue, which are far harder to detect. Nevertheless, in almost all cases, discovery of an abnormality leads rapidly to presentation to a GP. Those with certain signs and symptoms should be referred urgently (within 2 weeks) to secondary care: age ≥30 with an unexplained lump, and age ≥50 with discharge, retraction, or other change in one nipple only. Those with skin changes that suggest breast cancer or aged ≥30 with an unexplained lump in the axilla may be urgently referred, whilst those aged <30 with an unexplained breast lump may be non-urgently referred.

An important differential diagnosis is Paget's disease of the breast, which is characterized by a localized patch of eczema around the nipple/areola region. In over 90% of cases, it is indicative of an underlying ductal carcinoma *in situ* and is linked with a poor prognosis.

4.6.3 Diagnosis

Initial diagnoses are confirmed or refuted in secondary care with the help of the laboratory and imaging.

Blood tests

Serum tumour marker CA 15-3 (derived from the glycoprotein mucin-1 (MUC1), a product of *MUC1*) is often increased in breast cancer, but also in benign breast disease—so it may help in differential diagnosis.

Molecular genetics

Some 10–15% of cases of breast cancer have a known genetic mutation, and of these, 30% each are due to *BRCA1* or *BRCA2*, both very likely to have been inherited, the remaining 40% including *TP53* and *PTEN* at 10q23.31 (coding a tumour suppressor phosphatase). Each of these four represents a relative risk >10 fold, so that family history is crucial, not merely for diagnosis and management, but also for the benefit of sons and daughters. *BRCA1* is also a risk factor for ovarian cancer. Accordingly, variants of these genes will be sought by a regional genetics hub, and whilst there is a place for testing for individual genes, the value of multi-gene panels is established and may guide management, as described in section 4.6.5.

However, there are indications that other genetic markers may be useful, which may enter routine practice. Presence of a variant in miR-146a brings a fourfold odds ratio for breast cancer, whilst miR-27a is linked to a decreased (odds ratio 0.24) risk of this disease. Furthermore, miR-146a is strongly linked to nodal status and stage, whilst miR-196a2 is linked to tumour size, nodal status, and stage. In a further study, levels of miR-34a in combination with CA 15-3 improved the diagnostic value of each marker alone.

Imaging

First steps are X-ray and ultrasound, the latter to give enhanced views of potential lesions. A specialist form of X-ray, a mammogram, takes an image only of the breast, not of the entire chest. Later, there may be CT/MRI, the latter helping in determining disease in lymph nodes, primarily in the local axilla, and all imaging will help determine if the lump is in fact composed of fibrous tissue, fat, or is a cyst, or is likely to be a tumour. Similarly, staging of lymph node invasion, which with high-quality imaging can estimate the size of the tumour within a lymph node as being <0.2 mm, 0.2–2.0 mm, or >2.0 mm. This is a complex matter as several lymph nodes may carry micrometastases of different sizes undetectable by imaging, but the involvement of one to three lymph nodes confers N1 status, that of ≥4 N2 status. A clinical staging can then be formed (Table 4.11).

Histology

Should the MDT think it necessary, the next step is to obtain a sample of tissue (a biopsy) of the presumed malignancy. This is often with a 'fine-needle aspiration' (FNA, perhaps guided by ultrasound), as opposed to a tissue biopsy, which would require opening the skin, which is to be avoided, but may be needed. The products of the FNA go to the laboratory where the diagnosis is made, such as invasive breast cancer or DCIS. Part of this analysis will be for the expression of oestrogen receptors, progesterone receptors, and HER2, and for certain gene mutations, providing clues as to the cellular aetiology. Most tumours will express increased levels of these receptors, but perhaps 10–20% of tumours fail to express any, hence 'triple negative' tumours, pointing to other mechanisms. This is important because if the increased expression of these receptors is not present, then another aetiology must be present, and molecular genetics for *BRCA1* and *BRCA2* is required, as recommended by NICE NG101.

If there is lymphadenopathy, lymph nodes will be investigated by FNA, as there may be micrometastases. Lymph nodes (the primary being the sentinel lymph node) can be identified by following the course of blue dye as it passes along the lymphatics, or by an isotope. However, this procedure is generally reserved for invasive cancers, unless considered high risk (such as with a palpable mass and/or extensive microcalcifications), or if the patient will be having a mastectomy. MRI is not generally needed but may be used if there is a discrepancy in information from clinical examination and the mammogram, and to assess tumour size if breast-conserving surgery is indicated. Figure 4.15 shows normal and malignant breast tissues stained with dyes such as haematoxylin and eosin. Relevant special techniques include immunocytochemistry and fluorescence *in situ* hybridization (Figure 4.16).

Information to be collected should be age at diagnosis of any cancer in relatives (and their current age), current age of unaffected relatives, site of tumour(s), the presence of other cancers, and whether

TABLE 4.11 Breast cancer staging of the tumour

Finding	Stage
Tumour 1–5 mm in diameter	T1a
Tumour 5–10 mm in diameter	T1b
Tumour 10–20 mm in diameter	T1c
Tumour 20–50 mm in diameter	T2
Tumour >50 mm in diameter	T3
Tumour extends into the chest wall and/or skin	T4a
Marked involvement of the skin and/or a second nodule	T4b
Both of the above	T4c
Diffuse erythema and oedema involving a third of the breast (i.e. inflammatory cancer)	T4d

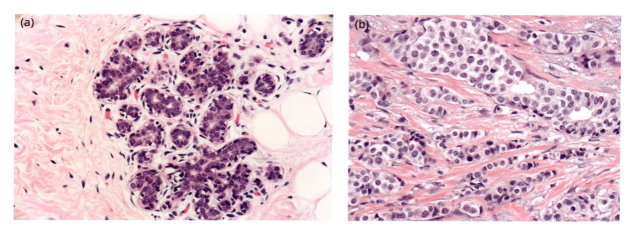

FIGURE 4.15

Normal and malignant breast tissue. (a) Normal breast tissue with typical lobular structure.
(b) Malignant breast tissue with loss of lobular structure.

(a) Shutterstock, (b) Science Photo Library.

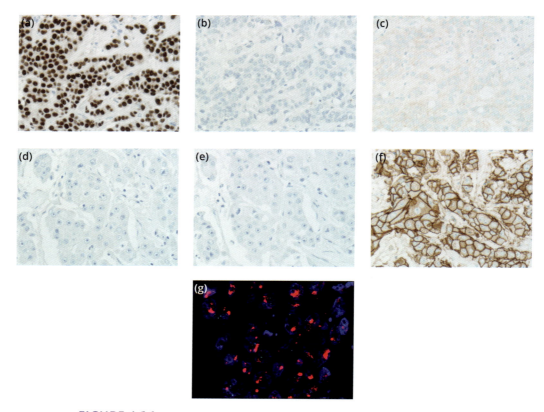

FIGURE 4.16

Immunocytochemistry and fluorescence *in situ* hybridization of breast cancer. Immunocytochemistry
for oestrogen receptor (a and d), progesterone receptor (b and e), and HER2 (c and f). In the
upper panel, the tissue is strongly positive for the oestrogen receptor, in the lower panel, there
is strong HER2 staining. (g) Marked amplification of HER2 (red signal), the nuclei stained blue by
4'6-diamidiino-2-phenylindole (DAPI). The weak green dots are for chromosome 17.

From Orchard & Nation, *Histopathology*, second edition, Oxford University Press, 2018.

BOX 4.3 Screening for breast cancer (mammography)

Current NHS practice is to invite all women aged 50–71 for mammographic screening every 3 years, but women outside this range may be screened if they are at high risk of disease (such as due to family history or target gene mutation). Like in all screening, issues include false positive and false negative results, and the potential for unnecessary treatment. The UK's leading cancer charity, Cancer Research UK, estimates that screening detects some 18,000 cancers (8 from 1,000 screened), and saves 1,300 lives a year. This is countered by an estimated 4,000 women who are over-diagnosed. However, an authoritative Cochrane meta-analysis in 2013 did not find an effect of breast screening on total cancer mortality, including breast cancer, after 10 years of follow-up. Nevertheless, as of 2018, the WHO recommends breast cancer screening by mammography. Mammography based on X-ray has been compared with magnetic resonance imaging. A systematic review by De Feo et al. (2022) reported sensitivities of 83.6% and 90.5% respectively, with specificities of 77.4% and 51.6%.

disease is bilateral (i.e. in both breasts). Once this information has been collected, and *BRCA1/BRCA2* status confirmed, a computer program (BOADICEA) can deliver an estimate of the risks of breast and ovarian cancer. If available, the program can also benefit from knowledge of mutations in *PALB*, *CHEK2*, and *ATM*. It is also the case that appropriate family members will benefit from knowledge of their risk and potential carrier status, and be advised on strategies to reduce their risk of disease.

SELF-CHECK 4.9

The primary tumour of woman A is 7 mm in diameter, that of woman B is 35 mm. Other lesions are also defined by imaging. What are their TNM stages?

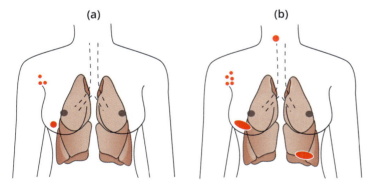

(a)   (b)

4.6.4 Management of high risk

Dependent on family history and genetics (*BRCA 1/2*), and so risk of disease, those free of cancer may be cared for in a primary or secondary setting. This is likely to include annual MRI or mammographic surveillance (but not ultrasound), dependent on their age, family history, and genetic status, and so moderate or high risk of cancer. The former is defined as a 17–30% risk of developing breast cancer, the latter as a >30% risk. Women with the highest risk of cancer (and some at moderate risk) may be offered chemoprevention with tamoxifen (e.g. 20 mg od), anastrozole (an aromatise inhibitor: 1 mg od), and/or (for osteoporosis) raloxifene (60 mg od), dependent on their menopause status and other factors, and also bilateral risk-reducing mastectomy (followed by reconstruction) and/or oophorectomy. Naturally, such procedures will involve the patient, her family, and a specialist MDT.

4.6.5 Management of early and locally advanced cancer

Once all the investigations have been completed, the MDT meets and will provide the patient with a synopsis and options. Although surgery is to the fore, some may first benefit from prior **neoadjuvant** therapy (NICE NC101). However, management decisions need to address molecular genetics in a number of panels. One such multi-target panel can test for *NTRK1*, *NTRK2*, and *NTRK3*, the results of which contribute to the decision to prescribe an NTKR inhibitor such as entrectinib, as according to NICE TA644. However, three other panels offer a more comprehensive view of key mutations in triple-negative disease: the Endopredict panel focuses on eight cancer-related genes and three reference genes, the Oncotype DX on 15 genes and five reference genes, whilst the Prosigna PAM50 panel includes 50 genes. NICE DG34 recommends the use of these panels as options for guiding adjuvant chemotherapy.

Neoadjuvant therapy

Also known as primary systemic therapies, these treatments hope to reduce tumour size, which will help breast-conserving therapy. Endocrine therapy and chemotherapy are offered to patients with ER-negative or ER-positive invasive breast cancer, whilst in those with triple-negative disease, treatment with platinum-based and/or anthracycline cytotoxics is likely. Patients with HER2 positive disease may benefit from a course of pertuzumab (a monoclonal antibody to HER2) with an initial loading dose of 840 mg followed by 420 mg every 3 weeks for three to six cycles (NICE TA424). Other neoadjuvant options include tamoxifen and **aromatase** inhibitors.

Surgery

The extent of surgery depends on the size of the tumour(s) on a spectrum from a simple small lumpectomy to a full mastectomy with radical clearance of all axillary lymph nodes, followed by immediate or delayed reconstruction. Midway between the two is a wide local excision, which removes the tumour and a small amount of normal breast tissues around the outside (the margin), the latter to ensure that all the tumour has been harvested. The most extensive, however, in those considering bilateral mastectomy, and with certain oncogenes, is oophorectomy to negate their risk of ovarian cancer.

Those with early or locally advanced disease may need adjuvant therapy after their surgery, and tools such as the PREDICT scoring system can help guide such treatment, although there are other calculators (Endopredict, Mammaprint, etc.), some of which need more extensive information about the tumour(s). This adjuvant treatment includes a combination of radiotherapy, chemotherapy, hormone therapy, and biological therapy, the choice of which depends on the major features of the case, such as history, grade, stage, genetics (ER, PR, HER2), and prognosis.

Axillary lymph node excision or damage brings a risk of lymphoedema, and if present, patients will need rapid referral to a specialist service which is likely to include physiotherapy, and recommend exercise and compression garments.

Chemotherapy

NICE recommends treatment with surgery and appropriate systemic chemotherapy, rather than endocrine therapy alone, unless significant co-morbidity precludes surgery. Endocrine therapy should also be considered after breast-conserving therapy for DCIS when radiotherapy is not recommended or received. The principal agents are tamoxifen and aromatase inhibitors which bring menopausal symptoms and many other side effects (increased risk of thrombosis, loss of bone density). This therapy is generally used for up to 5 years, but may be extended for women at high risk with a different agent.

Other regimes include both a taxane (docetaxel or paclitaxel) and an anthracycline. The benefit of the dual therapy is that a lower dose of the anthracycline can be used, so reducing potential side effects such as cardiac toxicity, nausea, and vomiting. Trastuzumab (herceptin, a monoclonal antibody to HER2, typically 4–8 mg/kg, dependent on stage) is an option in HER2-positive cancer, but should only be used with caution in those with certain heart conditions.

Radiotherapy

Usual clinical practice is to give external beam radiotherapy after successful breast-conserving surgery for early invasive breast cancer in 15 fractions over 3 weeks, but this may be extended in those at high risk. Radiotherapy is not needed in those with clear margins and at low risk (e.g. T1, ER-positive, HER2-negative) if they are prepared to take endocrine therapy for 5 years. This can be risky, as without radiotherapy, local recurrence occurs in 5% of women at 5 years, but at only 1% with radiotherapy, whilst overall survival at 10 years is the same.

Follow-up

All those with breast cancer, including DCIS, should have annual mammographs, which will be part of their care plan. This will also include details of further treatment, signs of symptoms they should be aware of, advice on a healthy lifestyle (weight, alcohol, exercise, smoking), and other specialist services. Although in principle, patients will be discharged from hospital back to the care of their GP, they would rapidly return to secondary care should it be needed.

4.6.6 Management of advanced disease

This section is based on NICE CG81. It refers to people with advanced (stage 4) disease. Major sites for these metastases are the bones, liver, and lungs.

Diagnosis and assessment

These rely on the same steps as in non-advanced disease, that is, a combination of radiography, ultrasound, CT, MRI, and PET/CT (where appropriate) to assess the location(s) and extent of disease. Care is taken to assess bone, and so determine any risk of fracture. Reassessing ER/PR/HER2 status is important and could lead to a change in management, as a new clone of abnormal cells may have developed.

Management

First-line treatment should be endocrine (aromatase inhibitor, tamoxifen, ovarian suppression) in ER-positive disease, but if the disease is imminently life-threatening or requires symptomatic relief, then chemotherapy is an option, as long as toxicity can be tolerated. However, if patients are already on chemotherapy, this should be completed before the patient is moved to endocrine therapy.

Chemotherapy options include doxorubicin (30–75 mg/m^2), docetaxel (75 mg/m^2 intravenously in combination with doxorubicin and cyclophosphamide, or 100 mg/m^2 in monotherapy), vinorelbine (25–30 mg/m^2 intravenously), or capecitabine (often in combination with docetaxel). Illustrating the non-specific action of this form of chemotherapy, these agents may also be used in other cancers, such as colorectal and gastric: doxorubicin may be used in 17 different cancers. CDK 4/6 inhibitors may have a place in hormone-receptor positive, HER2 negative advanced or metastatic disease, and include palbociclib (in cycles of 125 mg od for 21 days, then 7 days off, NICE TA495 and TA619), ribociclib (in cycles of 600 mg od for 21 days, then 7 days off, NICE TA593) and abemaciclib (150 mg bd, NICE TA579). These drugs are used alongside an aromatase inhibitor (such as anastrozole 1g od) or fulvestrant (an antagonist of the oestrogen receptor, 250 mg intramuscularly).

CASE STUDY 4.3 Breast cancer

Much to her concern, a 54-year-old woman is recalled for a second mammogram in further investigation of an abnormality. This confirms a suspicious lump in the left breast, and within 48 hours she is examined by her GP, who can feel the lump, and also some masses in the armpit. He immediately refers her to a breast cancer oncologist, and the following week both women meet for the first of many consultations. At this initial meeting, the patient is again examined, and a full medical history is taken, which is unremarkable. The patient considers that she has probably completed her menopause, has two children, both of whom were breastfed for at least 6 months, and that no relative has had breast, ovarian, or uterine cancer. She is a lifetime non-smoker, occasional drinker, and has a BMI of 28.5 kg/m². The oncologist fully explains to her patient the options, and both agree to routine blood and genetic tests (which include CA 15-3, and also *BRCA1*, *BRCA2*, and *TP53*) and an MRI to examine the lumps in the armpit. The latter will be scheduled within 2 weeks, after which they will meet again.

Meanwhile, the oncologist forms an MDT, which considers the options, referring to NICE NG101. The MRI finds a single breast mass of diameter 25 mm, with at least two abnormal lymph nodes in the same axillary cluster, and no marked bone lesions, pointing to cT2N2aM0. This is certainly sufficient to warrant fine-needle aspiration of the breast tumour and lymph nodes. This is put to the patient, and she agrees. The laboratory reports invasive malignant cells that express high levels of the oestrogen receptor (Figure 4.17) and Ki-67, but not the progesterone receptor or HER2 in the breast tissues, with abnormal cells in the lymph node tissues.

The CA 15-3 levels are 83 U/ml (reference range <30), whilst routine blood results are unremarkable, and the regional molecular genetics lab finds her to be *BRCA1*, *BRCA2*, and *TP53* negative. The MDT reconvenes, and the oncologist reports back to her patient that she has moderately advanced breast cancer that is amenable to surgery.

The patient is given options and meets with a clinical nurse specialist and Macmillan nurse, and women in a local breast cancer support group. Her choices are part-governed by experiences of a close friend and a distant friend, and she decides to by-pass neoadjuvant treatment with endocrine, chemotherapy, and radiotherapy and opt for breast-conserving wide local excision surgery without axillary node clearance, but with minimal surgery to remove only those lymph nodes shown to harbour micrometastases. Surgery proceeds without problems, and she has a course of whole-breast radiotherapy, which should reduce her risk of local recurrence from 5% to 1%, and to the axilla, to suppress possible disease in

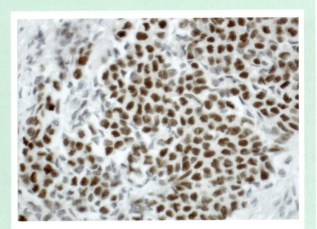

FIGURE 4.17

Breast cancer. Positive immunocytochemistry for the oestrogen receptor.

From Orchard & Nation, *Histopathology*, second edition, Oxford University Press, 2018.

the remaining lymph nodes. As hormonal therapy improves her 5-year survival from 75% to 81%, with the likelihood that she is, or shortly will be, post-menopausal, she starts a 5-year prescription of the aromatase inhibitor anastrazole 1 mg daily.

Her recovery is uneventful, and she tolerates the anastrazole well. The laboratory reports that the margins around the tumour are clear. Her care plan calls for regular follow-up with the hospital's breast team and annual mammograms, and recommends regular exercise to achieve and maintain a BMI <25 kg/m² and limiting alcohol intake to five units per week. She is in good general health, but after her fourth 'anniversary', she notes becoming increasingly tired, with non-specific aches and pains. The GP finds a small lump just above the left clavicle (collar bone), orders an urgent X-ray and routine bloods, and refers her on to the same oncologist, who confirms the findings and orders a round of ultrasound, CT, and MRI, focusing on the chest, axial skeleton, and proximal limb bones.

The imaging reports abnormalities in the remaining axillary lymph nodes and in some of the parasternal nodes. The collarbone lump is identified as a lymph node, and a fine-needle aspirate finds ER-positive cells typical of breast cancer. No major lesions are found in any bone, but her serum calcium of 2.6 mmol/l is right at the top of the reference range (2.2–2.6 mmol/l), suggestive of borderline lytic bone lesions. Similarly, her haemoglobin result of 118 g/l is at the bottom of the reference range (118–48). Although incomplete, the sum total of these results calls for an upgrading in treatment, with radiotherapy to the lymph nodes and chemotherapy. The MDT convenes and finds a consensus of advanced breast cancer best treated with a combination of docetaxel plus capecitabine, to which the patient reluctantly agrees. Although bone metastases have not been unequivocally demonstrated, she is also started on a bisphosphonate—zoledronic acid.

Despite the expected side effects of the cytotoxic drugs (alopecia, loss of appetite, and nausea), the patient completes the four cycles, lasting 12 weeks, and returns to endocrine therapy, but with tamoxifen and leuprorelin, an antagonist of gonadotropin-releasing hormone that acts to induce ovarian suppression. As the cytotoxic drugs wash out of the patient, her hair regrows, and imaging shows a reduction in the size of the lymph nodes. However, after 18 months, bone pain becomes more intense and is barely addressed with simple oral analgesics. Serum calcium has increased to 2.85 mmol/l. As CT and MRI findings are suspicious of, but do not confirm, bone metastases, bone scintillography with technetium-99 is performed, which shows multiple small metastases in certain ribs, vertebrae, and some long bones. As a consequence, a more aggressive chemotherapy regime of gemcitabine and paclitaxel is started, and those bones with the largest metastases are subjected to radiotherapy. Although the latter reduces some of the bone pain, in contrast to previous regimes, the patient tolerates these drugs badly, with a marked reduction in all indices in the full blood count (especially neutrophils) and increases in liver function tests and a fall in the estimated glomerular filtration rate, implying damage to the kidneys and potential renal failure. Doses of cytotoxic drugs are reduced to levels that are tolerable, and blood tests and symptoms improve. However, with few options remaining, her overall picture deteriorates, and she is eventually transferred to a hospice where she succumbs to the disease.

Supportive care

Naturally, the severity of advanced disease will call for more intense and broad support, addressing the patient's physical, psychological, rehabilitation, social, and financial needs. As the disease progresses, other pathology may be involved (such as anaemia, causing cancer-related fatigues), not merely driven by treatment. Ultimately, end-of-life care will be appropriate, with referrals to a hospice.

4.6.7 Bone

As bone is a common site for metastases (as it is in prostate cancer), bisphosphonates such as zoledronic acid (5 mg intravenously) or sodium clodronate (1,600 mg od) are often prescribed. The former may also be given to reduce the risk of osteoporosis caused by chemotherapy. Indeed, bone health should be checked by dual-energy X-ray absorptiometry (DEXA) in patients who are not on bisphosphonates but are about to start endocrine therapy, have treatment-induced menopause, or who are starting ovarian ablation/suppression therapy. If bone metastases are present, then external beam radiotherapy may help any pain and treat the tumour. The laboratory will expect requests for calcium, phosphates, and other bone markers.

SELF-CHECK 4.10

What types of drugs may be taken by the patient with breast cancer?

Key points

Although most breast cancer cells express high levels of oestrogen, progesterone, and epithelial growth factor receptors (probably explaining their development), around 15% fail to express these receptors, and so are 'triple negative', implying some other pathophysiology. There is considerable choice in chemotherapy.

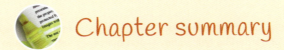

Chapter summary

- Cancer is the leading cause of death in the UK, the Western world, and globally. In 2021, in England and Wales, the number of male deaths exceeded that of females by some 14%, compared with 3.2% for all deaths.

- Many cancers present with general and non-specific features, such as weight loss, tiredness, and venous thrombosis, so that differential diagnoses must be considered.

- Common treatments include surgery, chemotherapy, and radiotherapy.

- Lung cancer is by far the leading cause of cancer deaths, and presents with features such as cough and problems with breathing.

- Colorectal cancer, the second most common cause of cancer death overall, may be detected with the faecal occult blood test and colonoscopy.

- Prostate cancer is the second most common cause of cancer deaths in men, and may first be detected with a PSA blood test.

- Breast cancer is the second most common cause of cancer deaths in women and is screened for in the general population.

Further reading

- Parkin DM, Boyd L, Walker LC. The fraction of cancer attributable to lifestyle and environmental factors in the UK in 2010. Br J Cancer 2011: 105; S77–S81.

- Yeo J, Crawford EL, Zhang X, et al. A lung cancer risk classifier comprising genome maintenance genes measured in normal bronchial epithelial cells. BMC Cancer 2017: 17; 301.

- Arvelo F, Sojo F, Cotte C. Biology of colorectal cancer. eCancer Med Sci 2015: 9; 520.

- Hasson SP, Menes T, Sonnenblick A. Comparison of patient susceptibility genes across breast cancer. Pharmacogenomics Person Med 2020: 13; 227–38.

- de Groot PM, Wu CC, Carter BW, Munden RF. The epidemiology of lung cancer. Transl Lung Cancer Res. 2018: 7; 220–33.

- Lange M, Begolli R, Giakountis A. Non-coding variants in cancer: Mechanistic insights and clinical potential for personalized medicine. Noncoding RNA. 2021: 7; 47.

- Lu RM, Hwang YC, Liu IJ, et al. Development of therapeutic antibodies for the treatment of diseases. J Biomed Sci 2020: 27; doi: 10.1186/s12929-021-00784-w.

Useful websites

- Statistics on lung cancer by the British Lung Foundation: **https://statistics.blf.org.uk/lung-cancer**

- Cancer Research UK: **https://www.cancerresearchuk.org**

- Predict breast cancer scoring system: **https://breast.predict.nhs.uk/tool**

- Predict prostate cancer scoring system: **https://prostate.predict.nhs.uk/tool**

- WHO press release on cigarette smoking: **https://www.who.int/news/item/29-05-2019-who-highlights-huge-scale-of-tobacco-related-lung-disease-deaths**

- National genomic test directory for England: **https://www.england.nhs.uk/publication/national-genomic-test-directories/**

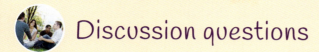 Discussion questions

4.1   What factors are to be avoided in order to minimize one's risk of cancer?

4.2   What are the similarities and differences in the epidemiology, presentation, diagnosis, and management of breast cancer and prostate cancer?

5

Clinical oncology 2: Further types of cancer

The previous chapter looked at four of the major cancers—those of the lung/bronchus, colon/rectum, prostate, and breast that, with blood cancer (Chapter 11), were linked to over half of cancer deaths in 2021 in England and Wales. This chapter will examine the next 12 most frequent cancers that together cause a third of deaths, and look briefly at those at a very low population frequency. As before, we will be introduced to each particular cancer, and then consider its aetiology, presentation, diagnosis, and management, in order of frequency. Once more, where relevant, we will refer to guidelines and other documents from the National Institute for Health and Care Excellence (NICE), the WHO, and elsewhere.

Learning objectives

After studying this chapter, you should confidently be able to:

■ describe the pathology of pancreatic, oesophageal, bladder, and liver cancer

■ explain the aetiology of stomach, brain, ovarian, and renal cancer

■ outline the diagnosis of oral and throat cancer, mesothelioma, melanoma, and cancer of the uterus

5.1 Pancreatic cancer

When considering pancreatic cancer, practitioners in the UK refer to NICE NG85 on 'Pancreatic cancer in adults: diagnosis and management'. The frequency of the disease in terms of all malignant neoplasms has increased by 14.8% year-on-year from 5.32% in 2013 to 6.11% in 2021, and it is found in proportionately more women than men.

5.1.1 Anatomy and aetiology

The pancreas is a loose collection of cells and tissues sandwiched between the diaphragm, stomach, and spleen with **endocrine** and exocrine function, the latter referring to digestive enzymes that pass via the pancreatic duct to the small intestine. Exocrine tissues in many organs can produce a malignant disease that carries a particularly poor prognosis, and as it results from an epithelial gland, it is an adenocarcinoma. The only other notable disease of this organ is **pancreatitis**, the aetiology of which is rarely microbiological, occasionally autoimmune, and more likely related to obesity and excessive alcohol consumption.

The leading lifestyle and environmental causes of this disease are tobacco (linked to 28% of cases) and overweight/obesity (12% of cases); the role of dietary animal fat and meat consumption may also be important. It may also arise from chronic pancreatitis and both types I and II diabetes mellitus. Several inherited genetic disorders increase the risk of pancreatic cancer, including Lynch syndrome (as described in section 4.4.1, a form of colorectal and other cancers), Li–Fraumeni syndrome (linked to several other cancers), Peutz–Jeghers syndrome (with intestinal polyps), and familial adenomatous polyposis (as discussed in Chapter 4, a further risk factor for colorectal cancer). These point to a familial/genetic aspect, and although there are few clues from molecular genetics regarding aetiology, *BRCA-2* is an established risk factor. Accordingly, family members must be made aware, and potential risks addressed, whilst in those at high risk (presence of gene mutations, family history of pancreatic disease, etc.) imaging may be considered, and they should remain under medical surveillance.

In common with many other cancers (Chapter 3), changes in proto-oncogenes *KRAS*, *p16*, (a tumour suppressor with a role in cell cycle regulation), and *TP53* (with roles in cell cycle regulation and apoptosis) are often found. NICE refers to *PRSS1* (at 7q3.4, coding for trypsin), *BCRA1* (at 17q21.31), and *PALB2* (at 16p12.2, coding for a protein that repairs breaks in double stranded DNA).

5.1.2 Presentation

Symptoms are all non-specific, and as yet there is no acceptable screening test. Common complaints are of nausea, vomiting, abdominal pain, and loss of appetite and so of weight. Should the tumour impact onto the bile duct and cause a blockage, there may well be jaundice and light-coloured faeces resulting from **obstructive cholestasis**. However, the latter may also be due to a gallstone in the common bile duct, and so unrelated to the cancer.

5.1.3 Diagnosis

Imaging

Ultrasound may be helpful, although with low sensitivity it may miss small masses, and may be combined in endoscopic ultrasound. However, the key method is CT (Figure 5.1), although MRI and fluorodeoxyglucose-PET scanning may also be used.

The laboratory

Diagnosis may be helped by cytological (perhaps by transdermal fine-needle aspiration), or histological sampling of suspect tissues (a biopsy), and if so, this will be via the intestinal route. These

FIGURE 5.1

CT scan of pancreatic cancer. The four red lines indicate a pancreatic tumour.

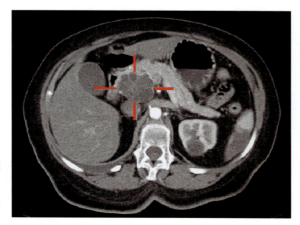

procedures may be guided by ultrasound. Any of these investigations may lead to an alternative diagnosis, such as a cyst, which can often be present.

Semi-specific blood tests **amylase** and **lipase** help differentiate pancreatitis, and liver function tests will also be useful. However, of most help is cancer marker CA19-9, also raised in other digestive system cancers (liver, oesophagus, colon, rectum), but also in pancreatitis. Measurement of CEA may also be helpful, although it too is cancer non-specific. Several potential markers in development include cell-free DNA and RNAs, such as oncogenes miR-21 and miR-155, and tumour suppressors miR-35 and miR-15a, as levels of the latter predict outcome and may therefore direct treatment.

5.1.4 Management

Neoadjuvant treatment (that which comes before a main treatment) is inappropriate, and if practicable, surgery should be considered as soon as possible, although this is feasible in only 15–20% of patients. MRI and CT are likely to help define staging and any lymph node involvement. Staging follows the pattern of several other cancers: stage 1 is a tumour ≤2cm, stage 2 a tumour >2cm. Tumours that have invaded nearby tissues are in stage 3, whilst stage 4 is reserved for local lymph node involvement or distant organ metastases (such as in the liver). Surgical options include the resection of selected lymph nodes, biliary drainage, placement of stents, and a bile duct bypass.

Adjuvant therapy should be started as soon as the patient is well enough to tolerate the chemotherapy, such as gemcitabine and capecitabine. Should the cancer be metastatic, a chemotherapy regime combines four drugs with different modes of action: irinotecan (a topoisomerase inhibitor), oxaliplatin (that cross-links DNA strands), fluorouracil (that inhibits thymidylate synthase), and folinic acid (to minimize adverse effect of fluorouracil). In one clinical trial, this combination extended mean survival from 6.8 to 11.1 months. Should this regime be unsuccessful, second-line treatments include oxaliplatin (if not a first-line agent (also used to treat colorectal cancer)) or gemcitabine (used to treat many cancers).

A problem with the disease and its management is that of the role of pancreatic secretions in digestion. One approach to help digestion is with oral enteric-coated pancreatin (a cocktail of digestive enzymes), another is with enteral feeding. The general population has a reasonably well-formed view of pancreatic cancer and is aware that it carries an especially poor prognosis. Accordingly, the patient and their family will need support to manage the psychological aspects of the disease, such as anxiety and depression. Figure 5.2 contrasts normal pancreatic histology with that of a pancreatic cancer.

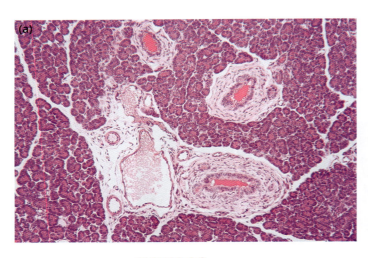
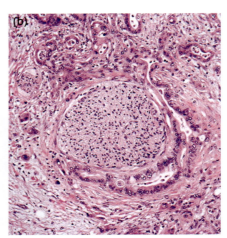

FIGURE 5.2

Histology of a normal and malignant pancreas. (a) Normal histology: three thick-walled blood vessels on the right contrast with a thin-walled exocrine duct on the left. (b) Perineural invasive cancer of the pancreas (same low-power magnification as above).

(a) iStock, (b) contributed by Wei Chen MD, PhD, from PathologyOutlines.com, Inc.

CASE STUDY 5.1 *Pancreatic cancer*

A 75-year-old man with type 2 diabetes mellitus and a body mass index of 29.5 kg/m² has become increasingly weak over a 3-month period, with weight loss (down from 32 kg/m²) despite a healthy appetite. More recently he complains to his GP of back pain, and the GP notices a degree of jaundice in the sclera of the patient's eyes. A full blood count, U&Es, HbA1c, and liver functions tests are ordered, which come back within 48 hours with a slight neutrophil leucocytosis with raised serum bilirubin and alkaline phosphatase. The GP organizes an urgent abdominal ultrasound and endoscopy, the former finding an enlarged gall bladder, no changes to the liver, but atypical changes to the pancreas. The endoscopy finds no pathological changes to the stomach or duodenum but a 'dry' **ampulla of Vater**. A further blood test reports mildly raised amylase (320 IU/l, reference range <300 IU/l), raised CA19-9 (67 U/ml, reference range <36 U/ml), raised CEA (15 ng/ml, reference range <2.5 ng/ml), but normal lipase, and the GP contacts the local oncologist with a request for an urgent MRI.

The patient and his spouse are stunned to learn that the MRI has identified changes that are consistent with pancreatic cancer, which the blood tests support. There are also suspicious changes to the liver, but these are equivocal. The GP then offers the couple the recommendations of the MDT, which are for surgery followed by chemotherapy. The operation proceeds with a little difficulty, and the patient spends several days in intensive care afterwards, where most of the pancreas (from the 'head', the section nearest the intestines) and several swollen lymph nodes are removed. The histology laboratory reports one 5 cm tumour, and two 2 cm tumours, the lymph nodes being infiltrated with the same cells as in the major pancreatic tumour, giving TNM stage 3. Molecular genetics in the excised tumour finds mutations in *KRAS* and in *TP53*. The patient then spends a further week on a post-surgical ward, where high serum urea and creatinine give fears of renal damage, but these eventually fall, as do the liver function tests, and the sclerotic jaundice resolves, as does serum bilirubin.

Two months after discharge (on oral anticoagulation), the patient begins the first of six rounds of chemotherapy with a regime consisting of folinic acid, fluorouracil, irinotecan, and oxaliplatin every 2 weeks. He tolerates this badly, with nausea, vomiting, increases in LFTs and U&Es, with falling haematology indices. These are partially reversed after three cycles when the patient is switched to two cycles of intravenous gemcitabine, at a dose of 1,000 mg/m² once weekly for 7 weeks with a 1-week rest period, then once a week for 3 weeks with a 1-week rest period.

However, after several months of improved health the patient's sclerotic jaundice returns, in parallel with bilirubin levels of 65 μmol/l (reference range <20). A repeat MRI finds several suspicious masses in the liver, and chemotherapy begins once more. However, this is also tolerated badly, and the doses are reduced. The patient is admitted to hospital, from where he is discharged to a hospice for terminal care, where he dies 18 months after diagnosis.

Key points

Long-term survival after a diagnosis of pancreatic cancer is particularly poor as it often fails to manifest itself until moderately advanced.

SELF-CHECK 5.1

Why does pancreatic cancer have such a poor prognosis?

5.2 Oesophageal cancer

The seventh most frequent malignant neoplasm, oesophageal cancer has consistently caused 4.7–4.9% of all cancer deaths in England and Wales since 2013, but with a marked twofold increase in both absolute number and proportion of deaths in men.

5.2.1 Anatomy and aetiology

The oesophagus is a simple tube connecting the throat and the stomach. The inner layer (mucosa) is several further layers of durable stratified squamous epithelia able to withstand physical abrasions, in common with the skin, mouth, and vagina, but also some mucus-secreting goblet cells. The lower end is bounded by a **sphincter muscle** to regulate food passing into the stomach.

There is no dedicated NICE document on oesophageal cancer alone, but NG83 deals with oesophageal-gastric cancer (defined as cancer of the oesophagus, the stomach, or the junction between the oesophagus and the stomach), and within this are separate notes on each individual disease. Major oesophageal diseases are cancer, inflammation (oesophagitis), Barrett's oesophagus (a form of inflammation, possibly due to reflux of stomach acid, with the transition of squamous to columnar epithelium), and the development of varices (distorted and swollen veins). Should the cancer arise from malignancy of the squamous epithelium (as 70% do), it is a carcinoma, whilst if it arises from glandular tissues, it would be an adenocarcinoma, and is allied to chromosomally unstable gastric cancer. In this respect, the distinction between the carcinogenesis of the adenocarcinoma and the aetiology of Barrett's oesophagus becomes blurred as both may arise from acid reflux—male sex and increased consumption of red meat are risk factors for both conditions. Obesity may compound this problem through elevated intra-abdominal pressure. The distinction between the cell biology of the squamous cell variant and the adenocarcinoma is important as it drives management.

There are no specific risk factors, but patients may have a history of smoking and excessive alcohol intake. The malignancy may develop from Barrett's oesophagus (possibly linked to the stepwise loss of tumour-suppressor gene function such as *CDKN2A* and *TP53*) and/or chronic gastric reflux, due to the effect of gastric acid. Some have suggested roles for the human papilloma virus, whilst large-scale studies in molecular genetics point to *PLCE1* (at 10q23.33, coding a phospholipase linked to growth, differentiation, and apoptosis) and *TMEM173* (at 5q31.2, a protein linked to the type 1 **interferon** response to microbial infection). The most commonly mutated genes in the squamous cell group include *TP53* (91%), *CDKN2a* (76%), and *SOX2/TP63* (48%).

5.2.2 Presentation

There are no physical signs, but symptoms include general pain, and pain on swallowing (dysphagia) and in trying to force food down the oesophagus (possibly past an obstructing tumour), which can be a cause of weight loss. Acid reflux causing heartburn may also be due to problems with the stomach and in particular the cardiac sphincter at the junction of the oesophagus and stomach. In advanced disease there may already be clinically apparent metastases in the supraclavicular lymph nodes and hepatomegaly.

5.2.3 Diagnosis

The number and severity of symptoms are likely to prompt an endoscopy (a camera on a flexible tube, allied to a colonoscopy) and/or a barium swallow. The oesophagus is clearly relatively easy to access by endoscopy, and this is given full consideration in early investigations to determine the extent and location of lesions and the general health of the organ. Should a suspicious lesion be found (Figure 5.3), it will be biopsied. A simpler device is the Cytosponge, a collection device on a string that is swallowed and then retrieved, ideally with representative luminal cells that can then be examined by standard cytological techniques and stained for features such as the lectin trefoil factor 3, and possibly molecular genetics for markers such as *TP53*.

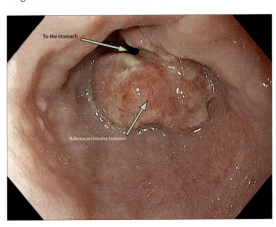

FIGURE 5.3

An oesophageal adenocarcinoma.

Lagergren J, Smyth E, Cunningham D, Lagergren P. Oesophageal cancer. Lancet 2017: 390; 2383–96. https://doi.org/10.1016/S0140-6736(17)31462-9.

The laboratory

Biopsies should be taken to determine the cellular basis of any lesion, and tissues may be stained by immunocytochemistry for periodic acid or cytokeratin 7/20 (for adenocarcinoma), for cytokeratin 5/6 and p63 (for squamous cell carcinoma), and to determine HER2 status (present in around 30%). Standard blood tests are unhelpful, with the possible exception of a full blood count for anaemia and liver function tests for liver metastases, although there may be a place for **non-coding RNAs**. Certain miRNAs, such as miR-106b and miR-1303, show differential expression in patients with or free of relapse after surgery, whilst high levels of miR-574-3b are linked to favourable outcome. Serum and tissue levels of miR-21 are elevated in oesophageal cancer, whilst miR-375 is downregulated. Time will tell if these prove to be useful in a full clinical setting.

Imaging

As the procedures described above are straightforward, in the first instance, imaging other than ultrasound is not immediately called for. However, should moderate or advanced disease be suspected, MRI/CT will be needed to determine disease outside the organ (lymphadenopathy, metastases), dependent on the biopsy and endoscopy findings. The former, if extensive, may help provide an early estimate of staging in common with other cancers, stage 1 being a small tumour localized within the inner layers and stage 4 being invasion through the oesophageal wall with metastases.

5.2.4 Management

Whole-body CT and PET–CT scanning may be offered to detect metastases before determining suitability for radical treatment. However, given the frequently aggressive nature of the cancer, and the poor overall survival rate, action will be urgent. In the UK, three-quarters of patients are diagnosed with either lymph-node or distant metastases, and many will not be amenable to curative therapy, hence the poor overall survival rate: that for patients with ≥T2 or N+ cancers (where most diagnoses are made) following surgery is around 50% at 10 years.

Neoadjuvant chemotherapy may be indicated in potentially curative cases, one regime being two cycles of cisplatin and fluorouracil, another being epirubicin, cisplatin, and capecitabine. The addition of radiotherapy is also effective. Surgical treatments may depend on the nature and extent of the malignancy. Where the cell biology is an adenocarcinoma, at stage T1bN0 surgery should be offered, whereas tumours other than T1N0 may be offered chemotherapy before, or before and after surgery, or may be offered chemoradiotherapy before surgery. Notably, the relatively low level of involvement of T1b still carries a 17–26% risk of lymph node metastases. Should the tumour be inoperable, in those with non-metastatic cancer, chemoradiotherapy and stenting should be considered. However, both types of cancers may present as **dysplastic** precursor lesions and can be treated locally without extensive surgery or chemotherapy. Unsurprisingly, minimally invasive (keyhole) oesophagectomy is preferred to a transthoracic approach.

Chemotherapy regimes for locally advanced or metastatic disease are based around cisplatin, 5-fluorouracil, oxaliplatin, epirubicin, and capecitabine. Should the tumour be HER2 positive, then trastuzumab (a mAb to HER2) may be added. As many tumours are unresectable, and over half of those having undergone curative treatment suffer recurrence, **palliative** treatment is common, most often with chemotherapy such as those based on platinum, irinitecan, and 5-flurouracil (as with pancreatic cancer). Such patients will also need help from dieticians, and non-medical help, giving support for emotional and social issues, most likely from a specialist care team and a peer support group.

Key points

The histology (and therefore oncology) of oesophageal cancer has much in common with that of gastric cancer, especially where the two organs meet.

What are the two different cell types in oesophageal cancer, and how can they be distinguished?

5.3 Liver cancer

This section includes disease of the **parenchyma** of the liver, and within the intrahepatic bile ducts: disease in parts of the biliary system outside the liver is considered alongside cancer of the gall bladder in 5.13.1. The eighth most frequent cause of cancer death, the relative proportion of liver cancer in terms of all malignant neoplasia has increased year-on-year from 2.94% in 2013 to 3.72% in 2021, a difference of 26.5%. This increase is widely believed to be due to a rise in the consumption of alcohol, and as with oesophageal cancer, the disease kills many more men than it does women.

5.3.1 Anatomy and aetiology

The liver is a large (~1.4 kg/3 lb) metabolically active organ of two lobes (separated by a ligament) on the right of the abdomen. The working cells of the liver are hepatocytes, other cells comprise the internal vessel system and micro-bile ducts (bile canaliculi), and there are also immunologically active Kupffer cells (modified macrophages) and stellate cells (pericytes that support and line the sinusoids and blood vessels). Almost all cases of liver cancer can be classified by their cellular basis: neoplastic transformation of hepatocytes leads to hepatocellular carcinoma (HCC), whereas cancer of the bile ducts within the liver is cholangiocarcinoma. With a large portfolio of functions, liver dysfunction is linked to many different diseases (Table 5.1).

Liver cancer can have many causes. Hepatitis viruses will invoke an immune response, and attack of infected cells by leukocytes could also involve DNA-damaging ROS (as can occur directly with ionizing radiation). However, hepatitis B virus, as a DNA retrovirus, can also insert itself directly into the genome and inactivate tumour suppressors *TP53* and *RB1*. The response of the body to infection with hepatitis B and C viruses leads to the replacement of functioning hepatocytes with non-functional fibrotic tissue, a process that is progressive and leads to cirrhosis and ultimately liver failure. Together, all infections have been reported to be linked to 16% of liver cancers in the UK. However, on the global stage, these viruses and liver parasites inflate liver cancer deaths to 8.2% of cancer deaths, compared to 3.5% in England and Wales. The mycotoxin aflatoxin (from the fungi *Aspergillus flavus* and *Aspergillus parasiticus*) has several subtypes, the most potent being B_1. Once within the hepatocyte, the aflatoxin is converted to a highly active epoxide, which forms a DNA adduct with guanine, the repair of which can lead to error and so oncogenesis. Tobacco is reputed to be responsible for 23% of liver cancer, far exceeding alcohol as a risk factor (9%).

TABLE 5.1 Liver disease

Aetiology	Examples	Text
Neoplastic	Hepatoma/hepatocellular carcinoma, metastases, cholangiocarcinoma	This chapter
Immunological	Hepatitis (viruses B and C), autoimmune disease, parasites	Chapter 10
Metabolic	Fatty liver disease, Wilson's disease, Gilbert's disease	Chapter 16
Toxic	Alcoholic liver disease, organic solvents, heavy metal (arsenic, cadmium), tobacco smoking, overdoses (e.g. paracetamol)	Chapter 16
Haematological	Haemochromatosis, haemostasis, some cases of Budd–Chiari syndrome	Chapter 11

The uniting feature, as with almost all other cancers, is abnormalities in certain genes, with ample evidence of mutations in a large number of genes, including *TP53*, *RB1*, *CDKN2A*, and others. There are also numerous reports implicating non-coding RNAs in HCC—miR-122 is the most abundant, comprising 70% of all liver miRNAs, whilst lncRNAs such as lncHUR1, which can bind to *TP53*, and so the p21/Bax promotor that regulates transcription of these genes. Similarly, many lncRNAs are upregulated in cholangiocarcinoma, including H19, CCAT1 (significantly linked to tumour stage and lymph node metastases), and CCAT2 (linked to TNM stage). Future potential diagnostic aids and/or options for treatment include miR-200b, miR-141, miR-214, and miR-21.

5.3.2 Presentation

The most common presenting symptom of any liver pathology is right-sided abdominal pain (perhaps with swelling—hepatomegaly), but there may also be semi-specific abdominal symptoms of fever, sweating, pruritus (itching), back pain, nausea, and vomiting. The leading sign is jaundice, noticeable in the sclera of the eye and, if extensive, the skin, although this may result independently from severe anaemia. It may also be noted macroscopically in a blood sample taken for another reason—the plasma being dark yellow/orange, indicating **hyperbilirubinaemia**. Should there be a tumour that blocks the outflow of bile (as may a gallstone in the common bile duct), then the faeces will no longer be brown, but grey. However, the latter may also be due to a gallstone in the common bile duct, and so unrelated to cancer.

A frequent intermediate stage in the malignant transformation is a step between the carcinogen and cancer—cirrhosis—the deposition of fibrous tissue in response to damage to the internal structure of the liver. This inevitably leads to a deterioration in liver function, and since the liver synthesizes coagulation proteins, a presenting sign may be unexplained haemorrhage.

5.3.3 Diagnosis

The laboratory

The first step is a full blood count, urea and electrolytes (for renal function), and liver function tests, the latter forming a major part of the investigation. Further blood tests include alpha-feto protein (AFP), carcinoembryonic antigen (CEA), and CA 19-9. The former is reasonably specific for the liver, and high levels have 100% specificity. Other indications for raised AFP include **germ cell** tumours (within the ovary or testis) and ataxia telangiectasia. CEA is less specific, with the potential to be raised in stomach, pancreatic, lung, and breast cancer, whilst CA 19-9 is also increased in colorectal and oesophageal cancer.

Imaging

In parallel with laboratory testing, imaging will be requested—ultrasound (the least sensitive), CT, and MRI (Figures 5.4 and 5.5). These tools will reveal the size and extent of any tumours, and also the status of the gall bladder and bile ducts, and with this information, the MDT can consider options. But, as in

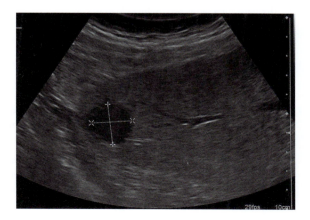

FIGURE 5.4

Ultrasound of the liver. Scan of the liver showing a large mass, the size of which can be determined by a cross-plot.

Danila M, Sporea I. Ultrasound screening for hepatocellular carcinoma in patients with advanced liver fibrosis. Med Ultrasonography 2014: 16; 139–44. DOI: 10.11152/mu.201.3.2066.162.md1is2.

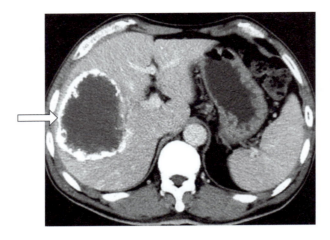

FIGURE 5.5
CT of the liver. A tumour is arrowed.

Sun J, Zhang Y, Nie C, et al. (2018). Effects of liver cirrhosis on portal vein embolization prior to right hepatectomy in patients with primary liver cancer. Oncology Letters 15: 1411–16. https://doi.org/10.3892/ol.2017.7530.

all cancers, imaging can only suggest changes *consistent* with HCC or cholangiocarcinoma. For a true diagnosis, a biopsy is required, as alternate diagnoses for an unexplained mass in the liver include granuloma, lipoma, a cyst, and **haemangioma**. However, although a biopsy provides definitive information, it is not without risk as it may lead to haemorrhage because the liver is highly vascular, and by definition liver disease is suspect, so the coagulation pathway may be impaired.

There are at least four potential indications of severe disease:

- Portal hypertension (high blood pressure in the hepatic portal vein) is most likely caused by liver fibrosis and may also cause an enlarged spleen (splenomegaly).

- This may be related to the common haematological finding of a low platelet count (thrombocytopenia), another reason to be cautious regarding biopsy.

- Ascites (fluid in the abdomen) may also be a consequence of portal hypertension.

- Encephalopathy (problems with cognitive and possible psychological function) could also be present.

5.3.4 Management

As with most cancers, management cannot proceed without a firm view of the stage of the disease.

Staging

Several models for staging exist:

- The Barcelona Clinic Liver Cancer system relies on the number and size of the tumours, a clinical performance status of the patient (day-to-day quality of life), and the presence of portal hypertension or increased serum bilirubin.

- The Child–Pugh system considers encephalopathy, ascites, and serum bilirubin, serum albumin, and the prolongation of the prothrombin time, and reports three classes (normal, mild/moderate, severe).

- The Model for End-stage Liver Disease (MELD) system requires serum creatinine, bilirubin and sodium, and the INR, and is generally preferred in assessing the severity of liver disease and the likelihood of a good outcome when considering transplantation.

Calculators for the Child–Pugh score and MELD are available online. As with several other cancers, the American Joint Committee on Cancer TNM system focuses on the size and number of the tumours (e.g. T1 = solitary tumour <5 cm, T2 = multiple tumours <5 cm, T3a = multiple tumours >5 cm, T3b = tumours involving a blood vessel, T4 = tumour invading adjacent organs and tissues), lymph node involvement (Nx, N1, N2), and distant metastases (M0, M1). Notably, none of these scoring systems considers the age, sex, co-morbidities, or the aetiology of the disease, although these will be considered by the MDT, and by the surgeon and anaesthetist, who will determine if the patient is fit for surgery. The purpose of these staging processes is to provide guidance as to the likelihood of the success of treatment options, the primary one being surgery.

Surgery

Although it is possible to resect only a part of the liver, or even a lobe (as can be done in live donor transplantation), removing the entire organ (hepatectomy) is by far the most common procedure. Although a very serious option, it may be curative—not only of the cancer, but also of underlying cirrhosis and related conditions (low platelet count, splenomegaly, encephalopathy, etc.). However, surgery does not guarantee a total cure, as there may already be metastatic disease elsewhere. Accordingly, a transplant is an option, and in many cases (such as in the face of deteriorating liver function), the only question is when—some centres have a cut-off point of a 5 cm tumour, others a 6.5 cm tumour, whilst some have a score of the total mass of tumours, sometimes adding the AFP value and MELD score to the decision.

Cholangiocarcinoma is frequently aggressive, with a median survival of 24 months, partly because the generally late presentation means that curative surgery can be offered to only 20–30% of patients. Consequently, 5-year survival rates are of the order of 5–30%, so that surgery is more urgent. Full transplantation is not usually offered because of the lack of clear effective selection criteria.

> **Cross reference**
>
> Full details of transplantation are presented in Chapter 10 on Clinical Immunology.

Non-surgical options

Should the patient be unfit for surgery, or the cancer advanced and/or with multiple metastases, there are other options. These include three-dimensional conformal radiotherapy to the tumour, ablation of the tumour with certain frequencies of the electromagnetic spectrum (e.g. microwave) that increases the temperature and so destroys the tissues, and the **percutaneous** ultrasound-guided injection of alcohol directly into the tumour. However, trans-arterial chemoembolization (TACE) is the leading option for intermediate-stage hepatocellular carcinoma. Recurrence is linked to an increased MELD score.

TACE requires identification of a viable blood vessel to deliver **embolic** microparticles carrying chemotherapy drugs or radioactivity (e.g. yttrium-90) directly to the tumour. These palliative treatments can also be combined (e.g. TACE and radiotherapy), used as bridging therapy until transplantation, or for downstaging to reduce tumour burden before surgery.

Portal hypertension may be part-relieved by the placement of a transjugular (that is, inserted via the jugular vein in the neck) intrahepatic portal vein shunt (or TIPS) wherein some of the blood normally destined for the liver bypasses this organ. **Ascites** may be treated with diuretics (renal function permitting) or direct abdominal taps. Treatment of encephalopathy is difficult and symptomatic, but anticonvulsants may help those having seizures, and in worst-case scenarios, dialysis will help reduce levels of potentially toxic bilirubin.

Chemotherapy

Chemotherapy may be considered in treating metastatic disease. The first choice is the tyrosine-kinase inhibitor (TKI) sorafenib (400 mg twice a day), possibly in combination with doxorubicin, which intercalates into DNA and so prevents the double helix from being repaired. Another potential drug is the TKI regorafenib. All such drugs are linked to the standard side effects of chemotherapy, but in addition, doxorubicin may be cardiotoxic.

Key points

Liver cancer has two forms—hepatocellular carcinoma (of hepatocytes) and cholangiocarcinoma (of the biliary system). Each has its own treatment pathway.

CASE STUDY 5.2 Liver cancer

A 40-year-old woman reads about the possibility of contracting an HIV or hepatitis C infection from transfused blood, and notes that she shares with those who have become infected a number of symptoms which include increasing tiredness, lethargy, and fevers without any clear cause or alternative diagnosis. She was transfused with two units of blood after the haemorrhaging delivery of her second child 5 years previously.

When she takes her story to her GP, he orders a full blood count, U&E, LFTs, and tests for hepatitis virus and HIV infection. The relief from the knowledge that she has not contracted HIV is countered by evidence of a hepatitis C infection. Further tests show an active infection, which is cleared by 2 months' treatment with antivirals ledipasvir and sofosbuvir. However, at the conclusion of the treatment blood tests reveal signs of liver damage (raised LFTs), a reduction in the platelet count (87×10^6/ml, reference range 150–400), and splenomegaly. Her hepatologist warns that the liver disease is likely to be progressive and that transplantation may be necessary, possibly within 5 years. This, and the 1- , 5-, and 10-year survival rates of 91%, 75%, and 60% respectively, are difficult to come to terms with as she does not feel particularly unwell.

All goes relatively well for a year until she coughs up blood, and a few days later vomits blood. Endoscopy reveals that she has oesophageal varices (section 5.2), which are found to be caused by portal hypertension that in turn is caused by liver cirrhosis. At this point the bilirubin has risen to 46 µmol/l (reference range <21). A routine of 6-monthly MRIs begins, the first two finding nothing abnormal, and parallel blood tests show reduction in liver markers, except the bilirubin, and that the low platelet count remains. At the next 6-monthly MRI a new 1.5 cm mass is noted, and the MRI scans are stepped up to 4-monthly, but at the next scan the mass has not grown. Nevertheless, the consultant hepatologist and pathologist consider the mass to be consistent with hepatocellular carcinoma, and the patient is further counselled for transplantation. She is likely to be a good candidate, with no other health issues and a MELD score of 10 (the cut-off point for surgery being ≥10).

The next MRI shows the mass is now 2 cm, and the MELD score has risen to 13, and 4 months later the mass has enlarged to 2.2 cm, a rate of growth that is slow. Indeed, the next scan still reports 2.2 cm, but the one after that finds that the mass has increased to 2.5 cm, which confers a stage of cT2. Blood results are stable, with a normal AFP, but the MELD score has increased to 15, so she is formally moved onto the waiting list. The call comes 95 days later, and after 6 hours in theatre receiving her new liver, she is transferred to intensive care for 2 days, then to a medical ward for a further 8 days. Recovery goes well, and on discharge all blood markers are within reference ranges.

She is passed back to the care of her GP, but has frequent outpatient appointments with her consultant hepatologist, primarily to monitor for early acute rejection within 90 days. Her drug regime is also regularly checked, immunosuppressive agents being steroids, azathioprine, and tacrolimus (details in section 10.4.5). Blood tests include full blood count, liver function tests, and urea and electrolytes. The latter is important as tacrolimus has a mildly damaging effect on the kidney. All drugs are well tolerated, doses are slowly lowered, and return check-up trips to hospital are becoming less and less frequent. Eventually, she is seen only once a year, but with 6-monthly blood tests, and 7 years after transplant she is well.

SELF-CHECK 5.3

What are the major causes of liver cancer?

5.4 Bladder cancer

The proportion of people dying from this disease in England and Wales is slowly rising, from 3.21% of all malignant cancers in 2013, to 3.42% (a 6.5% increase) in 2021. The ninth most common cause of a cancer death, it kills around twice as many men as it does women.

5.4.1 Anatomy and aetiology

The bladder is an expandable balloon of several layers of smooth muscle cells and connective tissue (the detrusor muscle) that receives (from the left and right ureters) and stores urine prior to urination. The outflow of urine is controlled by sphincter muscles at the base, where the bladder joins the urethra. In the female, this junction is uncomplicated, but in males the anatomy is more complex as the vas deferens, and ducts from the seminal vesicles and from the bulbourethral (Cowper's) glands, all empty into the urethra within the prostate. The innermost lining is of a transitional epithelium, resting on a basement membrane.

The major diseases of the bladder are cancer, overactivity, and inflammation. The bladder is prone to bacterial infection, and thus inflammation, and a urinary tract infection, although the site of the infection may be the urethra. An overactive bladder may be due to a number of factors, such as inappropriate neurological activity. As the inner layer of the bladder is epithelium, this cancer is a carcinoma, but a tumour arising from the smooth muscle cells will be a sarcoma. The site of the tumour is important: if near to the point of entry of the ureters it may block the entry of urine, which will back up the ureter and cause renal failure. Similarly, if the tumour is near the point of outflow to the urethra, urine will be retained, and in the worst instance, urine will again back up the ureters to damage both kidneys.

Tobacco smoking is the leading cause, one source suggesting that it is responsible for 30–40% of cancers of this organ, whilst occupation accounts for 5–10% (likely related to toxins as used in the chemical, dye, and petroleum industries) (Halaseh et al. 2022). The bladder is particularly sensitive to carcinogens because of the storage aspect, so that toxins have a relatively long time period to act on epithelium. Advances in molecular genetics point to the involvement of a large number of potential oncogenes and other genes, many of which we are aware of (*p21*, *TP53*, *Bcl-2*, *RB1*, etc.), and a large number of miRNAs (miR-9, miR-200, etc.) and lncRNAs (H19, MALAT-1, etc.) may be involved.

5.4.2 Presentation

Patients will complain of difficulty with urination (pain, increased frequency with reduced volume), but also of haematuria, some occasional, some permanent. Some of these symptoms may also be due to renal and/or prostate disease, so these must be eliminated, possibly by blood and microbiology testing, and by imaging.

5.4.3 Diagnosis

Traditional methods include urine cytology (which may detect shed malignant cells) and inspection of the bladder with a flexible cystoscopy. Routine blood tests are unhelpful, but certain urine biomarkers, such as that for nuclear matrix protein 22 (NMP22), ImmunoCyt, ADXBLADDER (an ELISA for minichromosome maintenance complex components 5) (see NICE MIB180), and UroVysion (using fluorescently labelled DNA probes that bind sections of chromosomes 3, 7, 9, and 17, the location of candidate genes) show promise. All of these markers have a varying degree of sensitivity and specificity, and although useful in their place, the gold standard is the transurethral (with local anaesthetic) cystoscopy (Figure 5.6). However, this procedure has limited sensitivity in that it can detect only very obvious lesions, but if such a lesion is found, it can be biopsied.

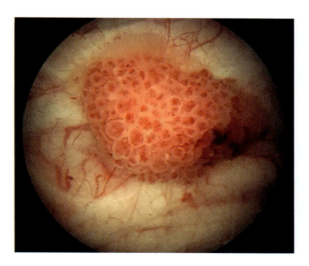

FIGURE 5.6

A bladder tumour revealed by cystoscopy.

Image Credit: Dr David Samadi.

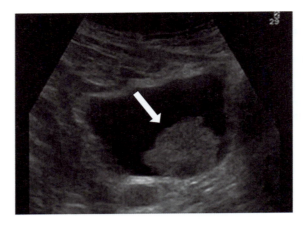

FIGURE 5.7
Ultrasound detection of a bladder tumour.

A more sensitive tool is blue-light cystoscopy, in which the bladder is first exposed to a precursor of a photoactive porphyrin, such as hexaminolevulinate, which accumulates in rapidly dividing cells (i.e. those that are malignant). When exposed to light of wavelength 360–450 nm, the cells that have taken up the porphyrin appear as if red. Given sufficient biopsy tissue, the laboratory can classify the tumour as G1 (well differentiated), G2 (moderately differentiated), or G3 (poorly differentiated). A further option is ultrasound, which although non-invasive, is of limited sensitivity, unless the tumour is large, as in Figure 5.7. Should muscle-invading cancer be suspected, this may be confirmed by CT or MRI.

5.4.4 Management

Once diagnosis has been clarified, the MDT will look at options such as transurethral resection of bladder tumour (TUBRT), which includes the underlying detrusor muscle, with use of a single dose of **intravesical** mitomycin C. Management depends on whether the tumour has invaded the underlying muscle: Ta refers to tumours restricted to the epithelium; T1 tumours have invaded the sub-epithelial connective tissue, and T2 tumours the muscle.

Non-muscle invading tumours

With the malignant tissue obtained by TUBRT, the laboratory provides additional information to permit staging. NICE NG2 classifies risk in three levels:

- Low risk is a single pTaG1 tumour <3cm in diameter, a single pTaG2 tumour <3 cm, or a **papillary** urothelial neoplasm of low malignant potential.

- Intermediate risk includes cancers of a solitary pTaG1 or pTaG2 mass >3 cm, multifocal pTaG1 or pTaG2 tumours, and any low-risk non-muscle invasive cancer that recurs within a year.

- High risk cases are those that are pTaG3, pT1G2, pT1G3, pT *in situ*, or aggressive variants, such as those that are micropapillary.

Patients at low risk need only cystoscopic follow-up 3 months and 12 months after diagnosis. Those at intermediate risk may be offered a course of at least six doses of intravesical mitomycin C, with cystoscopic follow-up at three, nine, and 18 months, and annually thereafter. High-risk patients may be offered intravesical BCG (Bacille Calmette–Guerin). If, following these treatments, the cancer recurs, then another TURBT or electrosurgery (also known as electrocautery or fulguration) may be called for. Cystoscopic follow-up happens every 3 months for 2 years, then every 6 months for 2 years, then it may be offered annually. The final option for high-risk patients is radical **cystectomy** (removal of the entire bladder and associated lymph nodes).

Muscle-invading tumours

Should there be clear evidence that the tumour extends across the sub-epithelial tissues to the muscle (that is, ≥stage T2), neoadjuvant treatment with a 3–4-week chemotherapy regime based on cisplatin should be offered. This should be followed by cystectomy or up to 4 weeks of radiotherapy with a radiosensitizer such as mitomycin C plus 5-fluorouracil. Adjuvant chemotherapy after cystectomy with a cisplatin combination should be considered.

Those who have opted for cystectomy will need a urinary **stoma**, or a continent urinary diversion (a bladder substitute or catheterizable external reservoir). Further post-cystectomy care, to be undertaken at least annually, includes monitoring for renal function with eGFR and imaging for **hydronephrosis**, stones, and cancer. There will also be a need for checks for local and distant recurrence by CT of abdomen, pelvis, and chest (six, 12, and 24 months after surgery). For men with a dysfunctional urethra, washing for cytology and/or urethroscopy are likely to be needed for 5 years. Should the bladder remain, then radiotherapy will be mandatory, and will be followed up with cystoscopy every 3 months for 2 years, then every 6 months for 2 years, then annually, with upper tract imaging for 5 years, and 6, 12, and 24 months CT monitoring for local and distant recurrence in the abdomen, pelvis, and chest.

Should there be locally advanced or metastatic muscle-invading disease, there are a number of chemotherapy options. First-line drugs include cisplatin in combination with gemcitabine, and the combination of high-dose methotrexate, vinblastine, doxorubicin, and cisplatin. As many of these regimes suppress the bone marrow, granulocyte-colony stimulating factor may be given. Second-line drugs include carboplatin or gemcitabine with paclitaxel, or the vinca alkaloid vinflunine (but see NICE TA272). The effects of these treatments will need to be assessed by regular clinical and radiological monitoring, and any toxicity will need to be addressed.

Palliative care

In common with all malignancies, should the cancer be incurable, the entire care team will be involved in offering support and options. This may include referral to a specialist palliative care team, in line with NICE CSG4 on 'Improving supportive and palliative care for adults with cancer'.

CASE STUDY 5.3 Bladder cancer

A 54-year-old man of average build, but a lifelong smoker, reports several months of changes in his pattern of urination, with occasional urine flecked with blood. On further questioning by his GP, he recalls odd twinges of pain at rest, when walking, and when urinating. The urine dipstick is unremarkable, so the GP arranges for a urine sample to be sent for analysis, and prepares the patient for a likely cystoscopy. The urine test comes back positive for tumour marker NMP22, indicating possible bladder cancer, and the request for cystoscopy is moved to urgent.

The cystoscopy reveals an oval 2 cm × 3 cm raised area with an irregular surface, which is biopsied. The lesion is on the anterior wall, far away from the junctions with the ureters and the exit sphincter to the urethra. Forty-eight hours later the lab reports a grade 2 carcinoma which is discussed at the next MDT. They recommend CT and MRI to assess the extent to which the tumour has invaded (or not) the muscle of the bladder, and whether there is any clear lymph node involvement, and the oncologist discusses treatment options with the patient and his partner.

The imaging is unable to make a firm commitment to the absence of extra-bladder disease, so the patient agrees to a transurethral resection of the bladder tumour. The histology report finds a tumour approximately 4 cm in diameter that has invaded the muscle wall, and the margins appear clear, pointing to stage 2 cancer with T2N0M0. With this new information,

the oncologist puts options to the patient, which are chemo-radiotherapy or cystectomy. The patient opts for the former, and starts a course of radiotherapy, BCG, and a combination of 5-fluorouracil and cisplatin, which is well tolerated. Regrettably, the patient continues to smoke 10–15 cigarettes daily and is very relaxed towards his disease. The first follow-up cystoscopy reports good healing of the bladder wall, and subsequent cystoscopies and CTs are also clear.

The 18-month blue-light cystoscopy shows two lesions—one near the site of the original lesion and a fresh lesion, and biopsies of both are taken, which both prove to be malignant. It is noted that the patient has lost 15% of his body weight over this time period. CT and MRI both concur that there is local lymphadenopathy, with probable lesions in the liver, and the oncologist recommends cystectomy. However, the patient is adamant that he does not want to lose his bladder, but will consent to keyhole surgery to remove the sentinel lymph nodes, and is stoically content to suffer the rigours of additional rounds of cytotoxic chemotherapy based on methotrexate and vinblastine. Over the 9 months that follow, there is growing pain, which is treated with oral opiates, and more weight loss. With pulmonary oedema and falling renal and bone marrow function, and the patient is hospitalized for specialist care. His condition deteriorates rapidly, and he is soon too ill to be transferred to a hospice. He dies 72 hours later.

Key points

Although seemingly easy to diagnose (urine problems), bladder cancer nevertheless carries a high risk of early mortality. As with prostate cancer, an early but bleak strategy is to remove the entire organ, but this brings major lifestyle changes.

SELF-CHECK 5.4

What are the major risk factors for bladder cancer?

5.5 Brain cancer

The relevant ONS/ICD section includes malignant neoplasms of the **meninges**, other parts of the central nervous system (spinal cord, **cranial nerves**, and other parts), and the brain, although the latter comprises >99% of the 4,000 deaths annually. The rare neuroblastoma is discussed in section 5.5.5. The brain is also a frequent site for metastases from up to 40% of cancers, principally the lung, breast, and melanoma, and 20–40% of patients with these cancers will suffer brain metastases. NICE NG99, entitled 'Brain tumours (primary) and brain metastases in adults' is pertinent. The frequency of deaths has been rising slowly from 2.51% in 2013 to 2.67% in 2021 (an increase of 6.4%), making it the tenth most mortal cancer.

5.5.1 Anatomy and aetiology

The brain is a dense and compact collection of neurological, vascular, and immunological cells, and a variety of other supporting cells, encased by bone (the skull). The nervous tissues of the spinal cord (itself protected by vertebrae) evolve into the three sections of the brain stem—the medulla oblongata, the pons, and the midbrain. To the rear (posterior) of the brain stem is the cerebellum. The brain stem merges with two parts of the diencephalon which in turn merge into the cerebrum. Between the brain and the skull are layers of connective tissue: the meninges, consisting of dura mater, arachnoid mater, and pia mater.

With a mass of some 3 lbs (1.3 kg), the neurological aspect of the brain is composed of around 100 billion neurons, each of which is estimated to be in contact with other neurons via an average of 100 synapses. The remaining cells within the brain may be considered in two parts—the cells of the lymphatics, arteries, veins, and capillaries, and glial cells, or **neuroglia**. Details of the four types of neuroglia are shown in Table 5.2.

As discussed in Chapter 3, the vast majority of cancers occur because of changes to the DNA during mitosis, so that cells that fail to divide—that is, are terminally differentiated—rarely transform into

TABLE 5.2 Neuroglia

Cell	Features and function
Astrocyte	Largest and most numerous of the neuroglia, providing strength and support, contribution to the blood–brain barrier, and help with homeostasis. Characterized by numerous extended cytoplasmic processes. Comprise 20–40% of all glial cells.
Ependymal cell	Line certain inner and external faces of the brain and generate cerebro-spinal fluid. Not found in the matrix of the brain.
Microglia	Modified macrophages with scavenging and phagocytic roles. Comprise 5–20% of all glial cells.
Oligodendrocyte	Form and maintain myelin sheaths around axons (as do Schwann cells in the peripheral nervous system).

cancerous cells, as is the case for the neuronal cells of the brain. However, all non-neuronal glial cells are capable of cell division and so can transform into cancers (e.g., a glioma), with a further classification based on the particular malignant cell, e.g. astrocytoma, ependymoma, and oligodendroglioma. A malignancy of cells of the arachnoid mater forms a meningioma, whilst the most common and aggressive tumour, a glioblastoma multiforme, arises from astrocytes.

There are few clear clues as to the risk factors for brain cancer: leading factors include exposure to radiation, toxins such as vinyl chloride (a possible risk factor for a glioma), and family history of brain tumours, pointing to a genetic link, such as neurofibromatosis types I (marked by a lesion at 17q11) and II (at 22q12). Li–Fraumeni syndrome is caused by a mutation (at 17p13) in tumour suppressor *TP53*, and one of its manifestations is a brain tumour. Similarly, von Hippel–Lindau syndrome, linked to a mutation in *VHL* at 3p25–26, may cause haemangioblastomas in the brain and spinal cord. Molecular genetics has pointed to roles for mutations in *IDH1* (coding for isocitrate dehydrogenase) and *IDH2*, and histone H3.3 K27M in gliomas and 1p/19q codeletion in oligodendrogliomas, whilst *BRAF V600E* mutations are often reported in astrocytomas. In certain circumstances, these genetics can have prognostic value, as wild-type *IDH* tumours have the worst prognosis regardless of treatment, whilst 1p/19q codeletion predicts response to chemotherapy.

Brain and other central nervous system tumours have an annual incidence of around 300 per million adults; 85–90% of these are of the brain and perhaps 30% are malignant. Meningiomas make up around 30%, but the dominant tumour is a glioblastoma, comprising some 60% of primary brain tumours. The latter is characterized by a poor survival rate of ~5% at 5 years, although this is strictly stage dependent, as low-grade tumours have a mean outcome survival of 7 years. However, the prognosis for an astrocytoma is much better, with reports of a 5-year survival of 90%. Some commentators group pituitary tumours and CNS lymphomas in with brain cancer. As regards the latter, there is growing evidence of lymphatics in the brain, but it is unclear if there are brain lymph nodes. Thus, it may be that a CNS lymphoma is a metastasis from a primary cancer elsewhere in the body. Spinal tumours are predominantly (79%) schwannomas, meningiomas, and ependymomas.

5.5.2 Presentation

Unsurprisingly, symptoms may be neurological, such as headache, seizures, problems with vision and balance, loss of motor function such as speech and sense of smell, and loss of sensation in a certain part of the head if a tumour presses on a cranial nerve. Non-specific symptoms include loss of appetite, nausea, and vomiting. There may also be cognitive changes such as loss of memory and concentration, and changes in mood, perceptions, and personality. All these also apply to metastases in the brain that arise from cancer elsewhere.

5.5.3 Diagnosis

In the absence of blood markers, imaging with CT and MRI is the first line of investigation (Figure 5.8), possibly with MR perfusion and MR spectroscopy. Imaging can show only a tumour, not its malignant nature, and if resection surgery is not possible, then a biopsy should be taken to determine the cell biology and molecular genetics. Differential diagnoses include an abscess or an arterio-venous malformation.

5.5.4 Management

The references for much of this section are NICE NG99 and the USA National Cancer Institute's 'Adult Brain Tumours Treatment' document. For most types of tumour, complete or near-complete resection is the goal, notwithstanding the patient's general health and potential neurological outcome. The latter may be minimized as in many cases the craniotomy (removing part of the skull) can proceed whilst the patient is awake and can be asked to count or read to determine the effect of the surgery on those parts of the brain.

Options for radiation therapy (which has a major role in high-grade gliomas) and chemotherapy depend on the nature of the tumour and its location. Brachytherapy may be used to deliver high doses of radiation directly into the tumour, whilst an option for low-grade gliomas is radiotherapy alone.

The **alkylating agent** carmustine may be used for systemic chemotherapy, but this has been largely replaced by temozolomide, although the former may be placed within the tumour base in a

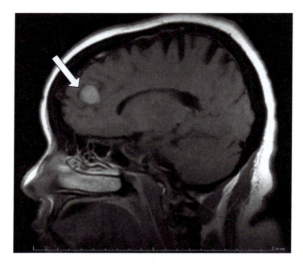

FIGURE 5.8
MRI of a meningioma. The tumour is arrowed.

Hain, TC. https://www.dizziness-and-hearing.com. 21 July 2022.

biodegradable disc. Other drugs for different tumours include procarbazine (a methylhydrazine derivative), lomustine (another alkylating agent), and vincristine (which inhibits microtubule formation), which may be used in combination. Oedema around the tumour can be treated with the corticosteroid dexamethasone, mannitol, and the **diuretic** furosemide, and those having seizures are treated with anticonvulsants. Each of the different types of brain cancer has its own semi-specific lines of management, emphasizing the crucial need for an accurate diagnosis. Long-term follow-up will then proceed, with regular imaging at intervals of 3 months, 6 months, and annually, depending on the grade of the tumour.

Investigation of suspected metastases should proceed in roughly the same manner as that of a primary tumour, initially with MRI. Systemic chemotherapy may be considered for those likely to respond efficiently, such as germ-cell and small-cell lung cancer metastases. Where there is a single brain metastasis, options include surgery, or radiosurgery/radiotherapy, with steroids taken in both instances. Other options include whole-brain radiotherapy. Unsurprisingly, people with brain cancers need a unique care package because (in addition to their physical disability), their disease will have effects on their behaviour, cognition, and personality, calling on several health and social care professionals.

5.5.5 Neuroblastoma

These tumours develop within renal and coeliac ganglia, and (mostly) the adrenal medulla of the sympathetic system. Arising from the embryonic neural crest, they account for 15% of paediatric cancer deaths, and being present at a frequency of 10 per million children, it is the most common malignancy of infancy. Aetiology can be sporadic or germline, and can arise from mutations in *ALK*, *PHOX2B*, *BARD2*, *LIN28B*, or *FLJ22536*, or from chromosomal abnormalities in 1p, 1q21, 11q, and 17q, whilst amplification of *MYCN* is present in 25% and carries a poor prognosis. Diagnosis is by imaging and biopsy, treatment by chemotherapy and surgery, and metastases are common.

Key points

There are several different types of brain tumour, but each has its own specific points of management, in terms of chemotherapy and radiotherapy.

SELF-CHECK 5.5

How do we classify brain tumours?

5.6 Renal cancer

There is no specific NICE guideline on renal cancer, but there are several TAs regarding chemotherapy, and guidance on several interventional procedures. The relative rate of renal cancer deaths has fluctuated between 2.33% of all cancers in 2013 and 2.47% in 2021, an overall 8.2% increase, that being over four times that of the increase in deaths due to all malignant cancers.

5.6.1 Anatomy and aetiology

The two kidneys lie either side of the vertebral column on the posterior aspect of the abdomen. Apart from cancer, these organs are also the site of inflammatory, cystic, autoimmune, diabetic, metabolic, hypertensive, and other forms of disease, each of which is discussed in other chapters. The leading pathology in renal cancer is renal cell carcinoma (RCC: 90–95% of cases), less frequent are transitional cell carcinoma (or urothelial carcinoma) and the very rare (2/million) Wilms' tumour (also known as a nephroblastoma), primarily a disease of children.

Leading lifestyle causes of renal cancer are tobacco smoking and overweight/obesity (both linked to 24% of cases), which together comprise 42% of all cases (as some smokers will also be obese). Four hereditary forms of renal cancer are recognized:

- most common are mutations in the von Hippel–Lindau (VHL) tumour-suppressor gene at 3p26, inactivated in ~85% of cases of the clear-cell variant of renal cell carcinoma

- hereditary papillary renal carcinoma caused by mutations in *MET*, a proto-oncogene at 7q34 coding for part of the hepatocyte growth factor receptor

- hereditary **leiomyomatosis** and renal cell cancer, caused by variants in *FH*, coding for the Krebs cycle's fumarate hydratase

- Birt–Hogg–Dubé syndrome, caused by changes to *FLCN* (both tumour suppressors at 1q42.1 and 17p11.2 respectively)

These hereditary diseases have many non-renal complications (such as spinal haemangioblastoma and lung cysts), so that poor survival may not necessarily reflect renal disease alone. Sporadic cases may be linked to abnormalities such as methylation of promotor CpG islands, and mutations in *SETD2*, *BAP1*, and *PBRM1*. One form results from a t(6;11)(p21;q12) translocation that places *TFEB*, coding a transcription factor, next to *MALAT1*, coding for a long non-coding RNA known to be over-expressed in malignancies.

5.6.2 Presentation

As with bladder cancer, the most common presenting sign is changes in urination habits, which may include haematuria. Non-specific symptoms include back pain (the most common) and abdominal pain, with the usual cancer symptoms of weight loss, tiredness, lethargy, etc.

5.6.3 Diagnosis

The GP will perform urine analysis, which may help with numerous differential diagnoses, such as an infection.

The laboratory

A blood sample will help in pointing to abnormal urea and electrolytes, and so a deterioration in overall renal function. Standard serum cancer biomarkers are unhelpful, but there may be a place for miRNAs as aids in diagnosis. Serum miRNA-378 is upregulated and miRNA-451 downregulated,

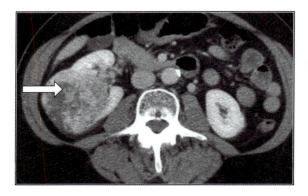

FIGURE 5.9

Renal cell carcinoma. A mass subsequently shown to be a renal cell carcinoma is detected (arrow). The normal kidney is on the right of the vertebra.

whilst expression of serum miRNA-210 is also raised, and levels fall after surgical treatment. Combined serum levels of miR-210 and miR-378 have superior sensitivity and specificity than either marker alone, and levels of both also fall after **nephrectomy**. Other methods include circulating tumour cells, which can be identified by epithelial cell adhesion molecule (CD326, although this is also present on colorectal cancer cells) and cytokeratin expression.

Imaging

Ultrasound and CT can provide essential information (Figure 5.9), but a definitive diagnosis can only be made by a biopsy (guided by ultrasound), which could be a fine-needle aspiration.

5.6.4 Management

Management is driven by the patient's general health, and by staging, itself informed largely by imaging. Smaller tumours may be treated by radiofrequency ablation and cryoablation, and chemotherapy may not be necessary if the disease is in early stages. Large tumours (>3 cm) can be managed by surgery, ideally to conserve as much of the larger nephron base as is possible, but total nephrectomy may be necessary.

In cases of advanced disease, and when metastases are present, there will be adjuvant chemotherapy. Treatment of advanced or metastatic disease with cytokines interferon and/or interleukin-2 (occasionally with cytotoxics such as vinblastine and 5-fluorouracil) has largely been superceded by mTOR inhibitors. NICE recommendations are shown in Table 5.3. It is clear from this section that practitioners pay close attention to guidance and trials, and the large number of drugs of different classes implies that the disease can be difficult to treat successfully.

Key points

Most cases of renal cancer are carcinomas. Unlike in many other cancers, chemotherapy is not cytotoxic.

SELF-CHECK 5.6

Explain why the general overall survival index is relatively good, but clinical trials generally have survival figures of less than 2 years.

TABLE 5.3 NICE chemotherapy recommendations for advanced renal carcinoma

Technology appraisal	Recommendation
169	Sunitinib for the first-line treatment of advanced and/or metastatic renal cell carcinoma
215	Pazopanib as a first-line agent in patients who had not had cytokine therapy
333	Axitinib as an option after failure of a first-line TKI or immunotherapy
417	Nivolumab (targeting CD274, the programmed death ligand 1)
432	mTOR inhibitor everolimus as an option where disease has progressed, or after anti-VEFG treatment
463	Cabozantinib as an option after VEGF-targeted therapy
498	Lenvatinib plus everolimus as an option in those who have had previous VEGF-targeted therapy
512	Tivozanib as an option in those who have had no previous treatment
542	Cabozantinib for untreated disease that is intermediate- or poor-risk
581	Nivolumab with ipilimumab (targeting CD152, the cytotoxic T-lymphocyte associated protein 4: CTLA-4) in disease that is intermediate- or poor-risk

Drugs with the suffix -nib are tyrosine kinase inhibitors (TKI); those with the suffix -mab are monoclonal antibodies.

5.7 Stomach cancer

NICE NG83 refers to oesophageal-gastric cancer, considering both organs and the junction between them, but there are also documents on procedures and the technology appraisal of drugs (such as TA208). Levels of this, the 12th most common cause of death from malignant cancer, have been falling almost consistently from 2.86% in 2013 to 2.3% in 2021, a reduction of almost 20%.

5.7.1 Anatomy and aetiology

The stomach is the first true digestive organ in that it stores and mixes food, and secretes into it enzymes, acid, and other molecules. It is formed from overlapping layers of muscles in which secretory glands are embedded. Other disease of the stomach besides cancer can be classified as inflammatory (gastritis, perhaps caused by excessive acid or autoimmune disease), mechanical (as in gastroparesis, caused by weakening of the muscles with prolonged emptying time, and failure of the sphincter muscles to close adequately), and **atrophic** (such as is caused by alcohol). These non-cancer diseases (and gastric anatomy) are considered in section 12.2.1. As with the greater part of intestinal cancer, the tumour is almost always an adenocarcinoma, arising from one of the many glands, and in advanced stages invades the muscle wall and lymph nodes. The tumours themselves may be subdivided into groups such as intestinal, papillary, diffuse, or non-diffuse. However, the stomach may also be the site of a gastro-intestinal **stromal** tumour of smooth muscle cells, as this can occur in any part of the intestines (see 5.7.5).

The leading lifestyle and environmental factors linked to stomach cancer are insufficient dietary fruit and vegetables (36% of cases), excessive dietary salt (24%), and tobacco smoking (22%). The role of infections in the cancer of this organ (32%, inevitably *H. pylori*) is second only to the role of HPV in cervical cancer (100%) and far outstrips links with infection in liver cancer (16%). Contrary to popular opinion, the stomach does not sterilize—with a pH <4, it merely suppresses bacterial overgrowth. Leaving aside common gastric infections, such as with *staphylococcus*, *shigella*, *salmonella*, *E. coli*, and

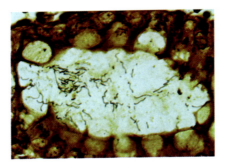

FIGURE 5.10

H. pylori in a gastric biopsy. Many *H. pylori* bacteria in the mucus near to epithelial cells.

From Orchard & Nation, *Histopathology*, second edition, Oxford University Press, 2018.

campylobacter species, the leading pathological organism is *H. pylori*, which also causes ulceration and gastritis (Figure 5.10). This link is so strong that Japan (where the incidence of gastric cancer is particularly high) has a policy of active eradication by antibiotics.

One possible neoplastic mechanism is the effect of ROS released by leukocytes as antibacterials in response to *H. pylori*. A further link is the presence of a certain virulence factor in this microbe—cytotoxin-associated gene a (*cagA*)—whose 120 kDa CagA product may interact with oncoprotein tyrosinase kinase within the cell.

There are no major genes linked strongly and directly to stomach cancer, although there is much evidence of alterations in *MYC, TP53, PIK3CA, RAS*, etc. Many receptors linked to tyrosine kinase, such as HER2 (amplified in ~15% of gastric cancers) and EGFR (= HER1) are relatively overexpressed in those cancers with a diffuse versus a non-diffuse histology, and this may be exploited in management. There may also be a place for mutations in genes coding for structural components such as adhesion molecule E-cadherin, and for chromosome or microsatellite instability. Notably, germline mutations *CHD1* at 5q15-21.1, coding for E-cadherin, may underlie hereditary diffuse gastric carcinoma, a variant causing ~2% of all gastric cancers. Indeed, the risk of advanced gastric cancer in a male 80-year-old with a *CDH1* mutation is ~80%. It is possible that these different carcinogenic processes may act on different anatomical sites of the cancer; for example, chromosome instability is more common in tumours in the cardiac region, whereas EBV pathology is more common in the body of the stomach. There is evidence that the degree of DNA methylation is important in prognosis, TNM stage, and metastatic potential, and there are possibly roles for mutations in genes coding for mucus production (i.e. *MUC-1, MUC5AC*, and *MUC6*), as these may permit *H. pylori* infection.

5.7.2 Presentation

Clearly, presentation is with complaints pertaining to the stomach—pain (not necessarily linked to eating), cramping, nausea, and vomiting (perhaps with blood). Alternative diagnoses include gastritis and food poisoning, whilst 'acid' indigestion may be treated with antacids. Quite possibly, in common with other cancers, there may be reports of loss of appetite, weight loss, and fatigue.

5.7.3 Diagnosis

Endoscopy and imaging

Once signs and symptoms have been collected, and disease is suspected, an endoscopy is inevitable. This has the advantage over CT scanning that the practitioner has a clear and (almost) unambiguous view of the stomach, and that a biopsy can be taken at the same time (Figure 5.10). However, more complex and high-resolution imaging is needed if lymph node involvement is to be assessed. NICE NG83 points to whole-body CT scanning, but suggests that fluorine-18 PET-CT scanning should only be considered if metastatic disease is suspected and it will help management.

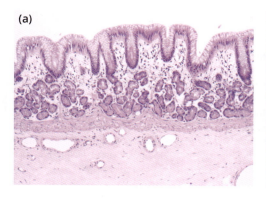
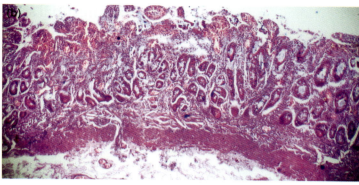

FIGURE 5.11
Routine histology of stomach cancer. (a) Normal structure of intact layers of tissues. (b) An intramucosal adenocarcinoma with loss of layers structure.

(a) Peter Takizawa, Director of Medical Studies, Department of Cell Biology, Yale University. Retrieved from http://medcell.org/histology/histology.php. (b) iStock.

The laboratory

Early in the disease, there are no particular changes to routine blood results, although patients with a **thrombocytosis** may have a worse stage and overall survival, and others suggest that high **IL-6** also predicts poor outcome. Both these changes may reflect a low-level acute phase response common in many malignancies. Detection and treatment of stomach cancer is hindered by a lack of high-quality serum markers, as CEA, AFP, CA19-9, and CA-125 all have poor sensitivity and specificity. Potential new markers include pepsinogens I and II, beta-catenin, **VEGF**, IgG antibodies to *H. pylori* and to a molecule called A Disintegrin and Metalloproteinase domain-containing protein 8 (ADAM8).

Some centres may assess levels of faecal antigens, gastrin, and urease (for *H. pylori*), although these may detect gastritis. Diagnosis is made by histological examination of the biopsy (Figure 5.11), and on immunocytochemistry for HER2.

Of the microRNAs, miR-21 is increased in the serum and peripheral blood mononuclear cells, out-performs CA199 and CEA, and can discriminate stages I and IV. High levels of miR-196a and 196b fall after the resection of the tumour and are linked to metastatic potential and survival. A panel of miRNAs may be more useful, as miR-501-3p, miR-143-3p, miR-451a, and miR-146a show potential as possible non-invasive biomarkers for prediction and prognosis of lymph node metastasis.

5.7.4 Management

The precise position of the tumour is important. NICE NG83 refers to stomach cancer, oesophageal cancer, and the junction between these organs, and many recommendations are the same. Once the diagnosis has been made, many patients opt for immediate surgery (if offered). Ideally, this will be laparoscopic, but if the tumour is extensive, and lymph nodes need to be removed, open surgery may be easier. Chemotherapy and radiotherapy may be offered before (to reduce tumour burden) and after surgery, and where the tumour is HER2-positive, and patients are naïve to chemotherapy, it should be cycles of intravenous trastuzumab (a mAb to HER2) in combination with cisplatin (which intercalates into DNA) and capecitabine (which is metabolised to 5-fluorouracil) or 5-fluorouracil (NICE TA208).

Possible combinations of first-line palliative chemotherapy for locally advanced or metastatic cancer include doublet treatment with 5-fluorouracil or capecitabine in combination with cisplatin or oxaliplatin, and triplet treatment consisting of 5-fluorouracil or capecitabine in combination with cisplatin or oxaliplatin plus epirubicin (an anthracycline). Adjuvant chemotherapy brings significant benefits in overall survival (hazard ratio (95% CI)- 0.82 (0.76–0.90)) and disease-free survival (0.82 (0.75–0.90)). Studies indicate one-, three-, five-, and 10-year overall survival rates in the order of 89%, 66%, 57%, and 20% respectively, but this varies with stage on diagnosis and treatment.

5.7.5 Gastrointestinal stromal tumours

These rare cancers (~900 new cases annually in the UK, 1–2% of all gastrointestinal neoplasia) can be found along the entire length of the intestines, from stomach to anus, although 60–70% arise in the stomach. The dominant cell biology is increased expression of CD117 a tyrosine kinase receptor for stem cell factor and coded for by *KIT* on 4q12. However, some 4% of GISTs are linked to *PDGFRA* and *BRAF*.

Once diagnosed by ultrasound/CT/MRI, primary treatment is laparoscopic surgery, but lymph node involvement may be found in ~30% of cases, calling for adjuvant chemotherapy. However, as only 50% of tumours are resectable, cytotoxic chemotherapy will be required, such as with the tyrosine kinase inhibitors imatinib (400 mg/day), and sunitinib (50 mg/day) for CD117 positive cases. Outcome survival is generally poor (median survival time 19 months), with a 4-year mortality of ~60%.

> ### Key points
>
> Compared to other cancers, the adenocarcinoma of stomach cancer is often positive for HER2, and this drives chemotherapy with monoclonal antibodies.

SELF-CHECK 5.7

Compare the lethality and frequency of stomach cancer with that of oesophageal cancer.

5.8 Ovarian cancer

NICE guideline CG122 'Ovarian cancer: recognition and initial management', and documents QS18, TA55, TA284, TA389, and TA528 are pertinent. The rate of deaths linked to this disease has fluctuated somewhat from 2.51% in 2013 to 2.30% in 2021, the lowest over this period (a reduction of around 8%). It is the sixth most common cause of cancer death in women. As with transgender women and prostate disease, transgender men who retain their ovaries—or their breasts, uterus, cervix, vagina, and/or vulva—remain at risk of cancer and other disease of these organs.

5.8.1 Anatomy and aetiology

The ovaries sit in the lower abdomen, linked to the uterus by a fibrous ligament. The two major types of disease of this organ are cysts, of which there may be many (hence polycystic ovary syndrome: 14.3.2), and cancer. In turn, by far the most frequent (95%) ovarian cancers are epithelial (and so are carcinomas); the remainder are stromal and germ cell tumours (the subject of section 5.8.5). Epithelial ovarian cancers may be classified histologically as serous (the most common), mucinous, endometrioid, and clear cell, and each has its subtypes. The International Federation of Gynaecology and Obstetrics (FIGO) is the leading authority and has published a staging system that absorbs the standard TNM classification (Table 5.4, which also shows 5-year survival rates for each stage). The T and M staging is as with other cancer—disease in lymph nodes and/or distant tissues and organs.

Lifestyle and environmental factors comprise 21% of cases of ovarian cancer, and the greatest within these is not having breastfed for at least 6 months, linked to 18%, and the implied pregnancy. This may be linked to reduced ovulation, as lack of pregnancy (being parous bringing a 30–60% reduced risk), early onset of menstruation, and late menopause are also risk factors. A further 20% of ovarian cancers are familial, hence the importance of family history, and may be linked to *BRCA1* and *BRCA2*.

Other risk factors include **endometriosis**, postmenopausal hormone replacement therapy (HRT), and obesity. The effect of *BRCA1* and *BRCA2* is so powerful that removal of the ovaries reduces the risk of ovarian cancer by 90% , and presence of *BRCA1* or *BRCA2* brings a life-time risk to an 80-year-old of 44% and 17% respectively (Box 5.1).

TABLE 5.4 FIGO/TMN classification of ovarian cancer

T stage	FIGO stage	Pathology	5-year survival
T0		No tumour present	
T1	I	Tumour limited to one or both ovaries	85–94%
T1a	IA	Tumour limited to one ovary, capsule intact, no tumour on the surface, no malignant cells in the peritoneum	94%
T1b	IB	As 1A but both ovaries involved	92%
T1c	IC	One or both ovaries involved, plus any of capsule not intact, tumour on the surface, malignant cells in the peritoneum	85%
T2	II	As I but with pelvic extensions	69–78%
T2a	IIA	As II but with uterus involvement but no malignant cells in the peritoneum	78%
T2b	IIB	As II but involving other pelvic tissues but no malignant cells in the peritoneum	73%
T2c	IIC	As IIA or IIB but with malignant cells in the peritoneum	69%
T3	III	Both ovaries involved plus metastasis outside the pelvis	17–59%
T3a	IIIA	As III but with microscopic metastasis	59%
T3b	IIIB	As III but tumour beyond the pelvis <2 cm in the greatest dimension	39%
T3c	IIIC	As III but tumour >2 cm and/or regional lymph node involvement	17%
T4	IV	Malignant cells in pleural effusion, metastases outside the abdomen, including lymph node involvement	12%

Modified from Testa U, Petrucci E, Pasquini L, Castelli G, Pelosi E. Ovarian Cancers: Genetic Abnormalities, Tumor Heterogeneity and Progression, Clonal Evolution and Cancer Stem Cells. Medicines 2018: 5; 16. https://doi.org/10.3390/medicines5010016.

BOX 5.1 Prophylactic removal of ovaries (oophorectomy)

The link between *BRCA1* and *BRCA2* and cancer is so strong that women with a marked family history (and other risk factors) are faced with the fear of ovarian cancer and a difficult decision—whether to have their ovaries removed. Evidence in favour of this approach comes from a study of 551 women with *BRCA1* (present in around 84%) and *BRCA2* (in 16%), of whom 259 had their ovaries removed (Rebbeck et al. 2002). After a follow-up 8 years later, there were two (<1%) cases of carcinoma in those who had surgery, and 58 (20%) in the women who retained their ovaries. This translates to a very significant hazard ratio (HR)(95% confidence interval) of 0.04 (0.01–0.16) for bilateral oophorectomy reducing the risk of this cancer. Other analyses showed that young age at first birth, any birth, and late age at menarche were also protective.

BRCA1 and *BRCA2* are also risk factors for breast cancer. Notably, women without ovaries had a reduced risk of breast cancer, with 21 cases (21%), compared to 60 cases (42%) in women with ovaries (HR 0.47 (0.29–0.77)). These advantages are in the face of the 20% of women with ovaries who have ever used HRT compared to the 48% of women with ovaries. The use of the hormones would have increased the risk of both cancers.

Other genes linked to ovarian cancer include those that methylate the DNA of certain tumour-suppressor genes, and mutations in *TP53*, *Notch*, *RAS/MEK*, and *PI3K* are common. Some mutations are semi-specific for different tumour types: those in *KRAS* and *TP53* are often found in mucinous tumours, whereas as *PIK3CA* and *ARDID1A* mutations are common in clear cell ovarian cancers. Overexpression of certain genes, such as those of the *CXCL* and *PSMB* family, is linked to infiltration of the tumour by T lymphocytes, and a better prognosis.

5.8.2 Presentation

Semi-specific signs and symptoms include persistent abdominal swelling (perhaps due to ascites), with or without tenderness and pain, changes in bowel habit, increase in urinary urgency and/or frequency, and possibly masses (not uterine fibroids) within the pelvic region. Differential diagnoses for the symptoms include inflammatory bowel disease (although this is unlikely as it appears in younger women than those with ovarian cancer) and colorectal cancer. There may be vaginal discharge of blood or fluids, and regrettably by the time these are recognized, the cancer is often advanced.

5.8.3 Diagnosis

The laboratory

Once a critical mass of signs and symptoms have been noted, blood will be taken for genetics and serum cancer markers. Principal among the latter are CA125 and CA19-9, whilst there is growing evidence of a place for Human Epididymis Protein 4, (a protease inhibitor) and osteopontin (an extracellular matrix protein). NICE recommends measuring AFP and **human chorionic gonadotropin (hCG)** as these are raised in germ cell tumours. As always, issues of sensitivity and specificity are to the fore, whilst in one study, 83% of women with ovarian cancer were identified by a CA125 level ≥35 units/ml and performs best in identifying advanced disease. Other potentially valuable tests include miRNAs such as miR-199, miR200, and miR-30a-5p (which are overexpressed), and miR-140, mir-145, and miR0125b1 (which are underexpressed) may one day become markers not only of the disease per se, but also of likelihood of metastasis and survival.

Imaging

Imaging includes transvaginal ultrasound, a technique that may also be useful in screening, especially in women who are at high risk. If this and CA125 suggest cancer, then for staging, abdominal and pelvic contrast-enhanced CT are required as the peritoneum and lymph nodes can be assessed, although MRI is preferred for determining bowel involvement. Figure 5.12 shows a CT of a large ovarian mass in the pelvis which histology showed to be an epithelial ovarian cancer.

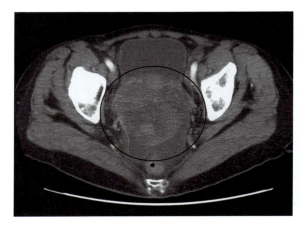

FIGURE 5.12
CT of ovarian cancer. The circle shows a large ovarian cancer.

5.8.4 Management

Several models for assessing the extent of disease may help diagnosis and management. The Overa test requires levels of follicle-stimulating hormone, HE4, apolipoprotein A-1, transferrin, and CA125, whilst the ADNEX score uses age, CA125, type of hospital centre, and six ultrasound-derived predictors (maximum diameter of lesion, ascites, etc.) to estimate the probability of the tumour being benign or malignant. The RMI score uses CA125, ultrasound imaging and menopausal status, whilst the ROMA score combines CA125 with HE4 and menopausal status. Others are described in NICE DG31.

The standard treatment is surgery followed by aggressive chemotherapy. The former allows for staging to be defined and is valued as a bilateral oophorectomy and salpingectomy debulking strategy, and a laparotomy will also allow for a hysterectomy, **omenectomy**, and (if indicated by imaging) **lymphadenectomy**. If surgery is not immediately planned, a laparoscopic biopsy of suspected tissue should be performed to define histology. In some cases surgery may be deferred in favour of neoadjuvant therapy, and the high rate of disease recurrence often calls for a second operation.

Chemotherapy should not (initially) be offered to women who have had optimal surgical staging and low-risk stage 1A and 1B disease (see Table 5.4: disease confined to ovaries only). However, at stage 1C, initial adjuvant therapy of six cycles of tubulin-inhibitor paclitaxel (a 3-hour intravenous infusion of 175 mg/m^2) with a platinum-based agent (cisplatin or carboplatin) at 3-week intervals should be offered. Most (~75%) patients respond, but most eventually also relapse (~62% within 2 years).

Second-line agents include alkylating agents chlorambucil, treosulfan, and altretamine, and topoisomerase inhibitors topotecan and liposomal doxorubicin. An alternative is paclitaxel with a platinum compound, possibly with a poly (ADP-ribose) polymerase inhibitor (PARPi, such as olaparib, rucaparib, niraparib, or talazoparib).

Emerging therapies include targeting the folate receptor (expressed by ~70% of primary epithelial ovarian cancers), and the promotion of cytotoxic T lymphocytes, as the presence of tumour-infiltrating lymphocytes brings an eight-fold improvement in 5-year survival.

5.8.5 Germ cell and stromal tumours

In embryology, a primordial germ cell gives rise to gametes, and so produces a malignancy that is fundamentally different from the epithelial cancers. Germ cell tumours are aggressive, rare (4/million females, ~2% of all ovarian malignancies), usually unilateral and almost always arise in teenagers and young women. Overall 5-year survival rates are ~87%, a very good rate compared to other ovarian cancers. The most common subtype is a dysgerminoma; others include a yolk sac tumour, embryonal carcinoma, choriocarcinoma, teratoma, and mixed germ cell tumour.

Presentation is as epithelial ovarian cancer—with abdominal pain, vaginal bleeding, etc.—but diagnosis rests on increased hCG (and so a positive pregnancy test) and AFP, and these are also useful in management. Lactate dehydrogenase and CA125 may also be elevated. Surgery is again advocated, but given the age of the patients, fertility-sparing surgery is emphasized, so egg harvest and storage may be required. However, in those whose families are complete, hysterectomy and bilateral salpingo-oophorectomy are called for, as may be lymphadenectomy (an option in ~23%). Most germ cell tumours are sensitive to chemotherapy, often with a cisplatin-based drug combined with bleomycin (that induces breaks in DNA), vinblastine (inhibits microtubule formation), and etoposide (topoisomerase inhibitor), and radiotherapy, and so are often curative.

Stromal tumours (some described as sex cord-stromal tumours) derive from non-germ cell, non-epithelial cells, these being fibroblasts, endothelial cells, thecal cells, and granulosa (the most frequent ~70%). Relative 5-year survival (~68%) may be described as 'intermediate' in terms of all ovarian cancers. Some stromal cells can produce sex hormones, including testosterone, which will be virilizing. Definitive diagnosis is by biopsy, and the rarity of these tumours is such that there is little evidence-based consensus of the correct management, which therefore reverts to that of epithelial ovarian cancer.

CASE STUDY 5.4   *Ovarian cancer*

An obese 75-year-old woman complains to her GP of increasing abdominal bloating, constipation, and occasional vaginal bleeding. Her GP orders CA-125, molecular genetics, and an ultrasound of the abdomen. These report back with CA-125 of 80 units/ml (reference range <35), positive for *BRCA1*, and changes in both ovaries consistent with ovarian cancer. The GP refers promptly to oncologists at the nearest teaching hospital, who apply the ADNEX scoring model, finding a high likelihood of malignancy. Ultrasound findings are confirmed by CT, which also notes multiple lymphadenopathy. The patient is offered surgery for removal of both ovaries and her uterus, omental fat, and several lymph nodes, to which she agrees.

At open laparotomy, the organs are removed and the peritoneum is washed out for cytology, which laboratory analysis shows to include high numbers of leukocytes and malignant cells staining positive for cytokeratins. The laboratory reports ovarian masses 3.5 cm and 4 cm in diameter, both having histology of moderately differentiated epithelial carcinoma, with invasion of local lymph nodes. The uterus appears to be uninvolved, and staging is set at T2cN1M0, FIGO stage IIC. Chemotherapy proceeds with paclitaxel, carboplatin, and bevacizumab, which is tolerated, and the CA125 level falls back to the reference range. Once the pelvic region has recovered from the surgery, CT reports a good overall picture, and after 18 weeks of cytotoxic therapy, the patient is maintained on bevacizumab for a further year.

All goes well, but 2 years later she returns to the GP with symptoms of tiredness, loss of appetite, and pain on the lower right side of her chest. The GP refers urgently to the oncologists, who order CT/MRI and blood tests, which point to metastatic disease in the liver and mesenteric lymph nodes, with a minor increase in pleural fluid. These prompt routine blood tests and a further round of chemotherapy, which is not tolerated as well. Within 3 weeks the patient's haemoglobin falls from 110 to 95 (reference range 118–48 g/l), the neutrophil count from 2.5 to 2.1 (reference range 2.0–6.5 × 10^9/l), and the serum creatinine rises from 75 to 100 (reference range 71–133 µmol/l), and she feels increasingly sick and short of breath. The chemotherapy is changed to liposomal doxorubicin, and after one round, paclitaxel is added. These too are poorly tolerated, and imaging finds no improvement, but an increase in pleural fluid. Routine blood tests show rising liver function tests (which may be in response to drugs and/or disease progression), so the drugs are again reduced in dose, but niraparib (an inhibitor of the enzyme poly-ADP-ribose polymerase: 300 mg once a day as she weighs 82 kg) is added.

Several months later the patient experiences increased shortness of breath, and imaging finds marked pleural effusions, and 150 ml of a cloudy fluid is aspirated by thoracentesis and sent to the laboratory. The procedure benefits the patient's breathing, but the cytology laboratory reports malignant cells. The patient is discharged, but breathing problems return, and a further thoracentesis aspirates 300 ml, which again carries malignant cells. The patient's symptoms increase as her general health deteriorates, and she dies at home in her sleep 2 months later.

Key points

Ovarian cancer has a strong genetic component, and a relatively poor prognosis (survival index 0.59) as it is often diagnosed in an advanced state. Accordingly, treatment must be rapid and aggressive.

SELF-CHECK 5.8

How can the laboratory help with the diagnosis of ovarian cancer?

5.9   Oral cavity and throat cancer

This cancer is often described as 'head and neck'. NICE NG36 focuses on cancer of the upper aerodigestive tract, and hospitals may have a dedicated ear, nose, and throat (ENT) department, with clear cross-over. The relative rate of death from cancer of these organs and tissues has increased year-on-year from 1.6% of all cancer deaths in 2013 to 2.04% in 2021, a difference of 25%—surprisingly large given the general slow but steady increase in the rate of smoking cessation.

5.9.1 Anatomy and aetiology

These organs and tissues are clearly diverse, being linked by anatomy and physiology, although it could be argued that the lip and mouth, the tonsils, the pharynx, and the larynx should all be considered separately. These organs and tissues are difficult to classify as most are part of the digestive system, but the larynx, being at the top of the trachea, is part of the respiratory system. However, the larynx and pharynx are clearly part of the neck, as is the thyroid, which is classified under endocrinology.

The surface layers of the aerodigestive system are epithelial, but underlying this layer are connective tissues, so that the cancers of these are carcinoma (ectodermal: generally squamous cell carcinoma, which make up 90% of head and neck cancers) and mesodermal-derived sarcoma. Other diseases of these tissues and organs are mostly inflammatory, and there may be lymphadenopathy, which may be a local metastasis or a lymphoma (Chapter 11) and melanoma (section 5.11).

The leading lifestyle and environmental factor is tobacco (in ~70%), followed by low dietary fruit and vegetables (~50%), high intake of alcohol (~27%), and infections (such as the human papilloma virus and Epstein–Barr viruses, ~12%), together making up >90% of all cases. Abnormalities in chromosomes 3p, 4q, 5p, 8p, 9p (a deletion of p21–22, location of the tumour suppressor *P16*), 11q, 13q (the location of tumour suppressor *RB1*), 14q, 17p (the location of tumour suppressor *TP53*), 18q, and 21q have been reported. All are potentially the result of carcinogens in tobacco smoke, although many abnormalities (including those in *Notch*, *Ras*, and *c-Myc*) can be present in lifelong non-smokers. The epidermal growth factor receptor (EGFR) is overexpressed in up to 90% of squamous cell carcinomas, giving a route to chemotherapy with monoclonal antibodies such as cetuximab and panitumumab. Changes in miRNAs have been reported, such as the absence of miR-199 and downregulation of miR-34a and members of the miR-29 family.

5.9.2 Presentation

As may be expected, symptoms include persistent pain in the mouth and throat, difficulty swallowing food and drink (dysphagia), nosebleeds, excess or insufficient saliva, bleeding, lumps in the neck below the jaw (these possibly being salivary gland tumours or lymphadenopathy), ear pain, and ulceration. The tongue may have red (erythroplakia) and white (leukoplakia) patches. Laryngeal disease may present with breathing problems, hoarseness, and cough.

5.9.3 Diagnosis

Being close to the surface, many cancers can be viewed directly, or with an endoscope, and endo-ultrasound is also valuable and can begin the staging process, to be completed by CT and/or MRI, which will also identify local invasion or lymph node invasion. If a suspicious lesion is found, it should be relatively easy to perform a biopsy, and this will give a diagnosis such as squamous cell carcinoma. There are no specific blood markers, routine tests remaining normal until the disease is advanced and possibly metastatic.

5.9.4 Management

Surgery is always to the fore, and ideally is curative, but should the tumour be inoperable, chemotherapy and radiotherapy are both required. In many cases, combined radiotherapy (e.g. 2 Gy daily, 5 days a week for 7 weeks) and chemotherapy (e.g. three cycles of cisplatin 100 mg/m^2 every 21 days) is an option, although another is cisplatin, 5-fluorouracil, and cetuximab (which targets the epidermal growth factor receptor, NICE TA145, TA473), or nivolumab (blocking programmed death-ligand 1 (PD-1), NICE TA490). Others suggest carboplatin, 5-fluorouracil, and docetaxel, and cetuximab (targeting the EGFR) which, in combination with radiotherapy, is more effective. Other potential treatments include other anti-EGFR agents, and the anti-PD-1 monoclonal antibody drugs pembrolizumab, durvalumab, and nivolumab. Other relevant NICE publications include QS146, NG36, and CSG6.

Key points

By its very nature, oral cavity and throat cancer (also known as head and neck cancer) is a diverse collection of diseases, each of which must be addressed according to its particular anatomical site.

SELF-CHECK 5.9

What are the most common types of oral cavity and throat cancer?

5.10 Cancer of the uterus

With a single exception, the frequency of uterine cancer has been rising year by year from 1.24% in 2013 to 1.57% in 2021, a notable increase of 26.6%. As with other organs of the reproductive system, transgender men remain at risk of cancer of the uterus whilst it is present.

5.10.1 Anatomy, pathology, aetiology, and epidemiology

The uterus lies in the lower region of the pelvis, between the bladder (on the anterior aspect) and the sacral sections of the vertebral column (on the posterior aspect), although it is in close contact with the colon and rectum. The two fallopian tubes merge with the superior (upper) aspect whilst the inferior (lower) aspect merges with the cervix. It is supplied with blood through two uterine arteries. Structurally, it has two components, a thick outer wall of smooth muscle cells (the myometrium), and an inner lining of epithelial cells (the endometrium) which, when reproductively active, thickens under the influence of steroid sex hormones, is shed during menstruation, and is restored post menstruation.

Both parts of the uterus may become malignant, but in addition the endometrium may grow outside the uterus, causing endometriosis, and there may be abnormal growth of the myometrium, causing fibroids. Cancer of the muscle of the uterus is a sarcoma (making up 4% of uterine cancers), and cancer of the endometrium a carcinoma. Accordingly, as the vast major of uterine cancers are endometrial, many sources merge the two diseases. A sarcoma may be further described as a leiomyosarcoma (leiomyo- referring to smooth muscle—rhabdomyo- refers to striated muscle). A third type is the endometrial stromal sarcoma, arising from the connective tissue that lies between the smooth muscles of the body of the uterus, and the endometrium. The uterus is rarely infected and there is essentially no other common or debilitating disease, except hormonal/reproductive (Chapter 14).

The leading lifestyle factor in uterine cancer is overweight/obesity, responsible for 34% of cases. Lack of physical exercise is linked to a further 4%, post-menopausal hormones to 1%. The most common likely aetiology of a sarcoma is pelvic radiation, another is long-term use of tamoxifen. The leading risk factor for endometrial cancer is overweight/obesity, linked to 50% of cases; others are the **metabolic syndrome**, **nulliparity**, use of tamoxifen, early menarche/late menopause (as with ovarian cancer, see section 5.8), and type 2 diabetes. Pre-menopausal disease seems likely to be linked to sex hormones (a hypothesis supported by the role of hormone-replacement therapy in carcinogenesis), but in post-menopausal cases this is less likely.

The heritable Lynch syndrome (as already described in the section on pancreatic cancer) brings a 60% lifetime risk of colorectal cancer, but an 80% risk of endometrial cancer, and is linked to *MHL1*, *MSH2*, *MSH6*, and *PMS2*. Other mutated genes linked to sporadic endometrial cancer include *ARID1A*, *PTEN*, *CTNNB1*, and *PIK3R1*. The t(7;17)(p15;q21) translocation is linked to endometrial stromal sarcomas. Absence of *PTEN*, mutation status of *TP53* and *PIK3CA*, aneuploidy, and HER2/neu overexpression all predict a poor prognosis.

5.10.2 Presentation

Leading semi-specific signs include vaginal discharge (mostly non-menstrual blood), whilst leading non-specific symptoms are pain, tenderness, and swelling in the lower abdomen, and frequent urination (as an enlarged uterus, regardless of pathology, will restrict full extension of the bladder). A leading differential diagnosis is fibroids.

5.10.3 Diagnosis

Following a physical examination, cells may be taken from the cervix to exclude cancer of this organ, and a transvaginal ultrasound ordered (likely to be more sensitive than transdermal ultrasound). There may be a case for hysteroscopy, whilst more invasive procedures are dilation and curettage, and biopsy, to harvest some of the endometrium which will go to the laboratory for analysis for grade (which part-predicts lymph node involvement) and hormone receptor status. CT and MRI can provide additional information (Figure 5.13).

Traditional blood markers are unhelpful in uterine cancer. However, levels of miR-152 and miR-24 are decreased, whilst those of miR-150, miR-205, and miR-222 are all increased in uterine sarcoma and predict the stage of the disease. Furthermore, the lowest levels of miR-24 and miR-152 bring superior 5-year survival.

5.10.4 Management

Surgery is the mainstay of all types of uterine cancer. However, staging after imaging must be performed before surgery as further resections may be necessary. Staging systems for the two most common cancers are very similar (Table 5.5), and options are therefore to offer hysterectomy and removal of relevant lymph nodes and other tissues (such as the ovaries), with adjuvant chemotherapy and/or radiotherapy (including vaginal brachytherapy), depending on stage.

Uterine sarcoma

Stage 1 disease is localized only within the uterus so that hysterectomy alone should be curative, although bilateral salpingo-oophorectomy and periaortic selective lymphadenectomy are options. In addition to surgery, in stages I and II there is evidence of the successful use of radiotherapy, cisplatin (which causes irreparable DNA damage), and doxorubicin (which interferes with topoisomerase). In stage III disease, alkylating agent ifostamide and tubulin inhibitor paclitaxel have also been used,

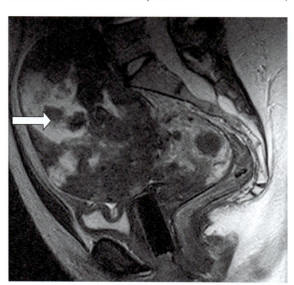

FIGURE 5.13
Abdominal MRI showing a tumour in the uterus. The intra-uterine cancer is arrowed.

Howell GA, MD. © Radsource 2022. All Rights Reserved.

TABLE 5.5 Staging of cancer of the uterus

Stage	Uterine cancer	Endometrial cancer
I	Tumour is present only in the uterus	Tumour is confined to the uterus
II	Tumour extends beyond the uterus, but within the pelvis, such as the adnexa	Tumour involves the uterus and cervical stroma
III	The tumour infiltrates abdominal tissues, perhaps regional lymph nodes	Tumour invades the serosa, adnexa, vaginal tissue, pelvic or para-aortic lymph nodes, or there is parametrial involvement
IV	Tumour invades the bladder or rectum, perhaps with distant metastasis to intra-abdominal or inguinal lymph nodes	Tumour invades the bladder and/or bowel mucosa, perhaps with abdominal metastases and/or inguinal lymph nodes.

Adapted from National Cancer Institute.

whilst in stage IV disease combinations of these drugs may be appropriate. As an example of the efficacy of these treatments, patients in some clinical trials have progression-free survivals of 6 to 8 months, and overall survival of 8 to 13 months. The same regimes may be used in recurrent disease, although alternatives include gemcitabine (which, when incorporated into DNA, leads to an irreparable error) and docetaxel (another tubulin inhibitor).

Endometrial cancer

The treatment broadly mirrors that of the sarcoma (surgery, then radiotherapy and/or chemotherapy), but there are differences depending on the subtype of the tumour (e.g. serous, clear cell, carcinosarcoma, undifferentiated). Grade 1 and 2 tumours may be considered low risk, and most patients do well with surgery alone. Grade 3 tumours may be considered high risk, and so lymph node dissection and adjuvant therapy with carboplatin plus paclitaxel, with radiotherapy, is likely. If possible, laparoscopic surgery is preferred to laparotomy, and vaginal brachytherapy is associated with less radiation-related morbidity than external-beam radiotherapy. Standard treatment options for stage III, stage IV, and recurrent endometrial cancer include the above plus hormone and biological therapy. Possible chemotherapy regimes include cisplatin, doxorubicin, ifosfamide, and paclitaxel, with added filgrastim (a granulocyte colony stimulating factor to treat potential neutropenia). Hormonal therapy includes hydroxyprogesterone, medroxyprogesterone, megestrol, letrozole (an aromatase inhibitor), and tamoxifen. Biological therapy includes mammalian target of rapamycin (mTOR) inhibitors such as everolimus and ridaforolimus (as many tumours have an altered AKT-PI3K pathway), and bevacizumab (which targets VEGF) with temsirolimus.

Key points

Uterine cancer includes two major groups—uterine sarcoma (of the smooth muscle cells of the wall) and endometrial cancer (of the epithelial lining cells).

SELF-CHECK 5.10

Vaginal brachytherapy is often preferred to standard external-beam radiotherapy—why is this?

5.11 Melanoma

The rates of deaths due to this cancer have also fluctuated, with levels in 2022 of 1.35% being 19% lower than those of 2015. There is an approximate 30% increased frequency of deaths in men.

5.11.1 Anatomy

Skin derives from embryonic ectoderm, invagination of the **gastrula** forming a continuous surface with the alimentary canal, respiratory system, and urogenital system at their respective orifices. It consists of a number of well-defined layers, the epidermis being the avascular outer superficial layer consisting of keratinized stratified squamous epithelial cells containing keratinocytes (90%, arranged in four or five layers), melanocytes (8%, which generate and transfer melanin to keratinocytes), immunologically active Langerhans cells, and Merkel cells (linked to the nervous system). At the base of the epidermis is a single layer of cuboidal or columnar keratinocytes that are fixed on a basement membrane of connective tissues. Below this is the dermis, a highly vascular and innervated region of dense connective tissue fibrils (mostly collagen and elastin) and a variety of cells and tissues, such as the hair follicles and sweat and sebaceous glands.

5.11.2 Pathology and aetiology

Pathology of the skin includes cancer, autoimmune disease (as in scleroderma), inflammation (contact hypersensitivity), dietary deficiencies (vitamin C), physical factors (frostbite, radiation), and infections (herpes simplex), many of these being investigated and treated by dermatologists. Specific conditions are described in other chapters. The melanotic lesion (which may develop from a mole) differs from many other skin lesions in that it can vary in colour and texture, and is asymmetrical.

The primary cause of melanoma is ultraviolet radiation (causing 86% of cases), and an immunological aspect is likely in some, as an increased frequency of the disease is noted in immunosuppressed transplant patients. The radiation (of wavelengths 275–320 nm) causes indiscriminate damage to DNA within melanocytes, molecular genetics pointing to *CDNK2A* and *TP53* as particular targets. Mutations in *BRAF* can also direct treatment.

5.11.3 Presentation and diagnosis

One of the reasons that the incidence of melanoma death is falling is because patients and their families are more aware, and so more likely to visit their GP early, when faced with any type of skin lesion. The most common differential diagnoses for melanoma are basal cell and squamous cell carcinomas, which can be differentiated by morphology. GPs are fully aware of skin lesions, and when presented with a likely melanoma (Figure 5.14) will refer promptly for a biopsy.

The laboratory

A number of histological stains are suitable, such as for S-100, vimentin, HMB-45 (human melanoma black), and melan A (Figure 5.15). To rule out a cutaneous lymphoma, tissue may be stained for CD45.

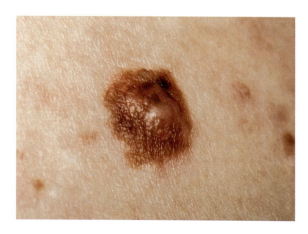

FIGURE 5.14
A malignant melanoma.

iStock.

(a)

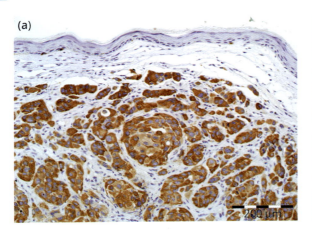

(b)

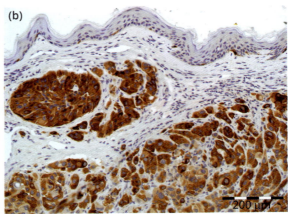

FIGURE 5.15

Tissue staining for melanoma. Immunocytochemistry with (a) S100 and (b) melan A, both supporting the diagnosis of malignant melanoma.

From Orchard & Nation, *Histopathology*, second edition, Oxford University Press, 2018.

Imaging

Further investigations will include imaging to determine stage, notably metastatic development to local and distant lymph nodes. These are important with regard to 5-year survival, which for stage 0 is 99.9%, stage I/II ~92%, stage II ~62%, stage III ~47%, and stage IV ~13%. The highest grade of disease, with distant metastases, perhaps in the lungs, is linked to high levels of serum lactate dehydrogenase. Should imaging find suspicious lymphadenopathy, this will also be biopsied as prognosis of a simple cutaneous lesion as 10-year survival with a negative biopsy is 30%, compared to 10% with a positive biopsy.

5.11.4 Management

Once diagnosis is complete, surgery to remove the lesion will proceed with urgency. The extent of surgery will depend on staging, as defined by biopsy and imaging. For example, therapeutic lymph node dissection should be offered in palpable stages IIIB–IIIC disease. A danger in not excising lymph nodes with micrometastases is that cancer will develop in remaining lymph nodes in 20% of cases. The surgeon will also excise a margin of healthy tissue, up to 2 cm around the primary lesion, to ensure all local malignant tissue has been harvested. The depth of the tumour from the surface of the skin (Breslow's depth) will be noted as it is linked to 5-year survival (<1 mm, 99–100%; 1–2 mm ~88%; .1–4 mm ~67%, >4 mm 50%). Topical treatment with imiquimod (which stimulates the immune system by activating Toll-like receptors) is recommended after excision of a stage 0 lesion. Genetic testing (e.g. for *BRAF* V600) is recommended for stages IIC and III disease, and imaging of the brain is recommended in those with actual or suspected stage IV disease, with whole-body MRI for those aged up to 24 years and with stages III–IV melanoma. Immunosuppressants should be minimized or avoided.

Depending on the stage, systemic adjuvant chemotherapy may be needed, but radiotherapy is not recommended in those with IIIB or IIIC cancer unless a reduction in the risk of local recurrence exceeds the risk of adverse effects. Options for stage IV disease include radiotherapy and radioembolization in addition to surgery. Targeted chemotherapy includes dabrafenib and vemurafenib for unresectable or metastatic *BRAF* V600 tumours (as these drugs inhibit the gene's product), ipilimumab (an mAb that activates the immune system by targeting CTLA-4), temozolomide (which alkylates or methylates guanine), and dacarbazine (which also methylates guanine, an option if the previous drugs are unsuitable).

CASE STUDY 5.5 *Melanoma*

A 56-year-old woman refers to her GP, showing him an irregular and mottled (brown/black) growth just below the left elbow. She says it wasn't there on her holiday photos taken 9 months ago. He passes her on to the local dermatologist, who immediately arranges a biopsy, the procedure occurring a week later in the outpatients department. When the biopsy comes back suggestive of a melanoma (positive straining for S-100 and HMB-45), the dermatologists informs the GP, and orders ultrasound and CT of the patient's left arm and chest, with attention to the axilla. Although there was no palpable lymphadenopathy, imaging suggests lymph node involvement, and an axillary sentinel lymph node is removed which is positive for metastatic deposits (Figure 5.16) and gives a proposed stage of pT2aN1M1.

The options are explained to the patient, who opts for wide and deep excision of the tumour and axillary lymph nodes, and subsequent chemotherapy. One of the excised lymph nodes has a very small melanotic infiltrate, three others are clear of any abnormal tissues. The molecular genetics laboratory reports a normal *BRAF*. The patient is started on an infusion of 3 mg/kg ipilimumab every 3 weeks for four cycles, and after a further 3 weeks she has an MRI, which fails to find any abnormalities. She continues with follow-up appointments, and repeated imaging, and 5 years after diagnosis is well with no signs of recurrence.

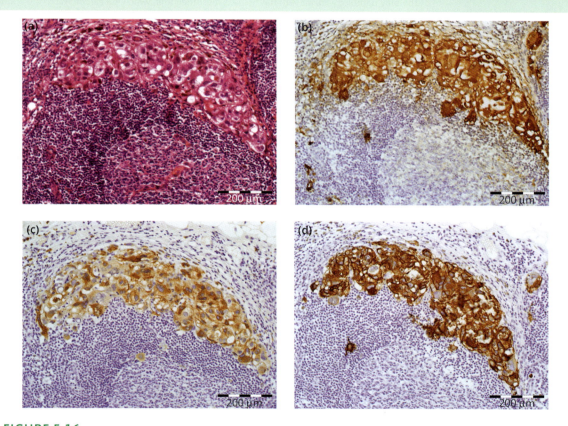

FIGURE 5.16

Sentinel lymph node staining. (a) Routine haematoxylin and eosin staining showing a suspicious arc-shaped infiltrate. (b–d) The same tissue stained by immunocytochemistry for S100, melan-A, and HMB-45 respectively, confirming the infiltrate as melanotic.

From Orchard & Nation, *Histopathology*, second edition, Oxford University Press, 2018.

SELF-CHECK 5.11

Why is immunology important in melanoma?

5.12 Mesothelioma

The annual rate of death from this cancer fluctuated somewhat between 2013 (1.59%) and 2021 (1.34%, a relative reduction of some 16%), but on the whole the trend is downward. This may be because of the reduction in use of the leading carcinogen, asbestos.

5.12.1 Anatomy and aetiology

The mesothelium is a sheet of squamous epithelial cells, supported by a basement membrane and loose connective tissue, that line the pericardial, pleural, and peritoneal cavities. Its purpose is to secrete a mucinous lubricating fluid to minimize friction between the heart, lungs, and ribs. The connective tissue carries fibroblasts and macrophages.

The only pathology linked with this tissue is mesothelioma, which is a carcinoma of the epithelial/mesothelial cells, although some will be sarcomatoid and others biphasic. There is no inflammatory or autoimmune disease, but alternative diagnoses include mesothelial hyperplasia and metastatic carcinoma from a primary tumour elsewhere. As the tissue is a sheet, the neoplasia rarely forms a spheroid mass, but more often spreads in two dimensions, and so is difficult to resect, in addition to the difficulties raised by its location and the crucial nature of nearby tissues—that is, the heart (the pericardium, <1% of cases) and lungs (the pleura, ~90% of cases). It is seen less often (~10%) in the abdominal mesothelium as the exposure route is through the lungs and so into the chest. At post-mortem, extra-thoracic metastases are found in over half of patients, mostly in the liver (32%), spleen (11%), and thyroid (7%).

Mesothelioma is unusual in that over 90% of cases have a single cause—asbestos—and accordingly it is top of the list of occupational hazards. This is likely linked to the considerable almost five-fold sex difference, with around 1,840 male deaths per year 2013–21, compared to around 380 female deaths. The effect is so strong that family members are also at risk from asbestos fibres brought into the home. A further unusual feature is the very long lag time (20–50 years) between exposure and symptoms, as demonstrated by men who were exposed for a short period as part of military service. Accordingly, it is likely that the incidence of this disease will increase as the dangers of asbestosis have been recognized and acted upon only in the last 20–30 years, it being banned in the UK in 1999. Other causes are the naturally occurring building material erionite, talc, simian virus 40, and radiation exposure. Familial mesothelioma may be linked to mutations in *BAP1* (at 3p21.1, encoding an enzyme that reverses ubiquitin) which are also present in some cases of melanoma. Other genes that may be involved include *MLH1*, *MLH2*, *CDKN2A*, *TP53*, and *BRCA2*. There is also evidence of homologous deletion of 9p21 (which includes *CDKN2A* and *CDKN2B*) in this disease.

5.12.2 Presentation and diagnosis

Symptoms include non-specific fatigue, sweating, and weight loss, although chest symptoms such as dry cough, chest pain, and dyspnoea are common, and many have stethoscopic 'chest sounds' suppressed by excessive pleural fluid. A good history to determine evidence of exposure to asbestos is step one, other differential diagnoses of chest disease excluded.

Imaging

Imaging with X-ray and CT follow (Figure 5.17), which can also help with staging, but MRI can also be used to determine the extent of disease.

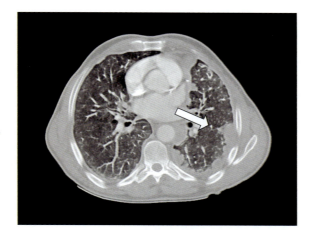

FIGURE 5.17

Chest CT showing mesothelioma. Contrast-enhanced CT scan showing increased nodular pleural thickening involving the mediastinal pleura, with a large mass on the right (arrowed).

Science Photo Library.

The laboratory

Blood tests and cancer markers are unhelpful, although there may be a place for mesothelin, a 40 kDa cell surface protein coded for by *MSLN* at 16p13.3, which is overexpressed by mesothelioma cells, and can also be detected in the plasma, but also in other cancers such as those of the ovary and pancreas. Ideally, a pleural biopsy will make the diagnosis. Histology and cytology (for cells harvested from a sample of pleural fluid: Figure 5.18) may be stained with Cam5.2, calretinin, epithelial membrane antigen, and cytokeratins (such as CK7 and AE1/AE3), although many of the markers are present in other cancers such as lung and breast adenocarcinomas.

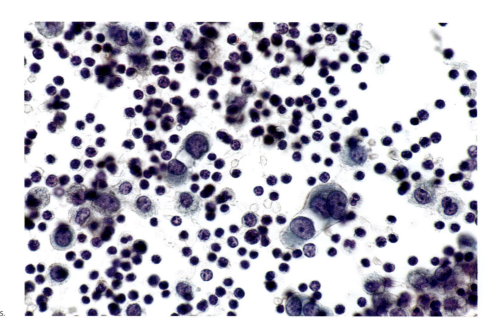

FIGURE 5.18

Cytopathology of pleural fluid in mesothelioma. There are two distinct populations of cells: most are small and resemble lymphocytes; others are large and display a malignant phenotype. Immunocytochemistry will further define the phenotype.

Nephron, CC BY-SA 3.0 <https://creativecommons.org/licenses/by-sa/3.0>, via Wikimedia Commons.

Precise diagnosis is helpful as it predicts outcome: median survival rates of epithelial, biphasic, and sarcomatoid tumours are in the region of 14, 10, and 44 months respectively. Once excised, the nature of the tissues can be determined with immunocytochemistry for markers such as mesothelin.

5.12.3 Management

Surgery is not indicated for the majority of patients, so that treatment aims to improve quality of life and survival, which includes psychosocial issues, bronchodilators, and draining excess pleural fluid. There is no standard chemotherapy, and although the combination pemetrexed (a folate antagonist) and cisplatin (which cross-links nucleotides) infusions is licensed, there is use of mitomycin C (and alkylating agent), and vinblastine and vinorelbine (both tubulin antagonists).

CASE STUDY 5.6 *Mesothelioma*

A 64-year-old ex-builder reports increasing breathlessness, a cough, and fatigue over a 6-month period. He reports a 30-year history of smoking but stopped 15 years ago, and that he had lost a stone in weight in the past year. The GP (with thoughts of lung cancer) orders routine blood tests and a chest X-ray, the latter reporting equivocal findings, with some pleural opacity, possible mediastinal lymphadenopathy, and a small pleural effusion. Blood results show a slightly low haemoglobin (130 g/l, reference range 133–67) and slightly raised CRP (12 mg/l, reference range <10). She calls the local chest consultant for an opinion, who recommends a CT of the chest and abdomen, which is performed a week later. This imaging is again imperfect, somewhat supporting the X-ray findings, but in addition points to a left-sided mediastinal shift, a thickened section of the pleural lining, indicating that a diagnosis of lung cancer is less likely.

Seeking a definitive diagnosis, the consultant orders an MRI and a PET scan, and these confirm all the other findings, but point to a mesothelioma, so that an MDT is called. This group concludes with a preliminary staging of cT2aN1M0, and seeks the patient's consent to biopsy, under ultrasound guidance, the largest section of abnormal tissue. With consent, the procedure harvests some fibro-fatty tissue and some pleural fluid—the former reports findings consistent with a sarcomatoid mesothelioma (positive for calretinin, CK7 and CK20) (Figure 5.19).

The MDT meets with an expert surgeon who considers the likelihood of curative surgery to be poor, and with a diagnosis of pT2aN1M0 the patient is started on pemetrexed (500 mg/m²) and cisplatin (75 mg/m²) infusions, with folic acid and vitamin B_{12}, on a 21-day cycle. As the enlarged lymph nodes are directly below the sternum, radiotherapy is thought inadvisable as it may damage the heart. The patient suffers nausea (controlled by an anti-emetic), vomiting, fatigue, a skin rash (which

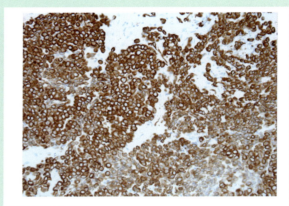

FIGURE 5.19

Immunocytochemistry. Staining pleural tissues with antibodies to CK7.

From Orchard & Nation, *Histopathology*, second edition, Oxford University Press, 2018.

responds to a corticosteroid), tinnitus, and minor mucositis. The laboratory reports a minor neutropenia and thrombocytopenia, rising creatinine, and hyperuricaemia: fortunately, these are not severe enough as to warrant a dose reduction.

After three cycles, the patient is rested and a PET scan is performed, which shows no changes to the high metabolic foci in the chest, so that the chemotherapy is continued. However, after two cycles, the neutrophil and platelet count fall, and the creatine rises so that a reduction in dose is required. The patient becomes increasingly unwell, loses weight rapidly, and requests that treatment is stopped and that he is allowed home, where he is becoming increasingly frail, and dies almost exactly a year after he first visited his GP.

Key points

Mesothelioma is a particularly malignant cancer that in most cases is inoperable.

SELF-CHECK 5.12

How may mesothelioma be contracted?

5.13 Cancers of undefined nature and low frequency

From a public health perspective, the cancers of defined organs and tissues described above have been focused upon simply because they are the most frequent and comprise 84% of all cancer deaths in 2021. Excluded from this focus on cancer of known specified organs are malignant neoplasms where the site is unspecified, that together were linked to 9,009 deaths in 2021, slightly more than those of pancreatic cancer. This illustrates the difficulties in diagnosis, many of which will be 'incidentalomas' (i.e. are discovered incidentally), so that progress in diagnostic pathology needs to be made. Indeed, NICE CG104 refers to metastatic malignant disease of unknown primary origin.

5.13.1 Organ-specific cancers

Cancers with a frequency of <1% to <0.1% are summarized in Table 5.6, together causing 4,633 deaths—more than brain cancer. The remaining specific malignant neoplasms (541 deaths, 0.4%) consist of 14 diverse cancers, each with a frequency of <0.1%. The most frequent in this group is adrenal cancer (0.09%, 128 deaths, and survival index 0.59). As a crucial endocrine organ (Chapter 15), manifestations of malignancy can include over- and underexpression of its products, although the former is dominant as deficiency in one gland can be restored by the other.

Kaposi's sarcoma gained prominence due to its association with the immunosuppression of HIV/ AIDS, but caused only four deaths in England and Wales in 2021 with an excellent survival index of 0.04, the global figure being over ten times worse at 0.44. The aetiology is dominated by a herpes virus infection, the most common presentations being of the skin and mouth. Accordingly, it is not a true sarcoma, and many cases are related to endothelial cells. If superficial, treatment is by surgery, but in many cases, lesions are present elsewhere, such as in lymph nodes, so that chemotherapy with drugs such as daunorubicin and paclitaxel may be needed.

5.13.2 Non-organ-specific cancers

Multiple endocrine neoplasia (MEN)

With a prevalence of 20/million, this cancer can be found in many different organs, and described in two types, both having a familial component.

- Type 1 MEN is characterized by a mutation in the tumour-suppressor gene *MEN1* on 11q13.1, which causes adenomas in the adrenals (40%), pancreas (50%), parathyroid (95%), pituitary (70%), and thyroid (20%), and accordingly the laboratory is often called upon for diagnosis. Dermal, gastric, and central nervous system tumours are rare, and women are at risk of breast cancer.

- Type 2 MEN is caused by mutations in proto-oncogene *RET* (at 10q11.21, encoding a tyrosine kinase) and leads to tumours in the adrenal (>95%), parathyroid (60%), and thyroid glands (>95%) (all type 2A), and intestinal and neural disease with a phenotype that resembles Marfan syndrome (type 2B, discussed in full in section 16.3.3). The leading adrenal tumour is a phaeochromocytoma which may lead to Cushing's disease. Wherever possible, tumours should be excised, and there is no place for chemotherapy or radiotherapy.

TABLE 5.6 Cancers of low frequency

Cancer	% of cancer deaths[1]	Aetiology/major risk factors	Symptoms	Diagnosis	Survival index (SI)[7]	Other features
Gall bladder[2]	0.75	Inflammation, gall stones	Abdominal pain, nausea, vomiting, jaundice	Imaging, liver function tests	0.52	Women dominate in presentation (two thirds/one third) and have a worse prognosis
Cervix	0.52	HPV, multiple sexual partners, smoking	Bleeding, other vaginal discharge, pain in sexual intercourse	Colposcopy, biopsy, cytopathology	0.26	5-year survival in stage 1 cervical cancer ~94%, stage 4 <10%
Vagina/vulva	0.37				0.36	
Small intestines	0.37	TP53, KRAS. Crohn's and coeliac diseases	Abdominal pain	Imaging, serum CEA and CA19.9, histology of biopsy	0.30	More cases are male, but no difference in outcomes
Anus and anal canal	0.27	HPV	Discharge, pain in defaecation	Imaging, biopsy	0.27	Almost two-fold increased frequency in women, and 50% more deaths.
Thyroid	0.25	Some RET, others BRAF (V600E)	Neck pain/nodules	Thyroid function tests, imaging, biopsy	0.10	Women comprise ~75% of new cases and 64% of deaths
Peritoneum[3]	0.21	Most undefined, some BRCA 1/2	Abdominal pain	Imaging, cytopathology	0.78	Markedly more severe in women: SI 0.85 v 0.46 in men
Ureters[4]	0.17	Smoking	Abdominal pain and on urination	Pyelogram, urinalysis	0.45	More men than women are diagnosed (67%) and die (57%)
Bone and articular cartilage	0.15	Most undefined	Bone/tissue pain, loss of flexure	Imaging, biopsy, bone markers[6]	0.67	Relatively good prognosis of >80% at 5 years, but 50% at higher stage
Male genitals[5]	0.14	Smoking, HIV, HPV	Pain in testes, groin. Abnormal growth on/in penis	Imaging, biopsy	Penis 0.21 Testes <0.1	Testicular cancer curable with surgery: 5-year survival 97%

[1] England and Wales, 2021.

[2] Includes unspecified parts of the biliary tree.

[3] Includes the retroperitoneum.

[4] Includes unspecified urinary organs.

[5] Includes penis and testes.

[6] Hydroxyproline, osteocalcin, procollagen, calcium, etc.

[7] Survival index = annual deaths/new cases: poor = 0.91(pancreatic), good = 0.13 (melanoma).

Carcinoid

Often described as a neuroendocrine tumour, this slow-growing tumour, with a frequency of 10/million, has a very low malignant and metastatic potential, and accordingly may be described as hyperplasia, not necessarily as a cancer. Indeed, it is not classified with neoplasia in the ICD system, and was linked to the death of one woman in England and Wales in 2021. With a number of semi-independent aspects, most arise in the **enterochromaffin** cells of the intestines, often the terminal ileum and appendix, with gastrointestinal symptoms. Some may secrete vasoactive molecules such as serotonin, histamine, and prostacyclin, leading to wheezing and other symptoms such as vasodilation and flushing. Carcinoid tumours may also occur in the lung (where it may grow to obstruct bronchioles) and the stomach. Should the tumour be secretory, then somatostatin analogues such as octretide and lanreotide should help symptoms. Surgery should proceed wherever possible, and adjuvant chemotherapy is unwarranted.

Chapter summary

- These twelve cancers together cause over a third of all cancer deaths, each linked to 1.4–5.7% of cancer deaths.

- Pancreatic cancer brings a particularly poor prognosis: blood tests amylase and CA19-9 may be helpful.

- Oesophageal cancer can vary with proximity to the throat or to the stomach, and acid reflux may be an important risk factor, alongside smoking and excessive consumption of alcohol.

- Tobacco smoking is the leading risk factor for cancer of the bladder, and more than twice as many men than women die from this disease.

- Liver cancer has several different aetiologies and brings a poor prognosis. The laboratory is particularly helpful in diagnosis and management.

- The leading causes of stomach cancer are poor diet, infections (such as *H. pylori*), and tobacco smoking.

- The malignant cells in brain cancer are not neurones, but supporting cells—the neuroglia—of which there are several types. The most common form of brain cancer is a glioblastoma.

- Some 20% of ovarian cancers are linked to reproductive factors, another 20% to *BRCA1* and *BRCA2*, which have a strong carcinogenetic effect.

- Over 40% of renal cancers are linked to tobacco smoking and overweight/obesity, and urinary symptoms generally lead to early intervention and so a relatively good prognosis.

- Cancer of the oral cavity and throat is diverse; it is most frequently caused by tobacco smoking, a poor diet, and excessive alcohol use.

- Over 90% of mesotheliomas have a single cause—asbestos—and most new cases (82%) and deaths (84%) are in men.

- Public perceptions of melanoma, and so early presentation, may be why this cancer has a good overall prognosis.

- About a third of uterine cancer can be accounted for by overweight/obesity, but this rises to 50% for endometrial cancer.

- Ten other cancers each have a low frequency (≤1%) of cancer death, these being gall bladder, cervix, vulva/vagina, peritoneum/retroperitoneum, small intestines, ureter, anus, thyroid, bone/cartilage, and penis/testes.

■ *RET* and *MEN1* are the major causes of multiple endocrine neoplasia, whilst a slow-growing carcinoid tumour may arise in different tissues.

Further reading

● Ilic M, Ilic I. Epidemiology of pancreatic cancer. World J Gastroenterol. 2016: 22; 9694–705.

● Yang FR, Li HJ, Li TT, et al. Prognostic Value of MicroRNA-15a in Human Cancers: A Meta-Analysis and Bioinformatics. Biomed Res Int. 2019: 2063823.

● Lagergren J, Smyth E, Cunningham D, Lagergren P. Oesophageal cancer. Lancet 2017: 390; 2383–96.

● Guidance on Cancer Services: Improving Outcomes in Urological Cancers: The Manual. NICE.

● Giulietti M, Occhipinti G, Righetti A. Emerging Biomarkers in Bladder Cancer Identified by Network Analysis of Transcriptomic Data. Front Oncol. 2018: 8; 450. doi: 10.3389/fonc.2018.00450.

● Dimitroulis D, Damaskos C, Valsami S, et al. From diagnosis to treatment of hepatocellular carcinoma: An epidemic problem for both developed and developing world. World J Gastroenterol. 2017: 23; 5282–94.

● Labib PL, Goodchild G, Pereira SP. Molecular Pathogenesis of Cholangiocarcinoma. BMC Cancer. 2019: 19; 185. doi: 10.1186/s12885-019-5391-0.

● Necula L, Matei L, Dragu D et al. Recent advances in gastric cancer early diagnosis. World J Gastroenterol. 2019;25:2029–2044. doi: 10.3748/wjg.v25.i17.2029.

● Lheureux S, Braunstein M, Oza AM. Epitheial ovarian cancer: Evolution of management in the era of precision medicine. Cancer J Clin 2019: 69; 280–304.

● Capitanio C, Montorsi F. Renal cancer. Lancet 2016: 387; 894–906.

● Alfouzan AF. Head and neck cancer pathology: Old world versus new world disease. Niger J Clin Pract. 2019: 22; 1–8.

● Carbone M, Adusumilli PS, Alexander HR, et al. Mesothelioma: Scientific clues for prevention, diagnosis, and therapy. Cancer J Clin. 2019: 10; 3322/caac.21572.

● Elsayed Z, Elhalawani H , Abdel-Rahman O. Gemcitabine-based chemotherapy for advanced biliary tract carcinomas. Cochrane Database Syst Rev. 2018: CD011746. doi: 10.1002/14651858.CD011746.pub2.

● Ma S, Wang J, Han Y, et al. Platinum single-agent vs. platinum-based doublet agent concurrent chemoradiotherapy for locally advanced cervical cancer. Gynecol Oncol. 2019: 154; 246–52.

● Barsouk A, Rawla P, Barsouk A, Thandra KC. Epidemiology of Cancers of the Small Intestine: Trends, Risk Factors, and Prevention. Med Sci. 2019: 7; p. ii: E46. doi: 10.3390/medsci7030046.

● Rusinek D, Chmielik E, Krajewska J, et al. Current Advances in Thyroid Cancer Management. Int J Mol Sci. 2017: 18; p. ii: E1817.

● Baird DC, Meyers GJ, Hu JS. Testicular Cancer: Diagnosis and Treatment. Am Fam Physician. 2018: 97; 261–8.

● Khatami F, Tavangar SM. Multiple Endocrine Neoplasia Syndromes from Genetic and Epigenetic Perspectives. Biomark Insights. 2018, Jul 2; 13:1177271918785129.

● Rubin de Celis Ferrari AC, Glasberg J, Riechelmann RP. Carcinoid syndrome: update on the pathophysiology and treatment. Clinics. 2018: 73(suppl. 1); e490s.

Useful websites

- National Institute for Health and Care Excellence: **https://www.nice.org.uk**
- Office for National Statistics: **https://www.ons.gov.uk**
- Cancer Research UK: **https://www.cancerresearchuk.org**
- International Agency for Research on Cancer: **http://gco.iarc.fr/today/home**
- National Cancer Institute of the United States: **https://www.cancer.gov/**

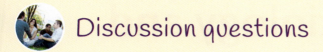

Discussion questions

5.1 Explain the difference in the survival profiles of mesothelioma and pancreatic cancer compared to melanoma and breast cancer.

5.2 The oesophagus and stomach are anatomically adjacent organs. Why is cancer of the oesophagus considerably more lethal?

6

The aetiology and cellular basis of cardiovascular disease

The second most frequent cause of death in England and Wales in 2021, of 134,174 people (22.9% of all deaths) comprising 72,498 men (24.4% of all men) and 61,676 women (21.4%), was disease of the circulation (generally synonymous with cardiovascular disease). At each decade, the proportion of men dying of this disease exceeds that of women, until aged 90 and over, where women comprise 64.5% of the deaths. The World Health Organization stated that the leading cause of death in 2019 globally was ischaemic heart disease and stroke, together linked to 27% of deaths.

The leading aetiology in cardiovascular disease is undoubtedly atherosclerosis, others are valve disease and cardiomyopathy, and there is often overlap. However, atherosclerosis is not so much a disease in its own right, causing only 60 deaths (0.04% of all cardiovascular deaths). Instead, it may be seen as the end point of four major traditional risk factors: hypertension, dyslipidaemia, smoking, and diabetes, whilst data from the last two decades have supported the hypothesis of a fifth—inflammation. A full understanding of the relationship between atherosclerosis and cardiovascular disease is not possible without an understanding of these risk factors, which is the subject of Chapter 7. A further aspect of this disease is thrombosis, details of which are presented in Chapter 11.

We begin our exploration of the cardiovascular system in section 6.1, looking at the different types of cells that make up this extended organ system. Section 6.2 discusses how these cells come together to form the different transport vessels and relate structure to function. We complete our revision of physiology in section 6.3 by looking at the cells that make up the heart, and how they ensure this most crucial organ provides the body with blood.

The second part of this chapter examines the diverse pathological processes that lead to cardiovascular disease. Section 6.4 will focus on atherosclerosis, and then, in section 6.5, on other cardiovascular disease: valve disease, cardiomyopathy, arrhythmia, and **heart failure**. The focus on the destructive processes of atherosclerosis, and the general view that the major risk factors are treatable, is countered by genetics. Numerous studies have shown that certain genes are linked to different aspects of cardiovascular disease, perhaps the best known being the mutation that causes familial hypercholesterolaemia. These issues are examined in section 6.6.

Learning objectives

After studying this chapter, you should confidently be able to:

- list the different cells and vessels of the cardiovascular system

- explain the structure and function of the cardiovascular system

- describe the pathophysiology of atherosclerosis

- recognize other forms of cardiovascular disease

- understand the relationship between genetics and cardiovascular disease

6.1 Cells of the cardiovascular system

A minimum requirement for all animal life, especially in the short term, is oxygen and a source of energy, such as glucose. The cells of the lungs are modified to absorb oxygen and facilitate the release of carbon dioxide, and red blood cells are highly specialized to carry and deliver oxygen to the tissues. In the mammalian cell, the key energy substrate is glucose, one of the many different sources of energy that are absorbed from the diet by the alimentary canal. Oxygen, glucose, and thousands of other molecules in the blood are moved around the body by the vascular system, a network of 60,000 miles of tubes consisting of arteries, arterioles, capillaries, venules, and veins. These blood vessels also remove the waste products of metabolism that are carried in the blood, such as carbon dioxide and urea, which are excreted by the lungs and the kidneys respectively.

Allied to the cardiovascular system is the lymphatic system, which has two functions. First, it facilitates the return of interstitial tissue fluid (lymph) to the heart, in which case it is acting as a drainage system. The second role of the **lymphatics** is to enable the movement of lymphocytes between different lymph nodes and other reticulo-endothelial tissue. In this respect the lymphatics are also important in immune responses and inflammation. The lymphatics are not a circulation—the content of the lymphatic system is not actively pumped in the way the blood system is driven by the heart. Instead, it is a closed system of capillary-like vessels that merge with the venous circulation near the heart.

The cells that make up the vessels and organs of the cardiovascular system include endothelial cells, smooth muscle cells, cardiomyocytes, pericytes, macrophages, fibroblasts, and adipocytes. The nervous system is also involved in the cardiovascular system, but as a partial regulator of the rate at which the heart beats and of blood pressure.

Key points

Cardiovascular disease is the second most frequent cause of death in England and Wales. The leading aetiology, atherosclerosis, is responsible for the majority of deaths from ischaemic heart disease (heart attack) and cerebrovascular disease (stroke). Other causes include heart failure and valve disease.

6.1.1 Endothelial cells

These flattened, orthogonal cells line all blood and lymphatic vessels, and the inside of the heart, forming a continuous layer, the endothelium. In the adult it consists of $1-6 \times 10^{13}$ cells, weighs approximately 1 kg and covers a surface area of approximately 4,000 to 7,000 m^2. Each cell is anchored to an underlying basic elastic lamina whilst individual cells are attached to their neighbours by specialized junctions in the gaps between the cells. These gaps regulate the passage of various cells and substances moving between the blood and the tissues. Endothelial cells are highly complex and have

BOX 6.1 Nitric oxide

Vasodilation is mediated through two distinct pathways, dependent or independent of the powerful vasodilator formerly called endothelium-derived relaxing factor and later discovered by Furchgott and Zawadzki to be nitric oxide. This molecule can be synthesized in response to shear stress and is produced by the activity of the enzyme nitric oxide synthase (eNOS) on its substrates L-arginine and oxygen. With a very short half-life, nitric acid must be continuously generated by eNOS. Nitric acid was first discovered in cardiovascular and neural systems as a vasodilator and a neurotransmitter, respectively.

Once generated, nitric acid diffuses to the smooth muscles, where it activates guanylate cyclase, which in turn produces cGMP. Increased cytoplasmic levels of this small cyclic nucleotide set off a chain of reactions that result in changes to the structure of the actin and myosin complex, and thus smooth muscle relaxation. By this pathway, the endothelium adapts to changes in cardiac output to maintain blood pressure and organ perfusion. Apart from changes in blood flow, several molecules, such as bradykinin, adenosine, vascular endothelial growth factor, and serotonin, can all induce the endothelium to upregulate eNOS and so levels of nitric oxide. Failure of the ability of the endothelium to generate nitric oxide may be a contributory factor in the development of hypertension and atherosclerosis.

many diverse roles, such as in blood pressure regulation, **haemostasis**, **leukocyte** migration, and with a barrier function.

Blood pressure

The role of the endothelium in blood pressure regulation involves the synthesis and release of factors that act on the smooth muscle cells of the middle part of arteries (the media), resulting in constriction or dilation. The balance between the molecules that mediate constriction and dilation of the vessel has a profound effect on blood pressure. **Vasodilators** include nitric oxide (see Box 6.1) and certain prostenoids (e.g. prostaglandin I_2 (prostacyclin) and prostaglandin E_2), whilst **vasoconstrictors** include endothelin, angiotensin, certain catecholamines, vasopressin, thromboxane A_2, and prostaglandin $F_{2\alpha}$. Indeed, high blood pressure (hypertension) can be the result of the failure of the endothelium to correctly regulate vascular tone, due either to overactivity of vasoconstrictors (too much endothelin) and/or underactivity of vasodilators (too little nitric acid). However, the endothelium is not the only factor influencing blood pressure. The sympathetic nervous system acts on the smooth muscle cells via nerves that feed from the outside of the blood vessel. Indeed, an entire class of drugs that interfere with this innervation (beta-blockers) are used to treat hypertension (section 7.2).

Haemostasis

A healthy endothelium resists clot formation as it is essentially anticoagulant. However, in certain circumstances, such as in acute blood loss where thrombosis is desirable, endothelial cells can become pro-coagulant and release molecules (such as von Willebrand factor, vWf) to promote clot formation and so minimize blood loss.

vWf is the major product of the endothelium that is released into the bloodstream. It is stored preformed in Weibel–Palade bodies (WPBs), and so large amounts can be released rapidly in response to an emergency, such as haemorrhage. A key constituent of the membrane of the WPBs is the adhesion molecule P-selectin. Thus, exocytosis of the WPBs results not only in release of vWf, but also in the appearance of P-selectin at the cell surface. This molecule promotes the adhesion of certain white blood cells and so possibly their passage into the vessel wall. vWf is also secreted into the vessel wall where it may help the adhesion of the endothelial cells to the basement membrane of the intima (the inner layer of the vessel, facing the lumen).

In addition to the release of vWf, endothelial cells also contribute to haemostasis by the expression of components of the coagulation pathway, described in Chapter 11. Endothelial cells also produce ectonucleotidases, enzymes that dephosphorylate ADP, via AMP to adenosine. As ADP activates platelets, the effects of these enzymes inhibit platelet aggregation and, in turn, thrombosis.

Leukocyte migration

An important response to inflammation or infection in the tissues is the movement of white blood cells such as monocytes and neutrophils from the blood into the tissues to combat the presumed or actual pathogen. The endothelium is strategically located at the blood–tissue interface and is essential to the immune and inflammatory response specifically by producing and reacting to a number of the low-molecular-weight protein mediators. These factors include bacterial endotoxin, colony stimulating factors, growth factors, and **cytokines**, examples of the latter being interleukins, tumour necrosis factor, and interferon. These stimulators produce various changes in the endothelial cell, one of which is the upregulation of adhesion molecules.

Endothelial cells promote the transfer of certain leukocytes (and possibly platelets) by increasing the expression of adhesion molecules on their cell membrane. Leukocytes cooperate in this process by expressing their own complimentary adhesion molecules and other receptors involved in cell–cell adhesion. These adhesion molecules include E-selectin, L-selectin, intercellular adhesion molecule, and vascular cell adhesion molecule. They are present constitutively at the endothelial surface at low levels, whilst P-selectin appears at the surface only after exocytosis of the WPBs.

Barrier function

Endothelial cells form a barrier to regulate interstitial fluid movement between the blood and the tissues. This may be directly through the cell itself, involving endocytosis and pinocytosis, but also by the regulation of the gaps between adjacent cells. Normally, cells have tight junctions that minimize the exposure of the blood to the sub-endothelium. If cells retract from each other, this can leave an intercellular space (i.e. a gap junction), that provides a route of access for blood constituents (including potentially atherogenic lipids) to the vessel wall. It may also allow the movement of leukocytes into the blood vessel wall and beyond independently of the endothelium.

Activation of the endothelial cell

A resting endothelial cell can become physiologically activated—that is, undertake a change in function—by a number of factors. These include the rate of blood flow and blood viscosity, both of which act by changes in shear rates at the endothelial surface. The consequences of endothelial cell activation include increased release of coagulation factors, WBP exocytosis (and so release of vWf and the appearance of P-selectin at the cell membrane), the increased expression of adhesion molecules, and possible changes in barrier function.

Immunological activation can be facilitated by the presence of certain inflammatory mediators (such as interleukins) which may be secreted by, for example, macrophages or certain lymphocytes, whilst haemostatic activation may be mediated by molecules such as histamine and thrombin. Histamine also increases vessel permeability to blood-borne products by rapidly and reversibly 'detaching' the endothelial cells from each other.

The endothelium also has the capacity to self-activate. Vascular endothelial growth factor (VEGF) can be released by endothelial cells, and has endocrine, paracrine, or autocrine (i.e. acting on itself) activity in promoting endothelial cell growth and the development of new blood vessels, a process known as angiogenesis. Indeed, endothelial cells about to participate in angiogenesis upregulate receptors for VEGF and other growth factors such as angiopoietin. An important aspect of endothelial cell activation is that it is reversible upon cessation of the stimulants. However, if the endothelium is activated for a long time period, it may not recover its normal function and so may become damaged.

SELF-CHECK 6.1

What is the molecular means by which the endothelium reduces blood pressure?

Damage to endothelial cells

The physiological processes we have just examined are sensitive to the health of the endothelium. Unfortunately, these cells are the target for the disease process in atherosclerosis and are attacked by the four risk factors for this disease (smoking, hypertension, dyslipidaemia, diabetes) and by pathogenic microbes. The consequence of this is endothelial dysfunction, which becomes manifest in the endothelium's failure to perform its physiological functions, such as in correctly regulating vascular tone and haemostasis. A damaged endothelium is less likely to be able to respond to those stimuli that control blood pressure; it may, for example, fail to upregulate eNOS and so generate and release NO.

It seems likely that a damaged endothelium will be less able to correctly regulate fluid passage into the vessel wall and tissues, possibly leading to oedema. A further consequence of damage to the endothelium is that cells will lose their ability to adhere to their basement membrane and so may be driven from the vessel wall by the blood flow. Indeed, circulating endothelial cells can be detected in the blood of patients with atherosclerosis and other conditions associated with endothelial damage, including inflammatory conditions such as vasculitis. Loss of endothelial cells exposes the subendothelium to flowing blood, which will permit the passage of fluids into the tissues, leading to **oedema**, and may also promote the adhesion of platelets and so the establishment of a thrombus.

There may also be organ-specific changes. Damage to endothelial cells of the cerebral circulation may lead to cerebral oedema, and possibly contribute to stroke, whilst damage to the endothelium of the glomerulus is likely to lead to renal failure and proteinuria. Loss of function of endothelial cells of the retina and its blood supply may cause visual disturbances and, ultimately, blindness.

Damage to the endothelium also results in the increased exocytosis of the WPBs and so increased plasma vWf that promotes thrombosis. The disease process may also cleave thrombomodulin from the cell surface, so that this molecule is no longer able to sequester thrombin, which therefore remains active. All these changes promote thrombosis. Further evidence of endothelial damage is the appearance of endothelial cells in the blood, and increased numbers of microparticles (essentially, cell membrane vesicles). Table 6.1 and Figure 6.1 summarize key features of endothelial cell biology. These will be revisited in section 6.4.2.

Key points

Endothelial cells have a number of functions, such as haemostasis and control of blood pressure, disruption of which leads to the initiation of atherosclerosis.

TABLE 6.1 Key features of endothelial cells

Feature	Physiology
Blood pressure regulation	Secretion of vasodilators (NO, prostacyclin) and vasoconstrictors (thromboxane, endothelin)
Haemostasis	Secretion of procoagulants (vWf, tissue plasminogen activator, factor V) and anticoagulants (plasminogen activator inhibitor)
Leukocyte migration	Expression of adhesion molecules such as E-selectin and vascular cell adhesion molecule
Barrier function	Maintains tight junctions to part-regulate tissue fluid movement through the vessel wall
Activation	Responsive to inflammatory cytokines that part-regulate other functions e.g. shift in the balance of haemostasis

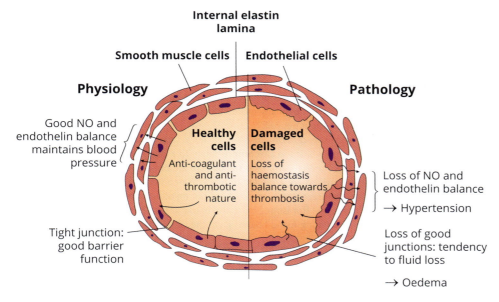

FIGURE 6.1
The endothelium in health and disease.

6.1.2 Smooth muscle cells

There are three types of muscle cells (myocytes): smooth muscle cells, skeletal muscle, and **cardiomyocytes**, the latter being the subject of section 6.1.3. Smooth muscle cells are so called because they lack the striations present in skeletal muscle and in cardiomyocytes. Other major features are that their contractions and relaxations are involuntary (meaning they have to be stimulated to do so), and that they are unbranched. Individual cells are spindle shaped, and, compared to skeletal and cardiac muscle, are slow to contract.

Smooth muscle cells are found in the walls of many organs, such as of the gastrointestinal tract (oesophagus, stomach, small and large intestines (where they drive peristalsis)) and genito-urinary system (ureters, bladder, urethra, and uterus). They are also found in the skin, where they are responsible for erecting hair. The media of arteries and (if present) of veins are composed mostly of (vascular) smooth muscle cells, and these cells provide physical strength to the vessel, and contract and relax to regulate blood pressure and the flow of blood.

The cytoplasm of the individual smooth muscle cell is dominated by two proteins: actin and myosin. Arranged in sheets and chains, they have the capacity to contract, according to the 'sliding-filament' theory. By fixing sections to the cell membrane (the sarcolemma), the entire cell will contract. Other proteins are calmodulin, tropomyosin, calponin, and caldesmon, and of these, calmodulin is involved in regulation of contraction. Instructions for the actin–myosin complex (and so the cell) to contract come from the autonomic nervous system, from catecholamines such as adrenaline (as smooth muscle cells have alpha-adrenergic receptors), and from the endothelium (endothelin). Upon receipt of these signals, intracellular second messengers such as calmodulin regulate levels of calcium. These in turn influence the action of kinase and phosphorylase enzymes that ultimately provide the ATP needed to enable the actin and myosin fibrils to contract.

Relaxation of smooth muscle cells can be mediated by factors such as nitric oxide and prostacyclin, which by different routes lead to a reduction in intracellular calcium. This leads, via a convoluted pathway, to the uncoupling of the actin–myosin filaments, which slide apart and so allow the cell to relax. Apart from their contractile properties, smooth muscle cells (like fibroblasts) have the capacity to secrete a large number of connective tissue components, including elastin, collagen, and proteoglycans. Fibres of these structural proteins also provide physical strength and elasticity to the media.

6.1.3 Cardiomyocytes

In common with smooth muscle cells, cardiomyocytes are involuntary, and in common with voluntary (skeletal) muscle, they have striations. However, cardiomyocytes have two unique features: they are branched, and they expand and contract spontaneously, even when isolated and grown in tissue culture. Unlike skeletal and smooth muscle cells, cardiomyocytes are permanently contracting and relaxing, and so demand considerable resources of oxygen and a source of energy, mostly fats and glucose. They are also characterized by many more mitochondria than are present in other muscle cells, and by having only one nucleus per cell.

The plasma membrane of the cardiomyocyte is called a sarcolemma, and the cytoplasm is referred to as the sarcoplasm, whilst the **sarcomere** refers to a basic unit of this type of muscle tissue. The major contractile material within the sarcoplasm are bundles of myofibrils, themselves composed of myofilaments, which are arranged in blocks (sarcomeres) and separated from each other by Z bands. Within each sarcomere, areas where the myofibrils are dense are called A (for anisotrophic) bands, whereas an I (for isotrophic) band marks areas where the myofibril density is low. This combination of alternating A and I bands gives the cardiomyocytes (and skeletal muscle) its striated appearance.

Myofibrils are composed of a complex of strands of proteins actin, myosin, tropomyosin, and troponin. The combination of these proteins produces a helical structure that allows them to slide over each other. Since the filaments are anchored to the Z lines, this has the effect of drawing the Z lines together, which produces the contraction of the cell. The importance of troponin in the diagnosis of myocardial infarction is discussed in section 8.2.2.

6.1.4 Other cells

The defining features of pericytes are still unclear, one principal reason being that they are present in small numbers in different tissues and are uniformly difficult to isolate and so characterize. They are thought to be derived from a **mesenchymal** cell precursor that has the capacity to differentiate into vascular smooth muscle cells, dependent on the direction of growth factors. Nevertheless, pericytes may be involved in different aspects of vascular physiology. For example, in the cerebral circulation, pericytes may be important in the integrity of the blood–brain barrier, and actively communicate with other cells of the neurovascular unit such as endothelial cells, astrocytes, and neurons. The presence of cell membrane receptors for molecules such as platelet-derived growth factor (PDGF), transforming growth factor-β (TGF-β), VEGF, and angiopoietin (another growth factor) on the surface of pericytes also implies a role or roles in angiogenesis. Furthermore, the presence of smooth muscle actin, non-muscle myosin, tropomyosin, and desmin in some pericytes suggests a structural role, perhaps an involvement in blood pressure control, and so a possible role in the regulation of vascular tone. There is also evidence that other pericytes can transform into fibroblasts or macrophages, if required, and it is suggested that pericytes have stem cell activity and can transform into muscle, bone, and skin cells, giving them possible roles in tissue regeneration after injury. One possible explanation for this multitude of functions is that pericytes should not be considered as a single cell type, but (as with macrophages) as cells that can have a number of diverse phenotypes.

Fibroblasts, arising from mesenchymal tissue, were once viewed as being uniform, with similar functions regardless of their particular anatomical site, be it the skin, bone marrow, lungs, or the heart. However, it is now clear that fibroblasts are phenotypically diverse and have a number of different properties dependent on the particular tissues and organs in which they are found. However, fibroblasts are often defined on morphological criteria—that is, that they are flat, spindle-shaped cells with multiple processes growing out of the main body of the cell in contact with neighbouring cells.

As regards the cardiovascular system, fibroblasts (like some macrophages) secrete a number of connective tissue molecules such as collagen, fibronectin, and elastin, all likely to provide support. Cardiac fibroblasts can be differentiated from other cells of the heart (myocytes, endothelial cells, vascular smooth muscle cells) by the heart-specific expression of a collagen receptor named DDR2. Other potential fibroblast markers include fibroblast-specific protein-1, fibroblast activation protein, and cadherin-11. In addition to cytoplasmic and cell-surface markers, genes highly expressed by cardiac fibroblasts include those for vimentin, beta-1 integrin, fibronectin, and connexins.

Adipocytes, often present near cells and organs which are metabolically very active, are able to both synthesize and store a variety of lipids. It seems likely that the fats and lipids stored within adipocytes are a source of energy. Indeed, there is increasing evidence that the heart draws a significant amount of energy from the fat stored in adipocytes.

Macrophages are often found within the matrix of the blood vessel, where they are assumed to perform tasks related to immunological surveillance and the removal of debris, and of dead and dying cells. They may also have a role in remodelling the connective-tissue aspects of the vessel. Further aspects of the function of these cells are discussed in Chapter 9.

SELF-CHECK 6.2

List major features of the other cells of the cardiovascular system.

Key points

The cardiovascular system is formed from several different types of cells (endothelial cells, smooth muscle cells, cardiomyocytes, pericytes, fibroblasts, adipocytes, and macrophages), each with specific functions.

Having looked at different types of cells, we now examine how these individual cells come together to form the functioning vessels and organs of the cardiovascular system.

6.2 Vessels of the cardiovascular system

Figure 6.2 shows a simple layout of the heart and circulation. The two types of vessels of the cardiovascular system are those that carry blood, and those that carry lymph. As blood passes through the kidneys, the proportions of different ions and pH are regulated, and waste material is excreted. This organ also regulates blood volume, which largely involves the amount of water in the blood. Deoxygenated blood returns from the lower part of the body and the upper part of the body and the head to the heart via veins—the inferior vena cava and superior vena cava respectively. Blood vessels carry blood around the body in two independent systems. In the pulmonary circulation, oxygenated blood leaves the lungs and passes via the pulmonary vein to the left side of the heart where it enters the left atrium. The heart then drives this oxygenated blood, in a continuous cycle of beats, around the body and the head via the principal artery, the aorta. Some blood passes into the digestive organs where nutrients are collected, whilst other blood is transported to the brain. The circulation is completed by blood moving from the right side of the heart to the lungs via the pulmonary artery. The vessels of the lymphatics do not form a circulation, but effectively drain the tissues of interstitial fluid. The lymphatics also act as a transport system for lymphocytes and link lymph nodes.

6.2.1 The arterial circulation

The aorta is the major artery, leaving the left ventricle and taking oxygenated blood away from the heart to the head and body (Figure 6.2). The first crucial branch of the aorta gives rise to the coronary arteries that feed the myocardium (not shown in Figure 6.2). Other arteries branch off the ascending part of the aorta, becoming the left and right carotid arteries (running up the neck to feed the brain) and the left and right sub-clavian arteries (feeding the shoulders and arms).

The descending part of the aorta moves down through the thoracic cavity and into the abdomen, where the left and right renal arteries feed the kidneys, and mesenteric arteries feed the intestines. In the lower abdomen the aorta divides to give the left and right iliac arteries. As each iliac artery feeds its particular leg, it becomes, in turn, the femoral, popliteal, tibial, and ultimately the small arteries of the foot.

In the pulmonary circulation, deoxygenated blood leaves the heart via the right ventricle and moves to the lungs via the left and right pulmonary arteries. As the major arteries move further away from

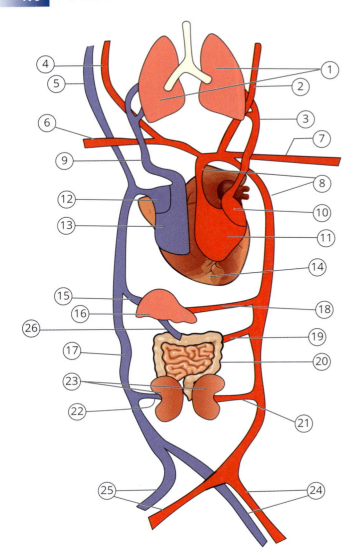

FIGURE 6.2

A simple representation of the circulation and selected organs. Red: oxygenated blood. Blue: deoxygenated blood. 1 = the lungs, 2 = left (L) carotid artery, 3 = pulmonary vein, 4 = right (R) carotid artery, 5 = superior vena cava, 6 = R subclavian artery, 7 = L subclavian artery, 8 = aorta, 9 = pulmonary artery, 10 = L atrium, 11 = L ventricle, 12 = R atrium, 13 = R ventricle, 14 = myocardium, 15 = hepatic vein, 16 = liver, 17 = inferior vena cava, 18 = hepatic artery, 19 = mesenteric artery, 20 = intestines, 21 = renal artery, 22 = renal vein, 23 = kidneys, 24 = L iliac and femoral arteries and veins, 25 = R iliac and femoral arteries and veins, 26 = hepatic portal vein.

the heart to deliver blood to the organs and tissues of the body, they divide many times and become progressively smaller, turning into arterioles.

The major driving force of the heartbeat is delivered by the left ventricle, and to enable this it has the greatest mass of cardiomyocytes of all the chambers. Blood leaving the left chamber through the aortic valve and into the aorta will be at a high velocity and pressure. This high pressure must be maintained if the blood is to be distributed throughout the body. Accordingly, the aorta and its branches have a thick muscular and elastic wall to enable this high pulse pressure to be maintained as the arteries become increasingly narrow and transform into arterioles. Blood leaving the heart for the lungs from the right ventricle is at lower pressure, but the relevant vessel, the pulmonary artery, still needs a thick wall. The largest arteries—those closest to the heart, such as the aorta—are likely to have a diameter of up to 20 mm. Arteries further away from the heart, such as the femoral arteries in the legs, may have diameters of 5–10 mm. The structure of a typical artery has three distinct layers. The innermost, the intima, interfaces with the blood. The outer layer is the adventitia, and between them lies the media (Figure 6.3).

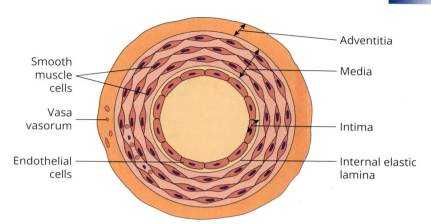

FIGURE 6.3
Structure of an artery. The microstructure of an artery consists of three concentric layers, the intima (endothelial cells), the media (mostly smooth muscle cells), and the adventitia (loose supportive connective tissue hosting the vasa vasorum (if present)).

The intima (or tunica intima)

The intima consists of a single layer of endothelial cells, beneath which is a basement membrane (the internal elastic membrane, also called a basal lamina), composed of connective tissue components such as collagen, actin, and elastin. The internal elastic lamina provides not only support but also a barrier to minimize the unregulated passage of fluids into the artery wall and (in capillaries) into the tissues. As the interface between the blood and the artery wall, as we have seen, the intima has key roles in haemostasis, leukocyte migration, and regulating blood pressure.

The media (or tunica media)

The thickest layer of the artery, the media, consists of smooth muscle cells interlaced with some elastic fibres such as collagen and elastin. The arrangement of these cells and fibres (some of which actually form sheets, or lamellae) is of concentric circles. These can lie perpendicular to the flow of blood, although some smooth muscle cells may lie parallel with the direction of blood flow. The degree to which these smooth muscle cells constrict and dilate, in response to vasoconstrictors and vasodilators respectively, and also sympathetic innervation, has a direct effect on blood pressure, a process known as 'tone'.

The smooth muscle cells and elastic fibres together confer strength, elasticity, and contractability, as arteries (and arterioles) need to be able to respond to the differences in blood pressure during the cardiac cycle and brought about by factors such as exercise. During systole, when blood is forced from the heart by the contraction of the ventricles, the extra pulse of blood must be accommodated by the arteries. This is enabled by the elastic properties of arteries, a feature known as compliance.

Vascular tone is controlled not only by the small molecular vasoconstrictors and vasodilators released by the endothelium, but also by the effects of sympathetic innervation by the autonomic nervous system. The interaction between the neurons and the blood vessels is mediated in most cases by synapses bearing beta-type receptors (alpha receptors are common in other tissues such as the ureter), the activation of which results in vasoconstriction and so the maintenance of vascular tone. An important class of drugs used after a myocardial infarction to reduce blood pressure are beta-blockers. These inhibit the messages passed from beta-receptors to the smooth muscle cells, so that the latter fail to constrict as firmly, which translates to lower blood pressure and so less stress on the heart. The external surface of the media of many arteries is bounded by an elastic membrane.

The adventitia (or tunica adventitia)

This outer layer is composed of a loose amalgam of connective tissue, such as elastic and collagenous fibres (possibly secreted by fibroblasts), which may bind to nearby tissues and so provide physical support. Adipocytes may also be present, and it is tempting to speculate that these serve as reservoirs for fat destined to be a source of energy for the smooth muscle cells of the media. Indeed, non-esterified fatty acids are the second most important energy substrate of the heart, but it is unclear if the same is true for arteries. The definition of the outer boundary of the adventitia is arbitrary.

In small arteries, the smooth muscle cells of the media are likely to be able to obtain oxygen and nutrients, and dispose of waste material, passively by diffusion. However, in the bigger arteries, the substantial mass of smooth muscle cells will not be able to obtain all its requirements or remove waste by simple diffusion alone. The problem is solved, curiously, by large arteries having their own blood supply. This consists of arterioles and venules found in the adventitia, termed vasa vasorum. These form branches that burrow into the media, and so deliver blood to the smooth muscles. The nervous supply to the smooth muscle cells of the media, which also part-regulate contraction and relaxation, and so vascular tone, is also found in the adventitia.

Arterioles

As arteries move further away from the heart, they become progressively smaller, and typically have a diameter of 0.1–0.5 mm. This is reflected by the more refined structure of arterioles, which have far fewer smooth muscle cells in the media, perhaps only two or three layers. However, these smooth muscle cells still contribute to the regulation of blood pressure. In the smallest arterioles there may only be a single layer of smooth muscle cells, and these are less likely to form the typical concentric layers of the larger arteries but may instead be arranged in a spiral pattern. There is also much less adventitia, although the endothelium remains present. Endothelial cells of arterioles retain the ability to influence vascular tone (perhaps under the control of sympathetic innervation) and to participate in haemostasis. The vasoconstriction of arterioles may be regulated, in part, by pericytes. As the arterioles subdivide, they lose the media of smooth muscle cells and ultimately give rise to capillaries.

SELF-CHECK 6.3

Name one feature of each of the three layers of an artery.

6.2.2 Capillaries

Capillaries are the smallest blood vessels, with a diameter of perhaps 4–15 μm, and consist only of a tube of endothelial cells on a basement membrane, with occasional pericytes. Lacking smooth muscle cells, they do not contribute to vascular tone, but are still able to express and/or release molecules important in haemostasis and in regulating leukocyte migration into the tissues. Given capillaries' thickness of only a single cell (i.e. the endothelial cell), nutrients are able to pass easily from the blood into the tissues. Similarly, waste products such as carbon dioxide can pass in the reverse direction, from the tissues into the blood.

The capillary endothelium regulates the passage of certain components of plasma from the blood into the tissues; the components form what is called tissue fluid, or interstitial fluid. Capillaries may not always be present as single vessels, as they often form large networks of vessels, and as such are described as a capillary bed. Sometimes described as microcirculation, the major capillary beds are found in the liver, skin, lungs, and brain. The capillary beds have site-specific functions; for instance, the cells in that of the lungs must allow the passage of oxygen and carbon dioxide.

We recognize three different patterns for the structure of a capillary (figure 6.4):

- Type 1 is the most frequent, the endothelium forming a continuous and complete layer of cells over the basement membrane. In this setting the capillary is essentially an arteriole free of a media. Adjacent endothelial cells often form 'tight junctions' between themselves to resist or perhaps control the flow of materials into the tissue.

- In type 2, there are occasional gaps or pores between individual endothelial cells (fenestrations) which allow the passage of materials into the tissues to be even easier. In these situations, the basement membrane is exposed to flowing blood. Fenestrations, which are perhaps

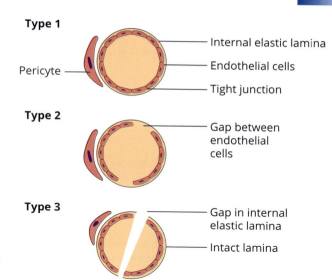

FIGURE 6.4
The structure of capillaries. The three types of capillary are characterized by the presence of pericytes and the extent to which there are gaps between endothelial cells and in the basic internal lamina.

70–100 nm in diameter, may therefore be a well-regulated method for controlling the movement of tissue fluid.

- A third structural variant, type 3, is of a further aid to the free passage of material—the lack of a basement membrane altogether. Thus, blood interfaces directly with the underling cells and there is no barrier to the transfer of nutrients or interaction with the cells of the blood.

All three classes of capillaries may be supported by pericytes, but the precise role of these latter cells (not withstanding section 6.1.4) is unclear. Nevertheless, the intimal relationship between endothelial cells and pericytes implies a role such as regulation of endothelial function, for example in angiogenesis. There is no evidence that pericytes are involved in any haemostatic activity of capillary endothelial cells.

6.2.3 The venous circulation

The venous circulation carries blood from the tissues and the lungs to the heart (Figure 6.2). In the pulmonary circulation, oxygenated blood leaves the lungs via the pulmonary vein and is delivered to the left atrium. In the peripheral circulation, deoxygenated blood from below the heart (the lower chest, abdomen, and legs) is delivered to the right atrium by the interior vena cava. Blood returning from above the level of the heart (the upper chest, arms, and head), is delivered by the superior vena cava. Having fed oxygen- and nutrient-rich blood to the heart via the coronary arteries, deoxygenated and nutrient-depleted blood returns from the myocardium via coronary veins, culminating in the coronary sinus, which also empties into the right atrium. In parallel to the arterial circulation, these large veins are fed by many small veins, which themselves are formed by the confluence of many even smaller veins—the venules.

Venules

Venules are formed when several capillaries merge. Those venules closest to the capillaries are composed of an intima of endothelial cells lying on an elastic internal membrane, with an external adventitia. The functional status of these endothelial cells in haemostasis and leukocyte migration is unknown. As several venules merge and the lumen increases, the larger size of the vessel needs to be supported by a media of smooth muscle cells.

   Having passed through a capillary bed, the force of blood pressure has been spent, and the flow of blood is weak. As additional venules join together, blood flow can be a problem as, in the absence of

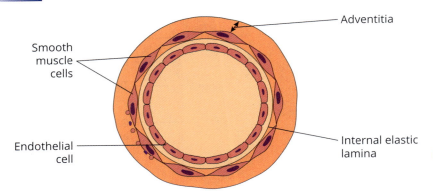

FIGURE 6.5

Cross-section of the structure of a typical vein.

blood pressure, blood may actually flow backwards (retrograde flow) from the venules to the capillaries. At the point at which this becomes haemodynamically important, retrograde flow is prevented by valves, and the large venules therefore become veins. Pericytes have been described in post-capillary venules, but their function needs to be clarified.

Veins

Veins are essentially large venules with valves and a thicker wall. As the vein increases in diameter, the media becomes thicker, with more smooth muscle cells. However, the media lacks elastic fibres that contribute to the strength of the arterial walls as venous blood pressure is low. With greater blood flow, the presence of valves becomes crucial to prevent retrograde flow of blood, especially in the veins below the level of the heart as gravity must be countered. Conversely, veins above the level of the heart (e.g. the jugular veins of the neck) have few valves (if any) as blood can return to the heart by gravity. The valves themselves are composed of fibrous tissue with modified smooth muscle cells covered by endothelial cells. Venous endothelium appears to have little influence on smooth muscle cells but does have a role in the haemostasis (Figure 6.5).

As with other large vessels, an adventitia holds the vein in place, fixed to nearby tissues. However, because there are fewer smooth muscle cells in the media, and those that are present are less meta-bolically active, then in many cases, vasa vasorum, if present, are less well developed than they are in arteries. With increasing age, and possibly because of mechanical factors, the valves become weak and possibly incompetent. This, accompanied by a reduction in the strength in the walls, leads to tor-tuous and dilated vessels (varicose veins), especially in the legs, and sluggish or even static blood flow. Indeed, deep-vein thromboses are often anchored near valves.

6.2.4 The lymphatic system

Blood consists of cells and plasma—the latter effectively water in which ions and molecules of vari-ous size and function are dissolved. In health, the cells (red blood cells, platelets, and most white blood cells) remain in the blood. However, a portion of the plasma moves across the capillary wall and into the tissues, often driven by the physics of high blood pressure. This filtered fluid, rich in nutrients and oxygen and lacking large proteins, is called interstitial fluid. Because of the relatively high pressure at the arteriolar ends of the capillaries, this tissue fluid is constantly being formed. Some fluid returns to the circulation via a specialized network of capillary-like vessels called the lymphatics. Once within the lymphatic system, the fluid is called lymph. Lymphatic vessels consist of endothelial cells on a basement membrane (as do capillaries), and the vessels may be anchored to nearby tissues by filaments of connective tissue such as collagen which are likely to be secreted by fibroblasts.

The lymphatics drain lymph from the tissues, and—in a manner analogous to capillaries, venules, and veins—converge to form larger vessels. In the lower part of the body, the lymphatics merge for a large vein-like vessel, the thoracic duct, which eventually empties into a vein near the heart.

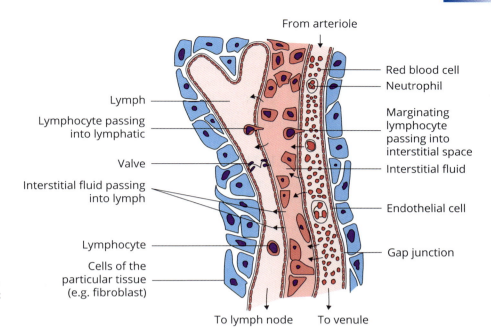

FIGURE 6.6
Structure of a lymphatic vessel.
Left: a lymphatic vessel. Centre:
interstitial tissue. Right: a
capillary.

Lymphatics draining the right side of the head, neck, and chest merge into the right lymphatic duct
that also empties into a vein near the heart.

In common with veins, lymphatics have valves to prevent retrograde flow. However, the lymphatics differ from blood vessels as the flow passes along a chain of small bodies known as lymph nodes, and these bodies are often fed by several lymphatics that drain a particular anatomical region. These masses are filled with lymphocytes and some macrophages and fibroblasts. Indeed, the lymphatics form not only a conduit for lymph but also a pathway for lymphocytes moving from lymph node to lymph node. Consequently, the lymphatics and lymph nodes are crucial in inflammation and immunity, as discussed in Chapter 9. In the gastrointestinal tract, lymphatics carry lymph rich in fats.

Figure 6.6 illustrates key features of the lymphatic system, and Table 6.2 summarizes key features of all the circulatory vessels.

SELF-CHECK 6.4

Table 6.2 shows differences between the lymphatics and blood vessels. How do these differences reflect their function?

TABLE 6.2 Key features of circulatory vessels

	Arteries	Veins	Capillaries	Lymphatics
Fluid pressure	High	Low	Low	Low
Structure of the wall	Thick	Moderate to thin	Thin	Thin
Direction of fluid flow	From the heart	To the heart	Intermediate	To the heart
Valves	Absent	Present	Absent	Present
Oxygenation	High	Low	Intermediate	Low

6.3 The heart

The heart is undoubtedly the most important organ in the body, certainly in an acute setting, and unlike almost all other organs, failure of the heart to perform its function can lead rapidly to death. A hollow, muscular organ lying in the thoracic cavity and protected by the sternum, it weighs perhaps 280–340 grammes in the adult male and 230–280 grammes in the adult female. The purpose of this organ is to provide the lungs, head, and body with over 3,500 litres of blood daily by the action of over 100,000 separate beats.

6.3.1 The structure of the heart

Almost all of the heart is composed of millions of highly specialized muscle cells, cardiomyocytes, which together make up the large mass of muscle—the myocardium—in which are scattered macrophages and fibroblasts. The external surface of the myocardium is bounded by a layer of fibrous connective tissue, the epicardium. This layer interfaces with the pericardium, a fibrous double-layered non-cellular sac, which also helps to keep the heart in place within the thoracic cavity. A small amount (~50 ml) of pericardial fluid minimizes friction between these fibrous layers as the heart beats. The internal surface of the myocardium, which interfaces with and so lines the chambers, is composed of a single layer of endothelial cells, and so is called the endocardium. These endothelial cells are continuous with the inner linings of the blood vessels that leave the heart, and also cover the valves and the tendons that hold the valves open.

The heart is of course a highly dynamic functioning organ in itself, and so needs to be supplied with blood. The high metabolic demands of a constant supply of oxygen and an energy substrate (such as glucose), and the removal of waste products (such as carbon dioxide), cannot be met by simple diffusion alone, so must be delivered by a separate coronary circulation. This blood is supplied by arteries, arising from the aorta, that run on the external surface (the epicardium), and then divide and burrow into the myocardium. The circulation is completed by cardiac veins on the epicardium, which carry the deoxygenated, nutrient-depleted, but waste product-rich blood back to the right atrium. A healthy heart also has modest fat deposits, formed from clusters of adipocytes, on the outside of the heart, through which the epicardial arteries and veins run. This adipose tissue may be a reservoir of fat that can serve as a source of energy but may also secrete **adipocytokines** that act on nearby vessels.

Although cardiomyocytes spontaneously contract individually or in bundles, when grouped together by the million in the myocardium, their contractions must be tightly regulated and coordinated in order for the heart to beat correctly. The stimulus for the heartbeat comes from two sets of highly specialized tissues located on the epicardium—the **sinoatrial node** and the atrioventricular node. From the cell biology viewpoint, these cells are difficult to accurately classify, but seem likely to have developed from nervous tissues. At the top of the heart are the left atrium and the right atrium (plural: atria), below which lie the left ventricle and right ventricle. The two ventricles are separated by a septum (Figure 6.7).

Key points

The heart is a highly complex organ whose dysfunction, in terms of rate, rhythm, and volume of blood pumped, can lead rapidly to serious disease and death.

6.3.2 The function of the heart

It is convenient to explain the function of the heart by looking at one complete cycle of a heartbeat, often called the cardiac cycle. The regulation of the beat is complex as the four chambers must act in a well-defined sequence. The stimulus for the first part of the cycle is the activation of the sinoatrial node. This activation is spontaneous and is associated with an electrical impulse that results in the

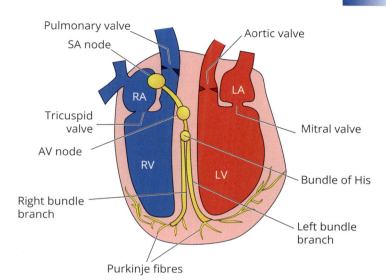

FIGURE 6.7

The heart, showing valves and conduction tissues. Valves separate the atria and ventricles, and the ventricles from the circulation. The conduction tissues control the correct beating of the four chambers, and part-regulate heart rate. SA = sinoarticular, AV = atrioventricular.

controlled contraction of muscles of the right and left atria. These chambers will have been provided with blood from the superior and inferior vena cavae, and from the pulmonary vein respectively. Blood then passes through valves (the tricuspid on the right and the bicuspid, or mitral, on the left) into the right and left ventricles.

The sinoatrial node acts not only on the myocardium of the atria, but also on a second specialized group of cells located near the junction of the right atrium and right ventricle. This second cluster is the **atrioventricular node**, and once activated, it in turn passes an impulse to the bundle of His. The impulse is then transmitted into the top of the septum, and the fibres then divide into the right bundle branch and the left bundle branch. Thus, the impulse message is transmitted down the septum that separates the left and right ventricles, to the muscles of the right and left ventricles.

The terminal sections of the bundle branches, those that actually contact the cardiomyocytes, are Purkinje fibres. These specially differentiated cardiac muscle fibres are not to be confused with cerebellar Purkinje cells. The atrioventricular node, the bundle, its left and right limbs, and the ventricular Purkinje fibres are surrounded by a connective tissue sheath which insulates them from the neighbouring myocardium.

Upon receipt of the impulse the cardiomyocytes of the right and left ventricles contract, forcing blood out of the right side of the heart through the pulmonary valve into the pulmonary artery, and from the left side of the heart through the aortic valve into the aorta, respectively. Once the ventricles have relaxed, the cardiac cycle is complete and another begins. Systole is the term given to the phase of the cardiac cycle when the chambers contract, diastole is the term used when the chambers relax. These two phases give us expressions for measuring blood pressure in arteries of the arm: systolic blood pressure and diastolic blood pressure. Sinus rhythm describes the normal sequence of the four chambers working together, the key being the microsecond delays between the impulse messages that cause the regulated contractions. Failure of this regulation can easily lead to life-threatening cardiac disease, such as when sinus rhythm is replaced by an abnormal pattern of beat. This is known as an arrhythmia, the most common being atrial fibrillation (AF), which is discussed in section 8.3.3.

6.3.3 Control of the heart rate

The sinoatrial node is the major pacemaker, which spontaneously delivers impulses at a rate of 60–100 per minute to control the rate at which the heart beats. However, this rate is also influenced by the action of nerves. A branch of the tenth cranial nerve (CN X, the vagus nerve), part of the parasympathetic nervous system, innervates the heart and acts on the sinoatrial and atrioventricular nodes to decrease the rate and strength of the contractions. A sympathetic nerve supply, the cardiac nerve acts to increase the rate and strength of the contractions. A further regulator is a group of cells embedded in the wall of the carotid artery which, when faced with increasing blood pressure, act to reduce the rate and force of the beating of the heart.

The combination of the heart rate and the volume of blood expelled from the left ventricle (perhaps 70 ml) gives the cardiac output. So if a heart beats 75 times a minute, then the cardiac output is 5.25 litres/minute.

The heart rate can also be influenced independently of the nervous system by other factors such as sex (the heart beats faster in women compared to men), temperature (faster when the temperature rises), age (slower with increasing age), thyroid hormones, adrenalin (as we experience during the 'fight or flight response', when a surge of adrenalin causes our heart rate to increase), and the concentrations of certain ions in the blood. If the heart rate is too high (tachycardia), the chambers may not have enough time to fill with sufficient blood, and so provide a poor blood supply to the circulation. Conversely, if the heart rate is too slow (bradycardia), although the chambers may be full of blood, the same insufficiency of supply may arise because of a small number of beats per minute. Both conditions need to be addressed.

SELF-CHECK 6.5

Describe the different types of cells that make up the heart.

6.4 The pathogenesis of atherosclerosis

The pathogenesis of atherosclerosis is complex and develops over years, possibly decades, as the coronary arteries of apparently healthy young men killed in armed conflicts demonstrate atherosclerotic changes. It is influenced by several independent pathological processes—the major risk factors of smoking, hypertension, dyslipidaemia, and diabetes. Although not all of these morbidities are required for the disease to develop or to progress, there is certainly a cumulative effect. This brings us to the concept that each individual has their own specific total cardiovascular risk, which has clear predictive value in determining major disease at some time in the future.

6.4.1 Total cardiovascular risk

The link between risk factors and actual disease is damage to the endothelium. The development of the risk factor hypothesis benefited from the clarity of early studies that focused on only one risk factor, to the exclusion of others. Subsequent studies showed that additional risk factors added mathematically to the likelihood of developing cardiovascular disease—that is, the more risk factors present, the greater the risk. Indeed, a significant proportion of patients have more than one risk factor, such as concurrent dyslipidaemia and diabetes, making management more complicated. Other risk factors, such as the presence of existing cardiovascular disease, and other clinical abnormalities such as left ventricular hypertrophy may be added to the refinement of general equations that are becoming increasingly successful at predicting an adverse event such as a **myocardial infarction** or stroke.

Risk factor equations

Much of our understanding of the risk factors, especially lipids and hypertension, has come from a lengthy epidemiology study from the town of Framingham in the USA (as outlined in section 2.7.2). This and other studies have led to the development of general tools for assessing any individual's personal risk of suffering a major cardiovascular event (such as a myocardial infarction or stroke) in the coming decade. Further refinement of the risk factor theory has allowed the evolution of this tool so that it can be applied to the general population, as they may be seen at their general practitioner. These tools are now widely applied and can be accessed at no cost online. The purpose of these models is to identify those at greatest risk, and so with the greatest need for targeted treatment. For example, a total cholesterol level that would not put a young man at risk, and so is unlikely to be treated, may certainly warrant treatment in his grandfather. This is because age is a major factor in these risk factor equations. Notably, none of these risk factor generators consider inflammation. Practical aspects of this topic are discussed in Chapters 7 and 8.

The product of risk factors

The consequence of risk factors is atherosclerosis. It has long been known that the walls of coronary arteries of many who die of a heart attack are infiltrated with lipids, so that the arteries are distended and narrowed. Early researchers found that most of this lipid resembled gruel, or, in Greek, *atherae*. This root combines with 'sclerosis', meaning thickening (that being thickening of the smooth muscle cells of the media), to give atherosclerosis. Recognizing the role of inappropriate platelet activity, the disease is often described as atherothrombosis. The mass of abnormal cells and other material is called an **atheroma**. However, the damaged arteries of those with heart disease represent only the final stages of the disease.

The dominant theory of the pathogenesis of atherosclerosis is of 'response to injury'. This theory states that the damage to, and/or dysfunction of, the endothelium (i.e. a lesion) is the first step from which all vascular pathology arises. Once this damage occurs, other steps follow that culminate in symptomatic, and then serious, and then possibly life-threatening disease. This powerful theory is supported by considerable evidence from a number of sources, many of which have their own specific animal models, and indicates that treatment of risk factors reduces the risk of a major cardiovascular event. The pathogenesis of atherosclerosis can be illustrated in a series of steps that, to some degree, follow an ordered pathway. However, it is likely, if not probable, that some steps are missed, bypassed, or develop at different times in different lesions.

6.4.2 The initial phase

The target organ for the disease process is the endothelium, and, as discussed in section 6.1.1, we can find clear evidence of damage to these cells in the absence of actual clinical cardiovascular disease. An unequivocal example of this is increased numbers of endothelial cells in the blood, presumed to have been damaged/destroyed by the disease process and so, having lost their ability to adhere to the substratum, driven from the vessel wall; another is increased plasma levels of endothelial-specific vWf. As we have seen, cigarette smoking is an excellent example of a noxious factor that is toxic to many cells, including the endothelium. Another injurious feature is high blood pressure, and there is evidence that the high blood glucose present in diabetes also damages endothelial cells. The extent to which inflammation is involved in the earliest stages, or whether it actively directly damages the endothelium (if at all), is unclear, but there is evidence that cytokines such as IL-1, **TNF**, and γ-interferon have many effects on these cells *in vitro*. However, whether this is also the case *in vivo* is unknown, especially early in the disease. In this situation the endothelium is said to be activated, one manifestation of which is the increased expression of leukocyte and platelet adhesion molecules which seem likely to promote the migration of these cells into the media.

A damaged endothelium is also unable to adequately participate in haemostasis and in the regulation of vascular tone, leading to the promotion of thrombosis and hypertension, respectively. Indeed, the endothelium may be the victim as well as the perpetrator of atherosclerosis. An example of this is its reduced ability to generate nitric oxide, which in health promotes low blood pressure (see Box 6.1). So the loss of nitric oxide leads to high blood pressure. Losing their normal anticoagulant characteristics, damaged endothelial cells promote thrombosis. Once the endothelium has been weakened, it becomes vulnerable to other pathology, such as losing its barrier function with loss of tight junctions between cells, which enables lipids to penetrate (or be driven into, by the high blood pressure) the basement membrane and enter the media. These initial steps are summarized in Figure 6.8.

6.4.3 The development of atheroma

With the barrier function of the endothelium disrupted (or absent), plasma and cells can enter, or are forced into, the media. Lipids—and especially low-density lipoprotein cholesterol (hence LDL)—can penetrate the basement membrane, where they accumulate. Animals fed a high-cholesterol diet develop characteristic lesions of fatty streaks that are very similar to those found in the arteries of

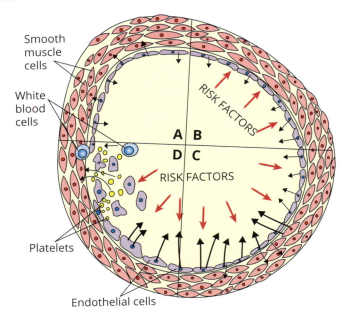

Smooth muscle cells

White blood cells

Platelets

Endothelial cells

FIGURE 6.8

Initial steps in atherogenesis. A: A healthy endothelium secreting vasodilators/vasoconstrictors into the smooth muscle layer of the media, and regulators of haemostasis into the lumen. B: Risk factors action on the endothelium. C: Changes in endothelial function into the lumen and media. D: Loss of endothelial integrity with detachment of cells from the basement membrane, exposing the subendothelial layers to platelets, white blood cells, and other components of the plasma, so enabling their movement into the media.

people in their 30s who have died from a non-medical cause such as a road traffic accident. These fatty streaks are found in the same anatomical positions as are more developed lesions in the late middle-aged and the elderly, suggesting that fatty streaks transform into juvenile atheroma, where the concentration of cholesterol can be so high that crystals are formed, which cause further cell damage. As the lesion develops, so too does the internal inflammatory and metabolic milieu. Activated immune and other cells (particularly macrophages) can produce inflammatory mediators, chemotactic factors, enzymes, and reactive oxygen species (ROS) such as the superoxide radical. Once the latter have overwhelmed local antioxidant defences, they can oxidize the LDL that has infiltrated the artery wall (hence oxLDL), and this may further damage the endothelial cell, a process that is accelerated in hypercholesterolaemia. Furthermore, oxLDLs stimulate endothelial cells and smooth muscle cells to secrete chemotactic factors to recruit leukocytes, whilst ROS can damage cell membranes.

A damaged or activated endothelium is more likely to promote the passage of white blood cells into the sub-endothelium by increasing the expression of adhesion molecules (such as E-selectin), and this may accelerate the growth of the atheroma. Monocytes are programmed to enter the tissues in order to scavenge for dead cells and damaged tissues—at which point they are described as macrophages—and are common in atherosclerotic plaques. Post-mortem studies show the presence of 'foam cells', which are macrophages laden with lipids, within fatty streaks. Once within the artery wall, macrophages can secrete a variety of factors, including inflammatory cytokines, growth factors, gelatinase, matrix metalloproteinases (MMPs, to digest the extracellular matrix and so promote cell invasion), and reactive oxygen species. These pro-atherogenic activities may also be undertaken at a later stage by invading neutrophils, although they are less likely to take up lipids. The infiltration of the artery wall by leukocytes is completed by the arrival of T, B, and NK lymphocytes, and these may contribute to an ongoing immunological lesion, with presumed humoral and cellular aspects (as fully explained in Chapters 9 and 10). The presence of dendritic antigen-presenting cells completes the requirement of a complete immunological response. In the midst of this collection of different cells, most of which will have some form of destructive capability, it is inevitable that there will be a great deal of apoptosis and necrosis, the products of which will further promote the general inflammatory situation.

Further changes include the proliferation of the smooth muscle cells of the media (driven by excess production of growth factors), a consequence of which is failure to control blood pressure (often described as hardening of the arteries). Myofibroblasts arising from smooth muscle cells accumulate in the intima and promote fibrosis by the secretion of collagens. There may be infiltration of calcium, which causes additional hardening, and which may have metabolic consequences. At this point the lipid- and platelet-rich lesion is in transition to an atherosclerotic plaque. These steps are summarized in Figure 6.9.

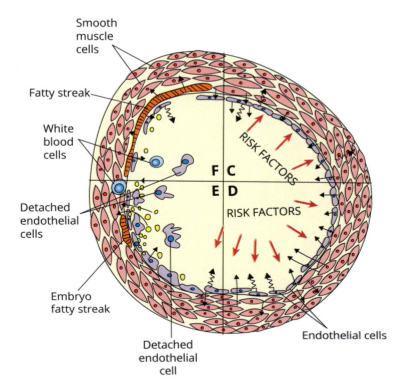

FIGURE 6.9

Intermediate steps in atherogenesis. C: Risk factors continue to act on the endothelium, and at D changes are becoming severe, with gaps between endothelial cells permitting infiltration. E: Endothelial cells are detaching, leaving spaces for platelets and lipids to enter the sub-endothelium, the lipid coalescing into a fatty streak. As this develops, at F the fatty streak extends and white blood cells, particularly monocytes, enter the lesion, take up lipids, and as such are described as foam cells.

6.4.4 The late stages

As the sub-endothelial fatty streak develops into an atherosclerotic plaque, there is an increased narrowing (stenosis) and irregularity of the lumen of the vessel. This leads to an increase in the velocity of the blood, which is turbulent, and also to locally increased blood pressure, all of which adds to endothelial dysfunction. This can also further promote thrombosis, and the fatty streak presents a barrier to endothelial cell products reaching the smooth muscle cells. As the plaque grows, its lipid-rich centre becomes necrotic, and this places physical strain on the outer margins (the shoulders) of the lesion.

SELF-CHECK 6.6

What are the roles of the fatty streak and the foam cell?

Risk factors continue to disrupt endothelial cells, which detach and expose more sub-endothelium, leading to more infiltration. This stage sees the **hyperplasia** of the smooth muscles of the media (potentially driven by platelet-derived growth factor and fibroblast growth factor), and the continued recruitment of a variety of white blood cells, although the macrophage and foam cells are prominent. When the mass of white blood cells reaches a critical mass, a local inflammatory response can develop, fed by the abnormal levels of lipid which can become oxidized. It is unclear whether the oxLDL found in the plasma originates in the plaque, or if it is generated in the blood and migrates or is driven into the vessel wall. Plaque leukocytes may also release local inflammatory cytokines and matrix-digesting enzymes, such as matrix metalloproteinase-9 (MMP-9), to further promote the general pathology of the lesion.

In most instances, the liver is the source of circulating inflammatory cytokines: it is unclear whether the inflamed plaque contributes to these messengers. Increased levels of C-reactive protein (CRP) are found in the blood, and arise in the liver, which is potentially stimulated to do so by IL-6 and TNF

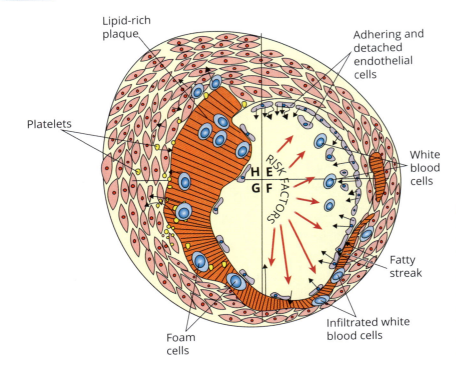

FIGURE 6.10

Late steps in atherogenesis. At G, the developing fatty streak evolves into a thick band of lipids that are detected by circulating leukocytes, which infiltrate. The smooth muscle cell layers become hyperplastic. H represents final stages, with thickened plaque and media. The luminal surface, where devoid of endothelium, can attract platelets which develop into a thrombus.

from plaque macrophages. There is a hypothesis that CRP is not merely an inflammatory marker, but (as it is present within plaque) may participate directly in **atherogenesis**, perhaps in binding to LDL and oxLDL. CRP may also activate the complement system (products of which include leukocyte chemotactins), inhibit nitric oxide, and promote thrombosis by inducing monocytes to increase tissue factor activity.

As the endothelium and the nervous system fail to control the contractions of the smooth muscle cells in the media, arterial spasm can occur, causing irregular and temporary narrowing of the blood vessel. This may further reduce blood flow down an already partially occluded artery, and stenosis is likely to lead to hypertension within the coronary arteries, further compromising their function. As the abnormality in the vessel wall continues to develop, it expands to a degree where it starts to narrow the lumen of the blood vessel, so that the blood is forced to flow more rapidly. This increase in blood flow, with increased shear stress, by itself changes the function of the endothelium to promote the steps we have already discussed. The late stages in atherogenesis are shown in Figure 6.10.

6.4.5 The mature plaque

Once mature plaque has been established, physiology responds to this site of damage by overlaying it with a protective fibrous cap of collagen and other connective tissue components, and in this state, it may not pose too many clinical problems. However, where this cap is thin and therefore weak, it may break open. This may result from physical pressure being placed on the weakened atherosclerotic plaque, perhaps due to increased blood flow and local hypertension, and/or to the underlying nature of the atheroma. This process may be promoted by proteolytic and lipolytic enzymes released by macrophages within the plaque. The nature of this fibrous cap is crucial: if sufficiently thick, it will resist disruption/digestion. However, if thin, it may be described as 'vulnerable', and more likely to rupture (perhaps precipitated by an arterial spasm, environmental factors, and/or extreme emotional disturbance), with spillage of atheroma and thrombotic material into the arterial blood stream. The pressure of blood flow will drive this atherothrombotic material downstream until it becomes lodged in arterioles and capillaries (Figure 6.11). Table 6.3 summarizes the steps in atherogenesis.

Cross reference

The devastating consequences of plaque rupture, such as a myocardial infarction or stroke, are discussed in Chapter 8.

FIGURE 6.11

A mature atherosclerotic plaque with a fibrous cap overlaying the atheroma. If thin, at the shoulders where it meets the rest of the intima/media, the vulnerable cap may rupture and attract platelets, and so a thrombus. In the worse-case scenario, the gruel-like atheroma may leave the plaque to be driven downstream by the circulation, and so become lodged in, and occlude, arterioles, precipitating an infarction.

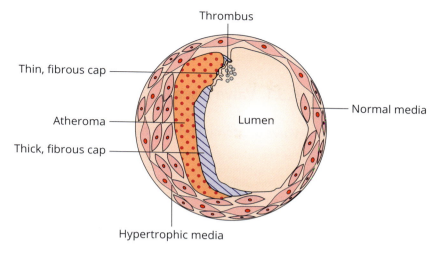

TABLE 6.3 A summary of atherogenesis

Stage	Key aspects
1: Early	Risk factors develop or are acquired: these damage the endothelium, which fails to maintain its role in blood pressure control, becomes pro-thrombotic and loses barrier function.
1: Late	The damaged endothelium creates pores that enable lipids to penetrate into the sub-endothelium and so the media. Dead or dying endothelial cells detach and appear in the blood.
2: Early	Lipids form fatty streaks, monocytes enter the vessel wall and become scavenging macrophages. Smooth muscle cell proliferation begins.
2: Late	Macrophages take up lipids and transform into foam cells. Secretion of inflammatory cytokines. Endothelial cell attack continues. Inflammatory focus develops, recruitment of lymphocytes.
3: Early	The fatty streak evolves into a lipid-rich atheroma: further recruitment of leukocytes and development of the inflammatory lesion. Potential for thrombosis over the surface of the atheroma.
3: Late	A fibrous cap forms over the atheroma, but if this is weak and/or eroded, it may rupture, spilling the contents of the lesion into the circulation.

Key points

The risk factor theory of atherosclerosis states that in addressing these factors, the risk of a major cardiovascular event will be reduced, as is the case. However, there is no evidence that any intervention can cure the disease. It is merely suppressed.

6.5 Other cardiovascular disease

Although atherosclerosis is the leading cause of cardiovascular disease, there are others, each with their own aetiology. We have already discussed atherosclerosis as a cause of the coronary artery disease that in turn inevitably causes angina and myocardial infarction. Other heart disease focuses on valves and cardiomyocytes, disease of the latter being cardiomyopathy. There is also disease caused by an irregular heartbeat (i.e. an arrhythmia), aneurysm (an abnormal bulging of the weakened wall of a blood vessel), and the failure of the heart to deliver enough blood—heart failure. The purpose of this chapter is to review only the major causes of each particular type of disease. Clinical presentation, diagnosis, and management of these conditions are discussed in Chapter 8.

6.5.1 Valve disease

Valve anatomy and function

The four valves of the heart, two on each side, regulate the movement of blood into and out of the four chambers. The aortic valve controls blood leaving from the left ventricle to the aorta, whilst the pulmonary valve regulates blood passing from the right ventricle to the lungs via the pulmonary artery. Movement of blood between the left atrium and left ventricle is regulated by the (bicuspid) mitral valve, and on the right the tricuspid valve links the right atrium and right ventricle (Figure 6.12). The mitral valve has two cusps, or leaflets, of fibrous tissue, and the other valves have three cusps. The major pump forces are from the ventricles, and to prevent blood from passing back into the atria, the integrity of the tricuspid and mitral valves is maintained by papillary muscles and **chordae tendineae**.

Valve pathology

If these valves fail to open and close correctly, or are leaky or narrowed, blood flow will be irregular, and this will lead to poor heart function. The major aetiology for most valve disease is rheumatic heart disease (RHD), but haemodynamic stress (such as hypertension) may also cause valves to fail. The greater part of valve diseases are stenosis and insufficiency and may be congenital or acquired (as following RHD). The opening of the valve may narrow (a stenosis), so that blood has difficulty passing from chamber to chamber, or from chamber to artery. Insufficiency of the valve can be seen in terms of failure to close correctly, so that blood can flow back from where it originated, the process of regurgitation. Hence the major diseases are aortic regurgitation and stenosis, mitral regurgitation and stenosis, pulmonary regurgitation and stenosis, and tricuspid regurgitation and stenosis:

- Aortic regurgitation may follow endocarditis or be present as part of Marfan's syndrome, whilst stenosis may be caused by calcification (the product of a low-grade inflammatory response), RHD, or renal disease.

- Mitral regurgitation may also follow RHD, endocarditis, and in Marfan's syndrome, but may also be associated with **ischaemic** heart disease and degeneration. Mitral stenosis may (once more) be the consequence of RHD, and valve function deteriorates after cusp thickening and calcification.

- Pulmonary regurgitation is most often a consequence of pulmonary hypertension (i.e. high blood pressure in the exit artery that the blood leaving the heart has to overcome), whilst stenosis is usually congenital but may also be caused by RHD and found in the carcinoid syndrome.

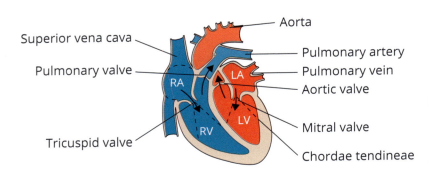

Superior vena cava

Pulmonary valve

Tricuspid valve

Aorta

Pulmonary artery

Pulmonary vein

Aortic valve

Mitral valve

Chordae tendineae

RA   LA

RV   LV

FIGURE 6.12

Structure of the heart showing valves. RA = right atrium, RV = right ventricle, LA = left atrium, LV = left ventricle. Red arrows indicate direction of blood flow.

- Tricuspid regurgitation may follow other heart disease: cor pulmonale, right ventricle dilatation (both of which may be caused by myocardial infarction), and pulmonary hypertension. Tricuspid stenosis is uncommon but is generally a consequence of RHD.

Rheumatic heart disease

As this disease (also known as scarlet fever) is mentioned several times in the above, discussion is warranted. The infective inflammatory disease rheumatic fever primarily attacks children and teenagers and is caused by infection with group A β-haemolytic *streptococcus* species such as *pyogenes*. Generally starting with a throat infection, the disease often spreads to the skin (causing a rash) and joints (causing **arthralgia** and often an **arthritis**), and, in 40% of cases, the heart. The immune response to the organism is believed to incorrectly identify 'self' components on the valves, and so a pseudo-autoimmune response, a process described as molecular mimicry. The immune response to this also causes thickening of the cusps and weakness in the papillary muscles and chordae tendineae. In severe cases there may be myocarditis, pericarditis, and arrhythmias with associated electrocardiogram (ECG) changes.

Endocarditis

Similarly, valve disease can follow endocarditis, this being inflammation of the endothelium lining the chambers. If infective, the most common culprits are *steptococcus* and *staphylococcus* species, requiring treatment with antibiotics, and septicaemia is likely. A build-up of organisms, and possibly white blood cells and platelets on the cusps, is described as a vegetation. By contrast, no organism is implicated in non-infective endocarditis, vegetations being much smaller. The most likely aetiology is thrombotic, but it may arise in certain cancers for reasons that are unclear.

Clinical aspects of valve disease

These abnormalities, if severe, may cause problems with breathing, angina, and damage to the left ventricle, requiring surgical repair or replacement of the particular valve. Replacement valves may be mechanical or bioprosthetic, and because of the increased risk of thrombotic stroke, lifelong anticoagulation is almost inevitable. Both natural and artificial valves are the target of bacteria, leading to infective endocarditis and treatment with antibiotics.

Many problems with valves can be diagnosed by an ECG, and by imaging the inside of the heart by ultrasound, termed echocardiography. This gives indices such as the velocity of the blood as it moves across the valves and (with colour **Doppler**) the degree of regurgitation.

Cross reference
Clinical aspects of valve disease are presented in section 8.3.1.

Key points

Valve disease may be classified as stenotic (a narrowing of the aperture) or insufficient (leading to regurgitation), and may be concurrent.

SELF-CHECK 6.7

What are the major risk factors for valve disease?

6.5.2 Cardiomyopathy

This is disease of the cardiomyocytes themselves and is not necessarily due to the poor oxygen supply that is the consequence of coronary artery disease. Causes can be primary (i.e. genetic), or secondary to a different pathology. An example of the latter is the ischaemic cardiomyopathy of coronary artery disease, where cardiomyocytes are starved of oxygen, and as such are at danger of necrosis and apoptosis. There are a number of different types, all of which lead to weakness of the heart as a whole, the consequences of which include arrhythmia and failure of the chambers to pump adequately and so deliver blood to the lungs and rest of the body.

Hypertrophic cardiomyopathy (HCM)

This variant is characterized by enlargement of parts of the myocardium, mostly of the wall of the left ventricle (typically 15 mm) and the septum. The **hypertrophy** can be so extensive that it prevents the outflow to blood, in which case it is described as obstructive. Present in one in 500 of the population, HCM is the most common cause of sudden cardiac death in young people and may be precipitated by exercise. The dominant aetiology is genetic (over 1,400 mutations have been described), and most cases (40–60%) are linked to mutations in one of nine genes coding for sarcomere proteins—secondary causes are very rare. Most mutations are **autosomal** dominant, demanding family studies (with possible screening) and awareness. The leading mutation, being present in 30–40% of patients, is *MYBPC3*, at 11p11.2, coding for myosin-binding protein C. Closely behind this, present in 20–30% of patients, is a mutation in *MYH7* at 14q11.2, which codes for β-myosin heavy chain. Mutations in genes coding for troponin T (*TNNT2* 1q32) and troponin I (*TNNI3* at 19q13) are present in 10% and 7% of patients respectively. Less frequent are mutations in genes coding for regulatory myosin light chain (*MYL2* at 12q23–24, 2–4%), essential myosin light chain (*MLY3* at 3p21.3, 1–2%), and for α-tropomyosin and α-cardiac actin (both <1%).

Non-sarcomere mutations (making up 5–10% of HCMs) include those in *LAMP2* (causing Danon disease—a glycogen storage disease), *GLA* (causing Fabry disease—a lysosomal storage disease of α-galactosidase), and *TTR* (coding for transthyretin—a protein carrying thyroxine—leading to **amyloidosis**). At the microscopic level, myocyte disarray exceeds 5–10%. Despite these associations, penetrance is generally weak: the development of perhaps the major clinical feature, left ventricular hypertrophy, rises from 55% at age 10–29, to 95% after age 50, implying other mechanisms such as epigenetics and/or a second gene. Some of these may be in one or more of the mitochondrial genes.

Dilated cardiomyopathy (DCM)

DCM is characterized by dilatation of the ventricles, generally on the left, but with normal or reduced ventricle wall thickness. With a frequency of 1/,2500 (but cited by some at 1/250–500), the principal consequence is systolic dysfunction—failure to adequately drive blood from the ventricle. As with HCM, most cases are genetic, and of these most are autosomal dominant with over 50 mutations identified. Some of these cross over with those of HCM (e.g. beta-myosin heavy chain, troponin), but other mutations most likely to be present in DCM include genes coding for actin, desmin, dystrophin, laminin, sarcoglycans, and vinculin. The most common (present in 12–24% of cases) is an abnormality in the protein titin, caused by a mutation in *TTN* (a shortening of titin), coded for at 2q31.2. Titin (also known as connectin) is a large (>1μm) protein, and is the third (after actin and myosin) most abundant muscle protein. The second most common mutation is in *MYH7* (as in HCM, and so named as a contraction of its product, myosin heavy chain beta) with a frequency of 4–7%, and the third is in *LMNA* (at 1q22), which codes for laminin. DCM is also a feature of Duchenne, Becker, and other forms of muscular dystrophy, all caused by mutations in the dystrophin gene at Xp21.2–21.1, as discussed in section 16.1.3 of Chapter 16.

Non-genetic causes include autoimmune disease; bacterial, parasitic (**Chagas disease**), and viral myocarditis; toxins such as alcohol, arsenic, heavy metals (cobalt, lead, mercury), and chemotherapy (especially cytotoxic, such as doxorubicin); thiamine deficiency; and tuberculosis. Rare causes of DCM are pregnancy and prolonged tachycardia.

Arrhythmogenic ventricular cardiomyopathy (AVC)

In contrast to HCM and CDM, this rare (200/million) condition primarily affects the right ventricle, although the left may also be targeted in 75% of patients, and if so, this biventricular involvement is more severe. The aetiology is of the replacement of cardiomyocytes by fibrous tissue and adipocytes, and so a dilatation. Genetic causes are linked to genes coding for the various desmosomal proteins that link cells together, and without which cells will fail to form a cohesive beating structure. These genes include *PKP2* (at 12p11.21) coding for plakophilin-2 (30–40% of cases), *DSP* (at 6p24.3) coding for desmoplakin (10–20%), and *DSG2* (18q12.1), coding for desmoglein-2 (5–20%). Less frequent are mutations in genes coding for desmocollin-2, junction plakoglobin, desmin, cadherin C, the ryanodine receptor (regulating calcium flow), and those of transforming growth factor-β.

Primary restrictive cardiomyopathy

This variant is characterized by enlargement of both atria with normal or decreased volumes of both ventricles, resulting in impaired ventricular filling. There is no hypertrophy, and valve function is normal. The aetiology is inevitably reactive to a number of diverse causes such as deposition of amyloid, **sarcoidosis**, and Loeffler's syndrome (caused by eosinophil infiltration and a consequent fibrosis and endocarditis). As described in section 8.3.2, cardiomyopathy brings a number of problems such as syncope, heart failure, and arrhythmia, all of which can be presenting features, and will be investigated by echocardiography and electrocardiogram.

> ## Key points
>
> Cardiomyopathy is characterized by damage to the cardiomyocytes, which can be primary (defective genes) or secondary (to factors such as endocarditis, alcohol abuse, and metabolic toxins such as chemotherapy).

6.5.3 Arrhythmia

The four chambers of the heart must beat in the correct order and are regulated by impulses from the sinoatrial (SA, the primary pacemaker) and atrioventricular (AV) nodes, hence sinus rhythm, as discussed in sections 6.4.2 and 6.4.3, and Figure 6.7. When normal sinus rhythm (i.e. rhythm directed by impulses from the SA node) is lost, the four chambers beat out of sequence, or faster or slower than normal, and an arrhythmia is present. Arrhythmias can be caused by a number of factors, and the patients themselves may complain of tiredness, dizziness, palpitations, and fainting spells (syncope), or they may be asymptomatic.

The heart rate can of course be measured by feeling the pulse in the wrist or neck, but determining the more complex arrhythmias requires an ECG. If the arrhythmia is profound, an external device can provide electrical signals that help the heart to beat in a correct manner. Such a device is called a pacemaker. There are several types of arrhythmia, which may be classified as changes in the rate of the heartbeat or changes in the rhythm of the heartbeat:

- A low heart rate, bradycardia, is generally <60 beats per minute during the day, <50 at night.

- An increased heart rate, tachycardia, is generally a heart rate of >100 beats per minute and can be further classified as that of the atria and that of the ventricles.

- Another factor is whether the rhythm is regular or irregular.

- Any combinations of the above may apply.

Bradycardias

Bradycardias can be further classified. Sinus bradycardia is generally caused by pathology of the SA node and may have cardiac (damage caused by nearby myocardial infarction or ischaemia, fibrosis, myocarditis, or cardiomyopathy), non-cardiac (jaundice, hypothermia, hypothyroidism, medications, overdose of cardiac medications), or other causes (aberrant nervous conduction).

Heart block refers to a number of conditions characterized by failure of the conduction system to correctly pass impulse messages to the myocardium, and again has subtypes. Atrioventricular block is due to issues with the AV node and/or the bundle of His (muscle cells adapted for electrical conduction), whilst blockage of the downstream conduction fibres leads to right bundle branch block and the more serious left bundle branch block. The causes of blockade are almost all cardiac or pulmonary in origin—aortic stenosis, arterial septal defect, myocardial infarction, pulmonary embolism or hypertension, coronary artery disease, and atrial septal defect. Perhaps the only external cause is Chagas disease.

Supraventricular tachycardias (SVTs)

Supraventricular tachycardias arise from problems with the atria (i.e. above or *supra* to the ventricles) and are a large and complex group that generally result in more serious clinical conditions. Relatively simple sinus tachycardia, in the absence of a clear disease process, may be due to intrinsic problems with the SA node or abnormal nerve regulation, or a number of other factors ranging from infections to acute heart failure. As implied, the lesion in atrioventricular junctional tachycardia lies at the meeting point of the right atrium and ventricle, the site of the AV node, and may result from a block in the conduction tissues between the two nodes. Consequently, the AV node becomes the dominant pacemaker, leading to myocardial confusion and so to the arrhythmia. These abnormalities are often described as re-entry or reciprocating pathways, reflecting the irregular passage of conduction impulses.

Atrial fibrillation (AF) is the most common of the arrhythmias, and it is very likely to be accompanied by a modest tachycardia. The population frequency is 2–3%, but this is strongly age dependent, as 10% of those aged over 80 are affected. From the physiological point of view, any factor that puts stress on the left atrium may lead to AF. Cardiac causes include heart failure, valve disease, myocardial infarction, surgery, and tumours. The leading extra-cardiac risk factor is hypertension; others include diabetes, pulmonary disease, excessive caffeine, and **thyrotoxicosis**. However, in around 25% of cases there is no clear precipitating factor.

In itself, AF brings a number of cardiac problems, but a further pathology lies with the turbulence within the heart that results from the abnormal heartbeat. This turbulence causes the formation of clots within the left atrium, which have the potential to run up the carotid arteries to the brain and there cause a stroke. Unfortunately, as hypertension and several other risk factors are often asymptomatic, both AF and its risk factor may be found only after this major event, and for others the arrhythmia may be discovered incidentally. With a high risk of stroke, almost all patients with AF (which in any case is mostly a disease of the elderly) should be anti-coagulated.

In the closely related atrial flutter, the atria contract at a rate of 250–300 beats per minute, whilst those of the ventricles at generally about half of this rate, i.e. 125–50 beats per minute. As with many other arrhythmias, it is associated with other cardiac disease such as infarction and cardiomyopathy, and non-cardiac causes such as hypertension and diabetes. The atrial disturbance leads to risk for thrombosis and stroke—as in AF, with which it has much in common, such as a strong age gradient and need for anticoagulation. Atrial tachycardia, with a heart rate of 125–50 beats per minute, is generally associated with structural heart defects.

Ventricular tachycardias (VTs)

Another type of arrhythmia, ventricular tachycardia, is characterized by a heart rate of 120 to over 200 beats per minute and can be life threatening as it may rapidly cause cardiac arrest. Most causes are cardiac, such as myocardial infarction, ischemia heart disease, aortic stenosis, and cardiomyopathy, but may follow low blood magnesium or potassium. In ventricular fibrillation the rate is so high that ventricles do not get the time to fill with blood and force it out into the arteries, and so effectively stop being able to pump blood. Accordingly, a pulse may be undetectable, so that it is a leading cause of cardiac arrest. Causes include cardiomyopathy, myocardial infarction, aortic stenosis, drug toxicity, and septicaemia.

The cycle of the heartbeat can be monitored by an ECG, and five prominent shapes are recognized and denoted PQRST. A recognized pathology is a prolongation of the interval between the Q wave and

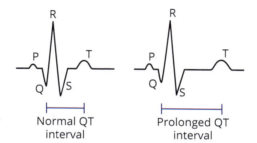

FIGURE 6.13
ECGs showing a normal and prolonged QT interval. The PQRST profile highlights the QT interval.

the T wave (Figure 6.13), which may be caused by numerous factors, such as low calcium, potassium, or magnesium, or certain drugs. But like the Brugada syndrome, long QT syndrome may also be the result of mutations in any one of ten genes coding for various parts of calcium and potassium channels, such as *KCNQ1* and *ANK2*.

In the absence of a structural heart defect, the Brugada syndrome is caused by a mutation in one to 20 genes involved in cation transport, present in 20% of cases. The leading genetic lesion, present in 20% of cases, is in *SCN5A* (at 3p22.2), which codes for a sodium channel, a mutation leading to low levels of the ion in cardiomyocytes. Mutations in several other genes with roles in sodium transport are causative, such as *GPD1L* (whose products include a glycerol-3-phosphate dehydrogenase motif) and *SCN1B* (coding a sodium channel subunit), whilst others, such as *CACNA1C* and *CACNB2*, code for components of the calcium channel, and so have roles in calcium transport. Figure 6.14 summarizes some of the features of arrhythmia.

Cross reference
Additional aspects of the ECG are discussed in Chapter 8.

Key points

Most cases of arrhythmia are caused by other diseases of the heart, or lungs, or hypertension. They can be asymptomatic (as in AF, which brings a risk of stroke) or carry a high risk of fatal cardiac arrest (as in most VTs).

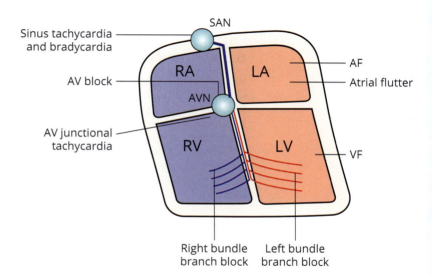

FIGURE 6.14
Selected features of arrhythmia. RA = right atrium, LA = left atrium, RV = right ventricle, LV = left ventricle, SAN = sino-atrial node, AVN = atrio-ventricular node, AF = atrial fibrillation, VF = ventricular fibrillation.

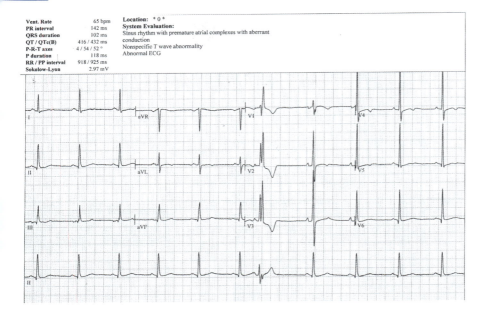

Vent. Rate	65 bpm
PR interval	142 ms
QRS duration	102 ms
QT / QTc(B)	416 / 432 ms
P-R-T axes	4 / 54 / 52 °
P duration	118 ms
RR / PP interval	918 / 925 ms
Sokolow-Lyon	2.97 mV

Location: * 0 *
System Evaluation:
Sinus rhythm with premature atrial complexes with aberrant conduction
Nonspecific T wave abnormality
Abnormal ECG

FIGURE 6.15

An ECG showing an ectopic beat. The key abnormality is present in V1, V2, and V3, notably the depressed S wave. The lower trace is a continuous feed from lead II, showing a small R wave, a depressed S wave, and an increased T wave. The software within the ECG machine has reported, in 'System Evaluation', that there is sinus rhythm with premature atrial complexes and aberrant conduction. An additional abnormality is the inverted T wave, most clearly seen in V4 on the top right of the printout. The patient was reassured by his GP and no further action was taken.

SELF-CHECK 6.8

Why do most patients with AF need to be anticoagulated?

Ectopics

Many of the middle aged and elderly are often concerned that their heart sometimes misses a beat, or that two or more beats come close together. These ectopic beats are entirely normal and may arise from a premature atrial or ventricular contraction, possibly driven by temporary stimuli such as exercise or emotional factors. The patient can be reassured that ectopics are not considered to be harbingers of more serious disease, and (generally) do not warrant investigation unless they persist and become more frequent. Figure 6.15 shows an example of a 12-lead ECG with an ectopic beat.

6.5.4 Aneurysm

Weakness of the media of arteries may lead to a distension and distortion of the shape of the vessel, which in turn leads to turbulence and the development of thrombus. This distension is called an aneurysm, generally defined as a 50% increase in diameter. Should it grow, the artery wall will become stretched, thinner, and will eventually burst (as would a balloon being blown up too much), often with catastrophic consequences. This pathology can (in theory) develop in any artery, but in practice it is restricted to only a few sites. Abdominal aortic aneurysms (AAA, defined by a lesion diameter >30 mm) generally form in the section beneath the kidneys but above the bifurcation of the aorta into the iliac arteries, whilst both the para-cardiac ascending and descending sections of the thoracic aorta can become aneurysmal. Rupture at all these sites is devastating: the mortality rate for thoracic aneurysm is in the region of 66%, for AAA the figure is 65–85%. Rupture of microaneurysms (0.8–1 mm diameter) in cerebral arteries (Charcot–Bouchard syndrome) can also cause massive and life-threatening haemorrhage. Aneurysms can also form in the iliac, carotid, femoral, and popliteal arteries.

All risk factors for atherosclerosis are also risk factors for aneurysm, but this does not necessarily imply that atherosclerosis causes aneurysm. Other risk factors include obesity, alcohol abuse, copper deficiency, polycystic kidney disease, and syphilis. From these, hypertension seems to have the greatest effect, but there may also be a place for structural changes, such as in smooth muscle cell/extra-cellular matrix dynamics, and the Marfan and Ehlers–Danlos type IV syndromes (see Chapter 16) are also risk factors. These support the hypothesis of a genetic component as family history is also a risk factor, derived from the note that 10–20% of AAA patients have at least one relative with the condition. Genome-wide association studies (GWAS) have identified 9 AAA risk loci, the strongest links being with *CDKN2BAS* (shortened from cyclin-dependent kinase 4 inhibitor B), *SORT1* (coding the protein Sortilin, involved in cytoplasmic transport), *LRP1* (coding the low-density lipoprotein receptor-related protein), *IL6R* (coding the receptor for interleukin-6), *MMP3* (coding for matrix metalloproteinase 3), *AGTR1* (coding angiotensin II receptor type 1), *ACE* (coding for angiotensin converting enzyme), and *APOA1* (the latter also involved in lipid metabolism).

6.5.5 Heart failure

Heart failure may be defined as the inability of the heart to deliver to the body the blood it needs, and has many potential and varied pathologies. Since the major driving force is the left ventricle, this chamber is often where the clinical problem lies (as it often is in cardiomyopathy). As it contracts, a healthy left ventricle is expected to eject perhaps 60–70% of the blood it holds, although this can exceed 75% or even 80% in top athletes. But even the most outstanding athlete cannot expel all the blood from their left ventricle (i.e. an ejection fraction of 100%). Heart failure becomes more and more likely as the ejection fraction falls, and is often defined clinically with an ejection fraction <40%.

The major causes of heart failure include damage to the myocardium of the left ventricle, perhaps after a myocardial infarction (present in 35–40% of cases), and/or excess strain on the left ventricle brought about by dilated cardiomyopathy (30–35%), and/or by hypertension (15–20%). We have already noted many of the other causes of heart failure, which underlines the complexity of heart disease. These include all other forms of cardiomyopathy, most forms of arrhythmia, most forms of valve disease, myocarditis, disease of the septum, several types of pulmonary disease, certain drugs (alcohol, chemotherapy), and inflammation (pericarditis, myocarditis). Clinical practitioners will be aware that heart failure may also be a consequence of cancer-direct chemotherapy.

6.5.6 Stroke

Stroke is the result of the irreversible loss of functional neural tissue of the brain. There are two major types of stroke: thrombotic (80% of all strokes) and haemorrhagic (17%), the remainder being due to venous thrombosis, **vasculitis**, and arterial dissection. Thrombotic stroke is inevitably the result of a clot that has blocked a cerebral arteriole and is essentially the product of atherosclerosis. As when a clot blocks a myocardial artery, the result for the brain is also deprivation of oxygen and glucose for the tissues downstream of the blockage, and so a cerebral infarction with loss of viable neuronal tissues. Similarly, the brain tissue downstream of a ruptured blood vessel will also be deprived of essentials, and so may also die. Figure 6.16 shows the fine structure of the three meningeal membranes in relation to the skull and brain tissues.

Transient ischaemic attack (TIA)

Whilst many strokes (in common with myocardial infarction) come as a complete surprise, a proportion could be predicted by a short-lived event called a transient ischaemic attack, or TIA. This is most often due to a reversible failure of the circulation to provide oxygenated blood to the brain (hence ischaemia), which in turn may be due to microemboli, arrhythmia, and/or stenosis

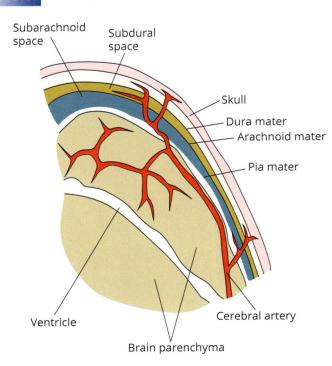

Subarachnoid space

Subdural space

Skull

Dura mater

Arachnoid mater

Pia mater

Cerebral artery

Ventricle

Brain parenchyma

FIGURE 6.16

Blood supply to the brain, skull, and meninges. Blood from the arterial cerebral circulation feeds the skull and brain tissues, and in doing so crosses the dura, arachnoid, and pia maters.

of the carotid arteries due to atheroma. To some extent, a TIA has parallels with the relationship between angina and myocardial infarction, the former often being (without treatment) a harbinger of the latter.

Haemorrhagic stroke

There are several types of haemorrhagic stroke, which make up some 17% of all strokes, this being the product of the rupture of a blood vessel and so a flood of blood into the tissues. This flood may be into the true brain (neuronal) tissues (which may be described as parenchyma); into the ventricles; and/or into the spaces between brain tissues and the membranes, mostly into the subarachnoid space (hence subarachnoid haemorrhage), but also between the dura mater and the skull, and between the dura mater and arachnoid membrane (see Figure 6.16).

Thrombotic stroke

In theory, a thrombotic stroke (which may also be described as ischaemic) may occlude any of the blood vessels shown in Figure 6.16, but in practice the more clinically important occlusions are to those arterioles that feed the parenchyma of the brain. The source of the thrombus is often (as we have seen in section 6.5.3) the product of AF, but may also arise from ruptured plaque from atheroma in the carotid arteries.

The causes of stroke

As described previously, clots may arise in the heart (due to arrhythmia, valve disease, or mural thrombosis) or atheroma in the carotid arteries or the aortic arch. The extra strain placed on arterioles by hypertension could be expected to cause a haemorrhagic stroke, but curiously this risk factor is also a major cause of thrombotic stroke, perhaps because it causes vulnerable atheroma to rupture. Other risk factors include haematological disease (high platelet count, **polycythaemia**, hyperviscosity—all promoting thrombosis) and autoimmune disease (anti-cardiolipin antibodies and lupus anticoagulants, rheumatoid arthritis, systemic lupus erythematosus (SLE), vasculitis, etc., all of which are pro-thrombotic, as discussed in Chapters 10 and 11).

6.5.7 Peripheral artery disease (PAD)

Atherosclerosis is a whole-body disease that may develop in any artery. We have already noted the role of this disease process in coronary arteries and cerebral arteries, and much is relevant to arteries elsewhere in the body. For example, atherothrombosis of arteries feeding the spine may lead to vertebral necrosis and collapse; atherosclerosis in a pudendal artery may lead to erectile dysfunction (particularly in people with diabetes, although another cause is neuropathy); whilst an occlusive thrombosis in a mesenteric artery may lead to ischaemia of a section of the intestines that may, in turn, lead to **peritonitis** and death. Whilst the four traditional risk factors are all implicated in PAD, diabetes and smoking do seem to predispose to disease of these arteries. The role of inflammation in the development of PAD is unclear, but it is certainly present in the terminal stage of lower limb disease. The arteries of the shoulder (subclavian) and arm (brachial) seem to be immune from this disease for reasons that are unclear. Section 6.5.6 has drawn our attention to carotid atherosclerosis as a risk factor for stroke and TIA. Similarly, it could be argued that abdominal and thoracic aortic aneurysms are the consequence of atherosclerosis, although it is clear that many aortae excised after an aneurysm repair are relatively free of this disease.

Arteries of the lower limb

Stenosis, obstruction, or occlusion of the iliac, femoral, and popliteal arteries will (depending on the site of the lesion) lead to downstream hypoxia of the thigh, calf, and foot. The former two present as intermittent claudication (pain upon walking, relieved by rest), the latter by a foot that is cold, pale, and with weak or absent pulses. Should intermittent claudication fail to be addressed, it will inevitably lead to critical limb ischaemia that, at best, will require debridement of dead and dying tissues, and at worse, amputation.

Figure 6.17 summarizes the anatomical distribution of the various forms of organ-based cardiovascular disease.

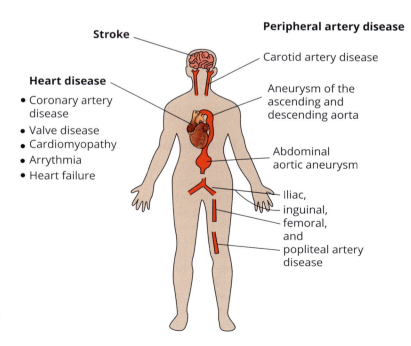

FIGURE 6.17
Anatomical distribution of the different forms of organ-based cardiovascular disease.

> ### Key points
>
> Aneurysm, heart failure, stroke, and peripheral artery disease often develop from risk factors or from defined genetic conditions, although heart failure may also arise from other cardiovascular disease, principally hypertension, myocardial infarction, and cardiomyopathy.

6.6 The genetics of cardiovascular disease

What is more important—nature or nurture? Or are they equally relevant? In this final section, we address an additional aspect of the aetiology of cardiovascular disease. Yes, we are all going to die, but of what, and when? Clearly, for some, death can be stalled, but in others, no adherence to 'good living' can delay morbidity and mortality. Commentators suggest that 30% to 60% of the variation in risk of coronary artery disease is genetic. It is well established that certain genes not only predispose to various diseases, but some can accelerate the process, and others confer an additive effect. In recent decades, GWAS on thousands of subjects have identified dozens of loci linked to various diseases, and so have revolutionized this aspect of clinical practice. This discussion continues in Chapter 16.

6.6.1 The major risk factors for atherosclerosis

Previous sections have emphasized the importance of the risk factors. Where applicable, the biochemistry and cell biology of each will be discussed in depth in Chapter 7, so that what follows will be limited to genetic aspects of these diseases.

Smoking

Tobacco smoking is clearly a psychosocial disease/lifestyle choice and should be addressed as such. However, molecular genetics has identified *CHRNA5*, at 15q25.1, which codes for a subunit of the neuronal acetylcholine receptor, mutations of which may help identify those in need of most help. A variant in this gene part-predicts delayed smoking cessation, possibly linked to nicotine addiction. An analysis of 76,972 people found that a single nucleotide polymorphism (SNP) in this gene was linked to the number of cigarettes smoked per day with a truly astonishing probability of 2.4×10^{-69} (this is not a typo; 69 zeros). The extent to which the toxic contents of tobacco smoke can be resisted may be partly under genetic control, such as variation in levels and function of genes coding for antioxidants, and in the ability to repair those genes damaged by carcinogens.

Hypertension

Decades of research at the end of the last century failed to identify any single gene that has a powerful and highly penetrant effect on high blood pressure, with a single exception (Liddle syndrome). This was frustrating as family history and twin studies suggest heritability estimates of 35–50%. The prevailing view was that it is likely that many different genes each act in a small way to control blood pressure, so that it is said to be under polygenic control. However, some analyses found evidence of roles for the genes for angiotensinogen and angiotensin converting enzyme (ACE) in blood pressure, which is relevant to the use of ACE inhibitors to treat high blood pressure. But these lines of evidence have been overshadowed by the power of molecular genetics, which, in analysing the genotypes of over 87,000 people of European ancestry, found powerful links between certain genes (*PDE1A, HLA-DQB1, VCL, H19, NUCB2, RELA, CDK6, FBN1, NFAT5, PRKAG2,* and *HOXC@*) with systolic and diastolic blood pressure, mean arterial pressure, and pulse pressure (Tragante et al. 2014). A subsequent study of over 1 million people reported 535 loci that influence blood pressure, although only 12 were associated with systolic blood pressure, diastolic blood pressure, and pulse pressure (Evangelou et al. 2018). Similar studies in African Americans have identified *SCNN1B*, *ARMC5*, *GRK4*, and *CACNA1D* as genes with roles in hypertension (Zilbermint et al. 2019). Genetic associations of blood pressure variants with daily fruit intake; urinary sodium and creatinine concentration; body mass index (BMI); weight; waist circumference; and intakes of water, caffeine, and tea have been reported, much of which is unsurprising. By merging data from 901 loci, a genetic risk score was linked to an increase in systolic blood pressure of 10 mmHg, with a probability of 1 $\times$ 10^{-300} (again, not a typo), also indicating a that high score predicts hypertension and incident stroke and myocardial infarction. Notably, GWAS has identified links with other risk factors, such as apoE and a variant of the LDL receptor-related protein (section 7.3.2). However, other links with hypertension are (at first sight) a little unclear, such as with an apoptosis regulator (*BCL*), the melatonin receptor (*MTNR1B*), keratinocyte differentiation factor 1 (*KDF1*), and carbomoyl-phosphate synthase-1 (*CPS1*).

Dyslipidaemia

This condition is dominated by high serum levels of total and low-density lipoprotein cholesterol (LDL) as a leading cause of atherosclerosis. The principal hereditary disease of this risk factor is familial hypercholesterolaemia (FH), and work on this disease and cholesterol metabolism led to the awarding of the 1985 Nobel Prize in Medicine to Goldstein and Brown. Three major forms of the disease are recognized, all based on the uptake of LDL by a specific surface receptor (LDLR, coded for by *LDLR* at 19p13.2):

- The classical form of FH arises from a mutation in LDLR generating a malformed receptor that fails to bind to LDL, circulating levels of which remain high and so contribute to atherogenesis.

- *ApoB* (located at 2p24.1) encodes a ~500 kDa molecule of around 100 amino acids (hence ApoB$_{100}$) that comprises the major protein part of LDL. A mutation in *ApoB* results in an alternate form (R3500Q: arginine to glutamine) of apoB$_{100}$ that fails to bind the LDLR, resulting in high levels of serum LDL. This mutation is the cause of 5–10% of FH cases.

- *PCSK9* at 1p32.3 codes for a chaperone enzyme (proprotein convertase subtilisin/kexin type 9) that conducts internalized LDLRs to the lysosome for degradation. A gain-in-function mutation in *PCSK9*, which results in low expression of the LDLR at the cell surface, and so high levels of serum LDL, is present in ~2% of FH cases.

A very rare cause of FH is mutations in *LDLRAP1* at 1p36.11, encoding the LDL receptor adapter protein 1. These associations are summarized in Figure 6.18.

Many other mutations in genes controlling lipoprotein metabolism leading to cardiovascular disease have been described. Familial lipoprotein lipase deficiency leads to a high concentration

FIGURE 6.18

Molecular causes of familial hypercholesterolaemia. Far left: The LDL particle binds to the LDL receptor (LDLR, shown in green) via apoB$_{100}$ and is internalized. The particle and the receptor dissociate and the LDLR returns to the cell membrane. Centre left: should PCSK9 bind the LDLR, the complex is still internalized, but upon dissociation the LDLR is conducted to the lysosome for destruction. Centre right: a mutated LDLR will fail to interact with a normal apoB$_{100}$. Far right: a normal LDLR will fail to bind to a mutated apoB$_{100}$. Both these latter two options result in the LDL remaining in the plasma, leading to FH.

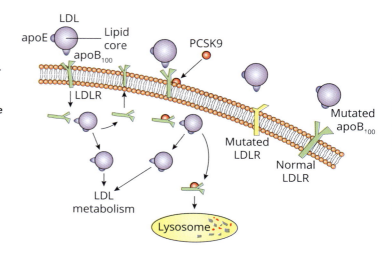

of **chylomicrons** and hypertriglyceridaemia, and may be caused by mutations in a number of genes such as *ApoC2* (at 19q13.32, coding for apolipoprotein C2) and *ApoA5* (at 11q23.3, coding for apolipoprotein A5). Familial hypertriglyceridaemia may also be caused by excessive production of VLDL by the liver. Similarly, deficiency in the enzyme lecithin-cholesterol acyl-transferase (LCAT), a consequence of loss-of-function mutations in *LCAT* at 16q21–22, leads to failure to convert cholesterol to a cholesterol ester form easily carried in HDL, so leading to hypercholesterolaemia.

APOE (at 19q13.32, close to *APOC-1*) is polymorphic, with three alleles (E2 5–10%, E3 65–70%, E4 15–20%) generating different phenotypes, of which E2/E2 is of note as E2 has only 1% of the binding affinity for the LDLR compared to the other forms, and so is associated with dysbetalipoproteinaemia and increased IDL. The E4 variant is associated with raised LDL, and is a risk factor for ischaemic stroke, with an odds ratio of 1.43 (95% CI 1.1–1.85) and may also be implicated in the pathogenesis of Alzheimer's disease. Familial combined hyperlipidaemia (increased cholesterol and triglycerides) may be linked to a haplotype cluster of *APOA1/C3/C4* and *APOA5* at 11q23.

Just as low LDL is desirable, so is high HDL. Twin studies suggest that around 70% of variability in HDL is under genetic control. Candidate genes include *APOA1*, *ABCA1*, *LCAT*, *CETP*, and lipases, whilst loss-of-function mutations in *ABCA1* lead to Tangier disease.

Diabetes

Type 1 diabetes mellitus (T1DM) certainly brings a risk of cardiovascular disease, but is markedly less fatal than type 2 disease (T2DM). In England and Wales in 2021, there were 505 (up 28% from 2019) and 4,428 (up 12%) deaths respectively, although there were also 2,396 unspecified diabetes deaths. Evidence in favour of a strong genetic component in T1DM includes a twin concordance rate of ~55%, a sibling risk of ~6.5%, and a parental risk of ~5%, with (unusually for an autoimmune disease) a significant male/female preponderance of 1.47. Of the numerous candidate genes linked to T1DM, changes in *VNTR*, *INS* (coding for insulin), IFIH1 (melanoma differentiation-associated protein 5), *PTPN22* (a protein tyrosine kinase), *ERBB3* (human epidermal growth factor receptor 3), *CTLA4* (CD152, cytotoxic T lymphocyte associated protein 4), and *IL2RA* (CD25, the IL-2 receptor alpha chain) are often cited. Combinations of these mutations can help with a diagnosis. However, the strongest links are with certain HLA class II types, notably the DRB1 alleles, where DRB1*04:05 brings an odds ratio for T1DM of 11, that of DRB1*04:01 being 8. Whereas only 2% of the population carry certain DR4-DQ8 types or DR3-DQ2, at least one of these is present in ~90% of T1DM patients, and 30% carry both, which together bring an odds ratio of 16. Several haplotypes are cited by the literature, including:

- DRB1*0301 + DQA1*0501 + DQB1*0201

- DRB1*0401 + DQA1*0301 + DQB1*0301

- DQA1*0301 + DQB1*0302

Not all links are pathogenic: the haplotype DRB1*1501 + DQA1*0102 + DQB1*0602 provides disease resistance, whilst HLA-B*57:01 is protective with an odds ratio of 0.19. Only 1% of children with T1DM express DQB1*06:02, a type that is present in 20% of the non-diabetic population. Despite these links, GWAS point to over 60 potential non-HLA susceptibility loci. Presence of a high-risk haplotype and more than one autoantibody predicts a 35% risk of T1DM within 5 years, 61% within 10 years, and 75% within 20 years. Combinations of some of these HLA genes and some of the non-HLA genes described here also give good discrimination between different forms of diabetes.

The fact that 40% of first-degree relatives of type 2 diabetes mellitus (T2DM) patients may develop the disease themselves (compared to 6% in the general population) implies a genetic component, as further supported by twin studies. Genome-wide association studies have identified some 75 candidate genes, some of which have staggeringly small p values, and so very strong probability of a true pathogenetic effect. These include *KCNQ1* at 11p15.4, coding an islet ATP-sensitive potassium channel subunit (p = 9.7×10^{-10} to 1.7×10^{-42}); *TCF7L2* at 10q25.2, coding a transcription factor regulating proglucagon (p = 2×10^{-31}); and *IRS1* at 2q36.2, coding an insulin receptor subunit

TABLE 6.4 T2DM candidate genes

Locus	Location	Product	Odds ratio (95% CI)	P-value
TMEM154	4q31.3	A transmembrane protein	1.08 (1.06–1.10)	4.1×10^{-14}
SSR1/RREB1	6p24.3	Subunit of an endoplasmic reticulum membrane receptor	1.06 (1.04–1.08)	1.4×10^{-9}
FAF1	1p32.3	FAS-associated factor 1	1.10 (1.07–1.14)	4.1×10^{-9}
POU5F1/TCF19	6p21.33	An octamer-binding transcription factor	1.07 (1.04–1.09)	4.2×10^{-9}
LPP	3q27.3	Lipoma-preferred partner	1.07 (1.04–1.09)	5.8×10^{-9}
ARL15	5q11.2	ADP-ribosylation factor-like 15	1.06 (1.04–1.08)	6.9×10^{-9}
ABCB9/MPHOSPH9	12q24.31	ATP-binding cassette sub-family B member 9	1.06 (1.04–1.08)	9.5×10^{-9}

Modified from DIAbetes Genetics Replication And Meta-analysis (DIAGRAM) Consortium, Asian Genetic Epidemiology Network Type 2 Diabetes (AGEN-T2D) Consortium, South Asian Type 2 Diabetes (SAT2D) Consortium, et al. Genome-wide trans-ancestry meta-analysis provides insight into the genetic architecture of type 2 diabetes susceptibility. Nat Genet 46: 234–44 (2014). https://doi.org/10.1038/ng.2897.

($p = 5.4 \times 10^{-20}$). Others include *SLC30A8* (at 8q24.11 coding zinc transporter 8), *SLC16A11* (at 17p13 coding a monocarboxylate transporter), *TM6SF2* (at 19p13.11 coding a transmembrane molecule of uncertain function), and several shared with other forms of diabetes, for example *HNF1A*, *HNF4A*, and *GCKR*. Table 6.4 shows data from a genome-wide meta-analysis of loci linked to susceptibility to T2DM.

miRNAs such as miR-7, miR-9, miR-29, and miR-34a are linked to insulin production and secretion, and so may be a therapeutic target. Numerous species, including miR-135a, miR-181a, miR-195 (all overexpressed), miR-25, miR-30, and miR-181a (all underexpressed) may be important in diabetic **nephropathy**, and miR15a is overexpressed in diabetic retinopathy. Similarly, there may be a place for lncRNAs such as CASC2 and GM1564, and circular non-coding RNAs such as circRNA_15698. It has also been suggested that different classes are linked, in that lncRNAs MALAT-1 may suppress miRNAs miR-23a and miR-449.

However, despite the links between these genes and T2DM, none are likely to help prevent the greater part of the development and treatment of this disease—that is, an imbalance between calorific intake and its expenditure as exercise, inevitably leading to overweight and obesity, which may take decades to develop.

6.6.2 Other potential risk factors

Almost all literature on molecular genetics and atherosclerosis focuses on the major risk factors, but there is much to be said on other potential factors. One such is oxidant stress, with some arguing that gene-driven variability in antioxidants (glutathione and related molecules, sirtuins, superoxide dismutase, etc.) has a role in atherogenesis. To some extent this may tie in with inflammation as leukocytes generate reactive oxygen species, and a further speculation suggests that variability in the release of matrix-degrading enzymes such as MMP-9 may have a role. The rs5498 polymorphism in the adhesion molecule *ICAM-1* at 19p13.2 decreases the risk of coronary artery disease, with potential causality, perhaps in failing to promote the passage of leukocytes into the artery wall. Increased levels of the amino acid homocysteine have long been promoted as a risk factor, as may be related to polymorphisms in a key metabolic enzyme 5,10-methylenetetrahydrofolate reductase gene (MTHFR). SNP rs1801133 in *MTHFR* at 1p36.22 (as discussed fully in section 16.6.4 of Chapter 16) is linked to risk of atherosclerosis, and interestingly, also to high levels of total cholesterol.

The power of GWAS brings intriguing data, but not all are easily explained in terms of our present understanding of atherosclerosis. For example, links between early-onset myocardial infarction and the phosphatase and actin regulator 1 gene (*PHACTR1*), the CDKN2B antisense RNA 1 gene (*CDKN2B-AS1*), stromal derived growth factor (*CXCL12*), and melanoma inhibitory activity protein 3 (*MIA3*) seem unclear, although associations with the LDL receptor (coded for by *LDLR*) and proprotein convertase subtillisin/kexin type 9 (*PCSK9*) with their links cholesterol cycling are more easily explained (Figure 6.18 and related text). The link with rs264 of the lipoprotein lipase gene (*LPL* at 8p21.3) is easily understood. Other coronary artery disease links are with an intron in the nitric oxide gene (*NOA1* at 4q12), voltage-dependent sodium and potassium channels (*REST*, also at 4q12), nitric oxide synthase (*NOS3* at 7q36.1), a signalling molecule involved in the regulation of filamentous-actin networks (*SWAP70* at 11p15.4), a downstream mediator of TGF-β (*SMAD3* at 15q22.33), a regulator of VEGF-dependent vascularization (*MFGE8* at 15q26.1), a protein (rudhira) that influences endothelial cells (*BCAS3* at 17q23.2), the melanocortin 4 receptor (*MC4R* at 18q21.32, with further links to obesity, high triglycerides, and low HDL), non-coding RNA POM121 (*POM121L9P* at 22q11.23), the kinase suppressor of RAS2 (*KSR2* at 12q24.23), and an uncharacterized non-coding RNA (*LOC400684* at 19q13.11) that is protective of coronary artery disease.

SELF-CHECK 6.9

Familial hypercholesterolaemia is the leading genetic cause of increased total and LDL-cholesterol—what are its causes?

6.6.3 The heart

Two SNPs in transcription factor *PITX2*, at 4q35, have been linked to atrial fibrillation, but clearly the penetrance is weak as these variants will be present from birth, and the arrhythmia may take decades to become apparent. The pathophysiological link may be that *PITX2* has a role in embryonic left/right axis development. Section 6.5.3 has pointed out that a long QT interval and the Brugada syndrome may have a genetic basis. A rapid increase in heart rate to an external stimulus, and its decline to normal, is physiological. Seventeen different genes are linked to heart rate increase and 26 to rhythm, with nine to both (van de Vegte et al. 2019). Some of these genes are linked to the development of the nervous system (such as *PAX2* at 10q24.31, coding for a transcription factor), prolongation of neuronal life span (such as *SYT10*, coding for a membrane-trafficking protein), and cardiac rhythm (*SCN10A* at 3p22.2, coding for a sodium channel subtype).

Section 6.5.2 described genes relevant to cardiomyopathy, and section 6.5.4 those that may be important in aneurysm. Genetics are also involved into a large number of congenital and/or structural conditions of the heart, and may involve other tissues and organs, such as in Marfan syndrome and Ehlers–Danlos syndrome (discussed in Chapter 16). Many chromosome defects are associated with heart disease, such as that of atrioventricular septal defects in trisomy 21 which are linked to a least nine genes.

6.6.4 Non-coding RNAs

Cross reference

The formation of miRNAs and lncRNAs, and their potential roles in cancer have been described in section 3.8.4, where Figures 3.11 and 3.12 are relevant.

Not all RNAs ultimately code for proteins—many are regulatory of DNA sequences and other RNA species and so are described as non-coding RNAs. Research in all ncRNAs is rapidly progressing and may bear practical fruit in diagnosis, as many are found in the blood, some in saliva and urine. They may also have a place in management, perhaps as a new class of therapeutic tools, or as targets, such

TABLE 6.5 Selected lncRNAs in cardiovascular disease

Short name	Full name	Cells involved	Function
GAS5	Growth arrest specific 5	SMCs, ECs, MACs	Regulates apoptosis, autophagy, SMC proliferation
HIF1a-AS1	Hypoxia-inducible factor 1-alpha Antisense RNA 1	SMCs	Regulation of apoptosis
CHROME	Cholesterol homeostasis regulator of miRNA expression	MACs, hepatocytes	Regulation of cholesterol homeostasis
CLDN10-AS1	Claudin 10 antisense transcript	EC	Regulation of EC signalling
STEEL	Spliced transcript endothelial-enriched lncRNA	EC	Angiogenesis
APOA1-AS	Apolipoprotein A1 antisense transcript	Hepatocyte	Regulation of APOA1 expression

SMCs: smooth muscle cells, ECs: endothelial cells, MACs: macrophages.

as in dyslipidaemia and blood pressure control, and are being patented. Most can be classified as microRNAs (miRNAs, 20–25 nucleotides long) or long-noncoding RNAs (lncRNAs, >200 nucleotides long and with a secondary structure). A third are small interfering RNAs (siRNAs), and a fourth are circular RNAs.

lncRNAs

One of the best known lncRNAs is ANRIL (antisense noncoding RNA in the INK4 locus, coded at 9p21.3), also known as CDKN2BAS, the transcription factor of cell cycle regulators CDKN2A and CDKN2B. ANRIL is expressed in cardiac tissues, which may affect development of CAD via regulating vascular cell proliferation and apoptosis. Some lncRNAs have very specific roles, such as that of MALAT1 as an activator of the antioxidant pathway and apoptosis via the p38 MAPK pathway, and could target miRNAs such as miR-145 to enhance the expression of VEGF. Potential biomarkers for myocardial infarction include MIAT, MIRT1/2, HIF1-AS2, and KCNQ1OT1, whilst meXis is atheroprotective in that it improves cholesterol cycling. SMILR is upregulated in atherosclerotic carotid plaque compared to paired normal tissues and correlated with CRP, and stimulation of vascular smooth muscle cells *in vitro* with PDGF and IL1α increases its expression, leading to a view that it may be a target for therapy. The list of lncRNAs continues to grow, and dozens, if not hundreds, are linked to cardiovascular disease: some are described in Table 6.5.

miRNAs

To some extent, advances in miRNA biology parallel those of lncRNAs. Once more, dozens or even hundreds of species have been reported with variable links with all aspects of cardiovascular disease, including lipid handling, inflammation, and cell biology. There is practical potential as animal and *in vitro* studies show considerable promise and may translate to the clinic and pharmacy. For example, miR-33 targets *ABCB1* (coding the ATP-binding cassette transporter-1), *CROT* (peroxisomal carnitine O-octanoyltransferase), and *CYP7A1* (cholesterol 7 alpha hydroxylase), and so has roles in HDL biogenesis and cholesterol biochemistry, fatty acid biosynthesis, and bile acid synthesis and secretion, whilst miR-301b, miR-130 b, miR-185, mir-128-1, miR-148a, and miR-27a/b are all linked to the LDL receptor. miR-146a and mir-146b are both significantly upregulated in unstable atherosclerotic plaques.

miRNAs may be classified into families depending on their function, such as in angiogenesis. Those promoting this process include miR-21, miR-27b, miR-31, miR-126, miR-130a, miR-210, miR-296, and miR-378, whilst inhibitors include miR-15, miR-16, miR-20a, miR-20b, miR-221/222, and miR-320.

TABLE 6.6 Selected miRNAs in cardiovascular disease

miRNA species	Mechanism	Link to pathology
miR-208a	Regulates the balance between α and β myosin heavy chains	Protective of heart failure
miR-145	Regulates expression of *KLF4* and *KLF5*	Prevents the development of pulmonary artery hypertension
miR-148a	*ABCD1*	Increased LDL
miR-30c	LPGAT1, MTP	Decreased lipid synthesis and LDL, protects from steatosis
miR-125b	Endothelin-1	Hypertension
miR-23b	*FOXO4*	Smooth muscle cell proliferation in restenosis

The target of both groups is VEGF. miR-21, miR-483, miR-23a, and miR-26a are all linked to arrhythmias. One practical aspect of these molecules is the observation that miR-499 has a superior diagnostic accuracy for acute myocardial infarction compared with troponin T. Serum miR-21, miR-24, miR-148a, miR-181a-5p, and miR-210-5a levels are increased in T1DM.

Table 6.6 shows selected groups of miRNAs and their roles in cardiovascular disease.

Key points

The split of nature versus nurture in the pathogenesis of cardiovascular disease is roughly 50/50 between risk factors and genetics. There are dozens of examples of diseases where genes have a profound and often intractable effect.

SELF-CHECK 6.10

What is the principal difference between the two major non-coding RNAs?

Chapter summary

- The cardiovascular system is composed principally of endothelial cells, smooth muscle cells, and cardiomyocytes, with smaller numbers of pericytes, fibroblasts, adipocytes, and macrophages.

- Blood is driven around the body in arteries, capillaries, and veins by the force of the pulse pressure of the heart.

- The heart is a complex modified muscle of four chambers separated by valves and the septum.

■ The pathogenesis of atherosclerosis starts with endothelial dysfunction and develops into a chronic infiltration of the artery wall by lipids and an inflammation.

■ Other cardiovascular diseases include valve disease, cardiomyopathy, arrhythmia, aneurysm, heart failure, stroke, and peripheral artery disease.

■ Perhaps half of the risk of cardiovascular disease can be ascribed to genetics and the remainder to environmental factors.

Further reading

● Wang D, Wang Z, Zhang L, Wang Y. Roles of Cells from the Arterial Vessel Wall in Athero-sclerosis. Mediators Inflamm. 2017: 8135934.

● Singh TP, Morris DR, Smith S, et al. Systematic Review and Meta-Analysis of the Association Between C-Reactive Protein and Major Cardiovascular Events in Patients with Peripheral Artery Disease. Eur J Vasc Endovasc Surg. 2017: 54; 220–33.

● Daiber A, Xia N, Steven S, et al. New Therapeutic Implications of Endothelial Nitric Oxide Synthase (eNOS) Function/Dysfunction in Cardiovascular Disease. Int J Mol Sci. 2019: 20, pii: E187.

● Zhu Y, Xian X, Wang Z, Bi Y, et al. Research Progress on the Relationship between Atherosclerosis and Inflammation. Biomolecules. 2018: 8(3), pii: E80.

● Poznyak AV, Nikiforov NG, Starodubova AV, et al. Macrophages and Foam Cells: Brief Overview of Their Role, Linkage, and Targeting Potential in Atherosclerosis. Biomedicines. 2021: 9; 1221. doi: 10.3390/biomedicines9091221.

● Tsimikas S, Fazio S, Ferdinand KC, et al. NHLBI Working Group Recommendations to Reduce Lipoprotein(a)-Mediated Risk of Cardiovascular Disease and Aortic Stenosis. J Am Coll Cardiol. 2018: 71; 177–92.

● Estrada-Luna D, Ortiz-Rodriguez MA, Medina-Briseño L, et al. Current Therapies Focused on High-Density Lipoproteins Associated with Cardiovascular Disease. Molecules. 2018: 23, pii: E2730.

● Ho CY, Charron P, Richard P, et al. Genetic advances in sarcomeric cardiomyopathies: state of the art. Cardiovasc Res. 2015: 105; 397–408.

● Bradley DT, Badger SA, McFarland M, Hughes AE. Abdominal Aortic Aneurysm Genetic Associations: Mostly False? A Systematic Review and Meta-analysis. Eur J Vasc Endovasc Surg. 2016: 51; 64–75.

● Ramsey AT, Chen LS, Hartz SM, et al. Toward the implementation of genomic applications for smoking cessation and smoking-related diseases. Transl Behav Med. 2018: 8; 7–17.

● Evangelou E, Warren HR, Mosen-Ansorena D. Genetic analysis of over 1 million people identifies 535 new loci associated with blood pressure traits. Nat Genet 2018: 50; 1412–25.

● Nasykhova YA, Barbitoff YA, Serebryakova EA, et al. Recent advances and perspectives in next generation sequencing application to the genetic research of type 2 diabetes. World J Diabetes. 2019: 10; 376–95.

● Mytilinaiou M, Kyrou I, Khan M, et al. Familial Hypercholesterolemia: New Horizons for Diagnosis and Effective Management. Front Pharmacol. 2018: Jul 12; 9: 707.

● Kessler T, Schunkert H. Coronary Artery Disease Genetics Enlightened by Genome-Wide Association Studies. JACC Basic Transl Sci. 2021: 6; 610–23.

● Turner AW, Wong D, Khan MD, et al. Multi-Omics Approaches to Study Long Non-coding RNA Function in Atherosclerosis. Front Cardiovasc Med. 2019: 6; 9.

 # Discussion questions

6.1 What are the anatomical and pathophysiological similarities and differences between endothelial cells and vascular smooth muscle cells?

6.2 Describe the various causes of arrhythmia.

The risk factors for cardiovascular disease

The previous chapter has provided a firm understanding of the aetiology of the various cardio-vascular diseases (CVDs), many of which arise as a result of well-defined risk factors and/or gene mutations. However, before discussing clinical aspects of the diseases themselves, we must first address the aetiology, pathophysiology, and management of these risk factors. Indeed, finding a CVD without any risk factors can be difficult. For each risk factor, we will discuss how it comes to be discovered, perhaps by the patient themselves, followed by diagnosis and management.

The dominant disease process in CVD is atherosclerosis, initiated by the action of the four major risk factors of tobacco smoking (section 7.1), hypertension (section 7.2), dyslipidaemia (section 7.3), and diabetes (section 7.4) on the endothelium, its target. However, there is more to CVD than atherosclerosis, as the risk factors may also cause non-atherosclerotic disease, such as heart failure. Furthermore, numerous diseases and syndromes have a genetic aetiology (such as certain forms of cardiomyopathy), and some an inflammatory basis (endocarditis may initiate valve dis-ease). It could be argued that, in fact, there are cases where a particular risk factor in itself does not cause disease—the related pathology comes from the effects of those risk factors on certain organs (the heart as in CVD and the lung as in cancer). Indeed, the ONS in 2021 reported no deaths linked to dyslipidaemia or smoking in England and Wales, but 7,342 due to diabetes, many of which would be linked to renal disease and peripheral artery disease.

Diabetes is generally defined as an endocrine disease (Chapter 15), and whilst the immediate biochemical consequence is hyperglycaemia, this later leads directly to atherosclerosis. This justi-fies the inclusion of this risk factor in the present chapter, not merely because it is part of the treat-ment profiles for both hypertension and dyslipidaemia, and, as described in Chapter 8, in those with clear CVD. Diabetes adds to the risk of CVD in both hypertension and dyslipidaemia in that the blood pressure and cholesterol targets are lower than in those free of diabetes.

Other risk factors, such as obesity and hyperhomocysteinaemia, have (as yet) failed to reach the level of importance of the 'big four', which dominate clinical practice. Obesity (BMI >30 kg/m^2) is certainly an independent risk factor for numerous cancers, deep-vein thrombosis, and osteoar-thritis, but in CVD this is less evident as its effect may in many cases be accounted for by diabetes, or it may simply be a marker of disease without actually participating in pathophysiology. There is also a growing consensus that renal failure may be an independent risk factor, although much renal disease is the consequence of other factors, as we have noted. Certainly, the most powerful risk factors for death are age and male sex, features that cannot be changed. Both of these com-pound all of the other risk factors, and in both sexes age is linked to many other disease processes, such as cancer. There is considerable evidence of roles for inflammation in atherosclerosis, such as leukocytes within atheroma, and an acute phase response. This is discussed in section 7.5.

Learning objectives

After studying this chapter, you should confidently be able to:

■ explain the role of risk factors in the development of CVD

■ describe the repercussions of tobacco smoking

■ outline the management of hypertension

■ list key features in the pathology of dyslipidaemia

■ identify key components of the pathophysiology of diabetes

■ summarize the evidence that inflammation is a risk factor for CVD

7.1 Tobacco smoking

No healthcare professional can be in any doubt that this is probably the leading most avoidable cause of mortality and morbidity in the world (Chapter 4, sections 4.1.1 and 4.3.1, and Chapter 6, section 6.6.1).

7.1.1 Epidemiology

One of the triumphs of epidemiology has been the classic work of Richard Doll and Austin Bradford Hill in discovering the association of cigarette smoking with lung cancer and heart disease, as described in section 2.4.2. The effect on heart disease is so strong that non-smoking wives of current or ex-smoking men (i.e. 'passive' smokers) have an increased risk of ischaemic heart disease, and young adults who as children were passive smokers demonstrate endothelial dysfunction. Now considered one of the major preventable causes of death globally, it is now abundantly clear that this association is not one of simple coincidence but is causative. However, the causes of smoking itself are of course complex and sociological, but notably, a leading UK institute for research into smoking and its cessation is based in the Maudsley Hospital in South London. Since the Maudsley is a mental health institute, it follows that smoking can be considered a psychiatric disease, akin to self-harm. Few of us can be unaware of the continuing public health campaign to prevent people from starting to smoke, and to help those who continue to smoke to stop doing so.

Data from the Global Burden of Disease Study from 2015 report that 18.1% of UK females and 19.9% of males considered themselves to be smokers, compared with 5.4% and 25.0% globally. Compared to never-smokers, current smoking brings a hazard ratio of the risk of any form of CVD of 2.1, falling to 1.4 in former smokers. Hazard ratios for acute coronary events are 2.0 for current smokers and 1.2 for former smokers, whilst for stroke these are 1.6 and 1.2 respectively. These increased risks fall steadily over 10–15 years after cessation to levels in never-smokers. Current global data from the WHO point to 8 million deaths due to smoking each year (11.9% of all deaths), with around 1.3 billion smokers (16.2% of the world's population).

7.1.2 Pathology

The link between smoking and CVD can be attributed to its effect on the endothelium, blood, and on the tissues of the lung. Notably, it is the combustion of tobacco that is important, as nicotine gum/patches/aerosols and chewing tobacco have no cardiovascular consequences (but the latter causes cancer of the mouth, lips, and throat). Precise damaging factors include the hot gas of the smoke, carbon monoxide, toxins (such as aromatic hydrocarbons), and oxidants. As is presented in Chapter 10, haemostasis is the balance between the factors promoting and preventing thrombosis. Evidence of the deleterious effect of smoking on the endothelium and links with thrombosis come from animal models, but also in the form of increased levels of pro-coagulant molecules such as von Willebrand factor and coagulation factor V. Haemostasis is further compounded by a smoker's reduced ability to prevent clots being formed (thrombogenesis) and to remove those clots that have formed (fibrinolysis).

A thrombus is composed of platelets and products of the coagulation pathway. In health, most platelets circulate in a resting state, but those in a smoker's blood are activated, and as such are more likely to participate in thrombogenesis. The laboratory has been crucial in demonstrating these changes to the coagulation pathway and to platelets, but all such methods are firmly research-based. There are no direct routine blood tests for smoking, but it may be determined by serum cotinine, the stable breakdown product of nicotine, and by carbon monoxide in exhaled breath.

A response of the body to any form of damage is inflammation, and the lung is no exception. Pulmonary tissues have a higher number of neutrophil and monocyte/macrophage phagocytes seeking to repair this damage and in response to linked inflammation, and these cells may also contribute to the pathology. At the cellular level, toxic constituents of tobacco smoke have the theoretical ability to promote reactive oxygen species, overwhelming local antioxidant defences. It seems plausible that these may contribute to damage to pulmonary endothelial and epithelial cells (and so cancer), but a direct or indirect effect on the cellular pathogenesis of CVD is difficult to prove. However, toxic metabolites leaving the lung arrive almost immediately at the heart where they may initiate damage and further promote atherogenesis.

In general, blood tests show a low-grade inflammatory picture in smoking, but whether this is an active promotor of atherosclerosis is unclear. Smoking also increases both whole blood and plasma viscosity, which may contribute to an increased risk of stroke (and possibly myocardial infarction). A further component of endothelial physiology is participation in blood pressure regulation by releasing nitric oxide which acts on the smooth muscles of the artery wall. There is indirect evidence that this process is impaired in smoking, which may contribute to the increased blood pressure (and possibly the higher pulse rate) in those who smoke, although the mechanism is unclear. These effects are summarized in Table 7.1.

7.1.3 Diagnosis

Many smokers readily admit to smoking, but for those who are in denial, expired breath carbon monoxide is a good indicator of smoking in the past 4–6 hours. Nicotine is relatively short-lived in the blood, but breaks down to a stable metabolite, cotinine, which can be determined, although this can only be used to confirm that someone is a smoker in those not taking nicotine as patches, gum, or aerosols.

Key points

Alongside obesity, smoking is the leading preventable cause of CVD and cancer. In the case of the former, the mechanism is by increased thrombosis, which itself may be due to endothelial damage.

TABLE 7.1   Deleterious effects of smoking

Cells/process/pathway/organ	Effect
Endothelial cells	Damage, leading to loss of anticoagulant nature, poor control of vascular tone, and the early stages of atherosclerosis
Platelets	Increased activation: more likely to promote thrombosis
Coagulation pathway	Increased activation: more likely to promote thrombosis
The lung	Damage to alveolar epithelium: likely to promote a general low-grade acute phase response and hypoxia. Major risk factor for many pulmonary diseases (cancer, asthma)
Blood and plasma viscosity	Both increased, putting additional strain on the heart, and a risk factor for stroke

7.1.4 Management

The importance of this risk factor is underlined by the widespread campaigns urging cessation, and the availability of stop-smoking clinics and other initiatives, many of which recommend nicotine replacement therapy (NRT). Naturally, there will be those who can give up with relative ease, but those having difficulty may be carrying the *CHRNA5* mutation (at 15q25.1, which codes for neuronal acetylcholine receptor subunit alpha-5), and so need additional support (see section 6.6.1). NICE offers a number of publications to help those highly dependent on nicotine, to support interventions on smoking prevention in schools, primary care, and in the community (NG92), in acute, maternity, mental health services, pregnancy, and after childbirth, and to support professionals and smokers. TA123 offers an evidence-based guideline on the use of varenicline (a partial agonist of the nicotinic acetylcholine receptor) for smoking cessation. A meta-analysis indicated that use of varenicline for 3 months is superior to both NRT and to bupropion (also a nicotine receptor antagonist, but one that is also used as an antidepressant), and that it is still superior to NRT and bupropion after 12 months.

SELF-CHECK 7.1

What are the pathophysiological links between smoking and CVD?

7.2 Hypertension

A complete understanding of hypertension is impossible without knowledge of the structure and function of arteries, as explained in sections 6.1 and 6.2. The importance of this risk factor is demonstrated by the fact that ICD has a dedicated section on deaths due to hypertensive diseases (of the heart, of the kidney, and secondary to endocrine disorders). Essential (primary) hypertension (i.e. not complicated by other disease) was linked to 1,743 deaths in England and Wales in 2021.

7.2.1 The physiology of blood pressure

The regulation of blood pressure (BP) relies on a complex relationship between nervous impulses and local hormones that act on the heart and on arteries. In health, blood pressure rises and falls as the physiological conditions demand. An example of cardiovascular conditioning is the observation that those taking regular exercise have lower resting BP and heart rate than those who rarely exercise. The most common method for assessing BP is to measure it in the brachial artery in the upper arm. The pioneers of this measurement used the ability of the BP to displace a column of mercury within a vacuum tube to develop a machine called a sphygmomanometer, hence units of millimetres of mercury (mm Hg). SBP and DBP are now measured by an electronic method, several types being available commercially.

Systolic and diastolic blood pressure

Two aspects of blood pressure in peripheral arteries are recognized—the systolic BP (SBP) and the diastolic BP (DBP). The former reflects the force of blood leaving the heart, known as systole, and so effectively assesses the force delivered by the contraction of the left ventricle. DBP reflects the resting or background pressure of the blood when the heart is not contracting—that is, when it is in diastole. SBP and DBP both rise with age, which may be due to factors such as increasing body weight, diet, and reduced exercise.

Assessment of blood pressure

In a clinical setting, the practitioner is likely to assess BP individually in each person, as other factors, such as diabetes or a history of CVD, will lead to a different target blood pressure. Therefore, a seemingly healthy individual with a SBP of 135 mmHg and a DBP of 85 mmHg (that is, 135/85), or possibly a little higher, is unlikely to cause too much concern to a practitioner. However, after a heart attack, the blood pressure target is <140/90, hopefully achieved. Furthermore, if the person has diabetes, the blood pressure target is lower—that is, <135/85. This is because diabetes carries a high risk of developing other CVD.

It follows that hypertension is not necessarily a fixed value of BP, such as >140/90. Indeed, starting from a baseline of 115 mmHg of SBP, the risk of CVD doubles with an increase in BP of 20 mmHg. Many practitioners use BP as part of an individual's total risk of developing CVD, and in the case of high BP, this is mostly likely to result in a stroke. It follows that if an individual is thought of as being at risk of CVD, especially a stroke, they may well benefit from treatment of their high BP. However, for many patients, the experience of having their BP taken in their general practitioner's office, or in hospital, is itself stressful and so may drive up an otherwise acceptable BP measurement. This effect, called 'white coat hypertension', can be detected by measuring BP away from the GP and hospital, and the best way of doing this is with 24-hour ambulatory BP monitoring. An alternative is home BP monitoring, which the patient can do themselves, but for which training is required.

7.2.2 The pathophysiology of hypertension

As the cause(s) of most cases of hypertension is/are unclear, it is often named 'essential hypertension'. Hypertension is more prevalent in African Americans and in Afro Caribbeans in the UK, possibly because of differences in genes linked to blood pressure regulation and sodium excretion, such as those of the renin, angiotensin, and aldosterone system. This is relevant because hypertension is a leading risk factor for stroke, the frequency of which is also increased in these groups (Messerli 1989; Lane et al. 2002). BP also rises with age. One problem in addressing these issues is that there are several forms of hypertension, as evidenced by the different classes of drugs that can reduce BP (section 7.2.6). However, there are a number of probable causes of hypertension:

- It has been suggested that hypertension is secondary to damage to the endothelium in that the latter may not release vasodilators (such as nitric oxide) in the media to act on the smooth muscle cells. It has also been suggested that endothelial damage is a consequence of hypertension, and so there may be a vicious cycle.

- A second possible pathological mechanism is of aberrant nervous stimulation of the smooth muscle cell of the media. Evidence in favour of this is the success of the beta-blocker class of drugs, although these also act on the heart to reduce BP and heart rate.

- There are few clear and major genetic causes of high BP, but those that have been reported suggest genes relating to smooth muscle cell function and renal function (as discussed in Chapter 14).

- Another factor that likely plays a role is a poor diet, especially one with a high level of salt or one that leads to obesity.

High BP may be secondary to many diseases and conditions. These include renal disease (possibly causing **hypernatraemia**), lack of exercise, endocrine disorders (such as Cushing's syndrome), thyroid disease, drugs (such as hormonal contraceptives), Conn's syndrome, and **phaeochromocytoma**.

Whatever the aetiology of hypertension, its effects are serious, with various forms of 'end organ damage' that include retinopathy, left ventricular hypertrophy (leading to left ventricular aneurysm and heart failure), and nephropathy. Curiously, it may have seemed reasonable to believe that the high BP of hypertension is a cause of haemorrhagic stroke, by simply bursting intra-cranial blood vessels. However, stroke in hypertension is mostly thrombotic, and linked to blood hypercoagulability. This could in theory be due to increased levels of von Willebrand factor from a stressed endothelium, but in fact non-endothelial coagulation molecules, such as factor VII, are also increased, for reasons that are unclear.

Hypotension

We know that high BP is dangerous, but the choice of a particular SBP/DBP result of 140/90 is somewhat arbitrary and does not imply that a BP of 139/89 is safe. Indeed, the risk of CVD in the band 130/70 to 139/79 is greater than in the band 120/60 to 129/69—so should we try for a very low BP? The answer is no, because we must maintain a basic level of BP to ensure all the organs are perfused with blood, as demonstrated by the consequences of hypotension. Postural hypotension can happen when someone who had been sitting or lying down suddenly stands up, the cardiovascular system then failing to act rapidly to deliver blood to the brain, hence the temporary loss of consciousness.

Key points

From a clinical perspective, almost all cases of hypertension have an unknown aetiology, although it is likely that a damaged endothelium can be either a cause or effect of the high BP.

7.2.3 Presentation

Hypertension can be asymptomatic well into the range that is considered to be treatable, so that patients can present when BP is dangerously high, often exceeding SBP/DBP >200/100 mmHg. However, the most common complaints are headaches, visual disturbances, tinnitus, and non-specific heart symptoms such as palpitations, and sensations of feeling the heart within the chest, and of the blood within the carotids. Fortunately, hypertension is now being actively sought in primary care, and this is where many cases are discovered, and it may also be found fortuitously in the work-up of other disease. Indeed, hypertension is now such a great problem that some consider it should be screened for in adults every 5 years.

7.2.4 Diagnosis

Every GP and clinic nurse will have a BP monitoring device on their desk (hence 'office' or 'clinic' measurement), and using these machines, BP can also often be determined by other healthcare professionals. BP is measured whilst the patient is sitting, with the cuff on the unclothed upper arm at the level of the heart, and the patient should remain still and silent whilst the measurement is being made. Some potential patients may well be anxious about having their BP recorded, which will therefore respond by being a little higher (and with a slightly higher pulse rate) than if the patient were not anxious. This anxiety-driven response may be described as white coat hypertension (WCH).

For this reason, for the most accurate 'office/clinic' determination BP should be measured three times, with a pause in between, as recorded levels are likely to fall as the patient becomes more relaxed. The reverse of WCH is masked hypertension, where BP is normal in the office/clinic but high elsewhere, such as at home.

Home blood pressure measurement

Consequently, current recommendations suggest that a diagnosis of hypertension should only be made on the basis of either repeated office BP measurements or out-of-office measurement, generally in the patient's home, often by themselves. This is popular as the patients will be more relaxed, with results being lower than when measured in the office or clinic (Figure 7.1).

Ambulatory blood pressure measurement

Should the practitioner still have concerns about the reliability of the office or home measurement, BP can be accurately measured by 24-hour ambulatory BP monitoring (ABPM). The patient is fitted with a modified cuff linked to a pump and recorder carried on a waistband or in a pocket (Figure 7.2). At semi-regular intervals (perhaps 15–30 minutes) over a 24-hour period, BP is measured in the usual way; the patient is warned a few seconds before the pump starts so that they can prepare by becoming calm and still.

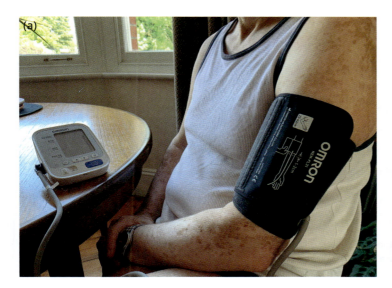

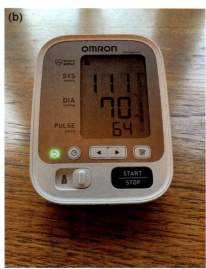

FIGURE 7.1

Home BP measurement. (a) The patient has a cuff on his arm, linked to a monitor on the table. (b) He will be pleased to see his result is good and shows a satisfactory pulse rate.

FIGURE 7.2

Ambulatory blood pressure monitoring. The patient is fitted with a cuff on the upper arm, which is linked to a power pack and recorder at the waist. He is thus able to relax, and ideally provide stress-free data.

Science Photo Library.

When data is downloaded, the serial measurements can be examined and outcomes discussed (Figure 7.3). The reduction in BP is such that hypertension may be present at ≥130/80 by ABPM, ≥135/85 by home measurement, and ≥140/90 by office/clinic measurement. Should sufficient good-quality readings be available for analysis, then daytime and night-time SBP and DBP can be averaged. The day/night aspect is important because of the established fall (dip) in BP when asleep.

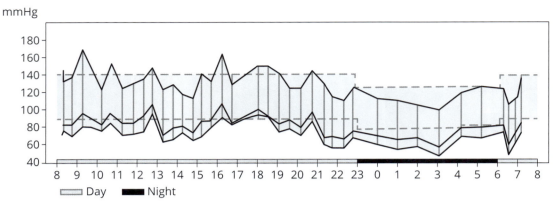

FIGURE 7.3
An ambulatory blood pressure monitoring trace. The graph plots BP on the *y*-axis and time on the *x*-axis. The trace links the SBP and DBP. On the right-hand side, on the *x*-axis, a black bar denotes night, and the BP traces fall, so the patient is a 'dipper'. There are a few peaks during the day, but overall the trace is normal.

Should the trace not fall, then we would describe the patient as a non-dipper, with consequences for treatment.

So, having obtained what is considered to be a reliable measurement, the next question is whether any action should follow. Guidelines for the definition of hypertension are shown in Table 7.2.

SELF-CHECK 7.2

What is the role of the endothelium in blood pressure control?

TABLE 7.2 European and US guideline categories

Category	Systolic blood pressure		Diastolic blood pressure
Optimal Eur. Normal USA	<120	and	<80
Normal Eur.	120–9	and/or	80–4
Elevated USA	120–9	and	<80
High normal Eur.	130–9	and/or	85–9
Stage 1 hypertension USA	130–9	or	80–9
Grade 1 hypertension Eur.	140–59	and/or	90–9
Stage 2 hypertension USA	≥140	or	≥90
Grade 2 hypertension Eur.	160–79	and/or	100–9
Grade 3 hypertension EUR	≥180	and/or	≥110
Isolated systolic hypertension Eur.	≥140	and	<90

Eur. = European guidelines (Williams et al. 2018), USA = US guidelines (Whelton et al. 2018).

7.2.5 General concepts of management

Precise management depends on the age, co-morbidities, and risk factors of the patient. Broadly speaking, the target is at least 140/90, but this will be lower in certain circumstances. Should the patient be relatively young and free of other issues, many would consider this target of 140/90 to be appropriate.

In the light of other factors, such as when cardiovascular risk is very high due to diabetes or exist-ing CVD, especially coronary artery disease, then the target may be reduced to 130–9/85–9, or lower in selected patient groups. Should there be a need for treatment, a full medical examination and other investigations (blood tests for lipids, diabetes, renal function etc., urine analysis, 12-lead elec-trocardiogram (ECG)) must be made to determine the likelihood of other issues, such as poor renal function, retinopathy, and left ventricular hypertrophy (LVH). A further useful tool is to assess an individual's overall cardiovascular score, one of the more pertinent being the Systematic COronary Risk Evaluation (SCORE) system, based on large cohorts of Europeans. Data required are age, sex, smoking status, cholesterol, and SBP. Referring to charts, this estimates the 10-year risk of a first fatal atherosclerotic event.

As with most diseases, the first step in treatment is often lifestyle interventions: attention to diet (e.g. reducing sodium, more fresh fruit and vegetables), weight (aim for BMI 20–25 kg/m^2), exercise (ide-ally 300 minutes of moderate exercise or 150 minutes of strenuous exercise a week), smoking (stop), alcohol (men <14 u/week, women <8), etc. However, although effective in reducing BP, it may not be enough, and a pharmacological approach may be necessary.

7.2.6 Classes of antihypertensives

Drugs are classified according to mode of action and cellular or molecular targets. The peptide hormone angiotensin is an important vasoconstrictor and is formed by cleavage of a part of its liver-derived precursor, angiotensinogen, by renin (from the kidney) and angiotensin-converting enzyme (ACE, on the pulmonary and renal endothelium). Thus inhibition of ACE will lead to failure to generate angiotensin, and so a reduction in vasoconstriction that translates to a fall in BP, hence ACE inhibitors (ACEi). Angiotensin (or more correctly, angiotensin II) exerts its effect on arteriolar smooth muscle cells and cells of the kidney, adrenal cortex, and posterior pituitary by docking into specific receptors. Preventing the interaction between the ligand and its receptor (hence angiotensin receptor blocker, ARB) will therefore prevent the constriction of the artery and so lead to a reduction in BP. Because ACEIs and ARBs operate on different parts of the same pathway, they may not be used together. Use of either is often referred to as renin–angiotensin system (RAS) blockade.

The **sympathetic nervous system** operates via adrenergic receptors, of which there are two types—alpha and beta. These receptors are present in various locations around the body, including heart, liver, bronchi, pancreas, and the arterial system where smooth muscle cells are prompted to constrict via beta-adrenergic innervation. Thus blocking this interaction will lead to a reduction in constriction and so a fall in BP, hence beta-blockers. A major issue is that as the bronchus has beta-receptors, certain blockers should be avoided in those with asthma. There are specific recommenda-tions for the use of these drugs in angina, myocardial infarction, arrhythmias, and heart failure. The alpha receptor may also be blocked, which leads to a reduction in BP, and these inhibitors are also prescribed for benign prostatic hyperplasia.

Calcium is a very important element in many ways (bone, coagulation, to name but two), but also participates crucially in the cation homeostasis within the cells, levels being regulated by calcium channels. As we have seen in Chapter 3, calcium is a key player in many second messenger pathways, and increased levels have consequences, such as in the contraction and relaxation of smooth muscles cells. Thus, blockade of calcium channels (hence CCB) will result in changes to intracellular levels of this cation, and so smooth muscle relaxation and thus reduction in BP.

Hypertension may be simply due to a large blood volume, one treatment for which is to induce the kidney to excrete water, using diuretics (i.e. making urine), although this may lead to elec-trolyte disturbances in the blood. To partly get around this problem, different classes of diuret-ics have been developed—some acting on the loop of Henle, others on the convoluted tubules

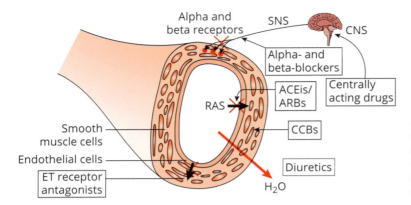

FIGURE 7.4

Mechanism of action of anti-hypertensive drugs. RAS = renin–angiotensin system, ET = endothelin, SNS = sympathetic nervous system, CNS = central nervous system. The major classes are ACEi/ARBs, CCBs, beta-blockers, and diuretics.

(the thiazides), whilst others have different effects on calcium and potassium. Use of these drugs requires care as too much diuresis will dehydrate, and some may lead to hyper- or **hypokalaemia**, which has consequences for cardiac function. Diuretics are also used to treat oedema (lung, ankle, cerebral). Other anti-hypertensive drugs include those acting directly on the central nervous system, mannitol (an osmotic diuretic), **mineralocorticoid** antagonists, hydralazine, minoxidil (both as an adjunct in resistant hypertension), and endothelin antagonists. Endothelin is a potent vasoactive peptide secreted from endothelial cells that acts on medial smooth muscle cells. Certain drugs (such as bosentan) are used only to treat pulmonary arterial hypertension.

A revolutionary treatment announced in 2023 was the subcutaneous delivery of a small interfering RNA molecule (zilebesiran, targeting hepatic angiotensin mRNA), which reduced blood pressure for 24 weeks.

Therefore, hypertension can be treated in various ways. Whilst some cases can be treated by only one class of drugs, most patients will need two or more classes to control their BP. The actions of different drug classes are summarized in Figure 7.4.

7.2.7 Management of major patient groups

With BPs >140/90, drug treatment should start immediately, depending on circumstances. BP reduction should be to target, or lower if possible, as long as the regime is tolerated. Notably, SBP should not fall below 120 mmHg in those with uncomplicated hypertension, or if complicated by diabetes, chronic kidney disease, coronary artery disease, or stroke/transient ischaemic attack, as this has been linked to increased risk of major cardiovascular events. All regimes may need to be modified in certain circumstances, such as in those who are both frail and living independently.

Uncomplicated hypertension

If there is no other major pathology, monotherapy may be considered in low-risk cases where the SBP is <150, or in those frail or age >80, but generally, a stepwise approach is taken:

- Step 1: An ACEi or ARB plus a CCB *or* a diuretic

- Step 2: An ACEi or ARB plus a CCB *and* a diuretic

- Step 3: As step 2 with added spironolactone, or other diuretic, alpha-blocker, or beta-blocker

- Step 4: If resistant: referral to a specialist centre in secondary or tertiary care

The patient with resistant hypertension is likely to be older (age >75), obese, diabetic, and/or with renal disease, and with high dietary sodium. Wherever possible, reinforcement of lifestyle measures (especially sodium restriction) is advised, and with the addition of low-dose spironolactone to existing treatment. If the patient is intolerant of spironolactone, further diuretics may be required, such as either eplerenone, amiloride, a higher dose of a thiazide or thiazide-like diuretic, or a loop diuretic.

BOX 7.1 The kidney and hypertension

These two subjects are worthy of further attention for a number of reasons as CKD can be both the cause and consequence of hypertension, although there is overlap and a chicken-and-egg situation may arise.

High BP (of whatever cause) acting on a healthy kidney can cause intimal thickening and sclerosis of intra-renal arterioles with fibrin deposition, tubular atrophy, and necrosis and so poor renal function, which may result in the retention of water and salt, and so further haemodynamic stress in a vicious cycle. It may also contribute to an atheroma causing renal artery stenosis that will reduce blood flow into the kidney, which will then become ischaemic and fail in its excretory and other functions. Should these pathological processes continue, there may well be glomerulonephritis.

An external pathology causing acute kidney injury (AKI) and/or CKD may lead to hypertension, the principal pathology being alterations in the renin–angiotensin system (RAS). In health, renal juxtaglomerular cells secrete renin that converts angiotensinogen to angiotensin I. Renal and pulmonary endothelial ACE then converts this angiotensin II, which mediates arteriolar constriction, sodium excretion, and aldosterone secretion. It follows that damage to the juxtaglomerular cells will activate the RAS and so promote increase in BP. Glomerulonephritis caused by another mechanism (such as an infection) may lead to water and sodium retention and so increased blood volume and so hypertension. Of course, in some patients, both mechanisms may be present at the same time—ischaemia may also activate the RAS. Other details of renal disease are presented in section 14.1.

An alternative is the addition of bisoprolol or doxazosin, and the patient is likely to be transferred to the care of a specialized centre.

Coronary artery disease

Should hypertension be present in patients with coronary artery disease, more care is required, although monotherapy is still an option in low-risk patients where SBP <150, or in those frail or aged >80. Therapy should be started when SBP ≥130 in these high-risk patients a common target being <135/85. The major difference from uncomplicated disease is the use of beta-blockers post myocardial infarction, to be added at all steps, whilst step 4 may be quadruple therapy.

Renal disease

Hypertension is common in chronic kidney disease (CKD, defined as eGFR <60 ml/min/1.72m^2 with or without proteinuria), as both a cause and a consequence of renal failure (Box 7.1). Patients in this group will have frequent tests of renal function—urea and **electrolytes** and the urinary albumin to creatinine ratio. Accordingly, the SBP target should be in the range 130–9. Loop diuretics should be used when the eGFR is <30 ml/min/1.72 m^2 because thiazide/thiazide-like diuretics are much less effective or are ineffective when eGFR is reduced to this level. Practitioners must be aware of the possibility of hyperkalaemia when using spironolactone, especially when the eGFR is <45 or potassium is ≥4.5 mmol/l.

SELF-CHECK 7.3

What are the four major classes of drugs for treating high blood pressure?

Heart failure

The leading risk factor for heart failure (with reduced or preserved ejection fraction) is hypertension, and this also causes LVH. Fortunately, the latter may be reversed by treatment. Treatment should be initiated (if not already prescribed) when the BP is >140/90, and then reduced further according to co-morbidities down to 120/70. Due to the high risk of a major cardiovascular event, and disease progression, step 1 is triple therapy (ACEi or ARB + thiazide or thiazide-like diuretic + beta-blocker) with addition of a further class in step 2 (a mineralocorticoid receptor antagonist: spironolactone or epelerone).

Atrial fibrillation (AF)

This, the most common arrhythmia, is a leading cause of stroke, and hypertension is a leading cause of AF and other arrhythmias. As tachycardia (high pulse rate) is common, beta-blockers or CCBs are helpful in reducing a high rate. Step 1 has two options, both calling for dual therapy: (a) ACEi or ARB + beta blocker or a CCB, (b) beta blocker + a dihydropyridine CCB. Should step 2 be necessary, a diuretic is added. The patient may be on oral anticoagulation, so practitioners will need to be aware of the risks of haemorrhage. The CCB must be a dihydropyridine, as non-dihydropyridines (e.g. verapamil or diltiazem) in this combination may cause a reduction in the heart rate.

> ### Key points
>
> **Management of hypertension in the face of other disease is complex and requires specialist attention: in all cases consideration of renal function is vital.**

Management of special cases

Hypertensive disorders are present in 5–10% of pregnancies, are part of the aetiology of pre-eclampsia (alongside oedema and proteinuria), and are linked to placental abruption, intrauterine growth retardation, and premature and intrauterine death. Women with hypertension who are considering pregnancy, and those already pregnant, should be treated with alpha methyl dopa, labetalol, or nifidipine. In neither instance should an ACEi, ARB, or a diuretic be used as they are linked to congenital malformations (section 14.4.2). Guidelines recommend treatment of all women with BP >150/95, but at >140/90 in **gestational** hypertension, which usually resolves by 6 weeks post-partum.

Hypertension is common in types 1 and 2 diabetes, and with their increased risk of cardiovascular and renal disease, target BP should be lower at 130/80 mmHg. However, if tolerated, BP can be reduced down to 120/80 in order to give greatest protection from stroke. Diabetic nephropathy may also be present, leading to a degree of renal dysfunction, so that changes in doses may be required (see Table 7.3 and related text). Hypertension is the most common co-morbidity in chronic obstructive pulmonary disease, and practitioners must be aware that this has an impact on choice of treatments.

Secondary hypertension

The hypertension we have described so far is 'essential'—that is, without an obvious organic cause. By contrast, secondary hypertension, which comprises 5–15% of cases, is that which is known to be caused by a precise factor, and is more likely in the young, in those with resistant hypertension, with hypertension-mediated organ damage (LVH, retinopathy, etc.), and in those whose BP control has become difficult.

Pathological conditions known to increase BP include obstructive sleep apnoea (snoring: present in 5–10% of cases of secondary hypertension), renal disease (parenchymal, fibromuscular dysplasia, or atherosclerotic, 3–20%), primary aldosteronism (5–15%), phaeochromocytoma (<1%), endocrine causes (Cushing's syndrome, thyroid disease, **hyperparathyroidism**: <1–2%), and coarctation of the aorta (<1%). Fortunately, many of these are relatively easy to diagnose by imaging and laboratory tests. Medication known to increase BP includes the oral contraceptive pill, diet pills, nasal decongestants, stimulant drugs, liquorice, immunosuppressants, certain anti-cancer drugs, anabolic steroids, non-steroidal anti-inflammatory drugs, and **erythropoietin**. As discussed in 6.6.1, Genetic causes of hypertension include the Liddle syndrome, an autosomal dominant disease caused by a loss-of-function mutation in *SCNN1B* or *SCNN1G* at 16p13–12 coding for epithelial sodium channel subunits.

7.2.8 Choice of drug

There are various drugs in each class (as shown in Table 7.3), allowing for personal preference and the exact clinical need of the patient (e.g. when the CCB should be a dihydropyridine). Some have short half-lives, requiring dosing twice a day, and there is variation in renal excretion, so doses may need to be adjusted in the face of CKD. Unsurprisingly, many patients have difficulty in juggling all their drugs (which may lead to non-compliance), and accordingly manufacturers have responded by combining

TABLE 7.3 Classes of selected anti-hypertensive drugs

Class		Drugs	Comments
Alpha-blocker		Doxazosin, Prazosin, Terazosin	May cause rapid BP fall with first doses: caution required
ACEi		Captopril, Enalapril, Lisinopril, Quinapril, Ramipril	May cause angioedema and a cough
ARB		Candesartan, Irbesartan, Losartan, Olmesartan, Telmisartan, Valsartan	Many properties similar to ACEis but do not cause cough.
Beta-blocker		Atenolol, Bisoprolol, Carvedilol, Labetolol, Metropolol, Nebivolol, Propranolol, Sotalol	Some also reduce the heart rate: some are lipid soluble, others water soluble, and so are less likely to enter the brain, but are excreted more readily
Calcium channel blocker		Amlodipine, *Diltiazem, Felodipine, Lercandipine, Nicardipine, Nifedipine, *Verapamil	Act on the myocardium and vascular smooth muscle cells. May also be used in heart failure (except*, as may be bradycardic) and angina
Centrally acting		Moxonidine	Agonist of the imidazoline receptor
		Methyldopa	May be used in pregnancy
		Clonidine	May cause rebound hypertension
Diuretic	Thiazides and related drugs	Bendroflumethiazide, Chlortalidone, Indapamide,	Inhibit sodium resorption at distal convoluted tubule.
	Loop diuretics	Bumetanide, Furosemide, Torasemide	Inhibit reabsorption from the ascending loop of Henle
	K-sparing and aldosterone antagonists	Amiloride	Retains K, so is given with a thiazide or loop diuretic (e.g. furosemide, see Case study 7.1)
		Spironolactone, Epelerone	Aldosterone antagonists: also beneficial in heart failure
	K-sparing diuretics combined with other drugs	Co-amilozide	Combination of amiloride and hydrochlorothiazide
		Co-amilofruse	Combination of amiloride with furosemide
		Co-triamterzide	Combination of triamterene with hydrochlorothiazide

Table intended to be neither exhaustive nor fully inclusive. Key references: British National Formulary and Williams et al. 2018.

commonly prescribed drugs of different classes into a single tablet. NICE NG136, based on evidence-based medicine (such as Helmer et al. 2018), recommends considering the use of an angiotensin II receptor blocker over an angiotensin-converting enzyme inhibitor in adults of Black African or African Caribbean descent. The most recent guidelines from the USA also recommend race-specific choices of anti-hypertensive therapy (Whelton et al. 2018), as do the International Society for Hypertension

CASE STUDY 7.1 *Blood pressure medication*

A 66-year-old man is brought to A&E, his wife having called 999 and reported that he collapsed and was poorly responsive. On further questioning she stated that he had been increasingly weak over the last 2 weeks with diarrhoea requiring several visits to the toilet each day and at night. On initial examination he is dehydrated with SBP/DBP 90/60, with a heart rate of 90 beats per minute. Routine blood tests are taken and he is immediately started on intravenous saline. Fortunately, he is known to the hospital, and to the cardiologists in particular. Past medical history is heart failure due to dilated cardiomyopathy (probably related to long-standing hypertension) and non-obstructive coronary artery disease. At his last outpatient visit, his left ventricular ejection fraction was 11% (criterion for heart failure, <30%), blood pressures 120/78, heart rate 58, blood results were consistent with stable CKD, and there was a ventricular tachycardia on ECG. His wife brought in his tablets, which were co-amilofruse (a combination of two diuretics: 40mg of amiloride and 5 mg of furosemide), perindopril (an ACE inhibitor, 4 mg), bisoprolol (a beta-blocker, 2.5 mg), spironolactone (an antagonist of the

mineralocorticoid receptor, 50 mg) and amiodarone (an anti-arrhythmic, 100 mg).

Whilst his history is being assessed, blood results are phoned through from the laboratory, staff being concerned to find exceedingly abnormal urea and creatinine, consistent with acute renal failure. All other blood tests are within reference range. He is immediately admitted, and over the following 36 hours is infused with 7.5 litres of fluids, followed by marked **diuresis**, and is discharged a week later. Blood results slowly normalize, as shown in Table 7.4. Once communicative, the patient reports his GP had recently started him on the spironolactone and, in the few days before his collapse, he noted that he was making increasingly less urine.

The likely pathophysiology is that the spironolactone caused the diarrhoea, leading to excessive fluid loss that precipitated the clinical picture of anuria, dehydration, tachycardia, acute renal failure, and so collapse. Fortunately, the clinical issues were resolved by the restoration of blood volume. The question that remains is why the GP started the patient on an additional blood pressure tablet in the face of apparently good vascular control.

TABLE 7.4 Serial blood results

	Urea	Creatinine	Bicarbonate	Sodium	Potassium
Reference range	3.3–6.7	71–133	24–9	135–45	3.8–5.0
Previous outpatient values	11.9	152	29	139	4.6
Day 1	70.0	896	19	138	6.9
Day 2	56.6	624	16	142	5.2
Day 3	38.5	352	19	137	5.6
Day 4	26.8	226	20	138	5.2
Day 5	23.5	202	19	140	5.5
Day 6	18.2	163	22	138	4.6
Day 7	14.0	155	21	137	3.7
1 month post-discharge	7.2	130	29	139	4.4

guidelines (Unger et al. 2020). However, the role of race in British hypertension guidelines has been questioned in a commentary (Gopal et al. 2022).

7.3 Dyslipidaemia

This section is crucial for a full understanding of the pathogenesis of atherosclerosis, as described in section 6.4.

7.3.1 Lipid biochemistry

Most lipids can be classed as triglycerides (also known as triacylglycerols) and cholesterol, and can combine with molecules of other classes, giving rise to lipoproteins, phospholipids, sphingolipids, and glycolipids. A fat is essentially a hydrophobic biological molecule of very low solubility in water, and perhaps is completely insoluble. However, the attachment of certain chemical groups makes fat more soluble, and so hydrophilic, and this can be achieved by the presence of other molecules, such as an acid group, hence fatty acid.

Triglycerides

There are several classes of triglycerides, all with the common structure of a backbone of glycerol, a three-carbon molecule, each atom of which is linked to a long chain fatty acid (an acyl group—typically 12–20 carbon atoms in length) (Figure 7.5).

Examples include lauric and oleic acid, which are the starting point for the synthesis of various molecules involved in platelet and vascular biology. Almost any three fatty acids may combine with glycerol, and name the molecule. For example, 2-stearodipalmitoylglycerol is composed of a stearic acid and two molecules palmitic acid. Triglycerides can be hydrolysed back to their component fatty acids and glycerol (i.e. lipolysis) by lipase enzymes whose synthesis may be part-regulated by hormones, notably insulin and glucagon.

Cholesterol

The structure of cholesterol is complex and markedly different from that of triglycerides. Although present in plants, in mammals it is synthesized in the liver by a series of complex biochemical steps based on a five-carbon isoprene unit and acetyl-coenzyme A. Several isoprene units fuse to give a ring structure (Figure 7.6).

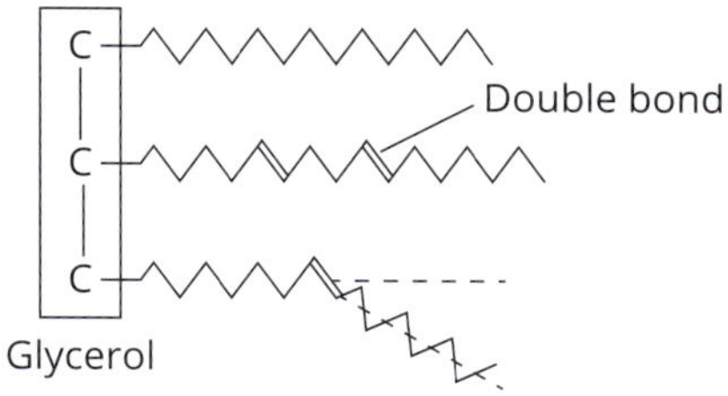

FIGURE 7.5

Structure of triglycerides. A triglyceride consisting of a three-carbon glycerol backbone and three long chain fatty acids. The topmost has only single carbon–carbon bonds, the middle two double bonds. Both are the trans-isomer form. The lower fatty acid is in the cis-isomer form, so the chain is angled.

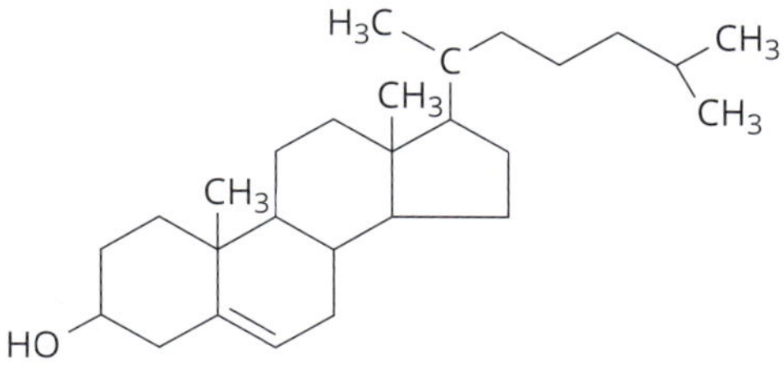

FIGURE 7.6

Structure of cholesterol. Cholesterol is formed of four carbon rings, five methyl groups, and an extended carbon chain (top right). The hydroxide group on the lower left can link with a fatty acid to give a cholesterol ester.

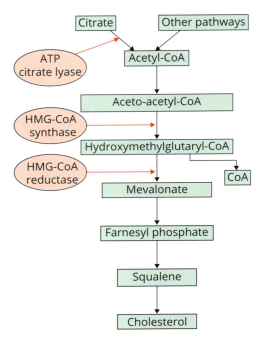

FIGURE 7.7

Key steps in the synthesis of cholesterol, showing the position of selected enzymes.

The greater proportion of cholesterol (perhaps 66–85%) is synthesized in the liver and exported into the plasma, where it combines with triglycerides and lipoproteins in several forms to enable it to be transported to other tissues and organs for use in metabolic processes. Key early steps include the synthesis of cholesterol from acetyl-coA (requiring the enzyme ATP citrate lyase), and hydroxymethylglutaryl-CoA (HMG CoA) from acetyl-CoA and acetoacetyl-CoA, followed by its reduction to form mevalonate. This step is regulated by the enzyme HMG CoA reductase (Figure 7.7). These two enzymes are the object of inhibitors, which can be taken orally, and which therefore reduce levels of cholesterol in the blood. Accordingly, they (and others) may be prescribed those with high cholesterol, but also those with CVD (post-myocardial infarction, stroke) regardless of their serum cholesterol, as discussed in section 7.3.6.

SELF-CHECK 7.4

From Figure 7.6, determine the approximate relative molecular mass of cholesterol (C = 12, H = 1, O = 16).

Cholesterol is the starting point for the synthesis of **corticosteroids** in the adrenal cortex and the sex hormones oestrogen, progesterone, and testosterone in the gonads, and it is also a component of the cell membrane. Various lipids and proteins combine to form microparticles, which can be classified by their size and density as chylomicrons, and very low-density (VLDL), low-density (LDL), intermediate-density (IDL), and high-density (HDL) lipoproteins.

Key points

Contrary to popular belief, fats and lipids are not always bad: several are essential to our wellbeing. The two major groups, triglycerides (as source of energy) and cholesterol (a component of the cell membrane), have markedly different structures and biochemistry.

Lipid conjugates

Lipids can combine with various other molecules, which bring their own particular chemistry and roles in pathophysiology.

- Phospholipids, making up ~half of the cell membrane, are related to triacylglycerols. Key members of this group are phosphatidylserine and phosphatidylcholine (lecithin).

- Sphingolipids, based on a serine molecule, include ceramide and sphingomyelin.

- Addition of galactose to ceramide generates a cerebroside, whilst addition of several carbohydrates makes it a ganglioside.

- Lipids may also become more soluble by combining with apoprotein (shortened to apo) of which there are five types, apoA to apoE, to form a lipoprotein. ApoB deserves special attention: $apoB_{100}$ (synthesized by **hepatocytes**) is the protein component of the LDL particle. $apoB_{48}$ (synthesized by **enterocytes**) is so named as it is composed of 48% of the larger molecule, being the product of a splice-edited and so truncated variant of *apoB* (located at 2p24.1).

Enzymes, transporters, and transfer molecules

A small number of other molecules complete our picture of the metabolism of lipids and lipoproteins.

- Lipoprotein lipase (LPL: EC 3.1.1.34) digests triglycerides back to their constituent fatty acids and glycerol (as happens in chylomicron metabolism).

- Cholesterol ester transfer protein (CETP), secreted primarily from the liver, promotes the transfer of cholesterol esters and triglycerides between various lipid species.

- Liver-derived lecithin cholesterol acyltransferase (LCAT), which apoA1 binds, mediates the transfer of a fatty acid from phosphatidylcholine (lecithin) to cholesterol.

- Phospholipid transfer protein (PTP) is also important in reverse cholesterol transport, transferring phospholipids between various lipoprotein particles.

- The collection of cholesterol and phospholipids from the tissues which are passed to apoA1 and apoE in HDL is mediated via ABC (ATP-binding cassette) transporter type 1 (ABCA1, coded for by *ABCA1* at 9q31.1).

Key points

Many lipids are non-polar, and in this form can be difficult to metabolize. Accordingly, they often conjugate to, or are closely associated with, non-lipids such as proteins and carbohydrates, which bring a non-polar chemistry, and so solubility.

7.3.2 Absorption and processing of lipids

Dietary fat consisting of coarse oil globules is digested by intestinal lipase and emulsified by bile salts, and so is converted into spherical **micelles**. These particles enter enterocytes where additional metabolism takes place, with the addition of apoproteins (such as apoE and $apoB_{48}$), converting some into lipoproteins. In the intestines, cholesterol and a fatty acid may be fused into a cholesterol ester by LCAT. Cholesterol in food is absorbed mainly in the upper part of the small intestines (the duodenum and jejunum) with other fats and bile salts, to form mixed micelles.

Chylomicrons

Cholesterol and triglycerides merge and subsequently enter the blood as large particles (chylomicrons) that transport the lipids to the tissues. These nascent chylomicrons are composed mostly of triglycerides with a minor component of cholesterol/cholesterol-esters, phospholipids, apoA1, and $apoB_{48}$, whilst apoE and apoC from HDL are soon added. In the capillaries of the heart, skeletal muscle, and adipose tissues, the triglycerides in these 'mature' chylomicrons are digested by lipoprotein lipase (coded for by *LPL* at 8p21.3) down to glycerol and fatty acids.

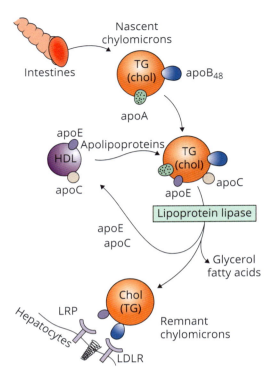

FIGURE 7.8

Chylomicron metabolism. Changes in chylomicrons in their journey from the intestines to the hepatocyte. For clarity, only directly relevant molecules and particles are shown (see Figure 7.10 for HDL metabolism).

With the loss of triglycerides, chylomicrons become progressively smaller, lose density, and become more cholesterol-rich, at which point they may be described as chylomicron or lipoprotein remnants. These remnants may then be taken up by hepatocytes (and to lesser extent by macrophages) via an interaction between apolipoproteins and two receptors. These are the LDL receptor (LDLR) and the LDL receptor-related protein (LRP or LRP1) (described in section 6.6.1) (Figure 7.8).

Intermediate and Very Low-Density Lipoproteins (IDL, VLDL)

VLDL is formed in the liver from triglycerides, cholesterol, $apoB_{100}$, apoC1, and apoE, and picks up apoCII and more apoE from HDL whilst in the blood. It transports lipids and lipoproteins from the liver to the tissues, primarily muscle and adipose tissues. The acquisition of these additional apoproteins and the digestion of triglycerides by lipoprotein lipase result in an increase in density and its transformation into intermediate density lipoprotein (IDL). Markedly smaller IDLs are characterized by little (if any) apoC1 and apoC2, but by high levels of apoE. The latter is recognized by receptors on hepatocytes, so that IDLs may be absorbed and recycled. As IDLs mature they lose their apoE, so that the dominant lipoprotein becomes $apoB_{100}$, at which point they are recognized as LDLs.

Low-density lipoprotein (LDL)

LDLs are the final particle in the lipoprotein pathway and are predominantly absorbed by the hepatocyte, less so by adipose tissues and the adrenals, via an interaction between $apoB_{100}$ and the LDLR, which also recognizes apoE. The consequences of loss-of-function mutations in the gene for the receptor include failure of the hepatocytes to absorb LDL, so that levels in the blood remain high (discussed in section 7.3.3).

Whilst in the blood, LDLs can be modified by oxidation (becoming oxidized LDL) or glycation, and so can be recognized by non-specific scavenger receptors (such as CD36) on the surface of hepatocytes, macrophages, and other cells. Once the LDLR has bound its ligand (i.e. LDL), it is internalized within the cell, gives up its LDL, and returns to the surface. This process is part-regulated by the activity of proprotein convertase subtilisin/kexin type 9 (PCSK9), which chaperones the receptor to lysosomes for digestion (Figure 6.18). Gain of function mutants of *PCSK9* (coded at 1p32.2) lead to high levels of this molecule, and so rapid removal of the LDLR, a consequence of which is high plasma LDL. Conversely, loss-of-function mutations lead to low levels of PCSK9, and so persistence of the receptor, which then is able to clear LDL all the more rapidly, leading to low plasma LDL. Overall, therefore, lipoprotein particles evolving from chylomicrons to LDL lose triacylglycerols and gain proteins. Figure 7.9 summarizes the metabolism of VLDL, IDL, LDL, and oxLDL.

High-density lipoprotein (HDL)

High-density lipoproteins (HDLs) are the smallest of the lipoproteins and composed almost entirely of cholesterol, phospholipids, and apoproteins; mostly apoA1 and apoA2, with smaller amounts of apoC2 and apoE. Their high density relative to other lipoproteins is due to their lack of triglycerides and the fact that almost half of them is composed of proteins. HDL is predominantly synthesized by the liver, but some is generated in enterocytes.

Nascent HDL is generated by apoA1 and cholesterol from the blood and tissues (macrophages, adipocytes, possibly others) via ABCA1, and further cholesterol is added via ABCG1/4, and takes a spherical shape, at which point it is termed HDL_3. It then exchanges cholesterol and triglycerides with IDLs and LDLs, mediated by cholesterol ester transport protein (CETP), and so becomes larger and less dense, being termed HDL_2. It also transfers and collects apolipoproteins to and from other lipid particles. Eventually HDL_2 particles pass into the liver, where hepatocytes bearing the scavenger receptor class B type 1 (SCARB1, coded for by *SCARB1* at 12q24.31) absorb the particle and so facilitate the digestion of the cholesterol. This pathway is summarized in Figure 7.10.

High levels of HDL confer a reduced risk of atherosclerosis. Originally thought to be due to its activity in clearing cholesterol from the blood, it is now clear that it has other features that can account for this protective effect. HDL contains antioxidants, can inhibit monocyte chemotaxis in response to oxLDL and prevent the upregulation of endothelial cell adhesion molecules, and has stores of lipoprotein-associated phospholipase A_2, which hydrolyses oxidized phospholipids.

SELF-CHECK 7.5

How does the composition of the various lipoprotein particles influence their density, and how does this occur?

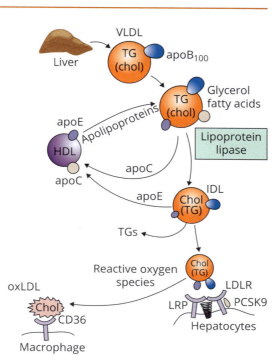

FIGURE 7.9

Metabolism of VLDL, IDL, and LDL. VLDL is transformed from a triglyceride-rich/cholesterol-poor particle, via IDL, to LDL, which is rich in cholesterol but poor in triglycerides. Reactive oxygen species acting on LDL convert it to oxLDL, at which point it may be taken up by macrophages via CD36. For clarity, only directly relevant molecules and particles are shown.

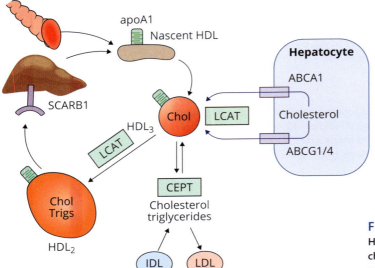

FIGURE 7.10

HDL metabolism. HDL interactions in reverse cholesterol transport. For clarity, only directly relevant molecules and particles are shown.

Lipoprotein (a)

Formed from the covalent union of a molecule of apo(a) with the apoB$_{100}$ of an LDL particle, Lp(a) is highly atherogenic, and also promotes the calcification of the aortic valve. Indeed, it has been estimated that up to 90% of oxidized lipoproteins are carried by this molecule, and levels rise in the acute phase of myocardial infarction and in rheumatoid disease.

With a density slightly greater than that of the 'standard' LDL, Lp(a) exists in a number of isoforms, and plasma levels vary considerably between individuals, being governed by the apo(a), generated by *LPA* at 6q25.3–q26, which is strongly expressed by hepatocytes, although it may be influenced by *APOE*. Levels of Lp(a) thought to be atherothrombotic (>75–125 nmol/l) are present in 20–30% of the population indicating a potentially unmet need.

Table 7.5 summarizes key aspects of the lipoproteins we have been discussing, Table 7.6 the apoproteins.

TABLE 7.5 Properties and composition of lipoproteins

	Diameter (nm)	Density (g/l)	Cholesterol esters (%)	Cholesterol (%)	Trigs (%)	Phospho-lipids (%)	Protein (%)
Chylomicrons	80–500	<950	1–3	1	85–95	3–7	1–2
VLDL	30–75	950–1006	12–15	7–9	45–65	12–20	7–12
IDL	20–30	1006–19	26–35	7–11	25–35	22–8	12–18
LDL	19–25	1019–<1063	36–48	7–10	7–11	19–25	19–23
Lp(a)	0–25	1052–<1063	30–6	8–10	3–4	20–5	30–5
HDL	6–11	1063–210	10–30	2–6	3–9	27–40	38–56

Trigs = triglycerides.

TABLE 7.6 Summary of key apolipoproteins

Apolipoprotein	Mass (kDa)	Links to particles	Function
apoA1	29	Chylomicrons, HDL	Major HDL protein, binds macrophage ABCA1, activates LCAT
apoB$_{48}$	241	Chylomicrons	Marker of chylomicrons
apoB$_{100}$	513	VLDL, IDL, LDL	Major LDL protein, binds LDLR
apoC1, apoC2	7.6, 8.9	Chylomicrons, HDL, VDL, IDL	Activate lipoprotein lipase
apoE	34	Chylomicron remnants, HDL, VLDL, IDL	Binds LDLR
apo(a)	300–800	LDL	Complexes with apoB$_{100}$ and LDL to form Lp(a)

Table not intended to be exhaustive: other apoproteins are known.

Key points

The metabolism of the numerous forms of plasma lipids is complex and, in the face of certain abnormalities, can lead eventually to atherosclerosis.

7.3.3 Lipids in the clinic and laboratory

Given the importance of lipids and atherosclerosis and other metabolic disease, practitioners need to be aware of levels of these molecules, and how they have come about, that being from the diet and by synthesis.

Hypertriglyceridaemia

Clearly, a high-fat diet is most likely to lead to high plasma triglycerides, and like cholesterol, is closely linked to obesity. Alcohol is a major independent influence on levels of triglycerides, and also VLDL. Primary hypertriglyceridaemia has a genetic component and so an inherited basis. The most common are familial endogenous hypertriglyceridaemia, lipoprotein lipase deficiency, and familial hypertriglyceridaemia (section 6.6.1). The leading causes of secondary hypertriglyceridaemia are (in order) pancreatitis, diabetes, alcohol, nephrotic syndrome, and hypothyroidism. As levels vary throughout the day (i.e. there is a diurnal variation), there may be high levels soon after a high-fat meal. Accordingly, for accurate results, the blood test must be performed on a fasting sample.

Hypercholesterolaemia

A poor diet, rich in cholesterol, is a clear cause of hypercholesterolaemia, but its resolution by diet alone may take months or even years to be effective. Even in a very low-fat diet, cholesterol can leach out of fat stored in **adipocytes** to keep blood levels high. Nevertheless, a low-fat diet is a key requirement for the treatment of obesity, and often of hypercholesterolaemia, and this is often the first line of treatment before drugs are prescribed. A modest intake of alcohol results in an increase in HDL, a fact often used to justify a small glass of wine with a meal.

Cholesterol is synthesized by the liver, and in many cases is regulated by several genes. As described in section 6.6.1, both loss-of-function and gain-of-function mutations exist which directly influence levels in the blood. The leading genetic disease is familial hypercholesterolaemia (FH), with three principal causes:

- 'Classical' FH is caused by failure of a mutated LDL receptor (LDLR) to remove LDL from the circulation, so that plasma levels remain high. Literally hundreds of mutations in *LDLR* have been described, and are present in one in 250–500 of the population (that is, are heterozygotes), leading to a serum cholesterol of perhaps 7 or 8 mmol/l. In homozygotes, the frequency is around 1/400,000, and as serum cholesterol is often in >10 mmol/l, aggressive treatment is required to counter premature atherosclerosis.

- The LDLR recognizes $apoB_{100}$ in LDL, so absorbing it into the hepatocyte. Familial defective apoprotein B_{100} (with a frequency of some 1/1,000) is so named because of an abnormality in *apoB*, coding a mutated $apoB_{100}$ product not recognized by the receptor. This results in raised serum LDL and a phenotype resembling classical FH.

- Binding of the LDLR to PCSK9 results in its destruction, so that high levels of the latter result in low LDLR levels and so increased LDL in the blood. Patients have a serum cholesterol level between that of heterozygous and homozygous classical FH.

In contrast to these single-gene-defect conditions (i.e. monogenic), there are many other genetic abnormalities leading to raised cholesterol. It is likely that several are present in an individual and act in concert with varying degrees of penetrance, leading to the concept of polygenic (i.e. many genes) hypercholesterolaemia. Secondary hypercholesterolaemia can be classified as being due to either disease or a drug effect. The former include anorexia nervosa, diabetes mellitus, chronic kidney disease, hypothyroidism, monoclonal gammopathies and paraproteinaemia, and **cholestasis**. Medications linked to hypercholesterolaemia include those used to treat hypertension (thiazide diuretics, beta-blockers) and acquired immunodeficiency syndrome, in addition to synthetic oestrogen, cyclosporine, glucocorticoids, and retinoic acid.

Increased HDL

Whilst increased total cholesterol and LDL are to be avoided, high HDL is to be encouraged. Where high HDL is present, this may be due to increased apoA1 or a reduced absorption of HDL from the blood by hepatocytes. Additional mechanisms include increased LCAT, reduced CETP activity, and reduced hepatic lipase activity.

Hypocholesterolaemia

Many cases of cancer are associated with hypocholesterolaemia, but this is a likely response to, or is caused by, the malignancy. An intensive low-fat diet and excessive treatment may of course lead to low cholesterol, but there are known biochemical syndromes. Hypobetalipoproteinaemia may be due to genetic defects that result in a familial form of the disease. Heterozygotes, who are generally asymptomatic, have lower total cholesterol than 'normal' (perhaps 2.5 mmol/l) but levels in homozygotes may be as low as 1.3 mmol/l. An extreme (and very rare) form of the above is a complete lack of apoB, described as abetalipoproteinaemia, leading to failure to produce VLDL and LDL. Tangier disease is caused by a loss of function mutation in *ABCA1*, whose product normally regulates apoA1, hence hypoalphalipoproteinaemia. Low levels of HDL may be genetic, secondary to other disease (such as diabetes and raised triacylglycerols), end-stage renal disease and severe inflammation.

Other conditions

Familial combined hyperlipidaemia (FCH) is estimated to be present in 0.5–2.0% of the population, is characterized by raised triglycerides and raised cholesterol, and there is also the likelihood of low HDL. Diagnosis is more certain if similar dyslipidaemia can be demonstrated in family members. Both hypercholesterolaemia and hypertriglyceridaemia can also arise at the same time from a poor (lipid-rich) diet, and especially if combined with other disease such as diabetes and often Cushing's syndrome.

The metabolic syndrome is defined by a clustering of factors that include high BP, central obesity, high triglycerides, low HDL, and high fasting glucose. It may also be associated with a low-grade inflammatory response with elevated CRP, IL-6, and TNF-α, and endothelial dysfunction. However, whilst epidemiological studies find the metabolic syndrome is a risk factor for diabetes and for CVD, it has failed to warrant a focused NICE guideline, possibly because most of the constituent parts are themselves already the object of clinical attention.

7.3.4 Why this is important

Much of Chapter 6 described the importance of lipids in the pathogenesis of atherosclerosis. Therefore, maintaining a low lipid profile in the blood is desirable if this disease (and others, such as a fatty liver), and its many manifestations, is to be resisted. The most effective method of ensuring low serum lipids is pharmaceutical intervention, although of course a low-fat diet in parallel is strongly encouraged. Hypercholesterolaemia is perhaps the most well-established laboratory marker of the presence of, and risk of, CVD, and the greater part of this risk can be accounted for by raised LDL. Unfortunately, LDL cannot be measured routinely, but can be estimated from an equation (the Friedewald formula) that calls for total cholesterol, HDL, and triglycerides. This equation is:

Equation 7.1

$$\text{LDL} = \text{total cholesterol} - \text{HDL cholesterol} - \frac{\text{triglycerides}}{2.19} \text{ mmol/l}$$

This equation is valid only with levels of triglycerides <4.5 mmol/l, although some consider an LDL result obtained with a triglyceride level of 2.5–4.5 mmol/l to be inaccurate. It is additionally unfortunate that the triglycerides must be measured in a fasting sample of blood: a random sample is unacceptable. Therefore, from the public health perspective, total cholesterol and HDL alone are often used as markers of CVD risk. An example of the risk of disease linked to these lipids and lipoproteins is the hazard ratio (with 95% confidence interval) for myocardial infarction brought by the particular molecule or particle. The strongest effect is with high $apoB_{100}$ (2.29 (1.92–2.74)), followed by LDL cholesterol (2.04 (1.72–2.43)) and total cholesterol (1.87 (1.78–2.11)). Conversely, reduction in risk is linked to both high HDL cholesterol (0.49 (0.41–0.59)) and (to a lesser extent) high apoA1 (0.60 (0.50–0.71)).

7.3.5 Presentation and diagnosis

In common with hypertension, most cases of dyslipidaemia are asymptomatic, lacking even non-specific signs and symptoms, and may only be found in the investigation of other disease. With a growing public awareness, members of the public may arrange their own testing, more so in the face of a relevant family history (such as major CVD in a first-degree relative aged <55 years). In those with longstanding very high cholesterol levels, subcutaneous deposits may form in the eyelids and xanthelasma (Figure 7.11). Other deposits (xanthoma) may be present in the knees, heel, elbows, creases in the palms, and elsewhere.

The clinical signs described above are useful, if not definitive, but are rare. Patients may bring to their GP a cholesterol result performed in the community or pharmacy. However, the ultimate diagnosis is with a full lipid screen, the traditional panel being triglycerides, and total cholesterol, LDL, and HDL. In some circumstances, the patient's Lp(a) result may be needed as well, such as in individuals with premature CVD, FH, a family history of premature CVD, recurrent CVD despite optimal lipid-lowering treatment, and a ≥5% 10-year risk of fatal CVD. The risk of CVD is considered significant when Lp(a) is >80th percentile, that being 50 mg/dl.

Key points

Despite the high level of awareness of xanthomas and xanthelasmas among practitioners, these are present in only a very small fraction of those with hypercholesterolaemia so that blood testing is the only reliable diagnostic tool.

7.3.6 Management of hypercholesterolaemia

Once serum lipids have been determined, the question that follows is whether treatment is necessary. Acknowledging the heterologous approach to CVD, guidelines consider the likelihood that an individual will suffer a major cardiovascular end-point in the coming decade. This risk score method, the same SCORE method described above for hypertension, relies not only on lipids but also on

FIGURE 7.11
Eyelid xanthelasma.

Shutterstock.

age, smoking, and blood pressure, and can be determined from a chart, giving the 10-year risk of a CVD event in per cent. NICE NG238 refers specifically to the QRISK3 assessment tool, which is valid only without a diagnosis of coronary artery disease, and requests (but does not demand) information beyond cardiovascular issues such as autoimmune disease and certain psychiatric diagnoses. These scores may be used to guide treatment.

Step one is to encourage patients to adopt a healthy lifestyle, with the avoidance of tobacco smoking, moderate intake of alcohol, a high intake of fresh fruit and vegetables, and regular exercise with adherence to a low-fat diet. Such a diet needs around 6–8 weeks for an effect (if any) to become noticeable, and should focus on a fat intake of <30% of total energy, reducing saturated fats to <7% of total energy intake, and increasing unsaturated fat. Patients may need to be referred to a dietician, and there is abundant literature. However, in many cases, although the diet is encouraged, drug treatment is likely to start in parallel. The risk-score-based holistic view of a patient described above will put them on a spectrum of risk of CVD and so guide management.

- Subjects at low to moderate risk should be treated to a target LDL of <3.0 mmol/l.

- Subjects at high risk should be treated to a target LDL of <2.6 mmol/l, or a reduction of >50% if the baseline LDL is 2.6–5.3 mmol/l.

- Subjects at very high risk should be treated to a target LDL of <1.8 mmol/l, or a reduction of <50% if the baseline LDL is 1.8–3.5 mmol/l.

If LDL is unavailable, secondary targets include non-HDL-cholesterol (i.e. total cholesterol minus HDL, such as <2.6 mmol/l in very high-risk patients and <3.4 mmol/l in high-risk patients).

Statins

Statins are now part of the pharmacopeia of almost every CVD strategy; the most frequently prescribed ones are simvastatin (where doses of 20 mg and 40 mg daily reduce serum LDL by 32% and 37% respectively) and atorvastatin (parallel data being reductions of 43% and 49% respectively) (Law et al. 2003). Each statin has slightly different effects, and side effects, the most well-known being **rhabdomyolysis**, which translates for the patients to muscle pain. Before starting statin treatment, a full range of blood tests is required, to include urea and electrolytes (U&Es), liver function tests (LFTs), and thyroid-stimulating hormone. NICE NG238 recommends 20 mg atorvastatin once daily (od) for those at low risk, rising to 80 mg od at high risk. If poorly tolerated, or if muscle damage is indicated, an alternative is simvastatin. However, failure of the maximum tolerated dose of a statin to bring LDL cholesterol to target will require adding a second class of drug.

SELF-CHECK 7.6

Why is a scoring system necessary—why not simply prescribe a statin for everyone with a raised cholesterol?

Bempedoic acid

This drug targets ATP citrate lyase, which part-regulates the production of cholesterol by inhibiting the production of acetyl-coA (Figure 7.7), at a point upstream of statins. Clinical trials indicated that the combination of this drug with ezetimibe (see below) was more effective in reducing LDL cholesterol than either drug alone. This, and other data, prompted NICE to release TA694, recommending this combination for treating primary hypercholesterolaemia or mixed dyslipidaemia in those for whom a statin is contraindicated, ineffective, or not tolerated. An appropriate starting dose is 180 mg bempedoic acid with 10 mg ezetimibe, once daily.

Cholesterol absorption inhibitors

The leading cholesterol absorption inhibitor, ezetimibe, acts by interacting with the Niemann–Pick C1-like protein on enterocytes, but also on hepatocytes. A dose of 10 mg once a day is effective in many, but not all patients, and may be prescribed alongside a statin. Ezetimibe is also recommended as an option for treating hypercholesterolaemia in the 0.5% to 3% of adults who cannot tolerate statins, or in whom these drugs are contraindicated. Common side effects include diarrhoea and arthralgia.

PCSK9 inhibitors

Sections 7.3.3 and 6.6.1, where Figure 6.18 applies, describe the role of PCSK9 in the regulation of the LDL receptor (LDLR). High levels of PCSK9 lead to low levels of the LDLR at cell surface, and so high plasma LDL. Conversely, low levels of PCSK9 lead to high levels of LDLR which in turn leads to low plasma LDL. This relationship is exploited by the generation of monoclonal antibodies (mAbs) that bind to PSCK9, and so reduce its presence at the cell membrane, and so high levels of the LDLR and in consequence, low LDL. The drugs are relatively well tolerated and are generally injected subcutaneously at a low dose every fortnight, or at a high dose monthly. Licensed PSCK9 inhibitors include evolocumab and alirocumab.

Other lipid-lowering agents

NICE does not recommend bile acid sequestrants, omega-3 fatty acids, **fibrates**, or nicotinic acid (niacin) for preventing CVD. However, other guidelines suggest that a combination of a statin and a bile acid sequestrant could be useful in achieving an LDL target, which it may reduce by an additional 10–20%. A bile acid sequestrant may also be added to ezetimibe when used for statin intolerance, should ezetimibe monotherapy be less than optimal.

Developing therapies

Non-coding RNAs offer new opportunities to treat dyslipidaemia. Inclisiran is long-acting small-interfering RNA that targets PCSK9, and so results in increased levels of the LDLR and so lower LDL. It is recommended by NICE TA733 for treating certain cases of primary hypercholesterolaemia or mixed dyslipidaemia. Other strategies include interfering with mRNA: mipomersen is an antisense 20-nucleotide RNA targeting the $apoB_{100}$ message, resulting in the latter's degradation. Other siRNAs targeting Lp(a) and apoC2 (ultimately targeting triglycerides) are also in development. Evinacumab is a monoclonal antibody to angiopoietin-like-3 (that is, a molecule that resembles angiopoietin) that targets triglycerides and cholesterol.

7.3.7 Management of hypertriglyceridaemia

The position of triglycerides as a risk factor for CVD has been debated for decades, not merely because of a close association with lipoproteins, but also because their role as constituents of chylomicrons and their remnants has recently become recognized and may part-drive atherosclerosis and CVD. Definition of a normal triglyceride level remains difficult as there are many physiological contributing factors, and accordingly definitions of hypertriglyceridaemia vary. Nevertheless, modern guidelines point to levels ≤1.7 mmol/l as desirable (others suggest <2.0 mmol/l), whilst levels >2.3 mmol/l call for treatment against a background of total CVD risk. A triglyceride result of 2–10 mmol has been described as mild to moderate, whilst there is consensus that very high levels of triglycerides are defined as ≥10 mmol/l and are a risk factor for pancreatitis and so warrant treatment.

Fibrates are effective in reducing levels of triglycerides. The mechanism is in activating receptors on the nuclear membrane—peroxisome proliferator-activated receptor-alpha (PPAR-α, coded for by *PPARA* at 22q13.31)—and so the induction of fatty-acid β-oxidation and lipoprotein lipase activity with a reduction in the activity of cholesterol ester transfer protein. Established fibrates include clofibrate, gemfibrozil, ciprofibrate, bezafibrate, and fenofibrate.

7.3.8 Summary of the treatment of dyslipidaemia

Treatment focuses on a reduction of LDL to a target dependent on the patient's risk of CVD. High cholesterol may need to be targeted by drugs that effect different aspects of the pathophysiology, these being the cholesterol-synthesis pathway (bempedoic acid, statins), the absorption and recycling of cholesterol (ezetimibe), promotion of the LDL receptors (PCSK9 inhibitors), with fibrates for hyper-triglyceridaemia. This model is reminiscent of that of hypertension, with increased doses of different classes of drugs required to achieve the BP target. With increasingly resistant hypercholesterolaemia, different classes of drugs are called for. Those with homozygous FH may require LDL-plasmapheresis, a procedure allied to renal dialysis. Figure 7.12 shows a possible algorithm for treatments with selected agents.

Key points

Dyslipidaemia is a leading cause of atherosclerosis and so potentially fatal coronary artery disease. Most disease is polygenic and is successfully treatable with a number of drugs of different classes.

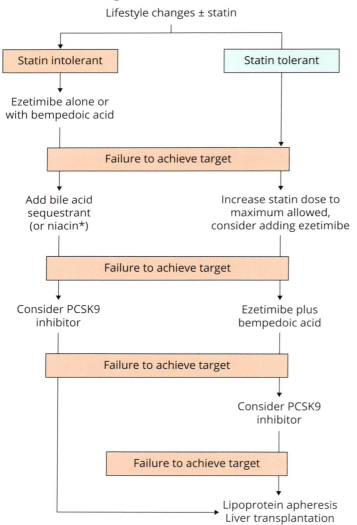

Diagnosis: Need to treat

Lifestyle changes ± statin

Statin intolerant

Statin tolerant

Ezetimibe alone or with bempedoic acid

Failure to achieve target

Add bile acid sequestrant (or niacin*)

Increase statin dose to maximum allowed, consider adding ezetimibe

Failure to achieve target

Consider PCSK9 inhibitor

Ezetimibe plus bempedoic acid

Failure to achieve target

Consider PCSK9 inhibitor

Failure to achieve target

Lipoprotein apheresis
Liver transplantation

FIGURE 7.12

Possible algorithm for treating hypercholesterolaemia. Practitioners will address their own regional, national, and international guidelines. The final stage, apheresis, is likely to be offered to those with resistant double-digit cholesterol.

* Use restricted to resistant FH.

CASE STUDY 7.2 *Hypercholesterolaemia*

At the funeral of the uncle of a 40-year-old man, the family discuss the observation that all the men in the family seem to die of a heart attack in their 60s, and the women in the family in their 70s. His children insist he gets checked out, and a high-street chemist finger-prick test reveals a whole blood cholesterol level of 8 mmol/l. A few weeks later he meets with his GP who takes a full medical and family history (the latter finding none of his father's family to have lived longer than 70, and another uncle having had a heart attack aged 55) and performs a physical examination, finding a BMI 27.5 kg/m², BPs 143/88 mm Hg, and normal urine analysis. The patient admits to taking around 15 units of alcohol a week and a lack of any exercise. Blood is taken for full blood count, U&Es, HbA1c, LFTs, thyroid-stimulating hormone, and a lipid screen.

All lab results come back normal, except that the raised cholesterol is confirmed at 7.7 mmol/l, with triglycerides 2.3 mmol/l and HDL 1.2 mmol/l, leading to a calculated LDL result of 5.4 mmol/l (Table 7.7). These data are fed into the QRISK3 cardiovascular risk score calculator which, for a non-smoker (the man has stopped 10 years ago) with a family history, gives a 10-year

risk of a heart attack or stroke of 4.8%. The GP follows NICE CG181 and begins lipid-lowering treatment, and suggests that his patient's relatives determine their own blood lipid levels.

The GP starts the patient on 40 mg of atorvastatin each day, and they agree on targets of a of BMI 25 kg/m² (aiming for a weight loss of around 7 kg/15 lb), alcohol at 10 units per week, and three sessions of exercise of at least 30 minutes each week, all to be achieved in 3 months. He leaves with leaflets and information on a low-fat diet and relevant websites, and an instruction to return in 6 weeks for LFTs, to check any possible effect of this statin on the liver, at which point all tests are normal.

At the 3-month review, the patient has lost some weight, is drinking less alcohol, and goes to the gym three times a week, all possibly related to a fall in blood pressure to 138/84 mm Hg. However, although better, the serum cholesterol remains high (Table 7.7). As the non-HDL cholesterol reduction of 20% is (at present) much less than NICE's target reduction of 40%, and the LDL is still far above NICE's moderate risk target of <2.6 mmol/l, his atorvastatin is increased to 80mg a day. Some blood is sent to the regional molecular genetics

TABLE 7.7 Cardiovascular risk profile

Metric	Month				
	0	3	6	9	12
Atorvastatin (mg/day)	0	40	80	80	80
Total cholesterol	7.7	6.5	6.0	5.0	4.4
HDL	1.2	1.3	1.2	1.3	1.3
Triglycerides	2.3	1.8	1.5	1.3	1.3
LDL	5.4	4.4	4.1	3.1	2.5
Cholesterol/HDL ratio	6.4	5.0	5.0	3.8	3.4
Weight (kg)	82.4	80.3	75.8	75.1	74.3
BMI (kg/m²)	27.5	26.8	25.3	25.1	24.8
Blood pressure (mm Hg)	143/88	138/84	134/80	132/78	131/77
QRISK score	4.8	3.5	3.6	2.8	2.7
Increased QRISK score compared to healthy peer	3.6	2.6	2.4	1.8	1.7
Healthy heart age (years)	53	50	50	47	46

Lipid indices in mmol/l.

(Continued)

CASE STUDY 7.2 Continued

laboratory with a request for confirmation of possible hetero-zygous familial hypercholesterolaemia. This increased atorv-astatin is tolerated well and LFTs are acceptable. The extended family is now very lipid aware, and an uncle and one of his two sons and a nephew also have raised cholesterol, and accordingly they have formed their own self-help group. At this point the very self-motivated patient aims to exercise almost daily and to adopt first a vegetarian and then a vegan diet.

Three months later the patient (now aged 41) has almost achieved his BMI target, blood pressure continues to fall (134/80), his exercise capacity has increased, and he is drinking less alcohol with an aim to stop completely. Despite this good news, the cholesterol results are still worrying; total cholesterol is still high.

The molecular genetics laboratory failed to find any abnormalities in *LDLR*, or in *ApoB*, but did find a gain-in-function mutation in *PSCK9*, and accordingly, measured serum levels, which were high at 600 ng/ml (reference range 120–420 ng/ml). This increased serum PCSK9 accounts for the high LDL as it will bind to the LDL receptor and so lead to its destruction, failure to recycle to the membrane, thus failing to take LDL into the cell. This makes the diagnosis of heterozygous familial hypercholesterolaemia (FH), and accordingly, the GP

passes her patient to the care of colleagues at the local university teaching hospital. With this diagnosis, NICE CG71 is applicable, and although this guideline does not recommend use of QRISK3 as those with FH are automatically at high risk, the metrics are continued as a record of progress.

With a 'normal' statin-resistant hypercholesterolaemia, the next step would be ezetimibe, an inhibitor of cholesterol absorption. However, in view of the high levels of serum PSCK9, which in preliminary reports are linked to coronary artery disease, the consultant pathologist recommends an injection of 140 mg of evolocumab every 2 weeks in addition to the statin. Three months later, his lipid profile has improved markedly (Table 7.7) and serum PSCK9 levels have fallen to 460 ng/ml. He moves to 420 mg of evolucumab once a month, and a year after presentation, all indices have improved and he is discharged back to this GP for continuing care. Notably, this regime has achieved the NICE CG71 target of a 50% reduction in LDL, but the patient nonetheless undertakes additional lifestyle changes to minimize further risk. Blood lipids will be checked regularly, and should the patient become resistant to evolocumab, other drugs such as ezetimibe can be used, or the small-interfering RNA agent inclisiran.

7.4 Diabetes

Although diabetes can affect many organs and organ systems (Chapter 15), this section will consider its place in cardiovascular disease.

Diabetes has a number of forms. Type 1 diabetes mellitus (T1DM) is characterized by beta-cell destruction (mostly immune-mediated) and absolute insulin deficiency, with an onset most common in childhood and early adulthood. Type 2 diabetes mellitus (T2DM) is summarized as various degrees of beta-cell dysfunction and insulin resistance that is commonly associated with overweight (BMI 25–30 kg/m^2) and obesity (BMI >30 kg/m^2). Both forms have much in common, including signs and symptoms, treatments, and the risk of microvascular and macrovascular disease. Many of these will be examined together, leaving type-specific details for later sections. NICE have stated that 90% of people with diabetes have T2DM, and that 370,000 people in the UK have T1DM (perhaps 0.5% of the population).

Diabetes may also be secondary to other disease and certain drugs, and less common types of the disease are linked to mutations in certain genes, many leading to maturity-onset diabetes of the young (MODY). Several other forms of **hyperglycaemic** disease are relevant. Impaired glucose tolerance (IGT) and impaired fasting glycaemia (IFG) are both pre-diabetes conditions that, without treatment, inevitably lead to T2DM. Gestational diabetes is described in section 14.4.4.

7.4.1 Anatomy and biochemistry

The pathology of diabetes is linked to both high (hyperglycaemia) and low (hypoglycaemia) blood levels of glucose: part of the biochemistry relevant to this section is described in section 12.4.1, where Figure 12.17 applies.

The hormones

Key to the understanding diabetes is an appreciation of the roles of the various molecules and cells that control levels of blood glucose and its storage form, glycogen.

- Insulin, a 51-amino-acid peptide of 5.8kDa, is coded for by *INS* at 11p15.5 and produced by the beta cells of the islets of Langerhans in the pancreas. Processing of pro-insulin also generates c-peptide.

- Glucagon, produced as a propeptide by *GCG* at 2q24.2, matures to a 29-amino-acid peptide of relative molecular mass 3.5kDa. It is produced by the alpha cells of the islets of Langerhans.

- Somatostatin, with two isoforms of 14 and 28 amino acids, is produced by 3q27.3 in the delta cells of the islets, the hypothalamus, and elsewhere. It inhibits the release of both insulin and glucagon.

- Two incretins (molecules that reduce blood glucose) are glucagon-like peptide-1 (GLP-1) and glucose-dependent insulinotropic peptide (GIP). Both act by enhancing the secretion of insulin and are inactivated by the membrane-bound enzyme di-peptidyl peptidase 4 (DPP-4).

The regulation of blood glucose

Acting in concert, the molecules described above ensure the correct homeostatic disposal of blood glucose. Increasing levels of glucose bind to the GLUT-2 receptor on beta islet cells, resulting in the production and secretion of insulin. At one of its many target cells (hepatocytes, myocytes, adipocytes), insulin binds to its receptor (coded for by *INSR* at 19p13.2) to promote the upregulation of GLUT-4 (coded for by *SCL2A4*) that in turn mediates the entry of glucose into the cell and so a fall in blood levels. The insulin receptor also binds two other ligands—insulin-like growth factors 1 (IGF-1, somatomedin C) and 2, that both have other non-glucose roles. IGF-1 is produced by the liver (and elsewhere) in response to growth hormone and has 50% homology with insulin.

A further action of insulin is in suppressing glycogenolysis (thus maintaining glycogen stores) and gluconeogenesis (the anabolic synthesis of glucose) by hepatocytes, and suppression of lipolysis in adipocytes. It also promotes the metabolism of ketones, and suppresses **ketogenesis**, an important aspect of the biochemistry of glucose and energy provision (Figure 7.13). However, high levels of ketones in the blood (hyperketonaemia) cause a metabolic acidosis (hence diabetic ketoacidosis, DKA) that if severe can cause considerable pathology and may be lethal (section 7.4.5).

Movement of glucose out of the blood produces a degree of relative hypoglycaemia that is detected by alpha islet cells, leading to the release of glucagon. At one of its many target cells, it binds its receptor (coded for by *GCRG* at 17q25.3), resulting in the liberation glucose-1-phophate from glycogen. However, other factors (hypoxia, hypothermia, high-protein meal, GIP, neurotransmitters) also

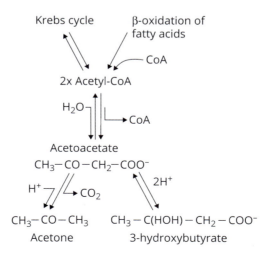

FIGURE 7.13

Ketogenesis. Simplified metabolic pathway for the generation of ketone bodies.

TABLE 7.8 Opposing actions of insulin and glucagon

Process	Insulin	Glucagon
Glycogenolysis	Suppressed	Promoted
Gluconeogenesis	Suppressed	Promoted
Glucose uptake	Promoted	Suppressed
Ketone bodies	Suppressed	Promoted
Blood glucose	Reduced	Increased

promote glucagon release, whereas somatostatin, GLP-1, high blood glucose, arginine, and ketones suppress secretion. These factors therefore work to maintain high levels of blood glucose, counter to the effects of insulin. Notably, whilst the metabolic effect of insulin is to promote ketone metabolism and suppress ketogenesis, glucagon has the reverse action, consistent with its role as a counter-regulator in this and other pathways (Table 7.8).

The consequences of hyperglycaemia

The metabolic and pathological consequences of high blood glucose can be summarized as effects of the following systems, cells, and molecules. Excess glucose moves easily into the red blood cell to increase the glycosylation of the haemoglobin, hence glycated haemoglobin (HbA1c) which is less efficient at carrying oxygen, thus contributing to tissue hypoxia/ischaemia. Hyperglycaemia (possibly as advanced glycation end-products (AGEs)) damages the endothelium (a primary target of the disease process), which loses its natural anticoagulant and anti-hypertensive character, so promoting thrombosis and hypertension.

Glucose binds plasma proteins (forming fructosamines) and lipids, generating AGEs that may be toxic and are cleared physiologically from the blood by specific receptors (hence RAGE, coded for by *RAGE* at 6p21.32) on endothelial cells and macrophages. There is extensive evidence of a low-grade acute phase response in diabetes whose origin is unclear, but may be the result of excessive adipocyte release of adipocytokines, such as leptin and adiponectin, and inflammatory cytokines. As discussed in section 6.4.3, inflammation is believed to contribute to atherosclerosis, as described in more detail in section 7.5. Considerable literature points to the loss of antioxidant protection as a disease mechanism, promoting cytotoxic reactive oxygen species that may oxidize LDL and damage the endothelium. These features partly explain the increased degree of renal disease, diabetic retinopathy, hypertension, and microvascular then macrovascular complications.

A further consequence of extreme hyperglycaemia (>30 mmol/l, up to 50 mmol/l is not unknown) is osmotic diuresis and dehydration, leading to the hyperglycaemic hyperosmolar state (HHS). This is defined as an **osmolality** of 320–40 mOsm/kg (reference range 280–95), in the absence of high levels of ketones. It can develop over several days and is often linked to high glucose intake and polyuria, symptoms including weakness, blurred vision, and a progressive fall in mental faculties. Precipitating factors include pneumonia, diuretics, infections, myocardial infarction, stroke, and steroids.

7.4.2 Pathophysiology of type 1 diabetes

The population frequency of the development of T1DM varies greatly with geography, from 600/million/year in Finland, 250/million/year in the UK, Ireland, and the USA, to 1/million/year in Papua New Guinea, possibly because of features described in the next subsection. However, the frequency is rising in all populations at around 2.5% per year, as is the frequency of T2DM.

Aetiology and pathology

Whilst various factors can lead to the destruction of the beta cells in T1DM, the leading cause is **autoimmunity**. Although it can present at any age, there are peaks at ages 5–7, and at ages 10–14, but 50% of cases arise in adults. An estimated 70–90% of cases are autoimmune, the remainder being linked to diet, pollutants (ozone, nitrates, heavy metals), lack of vitamin D, dyslipidaemia, obesity, viruses, and changes to the intestinal **microbiome** (Ogrotis et al. 2023). Around a third of T1DM patients are likely to suffer a second autoimmune disease (most commonly (in order) Hashimoto's thyroiditis, Graves' disease, and coeliac disease), with 25% suffering several diseases, although curiously T1DM protects from the development of inflammatory bowel disease.

The dominant leukocyte in the lesion is CD8+ve T cells, with macrophages, CD4+ve T cells, B cells, and plasma cells, but few FOXP3+ve Tregs or NK cells. A mutation in Treg locus *FOXP3* is **pathognomic** in the syndrome of X-linked polyendocrinopathy, immune dysfunction, and chronic diarrhoea (XPID), in which T1DM is inevitable.

Autoantibodies include those towards insulin, glutamic acid decarboxylase, islet antigen 2, tetraspannin-7, and zinc transporter 8, and have value in diagnosis and identifying those at risk, with the presence of two or more autoantibodies being linked to accelerated risk. The natural history generally describes a slow and steady decline in beta cell function, so that it may take decades for insulin levels to fall to zero. However, an important subgroup of T1DM may be described as 'latent autoimmune disease in adults' (LADA), and it may be misdiagnosed as T2DM, especially in the absence of obesity. These subjects are commonly positive for antibodies to glutamic acid decarboxylase.

The lack of insulin explains the signs, symptoms, and clinical features of the disease. Glucose denied entry to the cell (which therefore suffers from its own hypoglycaemia and so also suffers metabolically) remains in the blood, and has a number of consequences described above. The genetics of T1DM are described in section 6.6.1. Changes in non-coding RNAs may provide new opportunities for diagnosis and management: serum miR-21, miR-24, miR-148a, miR-181a-5p, and miR-210-5a levels are increased in T1DM.

Key points

Hyperglycaemia in T1DM is the product of T cell-mediated autoimmune destruction of B cells resulting in absolute insulin deficiency, with one or more autoantibodies and strong links to certain HLA DQA, DQB, and DRB alleles.

7.4.3 Pathophysiology of type 2 diabetes

The global frequency of T2DM is ~84,000/million, and varies in geography, ranging from 3.5% in sub-Saharan Africa and 8% in Europe to 12% in North America and the Caribbean. In contrast to T1DM, leading epidemiological features of T2DM are overweight/obesity and age: diabetes is present in 19.3% of global 65–99-year-olds, and again there is a strong geographical bias. A common feature of all groups is the increased incidence in women (in contrast to T1DM).

Aetiology and pathology

In addition to overweight/obesity and age, risk factors include physical inactivity, history of gestational diabetes, cardiovascular disease and its risk factors, and race and ethnicity (South Asian, African Caribbean, Hispanic) (for the latter, see Oldroyd et al. 2005). The extent to which excess adiposity is directly pathogenic is unclear, but it may contribute to endothelial dysfunction and certainly impairs physical activity and is a major risk factor for osteoarthritis. A further pathology is increased serum lipids, which can be converted to AGEs and oxidized LDLs, and so contribute to atherogenesis. Of note in the obese aspect of T2DM are adipocytokines leptin and adiponectin, both secreted by adipocytes. Obesity confers leptin resistance, with high serum levels, whilst adiponectin is low in obesity, insulin resistance, and T2DM. The genetics of T2DM are described in section 6.6.1.

7.4.4 MODY and other forms of diabetes

Numerous other infrequent forms of diabetes are linked to gene defects, other diseases, and to drugs.

Monogenic diabetes

The leading group of disorders in this group are those collected together as 'maturity onset diabetes of the young' (MODY) and represent 4–5% of diabetes cases. Several forms are recognized, often due to mutations in genes coding for glucokinase (*GKC*), present in ~45% of cases, hepatic nuclear factors (*HNF1α, HNF1β, HNF4α*), and ATP-binding cassette transported subfamily C member 8 (*ABCC8*) in the remainder. Mutations in *KCNJ11* (at 11p51.1, coding the ATP-sensitive potassium channel) can cause permanent neonatal diabetes (as can mutations in *INS*) and developmental delay, epilepsy, and neonatal diabetes. Monogenic defects in insulin action include those in the *INSR*, causing type A insulin resistance, Leprechaunism (the Donohue syndrome), and the Rabson–Mendenhall syndrome.

Non-genetic disease

Diseases of the exocrine pancreas linked to diabetes include pancreatitis, pancreatic cancer, cystic fibrosis, and **haemochromatosis**. Associated endocrine disorders include Cushing's syndrome, acromegaly, phaeochromocytoma, and hyperthyroidism. Drugs causing diabetes include glucocorticoids, thyroid hormones, thiazides, alpha-adrenergic agonists, and IFN-alpha. Diabetes may follow infections with rubella, coxsackie B, adenoviruses, and mumps.

SELF-CHECK 7.7

Describe the key features of the pathogenesis of T1DM and T2DM.

7.4.5 Presentation and diagnosis of diabetes

Signs and symptoms of the varying types of diabetes are broadly similar (although those of T1DM are generally more dramatic) and often follow the hyperglycaemia. For example, as excess sugar will be excreted, the laws of osmosis demand that it does not exceed a certain concentration, so that it must be diluted in water/urine, leading to excessive urination. Loss of water in urine leads to blood hypovolemia, which drives the subject to drink water (polydipsia). In the long term, other symptoms will develop and follow the glycation of tissues, with neuropathy, nephropathy, and potentially early atherosclerosis such as blurry vision, a sign of retinopathy.

Other signs and symptoms include weight loss (possibly resulting from dehydration and loss of muscle and fat secondary to insulin deficiency), sexual problems (erectile dysfunction, failure of vaginal lubrication, and vulva/vaginal infections), hunger, tiredness not explained by low red cell indices, and wounds that heal poorly. At the general practitioner's clinics, urine analysis is likely to show glucose and (possibly) ketones, and these will be enough to refer the patient to the local hospital.

The laboratory

Having established a preliminary diagnosis from signs and symptoms, the laboratory makes the final diagnosis with blood tests. For decades, the gold standard has been the oral glucose tolerance test, where the patient reports to the laboratory after an overnight fast (8–14 hours), and gives a sample of blood. They then drink a solution containing 75 g of glucose, and a second blood sample is taken 2 hours later. Diagnosis depends on the results of these two blood tests, which can also diagnose the two pre-diabetes conditions of IGT and IFG and gestational diabetes (Table 7.9).

As described in section 11.1, the erythrocyte membrane has many GLUT-1 transporters that mediate the entry of glucose. Once in the cytoplasm, some will bind to haemoglobin, generating glycosylated (or glycated) haemoglobin, HbA1c, which can be measured by high performance liquid chromatography, and reported in both percentage and (the preferred mode) as mmol of HbA1c per mol of Hb (mmol/mol). As the blood level of glucose increases, so too does the HbA1c, leading to diagnostic cut-off levels (Table 7.10).

TABLE 7.9   The oral glucose tolerance test

Condition	Fasting glucose		2-hour glucose
Healthy result	2.2–6.0*	and	<7.8
Impaired fasting glycaemia	6.1 –6.9	and	<7.8
Impaired glucose tolerance	≤ 7.0	and	≥7.8 and <11.1
Diabetes**	≥ 7.0	or	≥11.1
Gestational diabetes	5.1–6.9	or	8.5–11.0

Units are mmol/l. Considerable variation in the literature: refer to your local practice.

* Some describe the upper limit at 5.5 mmol/l.

** WHO definition.

TABLE 7.10   HbA1c in health and hyperglycaemia

Condition	Percentage	mmol/mol
Health	4.0–5.9	20–41
Pre-diabetes	6.0–6.4	42–7
Diabetes	≥ 6.5	≥ 48

The one-off HbA1c is clearly more convenient for both patient and laboratory. Furthermore, the HbA1c does not change upon fasting, and as it is irreversible, effectively gives a 3-month history of hyperglycaemia (that being the approximate lifespan of a red cell). Accordingly, HbA1c alone is becoming the tool of choice for the diagnosis of diabetes. Patients with T2DM are more likely to be overweight or obese, aged >10 years, have a strong family history of T2DM, and have **acanthosis nigricans**, undetectable islet autoantibodies, and elevated or normal C-peptide.

Diabetic ketoacidosis (DKA)

Figure 7.13 summarizes the generation of ketone bodies, and nearby text introduces the concept that should excessive ketones develop, they will lead to a metabolic acidosis, hence DKA. The root cause is generally insulin deficiency, leading to hyperglycaemia and lipolysis, the liberated free fatty acids being converted to acetoacetone and then acetone and β-hydroxybutrate in the mitochondrion. Although the vast majority of cases are seen in T1DM, it is not unknown in T2DM. The biochemical changes are easily detected—low pH, hyperglycaemia, low bicarbonate, total ketones >6 mmol/l, and possibly a hypokalaemia: <3.5 mmol/l. Signs and symptoms include dehydration, deep and forced breathing (Kussmaul respiration, attempting to resolve the acidosis), nausea, vomiting, low blood pressure, bradycardia (<60 beats per minute) or tachycardia (>100), and collapse. The hyperglycaemia should be treated with insulin. Around ~25% of the first diagnoses of T1DM are made in the presence of DKA, and if severe should be treated as an emergency with a bicarbonate/saline drip.

Hospitalization for DKA in T1DM occurs at a frequency of ~5 per 100 patient years in a paediatric population, and accounts for ~15% of T1DM deaths. The incidence of DKA in T1DM is greater in women, duration of diabetes, and is linked to a high HbA1c. DKA is biochemically linked to HHS (described in 7.4.1), the major difference being the higher degree of dehydration and hypotension and lack of ketosis in HHS and that it is more common in T2DM, in contrast to T1DM where DKA is more common. Alternative causes of ketoacidosis include excess alcohol (as ethanol), and ingestion of related molecules such as methanol, glycol, and acetone (in which case they would be treated as life-threatening toxins with the possible hepatotoxicity). A further variant is lactic acidosis, the product of high levels of lactic acid.

7.4.6 Management of diabetes

In summary of the proceeding sections, diabetes is a complex disease whose aetiology can vary by genetics, autoimmunity, overweight/obesity/lack of exercise, and external factors. Leading patho-physiological pathways include beta cell mass and function, the response of target cells to insulin, the secretion and action of glucagon, the secretion and action of incretins, and adipose tissue mass and distribution. As many of these as possible should be addressed in adults, NICE NG28 applies.

Initial steps

As in most diseases, step one is lifestyle changes, focusing on diet and exercise, with referral to a dietician, who will be part of the patient's multi-disciplinary team. In the obese, a weight loss of 1 kg has been linked to a reduction in the risk of T2DM of 16%, such that targets for weight loss (such as 5–10%) should be agreed upon. Whilst these options are pursued with determination by the professional, it is almost inevitable that drug treatment will be necessary. Whilst a low-carbohydrate diet can help, there are several families of pharmaceuticals.

Key points

Alongside hypoglycaemia, DKA can be a major metabolic issue for the person with diabetes, and may need in-patient treatment.

Assessment of the extent of the disease 1: Biochemistry

Relative levels of glucose and insulin can, alongside HbA1c and c-peptide, be useful in summarizing the patient's biochemistry. The key pathophysiological aspect of T2DM is the failure of target cells to respond to insulin, hence insulin resistance. The response of beta cells is to secrete more and more insulin in an attempt to overcome the insulin resistance, often leading to **hyperinsulinaemia**. The relationship between beta cell function (in secreting insulin) and insulin resistance (IR) can be quantified by measuring the fasting glucose (typical reference range 3.5–5.5 mmol/l) and fasting insulin (2–10 mIU/l, although others state 4.4–26 or 3–17, so refer to your local values). From these two measurements the HOMA (homeostatic model assessment) indices can be calculated:

- HOMA-IR (insulin resistance) is given by the product of glucose (in mmol/l) and insulin (in mIU/l) divided by 22.5.

- HOMA-B (beta cell function) is given by the insulin result multiplied by 20, and then divided by the glucose result minus 3.5, expressed as a percentage.

Since HOMA requires a fasting insulin result, and by definition T1DM patients lack this hormone, the HOMA is applicable only in cases of T2DM. Calculators are available online, and alternatives to the HOMA method for IR and beta-cell function include QUICKI. The insulin result for these calculators may need to be adjusted as some laboratories report as pmol/l (typical reference range 18–173). Reference ranges for HOMA indices also vary, one for HOMA-IR suggesting 0.5–1.4, another 0.7–2.0, where a result above the top of the reference range implies early IR and a result >2.9 suggests significant IR (Table 7.11).

HOMA-IR links strongly with overweight and obesity, with a mean of 1.2 at BMI <25, rising to 1.8 at BMI 25–30, and to 2.9 at BMI >30, whilst intake of fish, seafood, fruit, and vegetables are linked to lower HOMA-IR. Increased HOMA-IR in people with diabetes, alongside sex, age, smoking, HDL/total cholesterol and hypertension, is an independent risk factor for the prevalence and incidence of cardiovascular disease.

TABLE 7.11 HOMA indices

Theoretical condition	Insulin	Glucose	%B	%S	IR
Low normal	3	3.5	158.9	254.5	0.39
Medium normal	6	4.5	99.4	131.5	0.76
High normal	9	5.5	87.5	83.7	1.19
IGT	11	6.5	72.5	66.1	1.51
Diabetes in acceptable control	12	9.0	41.7	56.7	1.76
Diabetes in moderate control	14	10.0	38.7	47.7	2.10
Diabetes in poor control	20	14.0	29.7	29.8	3.36

%B = beta-cell function, %S = insulin sensitivity, IR = insulin resistance.

Data derived from the HOMA2 calculator: https://www.dtu.ox.ac.uk/homacalculator/.

SELF-CHECK 7.8

What can the laboratory offer in the assessment of diabetes?

Assessment of the extent of the disease 2: Target organ damage

The person with diabetes is at risk of damage to certain organs. Accordingly, patients will visit a number of specialists to determine the extent of any hypertension and dyslipidaemia, and pathology in (at least) their eyes, kidneys, heart and liver, investigations focusing on the laboratory and imaging. These examinations will be a regular part of the diabetic's care. Where necessary, a particular pathology will be discussed with the patient and options for treatment addressed. Clinical trials have shown that tight control of blood glucose and blood pressure are very effective in reducing both disease progression and the frequency of major cardiovascular events.

Pharmacological treatment 1: Glucose

The primary target of treatment is to reduce levels of serum glucose, of which HbA1c is an important surrogate. It should be monitored 3- to 6-monthly until the target has been achieved, then 6-monthly, but if use of HbA1c is invalid, plasma glucose, total glycated haemoglobin, or fructosamines may be used. Major classes of drugs are shown in Table 7.12, most focusing on promoting the effects of the body's own insulin to reduce levels of glucose. Since T1DM is characterized by the absence of insulin, most are irrelevant, treatment being insulin replacement therapy to mimic the function of the beta cells.

NICE has published an excellent patient decision aid accessible through NG28 to help them and their professional find the best way to treat their disease. Other agents include repaglinide (1–4 mg od), which stimulates the release of insulin from stores, and pramlintide (30 μg at mealtimes), an analogue of amylin which must be injected, which slows gastric emptying—thus promoting satiety—and inhibits the secretion of glucagon. Some preparations are combined, such as insulin degludec/liraglutide in single injection. Several NICE publications refer to the use of certain drugs in dual- (e.g. dapagliflozin and metformin) and triple-therapy (e.g. ertuglifloxin with metformin and a DPP4 inhibitor) combinations.

TABLE 7.12 Drug treatments for diabetes

Class	Mode of action	Drugs and typical doses
Biguanide	Complex, but ultimately decreases gluconeogenesis and enhances target cell sensitivity to insulin	Metformin, 500 mg–3 g daily. However, it may be poorly tolerated or have contraindications (e.g. acute or chronic disease that may cause tissue hypoxia)
Sulphonylurea (SU)	Act on the beta cell to promote insulin secretion, but require a functioning mass of beta cells	Tolbutamide (0.5–1.5 g daily in divided doses) Glibenclamide (2.5–15 mg od) Glipizide (2.5–15 mg) Glimepiride (1–4 mg od) Gliclazide (40–320 mg od)
Thiazolidinedione (TZD)	Reduces insulin resistance by interacting with peroxisome proliferator-activated receptor gamma (PPAR-γ)	Pioglitazone (15–40 mg od)
Di-peptidyl peptidase 4 inhibitor (DPP4i)	GIP and GLP-1 both reduce glucose by promoting insulin secretion. DPP4 degrades these incretins, thus inhibition of DPP4 increases incretins and so reduces glucose	Alogliptin (1.25–25 mg od) Linagliptin (5 mg od) Saxagliptin (5 mg od) Sitagliptin (100 mg od) Vildagliptin (50 mg bd)
Glucagon-like peptide-1 receptor agonist*	Long-acting analogues of GLP-1, and therefore enhance the release of insulin	Exenatide (5–10 μg bd) Liraglutide (0.6–1.8 mg od) Semaglutide (0.5–1 mg od) Lixisenatide (10–20 μg od) Dulaglutide (0.75–1.5 mg once weekly)
Sodium-glucose co-transporter inhibitor (SGLT-2i)	Renal SGLT-2 resorbs glucose, so inhibitors effectively promote glucose 'excretion'	Canagliflozin (100–300 mg od) Dapagliflozin (10 mg od) Empagliflozin (10 mg od) Ertugliflozin (5–15 mg od)
Insulins	Required for the transport of glucose into the cell	Several preparations with different half-lives thus short/intermediate/long action (Table 7.14)

Table not intended to be exhaustive: practitioners must refer to their local formulary.

* Must be injected.

Information based on NICE NG17, NG18, and NG28. Alogliptin, lixisenatide, and dulaglutide are subjects of ESNM20, 26, and 59, respectively. The SGLT-2 inhibitors are assessed in TAs 315, 336, 390, 418, and 572.

Treatment of T1DM

Treatment of T1DM focuses on injections of insulin at different times of the day and with respect to meals and their content, and exercise, and requires a high degree of patient education, such as in carbohydrate counting (NICE NG17) and use of a self-check device (Figure 7.14).

Many patients find the DAFNE (dose adjustment for normal eating) approach useful, which is accessible via diabetes charities and the NHS. The HbA1c target for T1DM is ≤48 mmol/l (6.5%) (≤58/7.5% in paediatric cases, and lower in those considering pregnancy), which should be measured every 3– 6 months, whilst patients should self-monitor their glucose at least four times a day (up to ten times in certain cases, such as during and after sport), including before each meal and before bed. Targets for levels of plasma glucose include 5–7 mmol/l on waking, 4–7 mmol/l before meals and at other times of the day, and 5–9 mmol/l at least 90 minutes after eating.

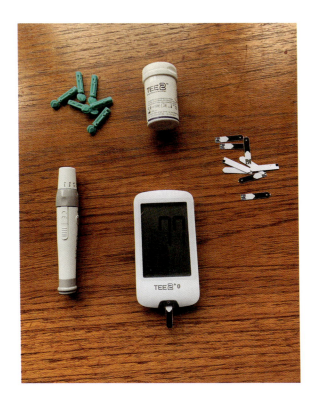

FIGURE 7.14
Self-monitoring blood glucose. A self-monitoring kit comes with disposable lancelets (top left), a pen-like device to prick a finger, the monitor device itself, the test strips (top right), and a container to hold these strips. A result is obtained within 5 seconds on blood being applied to the test strip.

The required dose of a bolus of insulin can be calculated from factors such as the blood glucose level and amount of carbohydrate estimated to be in a meal. Athletes should ensure their blood glucose is 7–12 mmol/l before their sport, and additional glucose intake during and after the activity may be required as periodic testing demands. Continuous glucose monitoring may be needed for those in fear of, or at risk of, hypoglycaemia, or who have persistent hyperglycaemia (HbA1c ≥75 mmol/mol (9%)) despite frequent testing and responses. The choice of type of insulin and timing of injection can be difficult to assimilate, and there are many preparations (Table 7.13). NICE NG17 recommends offering regimes that include multiple daily basal-bolus injections, rather than twice-daily mixed insulin regimes.

Basal insulin therapy could be with twice-daily long-acting detemir insulin (Levemir), an alternative being once-daily glargine insulin (Lantus). Rapid-acting insulin analogues (rather than human or animal forms) should be offered for injection before meals. An alternative such as a twice-daily mixed regime may be considered, as is the use of continuous subcutaneous insulin infusions from a micro-pump. In some cases, pramlintide may be useful as it has been shown to reduce postprandial

TABLE 7.13 Insulin formulations

Type of action	Insulins	Lag period	Length of action
Fast	Aspart, glulisine, lispro	5–15 minutes	3–4 hours
Short	Actrapid, humulin	30 minutes	5–8 hours
Intermediate	NPH insulin	1–3 hours	16–24 hours
Long	Glargine, detemir, degludec	1–2 hours	~24 hours

Source: British National Formulary, NICE. Other preparations include some supplemented with protamine to extend its duration of action, and a long-acting product mixed with zinc.

TABLE 7.14 Signs and symptoms of hypoglycaemia

Anxiety, confusion, feeling cold, headache, hunger, nausea, pallor, palpitations, personality changes, shivering, shortness of breath, sweating, tachycardia, tiredness, tremor/shakes.

hyperglycaemia and glycaemic variability when used alongside insulin. Metformin (as an insulin sensitizer), DPP4 inhibitors, and SGLT-2 inhibitors may also be used.

An important aspect of T1DM is hypoglycaemia, which, if severe, may be fatal. This may develop in those on certain drugs that predispose to hypoglycaemia, but most cases are due to incorrect insulin dosing in those who have over-exercised (sports, dancing), or have not had enough breakfast. In theory, hypoglycaemia may also develop in effective overdoses of oral drugs. Care is required as the biochemical definition of hypoglycaemia (blood glucose <3.9 mmol/l in people with diabetes, <2.8 mmol/l in adults) may differ from the clinical definition (according to signs and symptoms) in a particular patient (Table 7.14). But should either be severe, immediate treatment with oral (e.g. 0.3 g/kg of a fast-acting carbohydrate such as dextrose tablets or a sugary drink) and/or intravenous glucose (by a healthcare professional) is required. Severe hypoglycaemic events requiring treatment occur at a frequency of ~18 per 100 patient-years, those leading to unconsciousness or seizure at ~5 per 100 patient-years, and overall account for ~7% of T1DM deaths.

Treatment of T2DM

A typical target for a patient trying to control their disease with or without a single drug treatment not associated with hypoglycaemia is an HbA1c of 48 mmol/mol (6.5%), whereas a patient on a drug associated with hypoglycaemia would aim for 53 mmol/mol (7.0%). However, targets may be relaxed on a case-by-case basis for certain patients (the frail, the elderly, co-morbidities). The first-line drug treatment is standard-release metformin, but should this not be tolerated, a modified release form may be used. Should this fail, any one of a panel of second-line agents is needed. Should this also fail, dual- and triple-therapies are called for, and perhaps insulin (Figure 7.15).

Some advocate monotherapy for those with an entry HbA1c <7.5%, dual therapy for those with an entry HbA1c ≥7.5%, and triple therapy (which may include insulin) for those whose entry HbA1c is >9%. The choice of medications depend on clinical need and patient's preferences. The metformin dose will need to be reviewed in the light of poor renal function, and potentially replaced with a DPP4i, a sulphonyl-urea, or pioglitazone, although the latter has several contraindications.

SELF-CHECK 7.9

How are the different types of insulin classified?

Treatment of other forms of diabetes

The numerically minor number of non-T1 or -T2 cases of diabetes are generally treated at centres of expertise on an individual basis, depending on the precise aetiology. For example, should the disease follow monogenic failure of insulin production, treatments will resemble those of T1DM. On the whole, the aim is, as in T1DM and T2DM, to keep HbA1c as low as possible without causing any hypoglycaemia, and to address side issues as they arise.

Pharmacological treatment 2: Blood pressure and lipids

The disease process in diabetes inevitably leads to hypertension and dyslipidaemia. Diagnosis, treatment, and monitoring of hypertension is broadly the same in T2DM as without (NICE NG28 and NG136, which recommends annual blood pressure measurements). Because of the risk of target

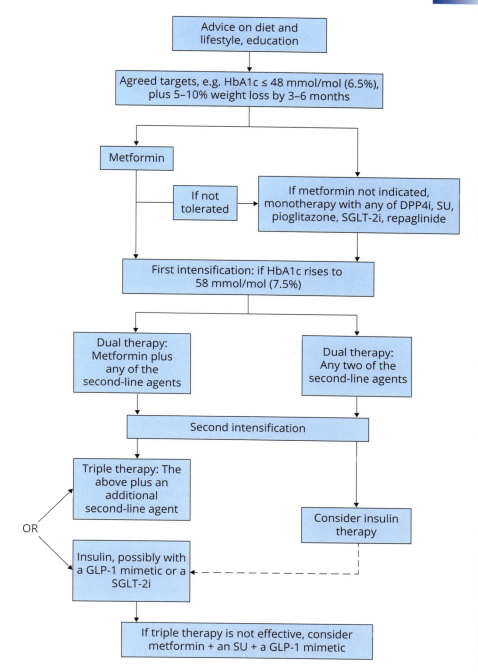

FIGURE 7.15

Schematic for the treatment of T2DM. Modified from NICE NG28. HbA1c = glycosylated haemoglobin, DPP4i = Di-peptidyl peptidase 4 inhibitor, SU = sulphonylurea, SGLT-2i = sodium-glucose co-transporter inhibitor, GLP-1 = glucagon-like peptide-1.

organ damage, the blood pressure target should be 130/80 mmHg, the first line of treatment being an ACEI or ARB. Should these be poorly tolerated or contra-indicated, a calcium channel blocker or thiazide-like diuretic are options. If this approach fails to achieve the target, dual and perhaps triple therapy may be needed. In the presence of clear cardiovascular disease, targets may be lowered to 125/75 mmHg.

In parallel, NICE guidance on dyslipidaemia follows roughly the same course: pharmacological intervention to reduce total and LDL cholesterol to a level of that of people without diabetes but with cardiovascular disease (NICE CG181). As diabetes figures highly in total risk-score calculators, the target LDL cholesterol level should be <2.6 mmol/l, the first line of treatment being a statin, which should be up-titrated until the target has been achieved. Despite focus on intensive blood glucose and blood pressure control, and the use of statins, the mortality rate in T1DM remains over three times that of the general population.

Other treatments

Surgical approaches to T1DM include islet cell or pancreatic transplantation, and for both forms of diabetes, obesity can be treated with gastric banding, sleeve gastrectomy, abdominal bypass, and general liposuction. Surgical intervention is effective in reducing the transition to diabetes and reducing markers of diabetes. Indications for transplantation include patients with severe and frequent hypoglycaemic episodes, despite optimum treatment.

7.4.7 Impaired glucose tolerance (IGT) and impaired fasting glycaemia (IFG)

The pre-diabetes conditions impaired glucose tolerance (IGT) and impaired fasting glycaemia (IFG) are important for a number of reasons. They indicate an increased risk of diabetes, open the door to awareness and potential lifestyle changes, and denote an increased risk of cardiovascular disease (marginally increased endothelial dysfunction is already present in IGT).

IGT is present in ~7.5% of the global population, almost half are aged under 50, with the frequency rising with age: 8% aged 45–9, 12% aged 60–4, 15% in those aged 75–9. The age-adjusted prevalence of IGT ranges from 4.4% in Europe to 12.3% in North America and the Caribbean. Five years after a diagnosis, the disease will have progressed to diabetes in 26% of those with IGT and in 50% of those with IFG.

IGT and IFG are generally not treated pharmacologically, but healthcare professionals to emphasize the need for a healthy lifestyle in resisting diabetes. Patients are often overweight, so may be referred to a hyperglycaemia/diabetes clinic or an overweight/obesity clinic and appointments with clinical nurse specialists and dieticians. Annual blood tests for glucose, HbA1c, U&Es, and for blood pressure are the norm, often in primary practice. There are no specific NICE documents for IFG and IGT, but NICE PH38 describes a fasting plasma glucose of 5.5–6.9 mmol/l (which would include all cases of IFG in Table 7.10) as being at high risk of progression to T2DM.

The common feature of the different forms of diabetes is hyperglycaemia, which, if not addressed fully, over decades slowly attacks and destroys various cells, tissues, and organs leading to disability, morbidity, and an estimated 8–15 years of life lost.

Key points

Whilst the objective of the treatment of hyperglycaemia is to reduce blood glucose, the danger lies in over-treatment, which could drive glucose levels down so low as to cause potentially life-threatening hypoglycaemia.

7.5 Inflammation

7.5.1 A historical perspective

Histological analyses of the components of the developing and mature atheroma in both animal and human models have long reported the presence of macrophages filled with lipids (foam cells, section 6.4.3). Subsequent work showed that these were circulating monocytes that had moved into the atheroma as macrophages, presumably to fulfill their roles as scavengers, but presence of T lymphocytes was not as easy to explain. The presence of the latter demanded a hypothesis of an active immune response within the atheroma, and so an antigen, which proved difficult to demonstrate, although oxidized LDL was thought to be the culprit. In parallel, slightly raised numbers of circulating leukocytes

CASE STUDY 7.3 Diabetes

A 56-year-old woman, living with her daughter, visits her GP with a history of increasing tiredness, occasional nausea, sweating, shortness of breath, and weight gain. She estimates a gain of 2 stones (13 kg) in a year, with a current BMI of 35.5 kg/m², and has been recently treated for **cystitis** and a vaginal infection of *C. albicans* with a topical antifungal. The GP notes an increasing frequency of clinic visits and performs a physical examination, finding a blood pressure of 142/85 mm Hg. When the clinic nurse reports the urine analysis that includes ++ sugar, she performs a random blood glucose on a finger-prick, which gives 10.9 mmol/l, and accordingly takes a venous blood sample for the local laboratory. On prompted questioning, the woman recognizes increasing thirst and the need to get more powerful reading glasses. They discuss weight-loss programmes, principally diet control and exercise, and agree a repeat consultation the following week.

The bloods are back, showing HbA1c of 58 mmol/mol (reference range 20–41)/7.5% (4.0–5.9), giving a firm diagnosis of diabetes. At this stage it may be type 1 or type 2, although the latter is far more likely. The GP arranges meetings with a diabetes nurse at the local hospital, and takes more blood for routine haematology and biochemistry indices, all of which come back within reference ranges. At the hospital, the patient and nurse (following NICE NG28) discuss a care plan, with initial targets of HbA1c 52 and 5% weight loss over 3 months. The patient receives a package of advice and support (principally, the DAFNE protocol for diet), an appointment for retinopathy screening, is put in touch with the local patient group, and is given a prescription for metformin, 500 mg to be taken twice a day during or after meals. The patient tolerates this regime well, so the diabetes nurse recommends increasing to three times a day.

At the 3-month review, the patient has lost 5 kg (6.25%), which she ascribes to swimming three or four times a week, participating in a weight-loss programme, using the bicycle, and diet changes, giving a BMI of 33.3 kg/m². She is delighted to learn that her HbA1c has fallen to 54 mmol/mol. However,

the nurse is not so pleased, and although offering congratulations and encouragement, recommends that she increase the metformin to 850 mg twice a day, retaining one 500 mg tablet. This increase is also well tolerated, and at the next meeting the patients' weight loss is now 7.5 kg (9.4% from baseline) and her HbA1c is down to 52 mmol/mol. The patient feels much better, as symptoms have mostly abated and blood pressure is down to 135/80. However, whilst the nurse is supportive, she checks with medical staff and advises that the patient go to 850 mg of metformin three times a day and starts a second-line treatment with pioglitazone, initially 15 mg a day, rising to 30 mg once a day.

At home, the patient's daughter moves out to live with her partner, and as their flat does not allow pets, it falls to the patient to walk her dog twice a day. Over the weeks and months that follow, the patient finds herself enjoying these walks, and extends them from 15 to 30 minutes and beyond. She notes further weight reduction, and at the next 3-month assessment is delighted that she is no longer obese, with a BMI of 29.8 kg/m², having lost 16.2% of her baseline weight, and with an HbA1c result of 48 mmol/l, enjoys the congratulations of all healthcare staff. However, wishing to continue the improvement in her health, she continues her rigorous pursuit of weight loss, and is rewarded a further 3 months later with a further fall in all indices—weight 65 kg (a reduction of 19% from baseline), BMI 28.9 kg/m², and HbA1c 46 mmol/mol. The question now arises as to the effects of the dog-walking and the pioglitazone, which started at around the same time. On discussion with her healthcare staff, the patient cuts down the pioglitazone to 15 mg daily, and then stops it altogether. This seems to have no effect on HbA1c, and as the patient is keen to self-manage completely, trials of reducing the metformin begin. In the year that follows, the patient's weight stabilizes at around 62–3 kg (still overweight, BMI >25), HbA1c around 43–4 (target <41) on metformin 500 mg tds, but nevertheless the hospital staff feel she is a success and discharge her back to her GP.

and an abnormal erythrocyte sedimentation rate and C-reactive protein (CRP) in patients with atherosclerosis were assumed to be epiphenomena, or 'genuine' subclinical asymptomatic disease, perhaps a low-grade acute phase response.

This view slowly eroded in the face of data from many different sources. For example, raised levels of inflammatory cytokines tumour necrosis factor-α (TNFα), interleukin 1 (IL-1), and IL-6, in addition to CRP, are predictive of major adverse cardiovascular events (MACEs) such as myocardial infarction and stroke. Others have showed that the circulating neutrophil count correlates with the mass of the infarcted myocardial tissue, whilst the neutrophil count predicts patients going on to suffer a MACE.

7.5.2 The need for evidence

Koch and Bradford Hill's criteria of causation can be applied to the risk factor hypothesis, and the weight of evidence indicating smoking, lipids, diabetes, and hypertension as causes is overwhelming.

Cross reference

A more in-depth discussion of causation, including Koch's postulates and Bradford Hill's criteria, can be found in section 2.4 of Chapter 2.

Based on the hypothesis that substance X causes a disease, removal or reduction of substance X must translate to a reduction in that disease. It has long been established that the well-known non-steroidal anti-inflammatory drug (NSAID) aspirin protects against cardiovascular disease, whilst aspirin resistance (by platelets) is a risk factor for MACE in patients who suffered an ischaemic stroke. However, the effect of aspirin could operate through several metabolic pathways, and the laboratory demonstration of aspirin resistance by platelets does not lend itself to an acute setting or to management. The success of another antiplatelet drug (clopidogrel) with no anti-inflammatory activity supports the view that the effect of aspirin on the platelet did not necessarily extend to a systemic anti-inflammatory response.

7.5.3 Proof of concept

The key to the hypothesis that inflammation is an independent risk factor for atherosclerosis is that steps to reduce inflammation will lead to a reduction in the frequency of the disease. Colchicine, a treatment for gout, inhibits neutrophil chemotaxis and cytokine production and is therefore an NSAID. In one study, giving the drug to patients with existing cardiovascular disease (and so already on aspirin) or with gout resulted in a reduction in the frequency of future MACEs. In others, colchicine was shown to significantly reduce both the volume of the atherosclerotic plaque in coronary arteries and the levels of CRP, IL-1, and IL-6. Use of hydroxychloroquine (with NSAID activity) in patients with rheumatoid arthritis (in which it is a standard treatment) results in a reduction in MACEs, and immunosuppressive therapy in RA and psoriasis leads to a reduction in blood pressure.

A third example of a specific anti-inflammatory intervention is a monoclonal antibody to IL-1 (canakinumab). Trialled in patients with existing cardiovascular disease and a raised CRP, the intravenous drug provided a dose–response reduction in CRP and a fall in the platelet count and neutrophil count, but no change in serum lipoproteins. The overall beneficial fall in MACEs, alongside the decreased incidence of arthritis, gout, and fatal cancer, is countered by the increased number of deaths due to infection or septicaemia.

Low-dose methotrexate (13–15 mg/week) is a successful anti-inflammatory treatment for rheumatoid arthritis, improving the disease activity score, IL-6, and CRP. It was trialled in atherosclerosis, but failed to have any effect on MACEs, CRP, IL-1, or IL-6. One reason for this is that rheumatoid arthritis is a far more inflammatory disease, and so a suppressive effect would be more noticeable. The last report aside, these studies, and other findings, confirm that inflammation is an independent risk factor for atherosclerosis. However, inflammation differs from the other risk factors as the question still remains as to whether it is linked causally to the disease or is a consequence of the disease. If the latter is the case, the precise antigenic stimuli need to be identified, and many presume that these are within the atherosclerotic plaque or atheroma. The fact that certain lipid-lowering treatments also reduce CRP implies that the inflammatory stimulus is linked to lipids. However, in one large study, around half of all patients who suffered an acute coronary syndrome (a large proportion of which would be caused by atherosclerosis) had normal levels of serum CRP. The role of inflammation will remain unclear until it can be shown that a simple specific treatment (as in the other risk factors) has a major effect on CVD outcomes.

SELF-CHECK 7.10

What evidence is there that atherosclerosis is an inflammatory disease?

Chapter summary

- The risk factor hypothesis is very strong, and, fortunately, many such factors can be treated. All factors must be considered together, and total CVD risk depends on the success of the treatment of all of them.

■ Tobacco smoking is a major risk factor for CVD and cancer. Patients must be urged to quit, perhaps with the support of nicotine replacement therapy.

■ Hypertension can be managed with a number of drugs of different classes, but is more complicated in the presence of other disease.

■ Treatment of dyslipidaemia focuses on reducing LDL cholesterol with statins and other drugs.

■ Diabetes in itself causes relatively few deaths (but considerable morbidity), but its major role is as a risk factor for atherosclerosis.

■ Inflammation seems to play some role in the development and/or progression of atherosclerosis, but evidence is weak compared to that of the major four factors.

 # Further reading

● Ridker PM, Everett BM, Thuren T, et al. Anti-inflammatory Therapy with Canakinumab for Atherosclerotic Disease. N Engl J Med. 2017: 377; 1119–31.

● Williams B, Mancia G, Spiering W, et al. 2018 Practice Guidelines for the management of arterial hypertension of the European Society of Hypertension and the European Society of Cardiology: ESH/ESC Task Force for the Management of Arterial Hypertension. J Hypertens. 2018: 36; 2284–309.

● Whelton PK, Carey RM, Aronow WS, et al. 2017 ACC/AHA/AAPA/ABC/ACPM/AGS/APhA/ASH/ASPC/NMA/PCNA Guideline for the prevention, detection, evaluation, and management of high blood pressure in adults. J Am Coll Cardiol. 2018: 71; 2199–269.

● Parsons RE, Liu X, Collister JA, et al. Independent external validation of the QRISK3 cardiovascular disease risk prediction model using UK Biobank. Heart 2023: 109; 1690–7.

● Hammersley D, Signy M. Ezetimibe: an update on its clinical usefulness in specific patient groups. Ther Adv Chronic Dis. 2017: 8; 4–11.

● Rosenson RS, Hegele RA, Fazio S, Cannon CP. The Evolving Future of PCSK9 Inhibitors. J Am Coll Cardiol. 2018: 72; 314–29.

● Ruscica M, Ferri N, Santos RD, et al. Lipid Lowering Drugs: Present Status and Future Developments. Curr Atheroscler Rep. 2021: 23; 17. doi: 10.1007/s11883-021-00918-3

● Grundy SN, Stone NJ, Bailey AL, et al. 2018 AHA/ACC/AACVPR/AAPA/ABC/ACPM/ADA/AGS/APhA/ASPC/NLA/PCNA Guideline on the management of blood cholesterol: Executive summary. J Amer Coll Cardiol 2019: 73; 3168–209.

● Santos RD, Gidding SS, Hegele RA, et al. Defining severe familial hypercholesterolaemia and the implications for clinical management. Lancet Diabetes-Endocrinol 2016: 4; 850–61.

● Katakami N. Mechanisms of development of atherosclerosis and cardiovascular disease in diabetes mellitus, J Atheroscler Thromb 2018: 25; 27–39.

● Todd JN, Srinivasan S, Pollin TI. Advances in the genetics of youth-onset type 2 diabetes. Curr Diab Rep 2019: 18; 57. doi:10.1007/s11892-018-1025-1.

● Wu Y, Ding Y, Tanaka Y, Zhang W. Risk factors contributing to type 2 disease and recent advances in the treatment and prevention. Int J Med Sci 2014: 11; 1185–200

● DiMeglio LA, Evans-Molina C, Oram RA. Type 1 diabetes. Lancet 2018: 391; 2449–62.

● Nashef, SAM et al. (2012). EuroSCORE II. Eur J Cardiothorac Surg. 2012: Apr; 41(4): 734–44; discussion 744–5. doi: 10.1093/ejcts/ezs043.

Useful websites

- National Institute for Health and Care Excellence: **www.nice.org.uk**
- The British National Formulary: **https://www.bnf.org/products/bnf-online/**
- The QRISK calculator: **https://qrisk.org**

 ## Discussion questions

7.1   Explain why inflammation is not a major risk factor for atherosclerosis.

7.2   In contrast to dyslipidaemia, explain why hypertension is not primarily a genetic disease.

Clinical cardiovascular disease

Chapter 6 has given us a firm understanding of the aetiology of the various cardiovascular diseases (CVDs), whilst Chapter 7 looked at clinical aspects of the risk factors for CVD. This chapter outlines the clinical consequences of these disease processes in three broad areas: the heart, the brain, and peripheral arteries of the abdomen, leg, and elsewhere. As in the chapters on clinical oncology, we will address each of the major clinical conditions, looking at presentation, diagnosis, and treatment.

Learning objectives

After studying this chapter, you should confidently be able to:

- describe the presentation, diagnosis, and management of the major types of coronary artery disease

- outline other forms of heart disease: valve disease, cardiomyopathy, arrhythmia, and heart failure

- appreciate the key clinical features of stroke and atrial fibrillation

- list the various types of peripheral artery disease and understand how they are defined and managed

8.1 Introduction

Similar to cancer, CVD is a large and complex group of conditions with a variety of aetiologies, many of which overlap. Although atherosclerosis is the leading disease mechanism that causes a stenosis (narrowing) of an artery, and then its occlusion, there are several others, many of which have a genetic basis (section 6.8). Table 8.1 summarizes the major causes of CVD deaths as defined by the International Classification of Disease (ICD), although the UK's Office for National Statistics (ONS) refers to these as diseases of the circulatory system.

The leading cause of CVD is heart disease, and within this group of diseases, ischaemic heart disease (IHD), which includes acute myocardial infarction (AMI), is the most common form. However, if we combine IHD deaths with those of 'Other heart diseases', these being pericardial disease (including pericarditis), myocarditis, endocarditis, diseases of the valves, cardiomyopathy, disease that follows the loss of the natural beating of the heart (arrhythmia), and heart failure (generally defined as the inability of the heart to eject more than 40% of the contents of the left ventricle), and rheumatic heart

TABLE 8.1 Leading cardiovascular deaths in England and Wales, 2021

Type of disease	Deaths	%
The entire circulatory system	134,174	100
Ischaemic heart diseases	56,829	42.3
Cerebrovascular diseases	28,998	21.6
Other heart diseases	28,918	21.5
Hypertensive diseases	7,954	5.9
Diseases of arteries, arterioles, and capillaries	7,750	5.8

© Office for National Statistics. Source: https://www.ons.gov.uk/. Licensed under the Open Government Licence 3.0.

disease, we come to a larger proportion of 63.8% of all CVD deaths of the circulation in England and Wales in 2021 (69.4% of the men and 57.4% of the women).

The second major grouping of CVD is cerebrovascular disease (blood vessels of the brain), major causes of which are carotid artery disease and hypertension. There are subgroups of intra-cranial haemorrhage (ICH, including sub-arachnoid haemorrhage), cerebral infarction, and stroke. Hypertensive diseases include essential and secondary hypertension, hypertensive heart and/or renal disease, and hypertensive heart failure. However, there can be overlap as some refer to hypertensive heart disease as that heart disease which is caused by hypertension, such as left ventricular hypertro-phy (LVH, a thickening in the wall of the left ventricle), heart failure, and coronary artery disease (CAD).

Diseases of the arteries, arterioles, and capillaries include atherosclerosis (60 deaths in England and Wales in 2021), aortic aneurysm and dissection (5,064), other aneurysm and dissection (208), other peripheral vascular disease (2,039), arterial embolism and thrombosis (154), other disorders of arter-ies and arterioles (210), and diseases of capillaries (15). Not shown in Table 8.1 are acute rheumatic fever (3 deaths), chronic rheumatic heart diseases (1,040; almost all being of the valves of the heart), diseases of the veins, lymphatic vessels, and lymph nodes not elsewhere classified (2,666), and other and unspecified disorders of the circulatory system (16).

8.2 Coronary artery disease

The heart is clearly your most crucial organ. The leading cellular aetiology is ischaemia (due to lack of oxygen), hence ischaemic heart disease (IHD) occurs as a consequence of CAD. The acute events of **angina** (chest pain of cardiac origin) and acute myocardial infarction (AMI) may lead to subsequent myocardial infarction (occurring within 4 weeks of an AMI) and, ultimately, chronic IHD. The presenta-tion of CAD runs a spectrum from stable angina and unstable angina to AMI, of which there are two types: non-ST-elevation myocardial infarction (NSTEMI, section 8.2.6) and ST-elevation myocardial infarction (STEMI, section 8.2.7). But before we look in detail at these conditions, we must briefly review pathophysiology (section 8.2.1) and investigations (section 8.2.2).

8.2.1 Pathophysiology

The purpose of arterial blood is to deliver oxygen, glucose, and other essentials to the tissues. Should it fail to do so, because the arteriole is blocked (occluded) by atheroma or its fragments, or by throm-bosis, then the tissue that would normally be fed with oxygen and glucose is denied these nutrients and, unless blood supply is rapidly restored, will die. This process, an infarction, can occur in any tissue or organ: in the heart, it is a myocardial infarction (MI), in the brain a cerebral infarction, and so on. Although infarction resulting from plaque rupture and/or thrombosis is certainly the most serious pathology, there can be other clinical consequences of atherosclerosis, such as reduced blood flow.

Should a particularly large artery be occluded, then the large mass of the myocardium that it feeds will be compromised, and so the damage is greater. A damaged myocardium will fail to contribute to good overall heartbeat, and this can lead to irregularities that can be fatal. Should the myocardial infarction be the result of the sudden rupture of an atherosclerotic plaque, then the symptoms of angina-like pain will come on rapidly, and if so, we can further qualify the process as an AMI.

At rest, the stenosis of a particular artery by a modestly sized atherosclerotic plaque will not result in functional ischaemia. However, at times of high muscle activity, such as strenuous exercise, the same stenosed artery will not be able to satisfy the increased demand for oxygenated blood. Reduced oxygen delivery will result in hypoxia (lack of oxygen), the consequences being ischaemia, hence myocardial ischaemia. When deprived of sufficient oxygen to satisfy their needs, muscles can still obtain energy from glucose via **anaerobic** respiration, which may lead to the build-up of a metabolic by-product, lactic acid, high levels of which cause cramp in the particular muscle bed. This happens when the demands of muscles exceed the ability of the cardiovascular system to supply these tissues with oxygen. Athletes may well feel cramp in their legs, and when the same process (generation of lactic acid) develops in the heart, that pain is angina. Objective definition of the different forms of heart disease relies on several investigations—blood tests, the electrocardiogram (ECG), echocardiography, and **angiography**.

8.2.2 Key investigations in heart disease

Blood tests

The heart is clearly a very metabolically active organ, requiring oxygen and glucose to generate ATP. One of the enzymes involved in these metabolic pathways is creatine kinase (CK), another is lactate dehydrogenase (LDH). However, LDH is also found in liver cells and red blood cells, and CK is also found in other muscle, as is myoglobin. Consequently, damage to the heart after a myocardial infarction can result in increased serum levels of both enzymes. Damage to the myocardium can be distinguished from damage to other muscles as a subtype of CK, that is, CK-MB, is found only in heart muscle. Thus, raised plasma CK-MB is specifically indicative of cardiomyocyte damage, and there is a clear relationship between increased CK and CK-MB and the mass of heart muscle that has been injured, so that this blood test effectively tells us of the severity of the heart attack.

The molecular troponin complex of three components is part of the musculature of the cardiomyocyte that includes the structural proteins actin and myosin, and (unlike CK) is specific for this cell. Two of these components, troponin T and troponin I (TropT, TropI), are often found in the blood after myocardial damage. The degree of increase in the markers varies with the time that has passed since the presumed acute event. The quickest markers to respond are CK and CK-MB, and the slowest are troponins, so that levels may not be abnormal on presentation to A&E, often peaking 3–4 hours after the infarction. Consequently, a second blood test is required several hours after the first to gauge the degree of increase (or not) of these markers, and the presumed extent of the damage to the myocardium.

Whilst troponins are specific for the heart, in themselves (i.e. without other clinical information) they have poor sensitivity. Conditions associated with raised cardiac troponins include cardiac amyloidosis and cardiomyopathy, such that it may be difficult to interpret if the presenting patient has one of these co-morbidities. There are also non-cardiac causes of a raised troponin, such as renal failure and stroke. The presenting chest pain may be due to pulmonary embolism, for which a semi-specific blood marker is levels of **D-dimers**, reflecting thrombosis. Unfortunately, atherosclerosis (with its burden of thrombosis) is also linked to raised D-dimers, compounding the problem. If there is indeed an acute coronary event in progress, the white blood cell count (and neutrophil count in particular) will start to rise, as will general and non-specific inflammatory markers erythrocyte sedimentation rate (ESR) and C-reactive protein (CRP), a process that will continue for the 24 hours that follow.

SELF-CHECK 8.1

What are the major blood tests of heart disease, and what do they tell us?

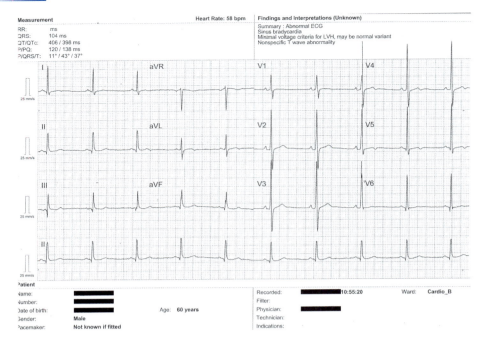

Measurement		Heart Rate: 58 bpm	Findings and Interpretations (Unknown)
RR:	ms		Summary : Abnormal ECG
QRS:	104 ms		Sinus bradycardia
QT/QTc:	406 / 398 ms		Minimal voltage criteria for LVH, may be normal variant
P/PQ:	120 / 138 ms		Nonspecific T wave abnormality
P/QRS/T:	11° / 43° / 37°		

Patient
Name:
Number:
Date of birth:
Gender: Male
Pacemaker: Not known if fitted

Age: 60 years

Recorded: 10:55:20   Ward: Cardio_B
Filter:
Physician:
Technician:
Indications:

FIGURE 8.1

An ECG printout. Data in the top left regard the different parts of the cycle, such as the time interval of the QRS complex. The heart rate (58 bpm) is given at the top centre, whilst the ECG machine automatically provides its assessment on the top right. In this case, it reports an abnormal ECG with a T wave abnormality. It also considers the heart rate to be a little slow (bradycardia).

The electrocardiogram (ECG)

This device detects changes in the electrical activity of the heart as the chambers contract and relax, and the valves open and close. It is expressed as a series of peaks and troughs that make up the **PQRST** pattern, and is printed out as an A4 sheet showing individual traces from each of the twelve leads, these being I, II, III, aVR, aVL, aVF, and V1–V6 (Figure 8.1), each of which refers to a different aspect of the heart along the bottom is a continuous trace of lead II, showing 12 complexes.

When the myocardium is damaged by an infarction, there are changes in the way that the chambers operate, and these changes can also be detected by the ECG. The left side of Figure 8.2 shows a typical trace of a normal ECG from one of twelve leads. On the right, the trace is typical of an ECG soon after an AMI, the section in red representing the abnormality.

The key difference is in the RST section. In health (the trace on the left), the 'peak' of the R wave is followed by a 'trough' of the S wave, which then rises to form a small peak, the T wave. The fundamental characteristic of an AMI (the trace on the right) is the loss of the normal trough of the S wave, which is replaced by an elevation in the S wave that merges into the T wave—the S wave does not have a trough. If present, this ST-elevation is a major contributor to a diagnosis of one of the more serious of the two types of myocardial infarction, a **STEMI**. However, an ST elevation is not specific for a heart attack; it may also be present in other conditions, such as inflammation of the pericardium (that is, pericarditis), or a clot in a vessel of the lung (a pulmonary embolus).

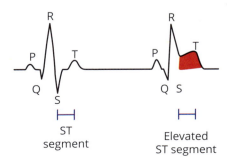

FIGURE 8.2

The ECG in ST-elevation myocardial infarction. The key abnormality in the increased ST segment is highlighted in red.

There may be other changes that point to damage to the myocardium near to the septal conduction tissues, leading to a left-bundle branch block (LBBB: the LBB being the tissue conducting instructions to contract to the left myocardium (see Figure 6.7)). In investigating angina, a number of changes on a resting 12-lead ECG are consistent with CAD and may indicate ischaemia or previous infarction. These include pathological Q waves, changes reflecting a LBBB, and ST-segment and T wave abnormalities.

The ECG is an extremely important tool for determining many different heart conditions, including LVH. The left ventricle delivers the major force to drive blood around the peripheral circulation. However, in hypertension, this force has to be stronger, and so the left ventricle becomes larger to counter the high blood pressure, and to drive blood out into the aorta. This enlargement (hypertrophy) is the basis of LVH, and over years the left ventricle eventually tires, becomes weak, and is unable to provide sufficient force to drive blood around the body. Illustrating the complexity of CVD, LVH may also be a long-term consequence of valve disease, an AMI and IHD, and can lead to heart failure (section 8.3.5).

Imaging

Echocardiography (Echo) is an important investigation that visualizes and quantifies crucial functions of the heart, and the integrity of structural components, such as the septum. A form of ultrasound, it is generally taken across the chest wall (i.e. trans-thoracic), but if this gives poor images (as potentially in the morbidly obese) the trans-oesophageal route will be required. With colour Doppler, Echo provides a host of complex haemodynamic data, most of which are interpreted by guidelines such as those of the British Society of Echocardiography. The value of Echo in valve disease (such as blood flow between chambers) will be discussed in section 8.3.1, in arrhythmia (the volume of the left atrium in AF) in section 8.3.3, and in heart failure (the fraction of the left ventricle that is ejected each heartbeat—the LVEF) in section 8.3.5. Several Echo metrics are also valuable in the management of CAD, an important measure being LVH.

The leading invasive investigation of CAD is a percutaneous coronary intervention, or PCI. Access to the coronary circulation is through the skin (hence percutaneous) and into a peripheral artery, the radial arteries at the wrist being the favoured access point. By running a special tube (a catheter: hence 'cath lab') gently up the brachial artery of the arm and then the sub-clavian, and finally the aorta, X-ray contrast medium can be infused into the coronary arteries, and any stenoses or occlusions can be seen in real time as an image on a monitor screen. Other more complex investigations include computerized tomography (CT), magnetic resonance imaging (MRI) (both techniques widely used elsewhere, such as in cancer (Chapters 4 and 5)), and those of nuclear medicine. These can be combined with angiography, and will be discussed in relation to the particular condition they are associated with in section 8.2.9.

Key points

Major tools for investigating coronary artery disease are signs and symptoms, blood tests, the ECG, Echo, and angiography.

8.2.3 Classification of coronary artery disease (CAD)

This disease can be classified in a number of stages with increasing severity. First, the clinical symptoms of CAD are often preceded by (and part-precipitated by) the 'pre-atherosclerosis' risk factors of Chapter 7. For many, the first clinical symptom of CAD is anginal chest pain, of which there are two types:

- stable angina, which resolves following rest as the demand for oxygen falls and the lactic acid is either washed out of the heart or is neutralized by biochemical pathways;

- unstable angina, the consequence of more severe CAD, which may be present at rest and take longer to resolve (if at all).

Neither of these forms of angina is associated with serious damage to the cardiomyocytes, and changes to the ECG are unusual. Without treatment, angina may lead to AMI and so long-term IHD. As with angina, there are two forms of AMI, each defined by changes in the traces of an ECG as described above.

- The presence of an elevation in the ST sections of the trace leads to the diagnosis of ST-elevation myocardial infarction—STEMI (Figure 8.2).

- However, if there is no elevation in the ST segment, the diagnosis is non-ST elevation myocardial infarction (NSTEMI).

Both are characterized by abnormal blood tests, which are higher in STEMIs. Although a progression, some may suffer an AMI without ever having had angina, whilst others may succumb to the consequences of coronary atherosclerosis (i.e. IHD) without having had an AMI. The sections that follow will address these stages in order of urgency and severity—stable angina, unstable angina, NSTEMI, STEMI. The latter three diagnoses are collectively described as acute coronary syndromes (ACS).

8.2.4 Stable angina

Stable angina is a chronic medical condition with a low but appreciable incidence within the general umbrella of acute coronary events that bring increased mortality (NICE CG126). NICE CG95 specifically refers to chest pain of recent onset, and whether it should be classified as angina, whilst NICE CG126 offers guidance on the management of stable angina.

Presentation

Whilst chest pain is the leading symptom, pain may also be present in the throat, jaw, shoulders, and arms, and may be brought about by emotional stress, physical exertion, or suddenly moving from a hot to a cold environment (as in winter). There may also be non-specific symptoms of nausea, vomiting, the urge to defaecate/urinate, **dyspnoea**, and sweating. The pain ,come and go, giving us stable and, if recurrent, unstable angina, the latter being more serious and often classed alongside AMI, as in NICE CG185, CG95, and CG126. However, the chest pain typical of angina may also be due to indigestion, a pulmonary problem, or musculoskeletal issues. Nevertheless, the presumption must be that the chest pain reflects genuine cardiac ischaemia, and so must be treated, as it may progress to AMI and IHD.

Diagnosis

Stable angina is effectively a diagnosis of exclusion—that is, it is not caused by the factors described above and it is not symptomatic (i.e. more severe, as in unstable angina) of an AMI or other pathology. There are no highly specific signs, symptoms, or investigations, but numerous indicators. Should the patient present to their GP with an appropriate history and symptoms, and differential diagnoses excluded, an immediate ambulance transfer to A&E will follow, as he or she may be having an ACS (effectively, a preliminary diagnosis). The term covers a range of conditions including unstable angina and different types of AMI and will be discussed in sections to come. On arrival, a brief history will be taken (family and personal medical history, extent and location of pain, precipitating factors, duration, etc.), blood will be taken, and a 12-lead ECG performed. Blood pressure will also be taken, and if very high (such as >200/100 mmHg) may demand urgent intravenous treatment (such as with nitrogylceride or labetolol) to bring the SBP to <140 mmHg. Blood will go to the laboratory where it will be tested principally for markers of cardiomyocyte damage, although some tests may be performed in A&E. A formal definition of anginal pain is:

- constricting discomfort in the front of the chest, or in the neck, shoulders, jaw, or arms, all more likely on the left side;

- precipitated by physical exertion;

- relieved by rest or glycerine trinitrate (GTN) within about 5 minutes. This drug breaks down to nitric oxide (NO), which acts to dilate arteries, thereby increasing blood flow that in turn delivers oxygen to relieve the hypoxia causing the pain.

The presence of all three defines typical angina, whilst any two is atypical angina, and one or none is non-anginal chest pain (NICE CG95). Factors making a diagnosis of stable angina more likely include age, male sex, risk factors (smoking, diabetes, hypertension, dyslipidaemia, and a family history of premature CAD), other CVD, and a history of established CAD, for example previous MI or coronary revascularization. Diagnosis also relies on blood tests and an ECG—the former will be normal on admission and 4–6 hours later, and the latter will show few (if any) serious abnormalities.

Management

The objective of the management of stable angina is to stop or minimize symptoms, improve quality of life, and reduce the risk of disease progression. Initial management of simple acute stable angina is with pain relief, such as 300 mg aspirin (which will also reduce the likelihood of thrombogenesis). But should the symptoms persist or get worse, other analgesia, such as opiates, would be appropriate. GTN is offered as a spray into the mouth, or a tablet under the tongue because, as a vasodilator, it should relax coronary arteries and so provide the myocardium with an improved blood flow (ideally carrying oxygen). An alternative to GTN is isosorbide mononitrate. Should mild to moderate symptoms and a favourable ECG call simply for observation, the patient will probably be kept in a side room in A&E and reviewed regularly, probably for 4–6 hours. After this interval, an ECG is repeated to determine any changes from the admission traces, and any new changes. If symptoms are severe and/or develop, it may be that the disease is more extensive, and it may be necessary to send the patient for an emergency percutaneous coronary intervention (PCI) or even for coronary bypass grafting (CABG).

If all investigations are normal (e.g. no increase in blood markers, no evidence of myocardial ischaemia on ECG), discharge will be likely, with plans for outpatient and GP appointments, and with drugs. For the secondary prevention of CVD, these will include (at least) a statin (NICE CG67), a blood pressure lowering drug (or drugs (NICE CG127), but an ACEI for people with diabetes) and aspirin, and for selected patients, a short-acting GTN spray. The patient will be trained in the latter and instructed to use it immediately before any planned exercise or exertion, and should there be further episodes of angina. Should the chest pain persist after 5 minutes, the patient will be advised to repeat the dose, and if this is not effective in relieving pain, to call an ambulance. The side effects (flushing, headache, light-headedness) will also be explained. Healthcare professionals will also discuss the precipitating events with the patient and review the risk factors which may need to be addressed. The patient will be directed to cardiac rehabilitation courses and self-help groups, both of which reduce the risk of recurrence or progression, and they will most likely return to hospital as an outpatient, and check in with their GP.

Later, there may also be a case for angiography, an Echo (to determine any abnormalities with the valves and the chambers, such as LVH), and an exercise ECG test on a treadmill (the Bruce protocol). Other investigations, such as myocardial perfusion scintigraphy, are unlikely at this stage (section 8.2.9).

SELF-CHECK 8.2

What are the acute coronary syndromes?

8.2.5 Unstable angina

By definition, this is a more severe disease than stable angina, from which it differs in a number of ways, principally in that the pain occurs at rest or after minimal exertion, may last for over 20 minutes, and is unpredictable. It may develop from stable angina, becoming increasingly painful and of longer duration over weeks or months. Should the patient have previously presented with stable angina, and been prescribed with a GTN spray, this would generally help resolve the immediate pain before they call for extra help. NICE CG94 applies, and a follow-on surveillance report refers to unstable angina and NSTEMI, indicating the close nature of the pathology of these conditions.

Acute coronary syndromes (ACS)

A further clinical definition we must now consider in more detail is an ACS, an umbrella term for any acutely presenting cardiac condition where the patient is in clear pain and distress that are consistent with an abrupt reduction in coronary blood flow. The spectrum of ACS runs from unstable angina via NSTEMI (accounting for ~59% of MIs) to STEMI (41%), with increasing loss of blood supply (and so myocardial ischaemia) in parallel with rising blood markers and ECG abnormalities. In unstable angina, there may be reversible coronary vasospasm that temporarily reduces blood flow, whereas NSTEMI is believed to be due to partial blockage and so a stenosis, and STEMI due to total blockage/occlusion.

Presentation

Whilst it is entirely possible that the patient will first go to their GP, the severe chest pain makes it more likely that 999 will be called or that they will arrive in A&E themselves. Triage will be prompt, as with stable angina, with history, bloods (which will include glucose and a full blood count), blood pressure measurement, and ECGs. The pain and general discomfort may lead to the use of analgesics and vasodilating nitrates.

Diagnosis

The question facing the staff in A&E concerns the extent of the disease. Is this a case of unstable angina, or a NSTEMI? These are defined by ECG and blood results. The former will give an immediate guide, but blood tests may not be abnormal until hours after the presumed ischaemic event. Consequently, refining the immediate diagnosis of ACS into unstable angina or NSTEMI will be delayed. However, in a 'worst possible outcome' scenario, it may be advisable to assume a NSTEMI until proven otherwise.

Management

Once more, the immediate comfort of the patient will be improved with analgesia, and any other urgent issues will be addressed, such as high blood pressure, dehydration, and nitrate vasodilators. A feature of unstable angina is the need for two classes of antiplatelet drugs: aspirin (300 mg: which the patient may have already taken) and a drug targeting the ADP receptor, such as clopidogrel (300 mg) or ticagrelor (a loading dose of 180 mg, then 90 mg twice a day). A further consideration is the use of an anticoagulant such as a low-molecular-weight heparin (LMWH), and the benefit of these agents will be countered by a fear of haemorrhage. The patient may be moved from A&E to a medical ward and so be formally admitted.

A key early step is the assessment of the risk of a future adverse cardiovascular event using a risk scoring system such as that of TIMI (thrombolysis in myocardial infarction) or the Global Registry of Acute Cardiac Events (GRACE), the latter being available online and favoured by NICE (Box 8.1). A further series of risk calculators for CAD, CVD, and other conditions derives from the Framingham Heart Study.

BOX 8.1 GRACE

GRACE is an algorithm using data pooled from hundreds of thousands of patients from hundreds of hospitals worldwide who were assessed on admission with an ACS and up to 6 months after admission. Much simpler than TIMI, it requires age, heart rate, SBP, creatinine, and the Killip class for chronic heart failure. Additional features are cardiac arrest at admission, ST-segment deviation, and elevated cardiac enzymes/markers. The model gives the probability (%) of death or MI, or simply death, both in-hospital and at 6 months. A second part is used at discharge, and requires age, heart rate, SBP, and congestive heart failure, but also in-hospital PCI, CABG, past history of MI, ST-segment depression, and elevated cardiac enzymes/markers. Use of his model to direct treatment has a profound effect on disease progression and overall survival.

Should the patient have a low GRACE score, perhaps 1.5%, no major ECG changes, and acceptable blood results, they will be treated conservatively according to local guidelines, referring to NICE and other documents. Once stable, patients may also be tested for inducible ischaemia on a treadmill (the Bruce protocol) and for left ventricular function (by Echo) before discharge, or at outpatients weeks later. However, should the acute patient have unequivocal ECG changes and bloods at the high end of the reference range and GRACE score >3%, other measures may be necessary, such as additional anti-thrombotic therapy (eptifibatide or tirofiban) and angiography. The latter will determine if there are indeed any major issues within the coronary arterial system, and if present, how to proceed. As with stable angina, patients with unstable angina will leave hospital with a host of medication and appointments for follow-up and for a cardiac rehabilitation course, and quite probably for further investigations such as Echo.

8.2.6 Non-ST elevation myocardial infarction

Non-ST elevation myocardial infarction (NSTEMI) is characterized by a severe stenosis and/or an occlusion of a coronary artery, with ECG changes and abnormal blood results. NICE CG94 considers unstable angina together with NSTEMI, considering a triad of symptoms, ECG changes, and abnormal blood markers.

Presentation and diagnosis

Presenting signs and symptoms are of ACS, as in unstable angina, and although the immediate treatment is the same (pain relief, antiplatelets), the ECG tells a different story. ECG changes typical of a NSTEMI include ST depression, and changes to the Q wave and T wave inversion, but there are several others that the experienced practitioner will recognize, such as those of an LBBB (as described in section 8.2.2). As AMI develops, the R wave loses its height and the Q wave often becomes deeper (described as pathological), both reflecting loss of viable myocardium. These Q waves may develop within an hour or two of the infarction but are generally late (12–24-hour) changes. Should there be an ST elevation, as in Figure 8.2, the diagnosis is a STEMI (section 8.2.7).

Once the blood results have returned from the laboratory, they may show an increase in CK, CK-MB, and/or troponin, although these may be normal (especially the troponin) if the precipitating pathological event is recent and has yet to cause material damage to the myocardium. However, some patients may have tolerated their chest pain for hours before calling for help, in which case there may be a raised troponin, justifying the need for an accurate history from the patient. The serum **creatinine** result will complete the GRACE score. Should the patient have a GRACE score >3.0%, they may need to undergo angiography within 96 hours, and if so, staff will consider the use of the anti-thrombotic drugs eptifibatide or tirofiban.

Management

Immediate management follows that of unstable angina (see section 8.2.5) but is led by consideration of the extent of the ACS and whether the patient is experiencing a NSTEMI, and if it will develop into STEMI. Once the ECG changes are established, and blood results are back, the next step must be addressed, which calls for objective knowledge of the extent of any issues with the coronary arteries to determine the exact state of stenoses and/or occlusions. This can be achieved with invasive and non-invasive methods, the former being angiography (Figure 8.3). Should a stenosis be present, the degree of narrowing will guide treatment—a small narrowing (generally <20%) may not prompt any immediate action, but a severe narrowing (generally >80%) could be deemed worthy of immediate intervention, perhaps by placement of a stent, or by **angioplasty**. This would be enhanced by coronary CT.

Stents and angioplasty

There are two options for treatment of a stenosis. Stents are small prestressed metal tubes that expand to open up a stenosis and so restore blood flow. They come in a variety of dimensions and lengths so that almost all stenoses can be addressed. Stent technology is now so advanced that there are few

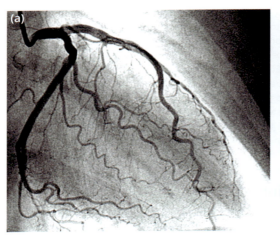
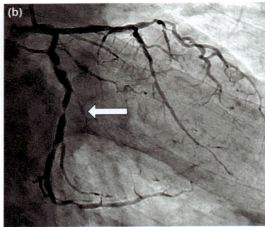

FIGURE 8.3

Coronary angiography. (a) Normal coronary arteries. (b) A stenotic artery (arrowed).

Science Photo Library.

lesions that cannot be stented, and the exceptional patient may well have a dozen or more individual stents. As with diagnostic angiography, the preferred approach is via a radial artery, and the precise placement of the stent within the target stenosis is monitored by bursts of contrast medium. The procedure is illustrated in Figure 8.4.

The stenosis in the artery in Figure 8.4 may well be amenable to treatment with a stent, and stents may be used in several other arteries, including the carotids, the aorta, and the renal and femoral arteries. Stents can be bare-metal, or coated with drugs such as the anti-proliferative agent sirolimus (i.e. drug-eluting stents), designed to prevent re-stenosis, which often have improved long-term patency.

The second option is angioplasty. A specialized catheter is again passed up the arm and shoulder arteries to the site of the stenosis, as in stent placement. However, the end of the angioplasty catheter is equipped with a deflated balloon, which is then inflated across the stenosis so that it presses the atheroma back into the artery wall. Once the deflated balloon is removed, blood flow will be improved. The trouble with this is that the atheroma is likely to reform, leading to problems in the years to come. Accordingly, stenting is by far the preferred option.

Treatment of an occlusion

It has long been known that intervention to restore blood flow (reperfusion) to the damaged tissues is a life-saver. This can be done by infusing a tissue plasminogen activator (such as alteplase, reteplase, or others) that mimics a natural molecule (streptokinase) to direct the natural enzyme plasmin to

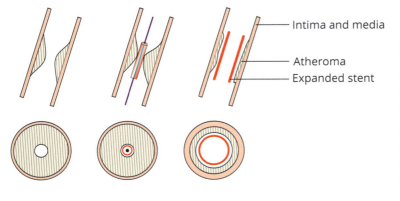

Intima and media

Atheroma
Expanded stent

FIGURE 8.4

Placement of a stent. On the left: the stenosis in cross-section and longitudinal aspects of an artery. Centre: placement of the prestressed stent via a guide wire over the stenosis. Right: the stent is then released and expands to its relaxed diameter, at which point it presses the atheroma back into the artery wall. The guide wire is removed, with a resultant improvement of blood flow.

induce the dissolution of the presumed clot that was theoretically blocking the artery. This process, **thrombolysis** (hence TIMI: thrombolysis in myocardial infarction, Box 8.1), is very effective in long-term follow-up studies of re-infarction, additional surgical interventions, and mortality.

Historically speaking, the occlusion in the diagonal artery in Figure 8.4 may have been treated with thrombolysis (as per TIMI), but despite its successes, in ~25% of patients thrombolysis fails to provide reperfusion, and in 1% it causes a haemorrhagic stroke. However, fibrinolysis/thrombolysis (clot removal) should be offered, especially in those situations where it is not possible to get to a PCI centre within 2 hours.

SELF-CHECK 8.3

What steps can be taken by the individual to reduce their risk of cardiovascular disease?

The time scale is important because, in the hours after its formation, the thrombus becomes consolidated and so increasingly difficult to remove. An alternative procedure to enzymatic digestion of a coronary thrombosis is atherectomy, where the atheroma is removed slice by slice by a rotating blade, and the carvings of atheroma are aspirated away. A thrombus may also be removed mechanically, perhaps by aspiration, although this may liberate particles that will flow downstream and cause micro-infarctions.

Figure 8.5 summarizes these interventions. Access to the coronary circulation is via the radial artery. A catheter is passed up the radial, brachial, and sub-clavian arteries and (in this illustration) into the left coronary artery. Thus X-ray contrast medium introduced at this point will pass into the downstream arteries and detect any abnormalities. In this example there is a stenosis midway down the left anterior descending artery that will reduce blood flow to the myocardium it feeds, causing ischaemia and so anaerobic respiration that will generate lactic acid and so the pain known as angina. This stenosis is amenable to angioplasty or stenting. On the right a thrombus has occluded the distal section of a diagonal artery, and so the downstream myocardium is infarcted and will release blood markers.

This damage will also cause changes to the ECG typical of a STEMI, and will be the object of thrombolysis or other techniques to ensure reperfusion. The puncture site of an angiogram catheter at the wrist leaves a small but distinct scar.

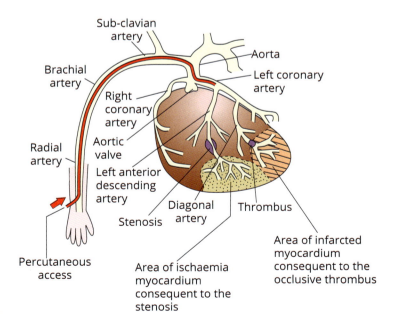

FIGURE 8.5
Angiography and CAD.

8.2.7 ST elevation myocardial infarction

A STEMI is the final and most serious form of an ACS. Almost half of the potentially damaged myocardium may be lost within an hour of occlusion, rising to two-thirds by 3 hours, so that diagnosis and management must be urgent and accurate.

Presentation and diagnosis

Management of any ACS depends on history, presentation, ECG changes, abnormal blood results (indicative of myocardial ischemia and necrosis), and a risk assessment such as the GRACE score or similar. AMI is defined by acute history and abnormal blood markers, and the type of AMI by ECG pattern. If there is an elevation in the ST segment (Figure 8.2) the diagnosis is STEMI, and there may also be other changes such as a pathological (deep) Q wave (Figure 8.6), and an indication of a new LBBB, as there may be with NSTEMI traces.

Immediate management

To a large extent, management has parallels to that of a NSTEMI, but once a STEMI has been established, and myocardial damage proven, steps must be taken to ensure that further damage does not develop, and ideally to reverse the ischaemia/hypoxia.

As described for NSTEMI, step one is to determine the extent of the damage, with standard diagnostic angiography. Step two is to act to restore blood flow (i.e. reperfusion), perhaps with removal of the occlusion by primary PCI or fibrinolysis, although this may be part of the diagnostic angiography. PCI via the radial artery is the preferred strategy if it can be offered within 12 hours of onset of symptoms and can be delivered within 120 minutes of the time when fibrinolysis could have been given. Patients will also be treated with unfractionated heparin or LMWH and prasugrel or ticagrelor. However, if PCI cannot be performed within 12 hours of the onset of symptoms, then fibrinolysis should be offered. The success of fibrinolysis can be assessed by an ECG 60–90 minutes after administration, and for those who have residual ST-elevation suggesting failed coronary reperfusion, there should be immediate coronary angiography, with follow-on PCI if indicated. After either procedure, the patient may well spend a day or two in hospital, being fully monitored, then discharged.

Follow-up management

As with other post-ACS discharge, the patient will have follow-up appointments for cardiac rehabilitation (to begin 10 days after discharge) and possibly other investigations, and will be provided with leaflets on risk prevention and recuperation. Medications as before will be a minimum of aspirin (or clopidogrel if aspirin-intolerant), a statin (most likely atorvastatin or simvastatin), and an ACEI (or ARB if ACEI-intolerant: Table 7.3), along with others depending on the individual. The ACEI and statin may need to be titrated or otherwise managed (perhaps in respect of renal function) to their respective targets. However, mandatory post-MI treatment also includes a beta-blocker (or diltiazem or verapamil if beta-blocker-intolerant: Table 7.3) and a second antiplatelet drug, such as ticagrelor, prasugrel, or clopidogrel.

Some patients may have an indication for anticoagulation, such as if they have a certain prosthetic heart valve, hypercoagulability as defined by unprovoked deep-vein thrombosis or pulmonary embolism, or AF. Accordingly, practitioners will need to balance bleeding risk, thromboembolic risk, and cardiovascular risk, and unless bleeding risk is high, anticoagulation should be continued alongside

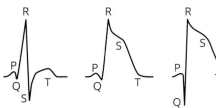
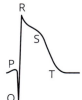

FIGURE 8.6
Lead V4 ECG changes in AMI. Left: normal trace with insignificant Q wave. Centre: ST elevation, also with insignificant Q wave. Right: ST elevation with pathological Q wave—note that the 'depth' of the Q wave exceeds that of the normal S wave on the left.

aspirin in those who have suffered an MI and have had balloon angioplasty or a CABG. NICE CG172 offers guidance on further strategies, such as in those with bare-metal or drug-eluting stents, combination of warfarin with prasugrel or ticagrelor, and the use of direct-acting anticoagulants rivaroxaban, apixaban, edoxaban, and dabigatran.

CASE STUDY 8.1 *Investigation of chest pain*

A 57-year-old man is woken at 4 a.m. by a dull aching pain in the left side of his chest, as if having slept in an awkward position. He takes 600 mg aspirin, but an hour later, the pain is still present (albeit a little better). He drives himself to the local hospital where, 10 minutes later, he is having an ECG and bloods are taken. Blood pressure is 135/75 mmHg, pulse rate 78 (the latter a little high: he is understandably anxious), body mass index is 24.8 kg/m² (target 18–25). The pain is subsiding, and no other analgesia is required. He is generally fit and well, but his father had five-vessel CABG when aged 63. A check on the hospital computer finds a routine 'well man clinic' lipid screen from 3 years previously, with total cholesterol 4.6 mmol/l (reference range 2.5–5.0), triglycerides 1.5 mmol/l (<2.3), and HDL 1.1 mmol/l (0.9–1.4), giving a calculated LDL of 2.5 mmol/l —all quite satisfactory.

The ECG shows only an absent T wave on lead V4 (so, crucially, no ST elevation as in Figure 8.2), and the first round of blood tests is normal: high-sensitivity Troponin-T 3 ng/l (reference range <14 ng/l) and creatine kinase 123 IU/l (<190), although CRP at 7 mg/l (<5) is slightly raised. The initial diagnosis is surprisingly difficult. He is not in severe pain, so a diagnosis of an ACS is unlikely. However, the ECG abnormality is clear, and there was pain on outset that woke him from sleep. He is moved to a side room, and as a second set of bloods taken 4 hours later is all normal (Troponin-T 4 ng/l, creatine kinase 120 IU/l, no abnormalities in the full blood count) and he has no other issues, he is prudently discharged on atorvastatin 40 mg, ramipril 2.5 mg, and aspirin 75 mg daily (as per NICE guidelines) with a diagnosis of an episode of transient stable angina.

The following day, bloods are still normal and the absent T wave abnormality remains. Several weeks later, at outpatient's, an ECG shows that the absent T wave on V4 has now become inverted (Figure 8.7), a change often found in NSTEMI. However,

as discussed, this is not a case of NSTEMI as blood results were normal. The consultant cardiologist recommends PCI to assess the extent of the presumed coronary atherosclerosis.

The PCI proceeds without problems, finding a 50%–60% stenosis near the origin of a diagonal artery arising from the circumflex artery, and a 25% stenosis in the mid-section of the left anterior descending artery. There is also a small area of hypokinesia (lack of movement) in the area of the left ventricle that is fed by the stenosed artery and close to the region where the ECG T wave abnormality was noted. This hypokinesia probably reflects the loss of active cardiomyocytes and their replacement by immobile fibrotic tissue, giving a diagnosis of mild CAD, as there is evidence of damage to the myocardium. This would normally call for a further PCI with a pressure wire (measuring blood flow either side of the stenosis to find the fractional flow reserve (FFR)) to determine whether the greater stenosis would benefit from the placement of a stent, but the patient declined (a FFR of <75–80% often leads to stent placement).

A month later he has an Echo, which reports a good LVEF of 67% and that all valves are working well. After a further 3 months, a blood test shows his total cholesterol is 2.8 mmol/l (reference range 2.5–5.0), HDL 1.2 mmol/l (>1.0), triglycerides 0.9 mmol/l (<2.3), giving a calculated LDL of 1.2 mmol/l. The standard screen for diabetes, HbA1c, is satisfactory at 38 mmol/mol (20–42). He has stopped taking the ramipril as his home BP machine was registering around 110/55, following which it increased to 125/65 (target <135/85). The chest pain has never recurred, and he continues to take active exercise every 2–3 days. After a further 2 years he undertakes a Bruce protocol on a treadmill, which is satisfactory except for an arrhythmia after 11 minutes of increasingly demanding work, which is of minimal pathological concern. He is discharged back to his GP until a full hospital revision at age 65.

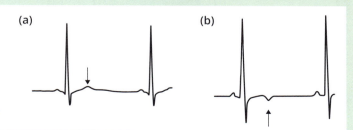

(a)   (b)

FIGURE 8.7
ECGs showing (a) normal and (b) inverted T waves, indicated by arrows.

Key points

Coronary artery disease becomes more serious as the pathology progresses from stable angina to unstable angina, NSTEMI, and STEMI.

NICE CG185 provides guidance on cardiac rehabilitation and the prevention of further CVD after an MI. Lifestyle changes that should be considered after an MI (and, indeed, in any form of CVD) include adoption of a Mediterranean-style diet with less meat, cheese, and butter and more bread, fresh fruit and vegetables, and fish, limiting alcohol intake to 21 units/week for men and 14 units/week for women, physical activity to the point of slight breathlessness for 20–30 minutes daily, cessation of tobacco smoking, and weight control to a body mass index <25 kg/m². Patients should be advised that there is no evidence that dietary supplements (such as omega-3 fatty acids and vitamins) are effective in reducing the risk of further CVD.

8.2.8 Ischaemic heart disease (IHD)

Myocardial infarctions cause irreversible injury to the heart: cardiomyocytes die and are replaced by inflexible fibrotic tissue that does not contribute to the pumping function of the ventricles. Over the years, the patient may suffer numerous infarctions, often ones that are silent, so that a once healthy myocardium is slowly degraded into isolated pockets of fibrotic tissues. Eventually, as the proportion of non-pumping tissues rises, the heart will start to fail, especially if there is untreated or poorly treated hypertension. Numerous stenoses will also contribute to poor cardiomyocyte function in failing to deliver sufficient oxygen for efficient myocardial action. These factors all lead to the chronic ischaemic heart disease (IHD), effectively the end consequence of a series of infarctions.

Technically, it could quite reasonably be argued that AMI is an ischaemic event, and so should be included in IHD, whereas others prefer the term to describe the long-term aspects of the disease. IHD may eventually lead to clinical heart failure (and other conditions), where the heart is unable to provide the lungs and rest of the body with enough oxygenated blood to meet its metabolic needs. This will be discussed in section 8.3.5. A 'route map' of the treatment of CAD is shown in Figure 8.8.

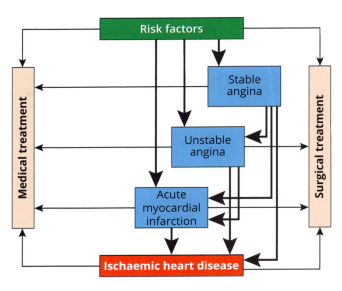

FIGURE 8.8

Treatments for coronary artery disease. The diagram shows a developing pathway of increasing disease severity. Medical treatments on the left include antiplatelets, statins, anti-hypertensives, etc. Surgical treatments on the right include PCI and CABG.

How is narrowing of a coronary artery assessed, and what is its common treatment?

8.2.9 Further investigations and treatments

Section 8.2 has so far focused on the most common features of the diagnosis and management of relatively straightforward CAD. Although ECG, blood tests, Echo, and angiography are very useful tools, medical technology continues to provide us with more complex tests that can give greater insights into the pathology of the heart. Whilst Echo is an excellent tool for assessing valve and chamber functions, it cannot easily determine the finer details of coronary atherosclerosis in the way that angiography can.

CT and MRI

By themselves, CT and MRI can provide images of the heart (Figures 8.9 and 8.10). However, CT combined with angiography (hence CTA, also described as coronary CT) provides better images of the entire heart pathophysiology of individual coronary arteries. The patient is injected with X-ray contrast media intravenously, and the heart is scanned whilst the contrast passes through the coronary circulation (as in standard angiography). Once the images have been collected, and analysed digitally, a three-dimensional model can be constructed to give more details of particular lesions. A version of this procedure

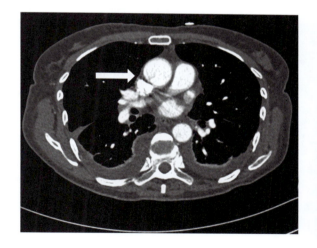

FIGURE 8.9

CT of the heart. The chambers appear as white bodies (one being arrowed).

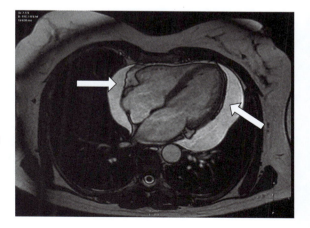

FIGURE 8.10

MRI of the heart, showing pericardial effusions (arrowed). (NB: the image is reversed—the heart is not normally on the right.)

Butz T, Faber L, Langer C, et al. Primary malignant pericardial mesothelioma—a rare cause of pericardial effusion and consecutive constrictive pericarditis: a case report. J Med Case Reports 3, 9256 (2009). https://doi.org/10.1186/1752-1947-0003-0000009256.

FIGURE 8.11

Computer-generated MRI image of the heart, aorta, pulmonary, and other vessels.

Shutterstock.

can assess the degree of calcification in the arteries. CTA is the preferred test in patients with a low likelihood of CAD, no previous diagnosis of CAD, and a high likelihood of good image quality. When combined with sophisticated software, these techniques can generate virtual three-dimensional images (Figure 8.11).

Nuclear perfusion studies

Radioactive emissions originating within the body can be detected and exploited in a number of isotope scans, and in cardiology a variant can be used to determine the degree to which the myocardium is perfused by blood—hence myocardial perfusion scan. This test can be performed at rest, but better information is provided when the heart is mildly stressed, such as on a treadmill or bicycle, or by certain drugs such as dipyridamole, adenosine, or dobutamine. The patient will be linked to an ECG throughout, to detect any irregularities that may develop during the stress test. Isotopes of thallium and technetium are preferred as they offer the best combination of signal and safety. The scan itself provides a coloured image of different aspects (cross-sectional, longitudinal) of the heart, and those areas of the myocardium that are not perfused by the isotope in the blood show a different pattern (Figure 8.12).

Other isotope tests using technetium include a multigated acquisition (MUGA) scan, which can be used to determine the left ventricle ejection fraction (LVEF) and other features of the left ventricle.

Cardiac surgery

Despite the widespread use of stents and other interventions, it may be that the damage to the coronary arteries is so extensive that it is effectively untreatable by these methods, so that the next step is CABG, where the occluded artery is literally bypassed by a section of healthy vein. In this procedure, the chest is opened (at the sternum: a sternotomy), blood is diverted into a heart-lung machine, and the heart stopped. Sections of saphenous veins from the patient's legs are harvested and used to circumvent those sections of the coronary circulation not amenable to stenting. Should the vein not be suitable, artificial tubes can instead be inserted. At the same time surgeons may divert the internal mammary artery (normally feeding the chest wall) to the coronary circulation (Figure 8.13). The patient is then taken off the heart-lung machine and their heart begins work anew. The chest is closed, and the patient moved to the intensive care unit.

FIGURE 8.12

A myocardial perfusion scan. The scan software is written so that colours (yellow, pink) indicate the degree of perfusion, so that changes in colour may be interpreted as areas where blood supply is poor. SA = short axis (cross-sectional), VLA = vertical long axis, HLA = horizontal long axis. Upper and lower panels are perfusions before and after application of attenuation and scatter correction.

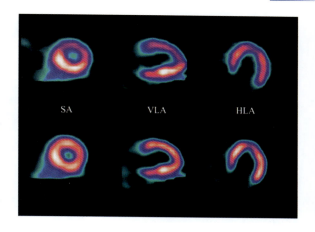

Hendel RC, Berman DS, Cullom SJ, et al. Multicenter clinical trial to evaluate the efficacy of correction for photon attenuation and scatter in SPECT myocardial perfusion imaging. *Circulation*. 1999: 99; 2742–9. https://doi.org/10.1161/01.CIR.99.21.2742. © American Heart Association. Wolters Kluwer Health, Inc.

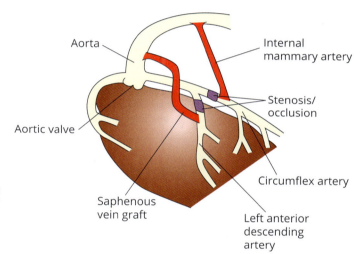

FIGURE 8.13

CABG. In this simplified illustration, a severe stenosis near the origin of the left anterior descending artery is bypassed by installing a section of a saphenous vein from the aorta to a point downstream of the lesion. A stenosis/occlusion in the circumflex artery is bypassed by an anastomosis of the internal mammary artery, whose ultimate origin is from the aorta.

Should CABG be inappropriate, the remaining option is heart transplantation (although increasing age may be a contraindication, and so exclude many with CVD), although CAD is not the most common indication, that falling to heart failure, cardiomyopathy, and other diseases.

8.2.10 Summary of coronary artery disease

CAD results from the effects of lipid-rich atheroma within the coronary arteries that cause narrowing (stenosis) and so ischaemia/hypoxia to the myocardium fed by those arteries. The clinical expression of this stenosis, if mild, is stable angina, whilst the more severe form is unstable angina. Should the atheroma rupture and generate emboli, these fragments may be driven down arterioles to become lodged and therefore occlude the vessel. This deprives the distal myocardium of oxygen and leads to such severe ischaemia that the tissues infarct, releasing cellular constituents (such as CK, CKMB, and troponins) into the bloodstream. The infarcted tissues fail to beat correctly, leading to changes that can be detected on an ECG. An increase in blood markers and certain changes on an ECG marks the condition as a NSTEMI, whereas other changes define the condition as a STEMI. Both conditions may lead to the PCI of a diagnostic angiogram, and further treatments include fibrinolysis and stent placement

Cross reference

Immunological aspects of heart transplantation are discussed in Chapter 10, section 10.4.3.

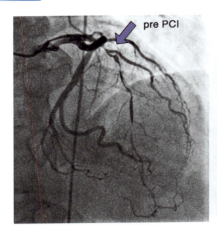
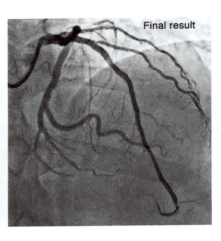

FIGURE 8.14

Successful restoration of blood flow by stenting. The site of the lesion is arrowed.

Basavarajaiah S, Qadir M, et al. Hybrid Strategy for Long Diffuse Coronary Lesion. J Am Coll Cardiol Intv. 2015: 8(11); 1518–21. https://doi.org/10.1016/j.jcin.2015.03.043. © 2015, American College of Cardiology Foundation. Published by Elsevier.

(Figure 8.14). If the disease is extensive, a CABG may be necessary. The patient will be discharged on antiplatelets, a statin, and blood pressure lowering drugs, and with appointments for follow-up and for cardiac rehabilitation.

Key points

Although the aetiology and management of coronary disease is reasonably well established, there will continue to be advances, examples being the use of stents and in computer-generated imaging.

8.3 Other heart disease

This diverse group was linked to over a fifth of all circulatory disease deaths in England and Wales in 2021, but this figure fails to account for the considerable morbidity they bring. The largest subgroup is arrhythmia (mostly atrial fibrillation and atrial flutter, 8,014 deaths), followed by heart failure (6,300), valve disease (4,575), and cardiomyopathy (1,579).

8.3.1 Valve disease

The anatomy, function, and pathology of the valves has been covered in section 6.5, where Figure 6.12 applies. To recapitulate, the four valves of the heart—aortic, pulmonary, mitral/bicuspid, and tricuspid—are formed of fibrous leaflets or cusps, with a surface layer of endothelial cells. Each valve links to the myocardium with a fibrous ring (an annulus). The tricuspid valve regulates blood flow from the right atrium to the right ventricle, the bicuspid/mitral valve that from the left atrium to the left ventricle. The cusps of both valves are linked to the internal apical aspects of their ventricles by *chordae tendineae*, themselves linked to the myocardium by papillary muscles to prevent the valves from inverting into their atria. The pulmonary valve and the aortic value (both with three cusps) regulate the flow of blood from the right ventricle to the pulmonary artery, and from the left ventricle to the aorta, respectively.

The principal pathologies of the valves are rheumatic heart disease, calcification (with thickening and inflexibility caused by calcium deposits), and infection (endocarditis), but there may also be congenital defects such as fusion of two of the three cusps. There is also growing evidence of the potential involvement of non-coding RNAs in the pathophysiology, and possible value in diagnosis and management (such as upregulated miR-466 in mitral stenosis). The end points of these pathological processes are two types of insufficiency: stenosis and regurgitation. The latter, which may be caused by prolapsed cusps, is characterized by the valve failing to prevent back-flow of blood from, for example, the right atrium back into the right ventricle through an incompetent mitral valve. Valve disease is pro-thrombotic: accordingly, pulmonary embolism and stroke are important consequences, although hypertension, often present, may be the cause.

Valve disease has a frequency of around 1.8% in the general population, rising steadily with age to around 12% in those aged 75 and over. The most common form is mitral valve regurgitation, with a frequency of ~63%, followed by aortic regurgitation (~18%), aortic stenosis (~15%), and mitral stenosis (4%). Being the end effect of perhaps decades of pathology, almost all severe valve disease requires surgery: other risk factors (for this disease and the risk factors for atherosclerosis) will be treated as necessary.

Presentation and diagnosis

The most common symptoms of valve disease are typically of the heart and lungs, and include breathlessness (which may interrupt sleep), cough, dizziness, fatigue, and palpitations. A cardiology examination with a stethoscope may find sounds such as murmurs, trills, and clicks, reflecting abnormal valve function—all crucial signs that take considerable experience to detect and interpret. These will be enough for a GP or A&E practitioner to send the patient for ECG, chest X-ray, and echocardiography. Serum marker **B-type natriuretic peptide** may be useful in assessing functional class, timing of interventions, and risk stratification.

The key investigation of valve disease is Echo. Metrics include cusp thickening, laminar or turbulent flow, annulus diameter, degree of calcification, valve area, peak velocity of blood across the valve, mean pressure drop, volume and fraction of regurgitant blood, vena contracta width, radius of the proximal isovelocity surface area (PISA: measuring the area of an orifice through which blood flows), and regurgitant jet area. From these, the practitioner can determine the extent of disease, and so what options may safely be offered to the patient.

Management

Management depends on a number of factors: severity, aetiology, co-morbidities (CAD, AF, etc.), symptoms, whether symptoms are related to disease, the expected benefits of intervention, mode of treatment (repair or replacement), and the wishes of the patient. In addition to the measures described above, the LVEF, the LV end-diastolic diameter, and the LV end-systolic diameter are important. Repair to a valve is generally considered when the cusps are not calcified, and can be pseudo-surgical (separating those cusps that have fused), or by forcing a larger opening. The EuroSCORE model of clinical and demographic factors calculates the risk of operative mortality and other outcomes and is available online.

Surgical intervention for valve disease provides the opportunity to address the AF, perhaps by surgical ablation or removal of the left atrial appendage. Replacement valves are of two types—mechanical or tissue (often called bioprosthetics, using human, pig, or cow material). Mechanical valves are more durable, common examples being Medtronic, Sorin Biocarbon, Carbometrics, Bjork–Shiley, and St Jude, but these come at the cost of thrombogenicity, so requiring lifelong oral anticoagulation (OAC: Chapter 11, section 11.6.3). The precise nature of the anti-thrombotic therapy will need to be tailored if the patient is to undergo PCI, comparing values of aspirin, clopidogrel, and OAC. The valve itself may attract an obstructive thrombus, which may adversely influence function. This may call for fibrinolysis, but in the worst case (such as a >10 mm thrombus), the valve will need to be replaced. Tissue valves generally require anticoagulation for 3 months or so, but all artificial valves carry a risk of infection and so endocarditis, so that antibiotics may be needed, and extra care will be needed to avoid valve infections, such as antibiotic prophylaxis in dental procedures.

SELF-CHECK 8.5

How do the symptoms of valve disease differ from those of coronary artery disease?

The aortic valve

Disease of this valve is linked to 91% of all valve disease deaths and has two major manifestations. Aortic regurgitation (AR) results in the flow of blood from the aorta back into the left ventricle as the latter relaxes (diastole) in order to fill with blood from the left atrium. This regurgitated blood therefore causes an unexpected increased mass of blood in the left ventricle, leading eventually to left ventricular enlargement (potentially noticeable on X-ray), LVH (also noticeable on ECG with tall R waves), and potential heart failure (HF).

Aortic stenosis (AS) obstructs blood outflow, therefore leading to blackouts (syncope), chest pain, dyspnoea, and fatigue, although these only appear when the disease is moderately severe. A normal aortic valve has an area of >2 cm^2, as determined by Echo, which defines severe disease as a diameter <1.0 cm^2, although other metrics (such as LVEF) are also important. The peak velocity of blood (a normal result being 1.1 m/s) increases (e.g. 3 m/s) as the LV forces the blood through a decreasingly effective area of the valve. Valves may be replaced by conventional surgery, but percutaneous replacement can be delivered by a catheter (hence transcatheter aortic valve implantation (TAVI)). The choice between TAVI and surgical aortic valve replacement may be assisted by the EuroSCORE model.

The bicuspid/mitral valve

With parallels to AR, mitral regurgitation (MR) results from a weak valve that fails to close effectively when the ventricle contracts to expel blood into the aorta, resulting in flow of blood from the LV back into the left atrium. MR is the second most frequent indication for valve surgery in Europe. Severity is gauged by regurgitant volume and fraction, as it is for AR: in mild MR these are <30 ml and <30% respectively, rising to ≥60 ml and ≥50% respectively in severe disease. Criteria for surgery include severe symptomatic MR and an LVEF <30%, and asymptotic MR with an LVEF <60%, whilst other disease such as AF and hypertension will be considered. The mitral valve may also prolapse—a condition where excessively large cusps, often thickened, are displaced into the left atrium, although there may also be an enlarged annulus and/or problems with the chordae tendineae. If severe, it may require repair.

Mitral stenosis (MS), the leading cause being rheumatic heart disease, is characterized by a reduction in the valve area from 4.0–6.0 cm^2 in health to <1.0 cm^2 in severe stenoses. The pathology, probably linked to inflammation, is of cusp fusion with thickening and calcification leading to immobility. An ECG can be informative, with an abnormal P wave (reflecting left atrial changes), but as with all valve disease, Echo is definitive. The leading invasive treatment is to force open the fused cusps by passing a balloon into the lumen of the valve, followed by its inflation, but this may also be done by open-heart surgery. Replacement of the entire valve with a mechanical or bioprosthetic valve may be necessary if there is also MR, if it is heavily calcified, and lifelong anticoagulation may be necessary.

The pulmonary valve

Pulmonary regurgitation (PR) most often arises from dilation of the annulus resulting from pulmonary hypertension: it is asymptomatic and rarely needs treatment. It is defined as mild if the regurgitant fraction is <40%, moderate if 40–60%, and severe if >60%. As with AR, pulmonary stenosis (PS) results in increased intra-ventricular pressure and so right ventricular and atrial hypertrophy, both of which can be demonstrated on an ECG, whilst an X-ray may show an enlarged pulmonary artery, although as always Echo is best. Mild PS (with a peak velocity <3 m/s, a normal result being 1 m/s) is likely to be asymptomatic, but severe PS (>4 m/s) will call for balloon valvotomy or surgery. NICE offers recommendations on the management of PV disease.

The tricuspid valve

Tricuspid regurgitation (TR) parallels MR, where blood flows back into the right atrium, perhaps due to valve/annulus dilatation, which increases the pressures within both chambers. Principal causes are, as before, rheumatic heart disease and endocarditis (primary disease), but may also be due to right ventricle dysfunction (secondary disease). In its most severe form, blood flow into the right atrium can be impaired, with associated high pressure in the inferior and superior vena cavae, giving rise to liver pathology and ascites, and to an engorged right jugular vein (hence increased jugular venous pressure, JVP), respectively. Tricuspid stenosis (TS) has much in common with TR in terms of causation and consequent cardiac pathophysiology (right-sided heart disease, liver pathology, raised JVP, and chamber hypertrophy). Indeed, TR and TS can coexist, and may well be present with left-sided disease, such as MS.

CASE STUDY 8.2 *Valve disease*

A 72-year-old woman complains to her GP about increasing tiredness whilst walking. She had been an active countryside walker with a group of friends, but had stopped around 6 months ago. Upon recent resumption, she was unable to keep up with her co-walkers: there was increasing shortness of breath but no chest pain or cough. The GP gives her an examination, finding a body mass index of 29.5 kg/m² (reference range 18–25) and hearing atypical heart sounds (a murmur) through a stethoscope. The patient gives a vague past medical history of colonic polyps, high blood pressure treated with a diuretic (furosemide 20 mg od) and a calcium channel blocker (amlodipine 5 mg od), and high cholesterol treated with atorvastatin (20 mg od), all started around 10 years ago. Finding a clinic SBP/DBP of 149/93 mm Hg, the GP recommends increasing both blood pressure tablets, to 40 mg and 10 mg respectively, and she orders routine bloods and refers the patient to see a cardiologist at the local district general hospital.

A month later, the cardiologist finds mild ankle oedema, and that the blood pressure has fallen marginally to 144/90. Based on this, she sends the patient for an ECG, chest and abdominal X-rays, and an Echo. The patient denies having had rheumatic fever. The ECG reports borderline left ventricular hypertrophy and the X-rays are unremarkable, but the Echo finds severe aortic stenosis (peak blood velocity 5.5 m/s, reference range <1.7 m/s, valve area 0.8 cm², reference range >2 cm²), moderate mitral valve regurgitation (regurgitant fraction 25%, reference range <10%), and mild tricuspid valve regurgitation, with severe left ventricular systolic dysfunction and an ejection fraction of 15–20% (reference range >40%). The patient is immediately admitted with a diagnosis of heart failure secondary to valve disease and undergoes cardiac CT/MRI, which confirms the Echo findings, but adds nothing to the picture. Consequently, she is discharged on ramipril 2.5 mg od, bisoprolol 5 mg od, eplerenone 25 mg od, and atorvastatin 20 mg od, giving blood pressure of 110/70 (target <135/85) and LDL-cholesterol of 2.8 mmol/l (target <3 mmol/l). She is offered aortic valve replacement by conventional open chest surgery or percutaneous transcatheter implantation, opting for the former.

Six months later she is admitted for elective aortic valve replacement with a bioprosthetic graft, which goes well, and is discharged 5 days later on the same drugs and with appointments for follow-up and for cardiac rehabilitation. When removed, the valve is moderately calcified, but the valve itself is not tricuspid, but bicuspid, accounting for the pathology. She will have been on short-term post-operative anticoagulation, most likely with an LMWH for a week, but no long-term anticoagulation is necessary. Three months after the operation, she has built up to gentle walking for 30 minutes on alternate days, increasing to daily after 6 months. She has lost weight (BMI now 27.5 kg/m²), feels fine, and is looking forward to joining a line-dancing group. At her 9-month post-operative assessment her ejection fraction has improved to 30% and the mitral value regurgitation reduced to 15%.

Key points

Valve disease is linked strongly to age and may resemble coronary artery disease. The leading diagnostic tool is echocardiography.

SELF-CHECK 8.6

Briefly describe the physiology and pathology of the valves of the heart.

8.3.2 Cardiomyopathy

Cell biology aspects of cardiomyocyte pathophysiology have been described in section 6.5.2 of Chapter 6. The two major forms are hypertrophic and dilated cardiomyopathy—HCM and DCM respectively.

Presentation

Both forms may present with any combination of all the semi-specific symptoms previously described—dyspnoea (especially after exertion), syncope, fatigue, palpitations, chest pain, etc. In the most severe cases (typical of HF) there may be pulmonary and peripheral oedema, the latter generally being of the ankles. Patients may be aware of their family history, prompting presentation, but certain cases of cardiomyopathy are late consequences of myocarditis, hypertension, CAD, valve disease, and certain arrhythmias.

Diagnosis

ECG can be suggestive, but diagnosis is made by Echo, focusing on LVEF and the thickness of the ventricular walls. There may also be a place for cardiac MR and genetic testing. In HCM, ECG abnormalities include those suggesting LVH and changes in ST, T, and Q waves (Figure 8.15), whereas DCM may be characterized by dispersed ST and T waves changes. In HCM the Echo-defined ventricle wall is thick (>1.1 cm; normal range 0.6–1.1 cm), whereas in DCM it is thin (<6 cm): in both conditions the LVEF is reduced (down from 55–75% to 35% or less).

Management

Wherever possible, the cause and risk factors should be addressed, such as valve disease and hypertension, the latter of which may respond to beta-blockers and calcium-channel blockers. Many of the treatment options are shared with that of HF (section 8.3.5), and options include implantable cardioverter-defibrillators (ICD: note that this is not the same as a pacemaker) for arrhythmias. Treatment will take account of co-morbidities such as CAD and AF. Ultimately, cardiac transplantation may be required.

8.3.3 Atrial fibrillation (AF)

A further strand of cardiac pathology considers the chambers failing to beat in concert, leading to arrhythmia, the leading condition being AF.

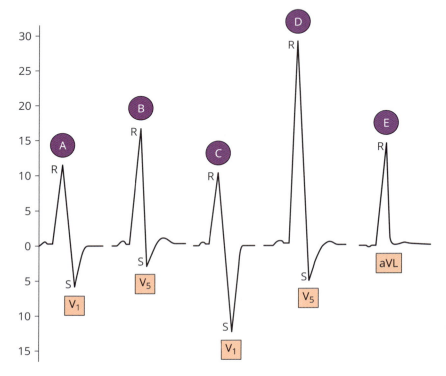

FIGURE 8.15
ECG voltage criteria for LVH. A and B: normal traces on leads V1 and V5. The S deflection on V1 is ~6 mm and the R deflection on V5 is 16 mm, giving a total of 23 mm. The S deflection on V1 of trace C is ~12 mm, the R deflection on V5 of trace D is 29 mm, summing to 41 mm, well over the Sokolow–Lyon cut-off of 35 mm. Trace E of an aVL lead shows the R wave to have a deflection of 14 mm, also exceeding a second cut-off point for LVH of ≥11 mm.

Aetiology

Section 6.3.2 describes the function of the heart, Figures 6.7 and 6.14 show the sites of the two cardiac pacemakers, the sinoatrial (SA) node and the atrioventricular (AV) node. These two nodes part-regulate the heartbeat by sending nervous impulses to the myocardium to induce contraction, and AF may arise from damage to these tissues. Although these two nodes may be damaged by other heart disease, such as in valve disease, the leading risk factor in the absence of clear heart disease is hypertension; others include increasing age, alcohol abuse, diabetes, and lung and thyroid disease.

Family studies point to genetic causes of AF, where heritability is around 22%, with rare familial forms mapped to 6q14–16 and 10q22–24, the leading mutation being the S140G mutation in *KCNQ1* at 11p15.4. The gene is one of a family that code for the α- and β-subunits of the voltage-gated potassium channel, there being 11 mutations, mostly gain-of-function. A second aetiology is genes coding for one of the β-subunits (these being *SCN1B* at 19q13.11, *SCN2B* and *SCN4B* at 11q23.3, and *SCN3B* at 11q24.1) and the α-subunit (*SCN5A* and *SCN10A* at 3p22.2) of the sodium channel Nav1.5. A third is in subunits of the calcium channel, notably *CACNB2* at 10p12.31 and *CACNA2D4* at 12p13.33. These ion channel mutations are believed to underlly inappropriate cation fluxes in cardiomyocytes and so irregularity in atrial function. Other gene aberrations include those involved in fibrosis and extracellular matrix remodelling (such as *COL12A1* at 6q13, coding for the α-chain of collagen type XII), and in cardiogenesis (such as *GATA4* at 8p23.1, coding for a transcription factor). There is also evidence of potential roles for ncRNAs, such as miR-1 targeting *KCNJ2* which is possibly linked to increased AF susceptibility, whilst lncRNA KCNQ1OT1 targets *CACNA1C* acting as a miR-384 sponge.

Epidemiology

AF is the most common sustained arrhythmia (present in 3% of the adult UK population, rising to 12% in the over-80s, and more prevalent in men) and is almost always of the left atrium. One in four of the middle aged in Europe and the USA will develop this condition, and it brings a five-fold increased risk of ischaemic stroke or TIA, a two-fold increased risk of mortality in women, and a 1.5-fold increased risk in men. It is present in ~40% of those suffering an ischaemic stroke, and there is associated vascular dementia and a decline in cognitive function, whilst 10–40% of AF patients are hospitalized annually. The leading co-morbidities of those admitted to hospital for acute AF include IHD (33%), hypertension (26%), and heart failure (24%). The Framingham Heart Study has an online calculator for risk of AF in the coming decade—it requires age, sex, height, weight, SBP, the PR interval from an ECG, and if there is treatment for hypertension, a heart murmur, and/or heart failure.

Presentation

One of the more unfortunate aspects of AF is that it is often first diagnosed after the patient has had a stroke, as the condition is mostly asymptomatic. Indeed, 20–30% of strokes are due to AF. Accordingly, there are now initiatives to detect this irregularity in primary care. Symptoms are mostly non-specific and include dyspnoea, chest pain, and dizziness alongside slightly more specific palpitations. However, as discussed above, AF may coexist with several other morbidities, and these may lead the practitioner to simply check the pulse. Having found an irregularity, the next step is an ECG. However, not all AF is present all the time: it can come and go, perhaps with severe symptoms that may last for hours or days and resolve spontaneously, in which case it is described as paroxysmal, and so may need a 24-hour ambulatory ECG. Rare AF in the absence of other disease is described as lone.

Diagnosis

The diagnosis of AF relies on the ECG. There are three factors: one is the loss of the very small P wave from the PQRST complex, and another is a tachycardia (<100 beats per minute (bpm)), although this is not always present. But the key feature is the irregular frequency of the heartbeat. This can also be seen on an ECG by the differing distances between adjacent R waves. In sinus rhythm, this distance is very regular (upper part of Figure 8.16), perhaps one every second. However, in AF, the distance between the peaks of the R wave varies (lower part of Figure 8.16). Note also the lack of P waves. In all patients, a trans-thoracic Echo will be required to identify other potential cardiac disease (unless already known), which may be followed up by a trans-oesophageal Echo. In this respect, left ventricular dysfunction is present in 20–30% of all AF patients.

Management—risk of stroke

A leading consequence of AF is thrombotic stroke, linked to increased plasma levels of pro-coagulant von Willebrand factor, and to the need for anticoagulation. NICE refers to AF in CG180, with several technology assessments of oral anticoagulants. Ideally, if present, an identifiable cause (hypertension (Table 7.5), smoking, excess alcohol, hyperthyroidism, obesity, habitual vigorous exercise) should be addressed and hopefully sinus rhythm will return, but this is rare. In most cases, the first step is to reduce the risk of stroke with anticoagulation. However, this therapy has potential drawbacks, and practitioners will balance the risk of stroke against the risk of haemorrhage caused by these drugs, and as the assessment of this balance may take some time, the patient should be covered with an LMWH. The risk of stroke is assessed by CHA_2DS_2-VASC, a score where one point is awarded for each of the following:

- Congestive heart failure history (e.g. reduced LVEF)
- Hypertension history
- Age 65–74
- Diabetes history
- Stroke/transient ischaemic attack history (scores two points)
- Vascular disease history (CAD, peripheral artery disease, aortic plaque)
- Age ≥75 (scores two points)
- Sex category—female scores one point

A score of zero suggests that the patient does not need anticoagulation, a score of 1 is low risk and so antiplatelet or anticoagulants may be considered, whilst a score of ≥2 suggests that the patient should

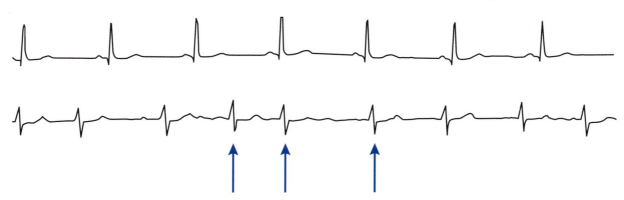

FIGURE 8.16

ECG trace of AF from lead II. The two traces are taken over the same time period of 7 seconds. The upper (normal) trace shows 7 regular beats at 60 bpm, the lower (AF) 9 irregular beats at a rate of 86 bpm—clearly higher, but not high enough for a formal diagnosis of tachycardia (e.g. >100 bpm). In the upper trace, the variation in the R–R interval is negligible, whereas in the lower, the longest R–R interval is over 70% longer than the shortest. Note the difference in the distance between the three adjacent arrowed PQRST patterns. In a normal sinus rhythm, this distance will be almost exactly the same. A further note is that the normal trace has no S wave, whereas the S wave in the AF trace is noticeable.

be on oral anticoagulation (OAC). The risk of haemorrhage can be quantified by HAS-BLED, a score derived from:

- Hypertension (uncontrolled SBP >160 mmHg)

- Abnormal renal/liver function (on dialysis, post-transplant, creatinine >200 µmol/l, cirrhosis, bilirubin >2× normal with AST/ALT/ALP >3× normal)

- Stroke/transient ischaemic attack history

- Bleeding history or predisposition

- Labile international normalized ratio (unstable or high INRs, time in therapeutic range <60%)

- Elderly (>65 years)

- Drugs/alcohol concomitantly (e.g. aspirin, clopidogrel, NSAIDs)

Each feature scores one point, and with the risk of haemorrhage being ~1% for score zero and 1–3.4% with a score of 1, an OAC should be considered in both scores. With two points the risk of bleeding is 1.9–4.1%, so that OAC may be considered, but the patients have moderate risk of bleeding. With a score of three points, the patient is at high risk of bleeding (3.7–5.8%) so that alternatives to OAC should be considered, such as closure or occlusion of the left atrial appendage (where thromboses may form, and if embolized, can run up the carotid artery to cause a stroke).

Thus with a high CHA_2DS_2-VASC score and a low HAS-BLED score, OAC is acceptable, whereas if the HAS-BLED score exceeds the CHA_2DS_2-VASC score, OAC is not advised. However, every patient should be considered individually, the decision being made by the lead clinician. Free online calculators for CHA_2DS_2-VASC and HAS-BLED are available, whilst alternatives to the latter include the $HEMORR_2HAGES$ and ATRIA risk scores, both of which consider many of the same features.

Once the risk of stroke has been assessed, and found to be sufficiently high, OAC will be considered with a vitamin K antagonist (almost inevitably warfarin), or a non-vitamin K antagonistic oral anticoagulant (NOAC, such as dabigatran, rivaroxaban, apixaban, or edoxaban). Each drug has different effects on risk of stroke and of haemorrhage, and has different pharmacokinetics and dynamics so that the best choice must be considered individually for each patient (NICE CG180). For example, doses of factor Xa inhibitors need to be adjusted for renal function—the direct thrombin inhibitor needs no adjustment. Antiplatelets are not recommended as monotherapy for stroke prevention.

Much of the above is applicable to lone AF. However, when there is concurrent disease such as of the valves, kidneys, and HF, the situation is clearly more complex and more consideration is required. For example, patients with CAD should be on aspirin, and those with AF should be on OAC, but both therapies increase the risk of haemorrhage. On balance, OAC is recommended for AF patients with stable CAD, but expert opinion may be sought. The patient in AF and suffering an ACS, or with stable CAD and elective stent placement, may be on antiplatelets and OACs for a month to reduce the risk of recurrent coronary and cerebral ischaemic events.

Management—the arrhythmia

A subsequent step in managing AF is to address two issues—the rate (as a tachycardia is likely) and the rhythm—both of which bring the same benefit in terms of symptoms and a reduction in the risk of stroke. Control of the heart rate should be offered to all AF patients except those whose disease is reversible, have HF caused by AF, have new-onset AF, have atrial flutter, and in those in whom rhythm control is preferred. Rate control is generally with drugs such as beta-blockers (except sotalol), CCBs (e.g. diltiazem), and digoxin (alone or in combination). Amiodarone is not recommended for long-term rate control. In an emergency setting, amiodarone, beta-blockers metropolol and esmolol, CCBs diltiazem and verapamil, and glycosides digoxin and digitoxin may be given intravenously. Exact choice of drugs and doses in emergency and long-term use depends on LVEF < or > 40%, aiming at a target rate of <80 bpm at rest and <100 during moderate exercise. Should these strategies fail, ablation of the atrioventricular node and an implanted pacemaker (set at 70–90 bpm) is an option, but this places the patient on pacing for life.

Rhythm control, which is preferred in younger patients (e.g. <65 years) may also be attempted with beta-blockers, but also with dronedarone, whilst for those with LV impairment due to HF, amiodarone is an option, and there are other drugs (such as quinidine, disopyramide, and sotalol). Patients with ischaemic or structural heart disease should not take flecainide or propafenone, but those with paroxysmal AF may benefit from the 'pill in the pocket' strategy (flecainide 200–300 mg or propafenone 450–600 mg) for when they are suffering from symptoms.

Management—electrical cardioversion

Use of the drugs described above to restore sinus rhythm may be seen as pharmacological cardioversion; another method is electrical cardioversion, which may be curative. This procedure seeks to shock the heart back to sinus rhythm with a 120–200-Joules bolus of energy delivered by paddle or pad electrodes on the front of the chest and on the back. Sedation is generally required, and there may be some post-cardioversion bradycardia. The procedure should be preceded with amiodarone, sotalol, or other drugs as these can improve the success, but as it brings an increased risk of stroke, the patient should be anticoagulated for 3 weeks before and 4 weeks after the cardioversion. However, although electrical cardioversion is often successful in the short term, the arrhythmia can return, leading to permanent AF, and so permanent OAC.

8.3.4 Other arrhythmias

Atrial flutter (rates of 250–300 bpm) is linked to AF and treatment is similar, diagnosis being by ECG (Figure 8.17). In atrial tachycardia, the beat rate is much lower (125–250). Treatments for both include pharmacological and electrical cardioversion, the latter requiring OAC before and after.

Ventricular tachycardias can be life-threatening, with dizziness, hypotension, and **syncopy** with a pulse rate of 120–220 bpm, and need to be distinguished from supraventricular tachycardia by their ECG patterns. Depending on severity, urgent direct current (DC) cardioversion may be needed, and pharmacotherapy comprises amiodarone and lidocaine. In ventricular fibrillation there is no effective LV contraction, and so no pulse, leading to loss of consciousness, respiratory arrest, and death. The ECG shows no identifiable pattern, and the only effective treatment is electrical defibrillation, after which survivors require a pacemaker (an implantable cardioverter-defibrillator, ICD, see below). Those with the Brugada syndrome (which can be heritable and carries a high risk of sudden death) also need an ICD. Other arrhythmias (which may be congenital or acquired) include short and long QT interval (such as in torsades de pointes) and all require specialized diagnosis and care (such as for devices and ablation of the left atrium). A long QT interval (present in ~150/million) may be due to hypocalcaemia (and so require measurement of parathyroid hormone), hypomagnesaemia (section 15.8.1), or hypokalaemia, and if life-threatening may require an infusion of the particular cation.

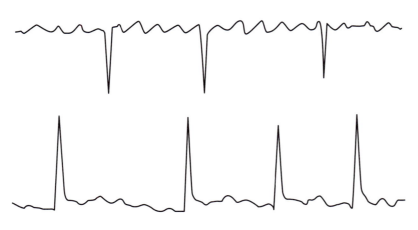

FIGURE 8.17

Atrial flutter. Upper trace: typical 'saw-tooth' pattern from V1. Lower trace from lead II showing irregular distances between R waves.

CASE STUDY 8.3 *Atrial fibrillation*

At a routine outpatient consultation at a cardiology clinic, a 76-year-old man with a history of high blood pressure and CAD brings a developing history of dizziness and loss of orientation, often at the same time as palpitations, in the past 6 months. Eight months ago his left ventricular ejection fraction was 62%, and no major valve disease was reported. Cardiovascular medications include perindopril 4 mg od and amlodipine 10mg od, which have until now been effective at keeping his blood pressure to <140/90, although today his pressures are 144/91. The practitioner sends him for an ECG, which provides no new information besides reporting a heart rate of 70 beats per minute, up from 58–62 a year ago. The patient is invited to keep a diary of his symptoms, perform his own daily blood pressure measurement on his home device, and return in 3 weeks with the diary and the device.

Upon review, the ECG is unchanged and clinic blood pressure is 146/93, which is in agreement with the patient's own device. The diary reveals no clear pattern, with several attacks on one day, and none for the next 48 hours, but when the palpitations are present his blood pressure temporarily rises to 151/98 and his heart rate to 95. The practitioner recommends he increase his perindopril to 8 mg od and arranges for him to be fitted with a Holter monitor to record his ECG over a 48-hour period. The following week the patient returns to learn that the downloaded data show numerous intermittent instances of an irregular heartbeat (Figure 8.16), some lasting up to 45 minutes, between a normal rhythm, which leads to the diagnosis of paroxysmal AF. In clinic, his blood pressure is still high at 143/92, and his pulse rate is 73 beats per minute.

The practitioner now focuses on treatment, the objectives being to address the high blood pressure, the heart rate, and the rhythm (NICE NG196), and talks the patient through the importance of this diagnosis as a risk factor for stroke, and options for treatment. The patient declines electrical conversion, preferring a continuous pharmaceutical approach (as opposed to the 'pill in the pocket' strategy, to be used only for immediate relief from a paroxysm), and as he is already taking two classes of drugs to reduce blood pressure, the beta-blocker bisoprolol is prescribed, starting at 2.5 mg od. Although this will help reduce blood pressure and heart rate, it will not address the arrhythmia and so the risk of stroke. Therefore, anticoagulation is necessary and the CHA_2DS_2-VASc and HAS-BLED scores must be determined. With acceptable renal and liver function, this patient has scores of 3 and 2 respectively, so as the risk of stroke exceeds the risk of haemorrhage, anticoagulation is indicated, and he starts on apixaban 5 mg bd.

Six weeks later the patient's ECG remains normal but the heart rate is a little slower at 64 beats per minute, and although his blood pressure is a little improved at 138/88 (target now <135/85), he still complains of palpitations, now a least every other day, which require him to lie down until they pass. Accordingly, his bisoprolol is increased to 5 mg od and (the practitioner acknowledging the fact that he has not had electrical cardioversion), he is started on dronedarone 400 mg bd. These drugs suit that patient well; after 6 weeks his symptoms have largely disappeared, blood pressure is 133/80 with a pulse rate of 58. With this degree of polypharmacy, he will continue to be seen in secondary care every 6 months.

Implantable devices

Serious arrhythmias that do not respond to medical management or electrical cardioversion must be treated by temporary or permanent pacing. Disc-like devices with a diameter of approximately 3.5 cm can be implanted in the chest, with wires passing from the control/battery unit to those parts of the myocardium that need support. We have already noted the use of ICDs in certain ventricular tachyarrhythmias, but they may also be valuable in CAD, and where there is reduced LVEF (i.e. HF).

Bradycardias (<60 bpm) may be treated with an ICD that can sense and pace either or both of the ventricles. NICE offers guidance on dual-chamber (i.e. atrium and ventricle) pacemakers for symptomatic bradycardia in cases of sick sinus syndrome (an irreversible dysfunction of the sinoatrial node) or atrioventricular block (failure of the conduction of electrical impulses from the atrial nodes to the ventricles).

Key points

The most common arrhythmia, AF, has a high prevalence in the elderly, and there is often confounding co-morbidity such as valve disease.

What are the treatment options in AF?

8.3.5 Heart failure (HF)

Heart failure may be defined as the inability of the heart to deliver to the body the blood it needs. Since the major driving force is the left ventricle, this chamber is often where the problem lies, hence left ventricular systolic dysfunction. As it contracts, a healthy left ventricle is expected to eject perhaps 60–70% of the blood it holds (more in athletes). Heart failure becomes more and more likely as the left ventricle ejection fraction (LVEF) falls and is often defined in the echocardiology laboratory at 40%. This is not to say that an ejection fraction of 41% is acceptable: some people have symptomatic heart failure with an ejection fraction of 50% or more.

Our examination of this condition follows naturally from our discussion of cardiomyopathy, valve disease, IHD, arrhythmias, and hypertension, as these may all cause this condition. It may also arise congenitally, following right heart failure, from alcohol abuse, pericardial disease, and infections. Incidence depends on definition, but it is likely that 1–2% of the UK population have heart failure, rising to ≥10% in those aged 70 and above, figures comparable to those of AF.

Presentation

It follows from the previous sections that presentation depends on the underlying aetiology, and accordingly practitioners will be aware of these. In the rare eventuality of the development of heart failure without a predisposing factor, the semi-specific symptoms of fatigue, breathlessness, etc. are likely. Slightly more specific signs include elevated jugular venous pressure, ascites, **paroxysmal nocturnal dyspnoea**, pulmonary crackles, the third heart sound (gallop rhythm), and peripheral oedema (typically of the ankles).

Diagnosis

As symptoms are far from specific, diagnosis rests on Echo-defined LVEF. However, further criteria are relevant structural heart diseases, namely LVH and/or left atrial enlargement, or diastolic dysfunction. A further criterion is high levels of one of the three natriuretic peptides (which promote renal excretion (hence -uretic) of sodium (hence natri-)), very likely in response to an excessive blood volume and the attempts of the body (in physiology) to relieve itself of this burden by promoting diuresis.

- Atrial natriuretic peptide (ANP) is secreted from the atrial myocytes in response to atrial stretching.

- Brain natriuretic peptide (BNP, so named from where it was discovered) is released in response to ventricular stretch, and is best assessed by the pro-BNF form, that has an N-terminal fragment (hence NT-proBNP).

- C-type natriuretic peptide (CNP) is released from endothelial cells in response to various stimuli and is linked to a number of diverse pathologies.

Increased serum BNP (>35 pg/ml) or NT-proBNP (>125 pg/ml) is of diagnostic value, but sensitivity and specificity profiles are such that use of the NPs is recommended for excluding HF, not to establish the diagnosis. Of further value is the classification of HF by quality of life criteria (those of the NYHA: Table 8.2), whilst the ECG may also have its place in that a normal ECG effectively excludes HF.

Management

Much of the preceding text has indicated directions for treatment, in that all risk factors should be addressed, with focus on hypertension and support for the heart. There is place for almost all classes of BP-lowering drugs (except CCBs in early stages), and Table 7.6 gives some pointers. Pharmacological treatment aims to address the pathophysiology but also the symptoms. Most patients would start on a

TABLE 8.2 New York Heart Association classification of heart failure

Class	Criteria
I	No limitation to general day-to-day quality of life. Normal physical exercise does not cause dyspnoea, fatigue, or palpitations
II	Mild limitation. Comfortable at rest but normal physical activity induces dyspnoea, fatigue, or palpitations
III	Marked limitation. Comfortable at rest but gentle physical activity induces dyspnoea, fatigue, or palpitations
IV	Dyspnoea, fatigue, or palpitations are present at rest and are exacerbated by any physical activity

White PD, Myers MM. The classification of cardiac diagnosis. JAMA. 1921: 77; 1414–15, cited in Criteria Committee, New York Heart Association, Inc. Diseases of the Heart and Blood Vessels. Nomenclature and Criteria for diagnosis, sixth edition, Boston, USA, Little, Brown and Co. 1964, p 114.

diuretic, an ACEi (or ARB if intolerant), and a beta-blocker, to which spironolactone or eplerenone can be added. If there is a need for further treatment, cardiac resynchronization therapy with an ICD (if in arrhythmia), an ACEI, ivabradine, hydralazine, isosorbide, and digoxin are options, in addition to a left ventricle assist device and transplantation. By definition, HF with relatively well-preserved ejection fraction (e.g. LVEF ≥50%) is less serious, but risk factors and other co-morbidities (e.g. AF, bradycardia) should still be addressed, and diuretics will usually improve congestion and symptoms.

Acute heart failure

Much of the above considers chronic heart failure (CHF), which may develop over years or decades. In contrast, acute heart failure (AHF) may present as an entirely new entity, or far more commonly as a sudden deterioration from CHF, perhaps triggered by factors such as an ACS, AF, a toxic insult (e.g. alcohol), pulmonary embolism, and infections. AHF can be very serious and life-threatening, with cardiogenic shock and respiratory failure, often requiring admission to intensive care or coronary care units, with transfer to a medical ward once haemodynamically stable. Other signs and symptoms include dyspnoea, peripheral oedema, oliguria (urine output <0.5 ml/kg/hour), hypotension (SBP <90 mmHg), bradycardia (<40 bpm), tachycardia (>120 bpm), low oxygen saturation (SaO$_2$ <90% by pulse oximetry), and acidosis (pH <7.35).

An Echo is mandatory, and chest X-ray can be useful, perhaps showing pulmonary congestion, and the ECG is rarely normal (and required to exclude AMI). Low levels of natriuretic peptides effectively exclude AHF, and a full panel of blood tests is required (FBC, U&Es, LFTs, troponin, thyroid-stimulating hormone, glucose, and D-dimers). One route to management is to consider the presence or absence of congestion (present in 95% of patients, marked by **orthopnoea**, oedema, ascites, hepatomegaly, etc.) and/or hypoperfusion (cold, sweaty extremities, oligouria, dizziness). Should the clinical picture, allied to the investigations, warrant treatment, this is likely to include a vasodilator (such as a nitrate), a diuretic (frusemide), an ionotrope (levosimemdan), ultrafiltration (removal of plasma water across a semi-permeable membrane), an anticoagulant (an LMWH), and mechanical circulatory support (intra-aortic balloon pump).

Patients will be fit for discharge when haemodynamically stable, established on medications, and with good renal function. They will be supplied with tailored education and advice regarding self-care, and have a written management plan. Best practice is to have appointments with their GP within a week, and for an outpatient visit within 2 weeks, and for other investigations, if necessary.

Mechanical circulatory support

Should the options described in the previous subsection fail to provide sufficient support of the ailing heart, in either AHF or CHF, mechanical measures will be needed. In severe AHF (such as cardiogenic shock, and so NYHA class III or IV), and terminal HF, perhaps with a LVEF of <25%, these may be extra-corporeal life support and extracorporeal membrane oxygenation, usually for a period of a few days to weeks. In the longer term, a left ventricle assistance device (LVAD) may be implanted within the chest. This device takes blood from the left ventricle and pumps it into the aorta, thereby relieving the ventricle of some of the pressure of maintaining the circulation. A second device is the intra-aortic balloon pump, placed just below the exit of the left subclavian artery and programmed to pulsate, and so to help drive blood down the aorta, which also relieves pressure on the left ventricle. These measures are intended to be bridges to heart transplantation.

8.3.6 The complexity of valve disease, arrhythmia, cardiomyopathy, and heart failure

It is clear from the preceding five sections that there is considerable overlap in these conditions, in both aetiology and management (Figure 8.18). This is evident in guidelines such as those of the European Cardiology Society, whose guidance on heart failure has sections on valve disease, CAD, and arrhythmias. Therefore it is incorrect to view each as a separate and discrete disease, and several of these conditions may be encountered in the same patient at the same time. Consequently, the practitioner will be thoroughly tested in juggling the different clinical aspects and treatments, some of which may be mutually confounding.

Key points

There is considerable overlap in clinical and pathological aspects of hypertension, valve disease, arrhythmia, cardiomyopathy, and heart failure, and any one of the diseases rarely exists by itself. With a high degree of mortality and morbidity, management is a challenge.

SELF-CHECK 8.8

What is the link between hypertension, valve disease, and heart failure?

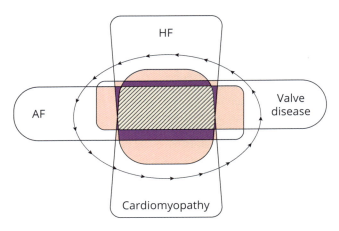

4 conditions

3 conditions

2 conditions

FIGURE 8.18

The complexity of valve disease, arrhythmia, cardiomyopathy, and heart failure. Overlapping Venn diagrams illustrating the complexity of these four conditions. Hypertension omitted for clarity (circular line with arrows). AF includes other arrhythmias. The size of overlaps does not imply a relationship based on population numbers or genuine interactions.

8.4 Stroke

8.4.1 Pathophysiology

The brain is another high-demand organ, receiving oxygenated and nutrient-rich blood via the left and right carotids and the vertebral arteries. Like the coronary arteries, these are also a target for atherosclerosis. Furthermore, the same disease process that occurs in the coronary circulation can also happen in the circulation of the brain, causing a cerebral infarction, or stroke. The source of the atheroma or thrombotic material causing this type of stroke may be a ruptured plaque in the carotid arteries, or a clot might arise from inside the heart. The most common risk factors for stroke include AF and hypertension. The latter, as in AF, is linked to increased levels of pro-coagulant von Willebrand factor, which is likely to promote thrombogenesis. However, a thrombus is not the only possible aetiology: stroke may also arise from a rupture of a cerebral arteriole, causing an intracranial haemorrhage (ICH).

Just as angina can be caused by a reduction in the ability of the coronary circulation to provide the myocardium with oxygen, so too can the carotids and cerebral circulation be unable to provide the brain with oxygen. In both cases this is likely to be due to a stenotic reduction in the lumen of the artery by atheroma. However, accurate assessment of cerebral artery atheroma remains a challenge. As regards cerebral blood supply, a reduction leads to short-lived symptoms such as loss of the power of speech or of control of the arms of legs. These events are called a transient ischaemic attack, or TIA. Just as angina is a warning sign of a possible heart attack, so too a TIA is a warning sign of a possible stroke.

As the disease process of atherosclerosis within the cerebral circulation is broadly the same as within the coronary circulation, treatments have much in common. In the short term, a thrombotic stroke or a severe stenosis causing many TIAs may also be treated, respectively, by drugs to dissolve the clot (fibrinolysis, or thrombolysis), or by angioplasty. In the latter, a guide wire with a deflated balloon is passed into the artery with the stenosis, and is then inflated, just as with a coronary artery stenosis (section 8.2.6). Long-term treatment of cerebrovascular disease consists of risk factor management (especially in reducing high blood pressure) and anti-thrombotics.

The strokes discussed above are all caused by an occlusion of a cerebral vessel by a thrombus. However, another type of stroke, a haemorrhagic stroke, is caused by the rupture of a cerebral blood vessel and the leaking of blood into the tissues of the brain. Haemorrhagic stroke can be further classified by anatomy—that is, by whether the ruptured vessel bleeds into the brain tissue itself (intracerebral haemorrhage, ICH) or into the spaces defined by the different membranes and the skull (such as subarachnoid haemorrhage, SAH, between the arachnoid mater and the pia mater).

The aetiology of stroke and TIA has been discussed in section 6.5.6, where Figure 6.16 applies. NICE guideline NG128 focuses on the diagnosis and initial management of stroke and TIA, whilst NG236 refers to stroke rehabilitation. The consequences of stroke can be devastating—just one example from NICE estimates that stroke leaves 30–70% of its victims with reduced or no use of one arm and hand.

BOX 8.2 Classification of arterial disease

The arteries feeding the myocardium are clearly easy to group together, as are those of the brain and skull. All other arteries may be described as peripheral, hence peripheral artery disease (PAD). Arteries of the leg are therefore a subset of these, and if they are affected, this is referred to as lower extremity arterial disease (LEAD). Likewise, disease in arteries of the arm and shoulders is referred to as upper extremity arterial disease (UEAD). As carotid artery disease is an essential component of stroke and TIA (10–15% of all strokes follow thrombo-embolism from a 50–99% stenosis of the internal carotid artery), it will be discussed in the current section.

8.4.2 Stroke and TIA

Presentation

Key features of stroke and TIA have been emphasized by a Public Health campaign in the UK summarized as 'FAST', which stands for face, arms, speech, time. The first three are clinical symptoms, while the *time* emphasizes the need to get the patient to a hospital urgently. As with the stable angina–ACS–STEMI spectrum, there is a spectrum between TIA and stroke. The former can be as brief as a few minutes in a single anatomical area, or last up to 23 hours 59 minutes in multiple areas, but at 24 hours the condition becomes a stroke. There is also the question of severity—some TIAs may bring serious but short-lived disabilities (fortunately reversible), whilst other strokes may be only mildly disabling with a duration of 36 hours with few after-effects.

A TIA may be defined as a brief episode of neurological dysfunction caused by lack of blood flow to the brain, spinal cord, or retina, without tissue death. The implied reversible ischaemia may have its origin in stenosed carotid arteries and/or the cerebral circulation. With the use of modern brain imaging, up to a third of patients with symptoms of <24 hours duration were found to have had an infarction. Nevertheless, the key points are recognition and action. Focal neurological signs and symptoms include the following:

- aphasia (loss or impairment of the ability to use and comprehend language)

- apraxia (difficulty in executing the voluntary movements needed for speech)

- dysarthria (difficulty in articulating words)

- dysphagia (difficulty in swallowing)

- dyspraxia (difficulty in planning and executing movement)

- hemianopia (blindness in one half of the visual field of one or more eyes), often described as amaurosis fugax

- paresis (weakness of voluntary muscles e.g. in arms, legs, or a combination is possible and may be unilateral or bilateral, and be present in one or all limbs, the latter being quadriparesis).

Some of the above signs and symptoms cross over with other clinical syndromes such as myasthenia (loss of strength in muscles) and ataxia (lack of coordination of muscles involved in gait, speech, and the eye). Several come together in those presenting with slurring and loss of speech, perhaps with uncontrolled salivation, whilst others may experience the arm and hand simply failing to follow instructions, as in inability to pick up an object. All of these symptoms are applicable to stroke, but are likely to be more severe and long lasting. Extensive and serious stroke brings irreversible physiological changes, and can deteriorate to death, even with the best care, and in this respect resembles ischaemic heart disease.

Diagnosis

The extent of the signs and symptoms described above, combined with a medical history and the events leading up to presentation, will give the practitioner enough for a preliminary, if not a full, diagnosis of TIA or stroke. But as with heart disease, other investigations include ultrasound of the carotid and cerebral arteries (as much as is possible) to determine stenosis due to atheroma, and cerebral angiography, which can also determine stenosis, and also any occlusion. Unlike heart disease, in stroke/TIA, an ECG will not demonstrate any specific changes, and there are no relevant specific blood tests. Several clinical scoring systems, such as ROSIER (recognition of stroke in the emergency room) and FAST (face, arm, speech, time) are available to help diagnose and differentiate TIA from stroke, and the possibility of development of stroke from TIA. Patients presenting to an A&E/emergency department with a suspected or confirmed TIA or stroke should have 300 mg of aspirin and be referred for a specialist opinion and imaging studies.

Imaging

Perhaps the key simplest investigation is carotid artery ultrasound to determine any atheroma, as this could be the cause of the TIA. A key component is measuring the diameters of the common, internal, and external carotid arteries, and of the lumen, so determining per cent of stenosis, if any. It can also help evaluate plaque morphology (echolucency, surface irregularity, and intra-plaque haemorrhage). A modification of ultrasound is to use duplex Doppler, a powerful tool for determining blood flow, which will accelerate as it is forced into a stenosis.

The velocity of blood is proportional to the stenosis, giving a further useful metric. In a healthy artery, blood flow is laminar, and without obstruction there is little or no turbulence. However, if there is an irregularity, blood flow will be turbulent, and this pattern is detected by ultrasound. Furthermore, changes in blood flow can be colour-coded to give additional clarity, and anatomical and pathological features can be characterized. This is illustrated in Figure 8.19.

In section 8.2.6, we looked at the value of coronary angiography (Figures 8.3, 8.4, and 8.7). The same process can give details of the cerebral circulation, and can also determine stenoses and occlusions. Angiography can be extensive, as in Figure 8.20, which shows the procedure imaging the arteries of the top of the chest, neck, and head.

A variant of the process is digital subtraction angiography (DSA) whereby software provides improved clarity, and where the light/dark aspects are reversed, so that arteries appear dark. Using this technique, focusing on selected regions of the carotids, any arterial disease pertinent to TIA and stroke can also be visualized, as in Figure 8.21.

More complex investigations in TIA are generally not called for, but MRI and CT scans may be informative, especially when combined with angiography (Figure 8.22). However, CT brain scanning is appropriate if there is clinical suspicion of an alternative diagnosis, whilst MRI may be used in stroke to determine the territory of ischaemia, or to detect haemorrhage or alternative pathologies. CT and MRI may also be used to determine silent cerebral infarctions, whilst spontaneous embolization can be detected with trans-cranial Doppler. Atherosclerosis in one carotid artery does not imply disease in the other (contralateral) artery, although it may be present, and both sides will be scanned.

It is likely that all imaging will be treated as urgent, so that tissue-saving treatments (if needed) can begin with a minimum of delay. For those suspected of having an acute stroke, imaging must be immediate.

There are several clinical features that mimic stroke/TIA and which therefore must be addressed. These include memory loss, generalized weakness, dizziness, loss of consciousness, tinnitus, headache, migraine, and blurred vision. More serious differential diagnoses are brain tumour, CNS infection (e.g. meningitis), hypoglycaemia, multiple sclerosis, vertigo, and some forms of epilepsy.

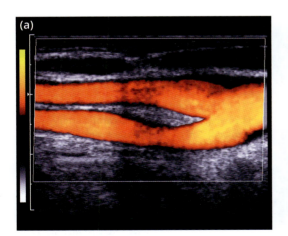
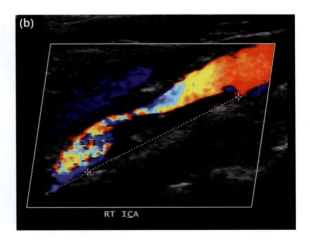

FIGURE 8.19

Carotid artery ultrasound. (a) shows a normal healthy common carotid artery (on the right) as it divides into the external (feeding the face and neck) and internal (feeding the brain) arteries. The colour is moderately uniform as blood flow is unimpeded. In contrast, (b) shows a large obstruction that has a clear effect on blood flow, as the normal red/orange colour is disrupted as the blood flow becomes turbulent, and there are many changes in the colour plot. The practitioner can determine the extent of the narrowing of the artery, and this will be a major factor in the decisions regarding further treatments.

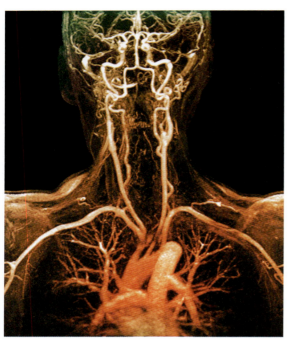

FIGURE 8.20

Angiography of the chest, neck, and head. The angiographic contrast medium has highlighted the ascending and descending aorta; the subclavian arteries; the common, internal, and external carotid arteries; and arteries of the cerebral circulation.

Science Photo Library.

(a)

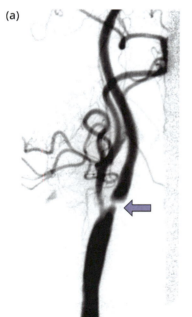

(b)

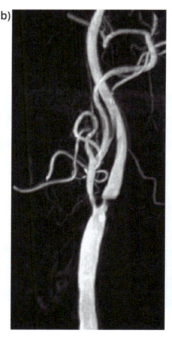

FIGURE 8.21

DSA of the carotid arteries. There is a clear narrowing of the left internal carotid artery (arrowed). Only a small passage remains, and the lesion is probably >90% stenosed, so being a candidate for stenting or carotid endarterectomy.

Hyde DE, Fox AJ, Gulka I, et al. Internal carotid artery stenosis measurement: comparison of 3D computed rotational angiography and conventional digital subtraction angiography. Stroke. 2004: 35; 2776–81. https://doi.org/10.1161/01.STR.0000147037.12223.d5. © 2004, Wolters Kluwer Health.

Management

Once more the parallel with angina/ACS can be applied, as the practitioner will need to determine the extent of any ischaemia, and whether it can be reversed. Best practice calls for dedicated stroke units to undertake all management and subsequent care, as these will have the required levels of expertise. In both TIA and suspected stroke, systolic blood pressure should be managed to 130–140 mmHg, and blood sugar to 4–11 mmol/l. Should ultrasound report no or minimal disease (stenosis <50% or <70%, depending on the precise anatomical location of the lesion and if the practitioner is using European

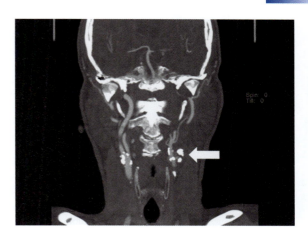

FIGURE 8.22
CT angiography of the carotids. CT angiography image showing heavy calcification of the carotid arteries (white nodules). The left internal artery is occluded at the origin (arrow).

Source: Evans NS, MD. All Rights Reserved.

or North American criteria), treatment is also likely to be relatively simple—as in stable angina—and involve addressing risk factors (hypertension, dyslipidaemia, hyperglycaemia, smoking, lack of exercise) and preventing progression or thrombosis, the latter using aspirin. Recent research has shown that serum vitamin D levels of 60 nmol/l bring the lowest risk of recurrent stroke, suggesting (where necessary) oral supplementation. NICE NG128 addresses diagnosis and initial management of stroke and TIA.

Carotid artery stenosis is a major risk factor for stroke and must be addressed. Criteria depend on symptoms and the degree of the stenosis. If asymptomatic, action is justified at 60–99%: if symptomatic at 50–99%, although other criteria include age, life expectancy, significant CAD, and pulmonary disease. One approach is to surgically open the artery, remove the atheroma, and then close the artery, a procedure known as carotid endarterectomy. One guideline recommends coronary angiography before elective carotid surgery. Although very effective, surgery is clearly very invasive and a high degree of care by the surgeon is required to ensure all minor particles of tissue are removed, lest they run up the artery into the brain and cause exactly the problem it is designed to prevent, that is, a stroke. An alternative is to place a stent across the lesion (as in the treatment of coronary artery stenosis), and this is often preferred as there is little chance of damage to nerves in the neck and the jugular veins.

Open surgery and stenting procedures bring the same degree of post-intervention stroke or death (cited in the range ≤1–4% after 30 days), although there is a lower risk of myocardial infarction in stenting. After carotid artery surgery or stenting, guidelines recommend dual antiplatelet therapy for at least a month (subject to risk of haemorrhage). Vertebral artery stenoses, responsible for up to 20% of ischaemic cerebrovascular events, are usually treated medically as they are often inaccessible, and so are likely to be considered by only the most experienced teams, who may insert stents.

As with CAD, brain tissue will die if deprived of oxygen by an occlusion, so that considerable effort is put towards restoring blood flow and so reducing the likelihood of a permanent stroke. With the benefit of more complex brain imaging (CT angiography and MR angiography), in acute ischaemic stroke the site of any occlusion can be determined, and thrombolysis or thrombectomy can proceed. If thrombectomy is an option, CT should be performed as soon as possible and within 24 hours of the onset of symptoms. Should thrombolysis be preferred, NICE NG128 and TA264 recommend alteplase as soon as possible within 4.5 hours of the onset of symptoms and once intracranial haemorrhage (ICH) has been excluded, and there are cases (such as an acute occlusion of the proximal anterior circulation) where thrombectomy with thrombolysis may be offered. Whilst in the peri-stroke period, oxygen should be given to maintain 95% saturation.

Treatment of ICH includes neurosurgery (craniotomy) to evacuate the haematoma (in which case high blood pressure should not be lowered), although exclusion criteria include hydrocephalus (excessive cerebrospinal fluid) and small deep haemorrhages. In some circumstances, depressive hemicraniotomy (to reduce intra-cranial pressure) may be applicable.

Aspirin (or alternative if intolerant) is the lead drug treatment after acute ischaemic stroke, for at least 2 weeks after onset by which time the plan for long-term antithrombotic therapy should be clear. To this is generally added dipyridamole, which has antiplatelet and vasodilator activity. However, low-dose rivaroxaban plus aspirin better protects against stroke in patients with stable CAD or PAD compared to aspirin alone. Should the stroke be in the venous circulation, heparin and then warfarin should be offered, and should an ICH have occurred whilst on warfarin, anticoagulation should be reversed urgently. Where a disabling stroke is linked to AF, aspirin should be given for 2 weeks, then oral anticoagulation (OAC), and where there is a prosthetic valve, OAC should be replaced with aspirin. Later, statin therapy should be initiated, aiming for LDL <1.8 mmol/l, or decreased by >50% if the initial LDL was 1.8–3.5 mmol/l.

Other aspects of medical care include attention to nutrition and hydration, with assessment of swallowing and, if problematic, this should be addressed with, for example, feeding by a nasogastric tube. Similarly, hydration will be monitored (and potential for aspiration pneumonia noted), and early mobilization encouraged (sit out of bed, stand, walk) as soon as possible. Advice on long-term management of stroke is offered in NICE CG162, with specific reference to specialist practitioners in physiotherapy, occupational therapy, speech and language therapy, and psychology. These will interface with professionals in social care in ensuring early supported discharge. Together, they will set goals for rehabilitation, such as at least 45 minutes of each targeted therapy 5 days a week, and monitor these.

Key points

There are similarities between angina/CAD and TIA/stroke in terms of pathogenesis, the role of imaging, and treatments.

SELF-CHECK 8.9

Why is imaging important in the investigation of TIA and stroke?

8.5 Peripheral artery disease

The same atherosclerotic disease process that can attack the coronary, carotid, and cerebral circulations may also attack other arteries, such as those of the abdomen, groin, and leg, leading to peripheral artery disease (PAD). Accordingly, there is a great deal of repetition as regards investigations (e.g. ultrasound, angiography) and treatment (e.g. stenting, antiplatelet therapy). In certain cases, carotid artery disease is included in PAD, whilst disease of the leg may be described as lower extremity arterial disease (LEAD). Notwithstanding the toxic effects of dyslipidaemia on the endothelium causing coronary atherosclerosis, both smoking and diabetes seem to be particularly linked with LEAD. Curiously, atherosclerosis of the arteries of the arms and shoulders (upper extremity arterial disease, UEAD) is rare. Aneurysmal disease primarily develops in the aorta (hence abdominal aortic aneurysm, AAA) and cerebral vessels. All patients with PAD (including carotid disease) should be screened for HF by NT-proBNP and echocardiography.

8.5.1 Arteries of the groin and legs

As the aorta passes through the diaphragm, the first major arteries it feeds are the left and right renal arteries, and a mesenteric artery. Below the kidneys, the lower part of the aorta bifurcates into the left and right common iliac arteries (about 4 cm long), which again bifurcate into the internal and external iliac arteries. The former (also 4 cm long) and its sub-arteries (such as the pudendal and gluteal) feed the pelvic floor—the urogenital systems, the rectum, the perineum, the buttock and the back of the thigh, the terminal vertebrae, and the pelvis itself. The larger external iliac passes through a fenestration in the pelvis and into the thigh, where it becomes the femoral artery. As it passes down the thigh, other arteries branch off to feed the thigh muscles, and behind the knee it becomes the popliteal artery. Moving down into the calf, the popliteal splits into the anterior and posterior tibial arteries, which eventually feed the ankle and foot. The pulse can be detected at a point slightly posterior to the internal meeting of the lateral malleolus of the fibula with the calcaneus (heel bone), and at a second point on the anterior (upper) surface of the foot, overlying the talus, navicular, and cuneiform bones. Detection of these two pulses is important in the diagnosis of peripheral artery disease.

Presentation

In parallel to angina/CAD, and TIA/stroke, intermittent claudication is the leading symptom of PAD. It appears as cramping pain in the muscle of the calf, the thigh, or the buttocks after exercise such as walking, causing limping (claudication) as a result of the build-up of lactic acid. This, in turn, is the consequence of ischaemia in the muscles due to the inability of the atherosclerotic arteries, such as the femoral and popliteal, to deliver sufficient oxygen to satisfy the metabolic demands of these tissues.

Once the muscles rest, and so demand less oxygen, there is less ischaemia, and the body can wash out the lactic acid so that the pain resolves. Naturally, the relief of pain will be taken by many to be a sign to resume walking. However, the atherosclerotic arteries will again be unable to satisfy the demands of the muscle for oxygen, so anaerobic respiration will resume with the build-up of lactic acid and the return of the claudication, hence intermittent claudication. If untreated, this symptom brings a 20% increased risk of myocardial infarction and stroke and a 10–15% risk of mortality within 5 years.

A further stage in pathology is failure of the peripheral arteries to feed sufficient blood to the lower leg, but more so to the foot, which may then become chronically ischaemic, perhaps with rest pain, which may be described as chronic limb-threatening ischaemia. Signs of this include a 'cold and white' foot, and weak or absent pedal pulses, with ulceration and poorly healing abrasions that are often infected. This is particularly noted in the 'diabetic foot', which is a prime target for the surgeon, with the need for debridement (removal of dead tissue) and amputation. Acute limb ischaemia, potentially arising from an arterial thrombus, is likely to present with features such as sudden pain, toe cyanosis, and with sensory and motor deficit, and must be managed promptly. The terminal stage of LEAD may be described as critical limb ischaemia, implying a need for urgent action if tissue is to be salvaged, and in its worst phase leads to debridement of necrotic tissues, gangrene of the toes, and, ultimately, the need for amputation of those toes, the foot, and ultimately the lower limb itself.

Diagnosis

A physical examination of the legs focuses on pulses and the colour and temperature of the toes (ideally pink and warm), which return with good colour once local capillary blood is driven out by squeezing. However, the key investigation is blood pressure measurement with a cuff at the extremity of the leg where it meets the foot. Ideally, the blood pressure at the arm will be the same as at the ankle, but in practice it is a little lower, leading to an ankle to brachial index (ABI) between 0.9 and 1.1. The use of an ultrasound probe over the arteries of the foot provides better sensitivity. Failure of the arteries to supply blood to the ankle will lead to a reduced ABI, with a result <0.9 implying vascular disease, and <0.5 severe disease, demanding urgent treatment. An ABI ≤0.9 is associated with a more than doubling of the 10-year rate of coronary events, cardiovascular mortality, and total mortality. However, calcification (hence stiffening) of peripheral arteries can increase the ABI, perhaps to >1.40, and this too is a risk factor for CVD and mortality. The procedure is so quick and simple that its use is being encouraged in all forms of cardiovascular disease management.

Ultrasound can also be used to estimate stenotic disease of the leg (femorals) and groin (iliacs) to pinpoint the exact position of the culprit lesion(s), and patients should be screened for AAA. A further test is angiography, which can demonstrate stenosis and occlusions with greater precision (Figure 8.23). The circulation of the leg can be challenged with the use of a treadmill, as in the Bruce protocol for the heart.

Management

Major strategies in low-grade intermittent claudication, as with uncomplicated TIA and stable angina, include risk factor management with blood pressure (to target <140/90 at a minimum) and lipids. In one study, statin use was associated with a 17% decrease in adverse cardiovascular events, so that the target LDL level is <1.8 mmol/l, or a decrease of >50% if baseline levels were 1.8–3.5 mmol/l (same

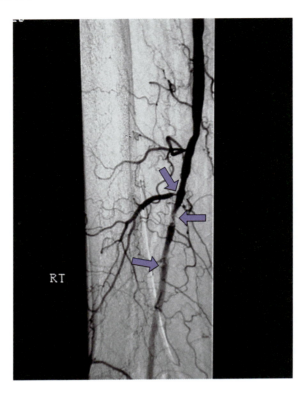

RT

FIGURE 8.23

Angiograph of atheromatous femoral artery. DSA of atheroma within a section of the femoral artery. The upper arrow shows a tight stenosis at the origin of a side artery, with a modest distal blood flow. The middle arrow shows atheroma that is minimally patent, as exhibited by the downstream blood flow. The lower arrow shows atheroma that all but occludes the artery.

© 2022 Vascular Society.

as with carotid artery disease). More specifically, patients should engage in structured and supervised exercise (simple walking will often suffice, an alternative being cycling) to promote leg artery circulation and general cardiovascular fitness, which can extend pain-free walking by around 100 m. A further initiative in PAD is antiplatelet therapy with aspirin or clopidogrel to reduce the risk of CAD. However, most patients on OAC should stay on their drug, an exception being those who have had a percutaneous intervention and with a low bleeding risk, who may be moved to dual antiplatelet therapy for at least a month. Perhaps unsurprisingly, PAD is a major risk factor for AMI and stroke, as all three are the consequence of the same disease process—atherosclerosis.

As the disease and its symptoms progresses towards critical limb ischaemia with imaging-proven stenosis and occlusions, balloon-driven angioplasty and stenting are considered (Figure 8.24).

Dual antiplatelet therapy is required for a month after stenting (as in all stent placements), followed by single antiplatelet therapy, whereas aspirin or clopidogrel is called for after surgery. Patients on OAC (e.g. because of AF or a prosthetic valve) should remain on this therapy. Should these prove ineffective or the disease be extensive, bypass is an option, such as diverting blood from the aorta or the iliac arteries to the femoral artery or arteries, from one femoral artery to the other, or from the femoral to the popliteal artery. The precise nature of the intervention depends on anatomy (common femoral v superficial femoral), age, surgical risk, and the extent of the stenotic/occlusive lesion (e.g. <5 cm stenting, </> 25 cm bypass). The patient's own veins are preferred bypass material (as in CABG), but options include prosthetics Teflon (made of polytetrafluoroethylene) and dacron (polyethylene terephthalate). Ideally, these will restore blood flow and salvage those tissues in danger of amputation (toes, the entire foot), the last resort.

Assessment of the risk of amputation may be made using the WIfI criteria, where W = wound (the extent of ulceration), I = ischaemia (determined by ABPI, ankle SBP, and Toe SBP or TpO_2), and fI = foot infection (no, local, or systemic). Each feature is scored on a scale of 0–3, so the maximum score is 9. A high risk of amputation is a score of 4, and the benefits are also high, depending on infection control.

Broadly speaking, the same criteria apply to the management of acute limb ischaemia as in chronic disease, but with the addition of heparin and analgesia and possible use of thrombus extraction or aspiration, and of catheter-directed thrombolysis.

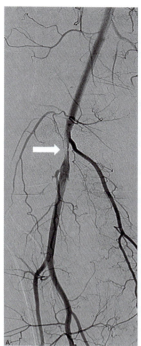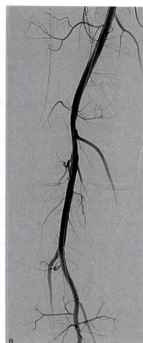

FIGURE 8.24

Successful treatment of a popliteal artery occlusion. The image on the left uses DSA to indicate an occlusion of the popliteal artery. The image on the right shows that flow has been restored by the intervention.

Jud, P MDa, Gary T Mda; Hafner F Mda, et al. Multiple arterial thromboses due to cystic medial degeneration Erdheim–Gsell: A case report. Medicine: Novr 2017: 96(47); p e8782, doi: 10.1097/ MD.0000000000008782. © 2017 the Authors. Wolters Kluwer Health, Inc. All rights reserved.

Key points

People with diabetes are at particular risk of LEAD, which can be delayed by optimum medical treatment with anti-glycaemics, anti-hypertensives and statins.

8.5.2 Aneurysm

Weakness of the media of arteries may lead to a distension and distortion of the shape of the vessel, which in turn leads to turbulence and the development of thrombus. This distension (an aneurysm) is most frequently found in the aorta as it passes through the abdomen (hence AAA – as introduced in section 6.5.4), the most common site being between the renal arteries and iliac arteries. Should this aneurysm grow, the artery wall will become stretched, thinner, and will eventually burst (as would a balloon being blown up too much). In the case of a ruptured AAA, this is likely to have catastrophic consequences with massive and rapid blood loss and carries a very high risk of mortality (65–85%). Cerebral artery aneurysms (CAA) are mostly saccular (berry) and are predominantly found in the anterior circulation. Rupture will lead to a subarachnoid haemorrhagic stroke, also with a high degree of mortality (sudden in 10–15%, 20–25% within 24 hours, and 50% at 3 months) and morbidity (30–40%).

In the UK, ultrasound scanning for AAA is routine practice for the over 65s, and depending on the result, surgery may be offered. Should the aneurysm be modest, ultrasound will be repeated at intervals until such time as surgery is required. Depending on the exact position, an aneurysmal section may be removed and replaced with a dacron tube graft in a lengthy open surgical session. However, a preferred option is to insert a modified stent via a femoral artery into the AAA, and anchor it at the level of the kidneys and at the origin of the iliac arteries, or within the iliac arteries. NICE NG156 offers guidance on the diagnosis and management of AAA.

CAAs are asymptomatic and cannot be detected by history or examination. The leading symptom of rupture is a sudden 'thunderclap' headache, although some report less debilitating headaches in the previous 3 weeks. Precise diagnosis relies on MRI, CT, and angiography, and possibly a lumbar puncture for increased CSF pressure and red cells and/or evidence of their degradation. Surgical repair involves removing a portion of the skull, placing a metal clip over the neck of the aneurysm to prevent more blood from entering, and then closing the skull. The aneurysm will eventually resolve, the clip will remain *in situ*. The second option is to place, with the help of a catheter, a platinum coil to promote local clot formation. Following either procedure, late effects such as vasospasm and seizures may occur, the patient recuperating in the intensive care unit.

8.5.3 Other arteries

Many other arteries are subject to atherosclerosis, but with tissue-specific consequences. For example, atherosclerosis of an artery feeding the vertebrae can lead to bony infarctions and ultimately destruction of that part of the spine, which is likely to lead to neurological symptoms. Atherosclerosis of the pudendal artery feeding the penis can lead to erectile dysfunction (particularly common in diabetes), but there is also a parallel with intermittent claudication. At physical rest, widespread stenosis of the iliac and pudendal arteries does not preclude an erection, but once engaged in sexual intercourse, atherosclerotic arteries may not be able to provide sufficient oxygenated blood to the respective muscles, leading to lactic acid build-up and pain, and also possible loss of erection. At rest, pain will subside and blood flow return to provide an erection, prompting a resumption of sexual intercourse which will once more be interrupted. A thrombus in an intestinal mesenteric artery can lead to death of that part of the bowel that relies on the artery for its nutrients, consequences including bowel gangrene, necrosis, and peritonitis. In patients with LEAD, over a quarter have a ≥50% stenosis in a mesenteric artery, symptoms being severe abdominal pain and vomiting/diarrhoea.

Atherosclerosis of the renal artery can lead to stenosis and so to insufficient blood entering the kidneys to be cleaned, leading to CKD, defined by reduced estimated glomerular filtration rate (eGFR). It is also a cause of renovascular hypertension, an important diagnostic sign. Fortunately, renal artery stenosis ≥60% may in certain circumstances be treated by angioplasty and/or stenting, medical treatment being ACEIs/ARBs etc. as discussed in section 7.2. Disease of arteries feeding the eye (notably the temporal artery, with temporal arteritis) may lead to problems with vision and perhaps, ultimately, blindness. Diabetic retinopathy, which can lead to blindness, is a serious consequence of long-term uncontrolled hyperglycaemia, and is described in sections 7.4.1 and 7.4.5. The part of the aorta that passes through the abdomen is also prone to atheroma, and so ruptured plaque can be shed, be driven down to the legs, and cause infarctions in the foot. Except for subclavian arteries, atherosclerosis is rare in the shoulder and arm arteries, and may be suggested by a ≥10% inter-arm SBP difference, and which would need confirmation with angiography. Estimated to be present in 2% of the general population of the UK, this rises to 9% in those with LEAD. Both stenting and surgery may be considered.

8.5.4 Multi-site arterial disease

As described above, atherosclerosis is a body-wide disease process attacking numerous arteries, so that many patients have disease at more than one site (Table 8.3). It follows that good practice requires that disease in all arteries be suspected, investigated, and considered in management. This reflects the

TABLE 8.3 Extent of multi-site arterial disease

Primary site	Additional carotid disease	Additional CAD	Additional LEAD	Additional RAS
CAD	5–7%	–	7–16%	4–15%
Carotid	–	39–61%	18–22%	n.a.
LEAD	14–19%	25–70%	–	10–23%

CAD: coronary artery disease, LEAD: lower extremity artery disease (ABI <0.90), RAS: renal artery stenosis (>75%). Carotid stenosis defined as >70%. n.a. = not available. Modified from Aboyans et al. 2018.

possibility of multiple-aetiology heart disease with cross-pathology of cardiomyopathy, valve disease, HF, and AF. Indeed, it is entirely possible that those with any of the latter conditions also have subclinical arterial disease.

Key points

Peripheral artery disease caused by atherosclerosis may attack the carotid, iliac, femoral, and renal arteries, treatments of which have certain techniques in common, such as stents.

SELF-CHECK 8.10

What are the key investigations in lower-extremity arterial disease?

Chapter summary

- Coronary artery disease runs a spectrum from stable angina to unstable angina, NSTEMI and STEMI: the latter three comprise the acute coronary syndromes. Key specific investigations are ECG and blood tests.

- Most heart valve disease can be classified as stenosis and regurgitation, and can lead to structural heart disease, such as left ventricular hypertrophy.

- The leading arrhythmia, atrial fibrillation (AF), is defined by ECG and is a leading risk factor for stroke, inevitably requiring anticoagulation.

- Heart failure may be secondary to valve disease, AF, CAD, and cardiomyopathy, and may call for mechanical support. The ultimate treatment for all heart disease is transplantation.

- Transient ischaemic attack is often the result of carotid artery disease, and both are a risk factors for stroke.

- Peripheral artery disease includes atherosclerosis of lower-extremity arterial disease (severity assessed by the specific ankle/brachial index) and aneurysmal disease, the rupture of which can be fatal.

Further reading

- Baumgartner H, Falk V, Bax JJ, et al. 2017 ESC/FACTS guidelines for the management of valvular heart disease. Eur Heart J 2017: 38; 2739–91.

- Thygesen K, Alpert JS, Jaffe AS, et al. Fourth universal definition of myocardial infarction. Eur Heart J 2019: 40; 237–69.

- Hindricks G, Potpara T, Dagres N, et al. 2020 ESC guidelines for the diagnosis and management of atrial fibrillation developed in collaboration with the European Association for Cardio-Thoracic Surgery (EACTS). Eur Heart J 2020: 00; 1–26.

- Aboyans V, Ricco JB, Bartelink ML, et al. 2017 ESC guidelines on the diagnosis and treatment of peripheral arterial diseases, in collaboration with the European Society for Vascular Surgery. Eur Heart J 2018: 39; 763–821.

- Lozano-Velasco E, Franco D, Aranega A, et al. Genetics and epigenetics of atrial fibrillation. Int J Mil Sci 2020: 21; 5717. Doi 10.3390/ijms21165717.

Useful websites

- National Institute for Health and Care Excellence: **www.nice.org.uk**
- The British National Formulary: **https://www.bnf.org/products/bnf-online/**
- The Framingham Heart Study: **https://www.framinghamheartstudy.org/**

Discussion questions

8.1 What are the major tools in the diagnosis of cardiovascular disease?

8.2 Defend the statement 'Atherosclerosis is present in all the arteries of the elderly—clinical disease appears depending on which artery is most diseased'.

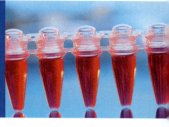

Basic and laboratory immunology

Immunology is all about recognition. At the level of the whole organism, we recognize the outside world by the senses, but at the level of the cell, recognition is of the shapes of certain molecules. Almost all nucleated cells express at their surface molecules unique to that individual which define 'self'. Having originally been described on white blood cells, these molecules take the name human leukocyte antigens (HLA). It follows that any different sets of HLA molecules that we encounter can't be 'self', and so are 'non-self'. This is important because we must recognize non-self pathogens that must be attacked and destroyed.

The immune system can be slow to become fully effective, and not all of our defences are finely tuned to the concept of self. Fortunately, natural selection has equipped us with a battery of other defence systems that are not reliant on the HLA recognition system. These innate defences are crudely effective, and buy time for a specific immune system to develop and adapt to the particular pathogen, and these often give rise to an inflammatory response. A key component of this is the acute phase response, likened to a general awakening and preparation of the body for a potentially dangerous infection.

This chapter is important because some degree of immunological and inflammatory process is present in almost all forms of disease: it has roles in renal, liver, endocrine, and reproductive disease, to name just a few. But before looking at pathologies of the immune system, we first need to consider its physiology. The function of the immune system is to defend us against pathogens, and if it fails to deal with these (and other issues), this results in immunopathology. The parts of a pathogen (a disease-causing entity) that stimulate the immune system are described as antigens.

The purpose of the first numbered section of this chapter is to identify and describe the pathogens against which the immune system defends us. We will then start our discussion of the immune system itself by reviewing its organs (bone marrow, liver, lymph nodes, etc.: section 9.2) and cells (neutrophils, lymphocytes, monocytes, eosinophils, basophils: section 9.3), before turning to humoral immunity, which describes immune responses mediated by soluble factors in the plasma (these being antibodies (IgG, IgM, IgD, IgE, IgA), complement, and cytokines: section 9.4). Section 9.5 describes precisely how these organs, cells, and molecules come together to defend us from pathogens in the innate immune system. Section 9.6 will look at the same features in the adaptive immune system, and in section 9.7 the practical qualities of these two aspects and how they come together to form inflammatory and immunological defence will be discussed. The chapter will conclude in section 9.8 with a look at the role of the laboratory in immunology. This will set us up for an in-depth examination of what happens when the immune system cannot cope, is otherwise malfunctioning, or can be manipulated—that is, clinical immunopathology—in Chapter 10.

Learning objectives

After studying this chapter, you should confidently be able to:

- appreciate that the immune system defends us from microscopic pathogens, much as the armed forces of a country defend it on the world stage
- describe the organs, cells, and humoral components of the immune system
- understand the importance of HLA in the concept of self, and in antigen recognition
- grasp the concepts of innate and adaptive immunity
- describe how the immunology laboratory can assess the function of the immune system

9.1 Defence

While we can usually identify and escape large organisms that would do us harm, such as carnivores, it can be more difficult to avoid harm caused by the smallest organisms—that is, microbes. The purpose of the immune system is to protect you—a biological body that can in this context be described as 'self'—from microbiological attack by recognizing anything that is not you (that is, 'non-self'), and then destroying it. We know several types of pathogens, which can be defined as factors that cause disease, and one method for classifying them is by size. While some atoms are toxic, such as arsenic, they are too small to be recognized by the immune system, and the same is true of certain molecules. The immune system also does not recognize prions (pathogenic and infectious proteins) as non-self and therefore these are not attacked. The immune system focuses on four types of pathogens that include millions of different parasitic microorganisms.

9.1.1 Viruses

Viruses consist of a segment of DNA or RNA enclosed within a protein coat, and they are only able to reproduce within a living cell. Most viruses are around 0.1 μm in size, which means that they can only be viewed with the help of electron microscopy. Examples of this type of pathogen include smallpox, polio, measles, influenza, hepatitis, and SARS-CoV-2 (which is addressed in Chapter 13). Some of these can be countered by vaccination, as described in Box 9.3 and in section 13.6.6, which specifically considers vaccination against SARS-CoV-2.

9.1.2 Bacteria

Cross reference

Further details on microbiology can be found in a dedicated text-book on *Medical Microbiology* in the Fundamentals of Biomedical Science series.

Bacteria show considerable diversity in size, but most of them are between 0.3 and 5 μm in length. Using light microscopy, they can be viewed either individually or as clusters or chains of spherical (cocci), rod-shaped (bacilli), or spiral (spirilla) organisms. Their outer layers consist of complex polysaccharides that are cross-linked by peptides, which defines them as 'non-self'. There is variation in the exact composition of the cell wall among different kinds of bacteria, which gives rise to different results when they are stained to identify them as Gram positive or negative—a useful classification tool.

9.1.3 Fungi

Notable species of fungi in the context of human disease include the hundreds of aspergillus species, which grow in characteristic long chains called hyphae. Other examples of these microbes are *Candida albicans* (yeast, which causes the infection thrush), histoplasmosis species, and *Pneumocystis jirovecci*, which can cause lung infections.

9.1.4 Parasites

Larger, more complex parasites can be classified into protozoa (single-celled organisms) and meta-zoa (multicellular organisms). Notable examples of protozoa include the various plasmodium species that cause malaria (*P. falciparum, vivax, ovale, knowlesi*, and *malariae*) and infect red cells, amoeba (*Entamoeba histolytica*, which often causes amoebic dysentery), leishmania, and trypanosomes. Metazoan parasites include filaria, schistosomes, and various worms (e.g., tapeworms and liver flukes). Many of these parasites enter the body of their host, but others, such as leeches and certain arthro-pods including fleas, ticks, and lice, stay on the outside of the body and can therefore be described as ectoparasites—that is, organisms that live on their host's skin.

9.2 Organs of the immune system

Different parts of the body provide specialized sites where immune cells arise and where they, and soluble factors, can perform their functions.

9.2.1 Bone marrow

Bone marrow is the organ where red blood cells, white blood cells, and platelets are produced. In the embryo and foetus, this happens in the yolk sac and liver, but during gestation, the stem cells (from which all blood cells derive) migrate to the bone marrow, principally of the long bones (femur, humerus) and ribs, but in adults the major sites are the sternum and the pelvis. Blood development (haemopoesis) can be broken down into cell-specific lineages, of red blood cells (erythropoiesis: section 11.1.1), of platelets (thrombopoiesis: section 11.2.1), and of white blood cells (leukopoiesis: section 9.3.1). Disease and damage to the bone marrow, perhaps by radiation or cytotoxic chemo-therapy, will result in a reduction or (in the worst possible situation) the cessation of production of white blood cells, which leaves the body open to infections. This situation, **immunodeficiency**, is the subject of section 10.1.

9.2.2 Skin

The skin is a very important first barrier—a tough, and ideally impenetrable, shield against the multi-tudes of pathogens encountered daily. The outer layers are effectively the dead keratinocytes of the epidermis, but below this the dermis houses specialist white blood cells. These sentinel Langerhans cells and histiocytes (modified monocytes) are often the first to the scene of a break or tear in the skin that may be an entry point for pathogens. These sentry cells recruit other white blood cells, notably neutrophils, to help with attacking potential invaders, principally by an ability to ingest and destroy the pathogen (the process of phagocytosis).

9.2.3 Lymph nodes

The lymph nodes comprise hundreds of micro-organs located across different parts of the body where white cells—primarily lymphocytes, monocytes, and dendritic cells—produce antibodies. However, they are not the only sites of antibody production, which can also occur in the spleen and the liver. Some anatomical sites feature specialized lymph nodes which have defined functions in counteract-ing pathogens from particular sources (Table 9.1).

Many lymph nodes in close proximity may be linked, as in the mucosal-associated lymphoid tissues (MALT, estimated to include half of all lymph nodes), a subgroup of which are the gut-associated lymphoid tissues (GALT) that include the Peyer's patches of the small intestines (section 12.3.1). A common feature of lymph nodes is their architecture (Figure 9.1), which brings togeth-er those white blood cells required to produce antibodies in germinal centres. Lymph nodes are arranged in chains, connected by lymphatic vessels that bring lymph-carrying potential pathogens, and also lymphocytes.

TABLE 9.1   Specialization of lymph nodes by anatomy

Lymph nodes	Location of pathogens
Tonsils	In food/particles in inspired air that lodge in the throat
Axilla	Those infecting the arms
Inguinal	Those infecting the legs
Bronchial	Airborne pathogens
Pulmonary	Airborne pathogens
Gastric	Pathogens in food
Abdominal	Pathogens in food

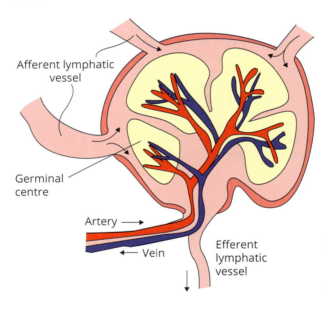

Afferent lymphatic vessel

Germinal centre

Artery →

← Vein

Efferent lymphatic vessel

FIGURE 9.1

Structure of a lymph node. The lymph node receives lymph from upstream lymph nodes or tissues via afferent lymphatic vessels, and with arterial blood. The germinal centres host the lymphocytes and antigen-presenting cells that together generate antibodies that leave the lymph nodes via venules. An efferent lymphatic vessel carries lymph and lymphocytes to the next lymph node, and ultimately to the bloodstream.

9.2.4 The liver

The liver produces numerous proteins, many of which (**complement**, certain cytokines) have important functions in immune responses. It also produces the clinically valuable C-reactive protein (CRP) and other inflammatory markers. The liver receives blood from the intestines, which may carry potential pathogens that have been ingested along with food. Accordingly, specialized Kupffer cells line certain hepatic vessels, ready to intercept and destroy pathogens, and this organ may also host those white blood cells that collaborate in generating antibodies. Digestive aspects of the liver are described in section 12.2.

9.2.5 The spleen

The spleen can be the site of antibody production, and so may be viewed as a lymph node. It also stores a large number of dormant neutrophils and red blood cells, which can enter the bloodstream rapidly. Further evidence of the value of this organ lies in those situations where it is removed (splenectomy), which leaves the patient prone to infections by certain types of bacteria, and so can lead to the requirement for antibiotics and vaccinations.

9.2.6 The thymus

The thymus is a crucial organ, lying between the heart and the sternum, where a subset of lymphocytes finally develop, the T lymphocytes, or T cells. Particularly large and active in the young, as they encounter foreign microorganisms for the first time, it shrinks with age. Its clinical importance is demonstrated in DiGeorge syndrome, a congenital condition in which the thymus may be rudimentary or even absent. This leads to T cell deficiency and the inability to mount a defence against a range of pathogens, or to produce certain antibodies, and so is another form of immunodeficiency (section 10.1).

Cross reference

See section 16.1.3 of Chapter 16 for an overview of DiGeorge syndrome.

9.2.7 The bursa equivalent

Just as T lymphocytes require the thymus to complete their development, so too does a second type of lymphocyte, that which is destined to make antibodies. In birds, this development occurs in the bursa of Fabricus, a small intestinal organ, hence the name B lymphocytes. No such organ has been found in mammals, and so it is assumed that B lymphocytes complete their development in any combination of the lymph nodes, bone marrow, and/or the spleen.

Key points

The immune system resides in a wide series of large and small organs, and consists of cellular and humoral components. Its defensive roles can be likened to those of the armed forces.

SELF-CHECK 9.1

Define the major anatomical sites of the immune system and their functions.

9.3 Cells of the immune system

Immunology and inflammation are primarily the province of white blood cells, present in the blood and in large numbers in the various anatomical sites described in section 9.2. Each of the five types of circulating white blood cells—neutrophils, lymphocytes, monocytes, eosinophils, and basophils—has a different function, although there are overlaps (as in phagocytosis and in binding certain classes of antibodies). As neutrophils, eosinophils, and basophils have intracytoplasmic granules, they may be collectively described as granulocytes (although monocytes may also have granules) and their irregular nucleus marks them as polymorphonuclear leukocytes. White blood cells develop in the bone marrow and elsewhere by the process of leukopoiesis.

9.3.1 Leukopoiesis

Whilst haemopoiesis is the process of the development of all blood cells, leukopoiesis refers particularly to the development of white blood cells. The formation of blood cells starts with a pluripotent stem cell that gives rise to two other stem cells, also known as colony-forming units (CFU). A unipotent CFU-L gives rise to cells that ultimately develop into lymphocytes, whilst a CFU-GEMM gives rise to all other blood cells. In turn, the CFU-GEMM generates a stem cell (CFU-EMk) for red cells (erythropoiesis) and megakaryocytes that then produce platelets (thrombopoiesis), and a CFU-GMo for granulocyte production. The CFU-GMo produces one final unipotent stem cell for each type of granulocyte: CFU-neutrophils, CFU-eosinophils, CFU-basophils, and CFU-monocytes.

Each unipotent stem cell produces immature forms of each cell type (blasts), which then mature into fully functioning blood cells. These immature cells are named according to their cell lineage, so, for example, a blast of the monocyte lineage is a monoblast. In health, blast cells should not appear in the blood, but they may do so in certain diseases and in acute inflammatory episodes.

Monocytes moving out of the blood and into tissues, such as those of the skin and lungs, become macrophages and may develop further with a number of different functions in different organs; for instance, they may turn into microglia in the brain and osteoclasts in bone. Similarly, when basophils move out of the blood to the tissues, they become mast cells. As discussed in sections 9.2.6 and 9.2.7, the final step in the maturation of lymphocytes is their processing through the thymus or the bursa equivalent, giving rise to T and B lymphocytes respectively, whilst a third type of lymphocyte is an NK (natural killer) cell (Figure 9.2).

The role of cytokines and growth factors

Stem cells and their products develop in response to cytokine growth factors, such as granulocyte colony-stimulating factor (G-CSF) and granulocyte-macrophage colony-stimulating factor (GM-CSF). There are also roles for small messenger molecules of the cytokine family, such as interleukins (ILs) and transforming growth factor beta (TGF-β) (section 9.4.3). Some act specifically, such as IL-7, which acts only on pre-B cells via the IL-7 receptor, and IL-6, which acts only on plasma cells (mature B cells that produce antibodies). Others have restricted activity, such as IL-4, acting on B cells, T cells, and basophils, and still others act non-specifically, such as IL-3, which acts on all non-lymphoid cells. As growth factors are necessary for cell growth and differentiation, it follows that lack of growth factors leads to cell death, and this is often by apoptosis.

> **Cross reference**
> For a detailed discussion of apoptosis, see section 3.4 of Chapter 3 on the aetiology of cancer.

Laboratory estimation of white cells

Subtypes of white cells can be distinguished by a number of methods. Once stained with certain dyes, one method is to examine the cell's morphology, which involves considering the proportions of the nucleus (and its shape) and cytoplasm, and the number of granules (if any). A further classification method considers the expression of molecules described as clusters of differentiation (CD) that serve as identifiable markers at the surface of the cell (Table 9.2).

These clusters can be detected with relative ease by fluorescence flow cytometry, using fluorescence-activated cell scanner (FACS, section 9.8). However, the true picture is complicated by the level of expression of the particular marker, and its specificity for a certain cell. For example, plasma cells are typically identified as $CD38^{high}$ $CD27^{high}$ $CD20^{low}$, the latter also being found on B lymphocytes.

9.3.2 Neutrophils

The most common type of white cell are neutrophils, which make up around 75% of the total amount of white cells. They have a diameter of 9–15 μm, and their circulating half-life may be as short as 6–8 hours. The characteristic features of their morphology are a nucleus in segments or lobes and the granular appearance of its cytoplasm (Figure 9.3). They normally feature 3 to 5 lobes, and a cell with only 2 lobes is described as hyposegmented, which is effectively one stage on from an almost mature **metamyelocyte**.

While many granules contain enzymes, others function as storage granules, which may, for instance, store glycogen. Neutrophils attack and destroy pathogenic microbes such as bacteria, often by phagocytosis, which makes them important mediators of inflammation. When someone has an infection with bacteria, we can therefore expect to see increased numbers of neutrophils (a neutrophilia, or neutrophil leukocytosis), which is a desirable effect under these circumstances. However, certain diseases (such as rheumatoid arthritis; section 10.3.2) also lead to an increased neutrophil count regardless of whether micropathogens are present. Neutrophils also play an important role in tissue repair after injury (as do macrophages).

9.3.3 Lymphocytes

Lymphocytes account for 20–40% of circulating leukocytes, which makes them the second largest group among them. However, lymphocytes do not only appear in blood: many more of them can be found within lymph nodes, and the spleen, liver, and bone marrow also contain considerable

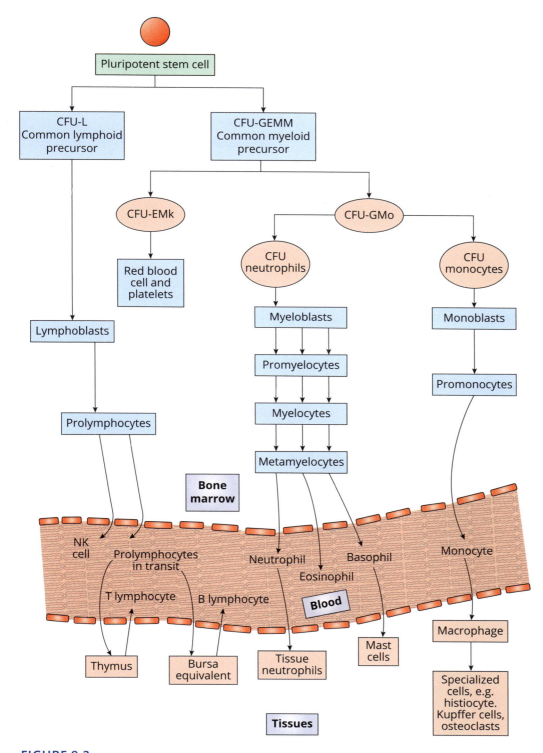

FIGURE 9.2

Leukopoiesis. Simplified diagram of the transition of immature white blood cells in the bone marrow to their mature forms in the blood and tissues.

TABLE 9.2   Selected CD molecules

CD marker	Distribution
CD3	T lymphocytes
CD4	T helper cells / regulatory
CD8	T cytotoxic cells
CD14, CD91	Monocytes
CD15	Granulocytes
CD16	NK cells, monocytes, neutrophils
CD19, CD20	B lymphocytes
CD30, CD56	NK cells
CD45	All white cells
CD114	Granulocytes, monocytes
CDw125	Eosinophils, basophils

CD = cluster of differentiation.

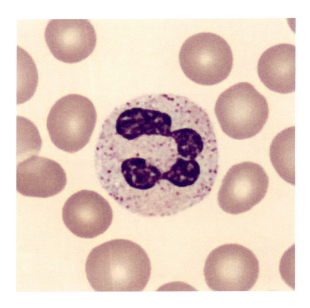

FIGURE 9.3

The neutrophil. A mature 4-lobed neutrophil with numerous purple/pink cytoplasmic granules.

Image courtesy of Dr Gordon Sinclair, London Metropolitan University.

numbers. Lymphocytes are characterized by a regular, circular nucleus that makes up about 90–98% of the cell and reduces the cytoplasm to a thin ring around the nucleus, as shown in Figure 9.4. They usually have a half-life of ~70 hours in blood, although some of them—probably those resident in lymph nodes—can live for decades. With a diameter of 9–12 μm, lymphocytes are the smallest type of white cell and only slightly larger than a red cell, which generally has a diameter of 7–8 μm. However, some lymphocytes can have a diameter of up to 15 μm. Especially when activated, these enlarged lymphocytes can have granules and may be described as large granular lymphocytes.

Lymphocytes are the primary cell involved in immunological processes, but (unlike granulocytes and monocytes) different subtypes have precise and non-overlapping functions. Several subtypes of lymphocytes that arise from a common lymphoid precursor are recognized, and although

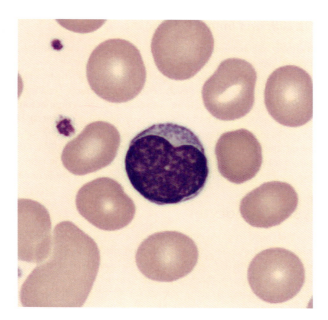

FIGURE 9.4

The lymphocyte. A lymphocyte whose nucleus occupies almost all of the cell.

Image courtesy of Dr Gordon Sinclair, London Metropolitan University.

indistinguishable by conventional staining (as in Figure 9.4), they may be defined by the presence of certain cell membrane CD molecules (Table 9.2) using immunofluorescence microscopy and FACS analysis.

T lymphocytes

T lymphocytes comprise 65–80% of circulating lymphocytes, making them the most frequent sub-type of lymphocytes, and are defined by the presence of CD3. They can be divided into several subtypes. Those carrying CD3 and CD8 seek and destroy cells they regard as foreign, such as those infected with viruses, or cells from another individual, such as a transplanted kidney, and so are called cytotoxic (hence Tc). These Tc cells are directed to their target by other cells, and (at the membrane level) by the T cell receptor (TcR). The most efficient production of certain antibodies by B lympho-cytes calls for the cooperation of dedicated T helper cells (T_h), which are characterized by CD3 and CD4. These specialist cells will be discussed in detail in section 9.6.2, and further Th subsets (e.g. T_{reg} Cells) introduced in section 9.6.3.

SELF-CHECK 9.2

Using defined examples, explain the value of CD markers.

B lymphocytes

B lymphocytes (10–30% of all lymphocytes) make antibodies. This is most unlikely to occur in the plasma, but happens in the lymph node, spleen, liver, and bone marrow, and generally requires den-dritic antigen presenting cells (APCs) and T_h cells. However, in some circumstances, B cells can make antibodies to certain bacteria without helper cells. The key recognition markers on mature B lym-phocytes are CD19 and CD20. However, those B cells already committed to the production of anti-bodies may also be identified by immunoglobulins on their cell surface. Such antibody-producing B cells, plasma cells, are identifiable by light microscopy using conventional dyes by a blue-stained cytoplasm. Finer details of the generation of antibodies by T cells, B cells, and APCs are presented in section 9.6.3.

NK cells and Innate lymphoid cells

The third and smallest group of lymphocytes (<5%) are defined by a lack of B or T cell markers, and include several subsets, the most well-established being NK cells. The latter arise from the common lymphoid precursor, with their own specific NK precursor. The first stages of maturation are linked to high expression of CD56 but low levels of CD16, but as the cell finally matures, CD56 expression wanes and CD16 expression increases. Those who have undergone **thymectomy** or have DiGeorge syndrome (and so lack T cells), or have had a splenectomy, all have a normal NK cell population, and NK cells can be generated *in vitro* from liver-derived NK precursors, suggesting that this organ may also have a role in NK cell development. Tissue-resident innate lymphoid cells (ILCs) are presumably sentinel cells. Many produce cytokines to help promote specific and non-specific immune responses, potentially in **helminth** infections and in recruiting eosinophils and basophils. Further discussion of ILCs is part of section 9.5.3.

9.3.4 Monocytes

The third most frequent circulating leukocyte, monocytes circulate in the blood for 2–3 days before moving into the tissues where they become macrophages and may live in an immunosurveillance role for several months (Figure 9.5). They are the largest type of leukocyte with a diameter of 15–25 μm. The nucleus is a single, roughly round shape (hence they are mononuclear, as are lymphocytes, in contrast to polymorphonuclear cells), but may be indented, with a C or U shape, and takes up 60–80% of the cells. Granules are often present. Monocytes can be shown to have several different functions *in vitro*, some of which, such as phagocytosis, also occur in the body. Key markers are CD14 (the receptor for bacterial lipopolysaccharide) and CD16 (the receptor for part of an IgG molecule).

Macrophages

As monocytes migrate into the tissues they are transformed into macrophages. Some undergo further changes in certain anatomical locations, such as dendritic cells (in lymph nodes), Kupffer cells (in the liver), Langerhans cells (various tissues, frequently the skin), histiocytes, and osteoclasts (in the bone). These specialized cells often lose classical monocyte/macrophage morphology, but some bear other markers—CD19, CD23, and CD83 on dendritic cells, CD1a, CD14, and CD83 on Langerhans cells, and CD53, which is shared between leukocytes, dendritic cells, and osteoclasts. Use of CD markers demands care because of poor sensitivity and specificity—CD19 is also found on B lymphocytes and C16 is also found on neutrophils.

Cross reference

Read more about sensitivity and specificity in section 1.3.5 of Chapter 1.

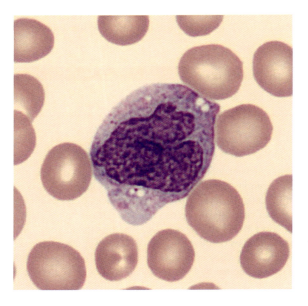

FIGURE 9.5

A monocyte with a characteristic C-shaped nucleus that occupies ~80% of the cell, and a small number of cytoplasmic granules. Note the contrast with the lymphocyte in Figure 9.4.

Image courtesy of Dr Gordon Sinclair, London Metropolitan University.

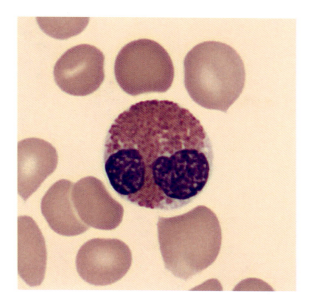

FIGURE 9.6

The eosinophil. A mature bi-lobed eosinophil with red/orange cytoplasmic granules.

Image courtesy of Dr Gordon Sinclair, London Metropolitan University.

Monocytes/macrophages also generate and release inflammatory cytokines such as tumour necrosis factor, interferons, and interleukins (see section 9.4.3) which promote the activation of other monocytes, but also lymphocytes, and endothelial cells. They also participate in haemostasis by generating tissue factor and coagulation factor V (Chapter 11).

9.3.5 Eosinophils

These granulocytes have a diameter of perhaps 12–17 μm. Almost all reside in the blood, with some in the lungs, and have a circulating half-life of ~6 hours. They are more long-lived in the tissues, where they may remain for around 10 days. The typical eosinophil has a bi-lobed nucleus with a cytoplasm dominated by reddish granules (Figure 9.6), and expresses CD9, CD15, CD23, CD35, and CDw125, although many other cells also bear these molecules.

The key to eosinophil function is its granules, dominated by those containing major basic protein, elastase, and peroxidase. However, there are also eosinophil-specific granules, such as that containing a cationic protein rich in arginine, another which has enzymes to degrade ribonucleic acid, a third containing a neurotoxin, and a fourth rich in histamine.

Besides having general 'housekeeping' anti-bacterial duties, eosinophils also have two more focused major roles. They are important mediators of allergic (hypersensitivity) reactions and are important in our defence primarily against helminth parasite infections (worms), such as filariasis, hookworm, schistosomiasis, and trichinosis. However, an eosinophilia (increased numbers of eosinophils in the blood) may be present in protozoal parasite infections.

9.3.6 Basophils

Basophils are the least frequent type of leukocyte, are slightly smaller than eosinophils at 10–14 μm, and transform into mast cells when they migrate into the tissues. The nucleus is bi-lobed and often obscured by the large number of purple or black granules (Figure 9.7). Basophils share CD9, CD68, and CDw123 with several other cells, and CDw125 with eosinophils, whilst mast cells bear CD49b. Like the eosinophil, the basophil shares several of the types of granules of its co-granulocytes (such as those that include elastase and other proteases). However, it has its own specific granules which include the anticoagulant heparin and the smooth muscle relaxant histamine. Other granule constituents include leukotrienes and cytokines.

Basophils have much in common with eosinophils, such as being active in defence against parasites and in **hypersensitivity** responses, but it is the latter that basophils focus upon. These include allergic and anaphylactic responses to factors such as certain pollens, and both basophils (in the

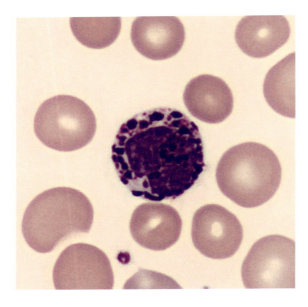

FIGURE 9.7
A mature basophil whose nucleus is obscured by numerous purple/black cytoplasmic granules.

Image courtesy of Dr Gordon Sinclair, London Metropolitan University.

circulation) and mast cells (in the tissues) degranulate in response to a variety of stimulants. A key activation pathway is the docking of an IgE molecule into a receptor on the surface of the basophil (to be discussed in section 9.4.1, where Table 9.5 applies). When basophil and/or mast cell activation and degranulation are excessive, consequences include several clinical conditions, as will be discussed in Chapter 10.

Table 9.3 summarizes major aspects of white blood cells.

TABLE 9.3 **Major physiological aspects of white blood cells**

Cell	Features
Neutrophil	The most numerous blood leukocyte. Phagocytosis, participation in inflammation
Lymphocyte	Production of antibodies, defence from viruses
Monocyte and macrophage	Phagocytosis, participation in inflammation, release of cytokines, subtypes present antigens to lymphocytes
Eosinophil	Defence from parasites, release of histamine
Basophil	Defence from parasites, release of histamine and heparin

Key points

Leukopoiesis produces five types of white blood cells, all with defined functions, and some of which move out of the blood into other organs where they have additional roles.

SELF-CHECK 9.3

Which of the white blood cells are phagocytes?

9.4 Humoral immunity

The three major aspects of humoral immunity that will be the focus of this section are antibodies, complement, and cytokines. However, it should be noted that there are several other molecules with roles in defending us from pathogens, such as some enzymes and mucopolysaccharides. The study of soluble molecules found in blood serum and other body fluids is called serology.

9.4.1 Antibodies

Antibodies are molecules generated by B lymphocytes, almost always with the aid of T lymphocytes and APCs (detailed in section 9.6.3), mostly in lymph nodes. From the protein perspective, antibodies are gamma-globulins (hence immunoglobulins: Ig), although they are also glycated. Antibodies are constructed from four polypeptide chains—two 'heavy' chains of ~50–5 kDa each, and two 'light' chains of ~25 kDa each—that form a 'Y' shape (Figure 9.8) with a total mass of around 150–160 kDa.

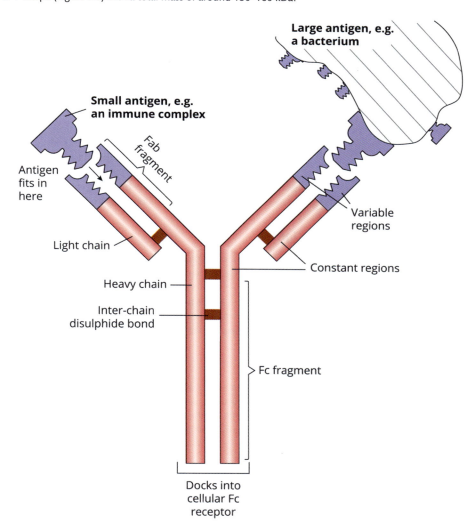

FIGURE 9.8

Simplified structure of an immunoglobulin molecule showing antigen binding. Antigens (which may be small soluble molecules (top left) or part of a much larger structure such as a bacterium or virus (top right)) are recognized by the binding site composed of the variable regions of the light and heavy chains.

The globular polypeptide chains (four in the heavy chain and two in the light chain) are linked by interchain disulphide bonds.

Antibody molecules recognize antigens by the combination of their shapes at the ends of their heavy and light chains, as the make-up of the particular amino acids is variable. In other parts of antigens, the amino acid sequences are (mostly) constant. This part of the antigen molecule is therefore the 'fraction antigen binding' region, or Fab. At the other end of the molecule, the two heavy chains form the Fc region, which can bind to specialized structures on certain white cells called 'Fc receptors', or FcR. There are five different types of antibodies, which are distinguished by the nature of their heavy chain and referred to as alpha (α), delta (δ), gamma (γ), epsilon (ε), and mu (μ). In addition, there are also two different types of light chains: kappa (κ) and lambda (λ). As the make-up of the heavy chain of the molecule is what determines the structure and function of the antibody, the heavy chains form the basis of the classification system for antibodies: if a gamma heavy chain is present, an antibody is classified as IgG, while the presence of an alpha, mu, delta, or epsilon heavy chain leads to classification as IgA, IgM, IgD, and IgE respectively. All five heavy chains can be associated with both kappa and lambda light chains. Together, the gamma-globulins make up around 20% of all serum proteins, with a concentration of 5.0–11.0 g/l.

IgG

The dominant antibody in plasma, IgG is also found in the tissues and is always a monomer of the basic 'Y' structure (Figure 9.8). There are four subclasses of the gamma heavy chain (IgG1–IgG4), which share around 90% homology, but each has different physicochemical and functional properties. IgG1 has the highest blood concentration (3.2–10.2 g/l) and is most effective in crossing the placenta and in binding to leukocytes via its Fc section. IgG2 has the second highest concentration (1.2–6.6 g/l), whilst IgG3 (0.2–1.9 g/l) is the most efficient at activating complement (section 9.4.2). IgG4 (0–1.3 g/l) is produced in response to multiple exposure to protein antigens and to extracellular parasites.

IgA

The antibody with the second highest concentration in the blood is IgA, which can activate the lectin pathway of complement. In the tissues, it is the most abundant antibody and appears in two isoforms, IgA1 and IgA2. IgA may be found in milk, **colostrum** (thereby protecting the infant), saliva, tears, nasal fluids, and sweat. It has important roles in defending mucosal surfaces, such as of the intestines and respiratory systems, where it is referred to as secretory IgA. It can occur as a monomer or as a dimer in which the two monomers are connected by an additional 15 kDa polypeptide—a J chain—coded for by *IGJ* at 4q13.3.

IgM

IgM circulates in the plasma as a pentomer of the basic Ig unit and five J chains (as in IgA dimers), which gives it a molecular weight of around 900–1,000 kDa. Accordingly, it is rarely found outside blood. However, monomeric IgM is found at the surface of the B lymphocyte as an integral part of the cell membrane. IgM is highly efficient at triggering the activation of complement, and because of its large size with multiple binding sites, it is able to cross-link separate antigens, for instance those present on two adjacent bacteria or other pathogens.

IgE

IgE is a monomer that is present at only very low levels in the plasma. It can bind to circulating basophils and their tissue version, mast cells, via an FcR at the surface of the cell. When binding to an antigen, this induces the cells to degranulate. In some cells, this happens too readily, which is the basis underlying many hypersensitivity reactions (Chapter 10).

IgD

While IgD is also present in small amounts in plasma, it is primarily expressed on the surface of B lymphocytes, where it may have a role in cell regulation.

Table 9.4 summarizes major aspects of the immunoglobulins.

TABLE 9.4 Major aspects of the immunoglobulins

Ig species	Adult reference range	Proportion of plasma Igs	Subclasses	Characteristics
IgG	6–16 g/l	75%	4	Major plasma antibody
IgA	0.8–4 g/l	15%	2	Major mucosal antibody: can be dimeric
IgM	0.5–2.0 g/l	10%	1	Pentameric: efficient in cross-linking
IgD	0–100 kU/l	<1%	1	Present on B lymphocytes
IgE	0–81 kU/l	<1%	1	Almost all bound to basophils/mast cells

Ig: immunoglobulin.

Fc receptors (FcR)

The process of antibodies binding to a cellular target via their Fab is called opsonization. Since the Fab binds the antigen, the exposed Fc region is free to bind with the FcR of a phagocyte, which is then able to kill the target cell. This act of killing is described as antibody-dependent cellular cytotoxicity (ADCC), or perhaps antibody-dependent cellular phagocytosis. As NK cells have FcRs they may also be active in ADCC, whilst the purpose of FcRs on B lymphocytes is unclear but may be part of a regulatory pathway (Table 9.5).

9.4.2 Complement

The complement system consists of a series of some 20 proteins that come together in an ordered pattern to perform a precise function, namely to punch a hole in a cell membrane. The system has some similarities with coagulation, most importantly the presence of different sub-pathways, the

TABLE 9.5 Fc Receptors

FcR	CD	Gene locus	Characteristics	Distribution
FcγR1	CD64	1q21	Binds IgG1 and IgG3 principally, with high affinity. Some IgG4 binding	Monocytes, macrophages, neutrophils, mast cells
FcγRII	CD32	1q23	Low affinity for monomeric IgG, more reactive towards immune complexes. Three isoforms: FcγRII a, b, and c	Monocytes, neutrophils, macrophages, B cells, mast cells
FcγRIII	CD16	1q23	Two isoforms: FcγRIIIa and FcγRIIIb, present on different leukocytes	Neutrophils, eosinophils, NK cells, some monocytes
FcεR	CD23	19q13	Principle interaction with basophils and mast cells, leading to degranulation. Excessive activation leads to hypersensitivity. Two isoforms.	Mostly mast cells and basophils, some expression on eosinophils and monocytes
FcαRI	CD89	19q13.42	Binding promotes ADCC and degranulation	All myeloid cells
FcRn		19	Neonatal FcR: Resembles MHC class 1 molecules—transports IgG from mother to infant	Syncytiotrophoblasts, duodenum, small intestine, and leukocytes
PIgR	CD351	1	Polymeric IgA and IgM receptor, binding IgM with high affinity and IgA with low affinity	B cells, macrophages, mucosal epithelium

ADCC: antibody-dependent cellular cytotoxicity.

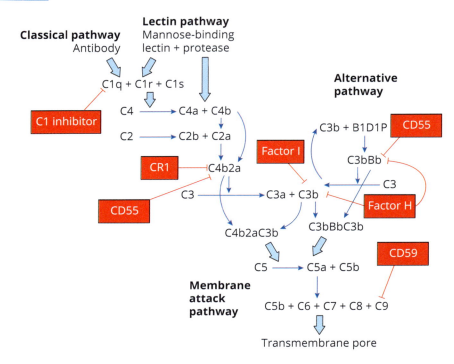

FIGURE 9.9

The complement pathway. The three generator pathways lead to the formation of C3 convertases that produce C3b. This combines with other components to form a C5 convertase to produce C5b that in turn leads to the formation of the membrane attack complex. Regulators are shown in rectangular red boxes.

amplification of the process, and the enzymatic activation of one molecule by another. The proteins that are part of the complement system are identified by the letter 'C' combined with a number or a name.

However, complement is not only about punching a hole in a membrane: small peptide breakdown products of the pathway can be functional. Some bind semi-specifically to pathogens such as bacteria, making them targets for phagocytes bearing complement receptors, and this enhances the phagocytosis. This is a further example of antibody-mediated opsonization, the complement components being described as opsonins.

There are three pathways for the generation of C3 convertase—the classical, lectin, and alternative pathways—and a fourth pathway known as the membrane attack pathway (Figure 9.9).

The classical pathway

In the classical pathway, three isoforms of C1 (C1q, C1r, and C1s) are activated by antibody–antigen complexes on the surface of cells and bacteria, and together bind to the antibody heavy chain, where they catalyse the formation of C4a and C4b from C4. Membrane-bound C4b provides the catalytic site for its substrate, C2, which is digested into C2a and C2b. The former combines with C4b to form a complex (C3 convertase) whose substrate is C3. This, the most abundant complement component (1-2 g/l in serum), is cleaved by the C4b2a complex into C3a and C3b, the latter forming a further complex with its 'parent', forming C4b2a3b. C3a is pro-inflammatory and induces the chemotaxis and activation of phagocytes. The final part is the cleavage of C5 by the C5 convertase supercomplex C4b2b3a to generate C5a and C5b.

The lectin pathway

This pathway also acts on C4, but in place of antibodies it is activated by mannose-binding lectin (MBL), which resembles C1q. MBL is associated with a serine protease, together acting on C4 in the same way as the classical pathway to generate C4a and C4b.

The alternative pathway

The alternative pathway is independent of antibodies and consists of three molecules, factors B, D, and P. It requires the cleavage of factor B into Ba and Bb. The pathway can be set in motion by C3b-binding bacteria and operates at a low level in the background to ensure that some C3 convertase is always available. A further supercomplex of C3bBbC3b is a second C5 convertase.

The membrane attack pathway

The two C5 convertase complexes (C4bC2aC3b and C3bBbC3b) act on C5 to generate C5a and C5b. Like C3a, C5a also has a variety of pro-inflammatory functions that include phagocyte chemoattraction, vasodilation, triggering the release of histamine by mast cells, stimulating an oxidative burst (with consumption of O_2) from neutrophils, and promoting cytokine release from hepatocytes and leukocytes. If these effects occur at an excessive level, they can be related to a series of pathological conditions (anaphylaxis), which explains why C3a and C5a are also described as anaphylatoxins (Chapter 10).

C5b functions as a key founder member of the last part of the pathway, which generates a membrane attack complex with factors C6 to C9, which form together in sequence. Their objective is to punch holes in cell membranes that are sufficiently large to facilitate the escape of cytoplasm from the cells and, inversely, the influx of tissue fluid. Both of these processes will lead to the death of the particular cell, which is referred to as complement-mediated lysis. However, a target cell may have methods for defending itself, such as cell-membrane components CD55 and CD59.

Regulators

Soluble regulators include C1-inhibitor (a serpin that binds C1r and C1s), factor I (a protease that degrades C3b to C3d, and C4b to C4d, cofactors for which include factor H), vitronectin, and carboxypeptidase N (which inactivates anaphylatoxins). The eukaryocyte cell membrane has its own defences against complement-mediated attack. Complement receptor 1 (CR1, also known as CD35, coded for by *CR1* at 1q32.2) accelerates the decay of both C4b2a and C3bBb. CD55, also known as decay-accelerating factor (coded for by *CD55*, also at 1q32.2), helps suppress C4b2a and C3bBb. CD59 (coded for by *CD59* at 11p13) blocks the incorporation of C9 into the membrane attack complex. The clinical consequences of the failure and/or absence of these regulators are discussed in Chapter 10.

9.4.3 Cytokines

Cytokines are small proteins that carry messages from cell to cell, and many have crucial roles in immunological and inflammatory responses. A number of families are recognized.

Interleukins

Originally thought to carry messages only between leukocytes (thus inter- and leuk-), interleukins (ILs) are known to also interact with, and can be produced by, several non-leukocytes, such as hepatocytes. The nature of the message is generally to activate the receiving cell via a specific receptor that will induce the cell to perform its particular function. Whilst around 40 ILs have been described, not all have been thoroughly characterized, and many share a common structure. Although all interleukins have important roles, several are prominent:

- IL-1 has two major isoforms (IL-1α and IL-1β, coded for at 2q12–14.2). Produced by activated macrophages and neutrophils, pro-inflammatory functions include promotion of the acute phase response, but it is also important in leukopoiesis.

- IL-2 (coded for at 4q27) is released mainly by activated CD4 T lymphocytes. It acts on other T cells, B cells and NK cells via specific IL-2 receptors to induce their further differentiation. IL-2 has been used in immunotherapy in certain malignancies.

- IL-3 (coded for at 5q31.1) is a haemopoietic growth factor, acting (with many other growth factors) to induce the differentiation of myeloid and other progenitor cells.

- IL-5 (coded for at 5q31.1) is a key mediator of eosinophil biology, both *in vivo* and in allergic diseases (discussed in Chapter 10).

- IL-6 (coded for at 7p15.3), as with IL-1, is produced by macrophages and, as a general pro-inflammatory cytokine, is linked to many aspects of the acute phase response. Production by adipocytes may explain why the obese have a low-grade inflammatory status with raised acute phase reactants such as CRP. Crucially, blockage of IL-6 is an effective therapy in rheumatoid arthritis (section 10.3.2).

Other cytokines

Secreted by macrophages and other cells, transforming growth factor-β (TGF-β), in common with IL-1 and IL-6, has a number of functions in the immune system, but also in other disease such as cancer. It may be involved in lymphocyte function and can induce apoptosis. Tumour necrosis factor-α (TNF-α), another pro-inflammatory cytokine, may be released by leukocytes, endothelial cells, and fibroblasts. It was discovered as a soluble substance that killed cancer cells *in vitro* and in animal models by apoptosis. Interferons (IFNs) were discovered by their role in anti-viral defence, the 20+ variants being classified by which receptors they bind. There are two principal forms. IFN-α, produced by plasmacytoid dendritic cells, acts on numerous cells to promote antiviral activity, such as increased macrophage, NK, and CD8 T cell cytolytic activity. IFN-γ is produced by NK and activated T cells, and acts on macrophages to contribute to defence from bacterial, protozoal, and viral infections.

SELF-CHECK 9.4

Which cells produce humoral immune factors?

Key points

Humoral immunity consists of antibodies, complement, and cytokines. The first two are directly involved in invader attacks whilst the latter promote and coordinate the defensive activity of leukocytes.

9.5 The innate immune system

Operating at numerous levels, a key component of the innate system is that it is relatively indiscriminate, lacking specificity. Its apparently crude effects buy time for a specific adaptive immune response to develop. It can also recruit defensive cells (of both innate and adaptive systems) to the site of an infection. The innate system can be divided into a number of independent but overlapping areas.

9.5.1 Physical defences

Our primary physical defence is the skin. Its barrier effect is supported by the effects of sweat, sebaceous secretions, and the shedding of keratinocytes. The nasal tract responds to perceived toxins by secreting mucus, which has anti-bacterial properties, supported by cilia to trap particles of dust and dirt. Similarly, tears protect the eyes, and over-salivation can help expel pathogens. Saliva also has antimicrobial properties. Oesophageal mucus may also have defensive properties, and although the stomach is strongly acidic, the presence of bacteria such as *Helicobacter pylori* shows that it is not entirely antiseptic. Indeed, the lower gastrointestinal system houses trillions of bacteria, although these may have invaded via the anus. The bronchus is also lined with cilia to trap airborne particles and pathogens. An acute infection is inevitably associated with a fever, as part of the acute phase response—presumably an attempt to stimulate the immune response and/or drive out the pathogen.

9.5.2 Humoral defences

Section 9.4.2 has introduced humoral innate defence—the lectin and alternative pathways of complement. Lysozyme (n-acetylmuramide glycanhydrolase) is a general enzyme found in many mucosal secretions with semi-specific activity against the cell wall of certain bacteria. Pentaxrins are a group of proteins characterized by five domains arranged in a circle, and levels rise markedly as part of the acute phase response:

- The principal pentaxrin (PTX1) is CRP, the major clinical laboratory marker of an acute phase response, coded for by *CRP* at 1q23.2, which generates a 25kDa protein. It is produced by the liver under direction of IL-6, and is named by its binding the capsular polysaccharide (hence C-) of bacteria such as *Streptococcus pneumoniae*.

- Serum amyloid P (SAP, PTX2, also coded for at 1q23.2) also binds debris and bacterial polysaccharides, but also apoptotic remnants, amyloid, and bacterial toxins, and in doing so is recognized by monocytes and macrophages, so promoting phagocytosis.

- PTX3, coded for at 3q25.32, is the product of monocytes and macrophages in response to cytokine signalling. Its roles include binding to complement component C1q and microorganisms such as the influenza virus, *E. coli*, and *aspergillus*, *pseudomonas*, and *shigella* species.

9.5.3 Cellular defence

The earliest step in haemopoiesis is the creation of two distinct lineages: that of the myeloid pathway and that of the lymphoid pathway (Figure 9.2). This distinction is evident in the innate response system and has repercussions for the adaptive response system.

Myeloid responses

Sentinels such as tissue macrophages of the skin, lungs, and mucosa, some of which are highly specialized (e.g. histiocytes and Langerhans cells), are first to interact with invaders and can call (with cytokines) for reinforcements, notably neutrophils. They are programmed to recognize and destroy any material considered to be foreign, generally by phagocytosis, but also by the release of their own toxic mediators to attack bacteria. These mediators include ROS, which include superoxide, the hydroxyl radical, etc., and enzymes (myeloperoxidase, lysozyme, etc.).

Central to the recognition of potential pathogens are **toll-like receptors (TLRs)**, some of which are expressed on the surface of dendritic cells, lymphocytes (mainly NK cells), monocytes, and macrophages, but also on epithelial cells, endothelial cells, and fibroblasts. Some TLRs focus on viral nucleic acids, and others on bacterial and fungal products such as lipopolysaccharide, zymosan, triacyl and diacyl lipopeptides, peptidoglycans, and flagellin. Once the TLR has bound its ligand, intracellular signalling results in pro-immunological and pro-inflammatory responses such as cytokine release.

Degranulation of tissue mast cells produces mediators that promote general defence, such as cytokines, to recruit other cells, and histamine and prostacyclin, which as vasodilators will aid the influx of other leukocytes. Once lymphocytes have been recruited, mast cell IL-4 helps induce the differentiation of naïve T_h cells. Once eosinophils are present, they too can degranulate and release many of the same mediators from mast cells. Cross-talk between humoral factors and myeloid cells includes enhanced phagocytosis of pathogens coated with complement components.

Innate lymphoid responses

Described briefly in section 9.3.3, innate lymphoid cells (ILCs) reside mainly within tissues. Three groups are recognized: ILC1 cells are involved in immune responses to tumour cells, and to viral and bacterial infections through the secretion of IFN-γ, TNF-α, and GM-CSF. ILC2 cells mediate responses

TABLE 9.6 Innate immunity

Aspect	Key features
Physical	Skin, cilia, mucus, sweat, tears, saliva, fever
Humoral	Lysozyme, gastric acid, defensins, alternative and lectin complement pathways
Cellular	Monocytes, macrophages, neutrophils, innate lymphoid cells, NK cells

to allergies and helminth infections and are an early source of IL-4, IL-5, and IL-13, recruiting eosinophils, basophils, and mast cells. ILC3 cells respond to monocyte/macrophage derived IL-23 and IL-1β by producing IL-17, IL-22, TNF-α, and GM-CSF that in turn act on other leukocytes.

NK cells

Unlike cytotoxic T lymphocytes, NK do not need specific instructions about which cells to kill—they lack a TcR, but instead express natural cytotoxicity receptors. The latter are a diverse and complex collection of transmembrane molecules, and include killer-cell immunoglobulin-like receptors (KIRs) (some being able to inhibit cytotoxic activity) and the NKG2 family (CD159, including NKG2D). Some of these receptors recognize invaders by their absence of self HLA molecules, others do so by presence of non-self HLA molecules (as in transplantation), and yet others by induced-self antigens. Indeed, the presence of self-HLA molecules inhibits NK activity.

The cytotoxic ability of NK cells is innate, and may be of use in recognizing and destroying cancer cells and virally infected cells, as certain NK cell subsets respond specifically to cytomegalovirus and the Epstein–Barr virus. NK cells are presumed to mature as do T cells but differ in that they do not express T cell markers (CD3, CD4, CD8), the most commonly used markers being CD16 (the IgG antibody FcR), CD30 (a TNF-α receptor), CD56 (neural cell adhesion molecule), CD91 (the apolipoprotein E receptor), and CD94 (a lectin receptor associated with NKG2). Some define NK cells as those bearing CD161. The presence of CD16, the FcγRIII, on the surface of NK cells brings the ability to kill targets bearing IgG and exposing its Fc segment, via the process of ADCC. Thus NK cells, like any cell bearing an FcR, may also be considered part of the adaptive immune system.

Although clearly important in certain circumstances, the clinical enumeration and functional assessment of NK cells is rarely called for, but if it is, then the presence of certain CD molecules will be sought. Some authorities refer to perhaps 10% of lymphocytes as large granular lymphocytes (LGLs), and this population includes many NK cells. Table 9.6 above summarizes innate immunity.

Key points

Innate immunity, consisting of physical, humoural, and cellular components, lacks much of the fine specificity of the adaptive immune system, but nonetheless provides excellent early defence against a variety of potential pathogens.

SELF-CHECK 9.5

How do non-T non-B lymphocytes recognize potential pathogens?

9.6 **The adaptive immune system**

Whilst the innate immune system is fighting a crude but efficient defence, behind the scenes the adaptive immune response is developing. A major part of this is processing the foreign pathogen (antigen) into small pieces that can be recognized by lymphocytes. This function is performed by APCs such as dendritic cells, Kupffer cells, and dermal Langerhans cells, and possibly by epithelial and endothelial

FIGURE 9.10

Map of the major histocompatibility complex. *The precise locations of E, F, G, DO, and DM are unclear. Each D locus set has a double bar representing A and B subregions, e.g. DPA and DPB.

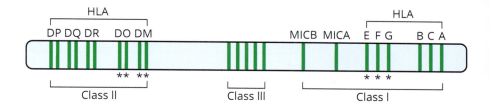

cells. Within the APC, the antigen is digested in proteasomes and fragments channelled to the endoplasmic reticulum where they form a complex with HLA molecules. This complex is then trafficked via the Golgi apparatus to the cell surface where it appears as altered-self.

The combination of HLA and a foreign antigen is presented to T and B cell receptors (TcR and BcR respectively). Once these lymphocytes have accepted the antigen, they progress to an activated and effector status to complete their respective functions in the pursuit and destruction of cells bearing that antigen, and the generation of antibodies specific for that antigen.

9.6.1 Human leucocyte antigens (HLA)

This complex series of molecules defines immunological self and, inversely, non-self. Pioneering work in transplantation pointed to a number of genes that induced a rapid and powerful rejection, and so the chromosomal region where they are located was designated the major histocompatibility complex (MHC). In our species the MHC consists of (at least) three major classes of molecules, coded for by genes at 6p21.3 (Figure 9.10).

Polymorphisms in immune response genes

Although we need almost all of our genes to be invariant (e.g. coding for two eyes, not three), the reverse is the case for genes involved in immune responses, which must be adaptable for the recognition of thousands of potential pathogens. Therefore it is positively desirable to have variability in our immune response genes that code for HLA molecules, antibodies, the TcR, and the BcR.

The structure of class I and class II HLA molecules resembles that of immunglobulins, where the protein chains form globular domains. This parallel extends to the presence of constant and variable regions, which are also present in the structure of the α, β, γ, and δ chains of the TcR and of the γ, μ, α, δ, ε, κ, and λ chains of BcR and the Fab of antibodies (discussed in section 9.6.2). The combination of a dozen or so different HLA genes confers a unique set of HLA molecules which is often described as an individual's HLA type, further described in section 9.8.5.

However, there is a major difference between the variability in antibody and T cell responses to antigens, and the variability in HLA molecules. Variability in the former is generally short-lived, can change, and involves differences in millions of different soluble molecules (antibodies) in the blood and on a small number of circulating cells (T lymphocytes). Conversely, the variability in HLA molecules is lifelong, stable, and unique to the individual, and whilst class I molecules are expressed on all nucleated cells of the body (and on platelets), class II ones are expressed only on activated lymphocytes and APCs (of whatever cellular background).

The HLA system is extremely polymorphic, which makes it reasonably likely that each person on the planet has a different HLA type. The genetically defined variations in these genes give rise to thousands of different HLA gene types coding for thousands of different proteins, with over 24,000 class I alleles and over 12,000 class II alleles (Table 9.7). The D locus is the most variable. Of the six major D locus genes, *DRB* is the most polymorphic (Table 9.8), and by far the greater part of the variation in DRB chains can be accounted for by isoforms of *DRB1* (3,094 alleles, 2,107 proteins, 103 null alleles) and *DRB3* (417, 310, 20 respectively). Practical aspects of HLA and COVID-19 are described in Box 9.1.

Cross reference

For more in-depth information on the relationship between HLA types and COVID-19, see the texts by Dobrijevic´ Z et al. (2023) and Blann AD and Heitmar R (2022), on which the information in Box 9.1 is based.

TABLE 9.7 HLA polymorphisms

HLA	Alleles	Proteins	Null alleles
A	7,354	4,302	369
B	8,756	5,287	302
C	7,307	4,042	313
D	12,790	8,331	547
E	298	118	7
F	48	7	0
G	94	30	5

Data Source: https://www.ebi.ac.uk/ipd/imgt/hla/about/statistics/. © EMBL 2022.

TABLE 9.8 HLA D locus polymorphisms

HLA	Major subtypes	Alleles	Proteins	Null alleles
DRA	DRA1	32	5	0
DRB	DRB1, DRB3, DRB4, DRB5	3,890	2,681	165
DQA	DQA1, DQA2	423	203	9
DQB	DQB1	2,193	1,386	96
DPA	DPA1, DPA2	378	158	10
DPB	DPB1, DPB2	1,915	1,198	101

Data Source: https://www.ebi.ac.uk/ipd/imgt/hla/about/statistics/. © EMBL 2022.

BOX 9.1 HLA and COVID-19

The millions of different HLA types ensure a neo-Darwinism spectrum of responses to a particular pathogen—at one end, some individuals will respond well (with no ill effects), while at the other, responses will be poor (with death possible, even certain). This has been reinforced in the COVID-19 pandemic of 2020–2, where a meta-analysis showed that, for example, the HLA-B*53 allele family was associated with a significantly increased risk of hospitalization with an odds ratio (OR) of 6.4 (95% confidence interval (CI) 2.25–18.2). Conversely, HLA-DRB1*11 was protective against hospitalization and clinical progression requiring intensive care with an OR (95% CI) of 0.51 (0.30–0.88). However, data of this nature can be criticized as they may not fully account for other risk factors such as age, obesity, and co-morbidities. Some associations may be valid in their host nations, such as HLA-DRB1*04:01 in the UK with severe disease, and HLA-A*03 and HLA-C*06 with death in Saudi Arabia. Other links have been reported in several locations: HLA-B*46:01 has been linked with a high risk of infection in 10 populations in China and Chinese populations nearby, and HLA-C*01:02 in different populations in Southeast Asia, Mexico, and New Zealand. HLA-A*02:02, HLA-B*15:03, and HLA-C*12.03 have been linked to a low risk of infection in many populations in Africa, Europe, and Asia.

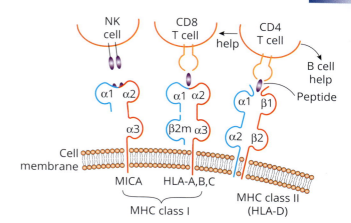

FIGURE 9.11

Structure/function aspects of HLA molecules. Class I molecules have three domains, class II molecules two. A single-domain-sized molecule of β-2-microglobulin (β2m) is associated with HLA-A, B, and C molecules. Class I and II molecules interact with T cell receptors (TcR). MICA resembles HLA-A/B/C molecules but does not associate with β2m.

Class I molecules

The physical structure of this group is of three globular domains, giving a combined relative molecular mass of 45 kDa. A crucial groove in the protein structure of the outer two domains enables the recognition of 8–10 amino acid peptides, which may be presented to a T cell or an NK cell (Figure 9.11). Three types of class I molecules are recognized:

- HLA-A, B, and C require the presence of a molecule of β-2-microglobulin (coded for by *B2M* at 15q21.1) to provide essential stability. Notably, it is neither anchored in the membrane nor attached covalently to the HLA molecule.

- HLA-E, F, and G are expressed on trophoblasts, implying roles in reproductive immunology, but there are other type-specific roles, such as with NK cells.

- MICA (around 388 alleles and 192 proteins) and MICB (237, 42), although not part of the HLA-system, are structurally similar to HLAs A–C, but do not associate with β-2-microglobulin, or bind peptide antigens. Described as stress-induced ligands, they may interact with NKG2B on NK and some T lymphocytes.

In addition, there are 12 pseudogenes (HLAs H–Y) that have around 147 alleles.

Class II molecules

The D locus is markedly more complex, coding for numerous genes that in turn code for thousands of different molecules at the cell surface. As with the components of the TcR, each HLA-D molecule is composed of two globular domains, and two molecules (α and β, of 31–4 kDa and 26–9 kDa respectively) form a dimer on the cell surface. The components of the three major groups are as follows:

- DPα and DPβ are coded for by *HLA-DPA* (with two isoforms, *DPA1* and *DPA2*) and by *HLA-DPB* (also with two isoforms *DPB1* and *DPB2*).

- Similarly, DQα and DQβ are coded for by *HLA-DQA* (with two isoforms *DQA1* and *DQA2*) and *HLA-DQB*, which has only one isoform—*DQB1*.

- DRα and DRβ are coded for by *HLA-DRA* (which has only one isoform—*DRA1*) and *HLA-DRB*.

The outer domains of the α and β chains come together to form a groove that can, much like the groove between the two outer domains of the class I molecules, bind a small peptide (Figure 9.11). Much less is known about HLA-DO and HLA-DM, although the evidence of roles for the latter in antigen presentation is growing. Although class II molecules are not linked to β-2-microglobulin, they are associated with an invariant CD74 molecule (coded for by *CD74* at 5q33.1), with a role in cytoplasmic transport, and so are described as the gamma chain. Individual HLA molecules are named according to an established method (Box 9.2).

BOX 9.2 Naming HLA genes

The highly polymorphic and complex nature of the whole HLA area demands a systematic naming procedure that must take account of the locus, gene, isotypes, allele group, and the specific protein product that is coded. For example, the meaning of HLA-DQB1*02:01 can be explained as follows:

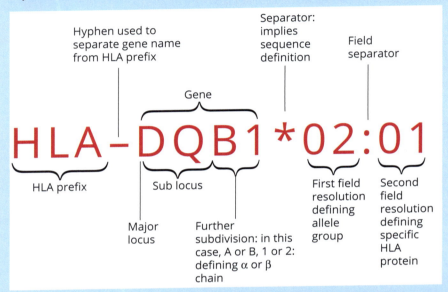

Hyphen used to separate gene name from HLA prefix

Separator: implies sequence definition

Field separator

Gene

HLA−DQB1*02:01

HLA prefix

Sub locus

First field resolution defining allele group

Second field resolution defining specific HLA protein

Major locus

Further subdivision: in this case, A or B, 1 or 2: defining α or β chain

However, this is an abbreviated version of a more complex full designation, which may be, for example, HLA-A*02:101:01:02N. In this system, the asterisk symbol (*) is used to distinguish a genetically defined type (e.g. HLA-A*02) from one that is (maybe historically) serologically defined (HLA-A2). Some place a 'w' after the C locus (e.g. HLA-Cw6) to differentiate it from a complement component. More information on naming conventions for HLA genes is available online at http://hla.alleles.org/nomenclature/naming.html.

Class III molecules

Class III genes do not code for HLA molecules, but for a variety of molecules, some of which are immunological, such as complement components C2, C4, and factor B, cytokines, and heat shock protein. Other genes include those for the receptor for advanced glycation end-products (possibly important in hyperglycaemia and diabetes), and *HFE*, coding for a molecule involved in iron metabolism (Chapter 11). Although there may be variations in the structure of class III molecules, they are rarely a major issue in immunopathology.

The function of HLA molecules

HLA-A, B and C molecules cooperate in the recognition of self and non-self, where non-self = self + peptide, the peptide deriving from the invading pathogen. In theory, an infinite variety of potentially foreign peptides in the domain cleft may be presented to the TcR, such as that of cytotoxic T lymphocytes bearing CD8, which is then licensed to attack those cells bearing the antigen, and to killer-cell immunoglobulin-like receptors. The importance of these molecules will be discussed shortly.

HLA-DP, DQ, and DR molecules cooperate in the recognition of self and non-self with T cells bearing CD4. These helper T lymphocytes are therefore primed to initiate the production of antibodies by B lymphocytes and to support the activity of cytotoxic T lymphocytes. As with CD8-bearing cells, CD4-bearing cells also co-associate with the TcR. However, there is a difference in peptide presentation in that the peptide groove is formed from a dimer of two HLA-D region monomers. In contrast to the wide expression of the major class II molecules on immunologically active cells, HLA-DO and DM have restricted expression, principally on antigen-presenting cells where they regulate and chaperone peptide processing.

Cross reference

Principal clinicopathological aspects of HLA in autoimmunity and transplantation are discussed in section 10.3.

9.6.2 Antigen recognition by lymphocytes

We have already discussed pathways by which NK cells and ILCs recognize foreign material. This subsection will focus on T and B cells.

T lymphocytes

The previous section has described the presentation of peptide antigens to T lymphocytes via the TcR. These structures consist of combinations of proteins coded for by genes *TRA*, *TRB*, *TRG*, and *TRD* located at 14q11.2, 7q34, 14q11.2, and 7p14 respectively, generating α, β, γ, and δ protein chains. The majority (90–95%) of T cells bear α and β chains, a minority γ and δ chains, hence $T_{\alpha\beta}$ and $T_{\gamma\delta}$. Each dimer forms a complex with, and is stabilized by, the six protein chains of CD3 (the γ, δ, and two ε chains on the external surface of the cell, and two ζ chains (CD247) on the cytoplasmic face).

B lymphocytes

The BcR for antigens is simply a monomeric antibody fixed into the cell membrane by the two terminal heavy chains. These can be any of the five classes (α, μ, γ, ε, and δ), linked to either of the two light chains (κ and λ), as discussed extensively in section 9.4.1. Just as CD3 supports the TcR, two copies of CD79 support the BcR. Figure 9.12 summarizes the structures of the two receptors.

Key points

Variation in genes coding for HLA molecules, the T cell receptor, and the B cell receptor (and thus antibodies) is the basis for the recognition of antigens.

SELF-CHECK 9.6

What the are similarities and differences between the T cell receptor and the B cell receptor?

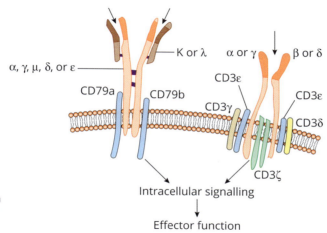

FIGURE 9.12

Structure of the BcR (left, essentially a membrane-bound antibody molecule) and TcR (right). Arrows indicate grooves for antigen binding. The darker shaded indicate variable regions.

9.6.3 Effector function of lymphocytes

We now move to examine processes leading to further activation of lymphocytes that lead to their final effector functions—killing cells they deem to be dangerous, and producing antibodies. To summarize previous sections:

- The MHC codes for HLA molecules that recognize peptide antigens.

- APCs of the innate immune system process alien material and offer peptide + HLA to lymphocytes.

- B and T lymphocytes express BcRs and TcRs to accept the presented antigens.

The complex association of the TcR, CD3, and HLA molecules ensures that T cells recognize antigens strictly in the context of self. Once a T cell is assured that whatever it has bound is non-self, the activation signal proceeds, and the cell starts those processes that ensure the destruction of the target cell bearing the particular antigen. The key step is the nature of the presentation of the antigen to the T lymphocyte. Presentation of the antigen as a part of an HLA class I or class II molecule to the TcR requires an additional signal by one of two other molecules: CD4 or CD8, which define different cell activity.

The CD4 T lymphocyte

The gene for CD4 at 12p13.31 codes for a further member of the immunoglobulin superfamily, and defines the T lymphocyte as a helper cell (T_h). The outer extracellular domain interacts with MHC class II molecules—HLA DP/DQ/DR—on the APC. These signals, the occupancy of the TcR by a peptide, and the recognition of an HLA class II molecule by CD4 together result in the activation of CD3ζ chains which initiate intracellular second messengers that ultimately move the cell to its effector function. This process is marked by the activation of a key intracellular enzyme, ZAP-70 (zeta-chain associated protein of 70 kDa), the absence of which results in a form of immunodeficiency.

The B lymphocyte

The activated T_h cell is now free to present the antigenic peptide to any B lymphocyte whose BcR recognizes that peptide. This interaction also requires the recognition by the CD40 molecule on the B cell with its ligand, CD40L, on the T cell, and very possibly other cytokine signalling. B cell activation also requires further signals from CD19, often described as a B cell coreceptor, and B-cell activating factor (BAFF). Once activated by binding to its antigen, and with additional help from a T cell, the B cell moves to its effector function via mobilization of second messengers (such as phosphoinositide-3-kinase) by the CD79 molecules. This culminates with clonal expansion and the 'industrial' production of antibodies, which are soluble versions of the BcR (but without the CD79). These antibodies recognize the same antigen that was presented to the B cell by the T_h cell. Negative regulators of BcR activation include CD22, CD72, and PECAM-1. During the process of antibody shedding, these proteins appear in large numbers in cytoplasm and on the cell surface, marking the lymphocyte as a plasmablast, which ultimately matures into a plasma cell.

This maturation process is accompanied by numerous changes in CD marker expression. Naïve B cells are CD19, CD20, CD45, and CD78 positive, but CD27 negative. However, once started on the pathway to antibody production, CD117 and CD27 appear, the latter being present in high levels on plasma cells. CD38 and CD126 (the IL6-receptor) appear, CD78 is upregulated, whilst expression of CD19, CD20, and CD45 falls as the cell matures. CD138 (syndecan-1) has also been described as a marker of antibody production.

The antibody response changes as the response to the antigen develops: there can be a switch of isotype. Individual B cells can modify the processed μ gene to a γ gene, generating an alternative mRNA so that an early IgM response can change to an IgG response. This processing generally retains the same variable regions so that the specificity for the antigen remains. However, there can also be variability in the affinity and avidity of subsequent antibodies so that the binding becomes stronger and more effective.

T helper cell subsets

The APC–T_h–B cell pathway described previously has long been established. However, this is only one type of T_h cell activity, described as T_h2, which focuses on B cells and is characterized by responsiveness to IL-4 and production of ILs 4, 5, 6, 9, and 10. In contrast, T_h1 cells, responsive to IL-12 and IFN-γ and secreting IFN-γ, TNF-β, IL-1, IL-2, and IL-10, are often involved in cell-mediated immunity, and are directed towards CD8-bearing cytotoxic cells, but also promote the cytotoxic activity of macrophages. T_h9 cells secrete IL-3, IL-9, and IL-27, whilst T_h17 cells are defined by their generation of IL-17, and in this role may be particularly active in defence from fungi and other extracellular pathogens, and in inflammatory responses in general.

CD4 T cells with regulatory function (T_{reg}) express markers Foxp3, CD25, CD28, CD45, and CD127, and receptors for IL-33 and TNF-α. With roles in the suppression of other immunological cells, they prevent autoimmune and excessive immunological activity, effectively limiting an inflammatory response. T_{reg} may also have a role in cancer immunotherapy, as cells with high expression of activation markers are often present within tumours, presumably suppressing host anti-tumour effector lymphocytes. There is evidence that the balance between T_{reg} and T_h17 cells is important in rheumatoid arthritis and other autoimmune disease (Chapter 10).

The CD8 T lymphocyte

The gene for CD8 at 2p12 codes for two 34 kDa globular molecules (α and β), both members of the immunoglobulin superfamily. Also present on NK cells and dendritic cells, CD8 defines the T lymphocyte as cytotoxic (T_c). The outer extracellular domain of CD8 interacts with MHC class I molecules—HLA A/B/C—on the APC.

The initial stages of the generation and activation of T_c lymphocytes have parallels with those of B cell activation. An APC processes potentially damaging molecules, perhaps from a virus, and places an antigenic peptide in the groove formed from parts of the outer two domains of an HLA class I molecule. This complex is recognized by the TcR, but a molecule of CD8 must recognize the self-aspect of the HLA molecule. As with the activation of T_h cells, cytokine signalling (such as IL-2) is important in generating T_c cells, and the activation of CD3ζ initiates the process of cell activation and final maturation. In further parallels with T_h cells, there are co-stimulatory signals between CD28 on the T cell and CD80 and/or CD86 on the APC that ensure correct signalling.

This process activates the T cell and transforms it into an effector cell capable of killing a target cell expressing the particular HLA molecule and the antigenic peptide. The activated T_c cell kills its targets directly by the release of cytotoxins such as perforin to punch holes in the cell membrane of the target. This leads, via various routes, to cell death, as we have already noted in the section on NK cells. Interestingly, perforin has structural similarities to C9, part of the final stage of the complement pathway that forms the membrane attack complex. An additional signal is that between CD95L (the Fas ligand; Fas) on the T cell and CD95 on the target cell. Crucially, the latter interaction may induce apoptosis. These processes are illustrated in Figure 9.13.

Memory lymphocytes

Reinfection with the same (or, as Jenner pioneered, as closely-related organism) pathogen induces far less serious responses, if any (box 3.1). We now know this is due to long-lasting B cell, CD4, and CD8 memory cells that can circulate or reside in various immunological (lymph nodes, liver) and non-immunological tissues. Memory T cells may also be important in protecting subsequent paternal-specific pregnancies (the immunology of pregnancy is discussed in Chapter 14, section 14.4.3). Different subgroups of memory cells may be identified by CD45RA, CD62L, CD127, CCR7, and others.

There is a developing view that there are NK memory cells, as exhibited by recall responses to cytomegalovirus, varicella-zoster, and other viruses, which appear to be long lived. This may be taken as further evidence that NK cells may be part of the adaptive immune system. A utopian view is that engineered antigen-specific memory cells will be used therapeutically.

Cross reference

Vaccination to counter COVID-19 is described in Chapter 13.

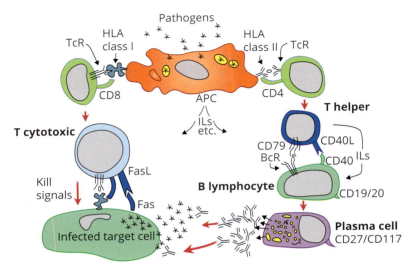

FIGURE 9.13

Lymphocyte effector function. Top central: an APC processes antigens and presents them in HLA class I or II molecules. On the left, class I molecules interact with the TcR and CD8 of a T lymphocyte, which is transformed into a cytotoxic cell primed to kill those cells bearing the class I/antigen complex. On the right, the APC presents a class II molecule bearing a peptide recognized by the TcR/CD3/CD4 complex on a naïve T cell, which transforms to a functional T helper cell. The latter instructs a naïve B cell bearing the BcR to transform into an antibody-producing plasma cell and produce antibodies to attack the antigen, be it soluble or cell-bound. Interactions are supported by cytokines and ligand–receptor binding.

BOX 9.3 *Manipulation of memory: vaccination*

The important procedure of vaccination relies entirely on the generation of memory cells following an encounter with an antigen. The technique of variolation, the immunization of susceptible individuals with tissue from an active smallpox (= *variola major* or *variola minor*) infection was brought to the UK by Mary Wortley Montague in 1721, and gained widespread acceptance, despite documented cases of full-blown and often fatal infections resulting directly from the procedure (one figure being a mortality rate of 2.5%) and major infections some time after the procedure.

In 1768, Surgeon John Fewster noted that, of two brothers who were variolated, one who previously contracted cowpox (*variolae vaccinae*, vacca being Latin for cow) failed to have an acute reaction, whereas the other brother, who had never suffered cowpox, did indeed have a reaction to the inoculation. He discussed this at a meeting attended by a young (age ~19) Edward Jenner. William Jesty, a farmer, noted in 1774 that two of his milkmaids (both of whom had previously been infected with cowpox) attended family members with smallpox, and did not to succumb to the disease. It is unclear whether he was aware of variolation, but he innoculated his wife and two sons with tissue from an infected cow, and all suffered minor inflammatory reactions. Fifteen years later, both boys were among a group that were variolated by a local surgeon—all the group had reactions, except the Jesty boys. In the 1780s, the observations of French cleric Jacques Rabaut-Pommier on the similarity of lesions caused by smallpox and by cowpox was passed via a merchant to a Dr Pugh, who duly passed them to his colleague Jenner. Thus Jenner was informed from several sources of the link between cowpox and smallpox several years before his famous inoculation of James Phipps. However, Jenner used material not from a cow (as did Jesty), but from the cowpox lesions of milkmaid Sarah Nelmes.

Vaccination has without doubt been a major triumph of immunology, and indeed of science itself. It has been used to eradicate smallpox and prevent millions of cases of other diseases caused by viral (hepatitis, yellow fever, rubella, polio, chickenpox) and bacterial (salmonella (causing typhoid fever), streptococcus, yersinia, mycobacterium (tuberculosis), tetanus, whooping cough) infections. In 1980 the global infection rate of measles was 945 per million, which by 2015 had fallen by 97% to 28 per million. However, due to widely publicised and potentially fraudulent bad science, so many parents are refusing to vaccinate their children that certain infections, principally measles, are becoming common again due to the loss of herd immunity, to the dismay and concern of many.

9.6.4 Other forms of antigen recognition

The explanations described previously must be tempered in the light of other forms of antigen recognition, some of which could be part of the innate system. For example, some monocytes, macrophages, and dendritic cells bear CD4, suggesting that they can provide T-independent help to B lymphocytes, and whilst CD19 is present on all B cells, it is also present on some lymph node dendritic cells. There are a number of situations where effector cells can respond to antigens in the absence of HLA and/or T_h cells.

$T\gamma\delta$ cells

Almost all of the T cell biology we have discussed has referred to $T\alpha\beta$ cells. Others, the $T\gamma\delta$ cells, share certain adaptive effector functions with their α/β cousins (such as VDJ gene rearrangement) and with NK cells, and early activation following recognition of stress-induced ligands does not always rely on the HLA system. Recognition of non-peptide phosphorylated antigens by the $\gamma\delta$ dimer is associated with many micropathogens (including *Mycobacterium tuberculosis* and HIV) and is a key feature of the innate immune system. In these circumstances, $T\gamma\delta$ cells rapidly secrete cytokines to induce other cells to participate in the local inflammatory response: they are a major initial source of IL-17. There is also evidence that they can adopt a cytotoxic profile, promote wound healing and tissue repair (like certain macrophages), and have a role in tumour cell killing.

Complement receptors (CR 1–4)

Several receptors for various complement components are recognized, some as mentioned in section 9.4.2. CR1 is widely present on white cells and red cells, and recognizes C3b and C4b. High expression of CR1 on myeloid cells implies that it assists phagocytosis of C3b/C4b-bearing targets. There is evidence that antibodies to certain antigens present on the surface of bacteria can be recognized via a second BcR, a complex of CD19, CD21, CD81, and CD225. The key molecule is CD21, which functions as CR2, recognizing complement components iC3b and C3d, products of C3 activation. Recognition of C3d on the surface of a bacterial or other potential target cell by CR2/CD21 is sufficient to induce the activation of lymphocytes via the CD79 dimer and its signalling properties. Notably, therefore, in this setting, antigen recognition/antibody generation is unusual in that it is independent of the HLA system and T cells.

CD11 has three isoforms that can combine with CD18. The CD11b/CD18 dimer is present on granulocytes (mostly neutrophils), macrophages, and NK cells, is markedly upregulated by IL-2, recognizes targets coated with iC3b and C3d, and is particularly active against *Mycobacterium tuberculosis*. CR4 is formed from CD11c and CD18, recognizes iC3b, and is also present on NK cells, leading to the concept of complement-dependent cell cytotoxicity.

Invariant NKT cells

Much of the discussion of the adaptive immune system has focused on peptide antigens. Invariant natural killer T cells are so named because they express CD161, a likely potential marker of classical NK cells, and a TcR whose α and β chains are drawn from a limited selection of germ line genes, hence invariant. This TcR is of interest because it recognizes self- or foreign lipids (such as those of mycobacteria) presented by CD1d on the surface of APCs and other cells.

SELF-CHECK 9.7

What are the minimum cell interactions required to produce an antibody?

> ## Key points
>
> As an immune response develops, recombination of genes for antibodies, the TcR, and the BcR generates increasingly specific and avid (and so effective) interactions with the pathogen.

9.7 Immunology and inflammation

It has been convenient to present the various parts of the jigsaw that is immunology and inflammation in separate sections. However, this is unsatisfactory as it fails to explain how the separate parts interact to form a recognizable picture. This will now be attempted.

9.7.1 The initial assault

Pathogens enter the body through various portals—breaches of the skin, the gastrointestinal system, the lungs, and the genitourinary system. Each of these parts of the body has sentry cells and systems ready to defend us—the innate immune system—and these generally perform well. However, should the invaders prove to be exceptionally virulent and/or in great numbers, and/or should the local innate system be impaired, the infection may be prolonged, requiring a sustained response.

9.7.2 Acute inflammation

When the innate immune system engages with an infection, a series of processes are initiated and messages (cytokines) sent to local and distant cells and organs, resulting in mobilization. The local environment will consequently be invaded with nearby resident sentinel leukocytes (macrophages, neutrophils, NK cells, and ILCs) and humoral factors that attack the invader. Release of inflammatory mediators from granulocytes contributes to local itching, increased temperature, irritation, redness, and swelling, much of which follows from increased blood flow. There may be complement activation, and if so, these contribute to white cell activity against the pathogens.

Acute inflammation is an entirely normal, short-lived, and healthy response to an invader (actual or perceived) or damage to tissues. The problem arises when the acute aspect becomes prolonged and/or misdirected, and so becomes chronic. When a particular organ or tissue becomes inflamed, this is expressed by adding the suffix -itis. Therefore, inflammation of the nephron is called nephritis or, in a more precise description, that of the glomerulus is glomerulonephritis, discussed in full in Chapter 10.

9.7.3 The acute phase response

Should local defences be at risk of failing to address the initial assault, then whole-body effects of the inflammation will follow, akin to an escalation of the conflict. This acute phase response includes a rise in blood pressure and heart rate, and may also affect other systems, depending on the extent of the infection. The acute phase response directs an increased production of a large number of plasma proteins with a wide variety of functions that are summarized in Table 9.9. Many of these plasma proteins are produced in the liver.

This response is likely driven by inflammatory cytokines such as interleukins (primarily IL-1 and IL-6) and TNF. The increased acute phase reactants are likely to contribute to an increased plasma viscosity and erythrocyte sedimentation rate (ESR), the latter being a very simple and quick marker. Conversely, levels of certain proteins including albumin and anti-thrombin are reduced. Many of these are directly defensive (such as complement components) and so clear functional members of the innate immune system, whilst others, such as CRP, are useful markers to define and monitor the acute phase.

The acute phase response also acts on the bone marrow, with the promotion of haemopoiesis resulting in an increase in the generation of all white cells, red cells, and platelets. Another consequence is that the neutrophil count and the total white cell count both rise as white cells that are

TABLE 9.9 Acute phase reactants

Reactant	Function
CRP	Leading serum marker of the acute phase, opsonin
Serum amyloid A	Unclear, but may part-regulate the acute phase reaction and/or promote leukocyte migration, induction of cytokines
C1s, C2, C3, C4, C9, factor B	Promote complement activity
Mannose binding lectin	Fixes complement, opsonin
Fibrinogen, prothrombin, factor XII	Promote coagulation
Ferritin, transferrin	Carriage and storage of iron
Haptoglobin	Carries free haemoglobin
Caeruloplasmin	Oxygen radical scavenger and carrier of copper
Orosomucoid/alpha-1-acid glycoprotein	Transport protein
Alpha-1-antitrypsin	Inhibits bacterial proteases
Alpha-1-antichymotrypsin	Inhibits bacterial proteases
Fibronectin	Cell–cell and cell–substratum adhesion

marginating—that is, temporarily located—on the wall of the blood vessel and those resident in the spleen are mobilized. This process leads to neutrophilia and leukocytosis.

All of these developments are generally desirable as they indicate that the body responds to the threat of—or an actual—microbial attack. However, if these responses are too powerful, and initial defences are overwhelmed, this may result in disease, as in septicaemia, which is also known as blood poisoning. Another potential complication is that if an acute inflammation persists for a long time, it may become transformed into chronic inflammation, which is an undesirable pathology. But even in the context of chronic inflammation, typified by rheumatoid arthritis (section 10.3.2), we still refer to the relevant markers as 'acute' phase reactants—and not as 'chronic' phase reactants.

The non-infective acute phase response

Over millions of years, survival programmes such as the acute phase response have evolved to defend us from micropathogens, which often infected via tears in the skin following escape from an encounter with a large carnivore. Fortunately, dangers from lions, tigers, and so on in the wild have been mostly eradicated, but the biological programme remains, and reappears after an encounter with the modern-day equivalent of a large carnivore that leaves tears in our skin—a surgeon. There are several consequences of this, such as an increased white blood cell count 24–48 hours after major surgery. This is indicative not necessarily of an infection, but of a healthy and active acute phase response to a presumed danger, namely the scalpel. However, a marked and prolonged increase in the white cell count is likely to be cause for concern.

A mild acute phase response is often present in cancer and reflects the body's response to an unspecified and potentially dangerous pathology. Similarly, a modestly raised CRP may be present in chronic and heavy cigarette smokers, wherein the tobacco smoke causes damage to the lungs, and quite possibly a local inflammation. The precise pathology explaining raised CRP in the obese is unclear, but this disease is linked with increased inflammatory cytokines, perhaps **adiponectin**, and adipocyte activity.

9.7.4 The developing adaptive response

Products of the initial encounter between the pathogen and the innate immune system can leave the original site via two routes.

The lymphatics

Tissue fluids drain via lymphatics to lymph nodes, and this is a route for pathogens, their degraded particles, and APCs bearing those antigens. Part-primed lymphocytes (possibly ILCs) may also pass down the lymphatics, and all meet in the lymph node to develop adaptive responses. APCs, T cells, and B cells all cooperate in generating antibodies, T_c cells are produced, and there will be generalized cytokine release to initiate other immune processes elsewhere. All these products leave the lymph node via efferent lymphatics and ultimately enter the blood, eventually reaching the site of the local infection and other organs such as the liver. An active lymph node will increase in size, and if on or near the surface of the body, can be noted. This increased size, lymphadenopathy, is reversible upon resolution of the immune response. However, as described in Chapter 11, lymphadenopathy is common in several forms of blood cancer.

Blood

The blood system itself may carry pathogens and part-primed cells into the general circulation, where the former will encounter circulating monocytes and neutrophils. Such blood will also perfuse the liver and spleen, where other sentinel cells are likely to be waiting, perhaps having been primed by pro-inflammatory cytokines.

9.7.5 The combined immune system in action

The fully functioning adaptive response generates antigen-specific antibodies and T_c lymphocytes which, together with the innate system, should clear the body of the pathogen. As the process develops, the precision of the antibodies and cytoxic effectors improves, and so the response becomes more efficient. Eventually, the combined systems overwhelm the invader, and recovery begins. The battleground is likely to bear scars, and so damaged cells and tissue will be removed by macrophages and neutrophils, and the tissues restored to their pre-infection state. Memory lymphocytes are long-term consequences of the infection, and respond rapidly should the same infection be repeated.

However, as with many conflicts, the battle may not always go our way. Should the invader overwhelm local innate defences, the adaptive response may not have had enough time to fully develop, and so a systemic infection may follow. This may develop into an attack on particular organs that may cause acute renal failure or acute pulmonary oedema, an extension of which is multi-system organ failure. Ultimately, there may be a life-threatening septicaemia or viraemia.

> **Cross reference**
>
> Clinical problems arising from an inefficent or aberrant immune system are discussed in Chapter 10.

SELF-CHECK 9.8

How does an immune response develop?

Key points

As an immune response develops, recombination of genes for antibodies, the TcR, and the BcR generates more specific and avid (and so effective) interactions with the pathogen.

9.8 The immunology laboratory

The laboratory is an indispensible part of the diagnosis and management of immunological and inflammatory disease, which can be divided into relatively discrete areas of immunodeficiency, hypersensitivity, and autoimmunity (Chapter 10).

9.8.1 Neutrophils

For leukocyte studies, blood must be taken into an anticoagulant, of which heparin is most often requested although EDTA may be acceptable. The practitioner must check with the laboratory, and analysis must be done as soon as possible after the blood has been drawn.

The leading disease of the failure of neutrophils to perform their function is chronic granulomatous disease (CGD), characterized by biochemical failure to generate a respiratory burst. The clinical consequences are recurrent bacterial infection, generally caused by the inability of the cell to kill these organisms. The molecular lesion can be in any one of several genes coding for enzymes generating the toxic ROSs (Box 9.4), the most common being in those for NADPH oxidase.

Molecular genetics are rarely called upon to confirm the diagnosis of CGD as good immunological and biochemical tests are available. In a common laboratory test, the normal granulocyte metabolizes nitro-blue tetrazolium (NBT), converting it to the insoluble blue product formazan. Those cells lacking the appropriate pathway fail to metabolize the NBT, so that no blue product is formed. For those laboratories with a flow cytometer, the dye dihydrorhodamine can be used as it is oxidized to rhodamine in normal cells. Consequently, failure of cells to generate rhodamine strongly suggests CGD.

The disease is tested for by challenging neutrophils with nitro-blue tetrazolium (which has a light yellow colour). In normal cells this chemical is acted on by ROSs to produce the blue product tetrazolium, easily identifiable by light microscopy. In CGD, there is no blue product, thus strongly indicating the diagnosis. Myeloperoxidase deficiency can also be determined with this test. An alternative

BOX 9.4 *Reactive oxygen species*

These key cytotoxic chemicals are produced by phagocytes as part of a respiratory burst involving enzymes such as nicotinamide dinucleotide phosphate (NADPH) oxidase, 5-lipooxygenase, and myeloperoxidase. A crucial pathway sees the generation of the superoxide anion from oxygen and the electron donor reduced NADPH:

$$2O_2 + NADPH \rightarrow 2O_2^- + NADP^+ + H^+$$

The superoxide radical participates in several other reactions that, alongside the chloride anion, generate hydrogen peroxide, the hydroxyl radical, and hypochlorous acid:

$$2H^+ + 2O_2^- \rightarrow O_2 + H_2O_2$$

$$O_2^- + O_2^- + 2H^+ \rightarrow H_2O_2 + O_2$$

$$Cl^- + H_2O_2 \rightarrow HOCl + H_2O$$

$$O_2^- + H_2O_2 \rightarrow OH + OH^- + O_2$$

Enzymes countering ROSs include superoxide dismutase and catalase. Loss-of-function mutations in genes for enzymes generating ROSs will lead to failure to generate these radicals (as in CGD). Similarly, gene mutations in anti-ROS enzyme malfunction and so permit the excessive activity of these cytotoxic chemicals.

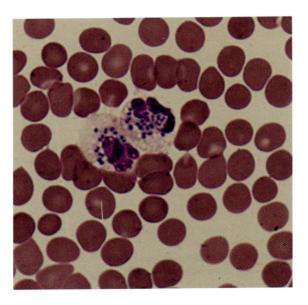

FIGURE 9.14
Phagocytosis of bacteria.
Numerous bacteria (blue dots)
within two phagocytes.

Source: van Kessel K. © 2022
Bacterial Infections and Immunity.
Retrieved from: https://www.evasio-
nutrecht.nl/.

test is to feed the cells with dihydrorhodamine and stimulate them with a chemical such as phorbol myristate acetate that in turn stimulates the production of enzymes to generate ROSs. If present, these oxygen groups oxidize the dihydrorhodamine to rhodamin in cells with normal function, which is also easily detected as it is fluorescent. Lack of fluorescence implies failure of the cell to generate ROSs, which effectively makes the diagnosis.

Clinical aspects of CGD are described in section 10.1.2. A practical but more complex biological test of phagocytosis is to feed live neutrophils or monocytes with yeast, which they should ingest. A blood film can be made and, using a light microscope, the yeast can be seen inside the neutrophils, just as bacteria can be seen within a phagocytic monocyte (Figure 9.14).

9.8.2 Lymphocytes

Almost all lymphocyte work is of defining subsets. Fluoresence flow cytometry (FFC) is the gold standard, often used to determine total T cells (CD3) and B lymphocytes (CD19 and CD20), and their ratio, and for the T cell subsets of helper cells and cytotoxic cells (CD4 and CD8 respectively). An absolute CD4 count is important in monitoring HIV infection and, ideally, a successful response to treatment. A positive HIV result demands FFC for CD4 and CD8, although a low lymphocyte count itself may trigger a request for T lymphocyte subsets which, if acted upon rapidly, could be performed on the same blood sample. A common result is low absolute numbers of CD4-positive cells, such as 325 cells/mm^3 (reference range 500–1,200). Should the CD8 count be normal, this will give a low CD4/CD8 ratio (such as 0.8; reference range generally 1.3–3.0). This is supportive of the diagnosis of an HIV infection (section 10.1.1), and possibly the early stages of the acquired immunodeficiency syndrome (AIDS). It may also prompt treatment with antiretroviral drugs (Figure 9.15).

Functional activity of lymphocytes can be assessed by the proliferative responses to stimulants such as pokeweed mitogen, phytohaemagglutinin, concanavalin A, purified protein derivative, tetanus toxin, and candida antigens. This testing requires considerable technical skill and is only performed in a reference laboratory.

9.8.3 Serology

The majority of the work of most routine immunology laboratories is serology and can be described in a precise number of areas: immunoglobulins, cryoglobulins, defined proteins, antibodies to defined antigens, and autoantibodies.

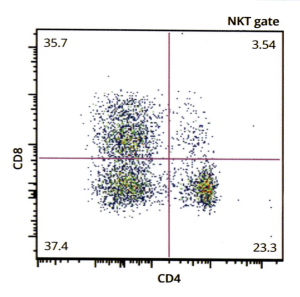

FIGURE 9.15

Fluorescence flow cytometry of CD4 and CD8 subsets. A typical four-quadrant result of CD3-positive cells, showing the number of cells positive for CD8 only (top left, 35.7% of all cells), for both CD4 and CD8 (top right, 3.54%), for neither CD4 nor CD8 (bottom left, 37.4%), and those positive for CD4 only (23.3%). Thus the CD4/CD8 ratio is 0.65. Notably, well over a third of all these T cells (as defined by CD3) fail to express either marker, so they are likely to include some NK cells that also bear CD3. The exact positions of the purple horizontal and vertical bars are defined according to other criteria to include/exclude certain groups of cells.

Department of Microbiology, Boston University, School of Medicine. Retrieved from: https://www.bumc.bu.edu/microbiology/research/.

Immunoglobulins

A key measurement of immunocompetence (and of certain malignancies) is of total immunoglobulin and the five isotypes. It is likely that total immunoglobulins will be measured by electrophoresis and densitometry, perhaps in the biochemistry or immunology section of the laboratory. Electrophoresis relies on the electrical properties of plasma proteins. A small amount of serum is absorbed onto a physical carrier (a gel or a membrane), or perhaps capillary zone electrophoresis, and an electric charge is applied. Each protein group in the serum sample migrates according to the sum of the electrical properties of its constituent amino acids. The resultant carrier can be identified with stains such as coomassie blue, amido black, and acid violet, which will allow a visualization of how the proteins have separated (Figure 9.16).

There are several major bands: from the top these are albumin (which also includes the lipoproteins), α_1-globulins (including, α_1-antitrypsin, thyroglobulin binding protein, transcortin, α_1-acid glycoprotein), α_2-globulins (α_2-macroglobulin, haptoglobulin, caeruloplasmin), β-globulins (transferrin, haemopexin, complement components 3 and 4, CRP), and, at the bottom. The somewhat diffuse band of γ-globulins, almost all of which are the immunoglobulins. Albumin is clearly the most heavily stained band, reflecting the high levels found in the blood, and this is reflected in the large peak on the left of the printout. Since the overall concentration of serum proteins can be determined (usually in the biochemistry lab), the amount of protein in each of the bands can be also determined. Many analysers automatically give this information, in the total protein results (Table 9.10).

The proportions and so concentration of the different fractions alter in certain conditions. In the acute phase response, and more so in frank infections, there will be increases in the α_1-globulin band, with normal, decreased, or increased β-globulins, and a broad increase in the γ-globulin band due to increased immunoglobulin antibodies. However, the most practical and powerful use of this technique is in the diagnosis of certain blood cancers, principally myeloma (section 11.4.4).

A further refinement of electrophoresis is to define the isotype nature of an abnormal immunoglobulin band, such as in a **gammopathy**, with immunofixation. Here, the serum sample is loaded onto six channels of an electrophoresis gel, which proceeds as normal. Once complete, antisera (perhaps of rabbit origin) specific for the major heavy and light chain classes are overloaded onto one of five of the channels. After incubation, excess proteins are washed out and the gel is stained (Figure 9.18). The binding of the specific antiserum with the separated proteins in the channels (if present) will stain with greater density, allowing identification. The method is essentially qualitative—the amount of protein stained is of no value, so densitometry is not required. Protein electrophoresis can also be performed on urine and on cerebro-spinal fluid. However, as regards the former, whole antibodies are not found

FIGURE 9.16

Normal serum protein electrophoresis.

From Hall, Scott, & Buckland, *Clinical Immunology*, second edition, Oxford University Press, 2016.

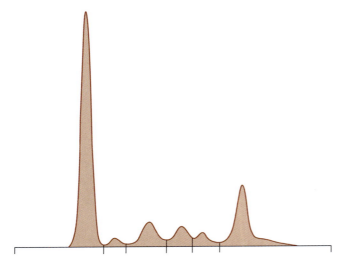

Serum protein electrophoresis

FIGURE 9.17

Densitometer analysis. An example of a densitometer scan. The large peak on the left is albumin; on the right is the gammaglobulin peak.

From Hall, Scott, & Buckland, *Clinical Immunology*, second edition, Oxford University Press, 2016.

FIGURE 9.18

Normal serum immunofixation. From left to right: total serum, G, A, M, κ, λ.

From Hall, Scott, & Buckland, *Clinical Immunology*, second edition, Oxford University Press, 2016.

TABLE 9.10 Quantification of serum protein by electrophoresis and densitometry

Fraction	%	Concentration* (g/l)	Reference range (g/l)
Albumin	65.1	45.6	35.0–50.0
Alpha-1	2.0	1.4	0.9–1.7
Alpha-2	9.1	6.4	5.2–7.0
Beta-1	6.9	4.8	4.1–6.0
Beta-2	4.1	2.9	1.3–3.3
Gamma	12.8	9.0	7.3–22.0

* Based on a total protein count of 70 g/l (54–90 g/l)

in urine, but there may be free light chains secreted by a gammopathy, as these molecules are small enough (22–24 kDa) to pass through the filter of the glomerulus. These urinary light chains are known as Bence-Jones protein. If there is glomerular damage, albumin may also be present in urine.

Immunoglobulin subgroups

The individual levels of the five heavy chain isotypes can also be quantified, but this is not by electrophoresis but by immunassay (IgG, IgA, and IgM perhaps using nephelometry and turbidimetry, IgD and IgE usually by ELISA), as will be levels of the four subtypes of IgG and two subtypes of IgA (section 9.4.1). Expected results for the five isotypes are shown in Table 9.11.

Cryoglobulins

Cryoglobulins are immunoglobulins that precipitate out of solution when below body temperature. The exact temperature varies, but as a far as the patient is concerned, it happens in the cold, generally in the winter months. The consequences are pain and parasthesia in the fingers, toes, and any

TABLE 9.11   Levels of immunoglobulins

Isotype	Adult reference range*
IgG	6.0–16.0 g/l
IgA	0.8–4.0 g/l
IgM	0.5–2.0 g/l
IgD	0–100 kU/l
IgE	0–81 kU/l

* Levels in those aged <18 years are likely to need age-band-specific ranges. From Hall, Scott, & Buckland, *Clinical Immunology*, second edition, Oxford University Press, 2016.

extremity exposed to low temperatures. The likely aetiology is the deposition of cryoglobulin in arterioles, capillaries, and venules, causing occlusion and so ischaemia. It is common in Raynaud's syndrome, although the latter may also be caused by arterial spasm and not necessarily by cryoglobulins.

In investigating cyroglobulins, the key aspect is to keep all the venepuncture equipment and vacutainer at 37°C, and keep it warm during transfer to the laboratory, where a warm centrifuge will be needed to separate the plasma or serum. Following separation, one aliquot is kept at 37°C and another at 4°C. The two are checked daily for the appearance of precipitate or turbidity in the latter. Once formed, cryoprecipitate becomes soluble on warming (Figure 9.19).

FIGURE 9.19
Cryoprecipitate, (a) at 4°C and (b) at 37°C.

Krishnaram AS, Geetha TP, Saigal A. Primary cryoglobulinemia with cutaneous features. © Indian J Dermatol Venereol Leprol 2013: 79; 427–30.

TABLE 9.12 Defined serum protein analyses

Test	Indication/comment (reference range)
Beta-2-microglobulin	Levels raised in renal failure and several B cell tumours, including myeloma (1.2–2.4 mg/l)
C1-inhibitor	Low levels in many cases of hereditary and acquired angioedema (0.15–0.35 g/l). A functional assay is possible (>70% normal)
Complement components C3 and C4	Low levels impy consumption (C3 0.75–1.75 g/l, C4 0.14–0.54 g/l)
Mast cell tryptase	Enzyme released from mast cells. High levels imply an ongoing allergic reaction or mastocytosis (2–14 µg/ml)
C-reactive protein	Leading marker of an acute phase response and actual infection (dependent on method, e.g. <5 mg/l, although high-sensitivity testing may give accurate results down to 1 mg/l)

The exact nature of the cryoglobulin can be determined and so classified. Type 1 (comprising 10% of all cryoglobulins) forms within hours, and is of a monoclonal gammopathy. Most are IgM and others IgG, whilst IgA is rare and may arise from a myeloma or related disease. Type 2 is the most common form (60% of cases) and a mixture of immunoglobulins, but often with a single monoclonal species, such as an IgM-κ whose antigen is on the Fc section of an IgG molecule. Type 3 (accounting for 30% of cases) is often seen in bacterial infections and autoimmune diseases, and is a mixture of polyclonal IgG and IgM antibodies.

As immune complexes, cryoglobulins can become deposited in the capillary beds, causing a type III hypersensitivity reaction (explained in Chapter 10, section 10.2.3). This can manifest itself as vasculitis with skin rashes and glomerulonephritis.

Defined protein analysis

Many other requests are for levels of defined plasma proteins (Table 9.12), where methods include enzyme immunoassay, ELISA, turbidimetry, and nephelometry.

C3 and C4 can be measured by standard immunoassays such as nephelometry and ELISA, and so can be measured in most routine laboratories. However, there is considerably more to complement than these two molecules. The CH50 (complement haemolysis 50) method is a bioassay and relies on the ability of the complement pathway to lyse 50% of a suspension of a target such as a sheep red blood cell that has been coated with antibodies. It is best performed in a microtitre plate, which allows a titration of the patient's sample against a standard suspension of red cells, the quantified end point being haemoglobin released from lysed red cells. A poor CH50 result may be due to the lack of one or more of the factors that constitute the pathway.

Similarly, the AP50 (alternative pathway 50) method tests the components that make up the alternative pathway. A common method uses rabbit red cells as these are extremely susceptible to non-antibody-dependent haemolysis. The end point, as in CH50, is the release of haemoglobin, which is easily assessed by a colourimeter. CH50 and AP50 are quantified in arbitrary units or per cent, and as with bioassay, extra attention must be paid to conditions, such as buffers and temperature, which must be standardized, and run with positive and negative controls. C3, C4, CH50, and AP50 testing can be used to determine whether the classical or alternative pathway is activated and whether this is reflective of an acute phase response.

Antibodies

The serology laboratory can also detect and quantify antibodies to defined non-self antigens (Table 9.13) and autoantibodies (Table 9.14). Key methods for determining these antibodies include indirect immunofluoresence (IIF) on cells and tissues captured on glass or plastic slides, enzyme-linked

TABLE 9.13 Antibodies to defined non-self antigens

Test	Indication
Aspergillus antibodies	Positive in allergic broncopulmonary aspergillosis (farmer's lung). Often IgE and IgG antibodies
Avian antibodies	May be present in extrinsic allergic alveolitis (bird-fancier's lung)
Allergy testing	IgE antibodies against peanut, hazelnut, egg, bee and wasp venom, latex, wheat, etc. (see section 9.8.4)
Antibacterial antibodies	To pneumococcal antigens, *Haemophilus influenzae* type b, tetanus toxin, and pneumococcal serotypes
Antiviral antibodies	To hepatitis viruses, influenza virus, HIV, cytomegalovirus

immunosorbent assays (ELISA), fluorescence-linked immunoassay, chemiluminescence, and latex immunoassay. Developing techniques such as robotics and multiplex technology are proving to be useful, if not essential, tools. The investigations outlined in Table 9.13 focus on antibodies to infective and possibly pathological antigens. The objective of such an investigation may be to determine a current or recent infection, or if a vaccination has been successful. Fine details of IgE antibodies and their role in allergy are presented in section 9.8.4 and in Chapter 10.

Autoantibodies

Autoantibodies may be classified as directed towards molecular, cytoplasmic, cell, and tissue antigens (Table 9.14), or towards components of the nucleus—that is, ANAs (Table 9.15)—and are found in many diverse autoimmune conditions.

ANCAs are detected on slides coated with neutrophils, which also allow determination of the pattern (Figure 9.20). Anti-mitochondrial antibodies may be detected on a liver cancer cell line called Hep G2, or on sections of a composite of liver, kidney, and stomach tissues (Figure 9.21). ANAs have a particularly complex nature, with several subtypes, being present in many inflammatory connective tissue diseases, principally systemic lupus erythematosus (SLE), Sjögren's syndrome, and systemic sclerosis (Table 9.15). The descriptor ANA itself can be a misnomer as nuclear antigens can be visualized by microscopy in the cytoplasm, to where they are presumed to have migrated, having leaked out of the nucleus or having been transported.

ANAs come in several different patterns, and further refinements of the ANA technique can determine with more precision links between the pattern and the exact antigen(s) in the nucleus (which may be the centromere, histones, riboproteins, or enzymes) and certain pathological situations. Examples of these patterns include coarse-speckled (binding to U1RNP and Sm, as found principally in SLE) and fine-speckled (binding to Ro (SS-A) and La (SS-B), principally in Sjögren's syndrome, less so in SLE) (Figure 9.22).

Historically, ANAs were defined on tissue section, but this has mostly given way to HepG2 cells. In general, most autoantibodies are reported as positive/negative, or with a titre, the latter often being used as an indicator of disease severity. Indeed, some ANAs (e.g. anti-dsDNA and SLE) are highly specific and are used to follow the effect of treatments. dsDNA antibodies are often first detected by their binding to the kinetoplast of *Crithidia luciliae*, a unicellular flagellate parasite of the common house fly (Figure 9.23). In some cases, methods involve defined recombinant antigens, some expressed by cells, giving exquisite specificity.

9.8.4 Allergy testing

The aetiology, cell biology, and clinical aspects of allergy and immediate (type 1) hypersensitivity are described in section 10.2.1, and as a generic is often described as atopy, the clinical consequence an atopic reaction. The gold-standard determination of allergy is to deliberately expose the patient to

TABLE 9.14 Autoantibodies to molecular, cytoplasmic, cell, and tissue antigens

Test	Indication	Method
Anti-neutrophil cytoplasmic antibodies (ANCAs)	Reported as positive or negative with three patterns: C, P and atypical (see below)	IIF on neutrophils
Anti-cardiolipin antibodies	Frequently found in SLE and thrombophilia. Can be IgG and/or IgM. Linked to anti-phospholipid antibodies	ELISA
Anti-cyclic citrullinated peptide antibodies	More specific for rheumatoid arthritis (RA) than rheumatoid factor	Enzyme immunoassay
Endomysial antibodies	Performed as a follow-up if IgA tissue transaminase antibody is positive	IIF on monkey oesophagus tissue
Epidermal antibodies	Antibodies to desmosomes or basement membrane differentiate pemphigus and pemphigoid	IIF on monkey oesophagus tissue
Gastric parietal cell antibodies	Linked to pernicious anaemia	IIF on rat liver, kidney and stomach
Glomerular basement membrane antibodies	Positive in Goodpasteur's syndrome	Enzyme immunoassay
Glutamic acid decarboxylase (GAD-65) antibodies	Can differentiate different types of diabetes: also present in stiff person syndrome	ELISA
Intrinsic factor antibodies	Present in most cases of pernicious anaemia	ELISA
Mitochondrial antibodies	Present in most cases of primary biliary cirrhosis	IIF on Hep G2 cells, rat liver, kidney and stomach
Myeloperoxidase antibodies (p-ANCA)	Present in certain inflammatory connective tissue disease: microscopic polyangiitis/eosinophilic granulomatosis with polyangiitis	ELISA, IIF
Proteinase 3 antibodies (c-ANCA)	Present in granulomatosis with polyangiitis	ELISA, IIF
Rheumatoid factor	Present in 80% of cases of RA	Nephelometry
Smooth muscle antibodies	Present in 75% of cases of auto-immune type 1 (chronic active) hepatitis	IIF on rat liver, kidney, and stomach
Thyroid peroxidase antibodies	Present in autoimmune thyroid disease	ELISA
Tissue transglutaminase (TTG) antibodies	TTG is the major antigen in coeliac disease	ELISA

ELISA = enzyme-linked immunosorbent assay, IIF = indirect immunofluoresence. Methods include quantifiable ELISAs and IIF, the latter often being on sections of rat and monkey liver, kidney, oesophagus, skin, and other tissues.

TABLE 9.15 Nuclear autoantibodies

Test		Indication	Method
Anti-nuclear antibodies (ANAs)		Reported as positive or negative, and with titre and pattern	IIF on Hep2 cells
Antibodies to single- or double-strand DNA		Positive or negative by screen, then numerical result by ELISA. Almost exclusively found in SLE	IIF on crithidia, then ELISA
Anti-centromere antibodies		Usually found in systemic sclerosis, dependent on clinical features. Possibly primary biliary sclerosis	IIF on Hep2 cells
Extractable nuclear antigens (ENAs)	Ro (SSA)	Sjögren's syndrome, SLE, congenital heart block, neonatal lupus, RA, primary biliary cirrhosis	ELISA
	La (SSB)	Sjögren's syndrome, SLE, RA	
	Ribonucleoprotein (RNP)	Mixed connective tissue disease, SLE	
	Sm (RNA-binding proteins)	SLE	
	Scl-70 (topoisomerase)	Systemic sclerosis	
	Jo-1	Myositis, Raynaud's disease Interstitial lung disease, arthritis	

ELISA = enzyme linked immunosorbent assay, IIF = indirect immunofluoresence.

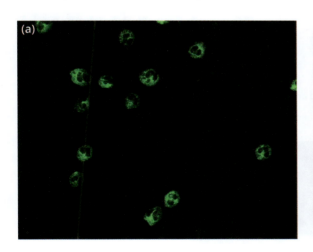
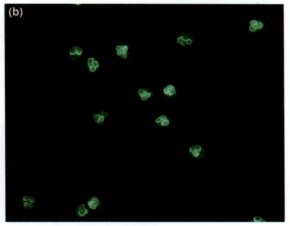

FIGURE 9.20

ANCA patterns by indirect immunofluoresence. (a) Cytoplasmic c-ANCA and (b) peripheral p-ANCA staining.

From Hall, Scott, & Buckland, *Clinical Immunology*, second edition, Oxford University Press, 2016.

the presumed allergen in a controlled setting, observing the result. In some cases the result may be available within minutes, whereas other reactions may take up to 48 hours to develop. The skin prick test pierces the skin to allow the potential allergen access to the subdermal tissues. Patch testing is supradermal—the potential allergen remains at the surface—and is useful in investigating contact dermatitis.

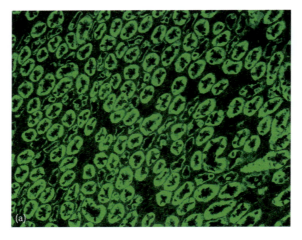

FIGURE 9.21
Anti-mitochondrial antibodies. High-intensity staining of rat kidney, positive at a titre of 1/10,240.

From Hall, Scott, & Buckland, *Clinical Immunology*, second edition, Oxford University Press, 2016.

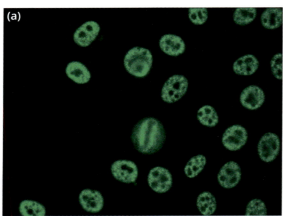

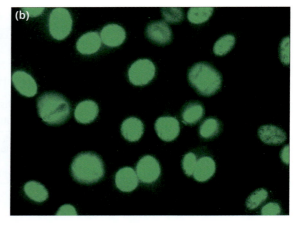

FIGURE 9.22
ANA patterns. (a) Coarse-speckled and (b) fine-speckled ANA patterns.

From Hall, Scott, & Buckland, *Clinical Immunology*, second edition, Oxford University Press, 2016.

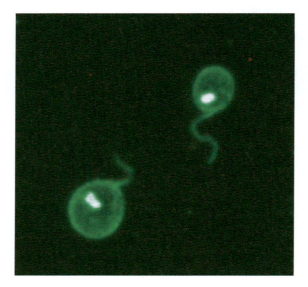

FIGURE 9.23
dsDNA antibodies detected in *Crithidea*. Antibodies have bound to the kinetoplast, a 'granule' of DNA.

From Hall, Scott, & Buckland, *Clinical Immunology*, second edition, Oxford University Press, 2016.

In testing for asthma, a formal clinical lung function setting is required. Challenge tests expose the patient to increasingly high doses of the allergen, and for all investigations a negative control (e.g. saline) is essential.

IgE

The overall level of serum IgE (perhaps by ELISA or chemiluminescence) and allergen-specific IgE can be provided. The latter can identify reactions towards certain pollens, fruit, egg, nuts etc., and so help determine which are to be avoided. Using micro-arrays, simultaneous testing for many allergens using a small sample of plasma is possible. When linked to a skin prick test, an IgE result of <0.35 kUA/l is regarded as negative, 0.35–0.70 as weak positive, 0.71–17.5 positive, and >17.5 as strong positive. Although focus is on IgE, there are IgG antibodies that cause hypersensitivity reactions, often manifesting with pulmonary symptoms, and may ultimately be diagnosed as farmer's lung, allergic alveolitis, and pigeon-fancier's lung.

Two systems offer multiplex allergen testing (NICE DG24). The ImmunoCAP chip can test for specific IgE antibodies to 112 components from 51 allergen sources simultaneously, whilst the microtest measures IgE antibodies to 22 allergen extracts and 4 allergen components simultaneously.

Basophils

The clinical consequences of an allergic reaction are caused by the micro-products of basophil and mast cell degranulation, itself caused by cross-linking cell-bound IgE. Consequently, an alternative approach is to test basophil reactivity towards putative allergens. When stimulated with an appropriate allergen (such as house dust mite allergen), the resting basophil upregulates expression of CD203c, and CD63 appears at the cell surface, and both can be quantified by fluorescence flow cytometry. As CD63 is a component of the membrane of intracullar granules, expression is proof of degranulation. Alternative markers include CD107a, CD107b, and CD164 (although the latter is also present on CD34-positive progenitor cells).

Quality control is essential, with unrelated allergens and a positive and negative control, and a sample from a non-allergic subject in parallel. Antibodies to chemokine receptor type 3 (CCR3) are also added to the sample, it being a stable marker for identification, whilst IL-3 can help enhance any reaction. Although a very powerful technique, the basophil activation test relies on blood cells and tells us nothing of mast cells in the tissues, and so may not be of value in assessing tissue mastocytosis.

Products of degranulation

The products of basophil and mast cell degranulation—histamine, heparin, cytokines, leukotienes, and prostaglandins—are difficult to measure. One that is stable and measurable is mast cell tryptase (coded for by *TPSAB1* at 16p13.3), which makes up >20% of the cell's total protein content. There are two isoforms, α and β, the latter being the dominant released variant. Measurement may be of use in two situations: in an acute allergic setting, as long as there is a reasonably firm knowledge of the time of the attack, and in a controlled setting, where the subject is deliberately challenged, to determine the effect of a known allergen.

There is a strong dose/response effect of allergens on tryptase release, so the timing of peak levels varies with the individual and dose. In one controlled test, baseline levels of the enzyme at ~40 ng/ml rose to ~70 ng/ml two hours after a challenge with 0.3 μg bee venom, returning to 43 ng/ml 16 hours after the challenge. As resting population levels generally lie between 35 and 105 ng/ml (others citing 2–14 μg/l), a second sample taken 24 hours after a presumed allergic attack will be needed. The half life of the enzyme is ~2–3 hours. One definition of a positive result is that the peak result must exceed the base/24hr result by >20% + 2 ng/ml (as it has above). Thus, in an acute setting, a peak result of 100 ng/ml (seemingly high) would not qualify if the 24 hour sample is 88 ng/ml—a peak result >108 ng/ml would have been positive. Other measures include urine methylhistamine (a stable histamine metabolite).

Although allergy testing focuses on basophils, mast cells, and IgE, there is also a role for IgG and the eosinophil. The granules of this cell contain cytotoxic proteins, enzymes, and leukotrienes, but also eosophil cationic protein. This molecule, which can stimulate the release of histamine from basophils and mast cells, may be measurable by fluorescence immunoassay, and may be of use as a marker of airways inflammation in patients with asthma. Allergen-specific IgGs may be present in chronic exposure to inhaled pathogens such as may be in mould and avian products, potentially leading to farmer's lung and bird-fancier's lung respectively.

9.8.5 Tissue typing, HLA, and transplantation

As described in section 9.6.1, the HLA system is a crucial part of our immune system. There are a small number of instances where the precise make-up of an individual's HLA type is required: in order to determine the potential response to a particular drug, in investigating autoimmune disease (section 10.3.6, such as HLA-B27 and ankylosing spondylitis) and in transplantation (section 10.4). The process of determining an HLA type for the purposes of transplantation (tissue typing) is performed only in dedicated laboratories, generally linked to regional transplant centres, although advances in technology are bringing some of this testing into the reference laboratory.

Serological typing for HLA type

Early in the development of tissue typing, it was noted that the serum of women multiparous to the same father often produced antibodies to his HLA type, so she could be a source of typing antisera analogous to ABO and Rh blood group typing. Rabbits immunized with leukocytes of known HLA type, and later with purified HLA molecules, were also used to raise polyclonal antisera. This approach proved a double-edged sword as the specificities of these antisera could be highly complex and of limited clinical value. For example, the original HLA-A28 type was subsequently split into A68 and A69, HLA-B12 into B44 and B45. Nevertheless, antisera proved to be of value in kick-starting the entire tissue typing process.

The principle relies on mixing antisera with presumed specificity (e.g. anti-HLA-A5) with the patient's mononuclear leukocytes (almost completely lymphocytes) to determine a possible reaction. In adding rabbit or guinea pig serum as a source of complement, cells binding the antibody would lose membrane integrity, and dyes such as trypan blue, propidium iodide, acridine orange, and ethidium bromide are used to detect positive or negative results. Thus the procedure is described as complement-dependent cytotoxicity (CDC) or perhaps the lymphocytoxicity test (LCT) (Figure 9.24). Complement-fixing monoclonal antibodies recognizing HLA protein domain structures subsequently replaced most polyclonal antisera.

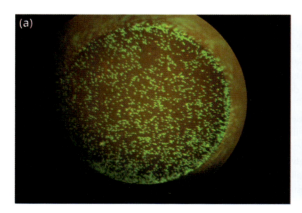
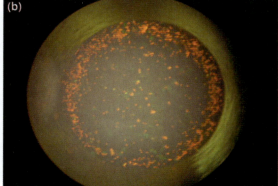

FIGURE 9.24

Complement-dependent cytotoxicity for determining antibody/cell HLA specificities. (a) negative result, (b) positive result.

From Hall, Scott, & Buckland, *Clinical Immunology*, second edition, Oxford University Press, 2016.

The procedure was refined with the development of a dedicated plastic tray (a Terasaki plate), consisting of 60 or 72 small wells where minute volumes (1–5 µl) of antisera, cells, and complement could react together. The result can be determined under an inverted microscope.

Molecular genetics for HLA type

Difficulties with serological typing can be largely abolished by determination of HLA types by direct DNA analysis. Although complex in theory, this testing is in fact relatively simple given advances in standard gene techniques such a polymerase chain reaction (PCR) and in cell biology, with a fluorescence flow cytometer. There are several forms of typing:

- PCR with sequence-specific primers (PCR-SSP) uses pairs of primers of a known sequence based on the DNA sequence of the particular HLA gene in question. Matches are determined by analysis of agarose gels.

- A variant method with oligonucleotide probes (PCR-SSOP) amplifies the gene in question by PCR, which is then hybridized to a solid matrix such as a nylon membrane. Primers conjugated to a fluorescent probe can detect their complementary sequence on the membrane.

- Luminex PCR-SSOP uses bead/primers conjugated to different fluorochromes, allowing up to a hundred specificities to be determined simultaneously on a fluorescence flow cytometer. This popular method provides better sensitivity than CDC or ELISA.

- Sequence-based PCR typing (PCR-SBT), rapidly becoming the gold-standard technique, is based on amplification of the exon(s) of interest as in PCR-SSOP, but amplicons are sequenced directly against a database of known sequences.

- Next-generation sequencing methods are being developed, which allow affordable (for the reference laboratory) high-throughput sequencing.

Detection of recipient antibodies

In transplantation, donor and recipient may well have very similar HLA types, but (as in blood tranfusion) this does not preclude the recipient having antibodies to one or more of the donor's antigens. Three major techniques for detecting these are known:

- Recipient antibodies may be detected by CDC against cells with a known HLA type. These may be transformed cell lines (often derived from a B cell malignancy, and so clonal) or from an individual whose HLA type is established.

- Advances in cell biology and protein chemistry allow purification of HLA proteins which can be coated on to the bottom of a 96-well microtitre plate. The recipient's serum can then be added and a modified ELISA performed to look for positive reactions.

- Another technique is an amended Luminex/immunoassay protocol where purified HLA proteins (as antigens) are bound onto coloured beads, recipient serum is added (and may bind to the beads), and a fluoresceinated anti-human antibody is used to detect binding. This method is generally performed using a fluorescence flow cytometer.

Transplantation cross-matching

Despite the accuracy and compatibility of HLA typing the donor and recipient, problems may still arise, and accordingly cross-matching is a prerequisite. Several methods are available, all of which are amendments for antibody detection above.

- CDC looks for recipient antibodies, but against donor T and B lymphocytes. However, by definition this does not pick up antibodies that do not fix complement and which may be dangerous in other settings, such as antibody-dependent cellular cytotoxicity.

- Flow cytometric crossmatch (FCXM) provides good data as it detects both complement fixing and non-fixing antibodies, different isotypes, a quantification of the extent of antibody binding, and leukocyte-specific binding (using a second antibody, such as directed towards CD3 for T cells and CD19 or CD20 for B cells).

- Solid phase cross-matching involves collecting donor cell surface molecules (antigens) and coating them onto beads, with subsequent recognition (or not) by recipient antibodies in a fluorescence flow cytometer.

Some of these techniques call for fresh blood from a potential donor, which assumes that they are alive, although some transplantable tissues from cadavers may be viable, and typing may be possible using lymph node or splenic lymphocytes. All methods have their advantages and disadvantages: genetic methods are certainly more definitive of the particular sequence, but are expensive and highly technical. However, genes are not necessarily expressed, so a better definition is whether a particular molecule is actually present at the cell membrane, which can be detected by antibody-based methods such as CDC and fluorescence flow cytometry. A further issue is that class II molecules are expressed only by APCs, so that low numbers in the peripheral blood may require an additional purification step. Ideally, all loci would be typed, but in practice the minimum is *HLA-A*, *HLA-B*, and *HLA-DRB2*, these being the most polymorphic and immunogenic loci, but *DQB1* may also be included.

Further details of transplantation are to be found in Chapter 10, section 10.4. A summary of tissue typing is presented in Table 9.16.

Key points

The laboratory offers a large panel of cellular and serological tests, most of the latter comprising autoantibodies. Most cellular tests are performed with FACS.

SELF-CHECK 9.9

Which autoantibodies are most commonly requested?

9.8.6 The bridge to clinical immunology

The purpose of the laboratory is to assist in the diagnosis and management of disease, each particular test providing a different part of the jigsaw puzzle. These tests, and the clues they provide, can be grouped together according to the clinical presentations.

- Repeated bacterial infections suggest antibody and/or complement deficiency, so that potentially useful initial investigations include immunoglobulins, functional antibodies, C3, C4, and CH50.

TABLE 9.16   The laboratory in tissue typing

Determination of HLA type	Serological
	Genetic
Detection of recipient antibodies	Complement-mediated cytotoxicity
	Modified ELISA with HLA proteins
	Modified Luminex with beads
Transplant cross-matching	Complement-mediated cytotoxicity
	Flow cytometric cross-match
	Solid phase cross-match

Hypogammaglobulinaemia (as is determined by serum protein electrophoresis) may be caused by a B cell defect such as a malignancy or cytotoxic chemotherapy.

- Viral and/or fungal infections suggest a T cell deficiency—an initial investigation would include full blood count for total lymphocyte numbers followed by flow cytometry for T cells and their subsets.

- Staphylococcal skin sepsis, and pulmonary and dermal fungal infections, suggest a neutrophil defect, so that functional studies of these cells would be considered.

- Confirmation of the diagnosis of SLE would be supported with results of ANA, RhF, and ESR tests. Those with a positive ANA on screening will be tested for antibodies to dsDNA, ENAs, and for C3 and C4 levels.

TABLE 9.17 **The immunology laboratory**

Section	Test(s)	Method(s)
Cellular immunity	Neutrophil function for CGD	Lack of ROS generation, phagocytosis
	Lymphocyte subsets	Fluorescence flow cytometry
Serology	Total immunoglobulins	Electrophoresis/densitometry
	Isotype determination	Immunofixation
	Immunoglobulin classes	Immunoassay
	Cryoglobulins	Warm/cold precipitation
	Defined protein analysis	Immunoassay, functional assay (CH50)
	Antibodies to foreign antigens	ELISA
	Autoantibodies	Indirect immunofluorescence, immunoassay
Allergy testing	Allergen-specific IgE	Immunoassay
	Basophil activation	Fluorescence flow cytometry
	Degranulation products	Enzyme assay, immunoassay
Tissue typing	Serological typing	CDC
	Molecular genetics	PCR, complimentary probe, Luminex
	Detection of recipient antibodies	CDC, ELISA, Luminex
Transplant cross-matching	Detection of recipient antibodies	CDC, ELISA, Luminex
	Flow cytometry cross-match	Fluorescence flow cytometry
	Solid phase cross-match	Fluorescence flow cytometry

ROS = reactive oxygen species. ELISA = enzyme-linked immunosorbent assay.

CDC = complement-dependent cytotoxicity. PCR = polymerase chain reaction.

- A suspected immunological aspect in renal disease (probably prompted by increased urea and electrolytes and/or reduced estimated glomerular filtration rate) would be investigated with ANAs, C3, C4, immunoglobulins, cryoglobulins, antibodies to the glomerular basement membrane, and C3 nephritic factor (only if C3 is absent).

These strands will be examined in depth in the next chapter. Table 9.17 summarizes analyses offered by the immunology laboratory.

Chapter summary

- The immune system is based in several widely distributed organs, and although all are important, the bone marrow and lymph nodes are crucial.

- All white blood cells are involved in immune and inflammatory responses, with specific and non-specific functions such as phagocytosis and antibody production.

- Humoral defences include antibodies and complement, whilst cytokines are key messengers.

- The innate immune response to an invading pathogen is local, rapid, and non-specific, and may lead to an acute inflammation.

- The adaptive immune response often develops away from the initial site of infection. It is slow but becomes increasingly more specific as it develops.

- The immunology laboratory provides an essential assessment of the capabilities of the immune system, the pathophysiology that arises, and how it can be treated.

Further reading

- Braza MS, Klein B. Anti-tumour immunotherapy with Vγ9Vδ2 T lymphocytes: from the bench to the bedside. Br J Haematol 2013:160; 123–32.

- D'Souza MP, Adams E, Altman JD, et al. Casting a wider net: immunosurveillance by nonclassical MHC molecules. PLOS Pathogens 2019: 15, e1007567.

- Robinette ML, Colonna M. Innate lymphoid cells and the MHC. HLA. 2016: 87; 5–11.

- Rakebrandt N, Littringer K, Joller N. Regulatory T cells: balancing protection versus pathology. Swiss Med Wkly. 2016: 146; w14343. doi: 10.4414/smw.2016.14343.

- Mathern DR, Heeger PS. Molecules Great and Small: The Complement System. Clin J Am Soc Nephrol. 2015: 10; 1636–50. doi: 10.2215/CJN.06230614.

- Kaufmann SHE. Immunology's Coming of Age. Front Immunol. 2019: 10; 684. doi: 10.3389/fimmu.2019.00684.

- Canè S, Ugel S, Trovato R, et al. The Endless Saga of Monocyte Diversity. Front Immunol. 2019: 10; 1786. doi: 10.3389/fimmu.2019.01786.

- Robinson J, Halliwell JA, Hayhurst JD, et al. The IPD and IMGT/HLA database: allele variant databases. Nucleic Acids Res. 2015: 43; D423–31. doi: 10.1093/nar/gku1161.

- Ka S, Lee S, Hong J, et al. HLAscan: genotyping of the HLA region using next-generation sequencing data. BMC Bioinformatics 2017: 18, 258 doi:10.1186/s12859-017-1671-3.

- Ma L, Li Q, Cai S, Peng H, et al. The role of NK cells in fighting the virus infection and sepsis. Int J Med Sci. 2021: 18; 3236–48.

Useful websites

- The IPD-IMGT/HLA Database for HLA typing: **https://www.ebi.ac.uk/ipd/imgt/hla/**

- Guidelines for managing allergies by the British Society for Allergy and Clinical Immunology: **https://www.bsaci.org/guidelines/**

- Lab tests online: **https://labtestsonline.org.uk**

 # Discussion questions

9.1 The immune system may be likened to a nation's armed forces. Using this metaphor, which equivalent role is undertaken by each of the major physiological parts of the body? For example, cytokines may be likened to messengers carrying orders between different parts of an army. Do the same for the bone marrow, white blood cells, neutrophils, antibodies, lymph nodes, the spleen, the liver, the skin, and the HLA system.

9.2 Compare and contrast the T cell receptor and the B cell receptor.

10

Clinical immunology

The previous chapter described the components and functions of the immune system, and the role of the laboratory. However, the immunological and inflammatory processes that have been introduced are not merely to do with fighting infections: they are present in many forms of disease, including renal, liver, endocrine, red blood cell, and reproductive disease, to name but a few. These diseases resulting from a failure on the part of the immune system to operate as it should to protect the body will be the subject of this chapter.

Immunological disease can be classified as underactivity and overactivity. The former is described as immunodeficiency, where the body has reduced ability to defend itself from pathogens, and will be examined in section 10.1. The polar extreme, overactivity, has two components: hypersensitivity and autoimmunity. An excessive response to an external stimulus—hypersensitivity—is discussed in section 10.2. An inappropriate immunological response to our own cells and tissues—autoimmunity, inevitably linked to the presence of damaging autoantibodies—is the subject of section 10.3. One treatment for autoimmunity is to manipulate the immune system with drugs to induce an immunosuppression. A more precise focus of this manipulation is in transplantation, which will be discussed in section 10.4.

Learning objectives

After studying this chapter, you should confidently be able to:

- discuss the causes and consequences of immunodeficiency

- recognize the key features of hypersensitivity

- explain the aetiology of autoimmune diseases

- describe the basis of inflammatory connective tissue diseases

- understand the processes of transplantation

10.1 Immunodeficiency

Immunodeficiency is a state in which the immune system fails to protect the body as it should and has two aspects: it is characterized by quantitative and qualitative changes in leukocytes and in the antibodies and complement of the humoral immune response. Immunodeficiency can be either primary—that is, caused by a genetic mutation that will often be congenital—or secondary, in which case it results from an outside factor such as an infectious agent, cancer, or radiation that compromises the immune system. Failure of the immune system can also be classified by whether disease affects only one part of the system (such as in T lymphocytes) or whether several components of the immune system are compromised, such as in severe combined immunodeficiency. In this section we will first look at the aetiology of the different types of immunodeficiency, and then consider their presentation, diagnosis, and management collectively.

10.1.1 Quantitative leukocyte deficiency

The causes of an insufficient number of circulating leukocytes (**leukopenia**) can be classified with relative ease. In most cases, the particular aetiology is known.

Bone marrow suppression

We start with bone marrow suppression as this organ is the site of production of all blood cells (Figure 9.2). Bone marrow leukopenia can have several causes:

- Suppression of **leukopoiesis**: The most common cause of the suppression of leukopoiesis is chemotherapy—typically either with cytotoxic drugs used to treat cancer or immunosuppressive drugs used in the context of severe autoimmune disease and transplantation. However, long-term use of certain 'everyday' drugs not usually thought of as dangerous—such as anticonvulsants, antibiotics, non-steroidal anti-inflammatory drugs (NSAIDs), and some antipsychotic and antihelminth drugs—may also suppress the bone marrow in rare cases.

- Cancer: If the bone marrow is invaded by a metastatic tumour that has spread from a primary site elsewhere in the body (such as the lung, breast, or prostate), the growth of malignant cells effectively takes over the bone marrow. As a result, those productive areas where **haemopoiesis** can proceed are reduced (Chapters 3, 4, and 5).

- Transformation of normal bone marrow cells into those with enhanced growth which also take over productive areas, e.g. B lymphocytes transforming into a myeloma (Chapter 11).

All of these processes may cause low levels of all types of white cells (panleukopenia), although some are specific to a particular class of white cells: for instance, carbimazole, used to treat **hyperthyroidism**, causes low numbers of neutrophils (neutropenia). Leukopenia that has been caused by drugs or a virus is typically reversible by stopping usage of the relevant drug or treating the viral infection. Likewise, leukopenia that results from malignancy is likely to improve after successful treatment of the disease.

Neutropenia

Several forms of severe congenital neutropenia (neutrophil count <0.5 × 10^9/l) are recognized, and most are due to the arrest of the maturation of precursors, mainly at the promyelocyte stage (Figure 9.2). These sporadic and mostly autosomal dominant disorders are linked to a variety of abnormal genes, such as ELA2 (encoding neutrophil elastase), CSF3R (encoding the granulocyte colony stimulating factor receptor), and HAX-1 (coding for a mitochondrial anti-apoptotic protein). Other causes include abnormalities in the Wiskott–Aldrich syndrome protein, which lead to decreased proliferation and increased apoptosis of myeloid progenitors. Children with the closely related autosomal-recessive Kostmann's syndrome have an even lower neutrophil count (<0.2 × 10^9/l). All diseases of this type are therefore examples of primary immunodeficiency. Table 10.1 summarizes causes of bone marrow suppression.

Human immunodeficiency virus-1 (HIV)

Pathology of this virus is due to the binding of protein gp120 to the CD4 molecule, which is present on T helper lymphocytes and several other cells. Infection of the cell also requires chemokine receptor CXCR4 to gain entry to lymphocytes, and CCR5 for entry into macrophages, and this triggers fusion of the viral envelope with the cell membrane. Viral RNA thus gains access to the host cell's cytoplasmic processes, which it hijacks in order to reproduce itself.

Shortly after infection anti-HIV IgM and IgG antibodies can be detected, but following this the disease can be latent for years. However, there will be a slow and steady reduction in CD4-positive lymphocytes, which can be plotted in the laboratory by a fall in the absolute number of CD4 bearing

TABLE 10.1   Bone marrow suppression

Causative agent	Examples
Infectious	Viruses (parvovirus, hepatitis)
Heavy metals	Cadmium, strontium, etc.
Cytotoxic drugs	Methotrexate, vincristine
Immunosuppressive drugs	Azathioprine, prednisolone
Cancer	Breast, prostate, colorectal, myeloma
Proliferative disease	Myelofibrosis*, myelodyplasia*

* Discussed in Chapter 11.

cells or a reversal of the CD4:CD8 ratio, which normally favours CD4 cells. Monitoring by fluorescence flow cytometry is the most common method, as explained in section 9.8.2, where figure 9.15 applies. CD4 T cells are the major lymphocyte subgroup, but a reduction will not show up in the total lymphocyte count from the white cell differential until levels are profoundly low.

The ultimate consequences of this infection for the CD4 cells are direct killing by the virus and an increased rate of apoptosis, which leads to falling numbers of the lymphocytes. Counter to pathophysiology, where low levels of antibodies would be expected due to a lack of T_h lymphocytes, there is often **hypergammaglobulinaemia**—that is, an elevated level of gamma-globulins. This may be the result of dysregulated B lymphocytes, which, lacking correct direction, synthesize abnormal antibodies that provide no protection. Falling numbers of CD4-positive T cells are a direct marker of the progression of the disease to AIDS, and with its pathophysiology, the clinical consequences of this progression are established. The loss of T lymphocytes and CD4-positive monocytes/macrophages leads to any number of a large series of opportunistic bacterial, viral, fungal, and parasitic infections.

DiGeorge syndrome

This genetic disease (section 16.1.3) is characterized by thymic **aplasia**. This can vary between a small thymus and no thymus at all, and because this organ is a requirement for T lymphocyte development (see Chapter 9), T cells may be reduced in number or absent altogether, leading to risk of viral infection. But an absence of T cells also leads to poor antibody production as B cells are not provided with helper cells. The laboratory method for confirming the diagnosis is fluorescence flow cytometry for lymphocyte subsets, although DiGeorge syndrome inevitably brings other clinical issues that aid diagnosis. The closely related Nezelof syndrome is also associated with thymic dysplasia and so T cell abnormalities.

Severe combined immunodeficiency (SCID)

This phenotype, which occurs with a frequency of ~10/million live births, is characterized predominantly by T lymphocyte abnormalities, and in several cases B cell function is unaffected. SCID has a number of causes. The X-chromosome-linked variant is caused by mutations in genes for receptors for cytokines, including IL-2, IL-4, IL-7, and IL-15. Accordingly, T cell (IL-7) and NK cell (IL-17) development that relies on these cytokines is impaired. A second major type of SCID is caused by a loss-of-function gene for the enzyme adenosine deaminase (ADA), resulting in increased levels of purines and dysregulation of nucleic acid metabolism and so the inhibition of lymphocyte proliferation. The rare 'bare lymphocyte syndrome' is due to loss-of-function mutations in transcription factors required for the expression of molecules of the major histocompatibility complex (MHC, the principles being those of the HLA system), and so failure of lymphocytes to recognize antigens.

Cross reference

Whilst organ transplantation is discussed later in this chapter (section 10.4.3), see section 11.4.6 in Chapter 11 to learn more about bone marrow transplantation.

Until such time as a bone marrow transplant can be arranged, the patient needs to be protected from infections that would normally be resisted, but which, in SCID, may be fatal.

10.1.2 Qualitative leukocyte deficiency

Diseases classified as qualitative leukocyte deficiencies are characterized by normal levels of leukocytes that fail to perform their functions. Of the few diseases in this group, one dominates.

Chronic granulomatous disease (CGD)

This condition is caused by a defect in the ability of the phagocyte to produce reactive oxygen species (ROSs), such as the superoxide anion, due to mutations in genes coding for NADPH (see Box 9.4). The biochemistry and laboratory testing is explained in section 9.8.1. Phagocytosis itself continues normally, and although a **phagolysosome** is formed, the ingested microbes are not killed. As ROSs are of great importance in the destruction of bacteria, failure to eradicate these organisms can lead to their consolidation in a small and localized inflammatory nodule called a granuloma, hence CGD. With an incidence of ~1/225,000 live births, and regardless of the aetiology, low numbers of neutrophils or dysfunctional neutrophils lead to bacterial infections. Often presenting as pneumonia, other clinical signs are lymphadenopathy, **hepatomegaly**, and/or **splenomegaly**.

Other neutrophil defects

A key feature of the Chediak–Higashi syndrome (alongside partial albinism) is the inability to correctly assemble lysosomes because of a loss-of-function mutation in a transport protein coded by *LYST* at 1q42.1–42.2. This leads to poor bacterial killing and so repeated infections, whilst in the laboratory, the blood film shows giant intracellular granules of acid hydrolases and myeloperoxidase. These are present in all granulocytes and monocytes, and platelets also have similar abnormalities, but in their dense granules.

Other leukocyte defects

Omenn's syndrome is caused by mutations in genes involved in the generation of B and T cell receptors (section 9.6). The syndrome is also characterized by an enlarged liver, spleen, and lymph nodes, raised IgE, and an **eosinophilia**. There is often an increased lymphocyte count (lymphocytosis), but this is due to the expansion of clones of abnormal T cells that leads to a form of SCID. The immunodeficiency of leukocyte malignancy is discussed in Chapter 11, section 11.4. Cell-mediated immunity is often compromised in **ataxia telangiectasia** and the Wiskott–Aldrich syndrome. In the former, it causes a lack of IgA and IgE responses, and so upper respiratory tract infections; the latter is associated with low IgM responses.

SELF-CHECK 10.1

How may bone marrow become suppressed?

10.1.3 Defects in humoral immunity

Cytokines

There are several examples of specific deficiency in cytokines and their receptors that lead to immunodeficiency. For example, mutations in *IL12RB1* in 19p13 encoding part of the receptor for IL-12 and IL-23 often result in **mycobacterial** disease, and clinical presentations with *Salmonella* species and **mucocutaneous candidiasis**. Signalling errors in TNFs and their receptors may lead to altered T_{reg} function and possible development of autoimmune disease (section 10.3). However, the vast majority of humoral defects are in antibodies and complement.

Antibodies

Being produced by the concerted action of T cells, B cells, and APCs (section 9.6.3), it follows that problems with any of these cells, or the lymph node microenvironment (such as lymphoma), will lead to low antibody levels (hypogammaglobulinaemia) or, in extreme circumstances, to their absence (agammaglobulinaemia). The precise definition depends on the total protein count and quantitative protein electrophoresis (section 9.8.3).

The most well-known agammaglobulinaemia is Bruton's disease, otherwise known as X-linked agammaglobulinaemia, and so most unlikely to be present in females. With a frequency of ~10/million, the cause is a loss-of-function mutation in *BTK* (located at Xq22.1), the gene for tyrosine kinase. This enzyme is required for the maturation of pre-B lymphocytes into mature B cells, leading to loss of the ability of the latter to generate antibodies.

Hyper-IgM syndrome is caused by the failure of immunoglobulin class switch, so that an IgM response is fixed and does not develop. This leads to low levels of IgG, IgA, and IgE antibodies. Therefore, extravascular tissues are undefended, leading to recurrent bacterial and opportunistic infections, such as that of *Pneumocystis jirovecii*. The size of IgM means it cannot leave the blood, and so there is increased viscosity with risk of thrombosis and stroke.

Complement

Deficiencies in many complement components are known: those due to decreased hepatic synthesis, increased loss (perhaps in renal failure), and increased consumption in inflammatory disease. An autoantibody (C3 nephritis/nephritic factor) to the C3 convertase complex can provide stability, leading to the consumption of complement components and so a deficiency in C3. Found at the centre of the complement system (Figure 9.9), C3 deficiency is clinically important. It may arise as above, by excessive consumption, or in a familial manner, inherited as an autosomal recessive mutation.

Deficiencies in several other complement components (e.g., C1, C2, C4) are recognized and may be inherited in an autosomal recessive manner. C2 deficiency is the most common, with the heterozygous frequency of 1/20,000 in Western countries (Sjöholm et al. 2006). Deficiency in C1 inhibitor is well described, leads to the unregulated activation of C1, and therefore reduced levels of C3 and C4. These abnormalities are often present in diseases associated with immune complexes, such as systemic lupus erythematosus (SLE), systemic vasculitis, and sub-acute bacterial **endocarditis**.

Genetic deficiencies in alternative pathway components factor D and properdin have been described but are exceedingly rare. Lack of inhibitor factors H and I are known and leads to excessive complement activity with the consumption of C3. The consequences of this include haemolytic–uraemic syndrome and proliferative **glomerulonephritis**.

There can be deficiency in all of the individual components of the terminal pathway (C5–C9). Lack of C5 has additional implications as there will also be reduced levels of C5a and so impaired chemotaxis. Loss-of-function mutations in genes coding for integrins and adhesion molecules lead to failure to express receptors such as CR3. Lack of C1 inhibitor, an essential regulator of the complement system, leads to the important clinical syndrome of **angioedema**, of which there are two types:

- With an estimated prevalence of ~20/million, hereditary angioedema (HAE) is caused by an autosomal dominant mutation in *SERPING1* (= serpin family G member 1) at 11q12.1, although ~ 20–25% of cases are spontaneous. In type I HAE (85% of all cases), C1 inhibitor is absent, or is present at low levels. In the remaining 15%, type II HAE, the molecule is present or even elevated.

- Angioedema may be acquired (hence AAE, meaning acquired angioedema) in lymphoproliferative (leukaemia, lymphoma) and autoimmune diseases. The most common aetiology is excessive complement consumption, possibly because an autoantibody renders it susceptible to degradation by serine proteases. Accordingly, AAE may be classified alongside non-organ-specific diseases such as RA and SLE.

In both cases of hereditary and acquired angioedema due to C1 inhibitor deficiency, the clinical presentation is similar. Affected individuals suffer from swelling of the soft tissues of the face and neck, but may also have painful gastrointestinal symptoms. At the cellular level, symptoms may be due to increased levels of bradykinin, which are normally part-regulated by C1 inhibitor. In HAE, symptoms may be controlled by infusions of C1 inhibitor, but in AAE this is less successful, so treatment is aimed at the underlying aetiology.

10.1.4 Presentation, diagnosis, and management

Presentation

The defining feature of immunodeficiency is repeated and prolonged infection with microbes that would normally be rapidly addressed. A reasonably simple classification is that T cell deficiency generally leads to viral and fungal infections, whereas B cell, macrophage/neutrophil, and complement deficiencies are most often associated with bacterial infections, although there are exceptions. Several immunodeficiency diseases have a genetic basis (SCID, CGD, DiGeorge syndrome), and so will be evident in neonates and small children. Acquired immunodeficiency (cancer, HIV) can be present at any age. Many cases of immunodeficiency present in the light of factors that suppress the bone marrow (radiotherapy, chemotherapy) and so may be a grudgingly acceptable side effect.

Although most cases of immunodeficiency present with common infections, certain particular pathogens dominate in some defined conditions. In HIV, these include the fungus *Pneumocystis jirovecii* (causing a pneumonia), the parasite *Toxoplasma gondii* (often leading to neurological disease), *Mycobacterium tuberculosis* (which also causes lung disease), and hepatitis, cytomegalovirus, herpes, and papilloma viruses. The latter two may be responsible for some of the increased risk of lymphoma and cervical cancer that is associated with AIDS. CGD often presents with infective organisms such as catalase-positive staphylococci.

Diagnosis

In most cases, the laboratory makes the diagnosis (section 9.8) with investigation of all leukocytes classes, both quantitative (e.g. full blood count, and fluorescence flow cytometry for CD4/CD8 T cells in HIV infection) and qualitative (neutrophil function in CGD). Bone marrow aspiration may be necessary, although if there is invasion of this organ, there will also be erythrocyte and platelet changes. Diagnosis of specific and total hypo- and agammaglobulinaemia will be made by protein electrophoresis and immunoelectrophoresis, supported by the absence of mature B lymphocytes, generally measured by fluorescence flow cytometry. A mutation in *BTK* (= Bruton's tyrosine kinase) and in other genes may be sought by molecular genetics.

Management

The standard treatment for bacterial infections is broad-spectrum antibiotics, which unfortunately feed anti-microbial resistance. Treatments for DiGeorge syndrome include thymus engrafting, and the loss of leukocyte function in SCID can be restored by bone marrow transplantation (section 11.4.6). Treatments for CGD include antibiotics for bacterial infections, typically *Staphylococcus aureus* (treated with trimethoprim, sulfamethoxazole, levofloxacin), whilst fungal infections with *Candida albicans* or *Aspergillus* species have their own specific treatments (clotrimazole, itraconazole, and amphotericin/voriconazole).

As defective cell-mediated immunity leads to bacterial infections, direct lack of antibodies leads to the same problem, and so requires antibiotics. However, another treatment is the infusion of purified gammaglobulins harvested from donated whole blood taken for transfusion. This is a requirement in agammaglobulinaemia, with its risk of infections with *staphylococcus*, *streptococcus*, *neisseria*, and *haemophilus* species, but needs to be repeated at 3–4-week intervals. Treatment of Hyper-IgM syndrome is to replace the IgG, IgA, and IgE in intravenous immunoglobulin infusions. In both cases the laboratory is called upon to monitor plasma levels of total and individual gammaglobulin classes. CD40 ligand deficiency has been successfully treated with bone marrow transplantation.

There is no replacement therapy for complement deficiencies: instead, preventative therapy with antibiotics will be required for the typical *Neisseria gonorrhoea* and *meningitidis* infections, common in situations where there is failure to form a functioning membrane attack complex. However, HAE caused by C1 esterase inhibitor deficiency may be treated with intravenous replacement therapy or an inhibitor of bradykinin B2 receptors.

Key points

Immunodeficiency may be present in almost any part of the immune system and leads to specific and non-specific infections. Deliberate immunodeficiency induced by immunosuppressive drugs is a key part of the treatment of autoimmune disease and in transplantation.

SELF-CHECK 10.2

What are the consequences of B lym phocyte immunodeficiency?

10.2 Hypersensitivity

As with immunodeficiency, we can break down an overactive/inappropriate immune response into a number of different aspects, the major areas being hypersensitivity and autoimmune disease. The former is generally an excessive and damaging response to a normal stimulation or situation, whilst the latter may be described as the reverse—effectively the normal response to an abnormal situation, as we will discuss in section 10.3. Hypersensitivity is classified into five types (I–V). Types I–III and V are linked to an excessive antibody response, type IV to abnormal white cell responses.

10.2.1 Type I hypersensitivity

This form of hypersensitivity, allergy, may be present in 10% of the UK population in its many forms. NICE estimates that 500,000 people in the UK have had a venom-induced anaphylactic response, and 220,000 under-44-year-olds a nut-induced reaction. Food allergy is one of the most common allergic conditions and a major paediatric issue: it has been estimated that around 1 in 6 children in the UK have two diagnosed allergies, and 1 in 40 have three allergies. Table 10.2 lists common allergens.

The pathology (section 9.8.4) involves IgE, basophils, and mast cells, which have receptors for the FcR of IgE, although there can be non-IgE-mediated responses. Complement components C3a and C5a can trigger mast cell degranulation independently of IgE.

Aetiology

The evolutionary basis of the physiology of most IgE responses is likely to have developed from a response to parasites or other external pathogens. These include those inhaled and therefore in the lung (particulate material such as pollen and fungal spores), as well as skin ectoparasites. An allergic response can be viewed in three stages:

1. An induction phase sees the allergen first contacting the immune system in the skin, lungs, and/or intestines. An antigen-presenting cell passes allergen particles to a Th2 T cell in conjunction with IL-5. The primed T cell then activates a B cell with antigen, IL-4, and IL-13 to produce circulating

TABLE 10.2   Common allergens

Foodstuffs	Peanuts, pecans, pine nuts, almonds, walnuts, shellfish, egg albumen, wheat, maize, kiwifruit, milk, soy, chestnut
Invertebrate material	House dust mites, storage mites (integument and faeces)
Plant material	Pollen (typically ragweed), poison ivy, certain oaks
Venoms	Insect and arachnid bites and stings

allergen-specific IgE that docks into the FcεR of basophils and mast cells. These cells are now primed.

2. Upon (re)exposure to the allergen, it binds and cross-links the IgE exposed on the basophils and mast cells, resulting in their degranulation. These products include heparin, histamine, elastase, leukotrienes, cytokines, platelet-activating factor, and leukocyte chemotactic factors. The latter recruit neutrophils and eosinophils that are likely to magnify the general inflammatory picture, also degranulating to release additional cytotoxic proteins, enzymes, leukotrienes, and other inflammatory mediators.

3. The third phase is characterized by local and systemic action of these inflammatory mediators. Histamine is perhaps the most active. The effects include constriction of the bronchus, dermal vasodilation, increased vessel permeability, pain, and itching. The latter may have developed as a method of scratching the skin to remove ectoparasites.

Most immune responses are measured and appropriate. However, the exaggerated and persistent allergic response of granulocytes results in pathological release of the inflammatory mediators, which go on to cause the signs and symptoms suffered by the patient. A key characteristic of type I hypersensitivity is that the response can develop rapidly, often within minutes of exposure to the allergen. The body's reaction to this can range from minor irritation to fatal. The cell biology of type 1 hypersensitivity is summarized in Figure 10.1.

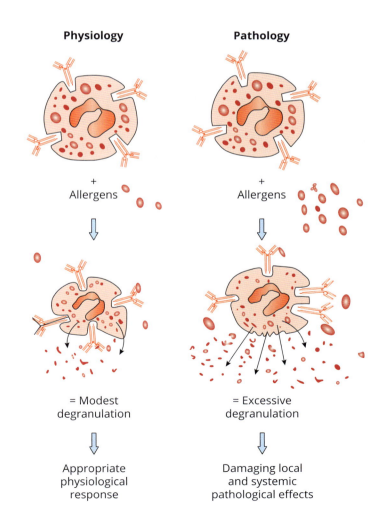

FIGURE 10.1

Cellular basis of type I hypersensitivity. Left: Modest physiology response of a basophil/mast cell with FcεR occupied by IgE that binds allergen, resulting in appropriate degranulation. Right: Excessive response by hypersensitive basophils/mast cells to an increased allergen stimulus resulting in excessive degranulation with pathological repercussions.

Presentation

The patient—or parent of a child—is likely to present to their GP with a number of signs and symptoms, all with varying degrees of severity. These can be local, such as irritation and itching in a certain area, but are just as likely to be systemic. Major clinical features are the following:

- Asthma is characterized by problems with breathing, and is by far the most common semi-permanent condition (section 13.5.2).

- Allergic conjunctivitis is similar to rhinitis, but an inflammation of the conjunctiva, with the same pain, irritation, and itching, and with tears. This is a modified response (like a runny nose) in trying to flush out the irritant.

- Eczema (also known as atopic or contact dermatitis—inflammation of the skin) is an umbrella term for several skin conditions that include rash, swelling, dryness, and itching. It may be caused by plants such as poison ivy, but also by exposure to industrial allergens such as nickel, latex, and even talcum powder.

- Intestinal problems, typically pain and cramps, are probably due to the action of histamine from mucosal mast cells, themselves triggered by allergens in the diet, such as ovalbumin (egg white).

- Urticaria, also known as hives, is a raised rash (weal) most often found on the arms, legs, and back (Figure 10.2), whilst facial urticaria is common with food allergies. However, there is a degree of cross-over with eczema and angioedema.

In many cases, type I hypersensitivity is tolerated with stoicism, and it is rarely of major clinical importance. However, two severe forms can be life-threatening:

- Anaphylaxis is perhaps the most serious reaction, and has been reported as being present in up to ~3,500/million per year, whilst NICE estimates that 1 in 1,333 of the population of England experience anaphylaxis at some point in their lives. It often develops rapidly (minutes after exposure), acts on different physiological systems, and may be fatal, causing ~20 deaths annually in the UK. One of its effects can be a drop in blood pressure resulting from vasodilation, and subjects are often described as being in **anaphylactic** 'shock'. Major issues are airway obstruction leading to breathing difficulties, often bringing a sense of impending doom.

- Angioedema: swelling of the skin and tissues resulting from increased blood vessel permeability. This can lead to breathing problems if throat tissues become swollen and the passage of air is impaired. In its worst manifestation, the patient may suffocate. An established non-immunological ADR is angioedema due to ACE inhibitors, and is due to the drug interfering with bradykinin metabolism.

Several of these symptoms can be present concurrently. For example, the most common manifestation of type 1 hypersensitivity is hay fever, likely to be a mixture of rhinitis, conjunctivitis, and often some pulmonary symptoms such as asthma (section 13.5.2).

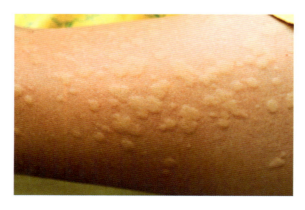

FIGURE 10.2
Urticarial rash.

iStock.

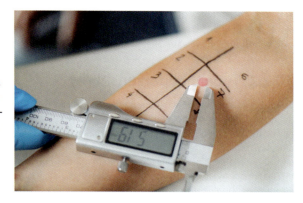

FIGURE 10.3
The skin prick test. Small amounts of liquified potential allergens are placed sub-dermally into the forearm. Reactions (redness, swelling) are noted until maximal and the diameter is measured to quantify effect.

Shutterstock.

Diagnosis

Diagnosis will generally be clinical. In most cases, and once the acute presentation has resolved, the precise allergen itself can be identified by challenging the patient by introducing low levels of a variety of potential allergens, increasing the dose, and then observing which ones evoke a response. This may be performed on the skin and is referred to as an intradermal (skin) prick test. The result is valuable as the patient is empowered to avoid the particular allergen (Figure 10.3).

The laboratory may be called upon for overall and allergen-specific IgE levels (section 9.8.4), increases in which may also be present in parasitic infections, hyper IgE syndrome, Wiskott–Aldrich syndrome, IgE myeloma, and eosinophilic granulomatosis with polyangiitis.

Management

Treatment depends on the nature of the stimulus and the frequency and severity of the attacks. In many cases, the first-line treatment is an anti-histamine to block the histamine receptor: chlorphenamine is commonly used. In case of severe allergy, patients will carry an EpiPen, a syringe loaded with adrenalin. An additional treatment is with drugs to stabilise the mast cell membrane and so suppress degranulation, the most common being sodium cromoglicate. Other targets of treatment include inflammatory mediators such as leukotrienes.

NICE CG134 offers guidance on the assessment and referral of anaphylaxis after emergency treatment, whilst TA246 refers to pharmalgen for the treatment of bee and wasp venom allergy. Food allergy should not be confused with food intolerance, as only 25–40% of self-reported food allergy is confirmed as a true clinical food allergy. Should corticosteroids fail to control the eczema (which by definition will be severe), topical tacrolimus and pimecrolimus may be offered. These two drugs are immunosuppressive calcineurin inhibitors; the former is commonly used to prevent transplant rejection.

> **Cross reference**
> The management of asthma is discussed in section 13.5.2 of Chapter 13 on respiratory diseases.

Molecular genetics

Numerous links between HLA types and allergic responses are known. These include ragweed pollen (*Artemisia artemisiifolia*) with DRB1*15; rye grass with DRB1*03 and DRB1*11; the house dust mite *Dermatophagoides* with DRB1, DRB3, and DRB5 variants; olive pollen allergen with DRB1*07 and DQB1*02; peanuts with DRB1*08 and DQB1*06; and aspergillosis with DRB1*15.

Genome-wide association studies (GWAS) have linked single-nucleotide polymorphisms (SNPs) in *TLR6* (coding for toll-like receptor 6), *STAT6* (a transcription factor), *SLC25A46* (a mitochondrial membrane transporter), *MYC* (a nuclear phosphoprotein), *IL2* (interleukin-2), and *HLA-DQB1* and *HLA-B* (both coding molecules involved in immune recognition) with allergic sensitization. Risk variants showed that ~25% of cases of allergic sensitization and allergic rhinitis could be accounted for by these ten loci. Examples of non-MHC gene involvement include allergic bronchopulmonary aspergillosis, which is linked to SNPs in *IL-4RA*, the promotor region of *IL-10*, and in *SP-A2*, coding for lung surfactant protein A2. Atopic dermatitis has been linked to SNPs in *IL-13* and *IL-4R*.

miR-126, miR-142, and miR-221/222 have been linked with increased mast cell degranulation and cytokine release, whilst miR-21, miR-132, and miR-302e have been linked to reduced mast cell degranulation and cytokine release. These changes may be exploitable in searches for new therapeutics.

10.2.2 Type II hypersensitivity

Chapter 9 describes antibody-dependent cell-mediated cytotoxicity (ADCC) and a subtype, antibody-dependent cellular phagocytosis—defence mechanisms that cross the innate/adaptive divide. They rely on the docking of the Fc section of (generally) IgG antibodies into their FcRs on the surface of granulocytes, monocytes, macrophages, and NK cells. The leukocyte is therefore primed to attack or ingest any other cell that is expressing the antigen that the antibodies recognize. It is likely that ADCC is effective in defence against pathogens, such as parasites, that are too large to be the object of phagocytosis.

However, should the antibody incorrectly recognize a structure on a healthy cell as abnormal, it will bind and so may be recognized by a leukocyte bearing the appropriate FcR, resulting in the activation of the leukocyte and attack of the healthy cell. In this instance the antibody is technically an autoantibody, and so part of section 10.3.3, where it presents a problem in SLE. A healthy cell may still be the target of ADCC should it bind a true antigen (such as bacterial product) on its surface. Figure 10.4 illustrates the cellular basis of type II hypersensitivity.

Some drugs bind to normal tissues, and so may form an unusual structure that the immune system regards as foreign and so worthy of destruction. Examples of this include chlorpromazine and phenacetin. In those very rare instances of an incompatible blood transfusion, and in other forms of transplantation, donor cells are recognized as being non-self and are destroyed.

Precise diagnosis, if needed, is complicated and requires considerable laboratory intrastructure and skills. Proof of an incompatible blood transfusion is simple by comparison, and will call for testing of

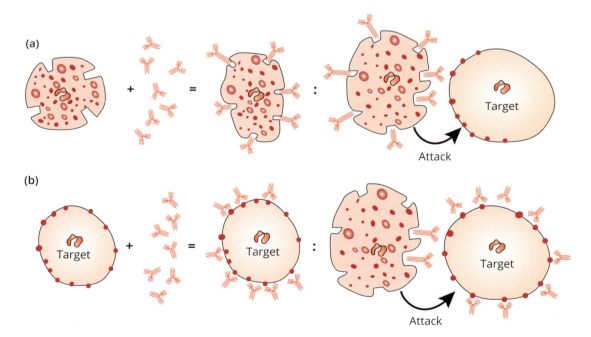

FIGURE 10.4

Cellular basis of type II hypersensitivity. In (a), (top left) a leukocyte with appropriate FcRs becomes primed by binding antibodies. To its right, the antibodies on the primed leukocyte recognize antigens on the surface of the target, which triggers attack by the leukocytes. In (b) (bottom left), antibodies recognize and bind to antigens on the target cells, exposing their Fc domains. On the right the latter dock into the FcR on the unprimed leukocyte, which then becomes activated and attack the target.

the donor blood pack and blood from the patient themselves. There is no direct treatment. When the antigen is a drug, the hypersensitivity should resolve itself upon removal, whilst autoantibodies may be treated by immunosuppression, although this non-specific treatment is often only modestly effective.

Haemolytic disease of the foetus and newborn (HDFN)

Given the trauma of vaginal delivery, some of her child's blood may enter the woman's circulation at birth. If there is a major blood group difference between the woman and her child, she may become immunologically sensitized to that difference, and so generate antibodies. Should she subsequently become pregnant with a foetus whose blood group is the same as that of the first, this will be recognized by her immune system, which will generate antibodies. Should these cross the placenta, they will be able to attack the red blood cells of the foetus, causing haemolytic disease of the foetus and neonate. If severe, this can cause late foetal death *in utero* and is a likely cause of early miscarriage.

Fortunately, this is recognized, and a common intervention immediately after delivery is to inject women who are blood group D negative with some antibodies against that blood group, this being the most common sensitizing agent. Since by definition the woman is group D negative, then the antibody will bind to, and so destroy, only those red cells bearing D which have entered her circulation. This form of ADCC ensures the woman is not sensitized to the D antigen, and so further D-positive foetuses should not be attacked.

10.2.3 Type III hypersensitivity

The concept of antibodies binding to antigens on large bodies such as bacteria, viruses, and cells is easy to grasp. However, antibodies can also bind to antigens that are present on much smaller soluble molecules and other non-cellular material in the blood, an excellent example being particles or remnants of the bacterial cell walls (perhaps the product of the action of antibiotics) and viruses. These combinations of soluble antibodies and their antigens are called immune complexes.

Normally, immune complexes should not pose too much of a problem, as they should be safely scavenged by phagocytes. However, if present in large amounts they may become insoluble, and fixed or lodged in certain anatomical sites where they are likely to attract the attention of phagocytes. The problem then arises as the phagocytes are programmed to respond to immune complexes, and if these are found on normal tissue, such as the endothelium, will inevitably cause a local inflammatory response and damage other healthy cells (Figure 10.5).

Presentation depends on the organ or tissue involved. Within blood vessels there may be dermatitis and vasculitis (inflammation of the blood vessels), with a skin rash, increased vessel permeability, and oedema. In the alveoli there will be breathing difficulty and possibly pneumonia. This may be precipitated by inhaled substances such as bacteria, fungal spores, and other material in house dust and other dust. Examples include 'farmer's lung' and 'pigeon-fancier's lung'. Deposition in the kidneys will cause immune-mediated glomerulonephritis, and so renal failure, with increased urea and creatinine.

Diagnosis is therefore dependent on the particular organ bed, but the gold standard is biopsy with immune complexes determined by, for example, indirect immunofluorescence. Clinically, type III disease is of greater importance than type I hypersensitivity (with the exception of potentially life-threatening anaphylaxis), but the general picture of antibodies causing tissue destruction is dominated by autoimmune disease (section 10.3). Accordingly, the only mode of treatment is, once more, immunosuppression.

10.2.4 Type IV hypersensitivity

This is the only form of hypersensitivity that is entirely cell mediated. The pathology is characterized by a slowly developing cell-mediated response and can be seen as a two-step process. At first, there is a normal physiological response to an antigen, with the generation of memory T lymphocytes primed to recognize their stimulating antigen. Perhaps weeks or months later, a second presentation of the same pathogen evokes a disproportionately over-active leukocyte response, that is, the hypersensitivity. Soluble mediators include inflammatory cytokines, such as IL-2, that recruit any combination of cytotoxic CD8 T cells, eosinophils, and macrophages (and possibly NK cells) which cause damage to the tissues surrounding the pathogen.

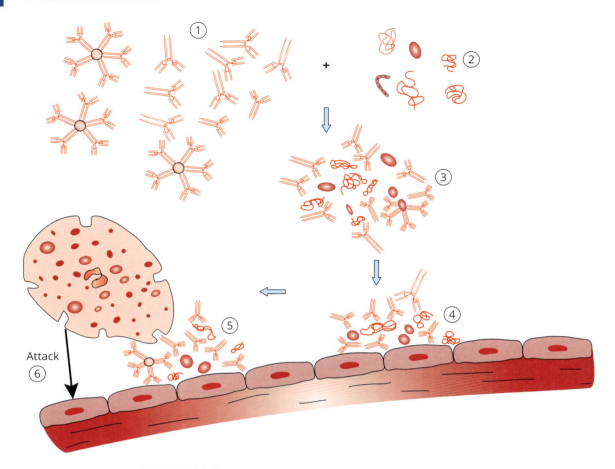

FIGURE 10.5

Cellular basis of type III hypersensitivity. IgM and IgG antibodies (1) react with the antigen(s) on small soluble molecules (2), forming an immune complex (3). Should this process develop, larger cross-linked complexes may become deposited on the internal wall of the blood vessel (the endothelium) (4). Once firmly deposited (5), the complexes may fix complement and attract the attention of cytotoxic leukocytes (6), leading to endothelial cell damage.

Type IV hypersensivity can be broadly classified as hypersensitivity that presents with symptoms involving internal organs and the skin. As to the former, precise diagnosis can again be difficult, but is not as important as acting on the clinical consequences of the disease, such as abdominal and pulmonary issues. Presentation and diagnosis of dermal disease is far easier, as it is a matter of distinguishing the skin lesion from other rash-like conditions. With varying clinical presentations, management depends on treating those which cause the greatest pathology. Broadly speaking, the condition has two aspects:

- The persistence of a bacterial infection can lead to the development of a chronic macrophage-led granuloma, such as occur in Crohn's disease (section 12.3), and in sarcoidosis (section 10.3.6), which manifests in the lung and skin.

- A second manifestation is contact hypersensitivity, causing a dermatitis. An example is effects of industrial chemicals (picryl-chloride and chromates) or metals such as nickel in jewellery clasps, the latter forming characteristic lesions at the back of the neck or wrist. A further cause is poison ivy, the active molecule being urushiol (Figure 10.6). Fortunately, the clinical symptoms resolve upon avoidance of the stimulants, but if persistent or clinically relevant, immunosuppression may be required.

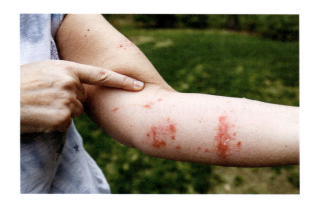

FIGURE 10.6
Rash caused by poison ivy.

Shutterstock.

This pathology is exploited in the Mantoux or Heaf tests for exposure (deliberate, as in a vaccination, or natural, as in an infection) to tuberculosis. The immune system is challenged by a small subdermal injection of tuberculin (an extract of the cell wall of *mycobacterium*). A positive result of small dermal lesions 48–72 hours later is evidence of primed cellular immunity.

10.2.5 Type V hypersensitivity

There is only one well-established syndrome in this group: the binding of an autoantibody to the receptor for thyroid-stimulating hormone (TSH) has the same effect on the target as the genuine ligand of TSH. The result is that the cell accepts the autoantibody as a true signal, and proceeds accordingly, which is to produce excess thyroid hormones, and so hyperthyroidism (Graves' disease). This therefore is an example of molecular mimicry. So presentation is with symptoms of thyroid disease, diagnosis by thyroid function tests and perhaps a biopsy, and treatment is with immunosuppression. It will be revisited in section 15.5.4.

Table 10.3 summarizes the leading forms of hypersensitivity.

Key points

Hypersensitivity can be classified as serological (types I, II, III, and V) and cell mediated (type IV). Treatments include immunosuppression and avoidance of the precipitating factor (the allergen).

TABLE 10.3 **Summary of hypersensitivity**

	Type I	Type II	Type III	Type IV
Nature of the pathology	Cross-linking IgE fixed into FcεR of basophils and mast cells over-reacts to allergens, causing mass degranulation	Antibody-dependent cellular cytotoxicity: autoantibodies link cytotoxic leukocytes to target cells(s)	Large insoluble immune complexes form and are deposited on capillary endothelial cells	Excessive secondary cellular responses to a previous antigen: no humoral component
Clinical consequences	Systemic anaphylactic shock, angioedema, asthma, conjunctivitis, eczema, rhinitis, urticaria	Destruction of target cells e.g. kidney, thyroid, skin	Generalized and/or organ-specific vasculitis, glomerulonephritis, pulmonary oedema, skin rash	Granuloma, dermatitis
Treatment	Avoidance of allergen, immuno-suppression	Immuno-suppression	Immuno-suppression	Avoidance of precipitant, immuno-suppression

Allergy is probably the leading public health issue in hypersensitivity—what are the major clinical signs, and what can we do about them?

10.3 Autoimmunity

Autoimmune disease is the third most important public health issue after cancer and cardiovascular disease. Although linked to only a relatively small annual number of deaths (in England and Wales in 2021, multiple sclerosis around 1,250; rheumatoid arthritis 830; vasculitis 200; systemic sclerosis 150; systemic lupus erythematosus (SLE) 75), it is responsible for considerable morbidity. Furthermore, deaths of people with autoimmune conditions are often classified otherwise—many with rheumatoid arthritis and SLE succumb to myocardial infarction and stroke, and so are listed as deaths from cardiovascular disease, not of auto-immune disease, although the latter may be listed as a contributing factor.

10.3.1 General aetiology of autoimmune disease

Autoimmune disease is effectively the result of a normal immune response to an abnormal situation. Normally we should not respond to our own tissues, but for a number of reasons 'self' becomes 'non-self', and so any altered normal cell or tissue is a legitimate target. An autoantibody is almost always present, inevitably contributing to destruction of cells and tissues. Thus, the real culprit is the abnormal B lymphocyte, which may develop via the failure of T lymphocyte regulation with increased levels of various pro-inflammatory cytokines. Should the suppressive activity of T_{reg} (expressing CD4, CD25, and FoxP3) be impaired, there may be failure to produce immunosuppressive cytokines TGF-β, IL-35, and IL-10, permitting potential autoreactive clones to develop, and so the autoantibodies.

Broad concepts in autoimmune disease

Although there is a great variability of the pathology and presentation of autoimmunity, several features are established. These are that women suffer three times as much as men do, that autoimmunity can be familial (so that there is a genetic mechanism), and that an infection can be a trigger, perhaps because microbial antigens resemble self antigens. The disease may strike a single or multiple organs. Almost all non-organ-specific diseases attack connective tissues (hence arthritis), and can present in any combination of bones, joints, ligaments and tendons, and skin, and may extend to non-connective tissues such as the eye, bone marrow, liver, and kidney. Nevertheless, the investigation and diagnosis of any autoimmune disease follows the common pattern of history and physical examination, laboratory investigations, and imaging (Table 10.4).

TABLE 10.4 The investigation and diagnosis of autoimmune disease

History and physical examination		Full personal and family medical history, medications, risk factors, social aspects, signs and symptoms
Imaging		X-ray, ultrasound, magnetic resonance imaging (MRI), computer-assisted tomography (CT)
Laboratory	Haematology	Full blood count, erythrocyte sedimentation rate (ESR), ferritin, coagulation screen
	Biochemistry	Liver function tests, urea and electrolytes, thyroid function tests, glucose metabolism
	Immunology	Autoantibodies, complement, C-reactive protein, immunoglobulins, leukocyte subgroups
	Molecular genetics	*HLA-B27, HLA-DRB1, HLA-DQB1*

Treatment of autoimmunity

Since a common feature of autoimmune disease is overactivity of the immune system, the leading treatment is immunosuppression, generally using drugs. Some modify the disease, hence disease-modifying anti-rheumatoid drug (DMARD), with prefixes c (for chemical, hence cDMARD) and b (biological, hence bDMARD). Several monoclonal antibodies are used, often targeting B lymphocytes via CD19 and CD20, or cytokines. More severe disease is treated with glucocorticoids (cortisone, prednisolone) and drugs also used in cancer chemotherapy (methotrexate, azathioprine).

The epidemiology and spectrum of autoimmunity

These diseases are reported to have a prevalence of ~1,100/million in Europe, which is likely an underestimate, as data from the USA suggest many more—60,000/million. As a group, autoimmunity is often overlooked as it does not feature amongst the major causes of death. For example, many with SLE succumb to chronic renal failure, heart disease, and stroke, which often take precedence in statistics and epidemiology. Where data are present, they are often classified by anatomy (e.g. autoimmune hepatitis is categorized with diseases of the liver) as opposed to aetiology, a model also adopted in this book. However, from a public health perspective, the leading autoimmunity is inflammatory connective tissue disease, and will be the focus of this section, other disease being discussed elsewhere (see Table 10.5). A further feature of autoimmunity and autoantibodies is their variability in specificity and presentation, both within and between different diseases. For example, autoantibodies to the basement membrane are common in Goodpasture's syndrome, and are often a cause of glomerulonephritis. They may also be present in SLE (being a cause of **lupus nephritis**), although the dominant autoantibodies in this disease are directed towards nucleic acids, as explained in section 10.3.3.

Multiple sclerosis

This disease is unusual as there is no clear autoantibody—the pathology arises from autoreactive CD4-positive T lymphocytes that recognize one or multiple antigen(s) in the myelin sheath of neural tissues within the central nervous system. The reasons for this are unclear, but genetic risk factors include certain alleles in genes coding the IL-2 receptor, the IL-7 receptor, and multiple SNPs in HLA-DRA,

TABLE 10.5 Autoantibodies and autoimmune diseases

Autoantibodies	Conditions were the autoantibody is often present	Primary tissues affected	Chapter
Anti-basement membrane	Goodpasture's syndrome, glomerulonephritis, SLE	Kidney, lungs, skin	14
Anti-TSH receptor Anti-thyroperoxidase Anti-thyroglobulin	Graves' disease, Hashimoto's thyroditis, hyper- and hypo-thyroidism	Thyroid	15
Anti-islet cell Anti-insulin	Type 1 diabetes	Pancreas	15
Anti-21-hydroxylase	Addison's disease	Adrenal cortex	15
Anti-mitochondria	Primary biliary cirrhosis	Liver	15
Anti-transglutaminase, anti-endomysium	Coeliac disease	Intestines	12
Anti-parietal cell Anti-intrinsic factor	Pernicious anaemia	Stomach	11

whilst miR-155 is upregulated. Roles for environmental factors are uncertain, but there is evidence of a role for the Epstein–Barr virus (Bjornevik et al. 2022).

The clinical course of the disease, which generally presents between 20 and 40 years of age and in three times as many women as men, may be alternating relapsing and remitting (85–90%), or a slow and steady deterioration—although the former may progess to the latter over time. Leading manifestations include bowel and bladder dysfunction, cognitive impairment, fatigue, gait dysfunction, incoordination, sensory loss, sexual dysfunction, and visual impairment. These are generally due to multifocal inflammatory lesions detectable by gadolinium-enhanced MRI. An estimated 100,000 people in the UK have this disease, which is linked to 1,241 deaths in England and Wales in 2021 (65% women).

Some patients benefit from baclofen (which helps with muscle spasms), gabapentin, tizanidine (anticonvulsants), amitryptyline (an antidepressant), amantadine (for fatigue), tetrahydrocannabinol (for pain and muscle spasms), and dantrolene (a muscle relaxant) (NICE CG186). Relapses may be treated with oral (0.5 g od for 5 days) or intravenous (1 g od for 3–5 days) methylprednisolone. Autologous haemopoietic stem cell transplantation is a viable option. Despite the focus on T cell pathogenicity, there may also be a role for B lymphocytes as oligoclonal immunoglobulin bands are found in the cerebrospinal fluid in 90% of cases. Developing therapies include anti-CD20 monoclonal antibodies, and one that targets VLA-4 integrin (CD49d), leading to the inhibition of leukocyte migration.

10.3.2 Rheumatoid arthritis (RA)

Rheumatoid arthritis (RA) is the leading autoimmune disease, present in ~7,500/million of the UK population (~420,000 people) with 21,000 new diagnoses annually. It is linked to smoking and overweight, strikes three to four times as many women as men, and has a strong familial component that increases the risk of the disease three- to nine-fold. Genetic factors point to a role for *HLA-DRB1*, bringing a three-fold increased risk, with over 100 other non-HLA susceptibility alleles described.

The two major autoantibodies in RA are rheumatoid factor (RhF) and anti-citrullinated protein antibodies (ACPA: proteins whose arginine groups have been changed to citrulline). Other changes, including elevated cytokines and chemokines, may be present 2–12 years before the onset of overt symptoms. Combining family genetics with serum markers provides additional power: an individual positive for ACPA and with two family members with RA has a 69% chance of developing RA within 5 years.

Synovitis (inflammation of the synovium) followed by joint destruction is a major hallmark of RA, but is also present in osteoarthritis (OA), which may cause as much morbidity and is often as severe. Differential diagnoses include septic arthritis, Lyme arthritis, and other arthritides, details of which follow. Generalized joint and muscle pain and arthritis are often present in other inflammatory and metabolic conditions such as viral infections, hepatitis, and **gout**. Accordingly, great care must be taken to ensure a correct diagnosis, and alternative diagnoses excluded.

Presentation

The patient will inevitably present with any combination of a wide variety of signs and symptoms, most non-specific but focusing on pain in the joints and muscles, of varying severity and duration. In themselves, these symptoms of arthralgia are not diagnostic, but may transform into RA. Other non-specific symptoms include a skin rash, unprovoked fevers and infections, and tiredness and lethargy. The latter may be due to anaemia, present in ~16% of newly diagnosed cases, and patients have on average one or two comorbidities such as hypertension (31%), depression (28%), fracture (24%), and chronic lung disease (21%).

Diagnosis

A full and accurate medical history is required, not merely to exclude other causes of the signs and symptoms, but also as part of the more precise diagnosis, focusing on number of tender and/or swollen joints, and other features such as a rash or flitting fevers, although these are rarely seen in early disease. Initial investigations are blood tests and X-rays, especially of those joints that are painful and/or swollen. These features can be scored, such as score of 1 for 2–10 asymmetric large joints affected, score 1.5 for 2–10 symmetric large joints affected, 2 for 1–3 small joints affected (with or without large joint involvement), 3 for 4–10 small joints affected (with or without large joint involvement), and 5 for more than

10 joints affected, including at least one small joint. Similarly, there is a role for autoantibodies: a low positive RhF or low positive ACPA scores 2, but high titres score 3.5. An abnormal ESR or CRP scores 0.5, and symptoms lasting more than six weeks score 1. A score of 6 or more (which can be made in primary care) is sufficient to confirm the diagnosis of RA, which will be followed by a referral to secondary care.

Historically, the defining autoantibody for RA is RhF, although it has poor specificity (85%) and sensitivity (69%). Indeed, it may be possible to have RA but be negative for the antibody, which is referred to as seronegative disease. The preferred test is ACPA as this has better specificity (95%) but also slightly worse sensitivity (67%) for RA. High levels are better predictors of adverse outcome such as joint erosions. Some may refer to cyclic citrullinated protein (CCP), of which the dominant species is mutated citrullinated vimentin. Changes in the full blood count are likely, with a **leukocytosis**, thrombocytosis, low haemoglobin and red cell count (the latter two of which may account for an anaemia).

In secondary care, additional investigations include X-rays of all joints, and ultrasound, with possibly MRI and CT, if other imaging is equivocal. These are important, providing a baseline, and as the joints are at risk of destructive erosions that bring (possibly considerable) pain and deformity (Figure 10.7).

The cause of the erosions are enzymes released by activated neutrophils that have found their way into the synovial fluid from the synovium. These enzymes (e.g. neutrophil elastase, collagenase, and synoviocyte matrix metalloproteinases) first digest the protective layer of collagen that covers the bone, and then the bone itself. Without the protective effects of the hydrostatic qualities of synovial fluid, bones can come into contact with each other, causing pain and additional damage.

The synovial fluid itself can be aspirated and analysed to confirm an inflammatory aetiology (low viscosity, and presence of white blood cells, red blood cells, proteins, enzymes, debris, etc.) as opposed to a traumatic synovitis or OA, where the fluid is clear. The pathology of an affected joint is summarized in Figure 10.8.

An alternative cause of a swollen joint is deposits of crystals (hence crystallopathy), as in sodium urate crystals in gout and calcium pyrophosphate in pseudogout, the presence of which can also be determined by light microscopy (which can also diagnose a septic arthropathy by the presence of bacteria).

SELF-CHECK 10.4

RA is described as a systemic disease—what does this mean?

Disease development

The clinical course of RA is variable, and should be managed in a specialist centre. The disease can develop in a number of directions. In some it may be relatively mild, and may even regress, often described as 'burnt out' (Figure 10.9).

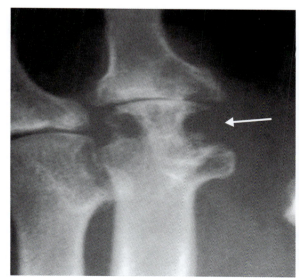

FIGURE 10.7

Joint destruction in rheumatoid arthritis. The arrow points to erosions of bone at the head of the tibia.

Reproduced by kind permission of Professor Peter Taylor, Nuffield Dept of Orthopaedics, Rheumatology and Musculoskeletal Sciences, University of Oxford.

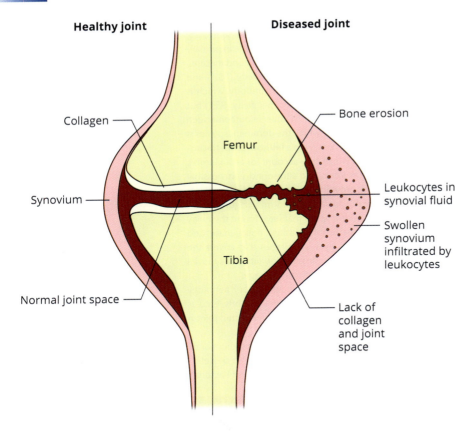

Healthy joint   **Diseased joint**

Collagen

Femur

Bone erosion

Synovium

Leukocytes in
synovial fluid

Swollen
synovium
infiltrated by
leukocytes

Tibia

Normal joint space

Lack of
collagen
and joint
space

FIGURE 10.8

Pathogenesis of joint destruction.
Left side: a healthy joint with a good
synovial space and bone covered
by collagen. Right side: Swollen
synovium infiltrated by leukocytes
feeds activated neutrophils into
the synovial fluid where enzymes
that digest the collagen layer and
then bone are released, leading to
reduced bone–bone space.

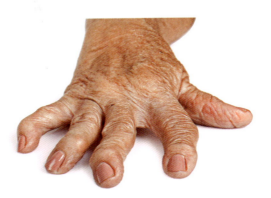

FIGURE 10.9

Swollen and deformed joints in
'burnt out' RA.

iStock.

In others it may attack few joints (1–3, hence pauci-articular disease), several (4–7, oligo-articular disease), or many (8 or more, being poly- or multi-articular disease). In early disease, the objective of management is to prevent disease progression, that is, (further) joint destruction. Prediction of potential progression to severe radiological damage may be guided by variants of *HLA-DRB1* and other markers. Disease progression can be assessed radiologically and with clinical scores, such as the disease activity score (DAS). This considers the tenderness and swelling of 28 joints, the erythrocyte sedimentation rate (ESR), and the patient's own view of their disease. On outset, a typical DAS28 score may be 4.0—a reduction by >1.2 is considered a good response to treatment, a reduction of ≤0.6 is considered ineffective. An ultimate objective is a DAS28 score of <2.6, as this generally implies low disease activity (remission), but in advanced disease, a more appropriate target may be 3.2. These (and other scores) allow an objective assessment of the rate of disease progression and the targets for testing the effectiveness of treatments.

The disease may spread from joints, causing extra-articular disease, such as in the lungs (resulting in fibrosis), skin (eczema, psoriasis, subcutaneous rheumatoid nodules), bone marrow (anaemia), eye (scleritis), nervous system (peripheral sensory and motor nephropathy), and liver (hepatitis). Further disease progression may lead to damage to the heart (angina, heart failure), chronic renal failure (risked by ~11% of patients), and rheumatoid vasculitis (section 10.3.5). The final stages are myocardial infarction and stroke: RA is the fifth risk factor for atherosclerosis, bringing a 50–60% increased risk of CVD compared to the general population. Investigations that may help assess disease progression include MRI and CT.

Pharmacological management

The first line of treatment in early stages aims to prevent disease progression with cDMARDs (Table 10.6). For many, methotrexate is the agent of choice for monotherapy, but if ineffective (as it may be in ~50%), a second cDMARD (e.g. leflunomide 10–20 mg od, sulphasalazine 500–2,000 mg od) or a low-dose glucocorticoid (≤10mg/day prednisolone) may be added. Use of certain cDMARDs calls for full blood counts, LFTs, and U&Es at 2/4 weeks after initiation, then at 8/12 weeks, and then at 3-monthly intervals.

Early intervention has been successful in reducing the rate of radiological progression. These drugs are effective in inhibiting radiologically defined disease progression. However, if ineffective, bDMARDs are the next step (Table 10.7).

Figure 10.10 summarizes treatment options for early RA (disease duration <6 months), where the target is low disease activity. Once early cDMARD options have been attempted, the disease duration is likely to be >6 months, and many consider the disease to be established. However, it is possible that the patient is still naïve to chemotherapy after 6 months, and if so, and the disease has low activity, methotrexate is the preferred initial therapy.

TABLE 10.6 Common cDMARDs

Group	Drugs
Non-specific	Penicillamine, gold, sulphasalazine, hydroxy-chloroquine
Targeted	Methotrexate (purine metabolism), leflunomide (pyrimidine metabolism), cyclosporine (T cells)

TABLE 10.7 bDMARDs

Class/Target	Drug	NICE document
IL-6 receptor	Sarilumab, tocilizumab	TA485, TA247
Janus kinase inhibitor	Tofacitinib, baricitinib, upadacitinib	TA480, TA466, ID1400
TNF-α	Adalimumab, etanercept, infliximab, golimimab, certolizumab pegol	TA195, TA225, TA415
CD20	Rituximab	TA195
CD80/CD86	Abatacept	TA195

TA375 refers to several agents.

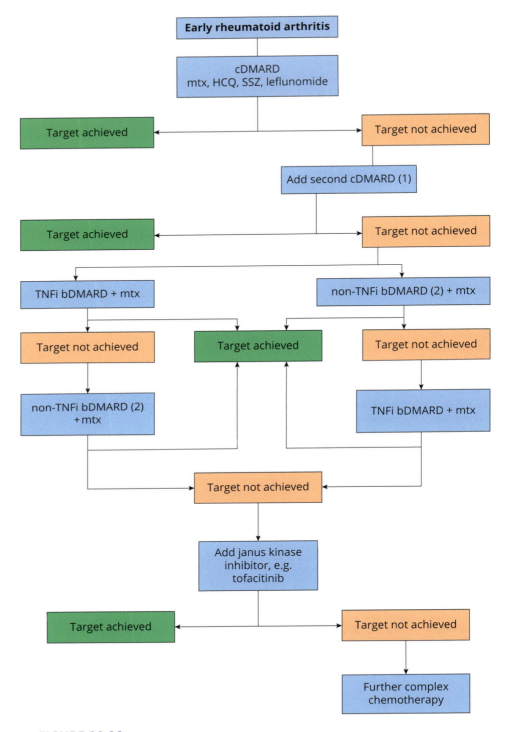

FIGURE 10.10

Treatment options for rheumatoid arthritis. Mtx = methotrexate, HCQ = hydroxychloroquin, SSZ = sulphasalazine. TNFi = TNF inhibitor, Non-TNFi = Non-TNF inhibitor. (1) a glucocorticoid could be used as a bridge, and may be added at any stage to treat an acute flare-up. (2) not anakira.

Figure modified from Singh et al. (2016), Smolen et al. (2020), and NICE NG100.

Treatment options for patients whose disease duration is >6 months and who have moderate or severe disease activity (e.g. DAS28 >5.1) include a TNF inhibitor or non-TNF inhibitor bDMARD with methotrexate. Should this be ineffective, a drug from an alternative bDMARD class may be an option, whilst further treatments include a janus kinase inhibitor. Should these fail to control the disease, other agents are available, such as high-dose steroids (prednisolone) and antimetabolites (azathioprine and cyclophosphamide). However, the high doses of these frequently cytotoxic drugs that are often required are likely to bring major side effects, as they do in cancer chemotherapy.

Non-pharmacological treatment

Other treatments include physiotherapy (to improve general fitness and encourage regular exercise) occupational therapy, and exercise programmes to strengthen the hands and wrists, and help reduce pain and dysfunction. Patients with foot problems should have access to a podiatrist. Surgery is an option in restoring function to those joints effectively destroyed by the disease process, for the relief of pain, and for restoration of flexibility. Knees, hips, shoulders, feet, wrists, and fingers are all established targets for the skill of surgeons. Some 10% of patients with a new diagnosis of RA undergo surgery within 3 years.

The molecular genetics of RA

Strong links with family history and twin studies point to a heritability of 40–60%. These imply a genetic component, the primary one being *HLA-DRB1*04*, although there may also be a place for *HLA-DRB1*01*. Genes outside the MHC with links to RA include *CTLA-4* (at 2q33.2, cytotoxic T lymphocyte associated 4, CD152), *FCRL3* (at 1q23.1, encoding an FcR-like molecule), and *CD244* (at 1q23.3, encoding natural killer cell receptor 2B4), although there are around a further hundred other susceptibility loci. As with cancer and cardiovascular disease, there are reports of scores of miRNAs that are either overexpressed/upregulated, or underexpressed/downregulated, and can be determined in leukocytes (e.g. miR-28, miR-125b, miR150, miR-451) and plasma (miR-125b, miR-130b-5p, miR-452, miR-579).

Osteoarthritis

Although not an autoimmune disease, discussion of OA is justified as a differential diagnosis to RA. The primary aetiology of OA is overweight/obesity, leading to the 'wear and tear' hypothesis, a view supported by evidence of inflammatory adipocytokines. However, this 'overweight' hypothesis cannot account for the link between *ALDH1A2* at 15q21.3, coding for aldehyde dehydrogenase, and hand OA. It is rarely associated with abnormal laboratory tests such as raised CRP and ESR and certainly not RhF or ACPA. There may be an increase in the volume of synovial fluid in OA, but (unlike in RA) it is clear and free of leukocytes and debris, and this too helps the differential diagnosis.

NICE CG177 suggests a diagnosis is made if the person is aged over 45, has activity-related joint pain, and <30 minutes morning stiffness. Treatments focus on exercise for local muscle strengthening, stretching, and general fitness. Pharmacological management includes pain relief with oral paracetamol and topical NSAIDs, escalating to oral NSAIDs/COX-2 inhibitors and opioid analgesics (and possibly a proton pump inhibitor). Intra-articular injections of steroids may help, and ultimately there will be a referral to a surgeon for arthroscopic knee washout, and hip/knee replacement. NICE NG226, MTG76 and QS87 are also relevant.

Key points

Rheumatoid arthritis is the most prevalent autoimmune disease, diagnosed by a scoring system that includes clinical (e.g. synovitis) and laboratory (e.g. RhF) features.

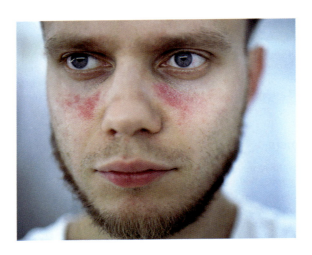

FIGURE 10.11
The facial rash of SLE.

Shutterstock.

10.3.3 Systemic lupus erythematosus (SLE)

This is the second most frequent well-described inflammatory connective tissue disease, and a major autoimmune disease, present at ~1,000/million in the UK (i.e. 10–15% that of RA). In England and Wales in 2021, SLE was the cause of 73 deaths (82% women)—most commonly between ages 70 and 74—compared to 830 deaths (70% women) due to RA—most commonly between ages 80 and 84. A major pathogenetic feature is increased levels of many cytokines, especially interleukins and IFNs, often described as a 'cytokine storm', an expression that resurfaces in the pathophysiology of COVID-19 (discussed in section 13.6). Like RA, SLE is also a multi-system disease, but it is often more aggressive, and so requires heavier immunosuppression, and also carries a worse prognosis (e.g. lupus nephritis). However, in perhaps 5% of patients, disease is limited to the skin, defining it as discoid SLE. Again, it attacks more women than men, but at a greater proportion (6:1, although some say 10:1), and the presenting age is generally a decade earlier (30–40) than in RA (40–50) with peak deaths also a decade earlier: 70–74 v 80–84. In men, the disease often has a more abrupt onset and is more severe (nephritis and serositis), and presents later in life (5th–7th decade) than in women. SLE is proportionally more common in Black and South/East Asian communities. With an estimated 80 HLA loci linked to susceptibility, heritability has been estimated as 44% genetic, 26% shared familial factor, and 30% environmental factors.

Presentation

The leading presenting symptom is of a facial rash (Figure 10.11, cutaneous lesions being present in 75%), with others on the hands, shoulders, abdomen, etc., often associated with **myalgia** and/or arthritis.

However, there are many other non-specific presenting features, reflecting the complexity of the disease, which include fatigue, fevers, infections, and coagulopathy (both thrombosis and haemorrhage), making diagnosis difficult. There may also be anxiety and psychiatric changes, including depression and psychosis, and the final diagnosis may be delayed for months or years. A thorough history may uncover musculoskeletal symptoms in the previous 5 years, present in 60%. These inculde hypertension (in 75%), dyslipidaemia and obesity (33%), gastrointestinal/hepatic signs and symptoms (50%), asthma (39%), and atopic dermatitis (20%).

Diagnosis

As with RA, the number of features provides a score that, when ≥4 are present, defines the disease. These include the following clinical features:

- A fixed **erythema** rash over the cheeks (e.g. Figure 10.11)

- An unusual reaction to sunlight (photosensitivity) causing a skin rash

- Ulcers in the mouth, throat, or pharynx

- Non-erosive arthritis involving two or more peripheral joints, characterized by tenderness, swelling, or effusion

- **Pleuritis** or **pericarditis**

- Neurological signs such as seizures, fits, and psychological changes.

The laboratory is important—there is a preponderance of autoantibodies to various components of the nucleus, both nucleic acid and protein. Antibodies to single- and double-stranded DNA (dsDNA) are almost exclusive to SLE, being present in >90% of patients with active disease. Others autobodies include SSA/Ro, SSB/La, RhF, the lupus anticoagulant, IgG and IgM anti-cardiolipin antibodies, antibodies to β-2 glycoprotein-1, and anti-phospholipid antibodies (aPLs). The latter are key components of the anti-phospholipid syndrome, are implicated in recurrent pregnancy loss and in unexplained arterial and venous thrombosis, and so should be tested for in all patients. Other haematological abnormalities pertinent to a diagnosis include haemolytic anaemia with **reticulocytosis**, leukopenia, **lymphopenia**, and **thrombocytopenia**. Urea and electrolytes must be checked regularly, as renal failure is a primary adverse end point. Abnormal LFTs may also provide evidence of early hepatitis, a further sign, although less common.

Management

As with RA, there may be limited disease with few clinical manifestations, or widespread disease with many organs involved, and the tendency to spread to the lungs, blood vessels, nervous tissue, and cardiovascular system. This may be due to the deposition of immune complexes, with the possible activation of complement, and so type III hypersensitivity, as described in section 10.2.3.

As SLE is certainly a more aggressive disease than RA, with many succumbing to renal failure (with lupus nephritis, present in up to 60% of patients), treatment must be more aggressive, aiming to minimize disease progression with immunosuppression, and to induce a remission. Treatment options are summarized in Table 10.8, although developing therapies such as anti-B cell agents rituximab (targets CD20) and belimumab (targets the cytokine B-cell activating factor, which binds to TNF receptors) are omitted. These options are clearly complex, perhaps reminiscent of other treatments (e.g. for hypertension) where several different classes of drugs are used in triple therapy. Therefore, patients must be under the care of only the most experienced rheumatologists. Moderate and severe disease may call for several rounds of pulses of high-dose therapy reminiscent of cancer chemotherapy, which share side effects (nausea, vomiting, bone marrow suppression, myalgia, fever, etc.).

One pulse regime calls for oral prednisolone at a dose of 0.5 mg/kg/day in parallel with three rounds of 500–750 mg methylprednisolone delivered intravenously, with mandatory U&Es to check renal function. An alternative in lupus nephritis is 500 mg of intravenous cyclophosphamide every fortnight for a total of 6 doses, followed by oral azathioprine (2 mg/kg/day). In the long term, patients should be monitored regularly for disease manifestations, drug toxicity, and co-morbidities with both clinical examination and laboratory investigations.

Active disease is likely to entail clinic visits every 1–3 months, whereas in stable disease or in remission, this may be 6–12 months, reflecting the control of the disease. These options are generally in steady state disease, but will needed to be amended, generally by increasing the doses of certain drugs, in the face of an acute flare-up of the disease, and patients may well self-refer to their care team. Laboratory indicators include increasing titres of antibodies to dsDNA and falling levels of complement, both being particularly useful in lupus nephritis. Co-morbidities of cardiovascular disease and its risk factors (including overweight/obesity), osteoporosis, avascular necrosis, and certain malignancies should be screened for at intervals and managed as appropriate. Women will need to be counselled regarding pregnancy, as certain antibodies may be directly pathological. Unsurprisingly, great care must be taken when choosing drug therapies during pregnancy, such as deciding whether to use hydroxychloroquine.

Molecular genetics of SLE

Over 40 gene loci have SLE links, especially in respect of IFN metabolism, but, as with RA, *HLA-DRB1* is to the fore, but with variants *DRB1*03:01*, *DRB1*15:01* (both conferring a two-fold relative risk), *DRB1*14:01*, and also certain *DQA2* and *DQB2* alleles. In common with many diseases, changes in

TABLE 10.8 Treatment options for SLE

Disease activity	Clinical and laboratory features	Initial drug therapy*	Maintenance drug therapy*
Mild	Platelets 50–149 × 10⁹/l, arthralgia, diffuse alopecia, fatigue, malar rash, mouth ulcers, myalgia	(a) A corticosteroid; topical preferred, but if not oral such as prednisolone ≤20 mg od for 1–2 weeks, *or* i.m. methyl-prednisolone 80–120 mg *and* (b) HCQ 200 mg/day *and/or* (c) methotrexate 7.5–15 mg/week *and/or* (d) NSAIDs (up to a few weeks only)	(a) Prednisolone ≤7.5mg/day *and* (b) HCQ 200 mg/day *and/or* (c) methotrexate 10 mg/week
Moderate	Platelets 25–49 × 10⁹/l, alopecia with scalp irritation, arthritis, cutaneous vasculitis, fever, hepatitis, lupus-related rash ≤2/9 body surface area, pleurisy	(a) Prednisolone ≤0.5mg/kg/day *or* i.v. methyl-prednisolone ≤250mg *or* i.m. methyl-prednisolone 80–200 mg *and* (b) azathioprine 1.5–2.0 mg/kg/day *or* methotrexate 10–25 mg/week *or* MMF 2–3 g/day *or* cyclosporin** ≤2.0 mg/kg/day, *and* (c) HCQ ≤6.5 mg/kg/day	(a) Prednisolone ≤7.5 mg/day *and* (b) azathioprine 50–100 mg/day *or* methotrexate 10 mg/week *or* MMF 1 g/day *or* cyclosporin 50–100 mg/day *and* (c) HCQ 200 mg/day
Severe	Platelets <25 × 10⁹/l, acute confusion, ascites, enteritis, myelopathy, myositis, optic neuritis, psychosis, rash involving >2/9 body surface area	(a) Prednisolone ≤0.5mg/kg/day *and/or* i.v. methyl-prednisolone 500mg *or* prednisolone ≤0.75–1 mg/kg/day *and* (b) azathioprine 2–3 mg/kg/day *or* MMF 2–3 g/day *or* cyclophosophamide 500 mg i.v. every two weeks or 500–750 mg monthly *or* cyclosporin ≤2.5 mg/kg/day *and* (c) HCQ <6.5 mg/kg/day	(a) Prednisolone ≤7.5 mg/day *and* (b) MMF 1.0–1.5 g/day *or* azathioprine 50–100 mg/day *or* cyclosporin 50–100 mg/day *and* (c) HCQ 200 mg/day

MMF = mycophenolate mofetil, HCQ = hydroxychloroquine (alternatives: chloroquine and mapacrine), iv = intravenous, im = intramuscular.

* Non-drug treatments include anti-UVA/UVB sun-block, sun avoidance, and protective clothing for cutaneous disease.

** An alternative to cyclosporin is tacrolimus.

Modified from Gordon et al., for the British Society for Rheumatology Standards, Audit and Guidelines Working Group, British Society for Rheumatology guideline for the management of systemic lupus erythematosus in adults: Executive Summary, Rheumatology, 57: 1; January 2018, 14–18, https://doi.org/10.1093/rheumatology/kex291.

many miRNAs have been reported, some in plasma, some in leukocyte nucleic acids. Some of these are increased, including miR-142-3p, miR-181a, and miR-551b, whilst and others, such as miR-124, miR-342-3p, and miR-17, are decreased. There may also be a place for some miRNAs in defined aspects of SLE, such as lupus nephritis, where miR-223, miR-20a, and miR-758-3p are decreased, whilst miR-21, miR-155, and miR-107-3p are increased.

SELF-CHECK 10.5

How can the laboratory help with the diagnosis and management of SLE?

Key points

SLE has a complex pathophysiology that brings a high risk of nephritis, cardiovascular disease, and infections, all of which can contribute to an increased risk of mortality.

CASE STUDY 10.1 Systemic lupus erythematosus

A 53-year-old man (a non-smoker) presented to his GP with a short (2 month) history of intermittent chest pain, tiredness, and general aches and pains. Following an examination, which finds a rash on the lower back, a BMI of 28.7 kg/m², SBP/DBP 148/89, but an unremarkable ECG, the GP orders routine bloods, which report total cholesterol 6 mmol/l and HDL 1.2 mmol/l, and an HbA1c of 44 mmol/mol. As these figures give a QRISK score of a 7.6% risk of heart attack or stroke within 10 years (section 7.3.6), the patient is started on 20 mg atorvastatin and 4 mg perindopril, both once daily, and is referred to a local cardiovascular risk factor clinic with advice to lose weight through diet and exercise. At clinic, bloods are repeated and confirmed, with new analyses performed (Table 10.9).

The serum creatinine translates to an estimated glomerular filtration rate (eGFR) of 58 ml/min/1.73m². The risk factor clinic refers the patient on to other colleagues with requests to investigate a moderately reduced renal function, mild liver disease, and a normocytic anaemia with pancytopenia.

Over the weeks that follow, the patient stoically visits the clinics, each repeating their particular blood tests, and is invited to return to this GP for an update, where he brings forward further complaints of aches and pains in the shoulders and wrists and some sweating fevers, and the GP notes that the rash has become more severe and more extensive. Considering the possibility of RA, she orders an autoantibody screen, which reports RhF 24 IU/ml (reference range <14) and ANAs at a titre of 1/320 (reference range <1/40) with a coarse speckled pattern. Upon receipt, the GP telephones the laboratory and requests testing for anti-dsDNA antibodies, and an urgent referral the local professor of rheumatology.

The following week, with the result of a high titre of anti-dsDNA antibodies, the professor considers all the results, and finding an asymptomatic nasal ulcer, makes a diagnosis of SLE. Considering the disease at present to be mild (Table 10.9), she starts her patient on 20 mg prednisolone and 200 mg hydroxychloroquine od, with a single dose of 10 mg methotrexate. The steroid is reduced to 15 mg/day after a week, then at week three to 7.5 mg/day (Table 10.9). At this point, having become accustomed to the minor side effects, the patient says he feels better, which is reflected by blood tests in week

TABLE 10.9 Case study 10.1—additional blood tests

Analyte (unit)	Result	Reference range
Urea (mmol/l)	6.0	3.3–6.7
Creatinine (μmol/l)	120	71–133
Bilirubin (μmol/l)	15	<20
Alanine aminotransferase (IU/l)	40	5–42
Aspartate aminotransferase (IU/l)	50	10–50
Alkaline phosphatase (IU/l)	145	20–130
Haemoglobin (g/l)	129	133–67
Mean cell volume (fl)	85	77–98
Haematocrit (L/l)	0.34	0.35–0.53
White blood cells (10⁹/l)	3.0	4.0–10.0
Neutrophils (10⁹/l)	2.0	2.0–7.0
Platelets (10⁹/l)	130	143–400
ESR (mm/Hr)	15	<10

four, as the previously abnormal results are all moving towards their reference range. The pharmaceutical regime continues, with reviews at increasingly longer intervals up to three months.

At a full 3-month review, blood pressure is 136/79, total cholesterol is 3.9 mmol/l on 40 mg atorvastatin daily, BMI is 26.5 kg/m², and all blood results, although not perfect, are acceptable. The professor is happy to reduce the steroids to 5 mg a day, the methotrexate to 5 mg/week, but with the same dose of hydroxychloroquine. After another satisfactory 3-month follow up, the patient is discharged back to his GP. His renal function is checked every 6 months, and he visits the hospital for annual check-ups.

10.3.4 Systemic sclerosis (SSc)

This complex disease (globally ~100/million) is characterized by the progressive fibrosis of a variety of organs, often most evident in the skin (described as scleroderma: scleros = thick, dermos = skin), and often a considerable vascular component. The pathology points to excessive cytokine activity (e.g. IL-1, IL-17, TNF-α, TGF-β, and IFN-γ) that induces fibroblasts to secrete the extracellular matrix components (collagens etc.), deposits of which cause the fibrosis in various organs and tissues.

The disease carries a particularly poor prognosis: in those with serious pulmonary or cardiac involvement, the three-year survival rate is around 50%. The sex ratio targets women, cited as anywhere between 4:1 to 10:1. In many cases, the aetiology and triggering factor(s) are incompletely understood, but there is evidence for a least three strands:

- Exposure to chemicals such as heavy metals, vinyl chloride and silica

- Myofibroblast infiltration of the arterial media, with endothelial damage and apoptosis, leading to thrombosis, hypertension, and vasculitis

- Autoantibodies.

Presentation

Patients may present with any number of a variety of non-specific signs and symptoms, of varying degree of severity and acute/chronic development, and also typical of RA and SLE (arthritis, myalgia), and of other diseases (dyspnoea (shortness of breath and/or difficulty in breathing), gastrointestinal upset, etc.). Following a thorough examination and history-taking, the astute practitioner will probably recognize many of these, and the more specific signs of changes to the skin and fingers. A common feature is Raynaud's phenomenon, caused by poor blood supply to the fingers, which subsequently become **hypoxic**, leading to painful 'white fingers' (Figure 10.12) and toes, often precipitated by cold, and both relieved and reversed by immersion in warm/hot water. This symptom can pre-date other symptoms by years. Older textbooks and guidelines refer to the acronym CREST, often seen as a precursor to systemic sclerosis, the R being Raynaud's. The remainder of the acronym are Calcinosis, oEsophagitis, Scleroderma, and Telangectasia.

The practitioner is likely to order routine blood tests and an auto-antibody screen, and if there are marked symptoms of oesophegeal and/or gastric disease, an endoscopy. At this point it is likely that the patient will be referred to a specialist centre.

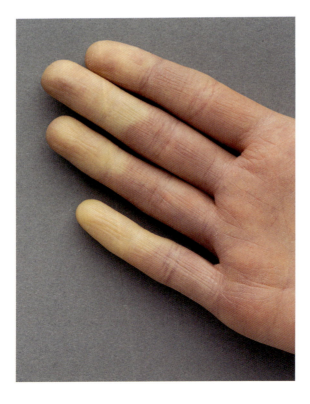

FIGURE 10.12
Raynaud's phenomenon.

Shutterstock.

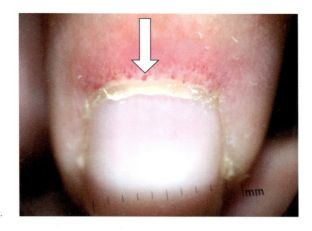

FIGURE 10.13
Abnormal nailfold capillaries.

© 2022 DermNet New Zealand Trust.
Retrieved from: https://dermnetnz.org.

Diagnosis

Whilst ANAs are present in almost all patients, semi-specific subtypes include anti-centromere (around two-thirds of patients), anti-topoisomerase I (Scl-70, ~10%), and anti-RNA polymerase III (least common, ~5%). Less frequent are autoantibodies to U3-RNP (fibrillarin), U1-RNP, and U11-U12 RNP. Anti-RNA polymerase III antibodies are linked to an increased risk of malignancy. However, two other investigations that go beyond a physical examination are also important. These are nailfold capillaroscopy, which requires a low-power microscope, abnormalities present in >75% of mild/early disease (Figure 10.13), and lung function, in view of potential pulmonary hypertension and fibrosis. From these, a points-based scoring system has been devised, based mostly on eight clinical and laboratory features (Table 10.10).

Once a primary diagnosis of SSc has been made, its subtype must be determined: limited cutaneous (lcSSc: few manifestations) or diffuse cutaneous (dcSSC: widespread disease) (Table 10.11).

TABLE 10.10 Diagnosis of SSc

Item	Specific feature	Score
Skin thickening of the fingers of both hands	Must extend proximally to the metacarpophalangeal joints	9
Skin thickening of the fingers	Puffy fingers	2*
	Sclerodactyly of the fingers	4*
Fingertip lesions	Digitial tip ulcers	2*
	Fingertip pitting scars	3*
Telangiectasia		2
Abnormal nailfold capillaries	(As in Figure 10.13)	2
Lung disease	Either pulmonary arterial hypertension or interstitial lung disease	2
Raynaud's phenomenon	(Need not be as severe as in Figure 10.12)	3
Autoantibodies	Anticentromere, anti-topoisomerase, or anti-RNA polymerase III antibodies	3

* Count only the higher score in this section.

Modified from van den Hoogen et al. 2013 Classification Criteria for Systemic Sclerosis: An American College of Rheumatology/European League Against Rheumatism Collaborative Initiative. Arthritis & Rheumatism, 65: 2737–47. https://doi.org/10.1002/art.38098.

TABLE 10.11   Clinical, demographic, and laboratory features of SSc

Features more common in dcSSc	Features with equal frequency	Features more common in lcSSc
Earlier age of presentation (typically 45 years), anti-Scl70 positive, active disease, digital ulcers, elevated acute phase response, synovitis, muscular disease, gastric disease, pulmonary disease, dyspnoea, palpitations, hypertensive renal crisis, proteinuria, short history of Raynaud's phenomenon	Anti-nuclear antibodies positive, presence of Raynaud's phenomenon, oesophageal disease, intestinal disease, pulmonary artery hypertension, heart conduction block, diastolic dysfunction, left ventricle ejection fraction, arterial hypertension	Later age of presentation (typically 48 years), female sex, anti-centromere positive (and more likely to be positive in women), long history of Raynaud's phenomenon

This classification is important because dcSSc carries a 10-year survival of around 75%, whilst the slow organ involvement in lcSSc brings a relatively good prognosis with a 10-year survival of >90%. A third type, Sine SSc (sSSc, 5% of patients), is characterized by Raynaud's phenomenon, typical SSc autoantibodies or capillaroscopic features, no skin thickening, but disease in certain organs or other vascular manifestation.

Management

Without a clear common pathophysiology, treatment is difficult and directed towards specific areas. Therefore, focus is on major organs, both in terms of screening for abnormalities, and treatment options.

- The lung is important because interstitial lung disease brings a mortality rate of 35% and pulmonary artery hypertension (PAH) a rate of 26%, and together, they are present in 80% of patients. Leading treatments include intravenous pulses of cyclophosphamide for 6–12 months, mycophenolate mofetil, and azathioprine, whilst endothelin receptor antagonists, epoprostenol analogues, and phosphodiesterase-5 inhibitors are treatments for PAH.

- Gastrointestinal symptoms, present in up to 90% of patients, may be investigated with endoscopy, barium swallow, gastric-emptying, and a breath test. Treatments include antacids, proton-pump inhibitors, and H2 blockers, whilst dietary adjustment may help. In severe cases, enteral or parenteral feeding may be necessary.

- Around a third of SSc patients have a cardiac pathology, which can progress to disease such as pericarditis and heart failure, and is linked to anti-centromere antibodies (present in 37%) and anti-Scl70 antibodies (24% of patients). Increased serum markers (NT-proBNP, troponin-T) are common and link to 20–30% of deaths. Thus, echocardiography and ECGs are recommended annually. Standard CVD treatments apply, such as ACEIs and diuretics, with immunosuppression for myocarditis.

- Renal disease may be screened for with urinalysis, U&Es, and eGFR. However, presence of anti-RNA polymerase III antibodies or early dcSSC are risk factors for renal disease.

- Skin disease is very common (except in sSSc), and although it often causes profound morbidity (ulcers, pruritus, loss of flexibility), it is rarely linked to mortality. Nevertheless, dermal disease may be treated with methotrexate (15–25 mg/week) or mycophenolate mofetil (≥3 g/day), with cyclophosphamide 2 mg/kg/day for severe lesions. Calcium channel blockers (such as nifidipine) are the primary choice for peripheral vasculopathy, followed by fluoxetine and angiotensin receptor blockers, with intravenous prostacyclin for serious Raynaud's phenomenon.

- Digital ulcers (present in 50% of patients at some time, and in 20% on a particular occasion) may be treated with phosphodiesterase-5 inhibitors. Subcutaneous calcinosis may respond to bisphosphonates and chelating agents, and may be treated by surgery.

- Arthritis and musculoskeletal disease may be treated with physiotherapy, NSAIDs, DMARDs and low-dose prednisolone (<10–25 mg/day).

A final option is autologous bone marrow transplantation and perhaps lung transplantation. Developing therapies include tocilizumab (anti-IL-6), nintedanib (inhibiting tyrosine kinase, for lung fibrosis), fresolimumab (a TGF-β inhibitor), tralokinumab (anti-IL-13), and abatacept (an antibody to CTLA). There is no dedicated NICE guideline, but ES7 and ESUOM42 are relevant.

Molecular genetics

Twin studies confirm a genetic factor, with *HLA-A3*, *HLA-B18*, and *HLAs DRB1*, *DQB1*, and *DPB1* being found more frequently. A GWAS of over 9,000 patients with SSc and over 17,500 controls found 28 risk loci, with strong links with *TNFSF4* (at 1q25.1, coding for OX40L/CD134), *STAT4* (2q32.2–q32.3, coding for a transcription factor), and *IRF5-TNPO3* (7q32.1, coding for interferon regulatory factor 5). The fibrosis found in early SSc and localized scleroderma may be linked to miR-485-5p and snRNA-U6, as the former may be a pro-fibrotic driver. In common with RA and SLE there is evidence of a place for miR-146, but in SSc low levels are linked to disease progression and lung fibrosis.

Key points

SSc may be seen as a triad of vascular disease (Raynaud's), autoantibodies (ANA, anti-Scl-70, anti-centromere), and fibrosis of major organs (the lungs, heart).

SELF-CHECK 10.6

What is the presumed immunological cause of the fibrosis in SSc?

CASE STUDY 10.2 Systemic sclerosis

A 50-year-old woman visits her GP, complaining of increasingly frequent acid reflux after eating, which is only part-managed by liquid antacids. The GP performs a full physical examination, and notes some irregular lumps under the skin of the fleshy parts of some fingers. Upon questioning, the patient says that these appeared a few months ago, but are not particularly painful. When the GP mentions the fingers look and feel a little swollen, she agrees, saying that she used to be able to rotate her wedding ring, but can no longer. She also says that she needs woolly gloves in even mildly chilly weather if she is to avoid painful fingers. There is no joint or muscle pain, no rash, and the patient denies fevers or more frequent infections.

The GP takes some blood for routine tests and an autoantibody screen, orders standard chest and abnormal X-rays, and arranges an endoscopy. Several days later, the full blood count, liver function tests, and U&Es are all within their reference ranges. As few days after this, the autoantibody screen results show moderately positive ANAs, and weakly positive anti-centromere and anti-Scl-70 antibodies, which, alongside the clinical signs, make the diagnosis of SSc. Further immediate lung function tests and cardiac assessments are all normal, although there are some nailfold capillary abnormalities. These make the diagnosis of limited systemic sclerosis.

Over the next few years, the patient complains of a series of additional signs and symptoms. These include not being able to open her mouth wide in order to eat, and that she is always the last to clear her plate, with 'stiff' cheeks (Figure 10.14) (indicating fibrosis, and leading to the prescription of methotrexate). She also reports the appearance of red blotches on her arms and legs.

The next symptoms to appear are Raynaud's phenomenon (prompting prescription of nifedipine), and becoming short of breath. The latter calls for lung function tests, which report poor function. The subcutaneous finger nodules are

(Continued)

CASE STUDY 10.2 *Continued*

getting large, are painful and more extensive, and are bursting through the skin, requiring local surgery and calling for bisphosphonates for the calcinosis. The titres of the autoantibodies are increasing, with evidence that the disease is transforming from the limited to the diffuse phenotype.

In the following years, the patient suffers weight loss, hair loss, and further difficulty in breathing, with pain in the chest. She is admitted to hospital for assessment, where pulmonary hypertension is diagnosed, but following a prolonged coughing spell, and an acute severe dyspnoea, she collapses and cannot be resusitated. The post-mortem finds a large pulmonary embolism.

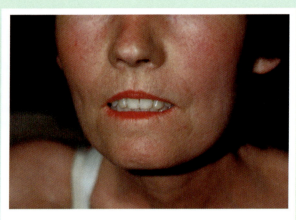

FIGURE 10.14
Systemic sclerosis. This image shows tightening of the skin of the cheeks. Note that the patient is wearing red lipstick.

Courtesy of Dr Dharam Ramnani, WebPathology LLC.

10.3.5 Vasculitis

Aetiology

There are numerous forms of this disease, literally, inflammation of blood vessels. Serology focuses on autoantibodies, whilst the histopathology points to infiltration of the vessel wall with leukocytes, fibrinoid necrosis, endothelial damage and granuloma formation. Formal diagnosis is generally by biopsy of a particular artery. Some arise from a background of RA or SLE (hence rheumatoid vasculitis and lupus vasculitis), or develop without any other obvious inflammatory disease.

The many variants have a number of non-specific clinical features in common that include anaemia, fevers, myalgia, arthritis and arthralgia, headache, and haemorrhage. Major organ diseases include those of the kidney (with glomerulonephritis and renal failure), the cardiovascular system, and the skin, the latter causing rash and ulcers, with purpura, haemorrhagic blisters, and digital gangrene (Figure 10.15). The disease may be classified by the size of the inflamed artery as large, medium, or small.

Large-vessel vasculitis (e.g. >150 mm)

This group consists of giant-cell arteritis (GCA) and Takayasu's arteritis—the histology of the vessels is indistinguishable (with the presence of granuloma) in these and so they may be the same disease. Morphological (MRI, CT, ultrasound) and metabolic (PET scanning) imaging is valuable to help diagnosis and to monitor treatment.

GCA, with an incidence of ~170/million, often (~50%) presents with fever; stiffness; painful and rheumatoid-like symptoms of the chest, lower neck, and shoulders referred to as polymyalgia rheumatica, with smoking being the only established risk factor. Cranial presentation, most likely as temporal arteritis, is often characterized by headache, pain in the muscle of the jaw, and ocular symptoms, and in the laboratory by an abnormal ESR (typically >60 mm/hour, reference range <10 mm/hour). Once a firm diagnosis has been made, treatment must be started promptly and aggressively if blindness is to be

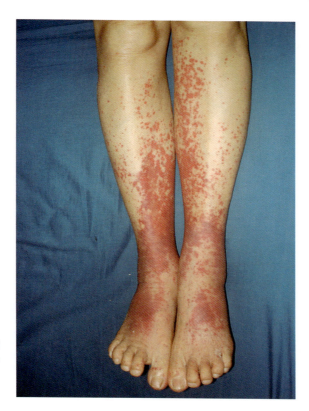

FIGURE 10.15

Cutaneous vasculitis. This form of vasculitis is associated with a rash of haemorrhagic lesions.

Shutterstock.

avoided as the temporal artery part-supplies the retina. The success of treatment with oral predniso-lone (often 40–60 mg a day) can be followed with a reduction in the ESR (and perhaps the CRP), which will in turn guide a reduction in the dose until a suitable maintenance dose is found.

Takayasu's arteritis, generally being diagnosed before the age of 50, primarily focuses on the aorta and its major branches, with imaging-defined arterial stiffness, stenoses, occlusion, and aneurysm. Diagnostic features include fever, absent or weak pulses, bruits, hypotension, weakness, visual loss, light-headedness, and stroke. Management is difficult due to its chronic and indolent pathology, whilst the laboratory (non-specific acute phase reactants CRP and ESR) and imaging have little con-crete to offer. Developing treatments for large-vessel vasculitis include bDMARDs such as janus kinase inhibitors and the anti-IL-6 agent tocilizumab.

Medium-vessel vasculitis (e.g. 50–150 mm)

This group also has two members: polyarteritis nodosa (PAN) and Kawasaki disease. The former is char-acterized by serial aneurysms and stenoses, often of the arteries feeding the internal organs, frequently with cutaneous involvement, but free of small-vessel disease and with no glomerulonephritis. Subgroups include virus-associated, cutaneous, and drug-induced PAN. The literature frequently describes PAN as necrotizing, implying necrosis of downstream tissues resulting from hypoxia and inflammation. Kawasaki disease is effectively paediatric vasculitis, and often attacks the coronary arteries.

Small-vessel vasculitis (e.g. <50 mm)

There are several diseases in this group, which focus on arterioles, capillaries, and venules. They are classifiable by autoantibodies to different components of the neutrophil cytoplasm (ANCAs, as described in section 9.8.3, where Figure 9.20 applies), hence ANCA-associated vasculitis (AAV). There may be immune complex deposition, and in all cases renal damage is a common problem (Table 10.12), also discussed in section 14.1.1.

TABLE 10.12 Small-vessel vasculitides

ANCA-positive	
Microscopic polyangiitis	A necrotizing vasculitis of small and medium-sized vessels, often of the kidneys (causing glomerulonephritis in ~95%) and lung (pulmonary capillaritis). pANCA dominates
Granulomatosis with polyangiitis (Wegener's granulomatosis)	Necrotizing granulomatosous inflammation, often of the respiratory tract and kidneys, where a necrotizing glomerulonephritis is common (~65%). cANCA dominates
Eosinophilic granulomatosis with polyangiitis (Churg–Strauss syndrome).	As above, but with eosinophil infiltration, peripheral neuropathy, and eosinophilia (>10%). pANCA dominates. Renal disease less likely (~40%). Tendency to asthma and allergic sinusitis
Immune complex related	
Cryoglobulinaemic vasculitis	Skin, glomeruli, and peripheral nerves often involved
IgA vasculitis (Henoch–Schonlein disease)	Often involves skin and gastrointestinal tract, and frequently causes arthritis and glomerulonephritis, which is indistinguishable from IgA nephropathy
Hypocomplementaemic urticarial vasculitis	Commonly causes ocular and pulmonary disease, arthritis, and glomerulonephritis
Anti-glomerular basement membrane (anti-GBM) disease	Chronic renal failure follows glomerulonephritis due to deposition of anti-GBM autoantibodies. May also cause pulmonary haemorrhage

Based on Jennette et al., 2012 Revised International Chapel Hill Consensus Conference Nomenclature of Vasculitides. Arthritis & Rheumatism 2013; 65; 1–11. https://doi.org/10.1002/art.37715.

ANCAs have two major forms: pANCA targets myeloperoxidase, and has a perinuclear distribution, whereas cANCA targets proteinase 3, and has a cytoplasmic distribution (Figure 9.20). These antibodies are not merely markers of the disease, but are active in the disease process, stimulating neutrophils to inappropriate activation and promoting complement activation with consequent endothelial damage.

Management

Treatment is dominated by immunosuppression, such as pulse intravenous or oral cyclophosphamide and glucocorticoids for inducing remission, although the high doses—such as prednisolone (0.75 mg/kg/day), cyclophosphamide (2.5 mg/kg/day), chlorambucil (0.2 mg/kg/day)—required to control the disease often lead to infections, infertility, cytopenias, end-stage renal disease, and malignancy. Choices of maintenance therapy include methotrexate (typically 10–15 mg/week), azathioprine (perhaps 50 mg/day), and mycophenolate mofetil (a starting dose may be 1 g/day). Unfortunately, patients on regimes of this nature often (50%) relapse within 3–5 years and in these, the 1-year mortality is around 15%. Accordingly, less toxic therapies such as anti-CD20 monoclonal antibodies, and those targeting BAFF, IL-6, and C5a have potential. Patients whose disease is linked to anti-glomerular basement membrane antibodies and ANCAs may benefit from plasma exchange.

In many cases, AAVs have relatively good outcomes, with 10-year survival of ~80%, although some, such as granulomatosis with polyangiitis, have a lower rate (65%, but 40% with renal involvement). Serious infections (often requiring hospitalization) are a common (25% of patients) problem, and with a high frequency of renal pathology and progression to end-stage disease, dialysis is common. AAV may also arise from drugs: cocaine, allopurinol, phenytoin, hydralazine, and propylthiouracil. Drug

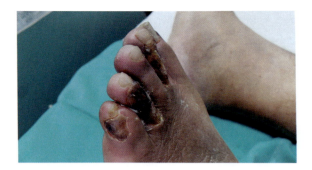

FIGURE 10.16
Digital gangrene in cryoglobulin vasculitis.

Shutterstock.

targets include B/plasma cells with rituximab (in combination with glucocorticoids), anti-BAFF blisibimod, and proteasome inhibitor bortezomib, T cells with anti-CTLA-4 abatacept, and cytokines with anti-IL-6R tocilizumab and IL-5-blocker mepolizumab (in eosinophilic disease).

Molecular genetics of vasculitis

GWASs have identified several genes involved in either susceptibility or resistance to AAV. In a European population, pANCA-positive microscopic polyangiitis is linked to *HLA-DQ*, whereas cANCA-positive granulomatosis with polyangitis is linked to *HLA-DP*. Non-MHC gene links include *PTPN22* (possibly related to IL-10), *SERPINA1* (coding for alpha-1-atiytypsin), *PRTN3* (coding protease 3, a component of neutrophil granules), and *SEMA6A* (coding for semaphorin-6). 'Immunological' gene links include *IRF5* (coding for interferon regulatory factor 5) and *TLR9* (coding toll-like receptor 9). The eosinophilic variant of vasculitis is linked to *HLA-DRB1*.

Secondary vasculitis

Vasculitis is a recognized as a late complication of SLE, sarcoid and RA, developed by ~3%, with the 5-year mortality rate being ~40%. Hepatitis C virus infection is associated with a vasculitis secondary to cryoglobulinaemia, often present in ~40–60% of patients. It is generally type 2 (mixed, see section 9.8.3), and may lead to renal disease and **digital gangrene** (Figure 10.16). Thus the practitioner may well be prepared for the use of cytotoxics such as pulse cyclophosphamide, prednisolone, and azathioprine—so much so that the outlook is often not as dark as it is when the vasculitis develops *de novo*. Other treatments are common with the primary vasculitides, that is, anti-cytokine and anti-B cell bDMARDs. The development of rheumatoid vasculitis may be linked to certain *HLA-DRB1* variants.

Key points

Vasculitis is a rare but complex disease with many variants, although renal disease is a common finding. The severity of the disease often calls for high-dose cytotoxic drugs to induce a remission, followed by a lower maintenance dose. However, several non-cytotoxic therapies are proving to be effective.

SELF-CHECK 10.7

Why are ANCAs important in vasculitis?

10.3.6 Other inflammatory connective tissue diseases

In this final section on autoimmune disease, we look at a number of diverse but reasonably well characterized conditions present in modest numbers that bring morbidity but are rarely fatal.

FIGURE 10.17
The Schirmer test for eye lubricants.

Shutterstock.

Sjögren's syndrome (SS)

This disease is characterized by lymphocyte invasion of glands that produce lubricating fluids. Cell pathology is based on Th1 and Th17 cells and dysregulated Foxp3 positive T_{reg} cells, with involvement of numerous interleukins, TGF-β, and IFN-γ. There are two characteristic autoantibodies to ribonucleoprotein complexes: Ro/SSA (present in ~75% of patients; there are two types: Ro-52 and Ro-60) and La/SSB (present in 45% of patients, of whom 95% also have anti-Ro/SSA antibodies). However, ANAs and RhF are also often present. The disease can arise *de novo*, described as primary Sjögren's syndrome (pSS), or in concert with RA (present in ~26% of patients), SLE (~25%), and SSc (17%), where it is secondary Sjögren's syndrome (sSS). pSS has a prevalence of ~600/million, and again, many more women are affected although men have more complex disease than women. The highest global disease activity index score was reported in Black/African Americans, followed by White, Asian, and Hispanic patients (6.7, 6.5, 5.4, and 4.8, respectively; P < 0.001) (Brito-Zerón et al. 2020).

The major symptoms are dry mouth (xerostomia) and dry eyes (keratoconjunctivitis sicca (sicca = dryness)) together present in 90% of patients, whilst arthralgia and arthritis is present in 50%, other pathology including interstitial lung disease, renal disease, skin lesions, and vasculitis. Defining investigations include a low rate of saliva production (<0.1 ml/min), the Schirmer test (assessing eye fluid absorption on to a strip of filter paper, <5 mm in 5 mins) (Figure 10.17), use of certain dyes (lissamine green, fluorescein) to detect ophthalmic lesions, and for the presence of autoantibodies to the Ro/SSA antigen.

As with other diseases, results of these investigations and clinical findings form a scoring system to make the diagnosis (Table 10.13), defined as ≥4, once inclusion and exclusion criteria have been addressed.

TABLE 10.13 Classification criteria for primary Sjögren's syndrome

Item	Weight/score
Labial salivary gland with focal lymphocytic sialadenitis	3
Positive for anti-SSA (Ro) antibodies	3
Ocular staining score ≥5 on at least one eye	1
Schirmer test ≤5 mm/5 min on at least one eye	1
Unstimulated whole saliva flow rate ≤0.1 ml/min	1

From Shiboski et al. American College of Rheumatology/European League Against Rheumatism Classification Criteria for Primary Sjögren's Syndrome: A Consensus and Data-Driven Methodology Involving Three International Patient Cohorts. *Arthritis & Rheumatology* 2016: 69(1); 35–45. https://onlinelibrary.wiley.com/doi/10.1002/art.39859.

TABLE 10.14 Diagnosis of SpA

(a) Axial SpA
Inflammatory back pain, current or past arthritis or psoriasis, buttock pain, enthesitis (inflammation of the entheses, the point where ligaments and tendon meet bone), uveitis (inflammation of the iris and nearby tissues), dactylitis (inflammation of an entire finger or toe), Crohn's disease ulcerative colitis, improvement within 48 hours of taking NSAIDs, family history of SpA, HLA-B27, elevated CRP, Age <45.

(b) Peripheral SpA
Arthritis, ethesitis, or dactylitis

Plus one or more of psoriasis, inflammatory bowel disease, HLA-B27 uveitis, sacroliitis on radiography or MRI imaging	or	Plus one or more of history of inflammatory back pain, family history of SpA

Based on Ward et al. (2019) and NICE QS170.

Treatment focuses on clinical issues. With eye disease, there are topical lubricants, **secretagogues**, and (with caution, for a short term) cyclosporin and steroids. Dry mouth may also be treated with lubricants and secretagogues, and topical fluorides to prevent dental disease. Other treatments include DMARDs (preferably hydroxychloroquine) for arthritis, NSAIDs and steroids for parotid gland swelling, and cyclophosphamide and steroids for lung disease. Rarely, cryoglobulins (section 9.8.3) may form and, if present, cause a vasculitis in a third of patients, which can be treated with methyl-prednisolone, **plasmapheresis**, and rituximab to suppress B cell activity. Emerging treatments include secukinimab (anti-IL-17), infliximab (anti-TNF), and belimimab (anti-BAFF). SS also brings a risk of non-Hodgkin's lymphoma (commonly mucosal), which accounts for a significant proportion of the increased mortality rate. Linked genetic loci include *HLA-B8* and *HLA-Dw3*.

Spondyloarthropathy (SpA)

The common feature of these diseases is pathology of the spine, and is dominated by HLA-B27. SpAs have a prevalence of ~1.2% in Europe and 1.4% in the USA, with a mean age at diagnosis of 30 years. Classification focuses on axial (vertebral) and peripheral involvement. Axial SpA may be diagnosed if there is (a) sacroiliitis on imaging (suggestive on MRI, definite on X-ray) plus ≥1 of the features in Table 10.14a, or (b) HLA-B27 plus ≥2 of the features in Table 10.14(a). Diagnosis of peripheral SpA calls for a different approach, but of mostly the same criteria (Table 10.14b). Correct diagnosis is important as >15% of the population complain of low back pain.

Specific SpAs include psoriatic arthritis, reactive arthritis, and an arthritis linked to inflammatory bowel disease (enteropathic spondyloarthritis). However, the dominant SpA is ankylosing spondylitis (AS), and is unusual among these inflammatory diseases as it is primarily found in young men (typical age at symptom onset being age 25), and is strongly linked to certain *HLA-B27* types and to the intestinal microbe population (the microbiome). However, as the national prevalence of HLA-B27 is 6.1% in the USA and 8% in the general UK population, care must be taken over the diagnosis as having this HLA type is not a sufficient criterion for a diagnosis with AS by itself. HLA-27 is present in 90–95% of UK patients with AS but is estimated to have a prevalence of 1% in a European population, such that presence of the HLA type brings a 16-fold risk of the illness. AS is three times more common in White Americans than Black Americans, possibly reflecting the frequency of HLA-B27 in these groups of 85.3% and 62.5% respectively. Although Latino ethnicity is largely self-reported and very heterologous, HLA-B27 is present in 86.7% of AS patients. There is also an ethnic difference in a functional impairment score (Black Americans 62.5, White Americans 27.8, Latinos 38.1), in a disease activity score (5.9, 3.5, 4.5 respectively), and erythrocyte sedimentation rate (median 27 mm/hour, 10, 17 respectively) (Jamalyaria et al. 2017).

The principal pathology is inappropriate bone growth and inflammation (osteitis, **enthesitis**) in the lower spine and sacro-iliac region, which brings pain and loss of flexibility with fusion of vertebrae. Accordingly, diagnosis is often confirmed by spinal X-ray or MRI, and there may also be other joint disease. Uveitis is present in ~35%, intestinal inflammation in ~55%, and 10% have inflammatory bowel disease.

Cell biology points to stimulation of CD4-positive T cells, γδT cells, and ILCs by IL-23 and TNF, which release IL-17, which in turn further promotes inflammation. *HLA–B27* subtypes **27.02, *27:03, *27:04, *27:05*, and **27:10* (but not subtypes **27:06* and **27:09*) are implicated, pointing to abnormal peptide presentation to T$_c$ cells. The *HLA-B27* link is particularly strong, with an odds ratio of ~100, but even so this accounts for only 30% of total heritability. Conversely, <2% of those with *HLA-B27* develop AS, pointing to other features. There are minor effects of *HLA-B40*01* and *HLA-B13*02*, and twin concordance is ~60%.

With few precisely targeted treatments, most patients are offered NSAIDs (ibuprofen etc.) as first-line treatments, followed by cDMARDs (sulphasalazine and methotrexate) and bDMARDs (anti-TNF, anti IL-12/23, and anti-IL-17 monoclonal antibodies), whilst active physiotherapy can slow bone fusion. Local spinal injections of corticosteroids may be effective, and in the worst cases, surgery can help. AS brings a 1.3-fold risk of acute coronary syndromes, 1.35 for stroke, and 1.6 for mortality, mostly related to co-morbidities.

Psoriatic arthritis

Psoriasis is a common (~3% of the general population) disease characterized by dry, scaly, and painful skin and nail lesions. Unfortunately, 25–30% will develop arthritis of peripheral and axial joints—psoriatic arthritis (PsA)—which in 20% will be irreversible, deforming, and with permanent loss of function, with the possibility of enthetitis (35–50%), and dactylitis in ~30% (Figure 10.18). The trigger for this transition is unclear, but may be due to a streptococcus infection. The sex ratio is roughly equal and the mean age of diagnosis between 40 and 50.

Around 17.5% are positive for ACPAs, and this subgroup is likely to have more swollen joints, dactylitis, and erosive change than seronegative patients. In addition to the established IL-17/IL-23 link in all SpAs, there is evidence for a role of the IL-9/IL-9R. These and other data strongly implicate T cells in general and T$_h$17 cells in particular, with possible roles for CD8-positive cells and NK cells in PsA synovium. The use of bDMARDs infliximab, certolizumab, etanercept, secukinumab, ustekinumab, and adalimumab (Table 10.8) has significantly changed the course of the disease for many patients. The working rule of 'treatment to target', defined as reduction in clinical indices such as tender and swollen joints, is adopted by many practitioners. CRP and ESR may respond, but other potential serum markers of disease activity include osteoprotegerin, Dickkopf-1, and MMP-3.

Of the various *HLA-B27* types, *HLA-B27*05:02* is linked to increased likelihood of symmetric sacroiliitis, enthesitis, and dactylitis. *HLA-B8*01:01* has been linked to the presence of peripheral joint erosions, axial joint fusion, extent of skin disease, dactylitis, asymmetrical sacroiliitis, and younger age at diagnosis.

Reactive arthritis

It has long been known that an arthritis closely follows genital infections with *Chlamydia trachomatosis* and *Neisseria gonorrhoeae*, hence reactive arthritis (ReA), previously also known as Reiter's syndrome. To these organisms, we now add a variety of **enteric** infections including *salmonella, yersinia,*

FIGURE 10.18
Dactylitis. Swollen fingers, often described as sausage-like, and marked dermal lesions.

Shutterstock.

shigella, and *campylobacter* species, alone or in concert, and viruses and parasites. Infection with a *Borrelia* bacteria (carried by a tick) can lead to Lyme arthritis. With an incidence of ~130/million, leading symptoms are those of most acute-onset arthritides (an asymmetric oligoarthritis, dactylitis, uveitis, urethritis, etc.), perhaps preceded by diarrhoea, but diagnosis rests on isolating and defining the causative organism.

Most cases of ReA are mild and self-limiting, but can develop into chronic arthritis in ~20%. However, it may transform into septic arthritis, wherein the infective agent (commonly *staphylococci* and *streptococci*) proliferates out of control, causing a very destructive arthritis, and may develop systemically into septicaemia, bringing mortality rates of up to 25%. Treatment is as with RA (i.e. DMARDs etc.), aiming for symptomatic relief and disease suppression (section 10.3.2).

SELF-CHECK 10.8

What are the major clinical signs of spondyloarthropathy?

Key points

Spondyloarthropathy is a disease characterized by arthritis of vertebral and peripheral joints, dominated by ankylosing spondylitis. Other spondyloarthropathies include psoriatic and reactive arthritis.

Sarcoid

This multisystem disease is characterized by a granuloma of macrophages, lymphocytes, and epithelial cells in numerous organs and tissues. The cell pathophysiology arises from T_h1 overactivity with increases in many cytokines. Leading HLA risk links defined by next-generation sequencing include *HLA-B08*01*, *-B8*04*, *-B3*01*, and *-B13*02*, whilst *DRB1*01/*04* has been associated with decreased risk. Extrapulmonary involvement has been linked to *DRB1*04:15* and disease progression to *DQB1*06:02*. Non-MHC genes involved in sarcoid include *IL1A* and *IL23R*. The incidence of sarcoid varies markedly with geography: 415/million in Iceland and Poland, 177/million in Sweden, 145/million in Norway, 114/million in Finland, 95/million in Canada, 87/million in Japan, 61/million in the USA, 42/million in Spain, and 11/million in Greece. Within the USA, the incidence and prevalence rates per million are, respectively, 178 and 1,414 for African Americans, 81 and 498 for White Americans, 43 and 217 for Hispanics, and 32 and 189 for Asians. In each group, prevalence was higher in women than in men (Baughman et al 2016). These data on variation within the United States and globally suggest a complex aetiology that involves sex, race, ethnicity, and the environment. Other risk factors include obesity, HLA-DRB1*03 (for acute onset disease), being a non-smoker, and exposure to silica (as with RA, SLE, and SSc). There is also an increased risk of malignancy: two-fold for haematologic, skin, and major organ cancers.

Around 90% of cases first present with pulmonary symptoms, where diagnosis can be difficult and so prolonged, and if in the lungs, diagnosis is reliant on non-specific symptoms such as dyspnoea, cough, and pain in the chest. In the absence of a biopsy, imaging can help the diagnosis. Other features include presentation in skin, which can take the form of soft subcutaneous nodules, ulcers, erythema, and other forms of rash, and a biopsy will make the diagnosis. There may also be sentinel lymphadenopathy. Ocular disease presents with uveitis and scleritis and is also easier to diagnose.

Should it be necessary, treatment is with glucocorticoids. In severe lung disease, this may be prednisolone 20–40 mg/day, reducing over months to a maintenance dose of 5–10 mg/day, whilst the most common cDMARD is 10–25 mg/week methotrexate. This regime may also apply to skin and ocular involvement at doses that reflect symptoms and severity (skin disease by lotions, ocular disease by eyedrops). As with other diseases, bDMARDs such as anti-TNFs may be useful, such as infliximab for pulmonary disease.

Myopathy

Inflammatory disease of the muscles (myositis) generally presents with features such as myalgia and muscle atrophy and weakness, and each has its own features. For a firm diagnosis, biopsy is required, and the laboratory may also report autoantibodies to Jo-1 (a tRNA-synthase), alongside raised serum enzymes—lactate dehydrogenase, AST, and ALT. The most common extra-muscular complication is interstitial lung disease, followed by cardiovascular disease. Dermatomyositis has an incidence of 14/million and prevalence of 58/million, with a female preponderance. Dermatological aspects include erythema and papules, whilst dysphagia, dysarthria, interstitial lung disease, and malignancies may all develop. Pathophysiology involves IFN-γ, IL-1, and TGF-β, with leukocyte invasion of the dermis and endothelial damage. Polymyositis has an incidence of 38/million and prevalence of 97/million, also with a female preponderance. It is characterized by markedly raised creatine kinase, possibly the result of cytotoxic CD8 T cells. IFN-γ, IL-1, and TGF-β are again implicated, as are TNF-α, IL-6, and IL-8.

Underlining the potential severity of the disease, the principal treatment is with pulsed intravenous 250–1,000 mg prednisolone/day for 3–5 days, then standard oral dosing of 1 mg/kg/day for at least 4 weeks, followed by a steady and tapered reduction to 2.5 mg/day, dependent on symptoms. Azathioprine, methotrexate, or mycophenolate mofetil may also be required in the initial immunosuppressive phase, or in maintenance, whilst cyclosporin and intravenous immunoglobulins may be additional or alternative treatments.

Behcet's disease

This rare and often debilitating chronic disease has no clear trigger, although intestinal pathology and infectious agents have been suggested. Principal clinical features are oral (97% of patients, Figure 10.19) and genital ulcers (~75%), others being skin, eye, urological, neurological, pulmonary, and musculoskeletal involvement. Vascular involvement brings risk of arterial and venous thrombosis. Prevalence per million also varies, from ~2,500 in Turkey and 680 in Iran to 60 in Europe. Presenting mainly at ages 20–40, and in 20% more men than women, the leading genetic link is with *HLA-B51*, present in 60% of cases, in line with the familial link of 30%. Other MHC links include *HLA-A30* and *HLA-A26*. There are links with *ERAP1* (as in AS), and with genes involved in the IL-10, IL-12, IL-23, and TLR pathways.

Treatment is directed to affected organs/tissues—for ulcers, corticosteroids (topical application of lotions, tablets, and a mouthwash), anti-bacterials, and local anaesthetics, whilst topical steroids may also be used for uveitis. For more severe disease, oral azathiprine (2–3 mg/kg/day), mycophenolate mofetil (2–3 g/day), and dapsone (2–3 mg/kg/day, best known as a leprosy treatment) are options. Symptoms such as arthritis may be treated with colchicine and cDMARDs, escalating to bDMARDs, and in some cases intravenous pulse methyprednisolone or cyclophosphamide may be required, as in the treatment of vasculitis.

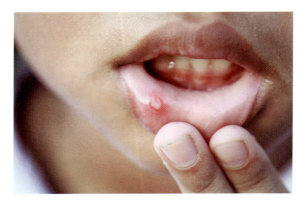

FIGURE 10.19

Mouth ulcer in Behcet's disease.

Shutterstock.

CASE STUDY 10.3 Reactive arthritis

A 21-year-old secretary complains to her GP of the recent (2–3 weeks) development of flitting aches and pains in various joints, back pain, a fever, and an increasingly sore elbow. The GP confirms all symptoms, finds normal urine analysis, takes routine screening bloods, sends his patient for X-rays, prescribes analgesics, and invites her to return in two weeks. Upon doing so, the patient reports the aches and pains and fever have improved but the elbow is now throbbingly painful, the GP feeling a tender and mobile mass along the outer part of the elbow, and she points out some pustules on her thigh that she failed to mention at the first visit. On reviewing her X-ray and lab results, which report a white blood cell count of 12.1 × 10⁹/l (reference range 4–11), an ESR of 25 mm/hour (reference range <10), and a CRP of 30 mg/dl (reference range <5), along with changes consistent with a large synovial fluid in the elbow joint in the X-ray, he telephones the local rheumatologist for an urgent appointment.

Upon arrival at the rheumatology outpatient clinic, staff confirm the diagnosis of an acute arthritis, a temperature of 38.5°C (reference <38.0), normal blood pressure but a pulse rate of 75 beats per minute (considered raised), and take more bloods. With the patient's consent, a needle is passed into the elbow joint and 12 ml of purulent turbid fluid is removed and sent to the microbiology laboratory for immediate microscopy and culture.

The cytopathology scientist telephones to report the synovial fluid has a marked white cell infiltration (50,000 cells/ml,

95% neutrophils), a few red cells, no crystals, but some cellular debris and bacteria whose precise identity is unclear. Shortly afterwards, the haematology scientist reports a leukocytosis of 15.5 with a neutrophilia of 9.5 × 10⁹/l (reference range 2–7). Concerned with the possibility of an actual or potential septic arthritis, the rheumatologist infuses some broad-spectrum antibiotics into the elbow joint, takes more blood for blood cultures, and prescribes a high 2-week dose of the same broad-spectrum antibiotics for her patient. A fine-needle aspiration of the larger of the pustules is performed and also sent for microbiology and cytopathology.

In the days that follow the results of the microbiological investigations find *Neisseria gonorrhoeae* in the synovial fluid but no microbes in the blood—results that are supported by nucleic acid amplification testing, and therefore make the diagnosis of a reactive arthritis. Rheumatoid factor, ACPA, and ANAs are negative. There are no microbiological abnormalities, but a leukocyte invasion in the pustule samples. Results are telephoned to the GP who calls in the patient for a single dose of 250 mg intramuscular ceftriaxone and 1 g of oral azithromycin, and referral to the local sexually transmitted diseases clinic. The patient confirms several acts of unprotected sexual intercourse during a holiday around a month ago. The fever and arthritis in the elbow slowly resolve, as do the pustules and blood abnormalities.

This case study is based on Wang and Lu 2019, Piazza and Gonzales-Zamora 2020, and Burns and Graf 2018.

SELF-CHECK 10.9

What is the basis of the classification of the inflammatory connective tissue diseases?

Key points

Inflammatory connective tissue diseases comprise the largest group of autoimmune conditions. They are characterized clinically by bone, joint, and skin involvement and in the laboratory by autoantibodies and increased markers of immunological activity.

10.3.7 Summary of the inflammatory connective tissue diseases

Table 10.15 summarizes key clinical and laboratory features of these diseases, many of which overlap, and which bring an additional burden of increased mortality (Table 10.16) and morbidity.

TABLE 10.15 Major inflammatory connective tissue diseases

Disease	Leading clinical features	Major laboratory features	Management
Rheumatoid arthritis	Multiple synovitis, progressing to extra-articular disease	Rheumatoid factor, anti-citrullinted protein antibodies, ESR, etc.	DMARDs (methotrexate, biologicals, etc.)
SLE	Malar rash, progressing to arthritis and lupus nephritis	Anti-nuclear antibodies, anti-dsDNA, ESR, etc.	Steroids, hydroxychloroquine, etc.
Systemic sclerosis	Calcinosis, Raynaud's syndrome, oesophagitis, scleroderma, lung disease, telangectasia	Anti-nuclear antibodies, anti-centromere antibodies, anti-topoisomerase antibodies	Methotrexate, prednisolone, calcium channel blockers, etc.
Vasculitis	Dependent on vessel(s) attacked: variously thrombosis, cutaneous lesions, nephritis	Leukocyte infiltration of blood vessels, ANCAs, immune complexes	Immunosuppression, e.g. pulse cyclophosphamide and steroids

ESR = erythrocyte sedimentation rate. ANCA = anti-neutrophil cytoplasmic antigen. DMARD = disease modifying anti-rheumatic drug.

TABLE 10.16 Standardized mortality ratios of certain inflammatory connective tissue diseases

Psoriatic arthritis	1.36 (1.13–1.64)
Rheumatoid arthritis	1.44 (1.23–1.69)
Ankylosing spondylitis	1.60 (1.44–1.77)
SLE	2.59 (1.95–3.44)
ANCA-positive vasculitis	2.83 (2.22–3.60)
Systemic sclerosis	3.51 (2.74–4.50)
Primary systemic vasculitis	4.80 (3.49–6.60)
All inflammatory connective tissue diseases	2.02 (1.79–2.29)

Data are standardized mortality ratios with 95% CIs.

10.4 Transplantation

This topic may be regarded as a major triumph of medical science, overcoming as it does a primary biological system. In this section we will address solid organ transplantation, leaving haemopoietic and blood transplantation to Chapter 11.

10.4.1 General introduction

In transplantation, tissues and organs—including the heart, lung (occasionally both at the same time), cornea, pancreas/islet cells, intestines, kidney, liver, and bone marrow—are transferred permanently from one person to another. Transplants between different people are allogeneic, those from one species to another are xenotransplants, whilst auto-transplantation (or autologous transplantation), where patients receive their own tissues, is often used in bone marrow transplantation, and occasionally in blood transfusion.

Clearly, an allogeneic transplant from a donor is non-self, and its long-term survival is naturally resisted by the immune system of the recipient (the host). As described in Chapter 9, the major recognition system defining self antigens is HLA. The major challenge is finding a donor with an HLA type that is as close as possible to that of the recipient, the general area being described as histocompatibility. It follows that in an ideal transplant the donor will carry exactly the same HLA molecules as those of the recipient, but in practice this is almost impossible, unless in transplants from one identical twin to another. The next best match is likely to be from a first-degree relative (sibling, parent), since by definition these share some of their HLA genes.

10.4.2 HLA revisited

The crucial immunological feature of transplantation is to prevent the recipient's leukocytes from reacting against the incoming tissues as alien (i.e. non-self), best achieved by the utopian perfect donor–recipient HLA match (section 9.6.1). However, not even a perfect HLA match guarantees full acceptance of the graft as there are hundreds of minor non-MHC differences that may prompt the recipient's T_c cells and NK cells to attack the donor tissues. Thus, drugs to suppress the latter are mandatory.

The need for transplantation and suitable donors

For certain diseases, transplantation is the final option, taken only when all other treatments have been exhausted. These are mostly single-organ transplants, but multi-organ transplants, such as heart–lung and kidney–pancreas are established. Most individuals admitted to the waiting list will be blood group ABO and Rh, and HLA typed and screened for any existing anti-HLA antibodies (section 9.8.5); then waiting for a suitable donor begins. If the patient's HLA and/or ABO type is unusual, a good match may not be possible, in which case, assuming the patient's disease is extremely serious and life-threatening, an imperfect 'best-case' match may be considered.

An ideal donor is a disease-free young adult who has been confirmed as brainstem dead, and from whom many organs can be harvested. Those who have died in acute setting, such as a road traffic accident or a cerebral haemorrhage, may be valuable so long as the interval between the heart no longer perfusing the tissues and the harvest is short, and the donor can be tissue typed. A key point is the ischaemic time (i.e. without oxygenation), which can be several hours, until the organ is transplanted and reperfused. Naturally, the organ starts to deteriorate as soon as it fails to be perfused, and the implanting surgeon may consider a retrieved organ too far gone (with necrosis) to transplant. Alternatively, it is possible that close family members will share enough HLA molecules to justify their consideration as live donors.

Inheritance

The chromosome and genes bearing HLA obey Mendelian genetics, which includes the possibility of cross-overs. Accordingly, sets of genes carried in gametes can be independently inherited by offspring. Thus, of the genes we have discussed thus far, an individual's HLA type may be A2/B6/C3/DPA1/DPB2/DQA2/DQB2/DRA1/DRB3 on one copy of chromosome 6, and A6/B15/C6/DPA2/DPB1/DQA1/DQB1/DRA1/DRB5 on the other. These spilt down to different haplotypes in the gametes. An attempt to illustrate the potential inheritance of all these genes would be untidy to say the least. So, for the sake of clarity, only variants of HLA-A, B, DRB1, and DQB1 are shown in Figure 10.20.

This set of four gene loci are the most polymorphic and most likely to induce a response, and so are the most important. So, by definition, each parent/offspring combination must share at least 50% of its HLA type (assuming no aberrant chromosome behaviour). Figure 10.20 shows that two siblings have a full match at the four loci, although they may also be fully mismatched at all the other loci.

10.4.3 Solid organ transplantation

Solid organ transplantation may include any large organ within the thorax and abdomen, which it is hoped will be removed from the donor and transported in haste to the recipient's hospital. In an ideal setting, the donor recipient(s) will be in adjacent or nearby theatres. A further aspect is the need for an intensive-care bed to be available immediately after the operation, and some perfectly acceptable transplants may not proceed if this is not the case.

Cross reference

The process of trying to match the donor and recipient HLA types in the laboratory, known as tissue typing, is described in section 9.8.5.

Cross reference

Learn more about the ABO and Rh blood group systems in section 11.1.2 of Chapter 11 on blood and blood diseases.

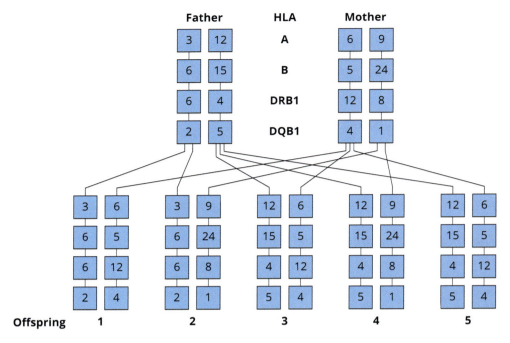

FIGURE 10.20
Inheritance of HLA types. A sperm carrying the haplotype HLA-A3, B6, DRB1*6, and DQB1*2, may fertilize an egg carrying HLA-A6, B5, DRB1*12, and DQB14. An offspring's cells may therefore express different combinations of HLA-A 3 and 6, B 6 and 5, DRB1 *6 and *12, and DQB1 *2 and *4.

COVID-19

Cross reference
Full details of COVID-19 are presented in section 13.6 of Chapter 13.

A further consequence of the COVID-19 pandemic in 2020 and 2021 was the reduced number of transplants of all organs. One of the many reasons for this is that those with the infection occupied proportionally more intensive-care beds that would otherwise have been used for the post-transplant period. Indeed, the increased length of stay of an infected patient effectively prevented 2.5 transplants. In addition, those dying with a SARS-CoV-2 infection were not considered as donors, thus reducing the number of available organs.

Organ retrieval

Organs are harvested from donors who are brainstem dead (DBD, where the heart is still beating, from whom ~68% of all organs were retrieved) or who are circulatory dead (DCD, i.e. heart not perfusing the organ, ~32% of all organs). However, merely removing an organ does not mean it will be transplanted: although overall retrievals are ~98% successful, only ~66% are transplanted, the median time between the two being ~10 hours. Predictably, DBD organs are more likely to be retrieved and transplanted. Table 10.17 shows ages of donors, and number and proportions of different organs retrieved and transplanted. The age difference between abdominal and thoracic donors (<10 years) is marked.

Data in this section are incomplete, as in several cases multiple organs are transplanted (e.g. heart and lung) so that, unless stated, figures refer to the organ in question only.

The kidney

Despite COVID-19, this organ remains the leader in terms of number of organs retrieved and transplanted (2,167 into adults and 96 into children in 2020/2021), because, of course, there are two per donor. Over the decades this organ has been the most widely transplanted, so that surgical and

TABLE 10.17 Organs retrieved and transplanted

	Donor brainstem dead	Donor circulatory dead
Abdominal transplants		
Abdominal donors attended (n)	791	587
Mean age (years)	48.8	48.9
Organs retrieved (n)	764	414
Organs retrieved/transplanted per donor (n)	2.5–2.9	2.0–2.5
Organs retrieved/transplanted		
Kidneys	90.7–91.8%	87.5–97.8%
Liver	86.2–90.1%	44.0–65.4%
Pancreas	20.8–54.7%	12.6–46.2%
Small bowel	1.8–85.7%	–
Cardiovascular transplants		
Cardiothoracic donors attended (n)	325	101
Mean age (years)	37.2	36.8
Organs retrieved/transplanted per donor (n)	1.5–1.5	1.6–1.2
Organs retrieved/transplanted		
Heart only	58.9–97.7%	49.0–75.9%
Lung only	26.0–90.3%	40.8–70.0%
Heart and lung	15.1%–*	10.2%–*

* Data not provided. Information taken from the NHS Blood and Transplant, Annual Report on the National Organ Retrieval Service (NORS) 2020/2021.

medical teams are far more experienced and therefore successful. Pre-pandemic, the overall picture was promising, with clear trends in those waiting and those transplanted (Table 10.18), and a resumption of the improvements in the metrics is expected in years to come. In 2021, those 4,710 patients suspended from the list exceed, for the first time, the number actively on the waiting list—a COVID-19 effect. This may also reflect the prolongation of the waiting time from 563 days in the period 2015–18.

The quality of the deceased donor kidney (in terms of its success when transplanted) can be scored by donor age, history of hypertension, donor weight, and days in hospital, ensuring only those organs with the highest likelihood of success are transplanted.

In 2020–1, of adults on the waiting list in the UK, 63% were men, and 72% were White, 15% were Asian, 9% were Black, and 4% were of other racial groups, with a mean age of ~53 years. Principal major indications were diabetes (42%), glomerulonephritis (17%), and hypertension/renovascular disease (9%). Risk-adjusted 1-, 5-, and 10-year adult survivals following deceased donor transplants were 98%, 87%, and 74% respectively. Other data are shown in Table 10.19.

TABLE 10.18   Renal patients waiting and transplanted

Year	Waiting*	Transplanted	%
2012	9,409	2,799	29.7
2013	9,237	3,001	32.4
2014	8,932	3,259	36.5
2015	8,945	3,131	35.0
2016	8,605	3,268	38.0
2017	8,453	3,351	39.4
2018	8,214	3,608	43.9
2019	8,101	3,597	44.4
2020	8,236	3,505	42.5
2021	8,229	2,358	28.6

* Includes those active and those suspended.

Information taken from the NHS Blood and Transplant, Annual Report on the National Organ Retrieval Service (NORS) 2020/2021.

TABLE 10.19   Selected transplant data

Organ	Waiting time (median days)	% died or transplanted after a year on the waiting list[a]	Cold ischaemia time (hours)	1-year survival	5-year survival
Kidney	813[b]	1–28	13.0[c], 3.9[d]	96%[e], 99%[f]	88%[e], 95%[f]
Liver	72[g], 2[a, h]	10–72	9.1[j], 7.1[k]	94%[g], 90%[h]	84%[g], 82%[h]
Pancreas and islet cells	363[l]	2–31	10.0[m]	99%[n], 100%[o]	89%[p], 84%[q]
Heart	Non-urgent 831, urgent 38, super-urgent 9	8–16	3.5	84%[n, j], 85%[k]	70%[j]
Lung	422	13–44	6.9	81%[n]	56%[p]
Intestines	113	–	5.3	88%[r], 70%[s]	74%[r], 39%[s]

[a] Data from 2019–20. [b] From start of dialysis (April 2020–March 2021). [c] Donor deceased. [d] Living donor. [e] Donor deceased, graft survival. [f] Living donor, patient survival. [g] April 2018–March 2020 (elective patients). [h] Super-urgent cases. [j] Donor brainstem death (DBD). [k] Donor circulatory death (DCD). [l] Data from 2015–19. [m] DBD and DCD. [n] Data from 2016–20, similtaneous pancrease and kidney (SPK). [o] Pancreas only. [p] Data from 2012–16, SPK 10-year survivial after SPK 77%. [q] Pancreas only. [r] Intestines only, data from 2011–21. [s] Intestines and liver, data from 2011–21.

Information taken from the NHS Blood and Transplant, Annual Report on the National Organ Retrieval Service (NORS) 2020/2021.

The liver

The caveats regarding HLA matching discussed in section 10.4.2 are irrelevant regarding this organ, because the only match required is for ABO and Rh blood groups. One of the possible reasons why HLA matching is not required is that this organ is the largest in the body, and so the rules of immunological recognition may be different, by sheer mass of non-self tissue. However, it is certainly the case that rejection may occur, and so immunosuppression is just as crucial. Liver transplantation is lagging behind that of the kidney, but is catching up, rising from 702 in 2009–10 to over 1,000 transplants each in 2018 and 2019. However, annual numbers then fell to 942 and 771, where 628 received a DBD and 121 a DCD. In 2020/2021, 85 patients were described as super-urgent, with a mean serum bilirubin level of 203 μmol/l (reference range <21), of whom 50 were transplanted.

Of the 535 adults transplanted in 2020–1 in the UK, 64% were men, the median age was 51, and 87% were White. Leading indications were alcoholic liver disease (29%), cancer (16%), metabolic liver disease (14%), primary sclerosing cholangitis (12%), autoimmune and cryptogenic disease (9%), and primary biliary cholangitis (10%). Encephalopathy was present in 68% and ascites in 55%, whilst 16% had a history of abdominal surgery. Unsurprisingly, recipients had hyperbilirubinaemia (median 55 μmol/l, reference <17), an **INR** of 1.4, and hypoalbuminaemia (31 g/l, reference 35–50). One-year outcomes differ little between indication, but at 5 years, that for cancer (~76%) is worse than that for primary biliary cholangitis (~90%). Donors were predominantly male (53%), White (88%), with a median age of 51 and an average body mass index (BMI) of 26 kg/m^2, and with the leading cause of death being intracranial (86%).

The pancreas and islet cells

This section is complicated by patients listed for pancreas only, islet cells only, or both simultaneously with a kidney. On 31 March 2021, 153 were active in awaiting a pancreas—a 27% reduction from the previous year, 262 being suspended. Most (92%) patients are listed for simultaneous pancreas and kidney transplant, and of these, 53% were male and 86% were White with a median age of 43 years. Those listed for islet cells transplantation are more likely to be female (58%), and with a mean age of 53. Biochemical success of islet cell transplants is shown by a fall in HbA1c from a median of ~62 mmol/mol before transplant to 48 mmol/mol at 1 year, and a fall in median daily insulin use from ~0.45 U/kg to 0.34 at 1 year. Just over a third achieved insulin independence at some time in their first year.

The heart

The active heart transplant list has been growing steadily, from 169 in 2010 to 340 in 2020, but fell to 311 in 2021 (243 active, 55 suspended, 16 urgent, 4 super-urgent). Patients are classified as non-urgent or urgent—waiting times being 559 days and 30 days respectively. Wait also depends on ABO blood group, varying from 82 days for group AB to 275 for group O, and use of a left-ventricular assist device (ever: 1,414 days, never: 259 days (data from 2018–19)).

Of the 217 adult heart transplants performed in 2020–1, the median age was 50, 70% were male, and 88% were White. Leading indications are cardiomyopathy (66%), coronary artery disease (20%), and congenital heart disease (7%). The typical donor was much younger (age 34) than for other organs, male (61% of donors), and White (88%), with a BMI of 26 kg/m^2. Patients transplanted with a DBD heart may expect a 30-day survival of 91% and 90-day survival of 88%. Survival with a DCD heart is better at 100% and 71% respectively. Survival rates with a DBD heart to 1 and 5 years are shown in Table 10.19, and are broadly similar by indication, although at 1 year, it is 91% in those transplanted for coronary artery disease and 72% for congenital heart disease. A meta-analysis of 860 combined heart and liver transplants (60% men) reported 1-year and 5-year survival rates of 85.3% and 71.4%.

The lung and heart/lung

The active lung and heart/lung transplant list has also been growing steadily, from 218 in 2010 to 367 in 2020, falling by 11% to 326 in 2021. In March 2021 demographics were 61% male, 88% White, with a median age of 56 years. Leading indications were fibrosing lung disease (48%), chronic obstructive pulmonary disease (COPD) and emphysema (20%), cystic fibrosis and bronchiectasis (9%), and primary pulmonary hypertension (6%). However, the leading indication in those actually transplanted was COPD/emphysema (37%). As with the heart, waiting time (Table 10.19) also depended on ABO blood group, varying from median 171 days for group AB to 732 days for group B.

Most donors were DBD (78%), female (54%), White (78%), and with a median age of 46. Most transplants were of both lungs (90%), with 8% having a single lung, and 2% heart and lung. Data from 2016–20 reported a 90-day survival of 90%, and whilst 1-year survival (81%) does not vary between indications, 5-year survival (56% overall) of those with fibrosing lung disease had a worse outcome (51% alive) than of those with cystic fibrosis and bronchiectasis (62% alive). Combined heart and lung survivals of 37 patients from 2012–20 are 81% at 90 days and 70% at 1 year.

Of the 27 children transplanted in 2020–1, 70% were urgent, 52% were girls, the median age was 10; most (59%) were already in hospital, and the leading indication was cardiomyopathy. The donors were mostly female (52%), White (93%), with a median age of 19, and a BMI of 19 kg/m^2.

Intestines

The small number of transplants of this organ limits accurate data. The active waiting list decreased from 19 in 2020 to 14 (8 adults, 6 children) in 2021. In the decade to March 2021 there were 177 intestine recipient registrations in England, of whom 141 were transplanted; 56 also received a liver. Leading indications were short bowel syndrome (36%), cancer (20%), mesenteric artery thrombosis (10%), and motility disorders (6%). Median age was 43, BMI was 22 kg/m^2, 86% had previous abdominal surgery, and 53% were men. The median age of the donor was 29, more likely to be female, and having died of a stroke (73%).

Discussion

Data in Table 10.19 tell several stories. The difference in waiting time is marked: that for liver may be accounted for by the small waiting list and the need to cross-match for ABO blood group only, that of the kidney by the large size of the waiting list. The heart, lung, and intestines have short cold-ischaemia times, driven by the need to transplant rapidly, perhaps reflecting the relative fragility of these organs. This may also be linked to the poor 1- and 5-year outcomes compared to liver and kidney transplants, organs that may be more robust. An alterative/additional reason for poor outcomes is that patients carried a greater burden of other disease.

10.4.4 Other organs and tissues

Other organ and tissue transplants include femoral heads, ground bone, the thymus, osteochondral bone, tendons, skin, amniotic membranes (for use in ophthalmic surgery), and aortic and pulmonary valves. The second most frequent transplant is of the cornea—around 3,800 in 2018/2019—for which HLA matching is not required. Three major types of corneal transplantation are recognized: endothelial keratoplasty, replacing the innermost layers of the cornea; deep anterior lamellar keratoplasty, replacing part of the front layers of the cornea; and penetrating keratoplasty, replacing all layers of the cornea. Some of the most ambitious transplants include the face and hands, whilst the most frequent and successful transplant (although not pathological) is discussed in section 14.4.3 of Chapter 14.

10.4.5 Rejection

As it is impossible to match every major and minor histocompatibility locus, rejection of the donor organ or tissue by the host receipient is always a possibility. Consequently, each transplant recipient has a panel of drugs to take, the effects of which may be assessed with regular blood tests and imaging. Notably, as described in the previous section, almost all of these drugs are also used in severe autoimmune disease.

Anti-rejection drugs

Ideally, these drugs would suppress only those T_c and NK cells attacking the graft, but in practice the drugs act on many different types of cell, not only leukocytes, causing adverse side effects such as tremor, hot flushes, and diarrhoea. In a worst-case scenario, high doses of these drugs may cause such profound immunosuppression that the patient is at risk of serious infections. Consequently, the transplant team must find the correct balance: sufficient immunosuppression to ensure the success of the transplant, but not so much that it causes other problems. An ideal drug regime will maintain a modicum of an anti-microbe defence whilst minimizing the risk of rejection. Some drugs have specific side effects: cyclosporin and tacrolimus may be nephtotoxic, and so require monitoring with U&Es. Accordingly, there will be variation in drugs and doses from organ to organ. A typical regime for immediate post-transplant care of the liver is:

- Prednisolone, starting at perhaps 20 mg/day, reducing to 15 mg/day at 3 weeks, to 10 mg/day at 6 weeks, 5 mg/day at 9 weeks, then no more after 12 weeks.

- Azathioprine 50 mg once daily for the first few weeks/months, then reducing.

- Tacrolimus, 3.5 mg twice a day. The dose will be adjusted depending on levels in the serum, aiming for 5–8 ng/dl, and also for patient's weight; the drug will be taken for life.

- Aspirin 75 mg daily for life.

- An antibiotic, such as co-trimoxazole (a combination of trimethoprim and sulfamethoxazole), 480 mg on alternate days for 3 months.

- An anti-viral, such as valganciclovir 900 mg daily (to target cytomegalovirus) for 3 months.

- An anti-fungal, such as nystatin mouthwash, 2 ml, not within 30 minutes of a meal.

- An antacid, such as lanzoprazole 30 mg bd or ranitidine 150 mg bd, possibly for life (dependent on symptoms).

Patients will also probably need self-directed analgesia with paracetamol and codeine in the few weeks after surgery. Other anti-rejection drugs include methotrexate, cyclophosphamide, mycophenolate mofetil, and basiliximab (targeting CD25, the α chain of the T cell IL-2 receptor).

Clinical rejection

Detection of a failing transplant can be difficult. In some cases there will be physical signs and symptoms (heart: chest pain; liver: jaundice; kidney: anuria), in others imaging and laboratory tests will be needed. As regards the latter, these are tropinins, anti-HLA antibodies, IL-2, and IFN-γ for the heart; liver function tests and interleukins for the liver; with U&Es and proteinuria for the kidney. In some cases, biopsy of the organ may be necessary: in the case of the heart this may show infiltration with leukocytes, mostly lymphocytes.

Rejection of solid organ transplants can be classified into three types:

- Hyperacute rejection, happening within hours, possibly minutes, is often caused by the recipient's pre-existing anti-ABO or anti-HLA antibodies that remained undetected in the cross-match. The principal cell attacked is the donor endothelium, and there is activation of the complement system, platelet activation, and thrombosis. This type of rejection is very rare as every attempt is made to ensure that there are no existing antibodies. If the organ is irreversibly damaged, a super-urgent salvage retransplant will probably be needed.

- The timescale of acute rejection depends on the organ: up to 3 months for the liver but a year for the heart. It is primarily the result of T cell mediated responses (inevitably cytotoxic) from the recipient that infiltrate and attack the donor tissues. There may also be some antibody-directed cellular cytotoxicity. It is generally treated by increasing the doses of immunosuppressive drugs.

- Chronic rejection is generally caused by multiple factors in antibody-, complement-, and cell-mediated responses. It develops months to years after the transplant and is marked by slow deterioration in the function of the particular organ. If detected, it may be resisted and possibly reversed by increasing doses of anti-rejection drugs, although most organs respond to rejection with fibrosis. A retransplant may be necessary.

Molecular genetics of rejection

Numerous SNPs have been linked to rejection. In the kidney these include SNPs in *ACE*, *CCR2*, *CCR5*, *CD28C*, *CTLA4*, many *ILs*, and others. SNPs in cytochrome genes *CYP3A5* and *CYP3A4* may be important in tacrolimus metabolism. As in other disease, there is much preliminary data on the value of leukocyte, serum, and urine miRNAs. There is literature on dozens of different ncRNAs in liver, lung, renal, and cardiac transplantation, with some being linked to several organs (e.g. miR-10a to heart, kidney, and lung, and miR-155 to heart, kidney, and liver). We await firm data of the value of this nucleic acid work in a routine clinical setting.

Key points

The natural desire of the host to reject the transplant can be minimized by matching for as many HLA molecules as possible, and by using immunosuppressive drugs.

SELF-CHECK 10.10

Which organ transplantations have the shortest waiting time? Which ones are the most common and the most successful?

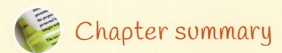

Chapter summary

- The immune system defends us from pathogenic viruses, bacteria, and parasites. The key mediators of an immune response are cellular (white blood cells) and humoral (antibodies, complement, and cytokines).

- Immunodeficiency is present if the body cannot defend itself from infections. Primary immunodeficiency diseases are genetic whilst secondary immunodeficiency may result from the suppression of the bone marrow, perhaps by drugs, an infection, or a malignancy.

- Most types of hypersensitivity are due to antibodies. The most prevalent is type I, of which a common symptom is asthma, characterized by an excessive response to pathogens in inspired air. Severe attacks can be life-threatening.

- Autoimmunity is characterized by an (auto)antibody to an antigen on the body's own tissues. The most prevalent conditions in this group are RA and SLE.

- Transplantation of organs such as the kidney, liver, heart, and lungs is an increasingly successful therapy for end-stage disease. Success is generally dependent on HLA matching and immunosuppressive drugs.

 # Further reading

- Fraenkel L, Bathon JM, England BR, et al. 2021 American College of Rheumatology guideline for the treatment of rheumatoid arthritis. Arthritis Care Res. 2021: 73; 924–39.

- Singh JA, Saag KG, Bridges SL, et al. 2015 American College of Rheumatology Guideline for the treatment of rheumatoid arthritis. Arthritis Rheum 2016: 68; 1–26.

- Kampstra ASB, Toes REM. HLA class II and rheumatoid arthritis: the bumpy road to revelation. Immunogenetics 2017: 69; 597–603.

- Moran-Moguel MC, Petarra-del Rio R, Mayorquin-Galvan EE, Zavala-Cerna MG. Rheumatoid arthritis and miRNAs: A critical review through a functional view. J Immunol Research 2018: Article ID 2474529://doi.org/10.1155/2018/2474529.

- Ghodke-Puranik Y, Niewold TB. Immunogenetics of systemic lupus erythematosus. J Autoimmun 2015: 64; 125–36.

- Márquez A, Kerick M, Zhernakova A, et al. Meta-analysis of Immunochip data of four auto-immune diseases reveals novel single disease and cross-phenotype associations. Genome Medicine 2018:10; 97 https://doi.org/10.1186/s13073-018-0604-8.

- Denton CP, Khanna D. Systemic sclerosis. Lancet 2017: http://dx.doi.org/10.1016/S0140-6736(17)30933-9.

- van den Hoogen F, Khanna D, Fransen J, et al. 2013 classification criteria for systemic sclerosis: an American College of Rheumatology/European League against Rheumatism collaborative initiative. Arthritis Rheum 2013: 72; 1747–55.

- Shiboski CH, Shiboski SC, Seror R, et al. 2016 ACR-EULAR Classification Criteria for primary Sjögren's Syndrome: A Consensus and Data-Driven Methodology Involving Three International Patient Cohorts. Arthritis Rheumatol 2017: 69; 35–45.

- Jennette JC, Falk RJ, Bacon PA, et al. 2012 Revised International Chapel Hill Consensus Conference Nomenclature of vasculitides. Arthritis Rhem 2013: 65; 1–11.

- Garcia-Montoya L, Gul H, Emery P. Recent advances in ankylosing spondylitis: understanding the disease and management. F1000Res. 2018: 7. pii: F1000 Faculty Rev-1512. doi: 10.12688/f1000research.14956.

- Hamdorf A, Kawakita S, Everly M. The potential of microRNAs as novel biomarkers for transplant rejection. J Immunol Res 2017: http://dx.doi.org/10.1155/2017/4072364.

- Jackson AM, Amato-Menker C, Bettinotti M. Cell-free DNA diagnostics in transplantation utilizing next generation sequencing. Hum Immunol. 2021: 82; 850–8.

- Aletaha D, Neogi T, Silman AJ, et al. 2010 rheumatoid arthritis classification criteria: an American College of Rheumatology/European League Against Rheumatism collaborative initiative. Ann Rheum Dis 2010: 69; 1580–8.

- Ward MM, Deodhar A, Gensler LS, et al. 2019 Update of the American College of Rheumatology/Spondylitis Association of America/Spondyloarthritis Research and Treatment Network Recommendations for the Treatment of Ankylosing Spondylitis and Nonradiographic Axial Spondyloarthritis. Arthritis Rheumatol 2019 Oct: 71(10); 1599–613.

Useful websites

■ Reports on organ donation and transplantation in the UK: **https://www.odt.nhs.uk/statistics-and-reports/**

 # Discussion questions

10.1 How do deficiencies in humoral immune factors lead to clinical disease?

10.2 Rheumatoid arthritis has been described as the ultimate challenge for a healthcare practitioner. Why so?

11

Blood and its diseases

If an organ is defined as a collection of diverse cells and tissues with a common aim or function, then blood is indeed an organ and, unlike any other organ, it is linked directly to all other tissues and organs via the vascular system. It is composed of three different types of cells (red cells, white cells, and platelets), suspended in a straw-coloured fluid called plasma: water in which ions and molecules are dissolved, including those concerned with haemostasis.

As Chapter 9 has already explained the functions of white cells, initial sections of this chapter will review the physiology of red cells (section 11.1), of platelets, plasma, and coagulation (11.2), and the role of the haematology laboratory (11.3). We then discuss the diseases of this organ. Excluding white cells in immunological disease, which have been discussed in Chapter 10, most other blood disease can be classified as cancer (11.4), anaemia (11.5), the consequences of the loss of haemostasis (11.6), and iron overload (11.7).

Learning objectives

After studying this chapter, you should confidently be able to:

- explain key aspects of the red blood cell, the platelet, plasma, and haemostasis
- appreciate the role of the haematology laboratory
- outline the aetiology, presentation, and treatment of blood cancer
- comment on aetiology, presentation, and treatment of anaemia
- describe the aetiology, presentation, and treatment of disorders of haemostasis
- understand the causes and consequences of iron overload

11.1 Red blood cells

The red blood cell, or erythrocyte, is one of the most abundant ($\sim$2.5 × 10^{13}) and most highly specialized cells in the body, each carrying $\sim$640 million haemoglobin (Hb) molecules accounting for $\sim$90% of its dry weight. The degree of specialization is so extreme that the cell has no nucleus, and so is unable to regenerate itself or produce proteins as efficiently as a nucleated cell. There is a marked sex difference: women have lower Hb, number of red cells, and the proportion of the blood that is cells (the haematocrit).

11.1.1 Erythropoiesis

This highly regulated process within the bone marrow produces $\sim$2 million cells/second (>100 billion/day), beginning with the common myeloid precursor, followed by the divergence of the erythroid lineage from the megakaryocyte lineage (which gives rise to platelets) (Figure 11.1). Subsequent

Cross reference
The different aspects of haematology and its laboratory investigations are covered in more detail in a dedicated volume on *Haematology* in this series.

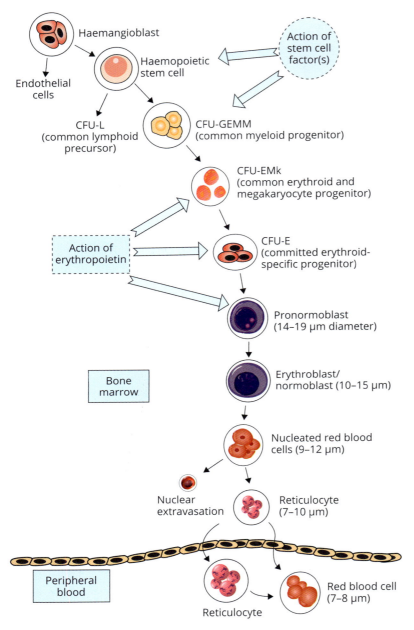

Haemangioblast

Endothelial cells

Haemopoietic stem cell

Action of stem cell factor(s)

CFU-L (common lymphoid precursor)

CFU-GEMM (common myeloid progenitor)

CFU-EMk (common erythroid and megakaryocyte progenitor)

Action of erythropoietin

CFU-E (committed erythroid-specific progenitor)

Pronormoblast (14–19 µm diameter)

Bone marrow

Erythroblast/ normoblast (10–15 µm)

Nucleated red blood cells (9–12 µm)

Nuclear extravasation

Reticulocyte (7–10 µm)

Peripheral blood

Reticulocyte

Red blood cell (7–8 µm)

FIGURE 11.1

Erythropoiesis. Stages in the development of the mature red blood cell (erythrocyte). A pathway of stem or precursor cells can be identified, commencing with the haemangioblast, and passing to various colony-forming units (CFUs) and ultimately to the mature cell. The 'early' stem cells such as the CFU-GEMM are stimulated primarily by stem cell growth factors, whereas 'late' stem cells such as the CFU-E are stimulated by Epo. As this pathway proceeds, cells become smaller and haemoglobin synthesis more prominent. A crucial stage is the extravasation of the nucleus from the nucleated red blood cell as it becomes a reticulocyte. Some reticulocytes remain in the bone marrow, where they mature into erythrocytes, but other reticulocytes pass from the bone marrow into the peripheral circulation, where their final maturation takes place.

From Moore, Knight, & Blann, *Haematology*, third edition, Oxford University Press, 2021.

cells become progressively smaller as they mature: the proerythroblast (diameter 14–19 µm), the **erythroblast** (or normoblast, 10–15 µm), the nucleated red blood cell (9–12 µm), the reticulocyte (8–10 µm, volume 120–50 fl) and finally the mature erythrocyte, with a diameter of 7–8 µm and a volume of 80–100 fl.

Key regulators of erythropoiesis include nuclear transcription factors such as GATA1 (so named as it recognizes a consensus nucleotide sequence ((T/A)GATA(A/G)) present in the regulatory elements of virtually all erythroid-expressed genes), which directs the common myeloid precursor stem cell down the erythrocyte/megakaryocyte pathway. Other transcription factors drive further specialization: Kruppel-like factor-1 to an erythroid fate, and FLI1 (= Friend leukaemia integration 1) and RUNX1

(= Runt-related transcription factor 1) towards megakaryocytopoiesis. There may also be roles for miRNAs: miR-221 and miR-222 inhibit erythropoiesis and erythroleukaemic cell growth.

Erythropoietin (Epo, coded for by *EPO* at 7q22.1) is an essential erythroid-specific factor for developing proerythroblasts and erythroblasts. With a mass of 30–40 kDa, it is produced mainly by the kidney, the key stimulus being hypoxia, major intracellular mediators being hypoxia inducible factor 2α and prolyl-4-hydrolase. Epo interacts with its target cells (erythroid progenitors) via the erythropoietin receptor (EpoR, coded for by *EPOR* at 19p13.2), a 56kDa protein membrane-spanning dimer, with ~1100 copies on an erythroid progenitor and ~300 on a late-stage erythroblast. Although Epo is the most important regulator of erythropoiesis, there are also roles for IL-3, thrombopoietin, and testosterone, the latter partially explaining why men have more red cells, Hb, and a higher haematocrit than women. Evidence of this is that men who have lost their testes to disease or trauma have a reduction in their red cell indices.

> ## Key points
>
> **Erythropoiesis is the process of the development of the red blood cell. The key growth factor is erythropoietin.**

11.1.2　The red blood cell membrane

Without a nucleus, and with a modified membrane, the red cell's flexibility allows it through the smallest capillaries—a normal 8 μm red cell can deform to pass through a 3 μm blood vessel lumen. The membrane has three components that can be classified by structure: first, a lipid bilayer of phospholipids and cholesterol; second, an internal cytoskeletal scaffold or skeleton that gives the red cell its characteristic round but flattened shape, a biconcave disc; and third, various proteins and glycoproteins embedded within the bilayer consisting of ~20 major and ~850 minor glycoproteins. Membrane components can also be classified by function as transporters of ions and molecules across the membrane (many supporting blood group antigens), as enzymes (such as those involved in glucose catabolism), and those with roles in adhesion and/or receptor and/or structural function. Of these molecules, band 3 (so named because of its migration in electrophoresis) is the largest (~102 kDa) and most abundant component, ~1.2 million copies per cell, and is stabilized in the membrane by binding to spectrins via protein 4.2, protein 4.1, and ankyrin. The importance of glucose is demonstrated by ~600,000 GLUT-1 (= glucose transporter 1) molecules per cell. The cytoskeleton is composed principally of spectrins, ankyrin, protein 4.1, band 3, and actin and its associated proteins.

The external sections of band 3, GLUT-1, RhAG (= Rh-associated glycoprotein), and aquaporin-1 express chains of carbohydrates. Variation in genes (located at 9q34.1–q34.2) for certain glycosyltransferases lead to variation in the composition of these carbohydrate chains, being a combination of L-fucose, N-acetyl D-galactosamine, and D-galactose. These differences are sufficiently strong as to lead to the generation of antibodies, and it is this combination, and the antibodies they induce, that gives us the ABO blood group system. The basic unit of the ABO system is an acetylgalactosamine-galactose-fucose trimer, this being the group 'O' chain. Addition of a further acetylgalactosamine residue to the galactose makes this an 'A' chain, whilst addition of a second galactose to the existing galactose molecule produces a 'B' chain. In most cases these are linked to the production of antibodies (Table 11.1). There are several variants of group A, the most common being A_1, A_2, and A_3, and of group B, these being B_1, B_3, and B_x.

The RhD blood group system is the second most important. Antigen 'D' is the principal member, with four others: 'C' and 'E', 'c' and 'e'. Mutations in two closely related genes, *RHD* and *RHCE*, located at 1p36-11, produce variations in the extracellular domains of these molecules that may be sufficiently non-self as to induce the formation of the alloantibodies. Those lacking the RhD molecules are D negative, generally due to a homozygous deletion of *RHD*.

The ABO and Rh systems are important because they may evoke powerful and possibly fatal consequences if there is a transfusion patient/donor mismatch. Furthermore, ABO blood group is linked to a risk of certain conditions, such as group AB with cognitive impairment; group O with cholera, tuberculosis, and mumps; and group A with cancer of the stomach, ovaries, cervix, uterus, and colon/rectum.

TABLE 11.1 ABO blood groups

Phenotype	Genotypes	UK frequency (%)	Antibodies
A	AA, AO	42	Anti-B
B	BB, BO	9	Anti-A
AB	AB	4	None
O	OO	45	Anti-A and anti-B

TABLE 11.2 Relationship between structure and function of red cells

Feature	Advantage	Disadvantage
Uncomplicated cell membrane	Simple passage of oxygen	Fragile, so relatively susceptible to damage
Anucleate, and lacking most enzymes and organelles present in a nucleated cell	Flexibility to penetrate fine capillaries	Unable to synthesize essential protective enzymes and other molecules, so diminished ability to maintain membrane integrity
Membrane lacks HLA molecules	Relatively easy to transplant (i.e. in blood transfusion)	None (?)
All features	Highly specialized	Short lifespan (~120 days)

The D antigen is a major cause of haemolytic disease of the foetus and newborn (HDFN), as discussed in section 10.2.2 of Chapter 10. Differences in many other structures give rise to minor blood groups, such as Kell, Lewis, Lutheran, and Duffy, so named because (in general) they are of less clinical importance. Like RhD, minor blood group mismatches can give rise to antibodies that can precipitate a transfusion reaction and cross the placenta to cause HDFN. Some people secrete Group A and B molecules into all body fluids, controlled by *Se* (*FUT2* at 19p13.3), leading to a value in forensic science.

Table 11.2 summarizes key aspects of the red blood cell.

11.1.3 Haemoglobin (Hb)

This molecule consists of protein (globin) and non-protein (haem, which contains iron) components.

Haem

Dietary ferrous iron (Fe^{2+}) is absorbed by intestinal epithelial cells and exported into the plasma via ferroportin. This process is controlled by the liver-derived 25-amino acid peptide hepcidin, coded for by *HAMP* at 19q13.1, whose regulation is crucial in the physiology and the pathology of iron metabolism. Once in the plasma, iron is carried principally by transferrin (a 76–80 kDa protein coded for by *TF* at 3q2.1). At the bone marrow, the iron–transferrin complex enters the erythroblast via transferrin receptors, which can be shed from the erythroblast surface into the circulation, hence the soluble transferrin receptor (sTfR), used as a surrogate laboratory marker for increased activity of the erythroblasts and possibly of erythropoiesis. Once inside the cell, the iron is released from the transferrin, and the resulting apotransferrin is returned to the plasma (Figure 11.2).

Almost all the imported iron is directed to mitochondria, where haem is synthesized, but it may be stored in the liver, pancreas, spleen, and bone marrow, within ferritin and haemosiderin. Apoferritin (iron-free ferritin), is a large (450–65 kDa) spherical molecule composed of heavy and light subunits coded for at 11q13 and 19q13.3–13.4. Haemosiderin is formed by the aggregation of partially digested ferritin and is found mostly in macrophages. It can be detected by light

FIGURE 11.2

Iron transport from the intestines to the erythroblast. Intestinal Fe^{3+} is converted to Fe^{2+} (1), entering the enterocyte (2) where some may be stored in ferritin (3). Export via ferroportin is regulated by hepcidin (4). Fe^{2+} is converted back to Fe^{3+}, is collected by apotransferrin (5), and is transported to the bone marrow (6), where it binds to the transferrin receptor on an erythroblast (7). Once internalized, the iron is released to form haemoglobin (Hb) (8) and the transferrin returns to the plasma.

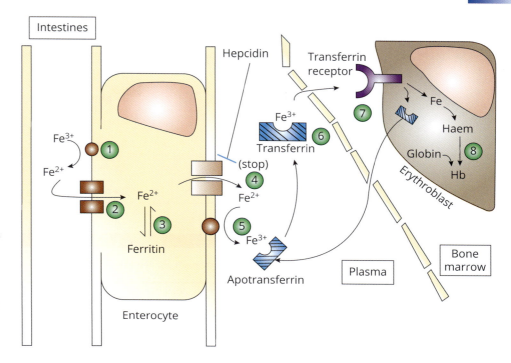

microscopy using Perls' Prussian blue stain, particularly valuable in assessing iron stores in the bone marrow and liver.

Haem is generated by a highly complex series of biochemical pathways in the cytoplasm and mitochondrion, requiring cofactors vitamins B_6, B_{12}, and folate. The initial step is the production of aminolaevulinic acid, which forms porphobilinogen, four molecules of which form coproporphyrinogen (Figure 11.3). This then moves into the mitochondrion and is converted to protoporphyrin. Iron is inserted, forming haem, which leaves the mitochondrion to merge with globin to give haemoglobin.

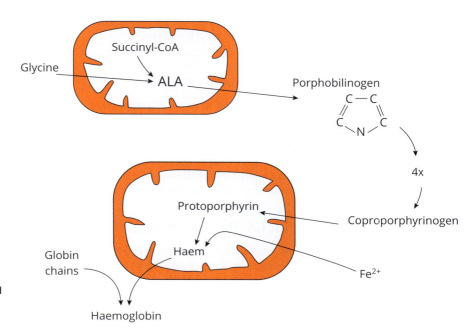

FIGURE 11.3

Synthesis of haemoglobin. Steps in the formation of haem and haemoglobin as they occur in cytoplasm and mitochondria of the erythroblast and nucleated red blood cell. ALA: aminolaevulinic acid.

TABLE 11.3   Changes in haemoglobin during development

	Chromosome 16 product	Chromosome 11 product	Haemoglobin species
Embryo	2 × zeta	2 × epsilon	Gower 1
	2 × zeta	2 × gamma	Portland
	2 × alpha	2 × epsilon	Gower 2
Foetus	2 × alpha	2 × gamma	HbF
Adult	2 × alpha	2 × delta	HbA$_2$
	2 × alpha	2 × beta	HbA

SELF-CHECK 11.1

What are the principal steps in red cell development?

Globin

This 16–17 kDa globular protein exists in a number of isoforms coded for by two alpha genes (*HBA1* and *HBA2*) and a single zeta gene (*HBZ*) at 16p13.3, and one epsilon (*HBE*), beta (*HBB*), and delta (*HBD*) gene and two gamma genes (*HBG1* and *HBG2*) at 11p15.5. The mature Hb molecule is a tetramer of individual globins and their haem ring, giving a relative molecular mass of ~64–68 kDa. Alternate combinations of globin monomers confer different oxygen-carrying properties on the Hb tetramer at different stages of the development, evolving from the embryo and fetus, via the neonate, to the adult (Table 11.3). The dominant Hb species in the adult, making up ~96–98%, is HbA, a minor form is HbA$_2$, making up ~2% of all Hb in the healthy adult. The remainder, HbF, comprises <1% of adult Hb.

> ## Key points
>
> Different variants of Hb are present during embryonic and foetal development to account for the growing need for oxygen.

Haemoglobin variants

Hb is susceptible to chemical changes, giving traces (<1%) of a number of variants, all with a reduced ability to carry oxygen. Carboxyhaemoglobin is formed by carbon monoxide-binding Hb at the same site where oxygen binds, thereby preventing the carriage of the latter. Most CO_2 is carried from the tissues to the lungs in plasma, but ~10% may be carried in a carbamino form, and can bind to Hb, giving carbaminohaemoglobin. Methaemoglobin (MetHb) is characterized by iron in the oxidized ferric state instead of its usual reduced ferrous state. Sulphaemoglobin, an inert non-toxic form of Hb, can be produced by the action of sulphur-containing drugs such as sulphonamides.

11.1.4 Red cell metabolism

As red cells do not contain mitochondria, they obtain energy enabling them to maintain shape and deformability, and to resist the toxicity of oxygen, by anaerobic respiration, principally the Embden–Meyerhof glycolytic pathway. The substrate for this pathway, glucose, is transported into the cell by membrane glycoprotein GLUT1. Figure 11.4 summarizes four aspects of interest, each having a clinical consequence, generally arising from partial or complete loss-of-function mutations leading to low or absent enzymes. These are (1) defence from oxidation, wherein the toxic effects of ROS are

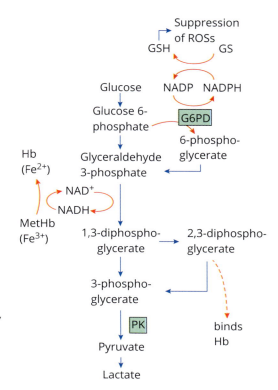

FIGURE 11.4

Selected red cell metabolic pathways. Simplified glycolytic pathway (e.g. ADP/ATP removed) showing biochemical aspects of interest. PK = pyruvate kinase, G6PD = glucose-6-phosphate dehydrogenase

addressed by antioxidants such as superoxide dismutase, catalase, and glutathione; (2) reduction of MetHb, which cannot carry oxygen, back to normal Hb; (3) pyruvate kinase (PK), an energy-producing enzyme generating pyruvate and ATP; and (4) 2,3-diphosphoglycerate (2,3-DPG), another substrate for PK which also binds to Hb.

11.1.5 Oxygen transport and the oxygen dissociation curve

Haemoglobin (or, more accurately, deoxyHb) collects oxygen at the lungs (where it is abundant) to become oxyHb, the oxygen molecule being held in place by a complex non-covalent interaction with iron at the centre of the haem ring. Subsequently, the oxyHb releases this oxygen where the gas is scarce (e.g. in the tissues), whereupon it reverts to deoxyHb. The key to understanding this process of uptake and release is the oxygen dissociation curve, which represents the relationship between the amount of oxygen in the blood and the tissues, and the amount of oxygen carried by Hb.

The amount of oxygen in the blood or tissues can be quantified in terms of partial pressure (denoted pO_2, quantified in units of mmHg). At a high pO_2 (such as 90 mmHg) at the lungs, all the Hb should be saturated with oxygen (i.e. oxygen saturation 100%). A convenient measure of ability to carry oxygen is P50, the particular pO_2 at which 50% of the Hb is oxygenated, so there are equal proportions of oxyHb and deoxyHb.

In health, this pO_2 is generally a little under 27 mmHg, but this can vary with disease or other conditions, and is associated with a left or right shift in the oxygen dissociation curve. Several factors influence the degree to which oxygen is absorbed by Hb at the lungs and given up at the tissues, and these include pH, carbon monoxide, carbon dioxide, MetHb, and 2,3-DPG.

11.2 Platelets, plasma, and coagulation

The third component of blood considers haemostasis: the balance between formation of a thrombus (composed of platelets, fibrin, and quite often, but not always, red blood cells) and haemorrhage.

The generation of fibrin is the end product of a complex series of interactions between various plasma proteins. We will address each of these three areas in turn.

11.2.1 Platelets

These small (7.5–11.5 fl) anucleate discoid (0.5 × 3.0 μm) bodies are generated by thrombopoiesis.

Thrombopoiesis

In a process similar to erythropoiesis, stem cells and precursors differentiate down increasingly specific pathways producing the megakaryoblast, which transforms into a pro-megakaryocyte and then the megakaryocyte. The platelet analogue of the red cell growth factor Epo is thrombopoietin (coded by *THPO* at 3q27.1), which acts on megakaryoblasts and megakaryocytes via receptor c-Mpl (CD110, coded for by *MPL* at 1p34.2). The megakaryocyte is a large cell (diameter 50–100 μm), with up to 32 sets of chromosomes, present in the bone marrow with a frequency of ~10,000 bone marrow cells. Some 2,000–5,000 platelets are produced by each cell by the budding-off of pseudopodia, the total megakaryocyte pool producing ~10^{11} platelets daily.

Platelet structure and biochemistry

The platelet membrane is a fluid mosaic of glycoproteins embedded in a phospholipid bilayer, as with other cells, and the principal components are listed in Table 11.4. Some of these mediate adhesion to other platelets (hence a role in thrombus growth), whilst others bind collagen, elastin, and von Willebrand factor (vWf) of the subendothelium.

There are two major organelles: first, dense bodies, about ~12 per cell, which contain ATP, ADP, calcium, and serotonin, the latter a powerful vasoconstrictor; and second, alpha granules, about

TABLE 11.4 Components of the platelet membrane

Adhesion molecule	CD62P (P selectin)
	GpIc-IIa
	GpIb-IX-V
	GpIIb/IIIa, also known as αIIbβ3
	GpVI
Binding partner	P-selectin glycoprotein ligand-1
	Fibronectin
	vWf
	Fibrinogen
	Collagen
Receptors	CD32a: IgG FcγRIIA
	P2Y1 & P2Y12 (purino-) receptors
	PAF receptor
	Thrombin receptor
	Thromboxane receptor
Ligands	Fc of IgG
	ADP*
	PAF (platelet activating factor)*
	Thrombin*
	Thromboxane*

* Binding results in marked activation.

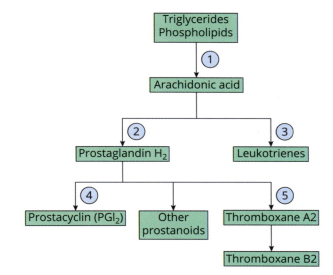

FIGURE 11.5
Key platelet metabolic
pathways. Enzymes:
1 = phospholipases A_2 and
C, 2 = cyclo-oxygenase
(Cox), 3 = lipo-oxygenase,
4 = prostacyclin synthase,
5 = thromboxane synthase.

~70–80 per cell, which contain platelet factor 4, β-thromboglobulin, tissue factor pathway inhibitor, Protein S, vWf, plasminogen activator inhibitor-1, plasminogen, platelet derived growth factor, and some coagulation factors. The membrane contains P selectin, so that upon degranulation, the activated platelet expresses this adhesion molecule, and so increased levels of soluble P select in mark platelet activation.. The cytoplasm also contains mitochondria, glycogen granules (providing energy), and lysosomes. It is also rich in eicosanoid constituents of several metabolic pathways that come into play once the platelet is activated and is participating in thrombogenesis (Figure 11.5).

Key points

The platelet is a complex ball of megakaryocyte cytoplasm that, when activated, releases a host of pro-coagulant enzymes and other mediators.

SELF-CHECK 11.2

How are platelets produced?

11.2.2 Plasma

Plasma is water in which ions and molecules, proportions of which are of crucial importance in the workings of the blood and therefore the entire body, are dissolved.

Plasma and serum

The greater part of the analysis of plasma is performed in the biochemistry (or perhaps clinical chemistry) laboratory, although some analyses may also be performed in haematology and immunology labs. However, removal of blood from the body results in the formation of a clot of the blood cells, which consumes some plasma proteins, the resulting fluid being serum. Therefore, analyses of blood cells and haemostasis must be in blood where the process of clot formation is prevented, achieved by anticoagulant chemicals, principally ethylene diamine tetra-acetic acid (EDTA), sodium citrate, and heparin. As these chemicals often interfere with biochemical analyses, the latter are generally performed on serum.

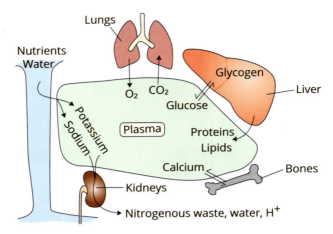

FIGURE 11.6

Key features of the composition of plasma. Diagram showing the major organs of relevance to the homeostasis of plasma.

Figure 11.6 summarizes certain features of the composition of plasma, it being convenient to discuss these in terms of the relevant organs, whose pathology is addressed in other chapters.

The kidneys

These organs are responsible for several homeostatic functions, principally maintaining levels of sodium and potassium, and the regulation of pH, which is partly achieved by the excretion of hydrogen ions. Failure to regulate the latter can lead to acidosis and alkalosis. Several of these come together in the most commonly requested biochemistry test of urea and electrolytes (U&Es: sodium, potassium, urea, and creatinine). These indices provide essential information regarding homeostasis and renal function, whilst serum creatinine is part of an equation for providing the estimated glomerular filtration rate (eGFR).

The lungs

These organs mediate the passage of oxygen and carbon dioxide in and out of the plasma, and almost all oxygen is carried in Hb. However, some carbon dioxide is carried in red cells as carbaminohaemoglobin, the remainder in conjunction with water as bicarbonic acid (H_2CO_3). This ionizes readily into a proton (H^+) and the bicarbonate ion (HCO_3^-), but at the alveoli, water and carbon dioxide are reformed, and the latter excreted in exhaled breath. These two ions have a marked influence on blood pH, so that pulmonary diseases (as renal disease) may also lead to acidosis and alkalosis.

The liver

The largest single solid organ in the body, the liver has several functions, one being the production of many plasma proteins and lipids. Albumin makes up around half of all the plasma proteins, and so has a considerable effect on the osmolarity of the plasma. The basic cellular unit of the liver, the hepatocyte, performs a large number of anabolic and catabolic functions, many of which involve enzymes such as alkaline phosphatase, gamma-glutamyl transferase, alanine aminotransferase, and aspartate aminotransferase. A further function is clearing the blood of a breakdown product of the red cell—bilirubin. This and the four enzymes are collectively the liver function tests (LFTs). The liver also produces several forms of lipids, principally cholesterol and triglycerides but also molecules required in haemostasis.

Bone

Over 99% of the body's calcium is in bone, complexed with inorganic phosphorus (measured as phosphate) in the mineral hydroxyapatite as $Ca_{10}(PO_4)_6(OH)_2$. Although once thought of as a specific LFT, alkaline phosphatase is also important in bone metabolism.

11.2.3 Coagulation

The coagulation pathway

Thirteen molecules (factors) come together in a complex pathway that, with platelets, can form a thrombus. These coagulation factors are synthesized in the liver, and several of them (prothrombin, FVII, FIX, and Fx) require vitamin K as an essential cofactor. This requirement is exploited in the use of warfarin as an anticoagulant, as described in section 11.6.3. Many of these molecules are named by roman numerals (e.g. FVIII), and together form a cascade to form fibrin by the interactions of highly regulated enzyme–substrate reactions, where the product of one reaction activates another (hence FXa). The cascade can be seen as a series of ordered but overlapping steps (Figure 11.7). The initiation stages are the generation of factor Xa from factor X—the so-called 'tenase' enzyme.

There are two routes to the formation of this super-enzyme. The intrinsic pathway involves high-molecular-weight kininogen, factor IXa and factor VIIIa. The extrinsic pathway calls for factor VIIa and tissue factor. Both pathways require calcium and phospholipids, which may be soluble or part of the platelet membrane.

Once sufficient factor Xa has been generated, it associates with factor Va, calcium, and phospholipids to generate a second complex enzyme that converts prothrombin into thrombin. In this amplification stage, thrombin generates factors Va and VIIIa (and so positive feedback), activation of platelets, and activation of factor XIII. The value placed on fibrinogen, a large molecule with mass of 340 kDa, is demonstrated by high plasma levels, around 6-fold more than the second most concentrated factor. Cleavage by thrombin generates fibrinopeptides A and B, and soluble fibrin, which is cross-linked by factor XIIIa, stabilizing the clot, making it mechanically stronger.

Inhibitors

Several regulators ensure that inappropriate coagulation does not occur. Antithrombin, the primary regulator, suppresses factors Va, IXa, Xa, and XIa. Its effect requires the binding of heparan sulphate, the basis of which leads to the use of purified soluble heparin as an anticoagulant. Tissue factor

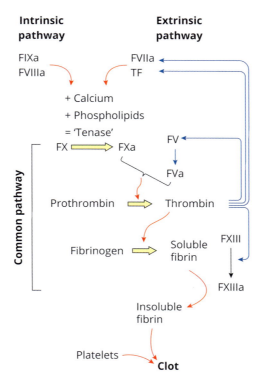

FIGURE 11.7
The coagulation cascade.

pathway inhibitor (TFPI) regulates the extrinsic tenase initiation phase of the coagulation cascade via an action factor VIIa/tissue factors, but it can act on factor Xa and thrombin. A second set of regulators are protein C and protein S that together inhibit factors Va, VIIa, and thrombin.

Platelet activation

Platelets circulate in a dormant state but respond rapidly to biochemical activation, vessel injury, and other stimuli. This may occur with any of the receptors described in Table 11.4 binding their ligands, and this sets in motion an ordered series of events:

1. The metabolic pathways are activated (Figure 11.5) with the generation of leukotrienes, prostacyclin (promoting vasodilation and opposing platelet aggregation), and thromboxane (promoting self-activation and/or the activation of nearby platelets). Leukotrienes can promote further activation, and outside the cell are inflammatory mediators, activating leukocytes.

2. This leads to alpha and dense body degranulation, and so release of numerous molecules that all promote further platelet and coagulation pathway activation.

3. Biochemical activation leads to cytoskeleton alteration, resulting in a shape change from discoid to spherical, with the outgrowth of pseudopodia, which increases its surface area and promotes interactions with other cells and tissues.

4. As a result of the shape change and pseudopodia, adhesion molecule interactions become important: P-selectin with its ligand, GpIb-IX-V with vWf, GpIc-IIa with fibronectin, and GpVI with collagen. These promote platelet adhesion to the sub-endothelium and possible platelet–platelet aggregation.

These changes allow the activated platelet to address potential blood loss through damage to blood vessels. Thromboxane and serotonin promote vasoconstriction, whilst the negatively charged lipid surface supports key reactions of coagulation and delivers additional vasoactive substances to the growing thrombus. This process also maintains the integrity of the endothelial cell–cell junctions, whilst vWf acts as a cellular glue to ensure that the platelet plug is effective.

Fibrinolysis

The final act of the coagulation play is the removal of the clot once it has performed its function—the process of fibrinolysis, or perhaps thrombolysis. The principal facilitator is the protease plasmin, which circulates as the liver-derived zymogen plasminogen (coded for by *PLG* at 6q26), which needs to be activated by endothelial product tissue plasminogen activator (tPA). tPA is itself regulated by plasminogen activator inhibitor (PAI, of which there are three isoforms).

The digestion of fibrin generates a number of degradation products (hence FDPs), the major form being D-dimers, measured with ease in the laboratory. Low levels of D-dimers are physiological, but high levels indicate increased fibrinolysis, which in turn is a surrogate for active thrombogenesis and the general body load of thrombus.

SELF-CHECK 11.3

What are the blood tests for liver and renal function?

11.3 The haematology laboratory

Much of the role of the haematology lab can be summarized in four main areas: the full blood count (FBC), the blood film, coagulation, and **haematinics**. Blood for a film and FBC is taken into a vacutainer with the anticoagulant ethylene diamine tetra-acetic acid (EDTA); coagulation tests require blood to be anticoagulated with sodium citrate, whereas haematinics (iron and micronutrients) are measured in serum. However, there are many other analyses that may be part of the haematology lab, or are found elsewhere, such as fluorescence flow cytometry in the immunology lab (Chapter 9), and ferritin or paraproteins in the biochemistry lab.

11.3.1 The Full Blood Count

The most requested and valued blood test consists of a package of information on red cells, white cells, and platelets provided by an autoanalyser. There are six red cell indices: haemoglobin (Hb), the mean cell volume (MCV), the red cell count, the mean cell haemoglobin (MCH), the mean cell haemoglobin concentration (MCHC), and the haematocrit (Hct: the proportion of whole blood that is taken up by all the blood cells). Reference ranges for Hb, red cell count, and haematocrit vary between the sexes, with lower levels in women (Table 11.5).

The FBC also provides a total white cell count, subdivided into each of the five major classes of cells, being referred to as the differential, or 'diff', with each cell having specific and non-specific features pertaining to immunology and inflammation. These indices (and the platelet count) do not vary with sex. Given this data, certain clinical conditions may be speculated upon. For example, a low platelet count (thrombocytopenia) may result from bone marrow suppression or excessive consumption.

Cross reference

For more information on the different types of white blood cells and their functions, see section 9.3 of Chapter 9 on basic and laboratory immunology.

TABLE 11.5 Refence range for a full blood count

Analyte (unit)	Reference range
Red cell indices	
Haemoglobin (g/l)	Men: 133–67. Women: 118–48*
Red blood cells (×10^{12}/l)	Men: 4.3–5.7. Women 3.9–5.0*
Haematocrit (l/l)	Men: 0.35–0.53. Women 0.33–0.47*
Mean cell volume (fl)	77–98
Mean cell haemoglobin (pg)	26–33
Mean cell haemoglobin concentration (g/l)	330–70
White cell indices	
White blood cells (×10^9/l)	4.0–11.0
Neutrophils (×10^9/l)	2.0–7.0
Lymphocytes (×10^9/l)	1.0–3.0
Monocytes (×10^9/l)	0.2–1.0
Eosinophils (×10^9/l)	0.05–0.5
Basophils (×10^9/l)	0.02–0.1
Blasts/atypicals (×10^9/l)	<0.02
Platelets (×10^9/l)	143–400

Note: Figures are likely to vary over time and with location. Always refer to your local reference ranges.

* Which reference range is most appropriate for transgender men and women depends on whether they receive gender-affirming treatment that affects hormone levels.

11.3.2 The blood film

This provides the opportunity to view blood cells under a microscope, and so confirm the proportions of the leukocytes provided by the FBC. However, the film also allows an assessment of other features, such as the morphology of the red cells, and any intra-cytoplasmic features, as may be important in certain types of anaemia (see section 11.5).

11.3.3 Haemostasis

The laboratory routinely offers two tests for haemostasis.

Prothrombin time (PT)

This assesses the ability of plasma to form a clot based on levels of prothrombin, fibrinogen, and factors V, VII, and X, and so the efficacy of the extrinsic and common pathways of the cascade (top right and centre of Figure 11.7). The PT is the time in seconds for the clot to form after the addition of certain reagents. Thus, if any of its constituents are suboptimal, the PT will be prolonged, for example from an expected ~11–14 seconds to ~18 seconds.

Activated partial thromboplastin time (APTT)

Similarly, and with partial duplication with the PT, the APTT measures the ability of prothrombin, fibrinogen, high-molecular-weight kininogen, and factors V, VIII, IX, X, and XII to form fibrin clot. It therefore assesses the intrinsic part of the cascade (top left of Figure 11.7). As with PT, deficiencies in constituents will prolong the APTT from 24 to 34 seconds to, for example, 50 seconds.

Other tests

Other tests that can be requested from most hospital path labs include fibrinogen, as may be needed in investigating haemorrhage (reference range 1.5–4.0 g/l). The thrombin time is effectively a fibrinogen time: thrombin is added to plasma, which converts all the fibrinogen to fibrin, and the time it takes for a clot to form is measured (9–11 seconds). D-dimers are the physiological products of fibrinolysis. Increased levels are present in venous thromboembolism (and many other conditions) and are an aid to diagnosis (section 11.6.3).

11.3.4 Other haematological investigations

Rheology and haematinics

Two tests of **rheology** stand out. The erythrocyte sedimentation rate (ESR) is a global score of physical aspects of the whole blood. The red cells in a column of anticoagulated blood will settle (sediment) over an hour, leaving a band of plasma at the top—hence ESR, measured in mm. An abnormal ESR (often >10 mm/hour, but may be higher) is present in a large number of conditions. These include inflammation, infection, the acute phase response, anaemia, leukaemia, almost all forms of cancer, and recovery from surgery. Because of this, most patients in hospital have an abnormal ESR. For these and other reasons, some prefer to measure whole blood and plasma viscosity. Haematinics include major micronutrients: iron, vitamins B_{12} and B_6, and folate, deficiencies in which lead to impaired production of red cells and therefore anaemia (section 11.5.3).

White cell studies

Subgroups of leukocytes can be determined by the chemical properties of contents of their cytoplasm and granules, as in neutrophils, eosinophils, and basophils, through cytochemistry. The allied process, immunocytochemistry, uses antibodies to detect antigens on cells and in tissues. However, these methods have been largely replaced by the more efficient and flexible use of fluorescence flow cytometry, especially in an immunology setting where the lymphocyte CD4/CD8 ratio is required (section 9.8.3, Figure 9.19), in blood cancer phenotyping (section 11.4), but also in investigations of red cell and platelet disease. Whilst many large hospitals will have sufficiently well-trained staff and equipment to undertake molecular genetics, such as fluorescence *in situ* hybridization (FISH) or detecting SNPs or deletions, most are undertaken by a reference lab (most likely a Genomic Hub), often allied to, or part of, a University Teaching Hospital.

Key points

The full blood count and biochemistry's U&Es are the most commonly requested and informative laboratory tests.

11.4 Blood cancer

Globally, blood cancer (defined officially as malignant neoplasms of lymphoid, haemopoietic, and related tissue) is the 5th most prevalent malignancy (3rd in England and Wales, Table 4.1), and the 5th most lethal (15th in England and Wales), almost all being lymphoma, leukaemia, and myeloma. Other forms of blood cancer include myelodysplasia, myelofibrosis, erythroleukaemia, and polycythaemia (section 11.4.5). Table 11.6 contrasts the epidemiology of blood cancer in England and Wales with that globally. In England and Wales, deaths due to lymphoma exceed those caused by leukaemia, but globally, the reverse is the case. The proportion of global deaths compared to the number of global new cases is equivalent for leukaemia and myeloma, but is lower in lymphoma, possibly reflecting less severe or malignant disease, and/or better treatments.

11.4.1 General features

We begin our exploration of blood cancer by looking at common aspects of these neoplasia, leaving disease-specific features for each particular disease.

TABLE 11.6 Epidemiology of malignant blood cancer

	All blood cancer	Lymphoma	Leukaemia	Myeloma*
England and Wales deaths**	11,634	4,542	4,128	2,830
England and Wales deaths %	100	39.0	35.5	24.3
Global deaths	711,840	283,169	311,594	117,077
Global deaths %	100	39.8	43.8	16.4
Global new cases	1,288,529	627,606	474,519	176,404
Global new cases %	100	25.4	48.7	13.7
Deaths/new cases	0.55	0.45	0.66	0.66

* Includes related neoplasms, discussed in 11.4.4.

** E&W = England and Wales 2021. Not all deaths are presented, hence sum does not = 100%.

Data sources: IARC, World Health Organization (WHO), https://gco.iarc.fr/ (global data); © Office for National Statistics. https://www.ons.gov.uk/, Licensed under the Open Government Licence 3.0.

Presentation

Presentation reflects loss of function of the three types of blood cell. Growing leukaemias and myelomas in the bone marrow inevitably suppress production of red cells and platelets, leading to common indicative signs and symptoms. The bone marrow is rarely involved in early lymphoma, but the disease may spread from its primary lymph node to others, and to the bone marrow. Signs and symptoms can be classified by blood cell and by organ involvement:

- The reduced ability of the marrow to maintain the production of normal leukocytes leads to infections—malignant white cells, although numerous, have no defensive capability.

- Anaemia, a fall in the red cell count and haemoglobin, due (as above) to the invasion of (or damage to) the bone marrow, leads to symptoms such as tiredness and lethargy (section 11.5).

- Haemorrhage may be due to thrombocytopenia, with bruising and bleeding (section 11.6).

- There may well be general non-specific symptoms such as malaise, weight loss/gain, night sweats, fevers, and a flu-like illness.

- As the blood cancer progresses, there may be hepatomegaly, lymphadenopathy, splenectomy, bone fractures, and renal failure.

A combination of these signs and symptoms is likely to prompt the practitioner to request a full blood count, often with an ESR.

Key points

Despite a relatively good 'cure' rate, malignancies of the blood are the third highest cause of cancer deaths (after bronchus/lung and colorectal), especially in the young.

Diagnosis

The full blood count is the principal test for diagnosing blood cancer. Almost all leukaemia is characterized by a high white cell count due to increased numbers of blast cells. An abnormal peripheral blood leukocyte picture is rare in the early stages of lymphoma and myeloma, but in advanced stages it is not unusual to find atypical cells in the blood (Table 11.7). The blood test is often enough for a diagnosis, in which case immediate referral will be required, and perhaps the laboratory itself will call the practitioner. Further investigations are required to characterize the malignancy.

The blood film is a simple, quick, and inexpensive tool that can be immensely instructive (section 11.3), and in leukaemia often points to the lineage of the disease. Armed with these results and the patient's signs and symptoms, immediate referral to secondary care is in order. An early investigation in secondary care is fluorescence flow cytometry to immunophenotype the cells. This technique is the gold standard for defining the key characteristics of malignant cells, and a panel of CD markers will be used. Molecular genetics has a place as much in the diagnosis and management of blood cancer as in any other disease, and in some cases is definitive, as in the diagnosis of chronic myeloid leukaemia.

Bone marrow

As this organ is where leukaemia and myeloma originate, it is very likely to be investigated. There are two types of analysis: an aspirate, where a large-gauge needle is driven into a bone (mostly the sternum or iliac crest) and some marrow extracted, and a trephine, where the structure of the bone marrow and its architecture is retained. Both analyses report different proportions of mature and precursor cells (Table 11.8). Fluorescence flow cytometry may also be used to search for malignant cells in a bone marrow aspirate.

TABLE 11.7 Full blood counts in blood cancer

	Normal profile (unit)*	Lymphoma	Leukaemia	Myeloma
Total white cell count	$5.8 \times 10^9/l$	9.9	27.1	13.6
Neutrophils	$3.3 \times 10^9/l$	4.5	9.5	3.9
Lymphocytes	$1.2 \times 10^9/l$	3.1	3.5	4.0
Monocytes	$0.9 \times 10^9/l$	0.6	2.0	1.2
Eosinophils	$0.3 \times 10^9/l$	0.2	0.8	0.5
Basophils	$0.1 \times 10^9/l$	0.1	0	0.4
Atypical cells	0	3.2	11.3	3.6
Red blood cells	$4.4 \times 10^{12}/l$	4.0	3.8	4.1
Haemoglobin	148 g/l	132	121	118
Mean cell volume	90.5 fl	88.5	82.5	85.2
Haematocrit	0.40 l/l	0.35	0.31	0.35
Mean cell haemoglobin	33.6 pg	33.0	0.32	0.29
Mean cell haemoglobin concentration	372 pg/l	388	386	338
Platelet count	$305 \times 10^9/l$	165	135	128

* Male (as men are 15% more likely to get cancer and 35% more likely to suffer blood cancer).

TABLE 11.8 Bone marrow cells in health and malignancy

Cell type	Normal bone marrow	An acute leukaemia	A malignant myeloma
Myeloblasts	0–3%	31.9%	4.2%
Promyelocytes	3.2–12.4%	18.1%	8.9%
Metamyelocytes	2.3–5.9%	9.5%	4.7%
Neutrophils	23.4–45%	17.5%	15.2%
Eosinophils	0.3–4.2%	1.0%	0.7%
Basophils	0–0.4%	1.5%	0.7%
Monocytes	0–2.6%	2.0%	1.5%
Erythroblasts	13.6–38.2%	10.9%	12.2%
Lymphocytes	6–20%	7.6%	22.5%
Plasma cells	0–1.2%	0%	29.4%
Myeloid:erythroid ratio	1.3–4.6:1	9:1	7:1

Note increased proportions of myeloid precursors in the acute leukaemia and increased lymphocytes and plasma cells in the myeloma at the expense of the erythrocyte precursors, reflected by abnormal myeloid:erythroid ratios.

Lymph nodes

Similarly, the lymph node may be aspirated or biopsied to determine the presence of a lymphoma or a differential diagnosis, such as inflammation. A needle is driven in the target node, and some of the contents are simply sucked out. Like the marrow trephine, a biopsy retains lymph node architecture, and must also be processed by standard histological methods.

Having introduced general aspects of the major white cell malignancies, we are now able to look at each in more detail.

What are the key clinical presenting features of leukaemia?

11.4.2 Leukaemia

The lineage of the leukaemia must be determined, as it provides clues to its natural history and treatment. The major forms are myeloid (60.1%) and lymphoid (28.0%), and both may be acute or chronic (hence AML, CML, ALL, and CLL). Monocytic leukaemias comprise the remaining 6.9%. Initial classification is based on morphology, subsequently by immunophenotyping for CD markers, and molecular genetics. Leukaemia is progressive: a serious event is 'blast transformation', wherein a relatively stable and slowly developing chronic disease changes to an acute and aggressive phenotype with increased numbers of blasts. In many cases this is associated with the development of a new genetic lesion, and demands a change in strategy.

Acute lymphoblastic leukaemia (ALL)

The key word in acute lymphoblastic leukaemia (ALL) is lymphoblastic, the dominant malignant cell being a lymphoblast (Figure 11.8). ALL differs from CLL as the genetic lesion causing the disease is closer to the stem cell, and so is more 'primitive', whereas the CLL lesion is further towards the final stages of differentiation, being more 'mature'.

The ALL lineage is principally defied by immunophenotyping with combinations of CD3, CD4, CD5, CD7, CD8, CD10, CD19, CD20, and CD79. Molecular genetics can define B and T ALLs by rearrangements in their BcR and TcR genes, although it is more likely to investigate the genetic lesion responsible for the malignant transformation. Examples include *t(9;22)(q34:q11.2)*, the Philadelphia chromosome, found in 30% of adult and 3% of paediatric ALLs. As discussed this lesion is present in CML (section 3.9.1); t(12;21)(p13;q22) brings together *RUNX1* at 21q22.12 and *ETV6* at 12p13.2 to form an oncogene and is present in around a quarter of all paediatric ALL cases.

Most ALLs have a relatively poor prognosis, and should standard chemotherapy with combinations of steroids and cytotoxic drugs (prednisolone, danorubicin, etc.) fail, transplantation is an option

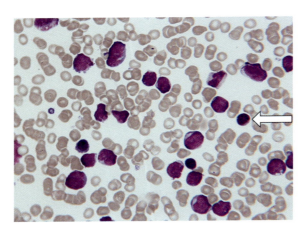

FIGURE 11.8

A blood film in ALL. Large numbers of large mononuclear cells (blasts). The arrowed cell is a normal lymphocyte.

BOX 11.1 Infectious mononucleosis

This acute condition, also known as glandular fever, is a self-limiting 7–10-day infection, generally with a fever, lymphadenopathy, and flu-like symptoms, and likely to be caused by the Epstein–Barr virus (EBV) or cytomegalovirus. The incidence is ~500/million/year, most being diagnosed between the ages of 15 and 25. The laboratory will report a lymphocytosis with atypical lymphocytes on the blood film, and a raised ESR, features in common with CLL. However, a CLL can be eliminated by the presence of heterophile antibodies detected by the monospot test, and by normal red cell and platelet indices. Treatment is bed rest and analgesics.

(see section 11.4.6). Publications from the NICE include use of ponatinib (TA451), tisagenlecleucel (the adoptive cell transfer of engineered T cells primed to attack cells bearing CD19, TA554), and pegaspargase (TA408). B-cell disease may also be treated with good success (80–90%) with monoclonal antibodies (mAbs) to CD19, CD20 (often rituximab), and CD22 (TAs 450, 541, 589), those with the Philadelphia chromosome with tyrosine kinase inhibitors such as irbitinib.

Chronic lymphocytic leukaemia (CLL)

Chronic lymphocytic leukaemia (CLL) is a disease of mature lymphoid cells that presents in around twice as many men as women. CLL and a lymphoma are often linked: some lymphomas are CLLs arising within a lymph node, and conversely 5–15% of CLLs metastasize to a lymph node. The phenotype of most CLL cells mimics a normal lymphocyte (Figure 11.9), and the film may resemble the lymphocytosis of infectious mononucleosis (Box 11.1). However, the CLL cell is fragile (as are many leukaemic cells), and may break up on the glass slide, making a smear or smudge cell.

Deletions are often (~80%) present; the most common (in ~40% of CLLs), del13q14.3, brings a moderately good prognosis (median 17 year survival). Del11q23, present in 20%, carries a poor prognosis, possibly because this region hosts the DNA damage response kinase *ATM*. Del17p13 brings a particularly poor prognosis to ~7% of CLLs (median survival 7 years), and may be linked to *TP53*. A further useful analysis is that of immunoglobulin heavy chain variable (IgHV) region genes: lack of rearrangement implies an earlier stage of differentiation and a more aggressive disease. Accordingly, routine work-up of CLL includes FISH for del11q, del13q, del17p, trisomy 12, and t(11;14), in addition to *TP53* and *IgHV* mutation analyses.

The typical CLL expresses CD5, CD19, CD20, CD23, CD79b, surface membrane immunoglobin (SMIg), and FMC7, whilst high levels of CD38 indicate a poor outcome. A minority of CLLs (~5%) are of T cells and NK cells: both are aggressive and bring a poor prognosis.

FIGURE 11.9

A blood film in CLL. Two neutrophils are on the right, a smear cell is arrowed. Most cells are of similar morphology, and are not much bigger than red cells (as is true of normal lymphocytes) and smaller than the ALL blasts in Figure 11.8.

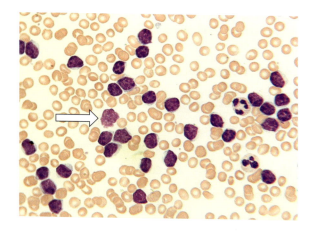

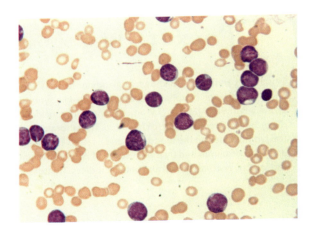

FIGURE 11.10

A blood film in AML. Most cells are of similar morphology, all much bigger than red cells.

Treatments are often guided by the genetics, such as normal or disrupted *TP53*. These include combined fludarabin, cyclophosphamide, and rituximab (anti-CD20); bendamustine (NICE TA216) and rituximab; and ibrutinib (TA429), and combinations of chlorambucil with anti-CD20s obinutuzumab or ofatumumab. Others include TA497 on the *BCL2* inhibitor venetoclax, and in combination with rituximab (TA561). Allogenic stem cell transplantation is recommended for those who have failed chemotherapy (see section 11.4.6).

Acute myeloid leukaemia (AML)

Acute myeloid leukaemia (AML) is generally characterized by large blasts (Figure 11.10), and there are many different genetic lesions. Immunophenotyping is required to establish the lineage (Table 11.9), but molecular genetics can define several groups and the oncogenic basis of the disease.

Mutations in *NPM1* (at 5q35.1, coding for a nucleolar phosphoprotein) are present in 30% of AMLs, whilst spliceosome mutations are found in 13%. *TP53* mutants with chromosomal aneuploidy are present in 10%. Acute promyelocytic leukaemias comprise 13% of all AMLs, and almost all (95%) demonstrate t(15;17)(q22;q12), resulting from the fusion of part of *PML* at 15q24.1 with that for the retinoic acid receptor-α (*RARA*, at 17q21.1). This generates a fused *PML–RARA* that results in a maturation arrest in myelocyte differentiation, which can be overcome with a cocktail of all-trans retinoic acid and arsenic, with a cure rate >80%.

Regimes for induction chemotherapy (and possibly consolidation) include an anthracycline such as daunorubicin (typically 45–90 mg/m^2 for 3 days) or idarubicin (around 12 mg/m^2 for 3–4 days) with 7 days of nucleoside antimetabolite cytarabine (100 mg/m^2 daily to 3 g/m^2 alternate days). Other options include midostaurin (TA523: an inhibitor of *FLT3* at 13q12.2 coding for cytokine receptor CD135), gemtuzumab ozogamicin (TA545: anti-CD33 linked to a toxin), and purine analogues such as cladribine and azacitidine (TA218 and TA399). Those with poor-prognosis genotypes are likely to undergo stem cell transplantation (section 11.4.6).

TABLE 11.9 CD markers in AML

Cell/leukaemia type	Marker
Precursors (blasts)	CD13, CD33, CD34, CD117, HLA-DR
Granulocytic (myeloid)	CD65, cytoplasmic myeloperoxidase
Monocytic	CD14, CD36, CD64
Mixed phenotype	Myeloperoxidase or monocyte differentiation (at least two of non-specific esterase, CD11c, CD14, CD64, lysozyme)

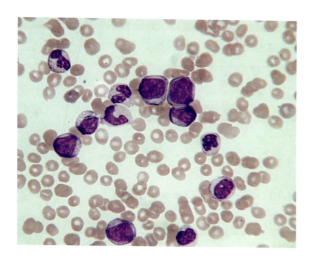

FIGURE 11.11

A blood film in CLL. Note cells of varying morphology, much bigger than the red cells.

Chronic myeloid leukaemia (CML)

The myeloid lineage of chronic myeloid leukaemia (CML) includes neutrophils, eosinophils, and basophils, but the former dominate. CML constitutes perhaps 15–20% of all cases of leukaemia with a frequency of ~10–15 per million per year and a median age at diagnosis of 67 years. The initial diagnosis is made by a full blood count and an examination of the blood film, the latter showing a varied collection of myeloid precursors (Figure 11.11). However, the ultimate diagnostic investigation is for the Philadelphia chromosome. This reciprocal translocation, t(9:22)(q34:q11), places *ABL* on chromosome 9 next to a breakpoint cluster region (*BCR*) on chromosome 22, forming the neo-oncogene *BCR-ABL*, whose product is a tyrosine kinase involved in cell signalling and proliferation.

Abnormally high levels of this enzyme can be targeted by therapeutic inhibitors (TKIs), such as imatinib (NICE TA70), bringing the usual side effects (nausea, vomiting, diarrhoea, etc.). Although some malignancies develop resistance, other TKIs (dasatinib, nilotinib, ponatinib: TA425, TA426, TA451) are better tolerated and less likely to induce resistance. Should TKIs fail, standard agents such as azacitidine (TA218), busulphan, hydroxyurea, and IFN-α may be effective. The final treatment is bone marrow transplantation (section 11.4.6). The effectiveness of treatment (i.e. measurable residual disease) can be determined by using molecular genetics such as FISH to probe for *BCR-ABL* (section 11.4.7).

Other types of leukaemia

The Office for National Statistics (ONS) reported 2,433 deaths in England and Wales in 2021 from monocytic leukaemia, which include acute and chronic myelomonocytic leukaemia (AMML, CMML), defined by CD13, CD33, and CD56, but generally with reduced levels of monocyte marker CD14. Often associated with abnormalities in *JAK2*, *RUNX1*, and *RAS*, CMML (by far the most common) has a median survival of 12–18 months. NICE TA218 recommends use of azacitidine.

Hairy cell leukaemia, accounting for ~2% of all leukaemias and ~8% of lymphoid leukaemias at a frequency of ~3/million, is named by the irregular pattern of the cell membrane. Diagnosed by morphology and B cell markers such as CD11c, CD25, CD103, and CD200, but lack of CD5 and CD23, the major genetic lesion is a gain in function in *BRAFV600E* at 7q34, present in almost 100% of cases, leading to oncogenic signalling via second messenger MEK-ERK. Treatments aim at BRAFV600E with vemurafenib and dabrafenib (targeting *BRAF*), or selumetinib and trametinib (targeting *MEK*).

Chronic neutrophil leukaemia is characterized by a high neutrophil count (>20 × 10^9/l) with many metamyelocytes, but few blasts, in the absence of a clear inflammatory response. The most common (~80%) aetiology is mutations in CSF3R at 1p34.3, coding for colony-stimulating factor 3 receptor (CD114). Eosinophil and basophil leukaemias are very rare.

CASE STUDY 11.1 *Chronic lymphocytic leukaemia*

A 65-year-old woman presents to her GP with a 6-month history of progressive tiredness and lethargy. The GP notes that she had two courses of antibiotics in the past four months. On examination the patient's left upper abdomen is firm and painful upon gentle pressure. Bloods are sent to the local hospital, requesting full blood count, U&Es, and LFTs (see Table 11.10 for the results).

There are numerous abnormalities in the full blood count: reduced haemoglobin and haematocrit, with the red cell count at the bottom of the reference range, suggesting a normocytic anaemia, which accounts for the tiredness and lethargy. The white cell count is high, with increased lymphocytes and blast/atypical cells. The laboratory reported smear cells on the blood film, and that the lymphocytes are small and with a uniform morphology. The tests of renal and liver function are normal. Together with the signs and symptoms, this makes a reasonably firm preliminary diagnosis of CLL. The issue of the left abdominal pain could be the spleen, as repeated use of antibiotics implies infection.

The patient is immediately referred to the consultant haematologist, who orders repeat bloods, a CT scan, immunophenotyping, and molecular genetics. The scan finds a large spleen, a normal liver, and no lymphadenopathy. Fluorescence flow cytometry finds a clonal expansion of small mononuclear cells bearing CD5, CD19, CD20, and CD23, but weak expression of surface immunoglobulin and negative for CD79b and FMC7. Molecular genetics reports rearrangement of *IgHV* genes and a normal *TP53*, and no deletions.

The multidisciplinary team concludes that she has early CLL, with such a low clinical score that treatment is not (yet) justified. She is reviewed with a full blood count every four months under 'watchful waiting', with the possibility of splenectomy. When the disease progresses, she may be started on fludarabine, cyclophosphamide, and rituximab, a regime recommended as initial therapy for previously untreated fit patients without *TP53* disruption.

TABLE 11.10   Blood tests in Case study 11.1

Analyte (unit)	Result	Reference range
Haemoglobin (g/l)	115	118–48
Mean cell volume (fl)	79	77–98
White blood cells (× 10⁹/l)	15.5	4.0–11.0
Platelets (× 10⁹/l)	166	143–400
Red blood cells (× 10¹²/l)	3.9	3.9–5.0
Haematocrit (l/l)	0.32	0.33–0.47
Mean cell haemoglobin (pg)	29	26–33
Mean cell haemoglobin concentration (g/l)	373	330–70
Neutrophils (× 10⁹/l)	6.6	2.0–7.0
Lymphocytes (× 10⁹/l)	6.0	1.0–3.0
Monocytes (× 10⁹/l)	1.1	0.2–1.0
Eosinophils (× 10⁹/l)	0.4	0.05–0.5
Basophils (× 10⁹/l)	0.1	0.02–0.1
Blasts/atypicals (× 10⁹/l)	1.3	<0.02
Erythrocyte sedimentation rate (mm/Hr)	20	1–14
Sodium (mmol/l)	139	135–45
Potassium (mmol/l)	4.2	3.8–5.0
Urea (mmol/l)	6.0	3.3–6.7
Creatinine (µmol/l)	95	71–133
eGFR (ml/min/1.73m²)	54	>90
Alanine aminotransferase (IU/l)	35	5–42
Aspartate aminotransferase (IU/l)	42	10–50
Gamma glutamyl transferase (IU/l)	46	5–55
Alkaline phosphatase (IU/l)	92	20–130
Bilirubin (µmol/l)	12	<17

eGFR = estimated glomerular filtration rate.

SELF-CHECK 11.6

How does the blood film in CML differ from that in AML?

11.4.3 Lymphoma

Lymphoma is the deadliest blood cancer in England and Wales, and signs and symptoms of this disease lead the practitioner to a general physical examination, which may reveal lymphadenopathy, and if present, further questioning may establish its history. Palpable non-inflammatory lymph nodes will lead to an immediate referral to secondary care. With the same symptoms, but no lymphadenopathy, diagnosis will be delayed until signs, symptoms, and other features (extra-nodal disease) prompt complex imaging with MRI and/or CT, and a biopsy will be necessary to make the diagnosis. The latter may also produce material for CD marker determination and for molecular genetics of the tumour. Clinical classification can describe localized or diffuse disease, or whether it is low (indolent) or high (active) grade. However, although the disease can be very heterogeneous, there are two major sub-diagnoses—Hodgkin lymphoma and non-Hodgkin lymphoma.

Hodgkin lymphoma

This disease comprises ~13% of all new cases of lymphoma but ~6% of deaths, and its diagnosis requires the identification of the large (30–50 μm) Reed–Sternberg cell, characterized by a multi- or bi-lobed nucleus. The leading genetic changes are to *STAT6* (30% of patients) and *TNFAIP3* (28%), which may be prognostic markers. New cases appear at a rate of ~27/million per year, with 30% of cases aged 20–35. Five-year survival is good at ~87%, but it is reduced to 70% if the disease is advanced on diagnosis.

Once diagnosed by biopsy, the disease can be staged, perhaps with local and distant metastases, and grouped for treatment. Most start on a combination of doxorubicin (25 mg/m^2), bleomycin (10 units/m^2), vinblastine (6 mg/m^2), and dacarbazine (375 mg/m^2) on days 1 and 15, followed by 20 Gy radiotherapy. If response is poor (defined by PET or CT scanning), the regime may be repeated. More advanced disease will need additional cycles, possibly with added etoposide (a topoisomerase inhibitor), cyclophosphamide (which cross-links DNA strands), procarbazine (an alkylating agent that methylates guanine), and prednisone (a steroid glucocorticoid that suppresses leukocyte function). Other options include monoclonal antibodies pembrolizumab and nivolumab (NICE TA540 and TA462: both targeting the programmed cell death receptor on lymphocytes), whereas brentuximab vedotin targets CD30 (TA524). As with other blood cancers, transplantation is an option.

Non-Hodgkin lymphoma (NHL)

The cancers in the large and diverse group of non-Hodgkin lymphomas (NHLs) have only a few features in common: most (85%) are B cell tumours and the majority of these have upregulated *BCL-2* (coding for an apoptosis regulator) and CD20. This disease is the subject of NICE NG52. Staging depends on local and distant metastases into other lymph nodes and organs (liver, central nervous system, spleen, bone marrow, blood) and this (alongside age >60) predicts outcome. Unfortunately, most present with more advanced disease, stage 1 being unusual. The two most common forms are:

1. Diffuse large B-cell lymphoma (DLBCL, the most common (~33%)), characterized by the malignant cell having a diameter >20μm, and generally high grade. More than 50% of these lymphomas can be cured. The leading genetic lesions are *BCL-2*, *BCL-6*, and *c-MYC* translocations (multiple lesions bringing worse prognosis), others being in *PTEN*, *CDKN2A*, and *NOTCH*.

2. Follicular lymphoma (20%), often low grade, and diagnosis relies on CD10, CD19, CD20 and a monoclonal gammopathy. Whilst 85% of cases have t(14;18), resulting in overexpression of Bcl-2, 27% have gain in function mutations in *EZH2* at 7q36.1, coding a methyltransferase. Bone marrow involvement is common (70%), and the median survival is ~10 years, with prognosis depending on age, haemoglobin, and stage.

Other forms include marginal zone lymphoma (~8%, with three types: splenic, nodal, and extranodal), mantle cell lymphoma (~7%), lymphoplasmacytoid lymphoma (also known as Waldenstrom's macroglobulinaemia, 1–2%), and Burkitt's lymphoma (<1%). The latter is of interest as it is most likely to be caused by EBV (which infects >90% of adults worldwide). Most cases are in malaria-endemic regions, hinting at a double-hit aetiology that first requires the immunosuppressive effects of the *plasmodium* parasite.

T cell lymphomas (15% of NHLs) can be classified as cutaneous or systemic, and generally have a very poor 5-year survival rate of <30%. Cutaneous disease comprises mycosis fungoides (60% of cases) and Sezary syndrome. Many have inactivating mutations in *TET2* at 4q24 (coding a methylcytosine dioxygenase), whilst others include t(2;5)(p23;q35), which brings together *NMP1* (at 5q35.1, coding for nucleophosmin, involved in ribosome biosynthesis) and *ALK* (at 2p23.2–23.1, coding for anaplastic lymphoma kinase). Malignant cells of both major groupings may express classic T cell markers of CD2, CD3, CD4, and CD5.

Treatments

Most B-NHLs are treated with chemotherapy alone, but in some, radiotherapy is added. A common regime is 6 cycles of R-CHOP, comprising anti-CD20 rituximab (NICE ESNM46: subcutaneous 1,400 mg, intravenous 375 mg/m^2), cyclophosphamide (400 mg/m^2), doxorubicin (25 mg/m^2), vincristine (1 mg capped-dose), and prednisone (~20mg), with 30 Gy adjuvant radiotherapy to bulky disease. In some, etoposide and bleomycin may be added. Knowledge of the precise genetic cause of the disease may inform a tailored regime. Mutations in *STAT3*, *KRAS*, *TP53*, and *JAK3* provide routes to treatment such as tofacitinib (a JAK inhibitor) and mAbs pembrolizumab, nivolumab (both anti-PD1), and daratumumab (anti-CD38).

CASE STUDY 11.2 Non–Hodgkin lymphoma

At a routine outpatient appointment, a 69-year-old overweight (BMI 29.1 kg/m^2) man with type II diabetes, reports a developing history of weight loss, fatigue, and intermittent night sweats. Diabetes control is acceptable. The practitioner takes blood for routine tests and X-rays, finding that red cell indices are towards the bottom end of their reference ranges, and that the ESR is slightly raised at 15 mm/hour (reference range <10); the X-ray is normal. The consultant calls the GP, and together they consider a CT to be valuable, which finds multiple small retroperitoneal lymphadenopathy. A positron emission tomography/CT scan finds enlarged ^{18}F-deoxyglucose positive lymph nodes either side of the diaphragm, and a biopsy of the largest and most active provides tissue consistent with DLBCL. Molecular genetics FISH showed rearrangement of *c-MYC* but not *BCL2* or *BCL6*. Immunohistochemistry showed CD19, CD20, BCL2, and MUM1, but CD10, BCL6, EBV RNA, and MYC protein were all negative. A sternal bone marrow aspirate is negative.

The patient's disease is defined as stage 3 DLBCL with activated B-cell phenotype. A cardiologist's opinion is that he is fit for chemotherapy, but as the largest tumours are deep, radiotherapy is not advised. He starts six cycles of R-CHOP, which he tolerates reasonably well, and following which a PET/CT scan suggests some shrinkage of some tumours. However, a further scan 3 months later suggests that the tumours are back to their original size, so he commences on a regime of two cycles of salvage chemotherapy of rituximab, gemcitabine, dexamethasone, and carboplatin, which causes some physical side effects. Blood tests indicate modest bone marrow suppression and slightly raised LDH. A repeat PET/CT scan shows reduction of the tumours' sizes, but shortly after this he suffers a myocardial infarction. Having recovered three months later, and rescanned, his tumours are as large as ever, and he is considered ineligible for transplantation, having lost two stone in weight. With constant monitoring, he is maintained on lower-dose chemotherapy, but the lymphoma transforms to a leukaemia, invading the bone marrow, to which he succumbs 18 months later.

Other agents include tyrosine kinase inhibitors (ibrutinib (TA502) and fostamatinib), mAbs obinutuzumab (targeting CD20, TA513, and with bendamustine, TA472), brentuximab vedotin (CD30, NICE TA577, TA478), alemtuzumab (CD52) and ofatumumab (CD20), proteasome inhibitor bortezomib (NICE TA370), thalidomide analogue lenalidomide, and idelalisib (a phosphoinositide-3-kinase inhibitor: TA604). A developing therapy is autologous engineered T cell treatment with a chimaeric antigen receptor to directly attack malignant cells (hence CAR-T: NICE TA567 and TA559).

Key points

Lymphomas are a complex group of diseases, generally with a relatively poor prognosis; they vary widely in genetic basis and methods of treatment.

11.4.4 Myeloma and related diseases

Unlike most blood cancer, the myeloma lesion is terminally differentiated bone marrow B lymphocytes (plasma cells, bearing CD27, coded for at 12p13.31). In many cases these produce nonfunctioning immunoglobulins (paraproteins). If small (<67 kDa), these aberrant molecules can be found in the urine as Bence–Jones proteins. The key diagnostic test is protein electrophoresis (section 9.8.3). There is a degree of progression of the disease through certain stages (see NICE NG35).

Monoclonal gammopathy of undetermined significance (MGUS)

Many myelomas originate in monoclonal gammopathy of undetermined significance (MGUS), present in ~2% of the over-50s, ~4% of the over-70s, and 10% of the over-80s. A paraproteinaemia (mostly IgG) <30 g/l, it is often benign, but as 1–2% transform to a myeloma or lymphoma annually, surveillance is justified, and (generally) there is no specific treatment. Progression is linked to the size and type of the paraprotein (IgM and IgA carrying a worse prognosis), and an abnormal serum kappa/lambda ratio (<0.2 or >2.0). Many (~50%) express *t(11;14)(q13:q32)*.

Smouldering multiple myeloma (SMM)

This Condition is an intermediate and asymptomatic stage between MGUS and malignant myeloma, characterized by a paraprotein of >30g/l and >10% plasma cells in a bone marrow aspirate: 10% progress to malignant disease within 5 years. However, SMM may be overlooked if MGUS transforms rapidly, and this risk is greater with the presence of t(4,14), t(14,16), 1q gain, and/or del13, so that those at high risk and with chromosome abnormalities have a 59% risk of 2-year progression.

Malignant myeloma

Frequently cited as comprising 10% of blood cancers, malignant myeloma has a poor prognosis (~50% at 5 years) and brings a far higher mortality rate of >20%. Median age of diagnosis is ~70, the male/female ratio being 1.5/1. A normal bone marrow aspirate should contain ~1% of plasma cells, but in myeloma this is can be up to 30%, dependent on stage. Malignant plasma cells generally express increased levels of normal plasma cell markers, such as heavy SMIg, CD27, CD28, CD33, CD38, CD56, CD117, CD126, and CD138 at the cell membrane. If present, the paraproteinaemia will cause a grossly elevated ESR and may lead to hyperviscosity (>1.72 mPa) that can precipitate chronic renal failure.

The disease typically metastasises to the skull, femur, vertebrae, and ribs, and may transform into plasma cell leukaemia. Osteolytic lesions may be noted on X-ray, leading to pain and spontaneous fractures, demonstrable in the laboratory with hypercalcaemia (and so low PTH) and raised alkaline phosphatase. Bone transition is linked to overexpression of genes such as the receptor activator of nuclear factor κB (*RANK*) and its ligand (*RANKL*) with increased cytokines such as TNFα, IL-1, and IL-6 that may activate osteoclasts. The dominant molecular genetic profiles are *c-MYC* activation (67% of cases), hyperploidy (55%), gain of 1q (50%), del(1p) (40%), MAPK activation (33%), and translocations involving 8q24 (20%) and t(11;14) (19%).

Treatments

Treatments for myeloma include 25 mg anti-angiogenic lenalidomide and 20 mg corticosteroid dexamethasone for various periods and cycles, perhaps adding 20–36 mg/m² of protease inhibitor carfilzomib, 200 mg/m² alkylating agent melphalan, or 16 mg/kg anti-CD38 daratumumab. NICE offers numerous technology assessments on suitable combinations of these drugs for first- and second-line treatments. NG35 gives recommendations as to the use of these agents in those being considered for, and those who are not candidates for, transplantation.

The paraprotein

A major laboratory aspect of these diseases is the abnormal serum monoclonal gammopathy present in almost all cases of myeloma (Figure 11.12), and in 80% there is Bence–Jones protein. Densitometry (section 9.8.3, Figures 9.17 and 9.18) will give an estimate as to the proportion of the gammaglobulin, and linking this with the total serum protein count will give the serum concentration of the paraprotein (Table 11.11).

The next step is determination of the isotypes of the gammopathy with immunofixation (section 9.8.3, Figure 9.19). Following a standard electrophoresis, sections of the gel are overlaid with antibodies targeting the major immunoglobulin classes. A small number of myelomas fail to produce a heavy chain paraprotein, but may generate an excess of light chains, which can be measured by ELISA.

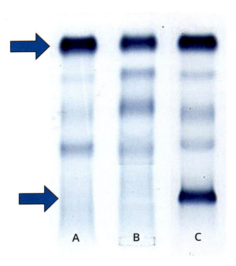

FIGURE 11.12

Serum protein electrophoresis. The top arrow shows the albumin band, strongly present in each column. The three weak bands below it are the alpha and beta globulins, the increased density of which in column B may represent an acute phase response. The lower arrow indicates the gamma globulin band; the heavy stain in column C indicates a myeloma.

TABLE 11.11   Protein subsets in health and myeloma

	Normal profile (g/l)	Normal profile (%)	Myeloma profile (g/l)	Myeloma protein (%)
Total proteins	70.0	100	92.5	100
Albumin	45.6	65.1	43.2	46.7
Alpha globulins	7.8	11.1	6.6	7.1
Beta globulins	7.7	11.0	6.1	6.6
Gamma globulins	8.9	12.7	36.6	39.6

Note the higher absolute protein count in the myeloma, and so the much higher estimate of the mass of gammaglobulins compared to normal serum.

As the normal amounts of each light chain in the serum are known, abnormal levels are relatively easy to recognize, and are generally reported as a kappa/lambda ratio, which in health is around 0.75 (reference range 0.2–2.0). A ratio far above/below this range defines a pathology, probably a myeloma. There are also instances where the ratio can be detected by electrophoresis of concentrated urine (i.e. Bence–Jones protein), and the light chain ratio in cerebrospinal fluid can also be useful in multiple sclerosis.

Key points

Myeloma is the end product of a disease process that starts with MGUS and is characterized by aberrant production of gammaglobulins by malignant B lymphocytes.

SELF-CHECK 11.7

What are the leading laboratory tools in myeloma?

11.4.5 Other blood cancer

Myelofibrosis (MF)

Myelofibrosis (MF) derives from dysregulated clonal haematopoiesis with increased megakaryocyte TGF-β bioavailability and fibroblast proliferation with deposition of reticulin, and so fibrosis. Leading presentations are anaemia and hepatosplenomegaly, with a median age at diagnosis of 67, and a median survival of 6 years. Mortality is most often due to transformation to AML (17%), progressive bone marrow failure and transfusion dependence (10%) and thrombosis (7%). Myelofibrosis may arise *de novo* or from PV or ET (see below), most cases being due to upregulation in *JAK-STAT* signalling (in 60%) or mutations in *CALR* (at 19p13.13), coding for calreticulin (25%). Chemotherapy focuses on *JAK*-inhibitors ruxolitinib ($\leq$20mg bd, NICE TA386), fedratinib, momelotinib, and pacritinib.

Polycythaemia vera (PV)

Defined by haemoglobin >185 g/l or Hct > 49% in men, and >165 or 0.48 in women, or increased red cell mass in either sex, an erythrocytosis (raised red cell count) is inevitable. Bone marrow biopsy will show panmyelotic hypercellularity, caused by overactivation of stem cells and so increased numbers of their progeny in the blood, leading to presenting signs such as bone pain, pruritis, headache, vertigo, dizziness, and night sweats. Presenting events may be arterial or venous thrombosis, haemorrhage, and hypertension caused by circulatory congestion.

In ~98% of cases the malignancy in myeloid stem cells is driven by *JAK2 V617F*, with a 20% risk of transformation to myelofibrosis 15 years after diagnosis, and a 10% risk of AML within 20 years. Treatments include venesection and low-dose aspirin (around 100 mg od), with either hydroxyurea or IFN-α in high-risk patients (platelets >1,500 or white cell count >15). Further causes include mutation in genes coding for hypoxia-inducible factors, such as in the von-Hippel–Lindau gene at 3p25–26, causing Chuvash polycythaemia.

Essential thrombocythaemia (ET)

Essential thrombocythaemia (ET) is diagnosed by a persistent platelet count >450 without other causes (e.g. inflammation), and a subsequent finding is increased numbers of enlarged mature megakaryocytes. *JAK2* mutations dominate (60% of cases), whilst those in *SRSF2* and *SF3B1* bring a poor prognosis. Outcome depends on age, leucocytosis, venous thrombosis, and an abnormal karyotype, leading to median survival of ~18 years, with risk of progression to MF (10% risk after 15 years) and AML (<5% after 20 years). Treatment, as in PV, focuses on avoiding thrombohaemorrhagic events, possibly by venesection and aspirin alone, but high-risk patients may need active cytoreduction with hydroxyurea, anagrelide, or *JAK* inhibitor ruxolitinib. Extreme thrombocytosis (>2,000) may be treated with 2 g hydroxyurea daily.

Myelodysplasia

Myelodysplasia is characterized by ineffective haemopoiesis due to stem cell aberrations and increased apoptosis, and consequences include peripheral blood cytopenias, often with macrocytic and other forms of anaemia (see section 11.5). Mean age at presentation is 74 with an estimated 3-year survival of <50%, to a significant extent due to the ~15% who develop AML. Mutations in *GATA2*, *RUNX1*, *CEBPA*, and *SRP72* are common. Chromosomal lesions can help prognosis: it is very good if -Y or del(11q) are present; good if del(5q), del(12p), or del (20q); intermediate if del(7q), +8, +19, or i(17q); and poor if -7, inv(3)/t(3q) or -7/del(7q). Treatment of individual problems is likely to be required (red cell transfusions, antibiotics, etc.). Potential treatments include erythroid stimulating agents, G-CSF, immunosuppression, and transplantation.

Erythroleukaemia

Erythroleukaemia is characterized by many of the non-specific symptoms outlined above, by erythroblasts (bearing E-cadherin and CD71) and nucleated red cells in the marrow, and by nucleated red cells in the blood and an erythrocytosis. It may arise *de novo* or from myelodysplasia, although (if there are excess myeloblasts in the bone marrow) some place it with AML, into which it may transform. Likely genetic lesions include *PU.1*, *KLF1*, and *GATA1*, whilst t(1;16)(p31;q24) and T(11;20)(p11;q11) have been reported. Without clear and specific treatment options, most regimes follow those of AML.

Table 11.12 summarizes major aspects of the three most common blood cancers.

11.4.6 Transplantation

Cross reference

A separate volume on *Transfusion and Transplantation Science* in this series covers haematopoietic stem cell transplantation and other topics related to transplantation in more detail.

Transplantation for blood cancer (haematopoietic stem cell transplantation, HSCT) differs from that of other organs (section 10.4) in that the incoming donor cells may attack the recipient's, causing graft versus host disease (GvHD). In many cases HSCT is a last resort after several rounds of failed chemotherapy, but in others it is considered soon after diagnosis. Around 20% of HSCTs are carried out in children and teenagers, and whilst most are for acute leukaemia, a third are for rare indications such as severe combined immunodeficiency, chronic granulomatous disease, Fanconi and Diamond–Blackfan anaemias, and syndromes of Wiskott–Aldrich, Hurler, Chediak–Higashi, and Kostmann.

TABLE 11.12 Major aspects of white cell malignancies

	Leukaemia	Lymphoma	Myeloma
Cellular basis	Can be any white cell	90% B lymphocytes 10% T lymphocytes	Always B lymphocytes
Basis of classification (in all cases, CD markers)	Cell lineage, proportion of blasts defines acute nature	Reed–Sternberg cells define Hodgkin's lymphoma, otherwise Non-Hodgkin's	Nature of the paraprotein (if present), Bence–Jones protein (if present)
Primary organ base	Bone marrow	Lymph nodes	Bone marrow
Leukocytosis	Always present (except in hairy cell leukaemia)	Uncommon, but often present in advanced disease where it may resemble CLL	Only present in advanced disease as plasma cell leukaemia
Bone involvement	Unusual	Rare	Common
Leading clinical aspects	Anaemia, thrombocytopenia, infections; all treatable with transplantation		

Autologous HSCT

GvHD can be avoided by using the patient's own (hence autologous) bone marrow or (after mobilization with granulocyte colony-stimulating factor or plerixafor, a CXCR4 antagonist) their peripheral blood. The procedure is reasonably straightforward: (1) extract as many CD34$^+$ bone marrow and/or peripheral blood stem cells from the patient as is possible; (2) destroy their remaining stem cell population (myeloablation) with high-dose chemotherapy (e.g. busulphan 16 mg/kg plus cyclophosphamide 120 mg/kg in AML, melphalan 200 mg/m^2 in myeloma, and busulphan 8 mg/kg plus fludarabine 180 mg/m^2 in myelofibrosis) and total body irradiation (12 Gy over 6 fractions); (3) eradicate all malignant cells from the harvested marrow in the laboratory; and (4) return the clean stem cell pool (free of all immunologically active leukocytes, especially lymphocytes) to the patient. This will hopefully restore normal and healthy haemopoiesis.

Step 1 is often preceded by prolonged chemotherapy aiming for complete remission, and there may also be some consolidating post-transplant and maintenance medications. Problems include ensuring that absolutely all of the malignant cells have been eliminated, and that sufficient stem cells are harvested and returned—ideally around 3–5 × 10^6 CD34$^+$ cells/kg. Disease relapse is a possibility up to 2 years after transplantation, and in AML, 5-, and 10-year probabilities of survival are 86% and 76% respectively.

Allogeneic HSCT

As with solid organ transplants, HLA types would ideally be fully matched in allogeneic HSCT, and this will also avoid GvHD, but in practice this is impossible. Accordingly, the minimum match will be for HLA-A, HLA-B, HLA-C, and HLA-DRB1 at 8 loci from the two chromosomes, preferably also HLA-DPB1, giving a 10/10 match; this is a challenge as there are ~25,000 HLA alleles (as discussed in sections 9.6.1 and 9.8.5). Broadly speaking, the regime for allo-HSCT is as that for auto-HSCT with eradication of the disease (e.g. etoposide 30–40 mg/kg with cyclophosphamide 120 mg/kg plus total body irradiation in Philadelphia-negative ALL), and infusion of replacement stem cells, but with more consolidating post-transplant immunosuppression to minimize GvHD. Allo-HSCT is generally the method of choice for AML, ALL, and the myeloproliferative neoplasia.

Post-transplant care

The patient will need a great deal of support until the new marrow produces effective mature cells, which is likely to take several weeks, possibly a month. This is likely to be blood and platelet transfusions, and cocktails of antimicrobials to counter the numerous potential pathogens. GvHD after allo-HSCT may be treated with agents such as calcineurin inhibitors (cyclosporine, tacrolimus), methotrexate, corticosteroids, and anti-thymocyte globulin, but some are aetiology specific, such as *JAK* inhibitors in myelofibrosis.

Undoubtedly the major disease of red cells, anaemia causes considerable morbidity and mortality, and the pathology comprises a range of diverse conditions of varied aetiology. As with blood cancer, some features are common, whilst others are specific for particular subtypes. Management depends on the aetiology of the particular anaemia.

11.4.7 Molecular pathology in blood cancer

The molecular pathology of solid organ tumours (lung, breast, etc.) is described in Chapter 4, whilst sections 11.4.2 to 11.4.6 have described instances where molecular genetics has proved valuable in the diagnosis and management of blood cancers. This is exemplified in the molecular pathology of *BCR-ABL* and CML, where knowledge of these genes and their products inform the production of highly specialized drugs, the tyrosine kinase inhibitors, as explained in section 3.9.1 of Chapter 3. It is very likely that advances in molecular genetics will result in more effective drug treatments for all cancers, including those of the blood. Regrettably, not all treatments for blood cancers are successful, which makes the ability to determine any measurable residual disease an important tool. Although the frequency of malignant cells can be determined by CD markers, the use of gene probes, such as

TABLE 11.13 Selected NGS panels in blood cancer

Clinical indication	Panel genes
Acute myeloid leukaemia	*NPM1, CEBPA, RUNX1, FLT3, IDH1, IDH2, KIT, WT1, ASXL1, SRSF2, STAG2, RAD21, TP53, KRAS, NRAS, PPM1D, DDX41, PHF6, CUX1*
Myelodysplasia	*TP53, SF3B1, IDH1, IDH2, NRAS, KRAS, TET2, SRSF2, ASXL1, DNMT3A, RUNX1, U2AF1, EZH2, BCOR, PTPN11, JAK2, SETBP1, PPM1D, DDX41, PHF6, CUX1, UBA1*
Myeloproliferative neoplasia	*KRAS, NRAS, TP53, JAK2, CALR, MPL, ASXL1, CBL, CSF3R, CUX1, DNMT3A, EZH2, IDH1, IDH2, IKZF1, KIT, NFE2, SF3B1, SH2B3, SRSF2, TET2, U2AF1, HRAS, RUNX1, SETBP1, ZRSR2*
Plasma cell dysplasia	*IGH-FGFR3, IGH-CCND3, IGH-CCND1, IGH-MAF, IGH-MAFB, MYC* rearrangement
Follicular lymphoma	*CARD11, CREBBP, EZH2, ARID1A, EP300, MEF2B, FOXO1*

Note that several genes appear more than once in different clinical indications.

for *RUNX1-RUNX1T1*, has much higher sensitivity, and so is better at assessing any relapse. The roles of *Bcl-2* in follicular lymphoma and of *c-Myc* in Burkitt lymphoma are described in detail in sections 3.9.4 and 3.9.5 of Chapter 3 respectively.

The practice of molecular pathology

The complex laboratory nature of this new discipline demands a small number of dedicated reference labs referred to as genomic gubs, situated in major university teaching hospitals. Using next-generation sequencing (NGS), the labs can determine multiple genetic lesions in the same sample, thus rapidly and unequivocally defining a particular abnormality, and so the disease. Examples of NGS panels of genes are shown in Table 11.13.

Non-coding RNAs

The potential value of these small molecules has been described in other cancers in section 3.8.4 of Chapter 3. However, in blood pathology they have not yet made an impact into guidelines regarding diagnosis and management, but may well do so in the future. Table 11.14 shows a selection of miRNAs linked to certain cancers.

TABLE 11.14 miRNAs and blood cancers

miRNA	Cancer	Relevance
miR-99a	Acute myeloid leukaemia	High expression may predict poor outcome in bone marrow transplantation
miR-150	Chronic myeloid leukaemia	Down regulated in cells—possible use in early disease diagnosis?
miR-29a	Chronic lymphocytic leukaemia	Plasma levels discriminate the disease from controls—value is rapid diagnosis?
miR-19a	Acute lymphoblastic leukaemia	Targets established T cell tumour-suppression genes such as *PTEN* and *CDKN1A*
miR-34a/b/c	Malignant myeloma	Hypermethylation attenuates the tumour-suppressor function of p53 via targeting *BCL-2* and *MYC*
miR-21	Diffuse large B-cell lymphoma	Increased expression linked to improved relapse-free survival
miR-221	Essential thrombocythaemia	Increased levels in leukocyte nucleic acid compared to controls

11.5 Anaemia

11.5.1 General features in diagnosis and classification

The diagnosis of anaemia considers presenting signs and symptoms, and red cell indices. Signs include pallor (especially of the conjunctiva), tachycardia (pulse rate >100 beats per minute), angular cheilitis (inflammation of the corners of the mouth), glossitis (a painful and swollen tongue), koilonychia (flattened, often spoon-like nails), dark urine, and jaundice. Symptoms include fatigue, lethargy, weakness, dizziness, palpitations, shortness of breath (especially on exertion), sleepiness, headaches, and tinnitus. Few of these are sufficiently sensitive or specific to be fully reliable. The second aspect is the full blood count (Table 11.15, partly copied from Table 11.5). The principles, which vary between the sexes, are haemoglobin, the red cell count, and the haematocrit, and low levels are indicative of anaemia. Both a sufficient number and severity of signs and symptoms, and the number and extent of the abnormal red cell indices, are considered before making a diagnosis. These criteria vary with age, pregnancy, racial group, and over time, but both parts are required.

It follows that someone whose haemoglobin is marginally below the reference range, but who is asymptomatic, is not necessarily anaemic. The reticulocyte count should be raised in most chronic anaemias as a physiological response to reduced red cell mass. However, in those anaemias where the bone marrow, and so erythropoiesis, is compromised, circulating reticulocytes may be low or absent, and grossly reduced in the bone marrow. The red cell distribution width (RDW) assesses the variation in size of the red cell population (anisocytosis) that, if high, points to a disease process.

Once diagnosed by symptoms and blood results, the nature of an anaemia must be determined before treatment is started. One classification of anaemia is based on the size of the red cell (i.e. the mean cell volume, MCV), so the MCV <77 defines microcytes and so microcytic anaemia, MCV 77–98 normocytes and normocytic anaemia, and MCV >98 macrocytes: macrocytic anaemia. Another definition is the amount of haemoglobin inside the cell (i.e. the MCH), which may be normochromic or hypochromic (the -chromic aspect referring to the stained colour of the cell on a blood film). These features tie in with the morphology of the cell (Figure 11.13).

TABLE 11.15 Red cell reference ranges

Index	Men*	Women*
Haemoglobin (g/l)	133–67	118–48
Red cell count ($\times 10^{12}$/l)	4.3–5.7	3.9–5.0
Haematocrit (Hct) (l/l)	0.35–0.53	0.33–0.47
	Both sexes	
Mean cell volume (MCV) (fl)	77–98	
Mean cell haemoglobin (MCH) (pg)	26–33	
Mean cell haemoglobin concentration (MCHC) (pg/l)	330–370	
Reticulocytes ($\times 10^9$/l, %)	25–125 : 0.5–2.5	
Red cell distribution width (RDW) (as CV (%) or SD (fl))	10.3–15.5 : 34.5–50.5	

These refer to adults, i.e. aged ≥18. CV = coefficient of variation; SD = standard deviation.

* Which reference range is most appropriate for transgender men and women depends on whether they receive gender-affirming treatment that affects hormone levels.

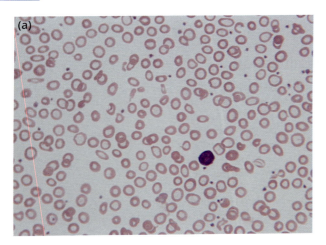

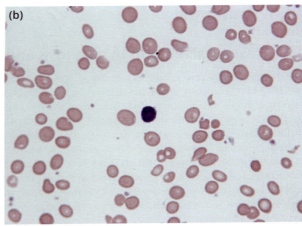

FIGURE 11.13

Red cell morphology. (a) shows red cells (many much smaller than the dark-stained lymphocyte) that appear as an empty ring, with little, if any, central colour: they are therefore not only microcytic but are also hypochromic. (b) the cell to the left of the dark-stained lymphocyte is of the same size (if not larger) and is therefore a macrocyte.

11.5.2 The bone marrow

Interference with erythropoiesis can lead to a reduction in red cell production, and in some cases there are no changes to white cells and platelets. Bone marrow aspiration may be needed to confirm a preliminary diagnosis made on clinical features and a full blood count. Section 11.4 examined anaemia in haemopoietic cancer, but other aetiologies include pure red cell aplasia and pancytopenia.

Pure red cell aplasia (PRCA)

The leading cause of pure red cell aplasia (PRCA) is Diamond–Blackfan anaemia, characterized by absent or reduced (<5%) erythroblasts, with an incidence of ~6.5/million live births. Almost all cases are congenital, and ~50% of patients also have **craniofacial**, cardiac, and renal abnormalities. There is a macrocytic anaemia (Hb 20–100 g/l, MCV >100 fl) with reticulocytopenia and a normocellular bone marrow. Most (60%) are caused by mutations in genes coding for ribosomal proteins, the most common (25% of cases) being in *RPS19* at 19q31.2.

Transient erythroblastopenia (i.e. lack of erythroblasts) usually develops in the second year of life. The anaemia is generally not severe (Hb 80–110 g/l) with mild macrocytosis and other morphological abnormalities. In congenital dyserythropoietic anaemia there is erythroid hyperplasia, and cells are often large (i.e. megaloblastic). Most cases (~90%) are autosomal recessive, linked to mutations in *CDAN1* at 15q15.2, or *SEC23B* at 20p11.23. Erythropoiesis is poor and the anaemia is generally moderate to severe (Hb often 70–95 g/l).

Pancytopenia

This condition is characterized by low levels of all three types of blood cell, and so aplastic anaemia. The principal congenital cause (~66% of cases) is Fanconi's anaemia, due to the autosomal recessive inheritance of any one of ~20 gene mutations leading to bone marrow failure. Androgens such as danazol 5 mg/kg/day can increase haemoglobin levels, but with a high risk of AML, transplantation may be considered. Other congenital causes include the Schwachman–Diamond syndrome and dyskeratosis congenita.

Miscellaneous causes

Bone marrow dysfunction may follow infections (hepatitis viruses, mumps, cytomegalovirus, HIV, tuberculosis, meningococcus, staphylococcus), autoimmune disease (SLE, rheumatoid arthritis, inflammatory bowel disease, Sjögren's syndrome: all described in Chapter 10, section 10.3), and chemotherapy (cytotoxics, immunosuppressives, gold, isoniazid, chloramphenicol, co-trimoxazole).

Management

Treatment depends on the aetiology and clinical severity. Where the cause of the anaemia is evident—perhaps due to a particular drug—the disease should be reversible upon its withdrawal. In other circumstances, treatment is palliative—red cell transfusion for incapacitating or life-threatening anaemia, platelet transfusion for severe thrombocytopenia, and prophylactic antibacterials and anti-virals if white cell numbers are profoundly low.

11.5.3 Micronutrients

Iron deficiency

The WHO estimates that 30% of the world's population is anaemic, the most common cause being iron-deficiency anaemia (IDA). Other causes, which can be found at each stage of the pathway of iron from the diet to the erythroblast mitosis (section 11.1.3), include stomach diseases (gastric atrophy, gastritis, alcoholism, and gastric carcinoma), intestinal disease (duodenitis, coeliac disease, ulceration, Crohn's disease), high levels of hepcidin (that prevent ferroportin from exporting iron into the plasma), and low transferrin levels (perhaps as a result of liver disease) that may not be able to carry iron from the enterocyte to the bone marrow.

Sideroblastic anaemia (SA)

Should iron fail to be incorporated into haem (Figure 11.3), it builds up in the mitochondrion, often as a complete or partial ring around the nucleus, forming a sideroblast. These can be detected in the bone marrow with Perls' stain, and define sideroblastic anaemia (SA). Erythropoiesis may still proceed, with microcytic red cells, some with iron granules (Pappenheimer bodies), these cells being sidero-cytes. The synthesis of haem is complex, governed by genes that may have loss-of-function mutations, such as in the synthesis of aminolaevulinic acid. Failure of the metabolic enzymes leads to build-up of intermediates (which can be toxic, bringing dermal, nervous system, and other disease) are detectable in the urine and faeces, and a diagnosis of porphyria.

SA may also be caused by drugs such as alcohol, isoniazid, and chloramphenicol, by a lack of cop-per, and by excess zinc. Lead poisoning is linked with SA: it interferes with several steps in proto-porphyrin synthesis and blocks the placement of iron into the centre of the protoporphyrin ring. Sideroblasts may also be present in myelodysplasia, which is described as refractory anaemia with ring sideroblasts, and where mutated *SF3B1* is found in 80% of patients. Severe cases may need to be transfused, although this may lead to iron overload, so that chelation therapy may be effective in reducing inappropriately high iron stores.

The laboratory

A major diagnostic point in IDA is microcytic anaemia, defined primarily by a low MCV, but possibly also by hypochromic microcytes on a blood film (Figure 11.13). However, a low MCV may be due to other pathology (such as **haemoglobinopathy**), so serum iron studies are required (Table 11.16), with low levels expected. Suspected SA can be confirmed with sideroblasts in the bone marrow, and although microcytes are prevalent, there may also be macrocytes, and so a high RDW. Iron stores may be normal or high.

TABLE 11.16 Iron reference ranges

Analyte	Men	Women
Total iron binding capacity (µmol/l)	54–72	55–81
Transferrin saturation (%)	18–40	13–37
Ferritin (µg/l)	25–380	28–365[a] : 7.5–224[b]

	Both sexes
Iron (µmol/l)	10–37
Transferrin (g/l)	2–4

[a] Post-menopausal,
[b] pre-menopausal.

Vitamin deficiency

Vitamins B_6, B_9 (folate), and B_{12} are essential cofactors in erythropoiesis and other aspects of physiology (e.g. neurology). Since animal products are the primary natural source of B_{12}, plant-based and especially vegan diets may not furnish B_{12} in sufficient quantities to avoid a B_{12} deficiency. Other causes of B_{12} deficiency parallel those of iron deficiency, including gastric disease and intestinal disease. The stomach is important as the acidic environment helps liberate B_{12} from food, and secretes an essential carrier, intrinsic factor (IF, coded by *GIF* at 11q12.1), without which the vitamin cannot be absorbed. The latter occurs in the intestines, where the B_{12}/IF complex is transported across the enterocyte membrane. Once in the blood, B_{12} is carried by transcobalamine-2 (coded by *TCN2* on 22q11.2). At its target cell, the complex binds a transcobalamine receptor (TCB1R/CD320, coded by *CD320* at 19p13.2) and is internalized, assisting in the synthesis of methionine and haem.

The leading (>90%) cause of B_{12} deficiency is autoimmune gastritis, with autoantibodies to the parietal cells that synthesize IF, commonly described as pernicious anaemia. The metabolic pathways reliant on B_{12} are widely distributed, so that there may be several pathological consequences. These include a swollen and painful tongue (glossitis), peripheral neuropathy/polyneuritis, with consciousness and personality changes, ranging from simple confusion and irritability to depression, loss of memory, and even psychosis.

Vitamin B_6 (pyridoxine) is an essential cofactor for the generation of haem, and so ultimately a sideroblastic anaemia. It is common in many foodstuffs (fruit, vegetables, and meat) and easily absorbed in the jejunum and ileum. Consequently, dietary deficiency is very rare, but may be caused by malabsorption. Similarly, vitamin B_9 (folate) is an essential requirement for the conversion of deoxyuridine to deoxythymidine, the latter forming nucleotides and so DNA, and both plasma and red cell levels may be needed.

The full blood count, signs, and symptoms will give a diagnosis of anaemia, a raised MCV (>100 fl) pointing to vitamin deficiency. The blood film will also show macrocytes and hypersegmented neutrophils (Figure 11.14), hence macrocytic anaemia.

However, there are numerous other causes of macrocytosis, including chemotherapy, metformin, myeloma, and liver disease. Similarly, there are other causes of vitamin deficiency, such as alcoholism, pregnancy, pancreatic disease, and tapeworm infestation. Accordingly, blood levels should be sought (Table 11.17). Should it be necessary, a bone marrow aspirate will show large erythroblasts, hence an alternative name—megaloblastic anaemia.

Whilst the cause of a deficiency should be determined and resolved, supplementation is a standard treatment (oral for B_6 and folate, intramuscular for B_{12}, as this bypasses the potential of intestinal disease). B_{12} is injected as a loading dose of 100–1,000 µg/day for a week, followed by a fortnightly dose, then a monthly or 3-monthly lower-maintenance dose. This should result in a burst of reticulocytes, followed by a slow normalization of red cell indices, but the autoimmune aspect of pernicious anaemia remains.

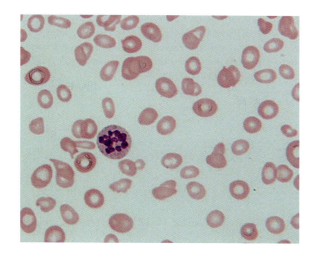

FIGURE 11.14

A hypersegmented neutrophil. This neutrophil has 6, possibly 7 lobes (normally 3–4). There are also some macrocytes.

From Moore, Knight, & Blann, *Haematology*, third edition, Oxford University Press, 2021.

TABLE 11.17 Vitamin reference ranges

Analyte	Moles	Grams
Vitamin B_{12}	120–680 pmol/l	160–925 ng/l
Vitamin B_6	17–279 nmol/l	2.9–47.0 µg/l
Red cell folate	>340 nmol/l	>150 µg/l
Serum folate	>7 nmol/l	>3 µg/l

SELF-CHECK 11.8

What are the major laboratory differences between iron-deficient and vitamin B_{12}-deficient anaemia?

Key points

Iron deficiency is the most common cause of anaemia, and whilst poor nutrition is a leading aetiology, there are several others.

11.5.4 Haemolytic anaemia

Anaemia may follow the destruction (haemolysis) of red cells, shortening their lifespan from ~120 days to perhaps 40–50 days. Certain extrinsic factors act on an otherwise healthy cell, whilst other, genetic factors are intrinsic to the cell itself. Damaged cells are removed by the reticuloendothelial systems of the liver and spleen (extravascular), or are destroyed in the blood (intravascular), perhaps by phagocytes and/or antibodies/complement. This can be detected in the laboratory with cell fragments on a blood film (schistocytes), free haemoglobin (often complexed with its carrier, haptoglobin, coded for by *HP* at 16q22.2) and haem (likewise, carried by haemopexin, coded for by *HPX* at 11p15.4).

Extrinsic haemolysis

Section 10.3 of Chapter 10 described autoimmunity, where the body makes antibodies to itself, and red cell autoantibodies result in autoimmune haemolytic anaemia (AIHA). With 10–15 cases per million per year, most (~80%) cases of AIHA have IgG autoantibodies reacting optimally at 37°C, and so are classified as warm AIHA. Some 10–15% have cold-reactive agglutinating autoantibodies (hence cold AIHA), generally of an IgM type, perhaps a gammopathy, which react at 30–32°C. A rare type of cold AIHA is paroxysmal cold haemoglobinuria, a transient condition mostly linked with viral infections and lymphomas and caused by a complement binding IgG autoantibody (the Donath–Landsteiner antibody) often with anti-P specificity.

CASE STUDY 11.3 Iron-deficiency anaemia

A middle-aged man is brought into A&E at lunchtime by ambulance semi-conscious, having been a passenger in a road traffic accident. With the possibility of concussion, blood is taken and he is sedated for head, chest, and abdominal X-rays, which find only old fractures, and he is retained for observation. Bloods (see Table 11.18 for the results) find low haemoglobin, MCV, MCH, and haematocrit, and slightly raised ESR, leading to a possible diagnosis of microcytic anaemia. There is also isolated raised gamma glutamyl transferase (GGT).

Late that evening he is discharged with a letter to his GP and an appointment for haematology outpatients, which he attends 3 months later. Bloods are again taken, to which iron studies are added. The routine bloods essentially mirror those taken in A&E, but there is low iron (8 μmol/l) with a raised total iron binding capacity (95 μmol/l), with ferritin at the low end of the reference range (40 μg/l). These are sufficient to extend the diagnosis to IDA. With a raised GGT the likely cause is high alcohol intake, and the patient admits to regular lunchtime drinking and most of a bottle of wine each evening, a probable weekly total of around 40–50 units (recommended intake ≤14). Details are passed back to his GP, and he is given support leaflets and referred on to a drugs and substance abuse service.

TABLE 11.18   Blood tests in Case study 11.3

Analyte (unit)	Result	Reference range
Haemoglobin (g/l)	125	133–67
Mean cell volume (fl)	72	77–98
White blood cells (× 10⁹/l)	9.6	4.0–11.0
Platelets (× 10⁹/l)	256	143–400
Red blood cells (× 10¹²/l)	4.0	4.3–5.7
Haematocrit (l/l)	0.29	0.35–0.53
Mean cell haemoglobin (pg)	31	26–33
Mean cell haemoglobin concentration (g/l)	434	330–70
Neutrophils (× 10⁹/l)	6.5	2.0–7.0
Lymphocytes (× 10⁹/l)	2.0	1.0–3.0
Monocytes (× 10⁹/l)	0.5	0.2–1.0
Eosinophils (× 10⁹/l)	0.5	0.05–0.5
Basophils (× 10⁹/l)	0.1	0.02–0.1
Blasts/atypicals (× 10⁹/l)	0.0	<0.02
Erythrocyte sedimentation rate (mm/Hr)	15	1–14
Sodium (mmol/l)	145	135–45
Potassium (mmol/l)	4.6	3.8–5.0
Urea (mmol/l)	4.5	3.3–6.7
Creatinine (μmol/l)	86	71–133
eGFR (ml/min/1.73m²)	82	>90
Alanine aminotransferase (IU/l)	22	5–42
Aspartate aminotransferase (IU/l)	36	10–50
Gamma glutamyl transferase (IU/l)	95	5–55
Alkaline phosphatase (IU/l)	74	20–130
Bilirubin (μmol/l)	15	<17

eGFR = estimated glomerular filtration rate.

FIGURE 11.15

Schistocytes and reticulocytes. Destruction of red cells in haemolytic anaemia generates fragments (schistocytes, left arrow), often with reticulocytes (larger, slightly more 'blue', right arrow). This slide also demonstrates variety in the colour of the cells (polychromasia) and anisocytosis. Three neutrophils are present.

From Moore, Knight, & Blann, *Haematology*, third edition, Oxford University Press, 2021.

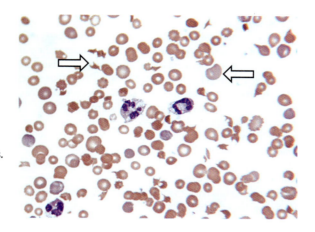

AIHAs may arise without a precipitant (i.e. are primary, or idiopathic), or with a clear cause, such as a blood or other cancer, other autoimmune disease such as SLE or ulcerative colitis, drugs such as α-methyldopa, or in infectious mononucleosis (Epstein–Barr virus).

The diagnostic test for AIHA is the presence of the autoantibody. This can be determined by direct and indirect antiglobulin tests, or by fluorescence flow cytometry and an anti-human globulin conjugated to a fluorochrome. The full blood count will show standard changes of anaemia, but a semi-specific blood film feature are schistocytes, often with increased reticulocytes (Figure 11.15). Intravascular red cell destruction liberates cell contents into the blood, so there will be free haemoglobin, haem, and increased lactate dehydrogenase.

Treatment for AIHA is immunosuppression, such as with prednisolone with anti-CD20 rituximab (NICE ESUOM39, 375 mg/m^2 weekly for 4 weeks), a regime that induces complete response in 75% of patients. If refractive, chlorambucil, cyclophosphamide, and/or fludarabine are options. A second-line therapy may be splenectomy, which can induce a response in 50% of patients. However, splenectomy brings increased risk from infection by encapsulated microorganisms, and patients should be vaccinated against pneumococcus, *Haemophilus influenzae*, and meningococcal serogroup C, and should be prescribed lifelong antibiotics.

Other extrinsic causes of haemolytic anaemia include haemolytic uraemic syndrome (HUS, a triad of acute renal failure, thrombocytopenia, and microangiopathic haemolytic anaemia, most (>60%) linked to an enteropathogenic strain of *E. coli*). Other causes include physical damage to red cells either on abnormal surfaces (such as an artificial heart valve) or caused by red cells travelling through fibrin strands deposited on capillary microvessels, and drug-induced haemolytic anaemia (accounting for 10–20% of cases).

Infections

Infection can lead to haemolytic anaemia via several routes, one being *Clostridium perfringens* septicaemia, whilst a high temperature, such as may be present in influenza, may also cause red cell destruction. Malaria is a major health and economic problem in many countries, and worldwide causes perhaps one million deaths annually from 300–500 million infected. The causative parasite, *Plasmodium*, precipitates a haemolytic anaemia, in addition to other laboratory (raised ESR) and clinical (fever) features. These occur partly because the parasite-loaded cell is detected as being abnormal, and so is eliminated. The parasite has driven the development of haemoglobinopathy (as those heterozygous for sickle haemoglobin are relatively protected from infection: see section 11.5.5) and the absence of Duffy blood group structures (CD234) (the receptor for *Plasmodium*) in malaria-endemic areas.

Membrane defects

Hereditary spherocytosis (HS) is the most common hereditary haemolytic anaemia in North Europeans, and in 75% of cases is inherited in an autosomal-dominant manner, with variable clinical presentations ranging from an asymptomatic to severe haemolytic anaemia. The most common causative mutations are in *ANK1* (coding for ankryn) and *SLC4A1* (Band 3) (see section 11.1.2). Parts of the membrane not supported by the cytoskeleton are lost, causing the cells to become spherical (Figure 11.16).

Being unable to pass through the splenic microcirculation, these cells are eliminated, resulting in a reduced lifespan (6–20 days) so that splenectomy is a common treatment. This results in the appearance of intracellular Howell–Jolly bodies (clusters of DNA). Other features of HS include an increased reticulocyte count (5–20%), low haemoglobin (perhaps 70 g/l), and reduced MCV with raised MCH, MCHC, and RDW (>14%) in most patients. Other membrane defects include hereditary elliptocytosis (most frequently found in Europeans and inherited as an autosomal-dominant characteristic), Southeast Asian ovalocytosis (common in Indonesia, the Philippines, and Malaysia, where the prevalence may be 30%), hereditary stomatocytosis (where red cells have an oblong bar of central pallor), hereditary pyropoikilocytosis (a rare autosomal-recessive condition, characterized by microcytes, poikilocytes, and schistocytes), and hereditary xerocytosis (caused by cation imbalance).

The simplest and most physiological test of a weak membrane is increased osmotic fragility, and lysis when hypotonically challenged. Alternatives are the acidified-glycerol lysis test, the cryohaemolysis test, and binding of the fluorescent probe eosin-5-maleimide (detectable by fluorescence flow cytometry), which binds less avidly if there are abnormalities in membrane components.

Cell membrane component glycosyl phosphatidylinositol (GPI, coded for by *PIGA* at Xp22.2) anchors a number of molecules, including CD55 and CD59. Mutations in *PIGA* lead to lack of GPI and so lack of its anchor-partners. The principal condition resulting from this is paroxysmal nocturnal haemoglobinuria (PNH), characterized by attack from the complement system, which CD55 and CD59 would normally resist. Fluorescence flow cytometry for the presence of cell-surface CD55 and CD59 is the method of choice for diagnosis and management. Treatment with a mAb (eculizumab, directed towards complement component C5) is effective, and almost eliminates intravascular haemolysis.

Enzyme defects

Enzyme biochemistry of glucose-6-phosphate dehydrogenase (G6PD) and pyruvate kinase (PK) is explained in section 11.1.4, where Figure 11.4 applies. Mutations in *G6PD* at Xq28 affect ~1% of the world's population, but are considerably higher (13%) in West Africans, showing marked clinical heterogeneity. The most common consequence of G6PD deficiency is drug-induced haemolysis (with haematuria and pain), such as with certain antimalarials, antibiotics, and antihelminths, but also ingestion of fava beans. In contrast to these acute crises, in steady-state G6PD deficiency there is chronic haemolytic anaemia and often jaundice. Heinz bodies (haemoglobin denatured by the high levels of oxidants) may be seen in reticulocytes. Commercial screening and assay kits are available.

Deficiency of PK (coded for by *PKLR* at 1q21) is inherited in an autosomal-recessive manner and is the most common enzyme deficiency in the Embden–Meyerhof pathway with a frequency varying

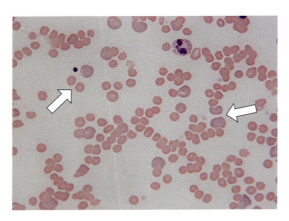

FIGURE 11.16

Spherocytosis. Numerous spherocytes (left arrow), slightly blue-ish reticulocytes (right arrow), and a neutrophil. There is also polychromasia and anisocytosis.

TABLE 11.19   Common membrane and enzyme defects leading to haemolysis

Condition	Nature of the pathology
Hereditary spherocytosis	Defects in the internal cytoskeleton leading to abnormal cell shape
Hereditary elliptocytosis	Defects in the internal cytoskeleton leading to abnormal cell shape
Paroxysmal nocturnal haemoglobinuria	Failure to anchor key protective molecules to the cell membrane
Glucose 6-phosphate dehydrogenase deficiency	Poor regeneration of reduced glutathione results in increased oxidant activity and failure to convert metHb to Hb
Pyruvate kinase deficiency	Reduced ability to generate ATP, build-up of 2,3 diphosphoglycerate

between 3.3 and ~50/million in white populations. There is generally a mild to moderate chronic non-spherocytic haemolytic anaemia (Hb typically 40–100 g/l) and an increased reticulocyte count. Reduced cell survival and chronic haemolysis result in increased iron turnover—increased ferritin and iron overload is found in 60% of untransfused PK-deficient patients and there is low hepcidin, which correlates with haemoglobin. Clinical consequences include jaundice, gallstones, iron overload, thrombosis, and osteopenia. Splenectomy is a common treatment (as in all haemolytic anaemias). Table 11.19 summarizes the pathology and laboratory aspects of membrane and enzyme defects.

SELF-CHECK 11.9

What is the molecular basis of the anaemia of mis-shaped red cells?

11.5.5 Haemoglobinopathy

Technically, haemoglobinopathy also causes a haemolytic anaemia, but it is of sufficient importance (present in 5% of the world's population) to warrant its own section. Section 11.1.3 provides details of haemoglobin genetics, section 16.7 describes chromosomal and mutation aspects. The two major manifestations, sickle cell disease and thalassaemia, have a variable microcytic hypochromic anaemia as the mutant haemoglobin fails to carry oxygen as efficiently (bringing the usual symptoms). The abnormal red cells are marked out and destroyed (thus haemolysis, contributing to the anaemia), which leads to a reticulocytosis and raised RDW. With a high level of general awareness, diagnosis is relatively straightforward as the subject's race, ethnicity, and family history will be known.

Abnormal haemoglobin variants can be detected with high-performance liquid chromatography (HPLC) (Figure 11.17) and electrophoresis, although molecular genetics is the ultimate test.

FIGURE 11.17

HPLC chromatograms. A composite chromatogram of different Hb species, the relative height of each indicating approximate blood levels.

From Moore, Knight, & Blann, *Haematology*, third edition, Oxford University Press, 2021.

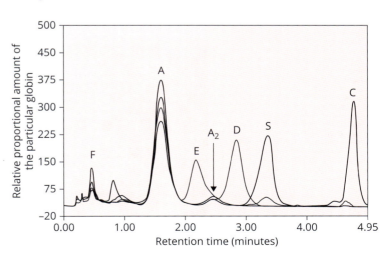

A common treatment of the anaemia is blood transfusion, bringing issues of iron overload and its problems (deposition in organs and their subsequent failure), often addressed by iron sequestration (section 11.7). Transfusions will become increasingly difficult because of the inevitable alloimmunization by minor antigens.

Sickle cell disease (SCD)

Sickle cell disease (SCD) is by far the most prevalent haemoglobinopathy, present in around 3.3% of the global population. Between 300,000 and 400,000 affected children are born annually, most of them (80%) in sub-Saharan Africa. The characteristic 'sickle' change in shape (Figure 11.18) occurs at low oxygen tension (i.e. in hypoxia), during dehydration, and with fever. In certain circumstances, reoxygenating blood can lead to the reversal of a newly formed sickle cell back to its normal state, but after repetitive cycles of oxygenation and deoxygenation, the otherwise reversible sickle cell is permanently damaged. Sickle red cells are very unstable and prone to lysis, with a lifespan reduced by 75% of that of normal red cells.

Mutations in the genes for β-globin lead to homozygous (i.e. HbSS) or heterozygous (HbAS) disease (section 16.7). In the latter, often described as a carrier, a residual amount of normal haemoglobin (such as HbF and HbA_2) moderates not only the extent of the polymerization, but also its reversal. Thus, sickle cell carrier status is generally asymptomatic, although painful crises may occur at times of physiological stress. SCD also brings inflammation and thrombosis, which with the anaemia lead to major clinical features:

- Painful vaso-occlusive crises caused by irreversibly sickled cells blocking blood vessels, thus tissue hypoxia and ischaemia, in turn leading to bone (hips, shoulders, and vertebrae) and muscle pain. These may be further enhanced by infection, the cold, strenuous exercise, dehydration, emotional disturbances, and pregnancy. The most serious vaso-occlusive crisis is of the brain (a stroke will occur in ~7% of patients).

- Visceral crises: Sickling of red cells in the liver and spleen can cause infarction, whilst sequestration in the lungs is partly responsible for the acute chest syndrome.

- Aplastic and haemolytic crises may occur, resulting from folic acid deficiency or parvovirus infection, and are characterized by an elevated rate of haemolysis and a sudden drop in haemoglobin level with the need for a blood transfusion.

Other clinical features include venous thromboembolism and ulceration resulting from hypercoagulability, vascular stasis, and local ischaemia. Further complications include retinopathy, **priapism**, gallstones, kidney damage, osteomyelitis, and leukaemia. A complex disease, the same SCD genotypes produce different phenotypes, as in the need for pain relief, which can vary from aspirin to opiates. Treatments include red cell transfusions, hydroxycarbamide/hydroxyurea (increasing HbF by ~10%), whilst the only cure is stem cell transplantation. The mAb crizanlizumab (targeting

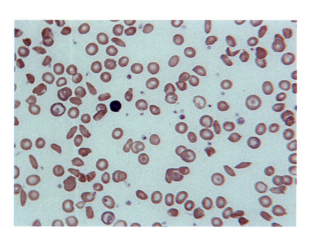

FIGURE 11.18
Numerous red cells are not round but are oblong or oval shaped. These are sickle cells: a striking example is in the top right corner. There are also several target cells, and a lymphocyte.

From Moore, Knight, & Blann, *Haematology*, third edition, Oxford University Press, 2021.

P-selectin, 2.5–5 mg/kg intravenously every 2 weeks) can be used to treat acute sickle crises, reducing their frequency by around 50%.

Thalassaemia

This disease is characterized by the reduced synthesis of either α or β globins, or both, and consequently a serious abnormality in the haemoglobin molecule. Present in around 1.7% of the global population, it is predominantly found in the eastern Mediterranean, the Middle East, the Indian subcontinent, and Southeast Asia. Notably, these are regions where sickle cell disease is also highly prevalent, reinforcing the view that the abnormal haemoglobin in carriers of haemoglobinopathy brings protection against malaria, which is endemic in these locations. Depending on which haemoglobin genes are mutated, the clinical extent of the disease may be described as minor, intermedia, or major.

The clinical consequences of the loss of α-globin function are sequential: loss of one gene is generally tolerated well with few clinical issues (i.e. thalassaemia minor), whilst two mutated genes lead to a mild or moderate anaemia, described as thalassaemia intermedia. With one remaining gene, clinical disease is inevitable with jaundice, hepatosplenomegaly, leg ulcers, gall stones, and folate deficiency, leading to thalassaemia major. Although missing α-globin can be compensated for by increased β- or γ-globin, this generally fails to address the poor oxygen-carrying aspect of the cell. Despite functional β-globin genes, and so β-globin, lack of any functional α-genes does not sustain life and the foetus dies either in utero or shortly after birth.

Mutations of the β-globin gene in the reduced or absent synthesis of β-globin lead to the increased synthesis of γ-globin, which pairs with α-globin chains to form HbF. As with α-thalassaemia, three broad clinical categories of disease are recognized. Thalassaemia major results from a mutation in each chromosome, hence little or no β-globin and so a profound transfusion-dependent microcytic anaemia, whilst thalassaemia intermedia may be caused by a variety of defects and so is often classified clinically. One form is homozygous β-thalassaemia, but effects of the anaemia are countered by increased HbF. The degree of anaemia varies and transfusions may be needed. Thalassaemia minor (or trait) are carriers of a mutant β gene.

Other haemoglobinopathy

Other qualitative β-globin gene disorders include haemoglobin E (HbE), commonly found in Southeast Asia with a frequency varying from 8% to 50%. Haemoglobin C (HbC) is found most frequently in West Africa, where the incidence may reach 20%. In the homozygous state (HbCC), splenomegaly and a mild haemolytic anaemia are common. As in HbE, the blood film is dominated by target cells. Haemoglobin D (HbD) disease may be described as a collection of different mutations in the β-globin gene and occurs with the highest frequency in the western region of India and in Pakistan.

As many heterozygous haemoglobinopathy states are asymptomatic, compound disease may arise. These include combinations of HbS and HbC (i.e. HbSC), and of HbS with HbD (HbSD), but also β-thalassaemia and sickle cell disease. In certain diseases, HbF is increased, giving at least some (limited) capacity to carry oxygen. Indeed, in homozygous δ-β thalassaemia, only HbF is synthesized, leading to 'hereditary persistence of foetal haemoglobin' (HPFH), which presents as thalassaemia intermedia. Accordingly, measurement of HbF can be useful.

Key points

The haemoglobinopathies are a major disease in several global regions that is treatable with blood transfusion, but curable only by bone marrow transplantation.

11.5.6 Anaemia of chronic disease (ACD)

Anaemia of chronic disease (ACD) is an umbrella term for anaemia arising from several different aetiologies, most of which focus on a particular organ (Table 11.20). Many ACDs are inflammatory, leading to aberrations in different processes. For example, in inflammatory bowel disease, there is likely to be poor absorption of micronutrients, whilst the systemic inflammation has the potential to increase hepcidin levels, leading to hypoferraemia. However, the precise aetiology of the well-established anaemia of rheumatoid arthritis is less clear, although it may be related to the suppression of erythropoiesis by cytokines.

11.5.7 Anaemia of the elderly

Even having excluded all the possible causes of anaemia described previously, it is often present in those aged over 60 years. The frequency of this unexplained anaemia of ageing in the over-65s ranges from 25% to 44%, with a mean of 35% (Guralnik et al. 2022). One possible explanation for this high figure is the accumulation of several otherwise subclinical aetiologies described in Table 11.20, such as nutrient deficiency, early malignancy, and low-grade inflammation, which act in concert to produce the effect.

Like IDA, many cases of the anaemia of chronic disease are characterized by low haemoglobin and MCV, whilst levels of CRP and the neutrophil count may be useful in differentiating those cases of the ACD that have an inflammatory component (e.g. a chronic infection or an autoimmune disease) from cases of (non-inflammatory) IDA.

11.6 Coagulopathy

Whilst section 11.2 explained the physiology of haemostasis, this section describes the pathology.

11.6.1 Overactive/excess platelets

Platelet numbers

A modest increase in the platelet count of >450 10^9/l (a thrombocytosis) is found in a wide range of conditions, such as smoking and rheumatoid arthritis (being part of the acute phase response). A platelet count >600 10^9/l defines the myeloproliferative neoplasm essential thrombocythaemia, which brings a risk of thrombosis (section 11.4.5).

Platelet activity

There is no major and recognized intrinsic defect in the platelet that leads to its overactivity—when this does happen it is the effect of an external factor acting on an 'innocent' platelet, an example

TABLE 11.20 The anaemia of chronic disease

Organ/system	Aetiopathology
Liver	Failure to store iron and synthesize carrier proteins (e.g. transferrin)
Kidney	Failure to produce erythropoietin
Gastrointestinal disease	Failure to absorb micronutrients: occult blood loss from bleeding tumours
Reproductive organs	Menorrhagia: failure of testes to produce testosterone
Endocrine disease	Anaemia associated with hypothyroidism
Systemic inflammatory disease	Suppression of erythropoiesis and hypoferraemia, both potentially due to inflammatory cytokines
Cancer	Suppression of the bone marrow by the metastatic process: effects of chemotherapy

being smoking. Nevertheless, the undoubted role of platelets in cardiovascular disease requires anti-platelet drugs in some primary conditions and in all cases of secondary prevention. A key metabolic enzyme, cyclo-oxygenase (Figure 11.5), is inhibitable by aspirin, and although the platelet may still be activated by other pathways, aspirin 75 mg daily remains the primary therapy in the treatment and prevention of secondary cardiovascular disease. Other NSAIDs such as ibuprofen and naproxen also inhibit cyclo-oxygenase, thus having anti-platelet activity, and, like aspirin, also have anti-pyrexial and analgesic activity.

Platelets may be activated by certain ligand–receptor combinations (Table 11.3), one being ADP with its P2Y12 purinoreceptors, an interaction that can be blocked by clopidogrel, prasugrel, and ticagrelor. These oral agents are cited by NICE in secondary prevention of cardiovascular disease. Clopidogrel may also be used alongside dipyridamole for the prevention of occlusive vascular events, and in those intolerant of aspirin.

The most abundant molecule on the platelet surface is GpIIb/IIIa, also known as integrin $\alpha_{IIb}\beta_3$, a receptor whose ligands are fibrinogen and vWf. Upon partial activation (e.g. by ADP binding its receptor), GpIIb/IIIa undergoes a conformational change increasing its affinity for its receptors, which bind to fully activate the cell and prompt adhesion and aggregation, early steps in thrombogenesis. The GpIIb/IIIa–fibrinogen interaction can be blocked by abciximab, integrellin, and tirofiban. All are given intravenously, and are NICE (CG94) and FDA approved for thrombosis prevention in percutaneous coronary intervention (PCI). An ADP receptor blocker such as prasugrel may be used in PCI for treating acute coronary syndromes in adults (NICE TA317, ESNM63).

11.6.2 Underactive/insufficient platelets

Investigation of unexplained haemorrhage starts with a full blood count, providing a platelet count, but this tells us nothing about the function of these platelets.

Platelet function

There are a small number of very rare (≤1/million) conditions of platelet hypofunction. Bernard–Soulier syndrome is caused by a lack of GpIb (a dimer of CD42b and CD42c, coded for at 17p13.2 and 22q11.21 respectively). The GpIb/V/IX complex (the receptor for vWF), fails to bind its ligand, which leads to weak or absent platelet–platelet and platelet–subendothelium adhesion. Glanzmann's thrombasthenia is caused by absent or low levels of GpIIb/IIIa that do not bind to fibrinogen, resulting in a lack of platelet activation with failure to contribute to haemostasis. This disease is caused by mutations in the genes for GpIIb (CD41) or for GpIIIa (CD61), both coded for at 17q21.32. Other diseases affecting platelet function are Scott syndrome (platelets exhibiting defective expression of phosphatidylserine); storage pool diseases, such as those of a deficiency in dense granules (such as in Chediak–Higashi syndrome); and grey platelet syndrome (caused by a lack of alpha granules). All these genetic disorders are inherited in an autosomal recessive manner.

Platelet function disorders may be diagnosed by fluorescence flow cytometry for the presence of relevant CD markers, but a simple test of function is to simulate platelet activation by exposing it to agonists such as thrombin. A normal platelet will degranulate and expose P selection (CD69P) on its surface, detectable by fluorescence flow cytometry. Other functional tests include:

- Light transmission aggregometry, where a beam of light passes through a suspension of platelet-rich plasma. Addition of an agonist (thrombin, ADP, collagen, epinephrine) causes the formation of micro- and then macro-thrombi, which clarifies the suspension, resulting in increased light passage.

- Thromboelastography, where whole blood aggregation is initiated by an activator of the coagulation pathway, such as kaolin. The clot thus forms from all the natural components of haemostasis. By allowing the process to develop over 30 minutes, the consolidation of the clot can be determined, and also its degradation by fibrinolysis.

- The Platelet Function Analyser (PFA-100), which assesses the ability of the platelets in a sample of whole blood to occlude an aperture once stimulated by an agonist. It is widely used as a first-line screening test in otherwise unexplained haemorrhage.

Thrombocytopenia (platelet count <150 × 10⁹/l)

The principal reason for thrombocytopenia is cancer and/or its treatment. Dysfunctional thrombogenesis is exceptionally rare (<1/million), an example being congenital amegakaryocytic thrombocytopenia. Causes of thrombocytopenia include (1) immune thrombocytopenia purpura (ITP, purpura = bruising), mostly caused by IgG autoantibodies to GpIIb/IIIa or GpIb/V/IX, presenting at a rate of ~75/million/year; (2) thrombotic thrombocytopenia purpura (TPP, 6/million/year), caused by lack of ADMTS13, an enzyme that digests vWf; (3) haemolytic uraemic syndrome (section 11.5.4); (4) thrombocytopenia as part of HELLP, a collection of features present in <1% of pregnancies, but in 10–20% of those suffering pre-eclampsia (section 14.4.2); and (5) heparin, an important anticoagulant, but in a small number of cases (~2.5% if unfractionated, ~0.5% of the low-molecular-weight form) it causes a thrombocytopenia 3–10 days after use—thus heparin-induced thrombocytopenia.

Treatments

The primary platelet disorders, Bernard–Soulier syndrome and Glanzmann's thrombasthenia, are treated with platelet transfusions as the clinical state requires, beyond which bone marrow transplantation is the next step. The secondary causes are treated according to aetiology: cessation of the particular drug, immunosuppression (steroids, cyclosporine, anti-B cell rituximab), or plasma exchange for a pathological (auto)antibody. In some cases, recombinant thrombopoietin can stimulate new platelet production, and therapeutic platelet transfusion is a widely accepted option in life-threatening thrombocytopenia and/or platelet dysfunction.

SELF-CHECK 11.10

How can we suppress overactive platelets?

11.6.3 Overactive coagulation

As explained in section 11.2.3, haemostasis is a balance between thrombogenesis and fibrinolysis. When the former exceeds the latter, there will be thrombosis, and in cases of the reverse, there will be haemorrhage. By far the greatest imbalance is in overactivity of the coagulation pathway (section 11.2.3), and the major clinical condition resulting from this misbalance is venous thromboembolism (VTE), mostly consisting of deep-vein thrombosis (DVT) and pulmonary embolism (PE), which together have an incidence of around 1,000/million/year.

Of the numerous well-established risk factors for VTE (Table 11.21), most are linked to existing disease and others to hospital procedures (surgery, chemotherapy). Furthermore, risk is additive: obesity brings a two-fold increased risk and use of the oral contraceptive pill a four-fold increased risk. Together they bring a risk over 20 times that of women free of both factors.

TABLE 11.21 Risk factors for VTE

Level of risk	Factors
Weak: increased <2 fold	Bed rest >3 days, immobility due to sitting, increasing age, laparoscopic surgery, obesity, pregnancy, varicose veins at age 60, non-O blood group, atrial fibrillation (AF), chronic kidney disease, raised CRP, hyperthyroidism, SLE
Moderate: increased 2–9 fold	Arthroscopic knee surgery, central venous lines, malignancy (2–4 fold, with chemotherapy 4–6 fold), heart failure, hormone replacement therapy, oral contraceptives, paralytic stroke, post-partum pregnancy, previous VTE, heterozygous factor V Leiden (and most other thrombophilias), varicose veins at age 45, nephrotic syndrome, inflammatory bowel disease
Strong: increased >10 fold	Hip, pelvis or leg fracture, hip or knee replacement, major general surgery, major trauma, spinal cord injury, hospital or nursing home confinement, homozygous factor V Leiden and deficiencies of inhibitors

BOX 11.2 Factor V Leiden (FVL)

The very high frequency (carried by 5% of a north-west European population: 1 in 20 people) of this abnormal protein known to cause an undesirable effect demands an explanation. The most persuasive is that blood loss at childbirth is reduced in those women carrying the gene, thus reducing the risk of potentially fatal postpartum haemorrhage, which provides a selective advantage. In support of this hypothesis is the observation that carriers of FVL have a higher haemoglobin level and reduced menstrual blood loss.

This model is similar in certain haemoglobinopathies and red blood cell membrane abnormalities, where heterozygosity brings protection from malaria. In both instances, homozygotes pay the price of an increased burden of disease whilst heterozygotes accrue an advantage (hence 'heterozygote advantage'). FVL is also an example of a founder effect, as levels are also high in areas of European immigration (North America, South Africa, Australia, and New Zealand).

Thrombophilia

This condition describes those at increased risk of thrombosis, brought about by several factors, many of which are genetic, and may account for ~60% of the risk of VTE. These include:

- A mutation in the non-coding region of the prothrombin gene—G2021A—causing increased levels of this protein and a 2–3-fold increased risk of VTE.

- Factor V Leiden (FVL), also known as F5-R506Q, present in 5% of those of north-west European descent and virtually non-existent in populations without some European ancestry. In its heterozygous and homozygous forms it brings a 6-fold and an 80-fold risk of VTE respectively (see Box 11.2).

- Increased serum homocysteine, which carries a 1.5–2.5-fold increased risk, possibly due to loss of function in methylene tetrahydrofolate reductase.

- Race/ethnicity: The rate of VTE in African Americans is ~5-fold that of Asian populations in East and Southeast Asia, with Europeans carrying an intermediate risk. The increased rate in African Americans and Black British people may be linked to lower levels of natural anticoagulants, compared with matched White people.

- Deficiencies of inhibitors protein C, protein S, and antithrombin, each conferring a relative risk of VTE of up to 15–20 times that of normal levels of each particular molecule.

- The anti-phospholipid syndrome (APS), an autoimmune disorder characterized by anti-phospholipid antibodies (APA), targets molecules beta-2 glycoprotein-1, prothrombin, or phospholipid complexes.

In testing for certain risk factors, the routine coagulation laboratory will determine all major coagulation molecules, and most special tests, such as APAs and the **lupus anticoagulant**. Most coagulation factors and inhibitors are assessed by functional assays (clot formation or chromogenic substrate digestion), others by immunoassay (e.g. ELISA). FVL, prothrombin G2021A, and other mutations will be determined by molecular genetics. Thrombophilia should be pursued in those with an unexplained VTE, if only to exclude certain factors.

Cross reference

For a discussion of the concept of the founder effect from the field of population genetics, see Box 16.2 in Chapter 16.

Diagnosis of VTE

Diagnosis of a VTE starts with a scoring system. Those of Wells and Geneva are most popular and rely on clinical features. Depending on this result, a D-dimer test may be required, and if negative, a VTE is excluded. Alternatively, a high score requires imaging: compression ultrasound of the leg for a DVT

TABLE 11.22 Diagnosis of deep-vein thrombosis (DVT)

Signs/symptoms	Score
Active cancer (treatment in <6 months or palliative); calf swelling ≥3 cm circumference compared to the other calf; swollen unilateral superficial varicose veins; unilateral pitting oedema; swelling of the entire leg; localized tenderness along the deep venous system; paralysis, paresis, or recent-case immobilization of the lower extremities; previous documented VTE; recently bedridden ≥3 days; major surgery requiring regional or general anaesthesia in <12 weeks: all score 1 point each. Plausible alternative diagnosis scores -2 points	0: Low risk 1–2: Moderate risk ≥3: High risk

and CT or pulmonary angiography for pulmonary embolism. Table 11.22 shows clinical features of a potential DVT; those for pulmonary embolism (several of which are in common with a DVT) are shown in Table 13.10 of section 13.5.4. Figure 11.19 summarizes this pathway.

NICE NG158 refers to diagnosis, management, and thrombophilia testing in VTE, and KTT16 focuses on anticoagulants, including direct-acting oral anticoagulants, whilst NG89 offers guidance on reducing the risk of hospital-acquired DVT or PE.

Key points

As advances are always being made, guidelines and recommendations are continuously revised. In the UK, NICE is the major authority, whilst local, national, and international guidelines may apply. The practitioner *must* refer to their own most recent up-to-date protocols. This applies to all such material in this book.

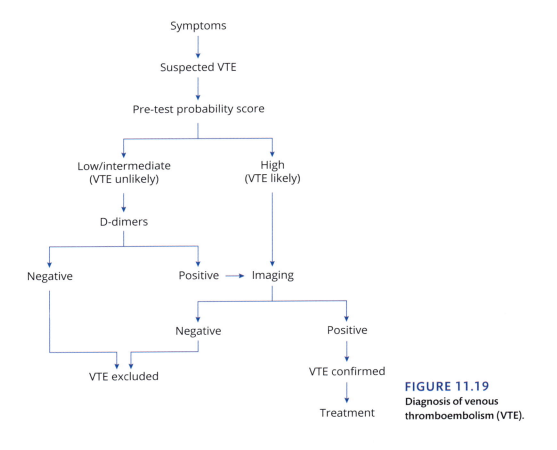

FIGURE 11.19
Diagnosis of venous thromboembolism (VTE).

Treatment and prophylaxis of VTE 1: Oral anticoagulants

Once diagnosed, treatment may begin. However, certain risk factors are so strong that prophylaxis is required to reduce the likelihood of the development of a VTE. These include AF, most cases of surgery (especially orthopaedic), and certain prosthetic heart valves. Prophylaxis is by anticoagulation, which may be given orally or transdermally, depending on circumstances.

Historically, the leading oral anticoagulant (OAC) has been warfarin, which inhibits hepatic synthesis of prothrombin and other coagulation factors. Although very effective, there is great variability in patient responses, and the dose requires monitoring and adjusting to extend the prothrombin to around 2.5 times its normal time, but 3 or 3.5 times for those at high risk. This translates to an international normalized ratio (INR) of 2.5, and although this is the target, any result between 2.0 and 3.0 is acceptable (Table 11.23, which also shows duration of treatment).

Four other direct-acting OACs (DOACs: NICE KTT16) are alternatives to warfarin and have several advantages, such as no need to routinely monitor their effect with a blood test. Dabigatran is a direct thrombin inhibitor; rivaroxaban, apixaban, and edoxaban are FXa inhibitors, and all may be used in DVT, PE, atrial fibrillation, and hip/knee replacement. Rivaroxaban may also be used in coronary or peripheral artery disease, and acute coronary syndromes. As is common practice with almost all anticoagulants, patients with renal impairment (GFR 30–50 ml/min) require reduced dosages, and liver function should be checked. The DOACs are also part of NICE documents CG180 and NG89.

Treatment and prophylaxis of VTE 2: Injectable anticoagulants

OACs are, on the whole, good drugs for stable and long-term use, but do not act rapidly. Where immediate anticoagulation is required, it can be provided by a number of injectable anticoagulants (IACs). Heparin, a natural product present on the cell membrane, can be purified and used as an anticoagulant in its whole, unfractionated form, but its effects are variable (as with warfarin) and need to be monitored with the APTT blood test. In practice, the dose aims to prolong the APTT by anything from 1.5 to 3 times that of normal, un-coagulated blood, that is, an APTT patient/control ratio of 1.5–3.0, analogous to the INR and warfarin. The lack of predictable effect of this unfractionated heparin has been overtaken by a formulation of only the low-molecular-weight parts of the crude heparin preparation (hence LMWH), which does not routinely need blood monitoring. LMWHs may now be given to appropriately tutored patients so they can self-dose daily at home. A third IAC, fondaparinux, is a selective and reversible Xa-inhibitor and so an alternative in certain types of surgery, such as orthopaedic (NICE NG158). Other IACs include idraparinux, danaparoid, lepirudin, bivalirudin, and argatroban.

TABLE 11.23 Target INRs and recommended duration of warfarin

Indication	INR target	Duration
Distal DVT due to temporary risk factors (such as pregnancy), cardiac mural thrombus	2.5	3 months
PE, proximal DVT, or DVT of unknown cause or those associated with ongoing risk factors, most surgical cases	2.5	6 months
PE/DVT associated with malignancy	2.5	6 months then review
Recurrence of PE/DVT whilst not on warfarin, rheumatic mitral valve disease, antiphospholipid syndrome (venous), AF, cardiomyopathy	2.5	Long term
Mechanical prosthetic heart valves (aortic)	3.0	Long term
Recurrence of PE/DVT whilst on warfarin, mechanical prosthetic heart valves (mitral), antiphospholipid syndrome (arterial)	3.5	Long term

TABLE 11.24 Properties of common anticoagulants

	Monitoring	Half-life	Target	Delivery
Unfractionated heparin	APTT	Short	Thrombin > factor Xa	Injected
LMWH	Generally not required	Short	Factor Xa > thrombin	Injected
Fondaparinux	Generally not required	Short	Factor Xa	Injected
Warfarin	INR	Long	Several factors	Oral
Dabigatran	Generally not required	Short	Thrombin	Oral
Rivaroxaban, apixaban, edoxaban	Generally not required	Short	Factor Xa	Oral

APTT = activated partial thromboplastin time; INR = international normalized ratio.

A key aspect of IACs is their almost immediate action, so that prophylaxis for those at risk of VTE can begin promptly. However, not all risk factors are equal, so that patients admitted to hospital must be scored as to their total risk of VTE and treated accordingly. These are: score 1: age >60, obesity, cardiovascular disease, blood cancer, nephrotic syndrome, inflammatory bowel disease; score 2: prescribed oestrogen, pregnancy, post-partum, thrombophilia, malignancy, sepsis, known family history in two 1st-degree relatives, immobile >3 days, history of VTE, active malignancy. A further aspect is the reason for admission. Should this be surgery, more risk factor points may be added: score 0: surgery lasting < 30 minutes; score 1: surgery lasting > 30 minutes; score 2: Intraperitoneal laparoscopic surgery lasting > 30 minutes, vascular surgery not intra-abdominal; score 3: thoracotomy or abdominal surgery involving mid-line laparotomy, total abdominal hysterectomy, including laparoscopic assisted; score 4: major joint replacement, surgery for fractured neck of femur, major trauma, e.g. lower-limb fractures. The sum of these factors gives a total score that can be used to guide anticoagulant management:

- Low risk (score 0–1): early ambulation: consider graduated elastic compression stockings (GECS).

- Medium risk (score 2–3): GECS plus low dose LMWH for surgical patients, high dose of LMWH for medical patients,

- High risk (score ≥4): GECS plus high-dose LMWH (not exceeding 14 days for medical inpatients). Surgical patients—consider intermittent pneumatic compression in theatre.

However, as with oral anticoagulants (and, in fact, any drug), there are a number of cautions and contradictions for the use of LMWHs. The former include severe hepatic or renal impairment, or major trauma or surgery to the brain, eye, or spinal cord. Contra-indications include known uncorrected bleeding disorders such as haemophilia, heparin allergy, heparin-induced thrombocytopenia, heparin-induced thrombosis, patients on existing anti-coagulation therapy, bleeding or potentially bleeding lesions, active peptic ulcer, recent intracranial haemorrhage, intracranial aneurysm, or vascular malformation. Table 11.24 summarizes properties of common anticoagulants.

11.6.4 Coagulation underactivity

The second aspect of the failure of haemostasis to form a clot is due to defects in coagulation factors, which can be classified as primary or secondary.

Primary hypocoagulation

Haemophilia is the leading severe genetic coagulation disease, caused by mutations in the gene for factor VIII (F8, coded at Xq2.8). Presenting with a frequency of ~65/million (1 in 10,000 males), ~70% are inherited. With two X chromosomes, females with an abnormal gene can generate sufficient factor

VIII for their coagulation needs from the normal gene, but they are nonetheless carriers of the disease and may pass the abnormal X chromosome to their sons.

A functioning tissue factor/factor VII pathway (Figure 11.7) which can activate factor X in the absence of factor VIII confers a degree of haemostasis. The disease can be described clinically as mild, moderate, or severe. Around 20% of cases are mild, where factor VIII levels are >5 IU/dl (reference range 50–150) and spontaneous haemorrhage is rare but may develop after trauma and surgery. Accounting for 30% of cases, with FVIII levels 1.0–5.0 IU/dl, moderate disease brings more spontaneous bleeds, and greater loss of blood after trauma. The severe disease, present in 50% of cases, is characterized by factor VIII <1.0 IU/dl, with frequent and spontaneous epistaxis (nosebleeds), bleeding into joints, muscles, and internal organs, and considerable haemorrhage after trauma.

The gene for vWf at 12p13.2 is large and so is subject to many different mutations that lead to variable penetrance into different clinical forms of von Willebrand disease (vWd), which may be present in 1% of the population. The mild form of vWd (type I) is the most common (~75% of cases) and in many it is asymptomatic, with only the occasional nosebleed. Levels of vWf are perhaps 20–50% of normal. Moderate disease (type II, of which there are four subtypes) is present in ~25% and characterized by normal or slightly reduced levels of vWf, but the molecule is dysfunctional (e.g. fails as a FVIII cofactor). Severe vWd (type III) is the rarest form (<1%) with very low or absent levels of vWf. It clinically resembles haemophilia, but is also found in females.

Deficiency of factor IX (haemophilia B or Christmas disease) arises at a frequency of 20/million (1 in 50,000 males). Mutations in the gene (coded at Xq27.1) lead to inability to transform to factor IXa and so participate in haemostasis. Deficiencies in fibrinogen and factors V, VII, X, XI, and XIII each present at ~3/million. The degree of deficiency of these molecules is variable, and therefore so is their haemorrhagic potential.

Replacement therapy with purified FVIII in haemophilia is frustrated by the development of auto- and allo-immune inhibitors. Should these be present, recombinant factor VII is an option. Purified vWf can be provided as cryoprecipitate of whole plasma, which includes some other factors. Desmopressin, a synthetic analogue of **vasopressin**, stimulates endothelial vWf release, and is used in both prophylaxis and in the treatment of haemorrhage. Whole fresh frozen plasma may also be given. If there is a great deal of blood loss in any condition, it may need to be replaced with packed red cell transfusion, and perhaps also plasma expanders. Curative treatment for the deficiencies includes whole liver and hepatocyte suspension transplantation, and possibly also of the use of gene therapy, but these are very far from routine practice.

Secondary hypocoagulation

The primary medical condition for haemorrhage is liver disease (as this organ produces the coagulation factors); others include a diet poor in vitamins C and K. Acquired vWd has been reported in leukocyte malignancies, autoimmunity (typically SLE), solid tumours, hypothyroidism, and some cardiovascular disease. Similarly, causes of acquired haemophilia include myelodysplasia, treatment with interferon or rituximab, essential thrombocythaemia, hepatitis, and anti-viral agents.

A major cause of acquired hypocoagulation is overdose with OACs and IACs. The clinical definition of haemorrhage is unclear, and if signs and symptoms are minor they may not warrant treatment. Those requiring treatment include persistent nosebleeds, haematuria, vomiting blood, and blood being passed *per vagina* or *per rectum*. Treatment is by cessation of the offending drug, which is usually sufficient for the IACs. However, warfarin has a prolonged effect on the liver, so that in actual or very high risk of a haemorrhage, it will be necessary to give some vitamin K orally and restart when the INR is in range. In the face of major bleeding, recommendations include to stop warfarin, give prothrombin complex concentrate (e.g. 50 units/kg or fresh frozen plasma 15ml/kg), then give 5 mg vitamin K IV, and repeat dose of vitamin K after 24 hours if the INR is still high. In the UK, all NHS hospitals will have their own specific guidelines which take precedence.

Disseminated intravascular coagulation (DIC)

Many of the concepts of this section can be distilled in this extremely dangerous and life-threatening condition, which may arise after protracted abdominal or thoracic surgery, obstetric problems, pneumonia, massive burns, septicaemia, and certain leukaemias. The aetiology is best summarized as coagulation out of control and semi-permanently activated, with vascular injury.

TABLE 11.25 An overview of major aspects of thrombosis and haemorrhage

	Overactive platelets and thrombocytosis	Overactive coagulation	Underactive platelets and thrombocytopenia	Underactive coagulation
Causes	(i) Risk factors for atherosclerosis (ii) High platelet count	(i) Risk factors for venous thrombosis (ii) Failure of inhibitors (often genetic)	(i) Intrinsic defect (ii) Over-treatment with anti-platelet drugs (iii) Low platelet count	(i) Intrinsic defect (ii) Over-treatment with anticoagulant drugs
Treatments	(i) Avoidance of risk factors (ii) Anti-platelets	(i) Avoidance of risk factors (ii) Anti-coagulants	(i) No specific treatment (ii) Correct treatment	(i) Replacement therapy (ii) Correct treatment
Laboratory assessments	(i) Monitoring risk factors (where possible) (ii) Full blood count	(i) Monitoring risk factors (if possible) (ii) Coagulation factor assays	(i) Platelet function studies (ii) Full blood count	(i) Coagulation factor assays

Excessive tissue factor and tenase generation lead to a thrombin burst, resulting in systemic fibrin generation and microthrombi production that the inhibitory pathways cannot address. Microthrombi may become lodged in capillaries, leading to tissue and organ damage. As platelets and coagulation factors are consumed, risk of haemorrhage rises as there are no components with which to form a constructive thrombus.

DIC is recognized in the laboratory with prolonged PT and APTT, low fibrinogen, thrombocytopenia, and raised D-dimers. Attempts to treat potential haemorrhage with infused coagulation factors and platelet transfusions often fail to restore haemostasis, but are themselves consumed. The underlying cause must be addressed, one immediate potential solution being the infusion of inhibitors, principally antithrombin and soluble thrombomodulin. Table 11.25 summarizes this section.

SELF-CHECK 11.11

How can we reduce the risk of venous thromboembolism?

11.7 Iron overload

Programmed by evolution to retain iron, we store it in ferritin and haemosiderin in the liver, bone marrow, and elsewhere. However, should stores be full, the iron will be stored in other tissues, causing the disease haemochromatosis, which has two major causes: hereditary haemochromatosis (HH) and blood transfusions.

11.7.1 Hereditary haemochromatosis (HH)

The most common genetic cause of iron overload, ~80% of autosomal recessive HH cases are caused by mutation C282Y in *HFE*, whilst 30 other variants are known. Heterozygous C282Y is present in ~9% of Europeans (i.e. 90,000/million), and although they have no clinically evident disease, iron levels may be raised in ~25%. In its homozygous form, HH inevitably has clinical consequences, and is present in 0.4% of Europeans (4,000/million, but 12,000/million in Ireland), whilst a haemochromatosis-associated *HFE* genotype and iron overload in African Americans is present in around 40/million (Barton et al. 2023). The reasons for this stark racial and geographical variation are unclear, but may relate to a need to save iron, although in Ireland it may be a founder effect.

The product of *HFE*, HFE is part of a complex that regulates *HAMP*, whose product is hepcidin. Although serum hepcidin is low in both C282Y heterozygotes and homozygotes, levels are strongly related to those of ferritin in homozygotes. Hepcidin levels in urine are negatively correlated with the severity of HH, adding further justification to the measurement of this small peptide.

11.7.2 Blood transfusions

Patients with any severe anaemia are likely to require regular blood transfusions and, as a consequence, develop iron overload. One unit of transfused blood contains ~200–250 mg of iron, so that a long-term consequence is iron overload, generally after 10–20 units. Consequently, iron chelation therapy may start after ~12 months of blood transfusion, but the figures vary markedly between patients. Transfusional iron overload can have serious clinical consequences and, unless body iron is controlled, patients may suffer significant morbidity and mortality.

11.7.3 Other reasons for iron overload

There are several other possible reasons for iron overload, which include:

- excess iron in the diet, perhaps taking decades to build up.

- mutations in *FPN1 (SLC40A1)*, coding for ferroportin, which (a) impair its iron-exporting function, particularly in macrophages, and (b) render it unresponsive to the inhibitory effects of hepcidin, so that it exports all absorbed iron into the plasma.

- acaeruloplasminaemia, a rare recessive disorder caused by a loss-of-function mutation in *CP* on 3q23–q24, which results in impaired ferroxidase activity, leading to excessive iron deposits, low transferrin saturation, and mild microcytic anaemia.

- atransferrinaemia, a recessive disorder caused by transferrin deficiency due to mutations in *TF* on 3q22.1; its characteristics include very low to undetectable plasma transferrin, leading to impaired erythropoiesis and a microcytic hypochromic anaemia.

11.7.4 Clinical consequences and treatments

Excess iron will become deposited in various organs. In the liver, there will be cirrhosis, fibrosis, and possibly carcinoma, whilst deposits in the heart will lead to myocarditis, arrhythmia, fibrosis, cardiomyopathy, and heart failure. Dermal iron leads to rash, hyperpigmentation, and dermatitis, and in the joints to arthralgia. Deposits in the pancreas will cause diabetes, and in the pituitary multiple issues may follow, such as hypogonadism.

Treatment should be tailored to each patient, although many begin when ferritin is >500 µg/l. Iron chelators are usually administered subcutaneously, the most widely used such agent being desferrioxamine, and the success of treatment is monitored through the assessment of ferritin levels: it also reduces the amount of iron in the liver and also labile plasma iron. An alternative to chelation is to simply bleed the patient as if they were a blood donor (venesection), and is the preferred method in HH. Venesection should proceed regularly (perhaps each month) until ferritin levels are at the lower end of the reference range (perhaps 50 µg/l). Furthermore, the response of the full blood count and iron indices to phlebotomy provides an assessment of the severity of the iron overload. However, these approaches do not reverse organ damage. Dietary advice is to minimize foodstuffs with a high iron content. Developing therapies include synthetic hepcidin analogues to inhibit further iron absorption, although these will do nothing for existing iron deposits.

SELF-CHECK 11.12

What are the major causes of iron overload and what are the treatments?

Chapter summary

- The red cell, produced by the bone marrow by erythropoiesis, has only one function—to transport oxygen to the tissues.

- Platelets combine with plasma proteins of the coagulation pathway (and others) to maintain haemostasis.

- The haematology laboratory offers a comprehensive panel of tests and procedures of great importance to the diagnosis and management of many diseases, not only those of the blood.

- Blood cancers include leukaemia, lymphoma, and myeloma: all have a clear genetic cause and are treatable with general cytotoxic drugs and others that are targeted at particular malignant cells.

- Anaemia, the major disease of red cells, may be caused by damage to the bone marrow, lack of micronutrients, and haemolysis. Examples of the latter include the haemoglobinopathies.

- Loss of haemostasis can result in thrombosis (due to excess platelets and/or an overactive coagulation pathway) or haemorrhage (insufficient platelets and/or an underactive coagulation pathway).

- The consequences of iron overload (haemochromatosis) include damage to organs such as the liver, heart, and pituitary.

Further reading

- Zivot A, Lipton JM, Narla A, Blanc L. Erythropoiesis: insights into pathophysiology and treatments in 2017. Molec Med 2018: 24; 11. doi: 10.1186/s10020-018-0011-z.

- Lux SE. Anatomy of the red cell membrane skeleton: unanswered questions. Blood 2016: 127; 187–99.

- Anderson GJ, Frazer DM. Current understanding of iron haemostasis. Am J Clin Nutr. 2017: 106(Suppl 6); 1559S–66S.

- Szymczyk A, Macheta A, Podhorecka M. Abnormal microRNA expression in the course of hematological malignancies. Cancer Manag Res 2018: 10; 4267–77.

- Kanakry CG, Fuchs EJ, Luznik L. Modern approaches to HLA-haploidentical blood or marrow transplantation. Nat Rev Clin Oncol. 2016: 13; 10–24.

- Riva G, Nasillo V, Ottomano AM, et al. Multiparametric Flow Cytometry for MRD Monitoring in Hematologic Malignancies: Clinical Applications and New Challenges. Cancers. 2021: 13; 4582. doi: 10.3390/cancers13184582.

- Madu AJ, Ughasoro MD. Anaemia of Chronic Disease: An In-Depth Review. Med Princ Pract 2017: 26; 1–9.

- Mohandas N. Inherited haemolytic anaemia: a possessive beginner's guide. Hematology Am Soc Hematol Educ Program. 2018: Nov 30; 377–81.

- Gurney D. Platelet function testing: from routine to specialist testing. Br J Biomed Sci 2016: 73; 10–20.

- Undas A, Zabczyk M. Antithrombotic medications and their impact on fibrin clot structure and function. J Physiol Pharmacol. 2018: 69. doi: 10.26402/jpp.2018.4.02.

- Lim W, Le Gal G, Bates SM, et al. American Society of Hematology 2018 guidelines for management of venous thromboembolism: diagnosis of venous thromboembolism. Blood Adv. 2018: 2; 3226–56.

- Pantopoulos K. Inherited disorders of iron overload. Frontiers Nutr 2018: doi:10.3389/nut.2018.00103.

- Nurden P, Stritt S, Favier R, Nurden AT. Inherited platelet diseases with normal platelet count: phenotypes, genotypes, and diagnostic strategy. Haematologica 2020: doi: 10.3324/haematol.2020.248153.

Useful websites

- Guidelines of the British Society of Haematology: **https://b-s-h.org.uk/guidelines**
- Overview on coagulation: **www.pathologyoutlines.com/coagulation.html**
- National Institute for Health and Care Excellence: **www.nice.org.uk/**
- United Kingdom Thalassaemia Society: **https://ukts.org/**

 # Discussion questions

11.1 Comparing the data in Table 11.7 with the reference values in Table 11.5, determine those which are abnormal, and discuss their meaning.

11.2 Summarize the causes and diagnoses of the various forms of anaemia.

11.3 How can we prevent and treat thromboses in the venous circulation?

Diseases of malnutrition and the digestive system

The Global Burden of Disease study reported almost 24 million deaths from digestive diseases in 2017. The UK's Office for National Statistics (ONS) reports that disease of the digestive system was the 7th most common cause of death in England and Wales in 2021, leading to the deaths of 28,470 people: 14,719 men and 13,751 women. However, to this may be added the 44,172 (25,073 men, 19,099 women) who died from cancer of digestive organs, and an unknown number of deaths from other inflammation of digestive organs. It could therefore be argued that diseases affecting this organ system are actually the 3rd most common cause of death, with a toll of over 72,000, exceeding those of COVID-19 (67,000) and those of the respiratory system (55,000).

We will work through the digestive system (literally) from top to bottom and consider three sections: the upper (section 12.1: above the diaphragm), middle (section 12.2: stomach, liver, biliary system, pancreas) and lower digestive tract (section 12.3: small and large intestines). Section 12.4 will describe the requirements of a good diet, and the disease that follows from a poor diet, that being malnutrition and its polar extremes of too much and too little.

Learning objectives

After studying this chapter, you should confidently be able to:

- explain diseases of the buccal cavity and throat

- understand the aetiology of oesophageal and gastric disease

- describe certain aspects of the pathology of the liver and pancreas

- outline the diagnosis and treatment of intestinal disease

- appreciate the importance of good diet and the pathology of a poor diet

12.1 The upper digestive tract

In this section we consider the lips to the base of the oesophagus. Certain parts of the upper section of the alimentary canal are often combined with other organs and tissues of the head under the medical speciality 'ears, nose, and throat', hence ENT, whilst oncologists have brought together head and

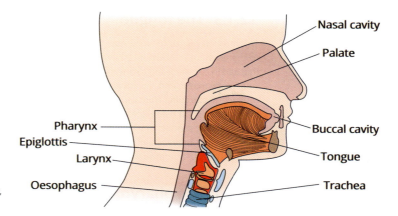

FIGURE 12.1
The upper digestive system.

Adapted from Fullick & Fullick, *Organs, Systems, and Surgery*, Oxford University Press, 2023.

neck cancer into its own speciality. The uppermost set of tissues (the mouth) consists of the lips, teeth, tongue, and palate (Figure 12.1). Above it is the nasal cavity, separated from the buccal cavity by the palate, a sheet of thin bone. The tissue at the base of the mouth is composed of connective tissues supported by the jawbone. The bones of the skull erupt through the soft tissues of the buccal cavity as teeth which will not be discussed in great detail. The throat houses the pharynx (which passes to the oesophagus) and larynx (merging with the trachea). However, the latter has no part in digestion, and so it will be discussed in Chapter 13 on respiratory disease. Similarly, the nose is a sense organ and respiratory conduit with no digestive role, so it also belongs in Chapter 13.

12.1.1 The lips

The lips are a pair of highly vascular soft tissues supported by connective tissues above and below the opening of the buccal cavity (the mouth). They are rich in sensory (taste and proprioception) receptors. Pathology is relatively rare, but there may be ulceration (as in Behcet's disease) and viral lesions (notably the herpes viruses). Inflammation of the corners of the lips, where they meet (angular cheilitis), is often caused by irritants, local infections (such as fungi, notably *Candida*), and mineral (deficiency such as iron), and is treated accordingly. Carcinoma may develop, especially in those smoking and chewing tobacco. The lips (and cheeks and the entire face) may suffer from angioedema, whether hereditary or drug-related.

12.1.2 The buccal cavity

The internal facings of the cavity are faced to the left and right by the cheeks, above by a sheet of thin bone (the palate), and lined with secretory epithelia generating mucus. Buried in the soft tissues are numerous salivary glands (parotid, submandibular, sublingual) secreting enzymes and other factors (e.g. **haptocorrin**) to begin the process of digestion. Mouth ulcers are surprisingly common, complained of by ~20% of the population, and are often idiopathic, although some are linked to vitamin deficiencies. Inflammation of the gums, gingivitis, is generally the result of the build-up of bacterial plaque following poor mouth hygiene. The uvula is a pea-sized nodule of soft tissues at the rear of the buccal cavity, hanging down from the hard palate.

Sjögren's syndrome (section 10.3.6) brings a dry mouth (xerostomia) resulting from autoimmune destruction of salivary glands, and so reduced salivation, and may be involved in sarcoidosis (also section 10.3.6). These glands may become enlarged, and may be mistaken for lymphadenopathy. On each side of the rear of the cavity, as it merges with the pharynx, are foci of immunologically active tissues—the tonsils—which have surveillance duties. When inflamed (tonsillitis), they may enlarge to narrow the exit passage so that food cannot pass. Accordingly, surgery (tonsillectomy) may be required, generally in the young, although this may be avoided if aggressive antibiotic therapy is successful.

12.1.3 The tongue

This muscular organ has a number of functions: to direct food to the top of the cavity for crushing and to the teeth for mastication (chewing), to inform on the quality of the food via taste buds/receptors (housed in papillae), and to fine-tune sounds arising from the larynx into speech. Once food has been sufficiently masticated, the tongue directs it down to the rear of the cavity, where it is swallowed. The tongue too can suffer from carcinoma, and very rarely, a sarcoma, and can become ulcerated. A leading clinical sign in haematology is glossitis (inflammation of the tongue, often red and sore) which is frequently the product of vitamin B_{12} deficiency. Other causes of tongue pathology include microbial infections as a brown staining of papillae. The highly vascular underside of the tongue is ideal for the rapid sublingual administration of acutely needed drugs such as nitrates for angina.

12.1.4 The pharynx

The pharynx sits at the back of the buccal cavity and merges with the oesophagus. A flap of skin with elastic cartilage (the epiglottis) prevents food and fluids from entering the larynx and so the trachea. Three sections are recognized: the upper nasopharynx, situated at the rear of the nasal cavity, with tonsil-like adenoids and the exit point of the **Eustachian tube**; the oropharynx, the exit point of the buccal cavity; and the lower laryngopharynx, which connects with the oesophagus and the larynx.

Effectively a muscular tube supported by cartilage, the pharynx may become inflamed (pharyngitis) with a local infection by a virus or bacteria such as a *Streptococcus* species. Such an infection may also cause a tonsillitis, and there may be a local reactive lymphadenopathy. A sore throat is also a symptom of infectious mononucleosis (glandular fever) (section 11.4.2).

Obstructive sleep apnoea

Apnoea is failure to breathe, and this may occur whilst sleeping, often associated with snoring. The dominant aetiology is obstruction of the airways, hence obstructive sleep apnoea (OSA), being defined by recurrent upper airways collapse that may lead to reduced oxygen saturation. It is highly unrecognized and underdiagnosed, likely to be present in up to half of adult men and a quarter of adult women, the rate in the latter rising after the menopause. At the anatomical level, collapse and poor positioning of the tongue and soft tissues at the rear of the buccal cavity/oropharynx impede the movement of air.

OSA is linked to a number of co-morbidities, almost all related to cardiovascular disease, including obesity. There may also be an increased neck circumference. However, a causation in either direction is unclear, but should OSA lead to night-time hypoxia, this may exacerbate existing disease.

12.1.5 The oesophagus

Anatomy and physiology

The oesophagus, like the trachea and bronchus, is simply a tube connecting two organs. Lacking the rigid structure (collagenous rings) of the airway tubes, it has several layers of circular, longitudinal and spiral smooth muscles, which provide strength and the ability to aid the passage of food to the stomach by peristalsis (Figure 12.2). In common with the bronchus, the inner layer (mucosa) is several layers of stratified squamous epithelium (protecting against abrasion, as in the skin, mouth, and vagina), and is defined by a sphincter muscle at the bottom. The passage of food is aided by mucus, secreted from glands within the muscular wall. Further details of epidemiology, molecular genetics, etc. are to be found in section 5.2.

SELF-CHECK 12.1

How does the buccal cavity prepare and initiate digestion?

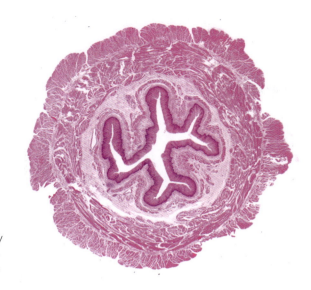

FIGURE 12.2
Histology of the oesophagus.
Note the multiple layers of
muscles and the dense epithelia
lining the lumen.

Takizawa P, Director of Medical
Studies, Department of Cell Biology,
Yale University. Retrieved from http://
medcell.org/histology/histology.php.
Kerr JB, *Functional Histology*, Elsevier,
2020.

Pathology

Cancer of the oesophagus (which kills around 8,000 people in the UK and over 500,000 globally per year) is described in section 5.2; other disease of this organ includes oesophagitis and Barrett's oesophagus, which is often described as a premalignant state. The latter is present in ~2% of the adult population, a frequency that increases with age, and is more common in men. Oesophageal varices (distended veins) may result from high blood pressure in the intestinal circulation, itself possibly due to liver cirrhosis.

Presentation and diagnosis

The leading symptom is pain as food passes to the stomach, and possibly when swallowing, although this may be a symptom of pharyngeal and/or laryngeal disease, and may be due to failure of peristalsis (achalasia). The primary diagnostic tool is endoscopy, which focuses on malignancy and inflammation. The nature of inflammatory disease (i.e. oesophagitis) changes from the upper section to the middle and lower where it is described as Barrett's oesophagus (NICE CG106), and where the pathology may be due to a low-level but persistent acid reflux from the stomach (CG184) (Figure 12.3).

   The Peptest determines pepsin in saliva or sputum, and if present, helps diagnosis of gastro-oesophageal reflux disease. Barium sulphate is an X-ray opaque radiocontrast medium which, taken as a drink (technically a swallow), can determine structural changes to the oesophagus, such as strictures and diverticula (intuckings and pouches). Rarely, eosinophils may infiltrate the mucosa, leading to a diagnosis of eosinophilic oesophagitis, possibly due to a food allergy. Oesophageal varices may be diagnosed by endoscopy, but more often by the patient vomiting copious amounts of blood due to variceal rupture, which may be fatal.

Treatment

Inflammatory disease will be treated by removal of the injurious stimulant—antibiotics for bacteria and antacids or radiofrequency ablation for acid reflux with surgery to correct the cause of the reflux (NICE IPG55, IPG461). A further option is electric stimulation of the lower oesophageal sphincter (NICE IPG540). Whilst oesophagectomy is standard treatment for high-grade dysplastic Barrett's oesophagus, NICE CG106 has guidance regarding radiofrequency ablative or photodynamic endoscopic therapy, with antacids for low-grade disease. Varices are treated by banding (ligation of selected high-risk veins

(a)

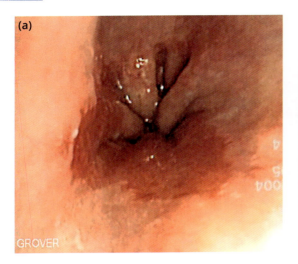

(b)

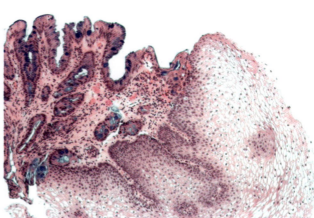

FIGURE 12.3

Barrett's oesophagus. (a) Endoscopy showing dark macroscopic changes consistent with Barrett's oesophagus. (b) Histological appearance of normal squamous epithelium (right) and the glandular epithelium of Barrett's oesophagus (left).

From Orchard & Nation, *Histopathology*, second edition, Oxford University Press, 2018.

to prevent bleeding), sclerotherapy (injecting veins with sclerosants such as foam of sodium tetradecyl sulphate), or with a stent (NICE IPG392), which can also be used to seal any perforations (as may be caused by endoscopy). A common cause of varices is portal vessel hypertension, which can itself be treated by placement of a transjugular intrahepatic-portal shunt, so relieving the hypertension.

Key points

Disease of the head, neck, and throat is an often overlooked but important aspect of clinical practice: few will escape some measure of inflammation, whilst over 3,000 die from cancer of these tissues in England and Wales every year.

12.2 **The middle digestive tract**

In this section we consider the stomach, pancreas, liver, and biliary system (Figure 12.4), disease of which covers a wide variety of aetiologies linked to over 14,000 non-malignant deaths in England and Wales in 2021.

12.2.1 **The stomach**

Anatomy and physiology

The J-shaped stomach lies at the top of the abdominal cavity directly under and on the left side of the diaphragm. It has two sets of sphincter (valve) muscles regulating food entering (the cardiac sphincter) from the oesophagus and leaving (the pyloric sphincter) to the small intestines (Figure 12.4). The former also prevents digested food from re-entering the oesophagus. Anatomical subregions, in order from the oesophageal/gastric junction, are the cardia, the fundus, the body, the antrum, and the pylorus. The stomach consists of a layer of specialized epithelial cells (the mucosa) encased by several

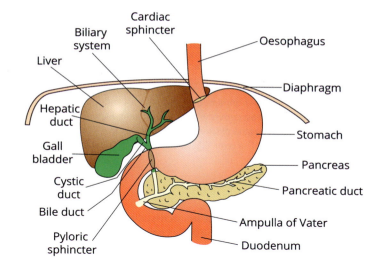

FIGURE 12.4
The middle digestive system.

layers of circular, transverse, and longitudinal smooth muscle, whose action is part-governed by a branch of the vagus nerve and by nerves arising from the spine, whilst nerves and hormones mediate the digestive aspects (Figure 12.5). The outer layers of the stomach are muscular, the inner are glandular.

As food is generally eaten at a faster rate than it can be digested, the stomach acts as a reservoir. Around 30 lymph nodes are present on the outer surface. The stomach has endocrine function as it secretes hormones such as ghrelin (which increases appetite: see section 16.5.3 on obesity), gastrin (which regulates gastric secretions), and cholecystokinin (which acts on the gall bladder and pancreas) (see Table 12.6, describing all intestinal hormones). The stomach has three major functions:

1. It physically mixes and breaks up food to promote digestion, enabled by the sheets of multiple layers of muscles. This process is under the control of **Auerbach's plexus** and local hormones such as gastrin, secreted from G cells. Slow and steady muscular activity helps maximize the exposure of the food to the secretions.

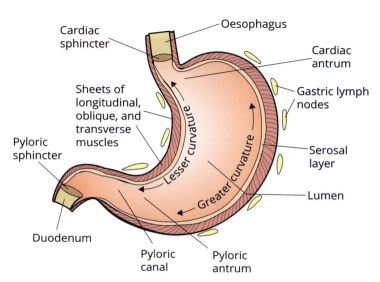

FIGURE 12.5
Structure of the stomach.

2. It chemically breaks down food, promoted by the secretion of pepsinogen from Chief cells, almost immediately activated to the digestive enzyme pepsin, and hydrochloric acid, arising from gastric glands, also part-regulated by gastrin.

3. It produces intrinsic factor, required for the absorption of vitamin B_{12}, by parietal cells, which also secrete hydrochloric acid.

The acidic environment may well have an antimicrobial effect, but this seems to have little effect on one of the most common gastric pathogens, *H pylori*. The secretion of supportive mucus by cells of the mucosa raises the pH near the luminal wall to 6–7, whilst the pH in the centre of the digesting mass is around 3–4. The semi-digested food leaves the stomach via the pyloric sphincter for the duodenum as chyme.

Pathology

Almost all disease of the stomach is inflammatory, malignant (responsible for over 3,000 in England and Wales and over 750,000 globally each year: section 5.7), or toxic (as exemplified by the effect of chronic alcohol intake). Disease of the muscular layers of the stomach is rare, the vast majority being of the mucosa and submucosa, generally leading to atrophy. Inflammation of the inner layer of the stomach (the mucosa) is gastritis (Figure 12.6) and may be caused by infection or autoimmunity.

Presentation and diagnosis

Symptoms of chronic gastric pathology are pain and discomfort in the upper abdomen, and vomiting. The latter may contain blood (haematemesis), which may reflect primary disease (gastritis, ulceration), but also other causes such as aspirin and NSAIDs. Gastric blood may also pass the length of the intestines, contributing to black and sticky faeces (melaena) which can be easily determined with the faecal occult blood test.

Lower intestinal haemorrhage will result in red-speckled faeces, and the degree of blood loss will be difficult to ascertain, so frequent haematological monitoring is required. Heartburn (generally acid reflux) is likely to be linked to a weak cardiac sphincter. Initially unrelated symptoms such as tiredness and lethargy may be tracked to vitamin B_{12} deficiency (which a blood test will define, section 11.5.3), possibly resulting from stomach disease.

Once there are sufficient symptoms, endoscopy will be performed, reporting the physical nature of the inner layer (the mucosa) of the stomach (Figure 12.6). Features include growth suggestive of tumours, ulceration (possibly due to excess acid and/or *H. pylori*: 70–80% of gastric ulcers are associated

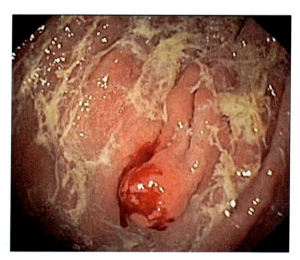

FIGURE 12.6

Gastritis. Note the blood around an inflamed mucosa.

Science Photo Library.

with this organism), and gastric atrophy, common causes being chronic alcoholism, excess dietary salt, or other poisoning, and leading to insufficient hydrochloric acid (achlorhydria). If required, a biopsy of the mucosa will help determine the nature of the disease. Should the biopsy find infiltration with leukocytes, the aetiology is inflammation, and so gastritis, which is almost always autoimmune in nature, although some inflammatory disease may be a response to *H. pylori* within the stomach wall, which may be detectable by immunohistochemistry. A biopsy may also show autoimmune attack of the parietal cells that produce intrinsic factor, and a serum autoantibody can be detected by indirect immunofluorescence which effectively makes the diagnosis.

Treatment

Management follows the aetiology: avoidance/eradication of toxins (alcohol, salt, *H. pylori*, etc.) in atrophy and inflammation. The presence of *H. pylori* can be confirmed by the ^{13}C urea breath. Eradication is generally with a 7–14-day course of a proton pump inhibitor (e.g. lansoprazole 30 mg bd) or a histamine H_2 receptor blocker (e.g. ranitidine 150 mg bd), with amoxicillin 1 g bd and either clarithromycin 500 mg bd or metronidazole 400 mg bd. Treatment of acidity includes over-the-counter alkaline preparations of sodium alginate, calcium carbonate, and sodium bicarbonate; proton pump inhibitors; and histamine H_2 receptor blockers (as above). Insufficient vitamin B_{12} is treated by injections (section 11.5.3).

Treatment of gastric cancer (linked mostly to insufficient dietary fruit and vegetables, infections (primarily *H. pylori*), excessive dietary salt, and tobacco smoking) is discussed in section 5.7. There is a place for radiotherapy, but most regimes consider chemotherapy and surgery, although if the latter is extensive there may not be sufficient healthy tissue remaining to enable the stomach to perform its functions.

SELF-CHECK 12.2

What are the broader consequences of stomach disease?

12.2.2 The pancreas

This organ has two distinct functions: digestion (apocrine) and endocrine. The latter is discussed in Chapter 15.

Anatomy and physiology

The pancreas is a loose collection of lobulated tissues, some 12–15 cm long, found on the left side of the abdomen, below the diaphragm and adjacent to the stomach and spleen. Cellular exocrine function comprises 99% of the organ, generating secretions that pass down the pancreatic duct (which merges with the common bile duct) and so into the intestines at the hepatopancreatic ampulla (the ampulla of Vater) (Figure 12.4). Pancreatic secretions include proteases (trypsinogen, convertible to trypsin), lipases (phospholipase A2, lysophospholipase, and cholesterol esterase), diastases (amylase), and sodium bicarbonate. The function of the latter is to neutralize the acidity of the chyme, providing the medium for neutral/alkaline pH enzymes. There is also a small accessory pancreatic duct that bypasses the bile duct, emptying directly into the duodenum. Almost all pancreatic disease is inflammatory (pancreatitis) or malignant (pancreatic cancer: section 5.1).

Pathology

The leading causes of non-malignant pancreatic disease include alcoholism, blockage of the exit of the pancreatic duct by gallstones, smoking, a diet poor in fruit and vegetables, obesity, and infections. At the cellular level, trypsinogen may be prematurely activated within the pancreas, attacking local connective tissues with autodigestion, and there may be calcification of the pancreatic duct system, impeding flow of secretions.

CASE STUDY 12.1 *Gastric disease*

A 42-year-old woman requests an early morning appointment with her GP, concerned about increasingly irregular menstruation periods, weight loss, tiredness, lethargy, and pains in her upper abdomen, which she presumes is the stomach. Considering the possibility of a premature menopause, the GP takes a history, performs an examination, and takes blood for routine biochemistry, haematology, and reproductive hormones. Urine analysis and blood pressure are normal. Results of the blood tests are in Table 12.1.

Levels of all reproductive hormones are within their reference ranges, but there is a borderline low haemoglobin and red cell count with a raised MCV. Gamma-glutamyl transferase is raised, and several others are at the top end of their normal range, but renal function is good. The GP calls his patient back for a 3 pm appointment, and when she arrives, he smells alcohol on her breath. She admits to a daily lunchtime drink or two, and often half a bottle of wine or more each evening, but is particularly vague on detail, so that the GP estimates her weekly total alcohol intake as 30–40 units, which is far in excess of guidelines (14 units per week). On further questioning, she admits to binge drinking at weekends, a habit that has been growing over the last 10 years since a death in the family. She also admits to stomach cramps and vomiting after some meals.

TABLE 12.1 Blood results for Case study 12.1

Analyte (unit)	Result	Reference range
Urea (mmol/l)	4.0	3.3–6.7
Creatinine (μmol/l)	93	71–133
Bilirubin (μmol/l)	18	<20
Alanine aminotransferase (IU/l)	38	5–42
Aspartate aminotransferase (IU/l)	45	10–50
Alkaline phosphatase (IU/l)	100	20–130
Gamma glutamyl transferase (IU/l)	270	0–100
Haemoglobin (g/l)	119	118–48
Red cell count (× 10^{12}/l)	4.0	3.9–5.0
Mean cell volume (fl)	105	77–98
Haematocrit (l/l)	0.34	0.33–0.47
White blood cells (× 10^9/l)	8.5	4.0–10.0
Neutrophils (× 10^9/l)	5.5	2.0–7.0
Lymphocytes (× 10^9/l)	2.0	1.0–3.0
Platelets (× 10^9/l)	255	143–400
ESR (mm/Hr)	10	<10
Luteinizing hormone (IU/l)*	56	14–96
Follicle-stimulating hormone (IU/l)*	18	5–22
Oestradiol (pmol/l)	750	100–1,500

* Mid-cycle.

The cause of the menstrual abnormalities is not endocrine, possibly the uterus, but may be related to the alcohol consumption. With the stomach pain, the GP sends his patient for an endoscopy, which reports around 50% of the surface in smooth atrophy, with loss of normal folding. This accounts for the stomach complaints, and possibly the haematology as an atrophic gastric lining will fail to generate the intrinsic factor required for the absorption of vitamin B_{12}. An additional sign of alcoholism is macrocytosis (as is vitamin B_{12} deficiency: section 11.5.3) and increased gamma glutamyl transferase, whilst the LFTs may be indicative of early liver disease.

The GP schedules the patient with a meeting with an alcohol addiction counsellor, who is working to documents such as NICE CG115 on alcohol-use disorders.

Rare causes include pancreatic head pseudocysts, autoimmunity, hypertriglyceridaemia, and a hereditary form, most commonly caused by an autosomal dominant gain-in-function mutation of *PRSS1*, found at 7q34 and coding for trypsinogen. Unsurprisingly, pancreatic disease is a risk factor for diabetes mellitus (Chapter 7, section 7.5), whilst long-term follow-up studies point to increasing pancreatic exocrine insufficiency. Around 20% of acute episodes of pancreatitis recur, and around a third transform to chronic disease. Mortality rates are estimated at 12/million/year and 1.5/million/year for acute and chronic disease respectively.

Presentation and diagnosis

Both acute and chronic pancreatitis present with severe abdominal pain, generally on the left side and radiating into the back, with nausea and vomiting. There may also be a fever and weight loss as a dysfunctional pancreas will fail to support digestion, and in its worst manifestation there may be shock and circulatory collapse. Diagnosis is relatively straightforward with high levels of serum amylase or lipase, and if equivocal, ultrasound, endoscopy, or abdominal CT may be needed. The integrity of the pancreatic and biliary ducts can be determined by MRI or endoscopic retrograde cholangiopancreatography (MRCP, ERCP), which may find a displaced gallstone (section 12.2.4) lodged in the ampulla of Vater.

Treatment

In acute disease, symptoms and major organ issues will be addressed (haemodynamics, risk of septicaemia, duct blockage, acid/base balance, etc.), perhaps with antibiotics. Should there be suspected infective pancreatic necrosis, this should be confirmed by endoscopy or a percutaneous investigation, and **debridement** considered. Nutritional support may be needed via a naso-gastric tube, which should end distal to the ampulla of Vater. Treatment of chronic disease again addresses aetiology, such as alcohol consumption. In view of the risk of diabetes, patients with chronic pancreatic disease should be tested for HbA1c every 6 months (section 7.4). NICE NG104 offers additional guidance.

SELF-CHECK 12.3

What are the major risk factors for pancreatic disease?

12.2.3 The liver

This section will focus on the parenchyma of the liver, composed of hepatocytes with a smaller number of immunosurveillant cells, leaving intrahepatic sections of the biliary tree and the gall bladder and its ducts for section 12.2.4.

Anatomy physiology

The liver is the largest single solid organ in the body, weighing around 1–1.7 kg in men and 0.75–1.6 kg in women. It is located at the right of the abdomen, under the diaphragm, and protected by ribs. It has many varied and complex metabolic, haematological, and immunological functions, one being

CASE STUDY 12.2 *Acute obstructive pancreatitis*

A 45-year-old obese (BMI 31.5 kg/m²) woman enjoys a night out with friends, overeating and taking considerably more than her usual consumption of alcohol. She wakes in the middle of the night to urinate and defaecate, but in doing so suffers an acute cramp-like pain in her upper abdomen, which she believes to be indigestion, with nausea but no desire to vomit. This dull pain has been developing during the past fortnight, but has been intermittent and more of a nuisance than an issue.

She feels unable to go to work and, as the morning proceeds, the pain becomes stronger and does not respond to analgesia. When it becomes unbearable in the middle of the afternoon, she calls a taxi to take her to hospital, fearing appendicitis. The admitting practitioner finds a painful and swollen abdomen, the pain radiating to the back; blood pressure is 136/78 mm Hg, pulse 77 beats per minute. She takes bloods for the standard routine analyses and requests serum amylase and lipase, and then sends her patient to the imaging department for an urgent chest and abdominal X-ray and ultrasound.

Whilst the practitioner is considering the blood results (Table 12.2), showing raised amylase, lipase, LFTs, ESR, and neutrophils, she is bleeped by the ultrasonographer. The report tells of an object consistent with a gallstone at the junction of the common bile duct and the pancreatic duct, and a moderately distended common bile duct, consistent

TABLE 12.2 Blood results for Case study 12.2

Analyte (unit)	Result	Reference range
Urea (mmol/l)	4.2	3.3–6.7
Creatinine (µmol/l)	88	71–133
Bilirubin (µmol/l)	28	<20
Alanine aminotransferase (IU/l)	40	5–42
Aspartate aminotransferase (IU/l)	55	10–50
Alkaline phosphatase (IU/l)	145	20–130
Gamma glutamyl transferase (IU/l)	150	0–100
Haemoglobin (g/l)	129	118–48
Red cell count (× 10¹²/l)	4.4	3.9–5.0
Mean cell volume (fl)	85	77–98
Haematocrit (l/l)	0.38	0.33–0.47
White blood cells (× 10⁹/l)	9.5	4.0–10.0
Neutrophils (× 10⁹/l)	8.0	2.0–7.0
Lymphocytes (× 10⁹/l)	1.1	1.0–3.0
Platelets (× 10⁹/l)	287	143–400
ESR (mm/Hr)	15	<10
Amylase (U/l)	450	28–100
Lipase (U/l)	300	0–200

with an increased volume of bile. There are also numerous other stones in the gall bladder (Figure 12.7).

In discussions with colleagues, the practitioner makes a diagnosis of acute obstructive pancreatitis and recommends to her patient that she undergo an endoscopic retrograde cholangiopancreatography (ERCP) with a view to removing the obstructing stone. The patient readily agrees, and within 2 hours is having her procedure, which is successful in retrieving a moderately sized stone approximately 0.5 cm in diameter.

The patient is moved to a medical ward where, over the next 48 hours, she makes a satisfactory recovery, and bloods slowly move towards their respective reference ranges. In follow-up appointments she is advised to lose weight, take regular exercise, and moderate her alcohol intake.

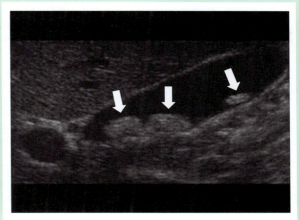

FIGURE 12.7
Ultrasound of the gall bladder. Three gallstones are arrowed.

to collect bilirubin and pass it, via the biliary system, to the gall bladder (to be discussed in section 12.2.4). It is fed oxygenated blood via the hepatic artery, and blood containing the products of digestion via the hepatic portal vein. Blood leaves the liver via the hepatic vein, which joins the interior vena cava. The working cells of the liver are hepatocytes, and other cells are those making up the internal blood vessel system and micro-bile ducts (bile canaliculi), immunologically active Kupffer cells (modified macrophages), and stellate cells (pericytes that support and line the sinusoids and blood vessels).

Pathology

Liver disease can be classified in a number of ways, including neoplastic (section 5.3) and haematological (e.g. iron overload: section 11.7) disease. This section will look at immunological, metabolic, and toxic aetiologies, which together were linked to over 10,000 deaths in England and Wales in 2021. Many of these bring several common and varied consequences, some of which may be fatal, including jaundice, loss of haemostasis with haemorrhage (sections 11.2 and 11.6), anaemia following failure to process iron (section 11.5), and ascites due to hypoalbuminaemia. A common feature of a damaged liver is fibrosis, leading to cirrhosis (Figure 12.8).

Hepatic disease can have effects outside the liver. Hepatic encephalopathy manifests itself as any number of symptoms related to consciousness—loss of concentration, tiredness, personality changes,

FIGURE 12.8
Severe liver fibrosis. Normal liver structure has been replaced by large streaks of fibrous connective tissues, shown here in blue.

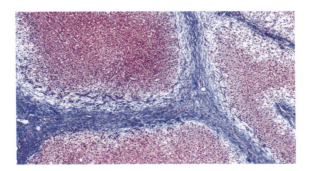

forgetfulness etc., developing into severe confusion. These are presumed to result from the disease process acting on the brain, perhaps by a combination of metabolites such as ammonia. Very high levels of bilirubin can become deposited in other organs, causing damage, such as dermal rashes and (via endothelial cell damage) peripheral vascular disease.

Immunological disease

The principal immunological disease, hepatitis—literally inflammation of the liver—is difficult to confirm definitively as it requires leukocyte infiltration on biopsy. Nevertheless, leading causes include an autoantibody, or the response to an infection with any one of five established hepatitis (A–E) or other viruses (such as herpes viruses), bacteria, or parasites (e.g. the liver fluke *Fasciola*), most of which can be demonstrated in the laboratory. Hepatitis viruses A, B, and C are a major public health issue in regions as far apart as Egypt and China.

The responses to the infections generally proceed as expected (section 9.7) with local inflammation as the infectious organisms and infected cells (hepatocytes) and tissues are attacked, and with all the systemic signs and symptoms of an acute phase response, such as fever. If the acute inflammation is successful, there should be minimal and repairable damage to the liver, but if the response is prolonged and chronic, extensive injury may occur, and damaged hepatocytes and tissues may be replaced not with functional cells but with inert fibrous tissue generated by invading fibroblasts and macrophages. Extensive fibrous tissue deposition leads to cirrhosis, with the liver losing flexibility and becoming stiff, a feature that can be detected clinically.

Hepatitis may also follow autoimmune damage to hepatocytes with anti-nuclear antibodies, anti-smooth muscle antibodies, and antibodies to liver and kidney microsomes. Together, these forms have an incidence of ~13/million/year with a prevalence ten times greater and a 10-year mortality rate ~8%; the female:male ratio is ~7:1. The disease progresses to fibrosis and cirrhosis, with ~4% eventually developing hepatocellular cancer. Predisposing factors include several *HLA-DRB1* and *DQB1* isoforms, whilst triggers include viral infections and certain drugs (minocycline, nitrofurantoin, melatonin, diclofenac, statins). Both serum miR-21 and miR-122 are increased, but their clinical significance is unclear.

Metabolic disease

Within the liver, hepatocytes perform a host of anabolic and catabolic procedures, and accordingly there are many inborn and acquired errors in metabolism (Chapter 16). A number of hepatic syndromes are well-characterized. Fatty liver disease (steatosis), the most common chronic liver disease in the developed world, is characterized by the deposition of lobules of fat (complex mixtures of cholesterol and triglyceride metabolites) (Figure 12.9). It must be distinguished from alcoholic liver disease where there are also fat deposits, hence non-alcoholic fatty liver disease (NAFLD). The strongest risk factor is obesity, but some (mostly lean) cases are linked to *PNPLA3* (at 22q13.31, coding for a

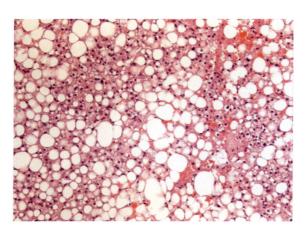

FIGURE 12.9
Non-alcoholic fatty liver disease. The circular white objects are deposits of fat.

Cabezas J, Mayorga M, & Crespo J, Nonalcoholic Fatty Liver Disease: A Pathological View. In (ed.), Liver Biopsy—Indications, Procedures, Results. IntechOpen. https://doi.org/10.5772/52622. © 2012 the Author(s). Licensee IntechOpen.

lipase), *TM6SF2* (at 19p13.11, coding for a product whose role is unclear but may be linked to fibrosis), and *APOB* at 2p24.1 (coding for apolipoprotein B). An additional hypothesis is of links with CD36, a general scavenger receptor coded for by *CD36* at 7q21.11 which may be involved in increased movement of free fatty acids into the cell.

Non-alcoholic steatohepatitis (NASH), a progression from NAFLD, is characterized by biopsy-proven leukocyte infiltration and, accordingly, is more severe. Fibrosis and cirrhosis are accelerated, with more rapid progression to hepatocellular carcinoma and therefore the need for transplantation. In both NAFLD and NASH, increased fatty acid oxidation may be responsible for the lipotoxicity (possibly responsible for reactive oxygen species) that drives the overloaded hepatocyte to **autophagy** and apoptosis.

Wilson's disease (present at a frequency of ~30/million) is an autosomal recessive condition caused by any one of a number of mutations in *ATP7B*, found on 13q14.3, which codes for a copper transporter. A loss-of-function mutation leads to failure to export copper and so its accumulation within the hepatocyte, with associated damage. Diagnosis is supported by biopsy (Figure 12.10) and the mass of copper in a sample of tissue (>250 µg/g) whilst serum copper and caeruloplasmin levels are low. In addition to liver disease, there may be neurological consequences (tremor, anxiety, depression, cognitive changes).

Gilbert's syndrome is a relatively mild disease, often asymptomatic, estimated to be present in 5–10% of the population. It is caused by an autosomal recessive mutation in *UCT1A1* at 2q37.1, coding for the enzyme UDP-glucuronosyltransferase-1, which conjugates bilirubin. A loss-of-function mutation leads to conjugation of bilirubin to a water-soluble carrier, so that levels of unconjugated bilirubin in the blood rise.

The metabolic syndrome is a complex cluster of a number of features, some of which are linked directly or indirectly to the liver, and with overlapping pathophysiology. It is defined by any three from abdominal obesity (inevitably with a waist/hip ratio >1), hypertension, hyperglycaemia, hypertriglyceridaemia, and low HDL, and accordingly is a risk factor for cardiovascular disease and diabetes. There may also be impaired fasting glucose and insulin resistance (section 7.4) linked to liver disease.

Other forms of liver disease can have consequences in other organs, an example being alpha-1-antitrypsin deficiency, often caused by loss-of-function mutations in *SERPINA1* at 14q32.13. Loss of function of this gene leads to reductions in alpha-1-antitrypsin, normally a general protease inhibitor, which leads to pulmonary as well as liver disease.

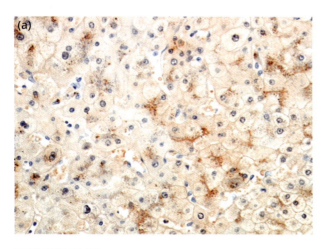
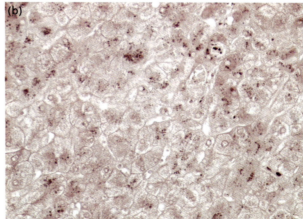

FIGURE 12.10

Copper deposits in Wilson's disease. Abnormal deposits of copper demonstrated by (a) rhodamine and (b) orcein staining.

From Orchard & Nation, *Histopathology*, second edition, Oxford University Press, 2018.

Toxic disease

The liver has been programmed by evolution to process and eliminate environmental toxins and can cope when presented with a modest mass of contemporary toxins, examples being complex synthesized pharmaceuticals never present in nature. However, should the mass of toxins exceed the metabolic capability of the liver, damage will follow, the classic example being alcohol abuse, where chronic excess leads inevitably to fibrosis, cirrhosis, and often cancer. Other toxins include industrial solvents, heavy metals, and drug overdoses. An example of acute toxicity is overdose with paracetamol (acetaminophen) generally of more than 15g in a single dose, which will absorb fully by 6 hours. If untreated, this will lead inevitably to measurable acute hepatic injury within 24–72 hours, followed by clinical symptoms (jaundice, abdominal pain, confusion) and death by 48–96 hours. Paracetamol overdosing accounts for ~15% of cases of acute liver failure in the USA and Europe.

Presentation and diagnosis

The leading sign of liver disease is jaundice, the result of hyperbilirubinaemia causing a yellowish tinge to the skin, most easily seen against the white ocular sclera, generally with a darkening in the urine as the excess bilirubin is excreted. Non-specific symptoms include nausea, vomiting, and abdominal discomfort (especially the upper right), whilst a semi-specific symptom is **pruritis**, possibly due to increased dermal bilirubin. However, an alternative cause of jaundice is marked intravascular haemolysis, as may be present in a fulminant malarial infection, where the liver is simply unable to rapidly clear the plasma of the high levels of bilirubin released from haemolysed red cells. If an immunological aetiology, there may well be symptoms of an infection with an acute phase response. In all cases, a recent history should be informative.

The initial diagnostic tool regardless of aetiology is liver function tests (LFTs) from the biochemistry laboratory, which can point to the severity of the disease with direct relevance for management. Historically consisting of five tests, many add albumin as it is a specific product of the liver, and some include prothrombin time or its linked international normalized ratio (INR). However, several of the markers are abnormal in different non-liver situations, requiring informed interpretation (Table 12.3).

The immunology laboratory will inform upon potential infective agents, particularly the hepatitis viruses, with antibodies and molecular genetics for viral nucleic acids, but also general acute phase reactants and autoantibodies. The haematology laboratory will show possible haemolysis, whilst a leucocytosis with raised ESR will support an inflammatory aetiology, and the biochemistry laboratory can offer a lipid panel for NAFLD and NASH. The toxicology laboratory can probe for common poisons and inform upon their clearance.

TABLE 12.3 Liver function tests

Test	Alternative interpretation of an abnormal result
Alkaline phosphatase (ALP)	Biliary disease (especially obstruction), bone disease e.g. myeloma, Paget's disease
Alanine aminotransferase* (ALT)	Diabetes, muscular damage e.g. myocardial infarction, myositis
Aspartate aminotransferase* (AST)	Diabetes, muscular damage e.g. myocardial infarction, myositis
Bilirubin	Intravascular haemolysis
Gamma glutamyltransferase (GGT)	Alcohol, phenytoin, barbiturates, heart failure
Albumin	Extensive burns, renal failure
Prothrombin time	Certain oral anticoagulants

* Also described as transaminase.

An additional important tool is ultrasound, especially valuable in investigating possible malignancy, vascular malformations, cystic disease, and occlusion of the biliary system. Complex imaging (CT, MRI, PET) may also be required should other investigations be equivocal. Despite the above, the ultimate diagnosis is by biopsy, which requires careful consideration as the liver may well be carrying considerable pathology. These tissues will unequivocally define a fatty liver, fibro-cirrhosis, and leukocyte infiltration. A key physical sign in Wilson's disease is the golden-brown corneal Kayser–Fleischer ring.

Treatment

Infections will be treated by standard microbiology methods: in the short term with support. Antivirals for hepatitis B virus (such as lamivudine and entecavir) can prevent replication, whilst chronic infections with hepatitis C virus can be cleared with 2 or 3 months of antivirals such as ledipasvir and sofosbuvir. Treatments for autoimmune hepatitis include immunosuppressants with steroids (initially 60 mg/day with a reducing dose) and azathioprine (50 mg/day). If this proves ineffective, tacrolimus and methotrexate are options to reduce levels of the autoantibodies. Monoclonal antibodies to CD20 (rituximab) and TNF-α (infliximab) are also potential treatments (as in many autoimmune diseases).

Gilbert's disease rarely calls for active treatment, but penicillamine is used in Wilson's disease to chelate copper and promote its excretion. Alpha-1-antitrypsin deficiency may be treated with intravenous replacement therapy, whilst the major consequence—lung disease—is initially treated symptomatically, for example with bronchodilators to help breathing. NAFLD, NASH, and metabolic syndrome are all treated by addressing risk factors, mostly poor diet and dyslipidaemia. Alcoholic liver disease is treated by abstinence, toxic disease by avoidance and by dialysis for the worst cases. Treatment of overdoses may, if the dose is not too high and addressed rapidly, be successful using gastric decontamination using activated charcoal and/or dialysis. If the dose is not too large, and again actioned urgently, acetylcysteine may reverse paracetamol-induced liver failure.

In some cases, notably alcoholic liver disease, hepatocellular carcinoma (perhaps a consequence of chronic hepatitis C infection), and Wilson's disease, liver transplantation may be considered (section 10.4.3). In this respect, the MELD (Model for End-stage Liver Disease) score may be helpful, being an equation based on the presence/absence of dialysis, serum bilirubin, creatinine, and the INR, whilst a modification also requires sodium. Although originally constructed as a score to predict mortality and placement of a transjugular intrahepatic-portal shunt, its use has extended to estimate the likelihood of mortality in the coming three months and so the urgency of the need for a transplant, and the likelihood of survival.

An additional score is that of Child–Pugh, which includes other features (Table 12.4), being more commonly used to estimate fibrosis. A score of 5 or 6 (class A) brings a 10% risk of death after abdominal surgery, a score of 7–9 (class B) brings this to 30%, whilst a score of 10–15 (class C) increases the risk to 70–80%. The score also gives a risk of any mortality within a year: 0–10% in class A, 20–30% in class B, 55–80% in class C. Given the commonality of units, unsurprisingly, the MELD and Child–Pugh scores

TABLE 12.4 The Child-Pugh score system

Feature	Scores 1	Scores 2	Scores 3
Total bilirubin*	<34 µmol/l	34–51 µmol/l	>51 µmol/l
Albumin**	>35 mg/l	28–35 mg/l	<28 mg/l
Prothrombin time prolongation or INR	<4 seconds, <1.7	4–6 seconds, 1.7–2.2	>6 seconds, >2.2
Ascites	None	Slight	Moderate
Encephalopathy	None	Grades 1 and 2	Grades 3 and 4

* Reference range <17 µmol/l, therefore 2 × and 3 × above the top of the reference range.

** Bottom of the reference range is 35 mg/l.

correlate well. A further score is that of the Barcelona Clinic Liver Cancer staging system, whilst fibrosis may also be estimated by scores using certain LFTs and other metrics, such as FIB 4 (a composite of AST, ALT, platelet count, and age).

Key points

With the organ's multitude of metabolic functions, liver disease inevitably has numerous consequences for other organs and for the whole body.

SELF-CHECK 12.4

Why do we need so many LFTs? Can't we just narrow them down to the two or three most important?

12.2.4 The biliary system

It is anatomically and pathophysiologically convenient to consider the biliary system apart from the liver and its hepatocytes.

Anatomy and physiology

The biliary system collects, stores, and delivers excretory by-products of the liver's metabolism (bile) to the intestines (Figure 12.11). Bile is mostly composed of three major groups of compounds: (1) bilirubin and related breakdown products of haem, the consequences of red cell destruction, and other toxic metabolites; (2) lipids: cholesterol, triglycerides, and related molecules; and (3) bile acids, being cholic acids and related molecules (such as ursodeoxycholic acid), becoming bile salts when conjugated to glycine or taurine. The latter two participate directly in digestion, alongside pancreatic secretions, in supporting the breakdown and absorption of fats.

The microanatomy of the liver can be viewed in orthogonal sections, with a branch of the draining hepatic vein at the centre and an arteriole and venule at each corner (Figure 12.12). The portal triad of vessels is completed by a bile ductule, merging with others into the hepatic duct, which leaves the liver

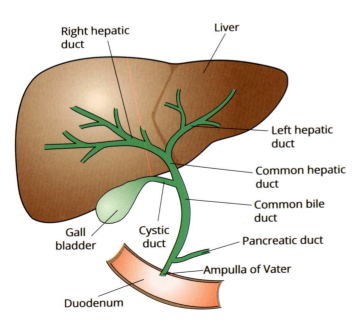

Right hepatic duct

Liver

Left hepatic duct

Common hepatic duct

Common bile duct

Pancreatic duct

Cystic duct

Gall bladder

Ampulla of Vater

Duodenum

FIGURE 12.11

The biliary system. The fine vessels of the biliary system within the liver merge into a single duct. Some bile passes into the gall bladder, and all ultimately passes along the common bile duct where it merges with the pancreatic ducts and passes into the intestines.

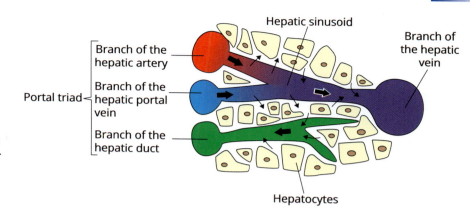

FIGURE 12.12

Microanatomy of the liver. Arrows indicate direction of fluid flow. For clarity, endothelial cells lining the vessels, Kupffer cells, and other cells are not shown.

at the **hilum**. Bile may continue its passage into the duodenum, or pass into the gall bladder, where it is concentrated, often described as bile sludge. It is periodically ejected down the cystic duct, which joins with the hepatic duct to form the common bile duct.

Pathology

Cancer of the biliary system is discussed in section 5.13.1. Almost inevitably present in each adult's gall bladder are gallstones, formed when bile sludge is hyperconcentrated so that the lipid and bile acid components come out of solution and form irregular crystals or stones. These may lie dormant and non-pathological, but too many can become an irritant and precipitate an inflammatory response: cholecystitis. The development of small gallstones (diameter < 3 mm) is described as microlithiasis (*lith* = stone). Leading risk factors for gallstones include obesity, diabetes, and increased age, whilst women have more and larger stones than men. Gallstones may leave the gall bladder and, if large enough, become lodged in the cystic duct or bile duct, causing an obstruction and so failure of the free flow of bile down the bile duct: cholestasis. Inflammation of the bile duct itself (cholangitis) may follow cholestasis and/or a bacterial infection, and as an inflammation of a tube narrows the lumen, obstruction is that much more possible.

There are two closely related conditions of inflammation of intra-hepatic biliary vessels: primary biliary cholangitis (PBC) and primary sclerosing cholangitis (PSC). The former, previously primary biliary sclerosis, is an autoimmune disease of the micro-bile ductlets within the liver with an incidence of ~33/million/year, prevalence of ~200–500/million and a female/male ratio of 10:1. The aetiology of PSC is not as clear, but involves inflammation and fibrosis causing ductlet thickening and narrowing, with a male/female ratio of 2:1. It has an incidence of ~11/million/year, and is strongly linked to inflammatory bowel disease (IBD). An important differential diagnosis to PSC is IgG4 sclerosing cholangitis (IgG4-SC). Neonates may suffer from biliary atresia (abnormal duct formation).

Presentation and diagnosis

Signs and symptoms generally resemble those of liver disease (right upper abdominal pain, nausea, vomiting, fatigue, pruritis, jaundice, fever, weight loss), but with certain differences, such as intermittent stabbing pain in gallbladder disease, and Sjögren's syndrome in PBC. Obstructive cholestasis of any aetiology results in bile backing up into the biliary system, with increased hydrostatic pressure, so that bilirubin will pass into the branch of the portal vein and thus the extra-hepatic circulation, so causing hyperbilirubinaemia and jaundice. This is also likely to cause damage to the hepatocytes, resulting in raised LFTs. A further consequence is that, as bilirubin normally colours our faeces brown, a sign of obstructive cholestasis is faeces being less brown and increasingly grey. The metabolic consequences of cholestasis and/or cholecystitis may lead to an acute phase response with a raised white cell count, ESR, and CRP, but this does not necessarily point to a bacterial aetiology. This may be confirmed by the culture of some recovered bile for common bacteria. The integrity of the lower biliary system (major ducts) can be determined by ERCP, whilst abdominal X-rays and ultrasound are standard practice.

PBC is likely to be linked to a sustained increase in ALP and GGT with antinuclear, anti-centromere, and anti-mitochondrial antibodies at a titre >1/40 (present in 95% of cases), some of which are PBC-specific. Biopsy tissue shows lymphocyte-mediated damage to, and destruction of, bile ductlets with granulomatous inflammation. Risk factors include infectious and chemical triggers, smoking, and recurrent urinary tract infection. Molecular genetics shows altered expression of many immune-related genes as in other autoimmune diseases. Further diagnosis of PSC relies on the laboratory (LFTs, serum perinuclear ANCAs, IgG4), but notably the disease may be detected in patients with newly diagnosed or established IBD. Ultrasound is rarely helpful in diagnosing PSC, but may exclude any lithiasis-based disease, whilst the gold standard is ERCP/MRCP. A biopsy is normally reserved for possible small-duct PSC and to determine the possibility of overlap syndromes: there may be a degree of autoimmune hepatitis in both PBC and PSC.

Treatment

Several forms of biliary system inflammation may be treated by antibiotics, the most common organisms being *E. coli* and *Bacteriodes*, *Clostridium*, *Klebsiella*, *Pseudomonas*, and *Streptococcus* species, a recommended first-line agent being ciprofloxacin 500 mg bd. A heavily loaded gallbladder can be removed by laparoscopic keyhole surgery, isolated obstructing stones by ERCP. Disease progression in PBC can be slowed with oral ursodeoxycholic acid (13–15 mg/kg/day, which solubilizes gallstones), with reductions in LFTs. Should this be ineffective (as defined by raised LFTs), obeticholic acid may be trialled (5 mg od, rising to 10 mg) (NICE TA443). Pruritis, present in up to 80% of patients with PBC and ~40% of PSC patients, may be treated with oral cholestyramine (4 g/day to a maximum of 16 g/day) or other resins, or rifampicin (300–600 mg od).

The clinical progression of PSC and PBC to fibrosis, portal hypertension, variceal bleeding and cirrhosis (as in most liver disease) is variable (the risk of cholangiocarcinoma in PSC being ~1% a year), and there may be a place for MELD and Child–Pugh in risk stratification in considering transplantation. Despite evidence of immune activation in PSC, immunosuppressants are not indicated, and ursodeoxycholic acid is ineffective in reducing the risk of malignant transformation, and accordingly MRCP/MRI/contrast enhanced CT should be used. Should biliary strictures be found, dilatation is preferable to stent placement. Reflecting the link with IBD, **colitis** should be screened for in PSC cases with colonoscopy and colonic biopsy, and these patients are also at risk of **osteopenia** and osteoporosis, and if confirmed and appropriate, should be treated according to NICE QS149, CG146, and TA464, for example with 400 IU (10 μg) vitamin D daily.

SELF-CHECK 12.5

What are the major features of the two forms of cholestasis?

12.3 The lower digestive tract

This third section includes the duodenum, jejunum, and ileum (the small intestines), and the appendix, caecum, colon, and rectum (the large intestines), the system being completed by the anus (Figure 12.13).

The intestines have essentially the same gross structure: long muscular tube, where sections differ by secretions and absorptions across the mucosal surface. As with the middle part of the digestive system, secretions, muscle action (peristalsis), and local blood supply are regulated by local hormones and peptides (such as secretin and somatostatin) and by an autonomic enteric nervous system consisting of a plexus within the intestinal wall. Evidence of an immunological role is provided by pockets of leukocytes forming gut-associated lymphoid tissues (GALT), one of the mucosal-associated lymphoid tissues (MALT) (section 9.2.3).

Malignancy of these tissues is described in sections 4.4 (colorectal cancer) and 5.13.1 (the small intestine), and in Table 5.6. Compared to the middle part of the digestive system, toxicity is rarely an

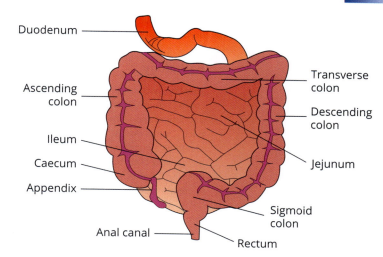

FIGURE 12.13

Anatomy of the lower digestive system.

Adapted from Fullick & Fullick, *Organs, Systems, and Surgery*, Oxford University Press, 2023.

issue. The consequences of atherosclerosis are rare, but ischaemia may result from atheromatous stenosis and a mesenteric artery occlusion resulting in infarction may lead to downstream bowel necrosis, rupture, and peritonitis that may be fatal. Consequently, focus will be on the remaining leading aetiology: inflammation and autoimmunity (Chapters 9 and 10). Genetic factors account for around 25% of the heritability of IBD, and there may be roles for long non-coding RNAs and miRNAs, as suggested by upregulation of DQ786243 in Crohn's disease, downregulation of CDNK2B-AS1 in IBD, downregulation of miR-141 in ulcerative colitis, and increased miR-199a-5p in active Crohn's disease.

A community of some 100 trillion microorganisms (the microbiome) in the intestines is a likely participant in pathophysiology, perhaps linked to a so-called leaky gut syndrome, a hypothesis suggesting that some microbes can escape the confines of the intestines and enter the circulation, with immunological consequences. Indeed, as the number of these organisms exceeds that of their host eukaryotic cells, it is perhaps not unsurprising that changes in the microbiome may be linked to ill-health. These may be caused by antibiotics targeting benign bacteria, permitting the overgrowth of pathogenic species such as C. *difficile*, being described as dysbiosis. To this end, 'allogeneic' faecal microbial transplantation, aiming to restore a normal microbiome, is an effective therapy in inducing remission in both ulcerative colitis and Crohn's disease.

Irritable bowel syndrome (IBS) is often defined by symptoms, not by bowel tissue pathology, although there is considerable cross-over of symptoms with inflammatory disease. An important discriminator of inflammatory or non-inflammatory disease is the presence of faecal calprotectin (NICE DG11) (present in neutrophils but whose concentration may be six-fold that in plasma) and/or lacto-ferrin (also present in neutrophil granules but also secreted by the pancreas) in the former. As in acute gastric disease, relatively severe and acute gastroenteritis (as may be caused by food poisoning with undercooked food and/or poor kitchen hygiene) must be excluded from all diagnoses that consider chronic disease.

12.3.1 The small intestines

Anatomy and physiology

Chyme leaving the stomach is almost immediately mixed with products of the biliary system and the pancreas in the duodenum (some 25 cm in length), where digestion continues. The wall of the duodenum contains Brunner's glands, secreting mucus and bicarbonate ions, in addition to enzyme-secreting glands, the latter continuing into the sections that follow. As the chyme progresses through the jejunum (~2.5 m, with extensive villi) and ileum (~3 m) additional digestion proceeds and the results are absorbed. The diameter of the small intestines is around 2.5 cm. The sub-mucosal wall of the ileum is characterized by foci of immunological cells (Peyer's patches, part of the GALT, and which appear to have similar functions as lymph nodes), whilst the mucosal layers are the site of the absorption of the vitamin B_{12}/intrinsic factor complex and of bile salts.

Pathology

There are two major inflammatory/autoimmune diseases of the small intestines—Crohn's disease and coeliac disease. The former is one of two IBDs (the other being ulcerative colitis) that can affect the duodenum and ileum, but also the large intestines, and in 13–16% of cases the upper gastrointestinal tract. Crohn's disease has an incidence of ~70/million/year, a prevalence of ~625/million. GWAS has identified a number of suspect genes, such as those coding for IL-17 and IL-23. However, any genetic basis is very likely to be polygenic, although HLA-DRB*0701 is linked to ileal disease. Crohn's disease may also be present in the oesophagus and stomach. Smoking is an established risk factor (as so often), as are use of NSAIDs and excessive domestic hygiene (being 'too' clean).

The affected areas can be patchy or extensive: inflammatory damage extends from the mucosa through the muscles to the serosa, and so a thicker structure, with ulcers, fissures, and granulomas. These may lead to strictures: narrowing of the lumen by fibrosis resulting from damage to the intestinal wall and which slows the process of digestion, often by dysregulated contraction of smooth muscles. An important differential diagnosis for Crohn's disease (and other IBD) is intestinal tuberculosis, investigations into which include the tuberculin skin test, isolating the whole organism or its DNA by standard PCR and molecular genetics.

Coeliac disease, an autoimmune condition, is a consequence of the inability of the intestines to process gluten protein (gliadins and glutenins) from wheat, barley, and rye: oats may be tolerated. Almost all (95–99%) of patients have HLA-DQ2 (resulting from the expression of DQA1*0501 and DQB1*02) or HLA-DQ8 (formed from DQB1*0302 and DQB1*03), and so with a population frequency of ≥1% (i.e. 10,000/million) this makes coeliac disease a leading genetic disorder. However, only 1/30 of those with these alleles will develop the disease, underlining the role of factors to be discovered, such as for the intestinal microbiome and/or toll-like receptors. A destructive autoimmune process is triggered once cells within the mucosa are sensitized to gluten epitopes, leading to mucosal destruction and the symptoms.

A third form of small bowel disease is lactose intolerance. In health, the disaccharide lactose is cleaved by lactase (coded for by *LAC* at 2q21) into glucose and galactose, and absorbed in the small intestines. Failure to do so resulting from low or absent intestinal lactase (where it is part of the brush border) leads to excess lactose being available as a food source for certain intestinal bacteria, which proliferate in response and cause standard enteropathic symptoms. Reduced levels of lactase may be secondary to mucosal damage caused by inflammatory/autoimmune disease. A useful tool to help diagnosis is the hydrogen breath test.

Merkel's **diverticulum** (of the distal ileum) is the most common anatomical intestinal abnormality, present in ~2.5% of the population. It is an embryonic remnant that may contain ectopic gastric or pancreatic tissue, which may adversely affect the local tissues, with blood loss and luminal obstruction. Duodenal ulcers may be present in Crohn's and coeliac disease, but 95% are associated with *H. pylori*.

Presentation and diagnosis

General intestinal disease presents with any combination of non- and semi-specific signs and symptoms of varying severity that include diarrhoea, nausea, constipation, loss of weight (and perhaps anorexia), steatorrhoea (pale, bulky, fatty faeces that often float), abdominal pain, and distension. These may lead to flatulence—excess gas production by an altered microbiome. However, some presentations can be with acute, localized pain, such as in diverticulitis, causing symptoms of appendicitis. Ectopic gastric tissues in Merkel's diverticulum can be detected with a [99]technetium scan. Non-enteropathic features include nutritional deficiencies (Table 12.5). Precise diagnosis requires endoscopy, complex imaging, and histological assessment of biopsied tissue.

Crohn's disease presents with the semi-specific signs described above, so that final diagnosis requires ileocolonoscopy, MRI/CT, and biopsy. Most diagnoses are made between the ages of 15 and 35, differential diagnoses being coeliac disease, chronic pancreatitis, diverticulitis, and small intestine lymphoma. Barium can be delivered directly into the duodenum via an extended naso-gastric tube, providing

TABLE 12.5 Nutritional deficiencies consequent to intestinal disease

Deficiency	Typical consequences
Vitamin B$_{12}$* and folate	Macrocytic anaemia
Iron**	Microcytic anaemia
Vitamin K	Haemorrhage
Calcium	Tetany
Iodine	Thyroid pathology
Vitamin D	Osteomalacia
Vitamin C	Myalgia, arthralgia
Protein	Oedema
Multiple vitamin deficiency	Glossitis, aphthamous ulcers, stomatitis

* Primarily in ileal disease.

** More likely in ulcerative colitis.

evidence of structural changes, and other preparations are useful for other intestinal investigations. Disease of the perianal region (the last 15 cm) manifests as **proctitis** (inflammation of this region), symptoms being perceiving the rectum to be full and so urge to defaecate, with pain on defaecation and the likely presence of fresh blood. Differential diagnoses for these symptoms include sexually transmitted diseases and excessive anal intercourse. At the cellular level, disease severity can be assessed by standard inflammatory markers, such as raised white cell count, ESR, CRP, etc., as in both IBDs.

Coeliac disease is likely to present with general symptoms as above, but as in many cases the disease may have been active for years, the BMI is likely to be low as a result of the malabsorption. It may lead to nutrient deficiency with anaemia, hypovitaminaemia, etc., and is likely to be responsible for failure to thrive in children, and short stature. In the laboratory the disease is characterized by IgA antibodies targeting tissue transglutamase (tTG), IgG antibodies targeting deaminated gliadin peptides, and IgA antibodies to the endomysium (connective tissue that ensheaths muscle fibres and which contains a transglutamase). Sufficiently high titres of anti-tTG antibodies and anti-endomysial antibodies are generally enough to make the diagnosis, but anti-gliadin antibodies are supportive in equivocal cases. Total IgA should also be measured to ensure a negative result is not due to IgA deficiency. However, the most conclusive evidence comes from duodenal biopsy material which will show blunted or atrophic villi, crypt hyperplasia, and leukocyte infiltration (such as Tγδ cells) into the lamina propria, the outer boundary of the mucosa. NICE NG20 recommends serological testing for those with appropriate symptoms, severe or persistent mouth ulcers, type 1 diabetes, autoimmune thyroid disease, and IBS, and for first-degree relative of patients with the disease. When investigating, probands should be eating a gluten-containing diet and should not start a gluten-free diet until diagnosis is confirmed. NICE QS134 also applies, and NICE CG116 refers to food allergy in those aged under 19.

Diverticular disease may be suspected in intermittent left lower quadrant abdominal pain with diarrhoea, rectal bleed, or constipation, most of which overlap with malignancy, IBS, and colitis. If the presentation is acute, with a fever, and a raised CRP is present, antibiotics may be justified, whilst MRI/CT and endoscopy will be needed to form a diagnosis. Guidance on the diagnosis and management of diverticular disease is offered in NICE NG147.

SELF-CHECK 12.6

There may still be vitamin B$_{12}$ deficiency in the face of a perfectly functioning ileum, where it is absorbed—why so?

BOX 12.1 Steroids

As is clear from this and other chapters, this class of drug is a very effective immunosuppressant, but brings many practical issues. Withdrawal must be gradual, for fear of acute adrenal insufficiency (section 15.3). In long-term use, steroids can bring about osteoporosis (so that some guidelines recommend vitamin D and calcium supplements), hyperglycaemia and diabetes, and hypertension (so that monitoring for both is required).

Treatment

Clinical management has two arms: chemotherapy and surgery. Mild to moderate Crohn's disease can be treated with 9 mg of the ileal-release steroid budesonide od for 8 weeks to induce remission. In severe disease, and to induce remission, leading treatments include intravenous or oral steroids (starting at 30–60mg/day, then tapering down; see Box 12.1) and mAbs to inflammatory cytokines (see below). Once in remission, other drugs such as azathioprine (50 mg od), methotrexate (7.5–15 mg once a week) are options, as are mycophenolate mofetil and 6-mercaptopurine. Other options include antibiotic therapy. Strictures (tight narrowings) may be treated by surgery or stricutureplasty (such as with a balloon, as in coronary artery angioplasty).

Diarrhoea (when derived from most aetiologies) may be treated with co-phenotrope, codeine, and loperamide (Imodium: maximum 8 mg a day). The latter binds to the gut wall opiate receptor, lengthening the transit time, reducing transit peristalsis (thereby inducing a form of constipation), and increasing the absorption of water and electrolytes. It also helps reduce faecal urgency and incontinence. The destructive progress of the disease, failure of the medical approach, and other issues lead to surgery in three-quarters of patients. Of course, resection is kept to a minimum, especially in the terminal ileum, but in many cases new disease develops in other locations. Sections of the small intestines can also be replaced by a transplant (section 10.4.3).

NICE has published several documents that are relevant in this context. NG129 refers to the management of Crohn's disease. Others include TA187 on the use of infliximab (targeting TNF-α) and adalimumab (TA556:TNF-α) on darvadstrocel for treating complex perianal **fistulas**, TA456 on ustekinumab (IL-12/IL-23), TA352 on vedolizumab (integrin $\alpha_4\beta_7$, an adhesion molecule on Peyer's patch lymphocytes), DG22 on therapeutic monitoring of TNF-α inhibitors, DG7 on the investigation of diarrhoea due to bile acid malabsorption (which also applies to IBS), and MIB178 on a method for predicting the likelihood of a flare-up.

The only currently effective form of treatment of coeliac disease is strict adherence to a gluten-free diet, as with lactose intolerance, which also requires abstinence. Patients will need bone density assessment as they are at risk of osteoporosis. In chronic diarrhoea, many practitioners consider performing faecal culture, especially for *Clostridium difficile* or other such enteropathic organisms such as *Shigella*. Treatment of Merkel's diverticulum is by surgery. In acute diverticulitis, some urgency may be in order as a rupture may lead to peritonitis (as in appendicitis), treatment being surgery, ideally laparoscopic.

In all intestinal disease, patients should be encouraged not to pursue self-directed exclusion diets, but to take a well-balanced diet rich in fresh fruit and vegetables and low in fat, sugar, and processed meats. More complex advice should be sought from dietitians whilst guidelines offer options. As a third of patients with IBD have iron deficiency anaemia, supplements may be necessary (Table 12.5).

Regulation of digestion

Food passes down different parts of the intestinal tract, and in doing so is mixed with enzymes and other factors. These processes are regulated by the nervous system (notably the vagus and spinal reflex arcs) and (as already described briefly above) by hormones (Table 12.6). Some of these also act on, or are released by, the hypothalamus. Tumours secreting gastrin (gastrinomas) have been reported in

TABLE 12.6 Intestinal hormones

Hormone	Gene/location	Leading source*	Principal function(s)
Ghrelin	GHRL, 3p25.3	Stomach, duodenum	Stimulates appetite
Gastrin	GAST, 17q21.2	Stomach G cells	Stimulates HCl production from parietal cells, pepsinogen from chief cells
Vasoactive intestinal peptide	VIP, 6q25.2	Intestines and pancreas	Stimulates pancreatic bicarbonate secretion, inhibits HCl production
Glucagon-like peptide-1*	GCG, 2q24.2	L cells of the small intestines and colon, the pancreas	Inhibits gastric emptying, acid secretion, and motility, promotes insulin secretion
Glucose-dependent insulinotrophic peptide*	GIP, 17q21.32	K cells of the jejunum and duodenum	Decreases gastric acid production, stimulates insulin production
Cholecystokinin	CCK, 3p22.1	Duodenum and jejunum	Prompts pancreatic and gall bladder secretions, suppresses hunger
Secretin	SCT, 11p15.5	S cells of the duodenum	Stimulates pancreatic bicarbonate secretion
Motilin	MLN, 6p21.31	Duodenum	Regulates peristaltic waves
Obestatin	GHRL, 3p25.3	Unclear	Opposes the actions of ghrelin
Pancreatic polypeptide	PPY, 17p11.1	Head of the pancreas and islet PP cells	Decreases gastric acid secretion, gastric emptying, and upper intestinal motility
Somatostatin**	GH1, 17q22--4	Stomach and pancreas	Decreases gastric emptying, suppresses pancreatic hormones

* See also section 7.5 on diabetes and insulin.

** See also section 15.3.1, as it is involved in hypothalamic/pituitary regulation of growth hormone.

the duodenum and pancreas, and cause excess gastric acidity and so ulceration. Malignancies such as these can be part of the Zollinger–Ellison syndrome and may be treated with surgery and proton-pump inhibitors. Vasoactive intestinal peptide (VIP) is of further interest as it is also a neurotransmitter, and enteric gangliomas that over-secrete VIP, causing flushing of the skin and a watery diarrhoea, have been described. Leptin and adiponectin are described in section 12.4.2.

12.3.2 The large intestines

Anatomy and physiology

The large intestines consist of the appendix, caecum, colon (ascending, transverse, descending, and sigmoid) and rectum (Figure 12.13). Continuing from the distal ileum, and separated by a sphincter valve, the caecum (around 6 cm long) merges with the ascending colon of the rear right of the abdomen, which in turn becomes the transverse and then descending colon on the left. The appendix contains lymphoid tissue, but its function is unknown. The final sigmoid colon (length ~40 cm) merges with the rectum (~20 cm) so that the large intestines are perhaps 1.5 m in length; the mean diameter is around 6.5 cm.

The histology is similar to that of the small intestine, with a mucosa, layers of muscle (for peristalsis), and a serosa. However, the large intestine has fewer villi and differences in glands: fewer secreting digestive enzymes and more goblet cells secreting mucus, both found together in the crypts of Lieberkühn. Consequently, there are marked structural differences as the function converts to absorption of water and the formation of faeces, composed of undigested material (fibre as roughage), mucus, unabsorbed ions and molecules, bacteria, and dead enterocytes. The microbiome is credited with producing vitamin K and biotin, and continuing digestion (potentially by fermentation), thus contributing to the diet. However, pathology may follow if the balance of the hundreds of different species of bacteria is disturbed.

Pathology

The leading (non-malignant) pathological and specific lower bowel disease is ulcerative colitis, with a prevalence of ~115/million. Whilst Crohn's disease proximal to the terminal ileum is present in ~16% of these patients, it has 'hot spots' in the ileocaecal region, the junction of the transverse and descending colon, and the perianal region. Inflammation of the rectum (proctitis) is less common but shares many features with colitis. Indeed, much of the pathology and presentation of different forms of lower IBD is similar, requiring an expert opinion (Table 12.7). However, ulcerative colitis has links with HLA-DRB*0103. A common finding on endoscopy are polyps, abnormal mucosal growths that part-occlude the lumen: being formed of glandular tissue defines a polyp as an adenoma, and they are generally benign, although some may transform (under genetic control) to a malignant phenotype (sections 3.7 and 4.4.1).

The third major form of large intestine disease, IBS, is diagnosed by symptoms, which appear in the absence of any clear dysfunction of part of the intestines described above (e.g. inflammation). However, many patients claim their IBS is post-infection, and accordingly a primary microbial aetiology cannot be excluded, as there are reports of increased serum TNFα, IL-1, and IL-6, with increased numbers of CD3/CD25 lymphocytes in the mucosa, suggesting an inflammatory aetiology. IBS is often described as a functional gastrointestinal disorder, with ~15% of the population reporting symptoms at some time in their lives, half of whom will self-refer to their GP, and of these a quarter will be referred to a hospital. One estimate puts the case load of IBS in a gastroenterology clinic at 40%, which makes it the single most prevalent intestinal disease. It is two to three times more common in women, the reverse of IBD.

Presentation and diagnosis

Patients present with the well-described semi-specific signs and symptoms, as in many diseases in this section. In contrast to Crohn's disease, where it is generally negative, pANCA may be positive in ulcerative colitis. The course of the disease is variable, with some patients experiencing few attacks, whilst in others the disease is always present at a low level. The faeces are often diarrhoeic with blood and mucus, and must be cultured to exclude C. *difficile* infection. Typical presentation of a severe attack or flare-up would include over 6 bloody defaecations per day (up to 20 is not uncommon), a fever (>37.8°C), tachycardia (>90 bpm), raised ESR (>25 mm/hour), anaemia (haemoglobin <105 g/l, resulting from loss of blood), raised CRP (>30 mg/l), and hypoalbuminaemia (<30 g/l, poor protein absorption). Diagnosis is by sigmoidoscopy with at least two mucosal biopsies, finding the mucosa to be ulcerated, inflamed, and haemorrhagic, and polyps are a likely finding, although endoscopy and histology cannot distinguish the two IBDs in 5–15% of cases (Figure 12.14).

TABLE 12.7   A comparison of Crohn's disease and ulcerative colitis

Crohn's disease	Ulcerative colitis
Attacks any part of the intestines	Found only in the colon
Disease affects the full thickness of the intestine	Disease only of the mucosa
Rectal bleeding uncommon	Rectal bleeding common
Vitamin B$_{12}$/folate/vitamin D deficiency often present	Iron deficiency common
Twin concordance ~44%	Twin concordance ~12.5%
Cigarette smoking is a risk factor	Non-smoking is a risk factor
Incidence ~70/million/year	Incidence ~10/million/year
Prevalence ~625/million	Prevalence ~115/million

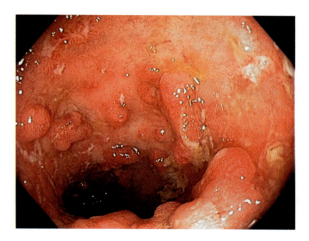

FIGURE 12.14

Colonoscopy for inflammatory bowel disease. Note the irregular surface, with inflamed and ulcerated areas.

Science Photo Library.

Any form of intestinal endoscopy will be undertaken with care, for fear of tearing or perforating the weakened bowel. Imaging has a place in determining dilatations and other anatomical and structural features. Barium enemas are also useful.

IBS is characterized by many of the symptoms already described, but may also be a part of a multisystem disorder that includes gynaecological, urinary, and connective tissue disorders. Accordingly, other disease must be excluded. There are numerous organic (intestinal infection (which may be present in ~25% of cases), surgery, antibiotics) and psychological/lifestyle (depression, anxiety, stress, change of diet, eating disorders) triggers. A negative faecal calprotectin (and serum inflammatory markers) excludes IBD. Notably, some linked genes (*TLR-9*, *TNFSF15*) are 'immunological', hinting at an aetiology.

Treatment

Treatment of IBD aims to reduce disease activity with the minimum of interventions. Disease activity may be quantified by a score derived from stool frequency, rectal bleeding, the state of the mucosa (e.g. erythema, erosions, ulceration), and a general global assessment. A first-line chemotherapy for mild and moderate ulcerative colitis (and possibly Crohn's disease) is 5-aminosalicyclic acid (5-ASA, as balsalazide (750 mg tds)), olsalazine (250 mg qd), or mesalazine (up to 2.4 g daily), to which 50% of patients should respond within 8 weeks. For colitis caused by Crohn's disease, it is 40 mg prednisolone od for 8 weeks to induce remission. The second-line agent is oral steroids, such as budesonide MMX 9 mg od or beclomethasone 5 mg od (possibly alongside 5-ASA), with 5 mg prednisolone for suppositories, if necessary.

Other chemotherapy options include azathioprine (50 mg od), vedolizumab (300 mg iv, 3 weeks apart, as for Crohn's disease), tofacitinib (a janus kinase inhibitor, 10 mg bd for 8 weeks for induction, then 5 mg bd for maintenance), a TNF-α inhibitor such as infliximab, adalimimab, or golimumab, or the IL-12/IL-23 inhibitor ustekinumab (as with Crohn's disease), all as per NICE guidance (NG130). Severe disease is likely to require oral prednisolone 40 mg od with weaning over 6–8 weeks. The most severe cases, possibly a flare-up (~20% of cases), will require admission, with intensive assessment, investigation (X-rays/CT, blood tests (CRP essential, FBC, U&Es, LFTs, magnesium), faecal culture), and treatment with steroids (such as 100 mg hydrocortisone or 40 mg methylprednisolone iv qds) and anticoagulation with a LMWH. Those failing this approach within 3 days may be eligible for iv cyclosporine (e.g. 2 mg/kg aiming for a plasma trough of 150–250 ng/ml) or infliximab (5 mg/kg).

This approach aims to avoid colectomy, the fate of ~20% of acute admissions, which may be required in the light of factors such as excessive haemorrhage, perforation, and dysplasia. However, if inevitable (as in ~33% of ulcerative colitis patients at some time), surgery (preferably laparoscopic) will be tissue-saving but, if extensive, may demand a complete colectomy, joining the terminal ileum to the rectum or anal canal, to create an ileal reservoir or pouch. Unfortunately, in ~50% of cases this can lead to a non-specific inflammation that may be treated with antibiotics (tinidazole, metronidazole, ciprofloxacin, or rifaximin (NICE ESUOM30)) or steroids (budesonide or beclomethasone). In the most challenging cases, there will be formation of an external stoma (colostomy) to bypass the anus altogether (the

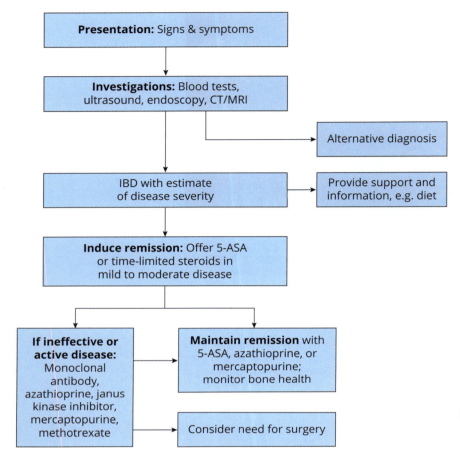

FIGURE 12.15

Management of inflammatory bowel disease.

faeces being collected in a bag), and a poor consolation is that such patients are absolved from the risk of malignancy associated with IBD. Figure 12.15 summarizes major points of the management of Crohn's disease and ulcerative colitis. Although the design is common for both IBDs, important differences are present, as in NICE NG129, NG130 and the guideline of Lamb and colleagues (2019).

Treatment of IBS (NICE CG61) focuses on diarrhoea and constipation, depending on whichever is present, and triggers (if identifiable), as described above (e.g. loperamide and laxatives, respectively). Other options include the antibiotic rifaximin and tailored diet, under the care of a dietician. In certain cases, restoration of a normal microbiome may be helpful, and be achieved by the introduction of probiotic microorganisms such as lactobacilli, bifidobacteria, and non-pathogenic strains of *E. coli*. This strategy may also be useful in other intestinal disease. If present, peritonitis is treated by addressing the cause, followed by sterile lavage and local and systemic antibiotics to prevent a possible septicaemia, largely with organisms such as *S. aureus*, *S. pneumoniae*, *E. coli*, *Klebsiella* species, and *P. aeruginosa*. Where the disease is dominated by diarrhoea, TA471 gives guidance on the use of the opioid receptor antagonist eluxadoline, NICE DG7 recommends tauroselcholic ([75]selenium) acid for diagnosing bile acid malabsorption, and if present, ESUOM22 summarizes the evidence for using colesevelam as a treatment. ESNM16 advises on the use of linaclotide in constipation.

Appendicitis

Appendicitis is the most common acute intestinal condition requiring hospitalization, with an incidence of ~2,250/million/year, predominantly in 10–20-year-olds, and is caused by obstruction of the opening, most often with a hardened bolus of faeces. It presents with pain in the right lower quadrant, developing over 24 hours, and is diagnosed by CT, which shows an enlarged appendix, although there may an acute phase reaction. Should the appendix rupture, the contents it releases may lead to inflammation of the peritoneal cavity and so a potential peritonitis. Accordingly, rapid keyhole surgery is likely.

12.3.3 The anus

A short canal links the distal rectum to the sphincter muscles of the anus, which are partly under voluntary control. Anal cancer is described in section 5.13.1. As described above, Crohn's disease may spread to this area, as may proctitis, and both may cause an anal fissure or fistula. The latter, possibly caused by an abscess, is treatable with anti-TNFαs and/or surgery, or the local transplantation of ~120 million allogeneic adipose tissue stem cells (NICE TA556).

Perianal disease with loss of the integrity of the anal sphincter and excess diarrhoea may lead to faecal incontinence, which may be treated with loperamide. NICE documents CG49 (faecal incontinence in adults), QS54, IPGs 210, 483, 393, 395, and 276, and MIBs 66 and 164 all apply. Conversely, constipation (very common, even in the absence of IBD or IBS) may be treated with high-fibre diet (wheat bran, methylcellulose), drugs to stimulate motility (e.g. prucalopride 2 mg od), lactulose, suppositories, and enemas. Haemorrhoids (anal and rectal blood vessels expressed outside the anus) are not generally pathological, but can bleed and cause distress. Risk factors include excessive anal intercourse, constipation, diarrhoea, and obesity, and NICE offers a number of documents regarding treatment (MIBs 75 and 201, IPGs 342, 525, and 589).

CASE STUDY 12.3 *Ulcerative colitis*

A 19-year-old woman presents to her GP with a 3-month history of lethargy, loss of appetite, photosensitivity, intermittent fevers, low-grade polyarticular pain, and weight loss. On examination, the GP fails to find any painful or swollen joints, and records an unremarkable history. Urine analysis is normal, and he takes blood for routine analyses, requesting rheumatoid factor (RhF), anti-citrullinated protein antibody (ACPA), and anti-nuclear antibodies (ANA). He recommends analgesia with aspirin and/or paracetamol, and invites her to return in 10 days to review the investigations. These find normal renal and liver function, a normocytic anaemia, and evidence of a low-grade inflammatory response (raised white blood cell count with neutrophil leucocytosis) with borderline increases in RhF, ACPA, and ANAs (Table 12.8).

The patient is relieved to learn that these, alongside history, signs, and symptoms, effectively exclude rheumatoid arthritis and systemic lupus erythematosus. As regards the anaemia, she denies heavy or irregular menstrual bleeding, and on further questioning on general health, reports intermittent abdominal pain with some diarrhoea. He takes additional blood for CRP and iron studies, and arranges for a faecal occult blood (FOB) test and faecal calprotectin to be done rapidly at the local hospital. In the meantime he recommends changing the drugs to an NSAID 200 mg four times a day, working up to 400 mg each time.

A few days later the results come back, with a raised CRP of 17 mg/dl (reference range <5 mg/dl), total iron 7 μmol/l (10–37), total iron binding capacity 45 μmol/l/l (55–81), transferrin saturation 10% (13–37%), FOB positive, and faecal calprotectin positive. These further confirm inflammation and change the anaemia to iron deficiency (despite a normal MCV). The reason for this is likely loss from the intestines and, alongside the positive calprotectin, this pushes the diagnosis in the direction of an inflammatory bowel disease. At this point the GP passes his patient to the local hospital where she meets with a consultant gastroenterologist, and agrees to have a colonoscopy to determine the likelihood of ulcerative colitis or Crohn's disease.

Several weeks later she has the colonoscopy which finds numerous shallow and often haemorrhagic ulcers in the colon and the rectum, making the diagnosis of ulcerative colitis (Figure 12.16). In view of the endoscopy finding, blood results, and signs/symptoms, she is thought to have mild to moderate disease, and is started on two 400 mg tablets of mesalazine three times a day (i.e. 2.4 grams daily) for two weeks, which are then reduced in steps to 1.6 grams daily. Three months later all her blood results and symptoms have improved, and by 6 months she has put on weight and is beginning to enjoy life once more. However, she has been counselled on the possibility of flare-ups with the return of the disease, the possible need for surgery, and the increased risk of colorectal cancer.

(Continued)

CASE STUDY 12.3 *Continued*

TABLE 12.8 Blood results for Case study 12.3

Analyte (unit)	Result	Reference range
Urea (mmol/l)	4.9	3.3–6.7
Creatinine (µmol/l)	98	71–133
Bilirubin (µmol/l)	12	<20
Alanine aminotransferase (IU/l)	16	5–42
Aspartate aminotransferase (IU/l)	22	10–50
Alkaline phosphatase (IU/l)	65	20–130
Gamma glutamyl transferase (IU/l)	49	0–100
Haemoglobin (g/l)	105	118–48
Red cell count ($\times 10^{12}$/l)	4.0	3.9–5.0
Mean cell volume (fl)	83	77–98
Haematocrit (l/l)	0.31	0.33–0.47
White blood cells ($\times 10^9$/l)	11.5	4.0–10.0
Neutrophils ($\times 10^9$/l)	9.5	2.0–7.0
Lymphocytes ($\times 10^9$/l)	1.8	1.0–3.0
Platelets ($\times 10^9$/l)	301	143–400
ESR (mm/Hr)	33	<10
Rheumatoid factor (U/ml)	18	<15
ACPA (U/ml)	13	<20
ANA titre	1/80	<1/60

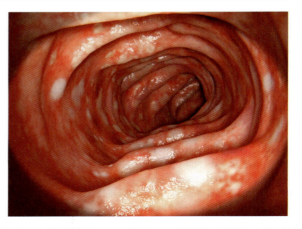

FIGURE 12.16
Colonoscopy for ulcerative colitis. Patches of white ischaemic tissue contrast with darker-red patches of inflamed tissues.

Shutterstock.

Key points

Although the IBDs Crohn's disease and ulcerative colitis grab one's attention because of their severity and the potential need for surgery, there is more morbidity in the non-IBDs of coeliac disease, lactose intolerance, and IBS.

SELF-CHECK 12.7

What are the differences between the two major diseases of the small intestines?

12.4 Nutrition

With the probable exception of obesity, much of nutrition is taken for granted. The Global Burden of Disease Study estimated that nutritional deficiency caused 270,000 deaths in 2017, 85% being due to protein energy malnutrition. A further 300,000 deaths were linked to eating disorders, of which 200,000 were of anorexia nervosa, the remainder to bulimia nervosa. Overweight and obesity are not included in these data (in common with the UK's Office for National Statistics), although both provide data on a leading consequence of obesity—diabetes (section 7.4).

12.4.1 Requirement of a healthy diet

A healthy diet has seven components—proteins, carbohydrates, fats, vitamins, minerals, roughage/fibre, and water—and all of these must be taken in the correct proportions. Too much or too little of any component will lead to disease (section 12.4.2), with most authorities recommending a daily adult intake of around 2,500 kilocalories by men and 2,000 by women, although these figures will vary according to energy expenditure by pregnancy, age, occupation, and exercise. Although the joule is the SI unit of energy, the calorie is widely used, where one kilocalorie is the energy required to raise the temperature of a kilogram of water by 1°C.

Whilst the liver is able to synthesize a broad range of proteins, it cannot provide all the protein needs of the body, required for plasma proteins and structural components (collagen, elastin, etc.). Metabolic pathways within the hepatocyte are able to synthesize certain amino acids and construct proteins from them, such as its own metabolic enzymes and those for export, including albumin and coagulation factors. However, as we lack the metabolic pathways to generate histidine, isoleucine, leucine, lysine, methionine, phenylalanine, threonine, tryptophan, and valine, these (literally) essential amino acids must be provided in the diet. Proteins are not stored, and at times of starvation can be used as energy sources.

The liver can also synthesize individual carbohydrate molecules (monosaccharides e.g. glucose, galactose, etc.) and build them into disaccharides (sucrose, maltose, etc.). However, the greater part of our glucose arises from the diet, and at times of plenty it may be polymerized into the storage molecule glycogen, the animal version of starch, from which it can be recovered. There are numerous metabolic pathways around glucose and glycogen, principally because the former is the main energy source for the body (Figure 12.17). These pathways are complex and interrelated: for example, the amino acid glutamine can enter the Krebs cycle, and alanine can be converted to pyruvate and so be part of the process of generating glucose.

Fats and lipids are also obtained in the diet, but can be synthesized from glucose, via glycerol and amino acids (the process of lipogenesis), whilst the Krebs cycle intermediate acetyl-coenzyme A can be part of the process of fatty acid synthesis. However, as with amino acids, we lack the metabolic pathways to enable the synthesis of certain omega-3 and omega-6 fatty acids. The former (such as α-linolenic acid, eicosapentaenoic acid, and docosahexaenoic acid) are present in fish, and the latter (such as linoleic acid and arachidonic acid) are widely present in grains, plant-based oils, poultry, and eggs. These are therefore also essential. Fat not immediately required is stored as triglycerides (also known as triacylglycerol) in adipocytes in various locations around the body, and can be an energy source when catabolized by β-oxidation (the process of lipolysis), often in adipose tissues and the liver,

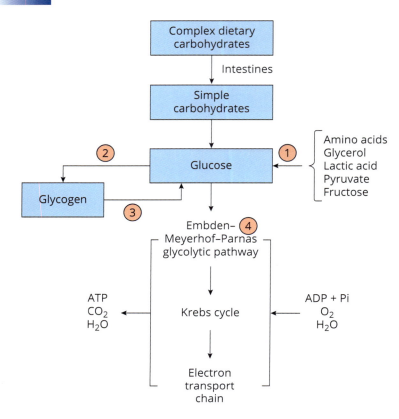

FIGURE 12.17

Glucose metabolism. Simplified metabolic pathways around glucose. 1 = glucogenesis, 2 = glycogenesis, 3 = glycogenolysis, 4 = glycolysis. Some pathways are part-controlled by insulin and glucagon (see section 7.4).

which generates ketone bodies. The liver synthesizes cholesterol and related molecules that serve as building blocks for steroid hormones and other molecules.

Vitamins are a heterologous group, and are indeed 'vital *amines*', being essential components of our diet as they cannot be synthesized, although the dietary precursor of vitamin D is moulded into its mature form by passing through the liver, skin, and kidney (Box 15.1). Variously named alphabetically, they are required in trace amounts as cofactors for certain metabolic processes and may be obtained from most food sources, although vitamin B_{12} is absent from the plant kingdom, requiring strict vegans (but not vegetarians) to take supplements. Small amounts of vitamins A, D, and B_{12} can be stored, mainly in the liver. Essential minerals include cobalt, calcium, copper, iodine, iron, manganese, molybdenum, selenium, and zinc. Numerous enzyme and other metabolic intermediates require certain minerals as cofactors (such as cobalt and vitamin B_{12}). The position of chromium as an essential trace mineral is unclear. Varying quantities of iron, copper, and iodine are stored in certain organs, the liver, and the thyroid, respectively.

Water is needed to maintain the blood volume, ensure adequate renal function, and replace insensible losses in sweat and through the lungs, whilst fibre/roughage (indigestible plant material) seems likely to aid the process of peristalsis and remove effete enterocytes from the intestinal villi.

12.4.2 An unhealthy diet

Few of us will be unaware of the advice regarding the necessity of a good balanced diet. Although there are some instances where pathology arises from an organic reason (Chapter 16), in many cases the aetiology is of poor choice of lifestyle, with insufficient exercise.

General nutrition

Malnutrition includes both an excessive intake and undernutrition, or perhaps undernourishment, and is a major global issue. In England and Wales, obesity was linked to 402 deaths in 2013, rising (with a single exception) year on year to 577 in 2021, an increase of 43.5%. The latter were 51.5% women,

48.5% men. WHO data indicate that 24.3% of the world's population are overweight or obese, whilst 5.9% are underweight. Undernutrition in the young leads to wasting and reduced growth (stunting). In all age groups, malnutrition may result from simple lack of calories in the diet, but other causes include chronic diseases such as parasitic infection and those linked to diarrhoea and micronutrient deficiency.

Leaving aside the need for essential amino acids, a diet poor in proteins has two extreme consequences, both characterized by muscle wastage and abdominal distension. Kwashiorkor is caused by sufficient calorie intake (such as mainly carbohydrates) but insufficient protein, whereas marasmus is primarily a disease of insufficient energy intake of all forms of nutrient. Treatment is by a high-quality diet, but this must be managed in stages, for fear of re-feeding syndrome, wherein metabolic disease follows from a sudden surge in calories. Cachexia is also characterized by weight loss and muscle wasting, but it is secondary to other underlying diseases such as cancer, AIDS, COPD, and congestive heart failure. There may well be other nutritional issues in those with chronic disease, stroke (with swallowing difficulties), or recovering from major surgery, but also in depression, dementia, alcoholism, poverty, and bereavement. A diet rich in fats and lipids is a risk factor for NAFLD, regardless of overweight/obesity.

SELF-CHECK 12.8

What are the seven essential constituents of a healthy diet?

With the exception of psychiatric conditions and 'fad' diets, a diet specifically deficient in carbohydrates and/or fats/lipids is exceedingly rare (at least in developed countries), but both food groups are lacking in marasmus. Adults whose diets are deficient in essential fatty acids generally have dermatological symptoms, and infants fed powdered milk low in linoleic acid demonstrate poor growth rates, susceptibility to certain bacterial infections, and histological alterations in the skin. Breast milk is enriched for linoleic acid.

Epidemiological evidence suggests that omega-3 polyunsaturated fatty acids present in fish oils provide benefits following a myocardial infarction, possibly by increasing levels of HDL-cholesterol and/or reducing the likelihood of arrhythmias. Evidence of this nature links to the advice of NICE CG172 that lifestyle changes after a myocardial infarction should include eating more fish, especially those deemed to be 'oily'. A diet high in indigestible fibre protects from ischaemic heart disease, stroke, type 2 diabetes, and colorectal and some other cancers. Those taking a high-fibre diet also benefit from lower blood pressure, body weight, and cholesterol.

Alcohol

Is alcohol a nutrient? It certainly has a calorific content (7 calories/gram, so around 200 kilocalories per 500 ml of beer), so for the sake of this chapter, the answer is 'yes', although others would disagree. The sociological and pathogenic effects of this molecule are evident, with a direct linear relationship between intake and liver disease and almost all forms of cancer, mostly gastric and hepatocellular carcinoma. However, cardiovascular epidemiology has shown a protective effect of a small intake of alcohol, leading to a J-shaped curve. The reason for this is unclear but has been linked to an increase in HDL-cholesterol. Therefore, it could be argued that the benefit of a small alcohol intake compared to abstinence in reducing the risk of myocardial infarction and stroke is balanced by the small increased risk of cancer. In contrast, chronic alcohol use leads to raised serum triglycerides and the LFT gamma-glutamyl transferase, and although increased levels of the latter generally reflect a degree of hepatic dysfunction, in this instance it may simply reflect a nonspecific metabolic reaction.

Body mass index (BMI)

BMI is simply an epidemiological and clinical tool for use in nutrition and pathology derived from the subject's weight (in kg) divided by the square of their height (in metres), also known as the Quetelet index. Several arbitrary categories have been defined: in adults, low BMI is considered to be <18.5 kg/m^2, normal 18.5–24.9, overweight 25–29.9, obese 30–39.9 and severely obese ≥40, or ≥35 if in combination with other disease such as diabetes. An alternative view of obesity describes three classes: Obesity class 1 being BMI 30–34.9, class 2 as 35–39.9, and class 3 as BMI 40 or more. Like hypertension (>140/90), these categories are not based on pure bioscience: overweight could just as easily be BMI 26–29.9. In fact, in different populations the spectrum of BMI can vary markedly, and alternative definitions of low/normal/overweight/obese may apply in different populations around the world.

NICE CG189 concludes that, compared to those from a White family background, people from certain ethnic/racial groups are at an equivalent risk of diabetes, other health conditions, or mortality at a lower BMI. Accordingly, overweight is considered to be BMI 23–27.4, and obesity to be BMI 27.5 and above in those with a South Asian, Chinese, other Asian, Middle Eastern, Black African, or African Caribbean family background. In 2020–1, in England, focusing on ethnicity, the Black ethnic group has the highest proportion of obese people (31%), the Chinese ethnic group the smallest proportion (7.3%).

Adipocytes are not simply storage vehicles: they are metabolically active, secreting cytokines and other molecules, including CRP, plasminogen activator inhibitor, leptin, adiponectin, angiotensinogen, omentin-1, and inflammatory cytokines such as IL-6. Indeed, adipose tissue has been described as one of the largest endocrine organs in the body. Two adipokines are of particular interest.

Under normal conditions, leptin (a 16 kDa protein coded for by *LEP*, also known as *Ob*, at 7q32.1) acts on specific receptors to suppress appetite, and is involved in blood pressure regulation and several other physiological processes. Obesity is characterized (paradoxically) by a hyperleptinaemia, but this may be a response to leptin resistance (a parallel to insulin resistance in type 2 diabetes mellitus—section 7.4) and so poor control of appetite. Adiponectin (present in isoforms of 200–1,000 kDa, coded for by *ADIPOQ* at 3q27.3) increases insulin sensitivity and has anti-inflammatory action. Serum levels are low in the obese, and in apparently healthy individuals predict the development of diabetes.

Increased BMI is linked to all major cancers and cardiovascular diseases, hyperglycaemic diseases (notably diabetes: section 15.2), venous thromboembolism (mostly deep-vein thrombosis: section 11.6.3), osteoarthritis (section 10.3.2), and complications of pregnancy (section 14.4.1). Overweight and obesity (the latter present in 30% of people in England) are linked to 22% of oesophageal cancers, 13% of colorectal cancers, 12% of pancreatic cancers, 18% of gall bladder cancers, 9% of breast cancers, 34% of uterine cancers, and 24% of renal cancers. Over half of all cases of endometriosis are found in obese women. In a prospective study of over 1 million Americans followed up for 14 years, BMI was a major influence on death from all causes, from cardiovascular disease, and from cancer, with a J-shaped curve. The BMI with the lowest risk was 23.5–24.9 in men and 22.0–23.4 in women. In England, for the period 2018–19, there were 11,117 hospital admissions with a primary diagnosis of obesity, most aged 45–54, up 4% from 2017–18, and 22% since 2014–15. Women made up 74% of these admissions. In 2018–19 there were 876,000 admissions (65% women) where obesity was a primary or secondary diagnosis, up 23% from 2017–18. Leading factors for admission were maternal care for other known or suspected foetal problems, arthrosis of the knee, gallstones, arthrosis of the hip, and chronic ischaemic heart disease.

The vast majority of instances of overweight and obesity result from an imbalance between calorie intake and expenditure (as exercise): true genetic causes include a genotype of *TFO* (at 16q12.2, coding for an enzyme that demethylates RA). Heterozygotes for the A allele of the *TFO* SNP are ~ 1.2 kg heavier whilst homozygotes are ~2.4 kg heavier. Prader–Willi syndrome (section 16.5.3) is characterized by chromosome 15 abnormalities, leading to constant hunger and so obesity. Following a 15-week weight-loss programme, miR-29a-3p and miR-29a-5p were upregulated and miR-20b-5p downregulated, whilst obesity is linked to upregulation of miR-221, itself associated with leptin-mediated fat metabolism.

NICE CG43 offers guidance on obesity prevention, whilst NICE PH42 focuses on working with local communities to prevent overweight and obesity. NICE CG189 covers the identification, assessment, and management of obesity, and points to bariatric surgery (e.g. duodenal–jejunal bypass sleeve (IPG471), duodeno–ileal bypass with sleeve gastrectomy (IPG569)) and very-low-calorie diets to help reduce weight. Surgical interventions are effective not only in weight reduction but also in reduced progression to diabetes and (where relevant) improvement in diabetic markers. Although the combination of naltrexone (an opioid) and bupropion (an antidepressant) is effective in reducing weight in the obese, it is not at present recommended for NHS use by NICE TA494 as it is not cost-effective. Other NICE publications regarding overweight and obesity include CG130 on hyperglycaemia in acute coronary syndromes, CG177 on osteoarthritis, and NG49 on non-alcoholic fatty liver disease.

The major diseases associated with a low BMI include anorexia nervosa and bulimia nervosa, both considered mental health conditions related to poor body image. The former is due to failure to eat, the latter to binge eating followed by induced vomiting (purging). A low BMI is often found in the terminal stages of cancer, where it is described as cachexia, in severe malnutrition/starvation, and in other conditions, as indicated above.

Enteral nutrition

Patients with digestion problems may be fed soup-like food via a tube passed via the nasal cavity down into the intestines (a nasogastric tube), the length of which can be tailored to terminate in the stomach or intestines. Very rarely, food may be introduced directly into the stomach or small intestines via a percutaneous catheter. The exact composition of the supplied foods must be determined with metabolic precision.

Parenteral nutrition

Those with more complex intestinal disease (e.g. Crohn's disease, ulcerative colitis, see section 12.3) and/or severe malnutrition (such as in cancer cachexia) may need to be fed intravenously with a complex mixture of all food groups except roughage. Should this be semi-permanent, it may be delivered by a central venous catheter (such as a Hickman line) into vessels such as the subclavian and jugular veins. Complications (as with dialysis) include infections and thrombosis, and in the long term there may be cholecystitis, fatty liver, and liver failure. The components of a parenteral diet must be calculated with even greater precision than those of enteral feeding, and its effects must be checked with regular blood tests, weighing, and imaging.

Vitamins and minerals

As they are an essential requirement of physiology, disease follows deficiency in vitamins and minerals, leading to the recommended reference intakes required for health (Table 12.9). The lag period between the absence of a particular component and the development of signs/symptoms depends on its biological half-life and the presence of stores, and ranges from a few weeks to several months.

Extreme vitamin D deficiency causes rickets in children and osteomalacia in adults. Although these are very rarely seen in the UK, NICE PH56 nonetheless offers guidance covering vitamin D supplements, which may be considered in a number of specific population groups: infants and children aged under 4; pregnant and breastfeeding women, particularly teenagers and young women; people over 65; people who have low or no exposure to the sun, for example those who cover their skin for cultural reasons, who are housebound or confined indoors for long periods; and people with darker skin, for example, people of African, African Caribbean, or South Asian descent.

People with particular dietary needs, such as people who avoid nuts, are vegan, or follow a halal or kosher diet, should also receive suitable supplements. It is important that supplements undergo quality control checks to ensure that they contain the correct dose of vitamin D. Of the essential minerals, almost all pathology follows from insufficient iron and iodine, which is easy to diagnose and treat. Consequences of a lack of iron (a microcytic anaemia) are described in section 11.5.3, the recommended daily allowance (RDA) being 8.7 mg (14.8 mg in pregnancy and in women who menstruate (e.g. ages 11–50)). Iodine is a component of the two thyroid hormones of tri-iodothyronine and tetra-iodothyronine (thyroxine), such that iodine deficiency leads to hypothyroidism, as described in section 15.5.3. The RDA of iodine is generally 150 μg in adults, but 200 μg if pregnant or breastfeeding.

TABLE 12.9 **Key aspects of vitamin biology**

Vitamin	Pseudonym	Stored	Consequence of deficiency	Reference intake
A*	–	Liver	Night blindness (xerophthalmia)	800 µg
B_1	Thiamine	–	Beriberi, polyneuritis	1.1 mg
B_2	Riboflavin	–	Dermatitis, anaemia, angular dermatitis	1.2 mg
B_5	Pantothenic acid	Liver and kidneys	Fatigue, muscle spasms, insomnia	6 mg
B_6	Pyridoxine	–	Dermatitis, anaemia, polyneuropathy	1.4 mg
B_{12}	Cyanocobalamin	Liver	Pernicious anaemia, neurological disorders	2.5 µg
Biotin	–	–	Depression, myalgia, dermatitis, fatigue	50 µg
C	Ascorbic acid	Glandular tissue	Scurvy, anaemia, haemorrhage, gingivitis	80 mg
D*	–	–	Rickets in children, osteomalacia in adults	5 µg
E*	Tocopherol	Liver	Pancytopenia, poor wound repair, ataxia	12 mg
Folic acid	Folate	–	Macrocytic anaemia, neonatal neural tube defects if mother deficient	200 µg
K*	–	Liver and spleen	Haemorrhage	75 µg
Niacin	Nicotinamide	–	Pellagra (dermatitis, diarrhoea, personality changes)	16 mg

* Fat soluble: Others are water soluble. See section 11.5.3 for precise roles of vitamins B_2, B_6, and B_{12}, and folic acid, and section 11.2.3 for vitamin K. Reference intake varies according to age, sex, and if pregnant or seeking pregnancy. Sources: Regulation (EU) No 1169/2011 of the European Parliament and of the Council, and British Nutrition Foundation.

Numerous other minerals are required as cofactors for enzymes, in maintaining bone, and for many other functions, and include calcium (RDA 700 mg), phosphorus (550 mg), magnesium (285 mg), sodium (1,600 mg), potassium (3,500 mg), chloride (2,500 mg), zinc (7 mg), copper (1.2 mg), cobalt (6.5 µg), and selenium (67.5 mg), but note some of these vary in relation to age and sex.

Generally, high dietary levels of vitamins and minerals do not cause major problems as excesses are excreted. Exceptions to this include those deliberately taking excessive levels of supplements, excessive intake of salt (leading to hypertension, stroke, and gastric carcinoma), and iron, causing haemochromatosis (section 11.7). But a little knowledge can be a dangerous thing. A case study reported a man who read that carrots are rich in vitamin A, and that this helps with vision in dim light. He bought a sack of carrots, and a vegetable blender, and drank 500 ml of carrot puree twice a day. He died of acute liver failure before the sack was empty.

Key points

Alongside cigarette smoking, excessive use of alcohol and overweight/obesity are the major lifestyle causes of morbidity and mortality.

SELF-CHECK 12.9

What are the medical consequences of obesity?

Chapter summary

■ Disease of the digestive system is the 7th most common cause of mortality in England and Wales.

■ Leading non-malignant diseases of the oesophagus include oesophagitis, Barrett's oesophagus, and oesophageal varices.

■ Diseases of the stomach include gastritis, ulceration, and atrophy, a leading cause being high consumption of alcohol.

■ Excess alcohol is also a cause of pancreatitis, although this disease may be due to obstruction of the pancreatic duct.

■ Leading diseases of the liver include hepatitis, cirrhosis and fibrosis, and fatty liver disease.

■ Almost all diseases of the small and large intestines are neoplastic or inflammatory, the latter including inflammatory bowel disease, colitis, coeliac disease, Crohn's disease, and lactose intolerance.

■ Malnutrition strikes the polar extremes of calorific intakes: undernutrition and obesity.

Further reading

● Hirschfield GM, Dyson JK, Alexander GJM, et al. The British Society for Gastroenterology/UK PBC primary biliary angitis treatment and management guidelines. Gut 2018: 67; 1568–94.

● Chapman MH, Thornburn D, Hirschfield GM, et al. The British Society for Gastroenterology and UK-PSC guidelines for the diagnosis and management of primary sclerosing cholangitis. Gut 2018: 67; 1568–94.

● Di Costanzo M, Canali RB. Lactose intolerance: common misunderstandings. Annals Nutr Metabol 2018: 73 (suppl 4); 30–37.

● Veauthier B, Hornecker JR. Crohn's disease: diagnosis and management. Am Fam Physician 2018: 98; 661–9.

● Chapman CG, Pekow J. The emerging role of miRNAs in inflammatory bowel disease: a review. Ther Adv Gastroentrol 2015: 8; 4–22.

● Lamb CA, Kennedy NA, Raine T, et al. British Society of Gastroenterology consensus guidelines on the management of inflammatory bowel disease in adults. Gut 2019: 68; s1–s106.

● Wang Y, Zheng F, Liu S, Lou H. Research progress in fecal microbiota transplantation as treatment for irritable bowel syndrome. Gastroenterol Research Pract 2019: doi. org/10.1155/2019/9759138.

● Byrne C, Targher G. NAFLD: A multisystem disease. J Hepatol 2015: 62; S47–S64.

● Iqbal U, Perumpail BJ, Akhtar D, et al. The Epidemiology, Risk Profiling and Diagnostic Challenges of Non-alcoholic Fatty Liver Disease. Medicines 2019: 6; 41; doi:10.3390/medicines6010041.

● Petrov MS, Yada D. Global epidemiology, and holistic prevention of pancreatitis. Nat Rev Gastroenterol Hepatol 2019: 16; 175–84.

● Peng Y, Qi X, Guo X. Child-Pugh versus MELD score of the assessment of progress in liver cirrhosis. Medicine 2016: 95; 1–29.

- Valitutti F, Cucchiara S, Fasano A. Coeliac disease and the microbiome. Nutrients 2019: 11; 2403; doi:10.3390/nu11102403.

- Bleszynski MS, Bressan AK, Joos E, et al. Acute care and emergency general surgery in patients with chronic liver disease: how can we optimize perioperative care? World J Emergency Surg 2018: 13; 32. doi.org/10.1186/s13017-018-0194-1.

- Di Costanzo M, Canani RB. Lactose Intolerance: Common Misunderstandings. Ann Nutr Metab 2018: 73 (suppl 4); 30–7.

- Rundo JV. Obstructive sleep apnoea basics. Cleve Clin J Med 2019: 86 (9 suppl 1) 2–9.

Useful websites

- International Agency for Research on Cancer: **http://gco.iarc.fr/today/home**
- National Institute for Health and Care Excellence: **www.nice.org.uk**
- British Society of Gastroenterology: **www.bsg.org.uk**
- Crohn's and Colitis UK: **www.crohnsandcolitis.org.uk**
- WHO overview on malnutrition: **https://www.who.int/health-topics/malnutrition**
- Statistics on obesity, physical activity, and diet in England: **https://digital.nhs.uk/data-and-information/publications/statistical/statistics-on-obesity-physical-activity-and-diet**
- UK Policy paper on 'Tackling obesity': **https://www.gov.uk/government/publications/tackling-obesity-government-strategy/tackling-obesity-empowering-adults-and-children-to-live-healthier-lives**

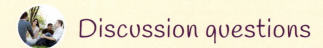 Discussion questions

12.1 How does the structure of the stomach reflect its function?

12.2 What are the principal causes of obesity, and what are its consequences?

13

Disease of the respiratory system

This system is composed of the nose and nasal cavity, the larynx, trachea, pharynx, the left and right bronchi, and the left and right lungs, each lung enclosed in its own pleural sac, and protected by the ribcage. The principal function of these tissues and organs is to provide the blood arriving in branches of the pulmonary artery with oxygen, and to remove waste products (principally carbon dioxide) in exhaled breath. They can be classified as the upper (section 13.2) and lower respiratory tracts, a distinction important in clinical practice as infections are referred to in these two parts—that is, upper and lower respiratory tract infections, the URTI and LRTI, respectively. As almost all of the burden of respiratory system disease is accounted for by that of the trachea, bronchus, and lung, they have their own separate sections (13.3–13.5). The full implications of the global COVID-19 pandemic that started in early 2020 will take years, perhaps a decade, to be fully understood. In England and Wales, SARS-CoV-2 killed 69,679 people in 2020 and 67,057 in 2021. It is discussed in section 13.6.

Learning objectives

After studying this chapter, you should confidently be able to:

- explain the anatomy of the airways

- understand inflammatory disease of the upper airways, trachea, and bronchi

- describe the pathology of lung disease

- outline the diagnosis and treatment of chronic obstructive pulmonary disease (COPD), asthma, and pulmonary embolism

- summarize the key pathological and epidemiological aspects of COVID-19

13.1 Introduction

As the biggest public health issue in a lifetime, it is natural that we should begin with COVID-19. Each year, the UK's Office for National Statistics (ONS) reports the causes of deaths in England and Wales (Table 13.1). From a baseline of 2013–2019, these show a reduction in deaths from all forms of respiratory disease, most notably in pneumonia and chronic lower respiratory disease, but also in cancer of the respiratory system. However, the relatively large number of COVID-19 deaths does not account for the 'missing' 12,048 respiratory system deaths in 2020, or the reduction by 19,175 in 2021, compared to the average during 7 pre-COVID-19 years. This disease will be discussed fully in section 13.6.

TABLE 13.1 Deaths from respiratory disease

Cause of death	2013–19*	2020	Difference From 2013–19	2021	Difference from 2013–19
The respiratory system	72,937	61,926	-15.1%	54,799	-24.9%
Chronic lower respiratory disease**	31,095	28,308	-9.0%	26,143	-15.9%
Pneumonia	26,945	19,620	-27.2%	16,197	-39.9%
Influenza	607	510	-16.0%	35	-94.2%
Other respiratory disease	5,861	5,769	-1.6%	6,076	+3.7%
Acute lower respiratory disease	4,493	4,286	-4.6%	3,508	-21.9
Malignant neoplasms of the lung, trachea, and bronchus	31,149	29,708	-4.6%	29,119	-6.5%
COVID-19	–	69,679	–	67,057	–

* Mean of these seven years.

** Includes asthma, bronchitis, emphysema, and other chronic obstructive pulmonary disease.

© Office for National Statistics, https://www.ons.gov.uk/. Licensed under the Open Government Licence 3.0.

According to the Global Burden of Disease study, the influenza mortality rate is of the order of 164 deaths/million, the figure in England and Wales having fallen from 8.9/million in 2013–19, to 7.5/million in 2020, to 0.6/million in 2021, and to 16.4/million in 2022. This remarkable fluctuation is likely to be linked to public health measures due to the COVID-19 pandemic, and the virus itself.

A leading UK charity, Asthma and Lung UK, estimates that 5 million people in the UK have asthma, although another, the British Lung Foundation, gives a figure of 8 million. This gives a prevalence of around 7,500–12,000/million, translating to the figure of around 0.02% for those who die of their illness annually. However, these data fail to consider the quality of life of those with the particular disease, which can be remarkably poor even if the condition is not fatal. The Global Burden of Disease Study provides data on chronic respiratory diseases, which in 2017 caused almost 56 million deaths, COPD being linked to 81.1% and asthma to 12.6% of fatalities.

The undoubted major pathogen in respiratory disease is cigarette smoking, with increasing evidence of the importance of industrial pollutants, microbes, and food, animal, and plant products. Atherosclerosis is rare, although there is pulmonary thrombosis, but both inflammatory and autoimmune aetiologies are common. As with digestive disease, we will start at the top and work our way down.

13.2 The upper respiratory tract

13.2.1 The nose and nasal cavity

The nose is constructed from bone at the upper part (where it meets the head) and cartilage at the lower (apical) aspect. The central septum separates the left and right nostrils, short connecting tubes that are lined with mucus-secreting epithelial and hair follicles. The inner aspect of the nostrils feed air into the nasal cavity (a sinus) that has four functions: to warm and filter incoming air, to host olfactory tissues that provide a sense of smell, to modify speech, and to provide a first line of defence against pathogenic microorganisms as part of the innate immune system. The rear of the nasal cavity merges with the superior (upper) section of the pharynx—the nasopharynx.

A common complaint is rhinitis (inflammation of the nose), which is mostly a response to agents causing the common cold, a physiological response being the secretion of mucus to help expel the pathogen (should there be one), as in sneezing. However, rhinitis and allied symptoms, present in ~20% of the population, may also be linked to IgE-mediated allergic hypersensitivity (as in to seasonal pollen, and moulds, house pets, and arthropods and their faeces, see section 10.2.1). Other symptoms include nasal obstruction, nasal itching, and conjunctivitis. If persistent, and with additional chest symptoms, asthma may be present. Treatment is by avoidance; antihistamines (e.g. loratadine 10 mg od); over-the-counter decongestants that thin mucus; anti-inflammatories (sodium cromoglycate as a topical preparation, spray, or inhaler); topical, oral, or spray corticosteroids (e.g. beclomethasone, prednisolone 5 mg od); and leukotriene antagonists (e.g. zafirlukast 20 mg od)—many of which are also treatments for asthma.

Sinusitis (inflammation of the nasal sinuses) often has a bacterial aetiology, perhaps *Streptococcus* and/or *Haemophilus* species, and is often accompanied by a purulent discharge. Should this fail to respond to broad-spectrum antibiotics, topical steroids may be effective, beyond which referral to a specialist is needed. The nose is relatively free of chronic disease, although there may be rare squamous cell carcinoma or granulomata as in vasculitic disease such as granulomatosis with polyangitis (formerly Wegener's granulomatosus—section 10.3.5).

13.2.2 The larynx

Sitting slightly anterior (in front) and inferior (below) the pharynx, the upper part of the larynx is bounded by the epiglottis. The organ itself links to the trachea and is responsible for sounds and speech. Formed from cartilage, muscles, and soft tissues, as air is forced out of the lungs, it is formed into noises by the combined action of the cartilage and muscles (the vocal cords, or folds), which is further influenced by the buccal cavity. These tissues are held in place by the hyoid bone. In adolescent males, the larynx falls and enlarges, resulting in low-frequency sounds.

Acute or chronic inflammation of the larynx (laryngitis) may be caused by infective (microbial) or environmental (smoke, excessive use of the voice) agents. If bacterial, it may be treated with antibiotics and soothing throat solutions/lozenges. The most common form is 'strep throat' (typically group A β-haemolytic *Streptococcus pyogenes*) and there may also be infection and inflammation of nearby tissue, such as the tonsils/tonsillitis and pharynx/pharyngitis.

13.3 The trachea

The airway passage continues from the pharynx to the larynx, which merges with the trachea. This tube, 10–12 cm long and 2–3 cm in diameter, lying anterior to (in front of) the oesophagus, is reinforced by ~18 cartilage rings fixed by connective tissues. At its distal end it merges with the bronchus. The internal layer is composed of mucus-secreting ciliary epithelia whose function is to collect and pass particulate material into the oesophagus. Inflammation of the trachea (tracheitis) is uncommon, and often presents alongside inflammation of those structures above and below. If bacterial, treatment is with antibiotics, but if severe, the trachea may narrow, and if inhibiting free gas exchange, will require steroids and possibly intubation. Cancer of the trachea killed 11 people in England and Wales in 2021.

13.4 The bronchi

At its lowest point (the carina) the trachea divides to form the left and right bronchi, which continue the transport of air to the lungs, where they repeatedly branch to deliver air to the alveoli for gas exchange. The bronchi retain almost all of the anatomy and histology of the trachea (i.e. cartilage rings and ciliated epithelia). On extending into the lungs, each bronchus repeatedly subdivides into increasingly narrow bronchioles with less cartilaginous support, which terminate with alveoli. In 2021 cancer of the main bronchus killed 15 people, the problem with statistics being that cancer in either the left or right bronchus is often considered to be part of cancer of the lung itself, as very little of the bronchus is outside the pleural lining that forms the boundary of the lung.

Inflammation of the bronchus—bronchitis—may be viral or bacterial, the latter being treated with antibiotics, such as amoxicillin 250 mg tds. The leading diagnostic differential from an URTI is chest pain (perhaps on deep inspiration) and an initially dry cough that may well deteriorate to producing yellow or even green globules of mucus (sputum) from which microorganisms may be cultured. There may also be wheezing, and chest sounds that can be determined with a stethoscope, such as crackles.

Should bronchitis be prolonged, perhaps due to a persist infection, it may develop into bronchiectasis, where the bronchi and/or bronchioles are permanently inflamed, enlarged, and dilated in response to organisms such as *P. aeruginosa* and *H. influenzae*. This results in the production of mucus, which is generally brought up as discoloured sputum, from which causative agents may be cultured. Treatment is with antibiotics (such as azithromycin 500–1,000 mg od, which may be in an aerosol), steroids (also oral or inhaled), and training the patient to expel their sputum by postural drainage. Bronchiolitis (inflammation of the bronchioles) is a very common condition of around a third of infants in their first year (NICE NG9), and in adults develops slowly with the same pathology as bronchitis. In England and Wales, acute bronchitis killed 96 people in 2020 and 55 in 2021.

Key points

The upper respiratory system is a series of tubes held open by connective tissue that act as conduits for the movement of air in and out of the lung parenchyma.

SELF-CHECK 13.1

Describe how the airways are adapted to fulfil their function.

13.5 The lungs

The lungs are paired organs located within the thoracic cavity, each protected with a pleural membrane of two parts—the parietal and visceral membranes. Between these two membranes is the pleural cavity, which has a small amount of lubricating fluid. The pulmonary artery takes deoxygenated blood from the right ventricle to both lungs via left and right pulmonary arteries. The terminal ends of the bronchioles are not supported by cartilage rings, which reduce to an alveolar duct and then to one of the ~300 million alveoli where gas exchange takes place (Figure 13.1) and which are true lung tissue.

The bronchi and bronchioles are lined with epithelial cilia and goblet cells that serve to clear the alveoli and airways of debris, dead cells, and inhaled particles by passing them up to the

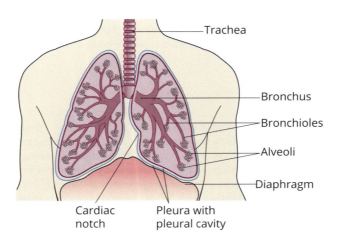

FIGURE 13.1

Anatomy of the trachea, bronchi, and lungs.

Adapted from Ford, *Medical Microbiology*, third edition, Oxford University Press, 2019.

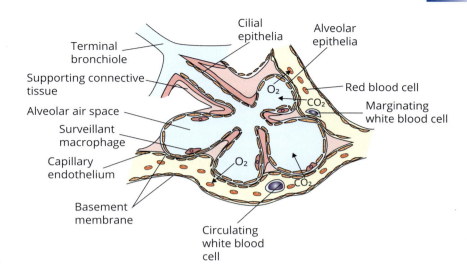

FIGURE 13.2
Microanatomy of the alveolus.
An alveolus consists of a group
of air sacs intertwined by blood
vessels and lymphatics, supported
by connective tissues. Gases cross
epithelial and endothelial cells
between the air spaces and the
circulation, where macrophages
provide immunosurveillant duties.

oesophagus and so into the digestive system. The alveoli also have macrophages for additional 'cleansing' activity (Figure 13.2), whilst pulmonary, bronchial, and tracheal lymph nodes provide further protection.

Lung cancer is described in section 4.3, mesothelioma of the pleural membranes in section 5.12. The pathophysiology and clinical aspects of autoimmune and inflammatory diseases are discussed in Chapters 9 and 10 respectively, thrombosis in Chapter 11. The most burdensome diseases by far, in terms of frequency within the population and severity, are COPD (including bronchitis), asthma, and pneumonia.

13.5.1 Chronic obstructive pulmonary disease (COPD)

COPD is a clinical umbrella term that includes conditions bringing about destruction of lung tissues that causes restricted air flow. It is the leading cause of fatal chronic lower respiratory system disease (89.4%), which is itself the leading cause of all deaths due to respiratory disease (47.7%), although these criteria fail to include respiratory death due to COVID-19. The key aetiological feature is poor gas exchange and so a degree of hypoxia and tissue ischaemia that can have further pathophysiological consequences. However, there is overlap with other respiratory system disease, such as asthma (1,146 deaths in England and Wales in 2021), bronchitis (35 deaths), and pneumonia (16,197 deaths). Estimates of the prevalence of adult COPD vary. Globally, it is reported as 4.9%, but other studies cite ~9% of the population. Leading risk factors include inhaled particulate matter (tobacco smoke, industrial dust) and infections, principally pneumonia.

Pathology

Although the pathophysiology can be complex, it can be reduced to features based on anatomy. Injury to the walls of the bronchi and bronchioles, with increased secretion of mucus by goblet cells, inflammation, narrowing, and fibrosis, results in reduced delivery of air to the alveoli. Damage to the alveoli themselves, with increased air spaces (bullae), leads to a retention of gases and a reduction in the total surface area available for gas exchange, and so emphysema. These factors overlap with other forms of pathology, such as scarring and fibrosis, inflammatory damage, increased alveolar mucus inhibiting gas diffusion, and loss of recoil elasticity in the ventilation cycle.

There are few established gene mutations associated with COPD, so it is widely seen as the end point of perhaps decades of insidious and multifactorial disease processes. However, there is a growing literature on miRNAs, such as increased miR-15b (targeting *SMAD7*, coding for an inhibitor of transforming growth factor β) in lung tissues, and increased serum miR-21 that correlates inversely with lung function. A further comparison with asthma is that eosinophils are components

Cross reference

Mesothelioma is discussed in detail in section 5.12 of Chapter 5.

of the pulmonary inflammation in both diseases, but in COPD there is no basophil activation. The effect of cigarette smoking on the risk of cardiovascular disease falls over time, eventually back to that of a non-smoker, but no such evidence exists with regard to lung disease. Some have speculated that a limited period of cigarette smoking or industrial exposure nevertheless causes irreversible damage that may not be clinically apparent for decades (as in exposure to asbestos and mesothelioma).

Although the disease is generally chronic, acute exacerbations present around 1–3 times a year, and are predicted by a rising CRP and white cell count (with an eosinophilia). Exacerbations can carry a mortality rate of ~25% in those with the most severe disease, the terminal events including myocardial infarction and pulmonary embolism. Leading causes (50–70%) are viral (e.g. respiratory syncytial virus, influenza virus, and rhinovirus) and/or bacterial (e.g. *S. pneumoniae*) infections, and a further 10% relate to environmental factors.

Presentation

As in many preliminary consultations, the practitioner will be faced with any number of semi-specific signs and symptoms of varying severity, so that a good medical history (personal and family) for the patient is essential in determining possible precipitants. NICE NG115 considers diagnostic factors and several other relevant features (Table 13.2). A rare and semi-specific sign is clubbed fingers, an extreme example of which is shown in Figure 13.3.

TABLE 13.2   Features to be considered in the diagnosis of COPD

Major diagnostic features	Allied features
Age over 35	Weight loss
Smoking or history of smoking	Reduced exercise tolerance
Exertional breathlessness	Waking at night with breathlessness
Chronic cough	Ankle swelling, fatigue
Regular sputum production	Occupational hazards
Frequent winter 'bronchitis'	Chest pain*
Wheeze	Haemoptysis*

* Uncommon in COPD, implying alternative diagnosis.

Source: NICE NG115.

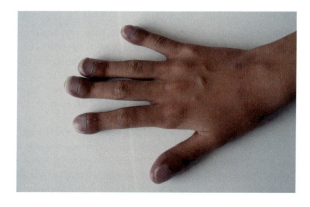

FIGURE 13.3
Finger clubbing.

Shutterstock.

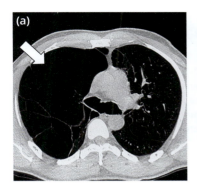 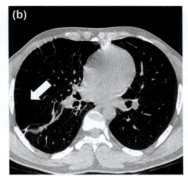 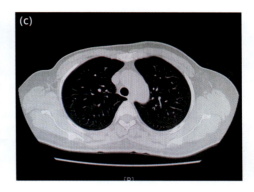

FIGURE 13.4

CT imaging for emphysema. High-resolution thoracic CT showing (a) extensive and (b) moderate emphysema (dense black space, arrowed). No such air spaces are present in (c), so emphysema can be excluded.

(a) and (b) © 2022 Stanford Medicine, Department of Radiology; (c) Shutterstock.

Diagnosis

A chest X-ray will be required as a screening tool to exclude other disease, such as cancer, cardiomegaly, pulmonary oedema, pneumonia, cysts, pneumatosis (localized pockets of air larger than alveoli), or a pneumothorax (air within the pleural layers). A common X-ray finding in COPD is a flattened diaphragm. Emphysema can be difficult to determine by X-ray, and for this and other features the preferred methods are complex imaging with CT and MRI (Figure 13.4). If there is still diagnostic uncertainty, measurement of the transfer factor for carbon monoxide may be helpful as it is closely linked to the degree of emphysema. This procedure compares carbon monoxide in inspired and expired air, a low transfer factor being associated with lung disease. An additional useful test is pulse oximetry to determine oxygen saturation (of haemoglobin), easily measured with a non-invasive probe attached to a finger, which ideally is close to 100%. If <92%, arterial blood gases should be assessed.

SELF-CHECK 13.2

What are the major presenting features of COPD?

Further differentiation of the different forms of lung disease can be made by referral to the patient's history and their age. In advanced COPD there may well be right-sided heart failure resulting from pulmonary hypertension as the right ventricle attempts to overcome the pulmonary congestion (cor pulmonale, also described as pulmonary heart disease). This, alongside poor alveolar gas exchange, could lead to generalized hypoxia (as indicated clinically with **cyanosis**), to which the kidneys may respond with increased erythropoietin, leading to an erythrocytosis (section 11.4.5), diagnosable with a full blood count.

The primary direct investigation is lung function testing, principally spirometry (see Box 13.1), where the test metric is the combination of the forced expiratory volume in 1 second (FEV$_1$, a single strong puff), and the forced volume vital capacity (FVC, as much of the lung volume as is possible, over perhaps 4–6 seconds), these being surrogates for breathing capability. These may be 4 litres and 5 litres respectively in healthy adults, giving a FEV$_1$/FVC ratio between the two of 0.8 (generally accepted reference range 0.75–0.85). A typical result in COPD would be 1.8 litres and 3.2 litres respectively, and so a ratio of 0.56, markedly less than the cut-off point of 0.75, although there may be a need for adjustment for age as the ratio falls in health with advancing years, according to sex, by body mass index, and race/ethnicity. There are many other measures of lung function, such as the tidal volume.

BOX 13.1 Spirometry

Use of this tool extends far beyond asthma and COPD: it is an essential technique in other lung disease, such as cystic fibrosis (where it can be used to help management, especially of exacerbations and the effects of physically clearing the lung of mucus); non-cystic fibrosis bronchiectasis; idiopathic pulmonary fibrosis; the pulmonary disease of rheumatoid arthritis; and systemic sclerosis, sarcoid, tuberculosis, and steroid-induced pulmonary fibrosis—to name but a few.

The investigations described above may be regarded as baseline. A more practical test (described by NICE NG115 as a reversibility test) is to challenge the airways with a stimulant, an aerosol of an adrenergic β_2-receptor agonist bronchodilator such as 400 μg salbutamol, or an anti-muscarinic receptor antagonist such as 40–80 μg ipratropium to relax smooth muscle cells and so dilate the airways. Ten to 15 minutes after aerosol inhalation, the spirometry is repeated: intervention should bronchodilate, relieve symptoms, and improve lung function tests, and if so, this completes the diagnosis of COPD. Bronchoalveolar lavage (examined in the laboratory for cells, bacteria, and other matter) is not routinely required but can be helpful in difficult cases with protracted disease.

Key points

An anatomical view of COPD is of disease of the airways (bronchi and bronchioles) and of gas exchanges (at the alveoli). Either can exist independently or they can be present at the same time.

Nitric oxide (NO) is a potent vasodilator, acting on smooth muscle cells in airways and blood vessels, and can be measured in exhaled breath. In COPD, the fraction of exhaled NO (hence FeNO) is reduced in smoking and more severe disease, is increased in asthma, COPD exacerbations, and in pulmonary inflammation (>50 parts per billion, reference range <25), and correlates with functional indices of respiration. Thus, monitoring FeNO may be a useful alternative to spirometry in managing COPD. NICE NG12 offers advice on devices to measure FeNO.

Although it was developed for pneumonia, the severity of COPD may be assessed by the CURB-65 score, where each of <u>C</u>onfusion of new onset, blood <u>U</u>rea >7 mmol/l, <u>R</u>espiratory rate >30/minute, systolic or diastolic <u>B</u>lood pressure <90 or <60 respectively, <u>A</u>ge ≥65 each score 1 point. The higher the score, the greater the likelihood of death.

Treatment

Lung tissue damage is irreversible, so that treatment is of symptoms and risk factors, severity of the airflow limitation following deterioration of the FEV_1 (mild, moderate, severe, very severe). The goals of treatment are relief of symptoms, improvement of exercise tolerance and health status, avoidance of hospital admission, preventing disease progression and exacerbations, and reduction in mortality. The dominant method for relief of symptoms (and other lung disease) is enlarging the diameter of the airways with a bronchodilator, the active drug often delivered as an aerosol from an inhaler device.

Bronchodilation can be achieved in a small number of ways by drugs of three different classes. Smooth muscle tone is influenced in part by stimulation of β_2-adrenergic receptors (that bind catecholamines such as adrenaline and noradrenaline) and by muscarinic/cholinergic receptors (that bind acetylcholine and muscarine). Antagonistic blockers for these receptors will therefore have physiological repercussions (as we have seen in Chapters 7 and 8 with the effects of beta-blockers on arterial smooth muscle cells to counter hypertension), which in this instance result in bronchial smooth muscle cell relaxation and so opening of the airways. There may also be an effect on goblet cells, with reduced secretion of mucus.

There are four types of agent: a short-acting (4–6 hours) β_2-receptor agonist (SABA), a short-acting muscarinic receptor antagonist (SAMA), a long-acting (~24 hours) β_2-receptor agonist (LABA), and a long-acting muscarinic antagonist (LAMA). An additional treatment is with an inhaled corticosteroid (ICS: as an adjunct only) which will, over several days or more, reduce inflammation, and so address airway constriction.

For most patients, long-term treatment is with a combined LABA and an ICS, a combination that is as effective as monotherapy with a LAMA in terms of lung inflammation and the number of acute exacerbations and deaths. However, several combinations of an ICS and a SABA, a LAMA, or a LABA are available (Table 13.8). Should these fail in the most intractable cases, triple therapy of a LABA, a LAMA, and an ICS are available, and these may be delivered by a nebulizer, especially in exacerbations (NICE ES17 and ES18). However, effectiveness of bronchodilators and other therapeutics should not be assessed by spirometry alone: clinical features must be included. Fenoterol and salbutamol are also available as a tablet and a syrup: oral steroids alone are not recommended, but in acute exacerbation 30 mg prednisolone for 5 days may be considered, followed by staggered reduction. Other combinations are also available, and other oral drugs include aminophylline, theophyline, and the phosphodiesterase-4 inhibitor roflumilast (which may be an adjunct to a long-acting bronchodilator). NICE TA461 recommends that it be used in those with severe COPD with a post-bronchodilator FEV_1 of <50% of their predicted normal value, and with two or more exacerbations in the past year despite triple therapy.

In those prone to exacerbations, a year-long treatment with antibiotics such as azithromycin (250 mg od or 500 mg three times a week) or erythromycin (500 mg bd) reduces the risk of further exacerbation, whilst NICE NG114 offers guidance on oral or intravenous microbial prescribing in an acute exacerbation, such as amoxicillin 500 mg tds for 3 days; doxycycline 200 mg on day one, then 100 mg od for the next 4 days; or clarithromycin 500 mg bd for 5 days. Alternatives include co-amoxiclav, cotrimoxazole, and levofloxacin. In selected patients, antioxidant mucolytics (based on cysteine) may be useful.

Most patients with COPD carry co-morbidities which should be addressed. The principal among these are hyperlipidaemia (46%), hypertension (43%), anxiety (38%), benign prostatic hyperplasia (30%), ischaemic heart disease (28%), sarcopenia (27%), obesity (24%), and pulmonary hypertension (22%), although the frequency does not reflect their impact on mortality. Unsurprisingly, as in many other chronic diseases, the number of co-morbidities predicts hospital admission (Table 13.3) and mortality.

Influenza and pneumococcal vaccination reduces the risk of LRTIs and so of COPD and should therefore be kept up to date. Those with most severe disease, and resting hypoxaemia (PaO$_2$ ≤7.3

TABLE 13.3 Factors predicting hospital admission for COPD

Arterial pH <7.35, arterial PaO$_2$ <7 pKa
Deterioration in existing symptoms
Development of new clinical signs (cyanosis, angina, oedema, hypertension, atrial fibrillation)
Exacerbations refractive to pharmacotherapy
Increasing age/frailty/polypharmacy
Increasing frequency of exacerbations
SaO$_2$ <90%, high CURB-65 score
Severe underlying COPD with falling FEV_1
Inability to cope at home
Worsening co-morbidities (falling left-ventricle ejection fraction, rising creatinine, falling haemoglobin)

BOX 13.2 *Acidosis and alkalosis*

The pH of the blood (7.35–7.45) is determined by a number of features, such as the ability of the kidneys to excrete hydrogen ions and the mass of the bicarbonate ions in the blood (ideally 22–26 mmol/l) (section 14.1.1). At the alveoli, carbon dioxide is reformed from bicarbonate and is excreted in exhaled breath. Lung disease can cause a respiratory acidosis due to the retention of CO_2 (hypercapnia: perhaps due to hypoventilation), and a respiratory alkalosis due to hyperventilation and excess removal of CO_2 (resulting in hypocapnia).

kPa (55 mg Hg) or oxygen saturation ≤88%), will benefit from continuous oxygen therapy, and this (and other factors) may also lead to hospital admission. Levels of oxygen and carbon dioxide are important as they have roles in the pH of the blood, and if abnormal can lead to acidosis and alkalosis (Box 13.2).

Surgical options include lung volume reduction surgery and bullectomy for severe bullous emphysema, the latter associated with decreased dyspnoea and improved lung function and exercise tolerance (NICE ILG114). Given the mean age of most patients with COPD, and the almost inevitable co-morbidities, transplantation is not usually an option, but if appropriate, it is likely to be of both lungs (section 10.4.3). Wherever possible, healthy living with a good diet and an appropriate level of exercise is recommended and tailored by dietitians (aiming for a BMI ≤25 kg/m²) and physiotherapists. Given the high burden of disease in COPD, many secondary care centres run pulmonary rehabilitation courses that parallel those of cardiovascular rehabilitation. Such rehabilitation reduces anxiety and depression, whilst other treatments include tricyclic antidepressants and anxiolytics, although care must be taken (especially in hypercapnia) as these drugs may depress the respiratory centre in the brainstem. Monoclonal antibodies to IL-5 may be helpful in eosinophilic COPD (which may be up to 40% of cases), as they are in eosinophilic asthma.

SELF-CHECK 13.3

What is the difference in mode of action between a LABA and a LAMA, and what does this mean for symptom relief?

13.5.2 Asthma

Various aspects of the pathology, presentation, diagnosis, and management of asthma overlap with, and/or are pertinent to, those of COPD, but there are major differences. Indeed, the asthma–COPD overlap is an established condition. A meta-analysis estimated the global prevalence of asthma alone to be 6.2%, that of COPD alone to be 4.9%, and the overlap to have a frequency of 2%, although these figures vary widely in different populations (asthma in Vietnam and China <2%, Western Europe >10%). Asthma is a heterogenous and genetically complex major public health issue commonly starting before the age of five and with a variable course over the decades that follow, although it may also present in adolescents and adults. Childhood asthma is more common in boys, but during puberty the burden of disease transfers to girls, a trend that continues into adult life.

Taking patients aged >50 whose primary diagnosis is asthma, 26.5% also have COPD, whilst 29.6% of those primarily diagnosed with COPD also have asthma. These data translate to an increasingly poor survival plot (Figure 13.5). The hazard ratio for mortality in asthma alone (group B) compared to neither asthma nor COPD (group A) is 1.04 and is not significant. However, for COPD (group C) the hazard ratio is 1.44, and for asthma + COPD (group D) the figure is 1.83, both sets being significant.

The Copenhagen City Heart Study of over 8,000 people found that people with asthma had a survival disadvantage of 3.3 years (smokers: 3.8 years, COPD: 10.1 years). Whilst the mortality rate of asthma is very low (1.8% of all respiratory disease deaths), it is nevertheless ranked 16th in the global scores of years lived with disability and 28th in that of causes of burden of disease.

CASE STUDY 13.1 Pneumonia

A 70-year-old retired steel worker is an outpatient at the local chest clinic where he is seen every 6 months for his COPD, which has been getting worse since he gave up cigarettes (20 a day) 12 years ago. He is overweight (BMI 29.5 kg/m^2), rapidly becomes short of breath on exertion, has a cough which occasionally brings up sputum, and suffered a myocardial infarction 7 years ago. His FEV$_1$/FVC ratio is borderline at 0.7. Medications are aspirin 75 mg od, atorvastatin 40 mg od (total cholesterol 3.5 mmol/l), perindopril 8 mg od and bisoprolol 2.5 mg od (holding blood pressure between 135–40 and 80–85), a salbutamol inhaler (delivering 200 µg) for when short of breath, and a GTN spray for occasional anginal chest pain. He was hospitalized for 3 weeks the previous winter with chronic bronchitis.

He requests an emergency appointment with his GP, complaining of increased breathlessness and coughing. The GP finds a temperature of 39.0°C, pulse rate of 90/minute, blood pressure 143/88, and a respiratory rate of 35/minute. Although this gives a CURB65 score (for predicting community-acquired pneumonia) of only 2, with the fever, tachycardia, hypertension, and medical history, the GP calls 999 to take him to the A&E unit of the local hospital. On arrival, the admitting team confirm the GP's findings, but blood tests also report a raised white cell count (of 11.5 × 10^6/l with neutrophilia), serum urea of 7.5 mmol/l (reference range 3.3–6.7), and CRP 50 mg/l (<5). A chest X-ray finds changes consistent with bilateral pneumonia, and bilateral ankle oedema is noted. An ECG is normal, except for tachycardia (pulse rate 95) and evidence of a previous infarction. Blood is taken for blood culture, and the patient is invited to provide a sputum sample, which he does with ease, and which is yellow/light green.

The uraemia increases the CURB65 score to 3, sufficient for admission to ICU, but as no beds are available, he is placed in a medical ward with a dedicated 1/1 nurse, where he is put on a saline drip and started on a diuretic, a low-molecular-weight heparin, and oral antibiotics, and his bisoprolol is increased to 5 mg. Blood gases are satisfactory with oxyhaemoglobin saturation of 97%, but as a precaution he is started on enriched nasal oxygen. He passes the night well, and at next morning's ward round his temperature has fallen to 38.5°C, blood pressure to 138/79, tachycardia to 80/minute, and respiratory rate to 25/minute, and although all previous blood tests are still abnormal, they have moved towards their respective reference range. However, he still complains of chest pain on inspiration, cannot fully inspire, is wheezing, and the registrar notes bilateral basal crackles with her stethoscope. His spouse and son visit after lunch, and as they leave the son has a word with the patient's nurse, expressing concern that his father is not thinking right, has poor memory, and is slow in conversation with long spells of silence, all of which are out of character.

At early evening observations, his nurse finds that the tachycardia has increased to 91/minute, respiratory rate 27/minute, temperature 39°C, but blood pressure is 135/85. He calls the registrar, who orders further blood tests, and the patient is finally moved to ICU. The patient is visibly concerned at the move, is uncooperative, wishes to go home, and attempts to remove his drip, so is given mild sedation. In ICU, blood gases are pO$_2$ 9.5 kPa (reference range 12–14.6), pCO$_2$ 6.5 kPa (reference range 4.7–6.0), and pH 7.2, consistent with respiratory acidosis, whilst oxygen saturation has fallen to 90%. Together, these bring an APACHE IV score (assessing the general state of the patient) of 43, and so an estimated mortality rate of 7.2% and an estimated length of stay of 4.3 days. Bicarbonate is added to the drip, and nasal oxygen concentration increased. Microbiology reports the presence of standard organisms in his sputum and a negative blood culture. Later that evening he becomes more agitated and passes in and out of consciousness. His temperature rises to 39.5°C (causing him to sweat in droplets), as do the tachycardia (95/minute) and the respiratory rate (32/minute, which begins to resemble panting), and oxygen saturation has fallen to 85%. Intravenous antibiotics are started, as results from the blood test taken earlier show a rising neutrophil leucocytosis, ESR, and CRP. The medical team starts to fear respiratory failure, leading them to deliver oxygen in a mask and prepare him for ventilation, but he suffers a cardiac arrest and cannot be resuscitated. The precise reason for the arrest is unclear, but it is likely due to a combination of factors that include poor cardiac oxygenation.

Pathology

The pathophysiology of almost all forms of asthma is understood: a defective type 1 (allergic) hypersensitivity response (section 10.2.1) that overreacts to a number of stimuli (allergens: Table 10.2) that would not normally elicit such a profound response. The leading cell biology aspect of excessive response of the immune system to the allergen focuses on cross-linking IgE antibodies whose Fc sections are locked into the Fc-receptors on circulating basophils and tissue mast cells. Activation of the leukocytes results

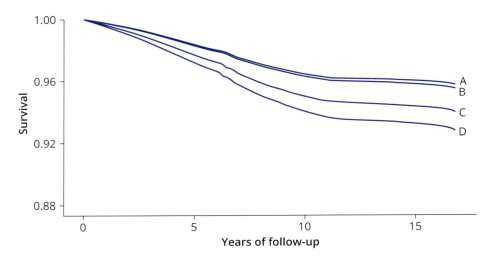

FIGURE 13.5

Survival plot for asthma, COPD, both, or neither. A = Neither asthma nor COPD, B = asthma alone, C = COPD alone, D = asthma plus COPD.

From Diaz-Guzman E, Khosravi M, & Mannino, DM (2011) Asthma, Chronic Obstructive Pulmonary Disease, and Mortality in the US Population, COPD: Journal of Chronic Obstructive Pulmonary Disease, 8: 6; 400–7, DOI: 10.3109/15412555.2011.611200. Reprinted by permission of the publisher (Taylor & Francis Ltd, http://www.tandfonline.com).

in degranulation with the release of histamine, heparin, cytokines, eicosanoids, leukotrienes, and other mediators. There is also evidence of the active participation (i.e. not merely as targets) of epithelial cells, fibroblasts, and endothelial cells, all of which can express IL-33. The inappropriate activation of granulocytes by these mechanisms has a number of consequences, the most clinically relevant being the degranulation of pulmonary and bronchial mast cells (and subsequently of eosinophils and neutrophils) that will cause constriction of the airways and other pathology that lead to the symptoms.

Common allergens more specific for asthma include inhaled house dust mites, storage mites, fungal spores, grass and tree pollen, pet dander (cat, dog, mouse, rabbit, and guinea pig), and food particles (such as from peanuts and tree nuts). Other precipitants include air with high density of smoke (including from cigarettes) and particulate matter (e.g. diesel fumes, ozone, SO_2, etc.). Chemical agents, such as heavy metals, bleaches, dyes, etc., but also aspirin, NSAIDs, and beta-blockers, may also cause a response (and also a COPD involvement). Some commentators refer to the latter group as causing non-allergic asthma, suggesting that 'classical' asthma is only present if an IgE response can be proven, which can be difficult for some stimuli, such as cigarette smoke.

Much of the above is described as environmental, to which others often add low-grade lung infections and the hygiene hypothesis. The latter proposes that children in an over-clean environment fail to be exposed to, and so generate a low level of resistance to, allergens and are thus overwhelmed when that allergen is presented in large amounts later in life, resulting in the overwhelming of the immune system and so the hypersensitivity. This is supported by considerable evidence of the lack of asthma in populations forced to live together in locations where complete cleanliness is difficult to achieve. There is also a developing view that asthma is a different disease when present with obesity (as each one predicts the other) where the excessive weight places greater stress on the respiratory system (as it does in COPD). Overweight (BMI 25–29.9) brings an odds ratio (OR) for asthma of 1.38, whilst for obesity (BMI >30) the OR is 1.92, although this has been challenged as some studies have failed to find a link. Caveats include the likelihood that people who are obese exhibit many of the symptoms of asthma as a result of their excess weight, and that their respiratory cycle is shallow and so the airways are likely to respond by adopting a smaller diameter, leading to breathlessness when exercising.

In many people with asthma, a two-hit model has been proposed, the first being environmental, the second being genetic. Twin concordance is around 50%, heritability being 0.40–0.85. Numerous

TABLE 13.4 Genetics and asthma

Risk loci identified by GWAS	*IL33, ERBB2, ORMDL3/GSDMB, IL1RL1/IL18R1, IL2RB, SMAD3, TSLP*
Loci related to allergen sensitization	*TLR6, C11orf30, STAT6, SLC25A46, IL1RL1, LPP, MYC, IL-2, HLA-B*
Genes methylated by particulate matter and related to asthma	*HLA-DPA1, CCL11, CD40LG, ECP, FCER1A, FCER1G, IL9, IL10, IL13, MBP*
Increased serum miRNAs	miR-16, miR-21, miR-125b, miR-126, miR-145, miR-146a, miR-148a
miRNAs with effects on bronchial smooth muscle cells in asthma	miR-140-3p, miR-708, miR-10, miR-142-3p, miR-25, miR-133

studies have identified risk loci and single-nucleotide polymorphisms (SNPs) with links to asthma (Table 13.4), but despite these associations, their effects are modest and predict <10% of heritability. At least 14 susceptibility genes can be traced to the 17q21 locus, whilst SNPs influencing response to treatment with a short-acting β_2-receptor agonist include those at 9q21 and 10q21, and those with effects on mepolizumab include those at 6q24. One of the strongest links is with an SNP in *ERBB2*, coding for Erb-B2 receptor kinase, which predicts asthma with a probability of 2.2×10^{-30}.

Unsurprisingly, there are many MHC links with asthma, including HLA-DQA1, HLA-DQA2, HLA-DQB1, HLA-DPB1, HLA-C/MICA, and HLA-DRA. As in other diseases, miRNAs may be useful markers of asthma and its pathophysiology, but also potential therapeutics as an animal model of allergic rhinitis was ameliorated by intra-nasal application of miR-133b, an intervention linked to a reduction in levels of IgE, IL-4, IL-5, and TNF-α.

Key points

Although rarely fatal, asthma may nonetheless bring decades of discomfort and morbidity.

Presentation and diagnosis

On the whole, people with asthma may present with the same symptoms as those with COPD (wheezing, coughing, chest tightness, shortness of breath, etc.), and should the cough be productive, it may also contain eosinophils. Accordingly, the primary clinical and demographic differentiators are age and medical history, and the diagnosis and severity are assessed by symptoms and objective tests of lung function. Several scores are available to help predict the likelihood of asthma in children: one factors in age, sex, wheezing frequency, parental history of asthma or allergy, eczema, activity disturbance, shortness of breath, exercise-related wheeze/cough, and aeroallergen-related wheeze/cough. Others use wheezing without colds, blood eosinophila, skin prick test, and specific IgE, although NICE NG80 suggest that skin prick testing, IgE, an eosinophil count, and exercise challenge (to adults and those 17 and over) should not be part of diagnosis. In certain cases, symptoms may be a response to pulmonary tuberculosis (see section 13.5.5).

Once clinical examination has been completed, diagnosis moves on to objective tests (NICE NG80). These include the FeNO test, where a result of ≥40 parts per billion (≥34 in children) is a positive (although this will be influenced by current smoking status) (NICE DG12). Spirometry should be offered (as above in COPD), a FEV_1/FVC <0.7 (or < the lower limit of the reference range) being regarded as positive. In those with a positive result, bronchodilatory reversibility should be determined, with an improvement of ≥12% together with an increase in volume of ≥200 ml being positive. In children aged 5–16, only the % increase is required. If there is still diagnostic uncertainty (e.g. positive spirometry with FeNO ≤39 parts per billion), peak flow variability should be monitored for 2–4 weeks, a

Cross reference

The role of the laboratory in testing for asthma is described in section 9.8.4.

>20% variability being diagnostic. If still uncertain, a direct bronchial challenge test with histamine or methacholine is an option, where a 20% fall in FEV_1 by 8 mg/ml metacholine is regarded as positive.

Treatment

The objective of treatment is to reduce the frequency and severity of the attacks. As a prototypical overactivity disease, the first-line treatment is avoidance of the precipitating allergen (where known or suspected), the second-line one being pharmacotherapy focusing on airways dilation and immunosuppression. Treatment varies markedly with age, as there is concern about the effect of the treatments on the growth of the young, and based on whether someone is newly diagnosed or with unstable disease. NICE NG10 has full details for those aged 5–16; what follows is a summary of treatment for those aged 17 or more, which effectively entails escalating the number and strength of different agents in the face of increasingly poor responses:

- A SABA alone should be offered to newly diagnosed patients.

- In those with more severe symptoms (e.g. waking at night) an inhaled corticosteroid (ICS) may be added.

- If the asthma is uncontrolled by SABA + ICS, a leukotriene receptor antagonist (LTRA) may be added, with review after 4–8 weeks.

- Should this fail to address symptoms, the SABA may be replaced by a long-acting β_2-receptor antagonist (LABA), and a role for the LTRA discussed.

- The next level is MART (= maintenance and reliever therapy), with a fast-acting LABA and a low-dose ICS (≤400 mg) combined that is used in daily maintenance and for the relief of symptoms as required.

- If this is not effective, the ICS could be increased to a moderate dose (400–800 mg), and if still ineffective, to a high dose (≥800 mg) or a long-acting muscarinic receptor antagonist or theophylline could be added.

Those patients whose asthma is severe, poorly controlled by inhaled drugs (bronchodilators and steroids), or potentially linked to a marked eosinophilia may benefit from treatment with transcutaneous monoclonal antibodies. These focus on inflammatory cytokines and their receptors (IL-4 and IL-5, the major source being T_h2 lymphocytes), but also, where the asthma has a proven specific allergic component, on IgE (Table 13.5).

TABLE 13.5 Monoclonal antibody treatments for severe asthma

Agent	NICE TA	Target	Indication for use
Benralizumab	565	IL-5R	Eosinophilia ≥300 cells/µl and ≥4 exacerbations requiring steroids or oral steroids over the past 6 months, OR eosinophilia ≥400 cells/µl and ≥3 exacerbations requiring steroids in the past 12 months
Reslizumab	479	IL-5	Eosinophilia ≥400 cells/µl and ≥3 severe asthma exacerbations requiring systemic steroids in the past 12 months
Mepolizumab	431	IL-5	Eosinophilia ≥300 cells/µl and ≥4 exacerbations requiring systemic steroids in the past 12 months OR has had continuous oral steroids over the previous 6 months
Dupilumab	751	IL-4R	Treatment with medium to high dose of inhaled glucocorticoid plus one or two other agents, e.g. LABA, LTRA, FEV_1 <80% of predicted normal value without bronchodilator, FEV_1 reversibility of 12% and 200 ml, etc.
Omalizumab	138	IgE	Severe persistent confirmed allergic IgE-mediated asthma as an adjunct to full standard therapy (i.e. LABA, high dose ICS, LTRA, theophylline and oral corticosteroids)

TA: technology appraisal.

Those whose asthma is compounded by rhinitis may benefit from antihistamines, but like all drugs in this section, there are numerous side effects such as dizziness and drowsiness that may be additive and so warrant discussion. In emergency situations (e.g. inability to complete a sentence in one breath, tachycardia >100, respiratory rate >30/minute), referral to hospital may be required for intensive treatment (oxygen, oral/IV steroids, ventilation).

How does the presentation of asthma differ from that of COPD?

As with almost all conditions, self-management is promoted wherever possible, with a personal plan and education programme regarding the side effects of steroids. With this in mind the possibility of decreasing the maintenance dose once symptoms have been controlled should be discussed, referring to risks and benefits. Should control of symptoms be suboptimal, adherence to therapy should be confirmed (NICE KTT5). A semi-invasive treatment for severe asthma is bronchial thermoplasty: the procedure reduces the mass of smooth muscles lining the airways, thereby reducing their ability to constrict (NICE IPG635).

Other NICE documents include NG149 on indoor air quality at home (ventilation, insulation, heating, etc.), CG183 on the diagnosis and management of drug allergy (noting exacerbations in asthma such as those due to NSAIDs), NG120 on prescribing guidelines for acute cough, and NG9 on bronchiolitis in children.

What are the targets for monoclonal antibody-based therapies, and why are these appropriate?

CASE STUDY 13.2 Asthma

During the spring of her first year at university, an undergraduate student experiences increasingly severe and longer-lasting breathing difficulties after going to the gym, and visits the campus health centre. She and her parents have no major health issues and no outstanding medical history, but she started on the oral contraceptive pill several months previously. Her body mass index is 22.5 kg/m², and she has never smoked, nor lived with a smoker, and her partner does not smoke.

The health centre practitioner discusses the possibility of asthma, orders a full blood count and IgE levels, and advises the young woman to keep a diary of events that may precipitate an attack that they will review in a fortnight. On her return, blood tests are normal, and the patient is unable to describe any major events linked to the breathing issues, except that they all occur shortly after exercise. The practitioner offers the student a salbutamol inhaler (delivering 100 µg per puff), and they agree to meet again to review its effect, at which point both patient and practitioner are satisfied, the latter being discharged with repeat prescriptions.

The student graduates, and a few years later moves to central London, but after only a few weeks notes her symptoms are getting worse, in terms of severity (with coughing),

duration, and frequency. Following NICE NG40, her GP offers an inhaler with a long-acting bronchodilator combined with a low-dose corticosteroid (200 µg, up to four times a day), which the patient finds effective, and over the months, symptoms become less severe and less frequent. She continues her medications during a pregnancy, finding the symptoms to be further reduced. However, 6 weeks post-partum the symptoms return to their previous intensity, and these are often accompanied by a runny nose, for which she takes over-the-counter antihistamines. She also notes a marked reduction in her exercise capacity. Following a particularly exhausting bout of breathlessness and coughing one evening, she returns to her GP, who refers her to the chest clinic at the local hospital.

The consultant chest physician orders spirometry and blood tests, and a referral to the allergy clinic. The former finds a baseline FEV_1/FVC of 0.65, with improvement to 0.78 after use of a bronchodilator, whilst the blood tests show her eosinophil count to be normal at $0.35 \times 10^9/l$ (reference range 0.02–0.5) and the IgE to be marginally raised at 90 KU/l (reference range 0–81). The allergy clinic reports negative skin prick tests for common allergens. The consultant recommends that she try two inhalers—one providing a maintenance dose of a long-acting

(Continued)

CASE STUDY 13.2 *Continued*

β2-receptor antagonist with a steroid, and a second of a short-acting muscarinic receptor antagonist to be used to give (it is hoped) immediate relief of acute symptoms, and which may be used in advance of the presumption of symptoms (such as before exercise). This combination works well for a number of years, but when her job begins to require business flights to the Indian subcontinent, with stays of 10–14 days, the patient starts to bring up phlegm during prolonged bouts of coughing. Back in the UK, she returns to her consultant for further tests, and brings samples of sputum, which reveal 40–60% eosinophils, 20–30% macrophages, and the remainder neutrophils. The spirometry tests have also deteriorated, and the full blood count finds an eosinophilia (1 ×10^9/l). The consultant suggests adding oral

prednisolone 5 mg od, but after 3 months this fails to control the symptoms, so it is increased to 10 mg od.

After a year, both consider the strategy to be unsuccessful, with persistent symptoms, an eosinophilia, and >50% of sputum leukocytes being eosinophils. The patient begins a course of the anti-IL-5 monoclonal antibody mepolizumab, 100 mg delivered subcutaneously once every 4 weeks. The patient suffers headaches and some nasal congestion in the initial months, and gains a little weight, but these subside as do the symptoms and eosinophils in the blood and sputum. After 6 months she tapers down the oral prednisolone, and at the annual review, the consultant and patient are happy and consider reducing the use of inhalers.

Asthma and COPD

We have already noted several links and overlaps in clinical and pathophysiological features between asthma and COPD (summarized in Table 13.6). Respirometry is an important tool in both conditions: typical data are shown in Table 13.7. This similarity extends to treatments, with several being applicable to both groups. Accordingly, the practitioner (in either condition) is faced with a range of potential agents as monotherapy and combinations (Table 13.8).

Key points

Asthma and COPD are important in the diagnosis and management of COVID-19 (section 13.6).

TABLE 13.6 Comparison of COPD and asthma

	COPD	Asthma
Onset	Midlife/latelife	Early in life (often childhood)
Symptoms	Generally constant from day to day, worse as the day progresses	Vary widely from day to day, worse in early morning/night
Pathophysiology	Chronic inflammation	Autoimmune type 1 hypersensitivity
Smoking or ex-smoker	Nearly all	Possibly
Chronic productive cough	Common	Uncommon
Breathlessness	Persistent and progressive	Variable
Symptoms under age 35	Rare	Common
Genetic links	None established	Numerous
Percentage of respiratory system deaths	38.9	1.8

Many of the features of COPD are shared with idiopathic pulmonary fibrosis.

TABLE 13.7 Respirometry in asthma and COPD

Characteristics	Controls (n = 220)	Asthma (n = 150)	COPD (n = 194)
Smokers	51.3%	15.0%	89.2%
Pre FVC (L)	2.9 (0.8)	2.3 (0.8)	2.1 (0.7)
Post FVC (L)	2.9 (0.8)	2.4 (0.9)	2.2 (0.7)
Pre FEV1 (L)	2.4 (0.7)	1.5 (0.5)	1.2 (0.5)
Post FEV1(L)	2.5 (0.7)	1.8 (0.4)	1.3 (0.5)
Post FEV1/FVC (%)	95.1 (9.6)	71.4 (10.6)	56.0 (8.9)

Data are mean (SD), groups matched for age and BMI. Pre/Post refers to use of bronchodilators. All differences statisically significant ($p < 0.05$). The bronchodilator has little effect on the FVC in the controls, but has improved the low levels in both lung diseases. The FEV1 and post FEV1/FVC ratio follow a trend of health > asthma > COPD. Note also the high % of smokers in the healthy and (more so) in the COPD groups.

From Sahu A, Swaroop S, Kant S, et al. (2021) Signatures for chronic obstructive pulmonary disease (COPD) and asthma: a comparative genetic analysis, British Journal of Biomedical Science, 78: 4; 177–83, DOI: 10.1080/09674845.2021.1905988.

TABLE 13.8 Named treatment options in asthma and COPD

Class		Agents	COPD (NG115)	Asthma (NG80)
Short-acting bronchodilators (SABAs and SAMAs)	β_2-receptor agonists	Salbutamol	✓ (BNF)	✓ (BNF)
		Fenoterol, levalbuterol	✓	✓
	muscarinic receptor antagonists	Ipratropium	✓ (BNF)	✓ (BNF)
		Oxitropium	✓	✓
Long-acting bronchodilators (LABAs)		Formoterol	✓ (BNF)	✓ (BNF)
		Indacaterol	✓ (BNF)	✓
		Vilanterol	✓	✓
		Olodaterol	✓ (ESNM54)	✓
		Salmeterol	✓ (BNF)	✓ (BNF)
Long-acting muscarinic receptor antagonists (LAMAs)		Aclidinium	✓ (ESNM8)	X
		Glycopyrronium	✓ (ESNM9)	X
		Tiotropium	✓ (BNF)	✓ (ESNM55)
		Umeclidinium	✓ (ESNM52)	X
Inhaled corticosteroids (ICS)		Beclomethasone, Fluticasone, budesonide Mometasone, fluticasone	✓	✓ (TA138) (TA131)

(Continued)

TABLE 13.8 Continued

Class	Agents	COPD (NG115)	Asthma (NG80)
Combined LABA and ICS	Vilanterol/bluticasone	✓ (ESNM21)	✓ (ESNM34)
	Formoterol/beclametasone	✓ (ESNM47)	✓ (ESNM22) (ESNM53)
	Formoterol/fluticasone	✓	✓ (ESNM3)
Combined LABA and LAMA*	Indaceterol/glycopyrronium	✓ (ESNM33)	X
	Formoterol/aclidinium	✓ (ESNM57)	X
	Olodaterol/tiotropium	✓ (ESNM72)	X
	Umeclidinium/vilanterol	✓ (ESNM49)	X

X = not described by either source.

Source: NICE, https://www.nice.org.uk/ (reference in brackets) and the British National Formulary (BNF). Other national guidelines and formularies outside the UK may have different recommendations.

* NICE NG80 recommends the class, but does not name specific agents.

13.5.3 Pneumonia

Pneumonia is an infection of the parenchyma of the lungs, mostly of the alveoli and their ducts, and the principal pathology is COVID-19 (see section 13.6, where COVID-19 pneumonia is discussed). If acquired in the community, pneumonia is described as community acquired (CAP), to distinguish it from that acquired in hospitals (HAP), or more specifically by ventilation (VAP). According to NICE CG191, ~8.5% of those presenting to their GP with symptoms of LRTI are diagnosed with CAP, and of these around a third will be admitted to hospital, where ~9.5% are managed in an intensive care unit (ICU). With a median age of presentation of 75, the disease has a particularly high ICU mortality rate (>30%) as patients are likely to be carrying significant other morbidity, with >50% of ICU deaths being in those aged >84. Indeed, a major risk factor for CAP is other lung disease.

Risk factors for mortality in ICU include aspiration, shock, cachexia, metastases (odds ratio around 3.0), age >75, sepsis, MRSA, lung cancer (~2.4), *Pseudomonas* infection, readmission, poor mobility, dementia, and cardiovascular disease (~1.6). The incidence of CAP varies with age; 310/million aged <19, 800–2,500/million in 18–64-year-olds, but 6,500–32,000/million in those aged 65 or more. Around 0.8% of inpatients are diagnosed with HAP, which is the most common infection in ICU and brings mortality rate of around 50%.

Pathology

The pathological basis of pneumonia is of the alveoli (as opposed to the airways), which become filed with fluid, inflammatory cells, and microbes, so that gas exchange is inhibited. However, if bronchi and bronchioles are involed, it is described as bronchopneumonia. The aetiology may be bacterial, viral, or both (or neither), which has major implications for treatment. The most common bacterial cause are pneumococcus species, in one study reponsible for 41% of bacterial cases, others including *Streptococcus pneumoniae*, *Pseudomonas aeruginosa*, *Acinetobacter baumannii*, *Klebsiella* species, *E. coli*, and *Staphylococcus aureus*, whilst 25% of cases are polymicrobial. Viruses causing pneumonia include rhinoviruses, influenza, coronaviruses, SARS-CoV-2, and respiratory syncytial virus. At the cellular level, these pathogens, and the immune response to them, induce in the alveoli an inflammatory state with increased alveolar endothelial cell permeability, leading to increased tissue fluid infiltration and so poor gas exchange. However, pneumonia may

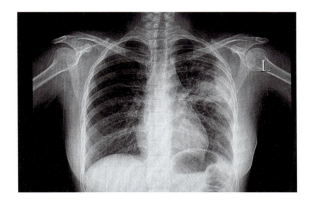

FIGURE 13.6
X-ray of pneumonia. Opacity in the right lower quadrant consistent with pneumonia.

Shutterstock.

also arise from fungal infections and opportunistic parasites, as in those severely immunocompromised. Numerous rare forms of pneumonia without a clear infective agent have been described.

Presentation and diagnosis

The standard symptoms of most respiratory disease, and LRTIs in particular, may be expected in pneumonia, these being breathlessness, chest pain, cough, fever, production of sputum, and wheeze. Based on these symptoms, major differential diagnoses include COPD, asthma (as above), pulmonary embolism (to follow), and acute coronary syndrome (section 8.2.3). However, most cases of pneumonia present in a relatively acute manner (7–10 days), so that asthma and COPD may be excluded, although pneumonia may arise from these chronic conditions.

Should the case strongly point to acute pneumonia, measurement of CRP is important, which may be performed in primary care with a point-of-care device. One study reported a mean CRP of 32 mg/l in bacterial infections compared to 9 mg/l in viral infections, whilst procalcitonin may also be a useful marker (≥25 µg/l in bacterial infections, ≤0.1 µg/l in viral infections), although a false negative rate of 23% has been cited. The white cell count is more likely to be raised in bacterial than in viral infections. NICE CG191 recommends that antibiotics should not be offered if the CRP result is <20 mg/l, that their use should be delayed if the CRP is 20–100 mg/l, but that they should used when >100 mg/l.

As described in the section on COPD, the CURB-65 score is helpful. With a low score (0,1) home care should be considered (risk of mortality <3.2% within 30 days), a score of 2 is suggestive of need for hospital care (13%), whilst score 3 (17%) calls for admission to an ICU. Scores of 4 and 5 bring risk of death to 41.5% and 57% respectively. A chest X-ray is essential and in pneumonia will show new shadowing, although alternative diagnoses are pulmonary oedema and myocardial infarction (Figure 13.6), and to distinguish it from UTRI.

Although the diagnostic accuracy of an X-ray for a viral or bacterial infection is ~66%, different organisms may be linked with certain patterns. Focal non-segmental or lobular pneumonia is often seen in *Streptococcus pneumoniae* infections, whilst multi-focal bronchopneumonia or lobular pneumonia is common in *Staphylococcus aureus*, *Haemophilus influenzae*, and fungal infections. Focal or diffuse interstitial pneumonia may be present in atypical bacterial organisms including *Legionella pneumophila*, *Mycoplasma pneumoniae*, and *Chlamydophila pneumoniae*, and diffuse, patchy, bilateral lung involvement is often seen in viral infections.

Blood and sputum cultures should be obtained in patients at moderate and high CURB-65 risk in an attempt to determine a treatable causative agent. However, these measures can be problematical as they assume the infection has spread to the blood and that any organism in sputum reflects those in the airways or upper respiratory tract. Nevertheless, those at moderate and risk should have urinary antigen tests for *L. pneumophila* and *S. pneumoniae*. A further scoring system can provide improved sensitivity, based on a more comprehensive series of clinical and laboratory features (Table 13.9). A score is constructed from each risk factor, to which is added age (age minus 10 for women), such that <51 points points to risk class I, 51–70 class II (both suggesting outpatient management), 71–90 class III (outpatient management with support), 91–130 class IV, and >130 class V (both inpatient management).

TABLE 13.9 Pneumonia severity index

Risk factor	Score
Nursing home resident, heart failure, stroke, renal failure, tachycardia >125 beats per minute, glucose >13.9 mmol/l, haematocrit <0.30, arterial pO_2 <60 mm Hg, pleural effusion	10
Liver disease, altered mental state, respiratory rate >30 breaths/minute, serum urea >5 mmol/l,	20
Cancer, arterial pH <7.35	30

Modified from Modi and Kovacs (2020) and Rider and Frazee (2018).

Treatment

Once CAP has been diagnosed, treatment should proceed urgently. NICE NG138 offers guidance on antimicrobial prescribing in CAP, dependent on severity. For low-severity CAP, clarithromycin (e.g. 500 mg bd), amoxicillin (500 mg tds), or doxycycline (100 mg od after a load dose of 200 mg) are appropriate, whilst erythromycin, co-amoxiclav, levofloxacin, a cephalosporin, and ceftriaxone may also be considered, generally for 7 to 14 days. Fluoroquinolones should not be offered. For moderate- to high-severity CAP, treatment should start within 4 hours of presentation to hospital. There is no difference between a macrolide (where the antibiotic has the suffix -romycin) compared to penicillins, beta-lactam inhibitors, cephalosporins, and carbapenems, or between a fluoroquinolone and penicillins, beta-lactam inhibitors, cephalosporins, and carbapenems, or between levofloxacin, tigecycline, doxycycline, ofloxacin, erythromycin, moxifloxicin, ertepenum, and ceftriaxone. Doses and frequencies of these drugs are generally at modest levels when used in cases of low severity, but in most instances these will need to be increased in the face of more severe infections, and some used intravenously. Clearly there are multiple options. Other guidelines suggest dual combination antibiotics of amoxicillin and a macrolide in moderate-severity CAP, with a β-lactamase stable β-lactam (such as co-amoxiclav) and a macrolide for those with high-severity CAP, who should also be investigated for *Legionella*.

NICE NG139 offers guidance on antimicrobial prescribing in HAP, describing many of the same antibiotics and others such as vancomycin, teicoplanin, and linezolid, but also details of doses and durations, such as doxycycline 200 mg on day one, then 100 mg daily for a further 4 days. VAP, a major cause of in-ICU pneumonia, may be avoided by providing non-invasive ventilation. Practitioners are likely to refer to their own local guielines.

SELF-CHECK 13.6

How does pneumonia differ from COPD?

13.5.4 Pulmonary embolism (PE)

A crucial aspect of blood disease is venous thromboembolism (VTE), almost all instances of which are deep-vein thrombosis (DVT) or pulmonary embolism (PE). The latter may form within the pulmonary circulation or are formed from clot that has broken away from an existing DVT in the calf, thigh, or groin and come to rest in a pulmonary arteriole (section 11.6.3).

Pathology

Whilst a single large embolus can indeed be life-threatening, a large number of small emboli scattered throughout the pulmonary vascular bed can together impede blood flow and cause pulmonary hypertension. Although atheroma is rarely found in these arteries, abnormal pulmonary haemodynamics resulting from pulmonary emboli can cause right-sided heart failure (section 8.2.1). PEs may develop in ~15% of patients with acute exacerbation of COPD and are present in over a third of COPD lungs at post-mortem. Moderate and severe COPD brings a two-fold risk of PE compared to those with normal spirometry.

Section 11.6.3 considers the aetiology and pathophysiology of overactive coagulation, focusing on both PE and DVT, with Table 11.21 showing the dozens of risk factors for both VTE variants. Briefly, these include obesity, atrial fibrillation, heart failure, cancer, and stroke, in addition to most forms of surgery, especially orthopaedic.

Presentation and diagnosis

PE most likely presents with acute chest pain, dysponea and hyperventilation (each in >50% of cases), tachycardia, anxiety, fever, hypotension, cough, syncope, and leg pain/swelling, for which the major alternative diagnosis is an acute coronary syndrome. An alternative differential diagnosis is pleurisy, inflammation of the pleural membranes often caused by a viral infection, and which presents with a sharp stabbing chest pain whilst breathing. Diagnosis of PE relies on clinical signs and symptoms, the laboratory, and imaging. Unfortunately, each of the former generally have poor sensitivity and specificity, but together they have value, whilst the two scoring systems of Wells and Geneva bring many clinical features together to estimate the probability of PE (Table 13.10). If the probability of PE is intermediate or likely, the next step is to measure D-dimers (perhaps by a point-of-contact device), and if negative, PE is excluded.

Other scoring systems consider oxygen saturation <94% and D-dimer levels ≥500 ng/ml or ≥1,000 ng/ml. Blood gases are generally normal and are rarely requested, and an ECG will be useful as it can exclude serious cardiac pathology. If the probability of PE is high, the final step is complex imaging. Ideally, this would be computed tomography pulmonary angiography, but other techniques include pulmonary angiography, magnetic resonance angiography, and ventilation/perfusion (V/Q) scanning.

A VQ scan consists of a comparison of the lung spaces after the inhalation of 25–30 MBq of xenon-133, krypton-81, or technetium-99, with the pattern of the pulmonary circulation when perfused with 100–120 MBq of technetium 99 radiolabelled albumin given intravenously. In Figure 13.7, the

TABLE 13.10 Diagnosis of PE

System	Signs/symptoms	Score for PE probability
Geneva	Age >65 years, previous DVT/PE, surgery or fracture within a month, active malignancy, unilateral lower-limb pain, **haemoptysis**, heart rate 75–94 beats per minute, and pain on lower limb deep-vein palpation and unilateral oedema all score 1 point: heart rate >95 scores 2 points	0–1: low 2–4: intermediate ≥5: high 0–2: PE Unlikely ≥3: PE Likely.
Wells	Previous DVT/PE, surgery or immobilization in the past 4 weeks, cancer, haemoptysis, heart rate >100, clinical signs of DVT, and an alternative diagnosis less likely than PE all score 1 point.	0–1: PE unlikely ≥2: PE likely

PE = pulmonary embolism, DVT = deep-vein thrombosis

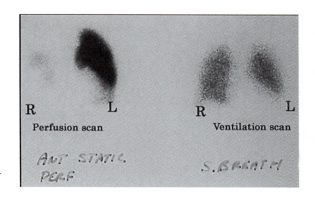

FIGURE 13.7

A positive VQ scan. Left: image formed by the radioactive albumin in the circulation of the left lung only. Right: image formed by radioactive gas in both lungs.

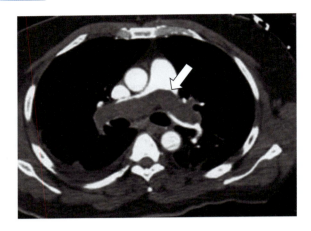

FIGURE 13.8

Pulmonary embolism on CT with pulmonary angiography. A large inverted V- shaped PE (also described as 'saddle') is arrowed, and which straddles the left and right pulmonary arteries.

Samra SR, Gomaa A, & Shaalan A, Assessment of acute pulmonary embolism outcome in hospital through Tricuspid Annular Plane Systolic Excursion versus Pulmonary Embolism Severity Index score, Egypt J Chest Dis Tuberc, 2017: Vol 66(4); , 663–9. http://dx.doi.org/10.1016/j.ejcdt.2017.06.001.

ventilation scan on the right shows radioactivity in the air spaces of both lungs. However, the plot on the left shows radioactivity only over the left lung but none over the right lung. The implication is that a PE is preventing blood carrying radiolabelled albumin from entering the circulation of the right lung. In this instance the PE must be very large, and attempts would have been made to remove the blockage.

A V/Q scan has practical value in demonstrating lack of blood flow, both CT and MRI scanning can provide a visual demonstration of a clot, whilst colour Doppler ultrasound can show changes in blood flow. Further diagnostic tools are angiography and a combination of CT with pulmonary angiography (Figure 13.8).

Treatment

If life-threatening, a PE may be removed by systemic or (preferably) catheter-directed ultrasound-guided thrombolysis with streptokinase, urokinase, tenecteplase, or alteplase. However, there are a number of contraindications, such as structural intracranial disease and recent ischaemic stroke, and all thrombolytics carry a risk of haemorrhage. Alternatives are the use of extraction catheters to capture and remove the embolus, and surgical embolectomy. The PE shown in the case described in Figure 13.9 was removed by open-chest surgery.

Further treatment of PE is to prevent its recurrence. Wherever possible, this is by addressing modifiable risk factors (smoking, obesity, atrial fibrillation, etc.), but almost all cases will be prescribed oral anticoagulation (section 11.6.3). This may be with a vitamin K antagonist (e.g. warfarin: Table 11.23) or a direct-acting oral anticoagulant (dabigatran, rivaroxaban, apixaban, or edoxaban: Table 11.24) as recommended by NICE CG144 and KTT14 (which also refers to the technology appraisal for each DOAC).

SELF-CHECK 13.7

What is the most important differential diagnosis for pulmonary embolism, and how can it be excluded?

Key points

Both forms of venous thromboembolism (DVT, PE) are a major risk in several conditions and after certain forms of surgery, so that anticoagulation is mandatory.

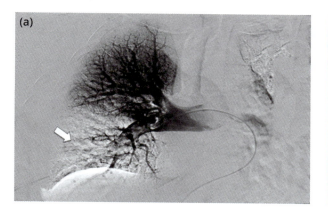

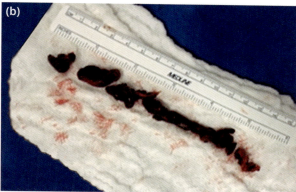

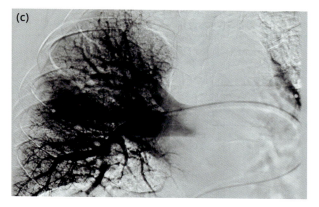

FIGURE 13.9

(a) Pre-intervention angiogram. The arrow indicates the region where the contrast medium fails to perfuse the pulmonary circulation, and so the likely location of the embolism. (b) Retrieved emboli. (c) Post-intervention angiogram showing full restoration of the pulmonary circulation.

© 2022, The Trustees of the University of Pennsylvania.

13.5.5 Other pulmonary disease

α_1-antitrypsin deficiency

The only major metabolic/genetic disease in lung pathology is deficiency of α_1-antitrypsin (A1AT). A product of *SERPINA1* at 14q32.13, this general protease inhibitor (not only of trypsin) protects tissues from excessive enzyme activity. Although there are numerous error-causing SNPs, the most common is present in its heterozgous form in ~3.5% of European-descent populations, and in ~0.04% as the homozygous form. In the lung A1AT inhibits elastase and other enzymes following granulocyte degranulation. Thus loss of function mutations in *SERPINA1* lead to low plasma A1AT and so unfettered protease activity that in the lung leads to tissue digestion, inflammation, and leukocyte degranulation, and ultimately emphysema and COPD. Liver pathology is also likely, this organ being the source of the molecule. Replacement therapy is not recommendend (NICE NG115).

Acute respiratory distress syndrome (ARDS)

This potentially life-threatening condition is characterized by the sudden onset of severe pulmonary inflammation and oedema leading to compromised oxygenation that develops in ~550/million/year. It is a major pathology in COVID-19 (discussed below). In its most severe manifestation,

ARDS is described as respiratory failure, and is a final common pathway of many pulmonary injuries. Diagnosis and staging rely on the ratio between arterial partial pressure of oxygen (PaO_2) and the fraction of inspired oxygen (FiO_2). Although heterologous, the pathophysiology has much in common with COPD and pneumonia, from which it may develop (and from which it must be distinguished). Triggers include viral (particularly coronaviruses—see section 13.6) and bacterial infection, malignancy, aspiration (such as vomitus), burns, smoke inhalation, blood transfusion, pancreatitis, sepsis, and trauma (including surgery). There is no specific cure; treatment (most often in ICU, where mortality is commonly >50%) addresses symptoms and aetiology, whilst poor oxygenation may require ventilation and extra-corporeal oxygenation (NICE IPG391) and removal of carbon dioxide (IPG564). Use of corticosteroids can reduce the duration of ventilation and reduce length of stay, probably because of anti-inflammatory actions.

Tuberculosis

The lungs are the principal site of infection by *M. tuberculosis*, causing tuberculosis (TB), a major extra-pulmonary form causing meningitis (treatable with dexamethasone (oral or intravenous: NICE NG33) or prednisolone), although bone TB is also important. A global disease, the epidemiological impact of TB can be measured in prevalence, in years lived with disability, and in deaths (Table 13.11).

By far the most prevalent respiratory infection is TB, accounting for eight times as many cases as LRTIs (COPD, pneumonia, etc.) and many more years of living with disability. However, around twice as many die of LRTIs than from TB. Although the WHO estimates that 1.7 billion people (22% of the global population) are infected with *M. tuberculosis*, only 5–10% will develop clinically active TB, pointing to a great deal of latent disease. Further estimates are the number of new cases annually—10 million—and the number of deaths—1.2 million, making it the leading cause of death from an infectious agent before the advent of COVID-19. Without treatment, some 70% of people with smear-positive pulmonary TB die within a decade, whilst HIV infection brings a 20–30-fold increased risk of the development of active TB. Other risk factors include immunosuppression, chronic lung disease, malnutrition, diabetes, use of > 40 g of alcohol a day, intravenous drug abuse, indoor pollution, and silicosis. The frequency of TB varies globally (Table 13.12); in England and Wales in 2021, 151 people died from TB.

TABLE 13.11   Epidemiology of certain respiratory infections

	Prevalence (million, %)	Years lived with disability (million, %)	Deaths (million, %)
All respiratory infections	2,187 (100)	11.7 (100)	3.75 (100)
Tuberculosis	1,929 (88.2)	3.1 (26.5)	1.12 (29.9)
Lower respiratory tract infections	10.6 (0.5)	0.65 (5.6)	2.5 (66.7)

GBD 2016 Causes of Death Collaborators. Global, regional, and national age–sex specific mortality for 264 causes of death, 1980–2016: a systematic analysis for the Global Burden of Disease Study 2016. Lancet 2017: 390; 1151–210. Table 2. © Elsevier.

TABLE 13.12   Global occurrence of tuberculosis

Region	Frequency/million
Indian subcontinent, sub-Saharan Africa, Micronesia, archipelago of Southeast Asia	1,000
China, Eastern Europe, Central and South America, North Africa	260–1,000
North America, Japan, Western Europe, Australia	<250

Following signs and symptoms of respiratory disease (in particular haemoptyosis, rarely encountered in COPD, and less so in asthma), diagnostic tests include sputum smear microscopy (for acid-fast bacilli, the Ziehl–Neelsen stain), microbial culture (problematic and slow), molecular genetics for *M. tuberculosis* nucleic acids (effective but relatively expensive), and a chest X-ray. An infection tends to form pseudo-granuloma (tubercles), detectable on radiography and by CT. For proof of immune recognition, the Mantoux test injects a small amount of purified protein derivative of *M. tuberculosis*, which, if positive, will produce an erythematous wheal 24–48 hours later. A more complex test is that of the interferon-gamma release assay, which assess the release of the cytokine by T lymphocytes when stimulated by *M. tuberculosis* extracts. The WHO refers to treatment for a 6-month regime of isoniazid (NICE NG33 recommending alongside pyridoxine), rifampicin, ethambutol, and pyrazinamide, which has an 85% success rate, a problem arising from drug-resistant strains. Prevention is by vaccination with the bacillus Calmette–Guerin (BCG), proof of effect being the Mantoux test.

Idiopathic pulmonary fibrosis

The British Lung Foundation estimates this condition to be present in 500/million of the UK population. By definition, the precise aetiology of this condition is unknown (although it may be linked to smoking and viral infections), so in many cases it is a diagnosis of exclusion. Symptoms are as other lung disease—cough, shortness of breath, etc. There are generally few changes on X-ray, but MRI and CT can define scarring with strands or sheets of fibrosis that are presumed to have been formed following damage (as in liver fibrosis and cirrhosis). It is likely that the fibrosis is the consequence of fibroblast activity with generation of proinflammatory cytokines and growth factors (IL-6, TNF-α, TGF-β, PDGF). A common treatment for mild to moderate disease is pirfenidone (with anti-fibrotic and anti-inflammatory properties), on an escalating scale of 267 mg three times a day to 801 mg three times a day. An alternative agent is ninteanib (a tyrosine kinase inhibitor), 150 mg taken orally twice a day, which in trials also markedly slowed the rate of decline in FVC, and reduced the incidence of death, compared to placebo.

Other conditions

The effect of cystic fibrosis on the lung is described in Chapter 15. Although obstructive sleep apnoea is primarily a disease of the pharynx (section 12.1.4), if linked to low oxygen saturation it may compound the pathophysiology of COPD and asthma. Other conditions that may be lung pathology include sarcoid, granulomatosis with polyangiitis, Goodpasture's syndrome, and rheumatoid arthritis (with pulmonary nodules) (Chapter 10). Certain drugs (especially cancer chemotherapy) and radiotherapy lead to pulmonary fibrosis.

13.6 COVID-19

Quite possibly, more has been written on this virus than any other, including HIV, and will continue to be written for years to come. In 2020 it was linked to 69,679 deaths in England and Wales (11.5% of all deaths), in 2021 to 67,057 (also 11.5%), despite the vaccination programme, and was the third most frequent cause of death. In 2022 there were 22,396 deaths (3.9%).

13.6.1 Coronaviruses

Classification

A coronavirus is named on morphological grounds as it expresses a variable number of spike-like protein protrusions, and so is said to resemble the corona of the sun. These spherical viruses are composed of a short section of RNA enclosed in a protein coat (the nucleocapsid) and infect many mammals, including bats, mink, rats, pigs, hedgehogs, mice, and our own species. They bring a number of typical signs and symptoms of a viral infection, including fever, respiratory distress, and diarrhoea.

Of the four coronavirus families, the alpha-coronaviruses are common in the general population and cause a mild and self-limiting respiratory infection, the common cold.

One member of the beta-coronavirus family emerged as a major public health issue in China in 2002, causing severe acute respiratory symptoms (SARS), hence the name SARS-CoV. Of 8,098 confirmed cases, there were 774 deaths (9.7%), and the virus was seemingly eliminated. A decade later, a second beta-coronavirus, also causing SARS, appeared in the Middle East, hence its name MERS-CoV. Of 2,494 laboratory-confirmed cases, it caused 858 deaths (34.4%). The current manifestation, SARS-CoV-2, has much in common (genetically and structurally) with SARS-CoV and MERS-CoV, and causes the infection and pandemic as COVID-19. With a broad case fatality rate of ~1.2% of those infected that varies greatly with clinical and demographic factors (although it was much higher in the early stages of the pandemic), SARS-CoV-2 is therefore far less deadly, but is far more transmissible, making it much more of a global health issue. An as yet unknown quantity is the short- and long-term morbidity in the survivors.

SARS-CoV-2 and its pathophysiology

At the centre of the virus is a ~31 kb RNA molecule whose nucleotide sequence is 96% homologous to that of the virus BatVoVRaTG13, previously detected in the bat *Rhinolophus affinis*. The RNA is enclosed within a spherical, non-segmented capsule with a diameter of ~125 nm expressing a heavily glycosylated protruding spike protein. This has a domain (S1) that binds to angiotensin-converting enzyme type 2 (ACE2, coded for by *ACE2* at Xp22.2), and so may be considered a receptor for the virus, and another that anchors it into the viral membrane that is involved in viral penetration of the host cell. ACE2 is present on the surface of vascular endothelial cells, but is also expressed by epithelial cells of the alveoli (see Figure 13.2), facilitating viral entry to the cell. Notably, those with diabetes and hypertension exhibit increased expression of ACE2, which provides a rationale for the role of these conditions as COVID-19 risk factors.

This provides the link with pulmonary disease as endothelial cells form the border between alveolar epithelial cells and the blood, and helps explain why those with existing pulmonary disease are at increased risk of the development of additional lung disease. A second crucial target cell molecule is the enzyme transmembrane protease serine 2 (TMPRSS2, coded for by *TMPRSS2* at 21q22.3), which modifies part of the S2 section, so facilitating entry of the virus (Figure 13.10). A cell infected with a virus and expressing non-self viral components will attract the attention of the immune system, first with an acute inflammation (increased neutrophils, acute phase response, etc.). In the lung, these are epithelial and endothelial cells of the alveoli and capillary network (Figure 13.2).

A virally damaged endothelium loses its barrier function, and this may explain the increased infiltration of tissue fluid exudates in air spaces that are characteristic of pulmonary oedema and pneumonia.

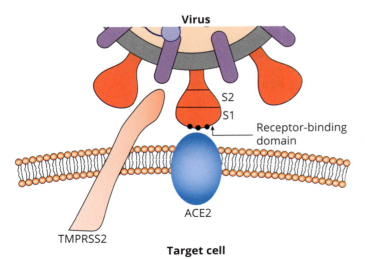

FIGURE 13.10

The SARS-CoV-2 spike protein (in red) interacting with its target cell. The S1 section binds ACE2 in blue via a receptor binding domain, whose precise amino acid structure is crucial in determining affinity and avidity of the interaction, and which represents a target for defensive antibodies. The interaction between S2 and TMPRSS2 facilitates the entry of the virus into the target cell.

In addition, a damaged endothelium loses its anticoagulant nature, becoming pro-coagulant, and so promoting thrombosis. Accordingly, COVID-19 may also be described as vascular disease. Further analyses point to high serum levels of cytokines such as IL-1β, IL-6, IL-10, IL-12B, TNFα, etc, often described as cytokine storm, a phrase borrowed from several other inflammatory conditions, including graft-versus-host disease, pancreatitis, septicaemia, and multiple sclerosis.

13.6.2 Global dissemination and epidemiology

The World Health Organization (WHO) reported unexplained cases of pneumonia in Wuhan, China, on 5 January 2020. The causative agent was subsequently named SARS-CoV-2, and the disease COVID-19. Behaving as mathematical models predicted, the virus spread from China in a southerly and westerly direction to all parts of the globe, initially into Southeast Asia, but at a far greater speed than expected. From there it jumped (temporarily) over the Indian subcontinent to Western Europe (initially Italy and Spain), then North and South America, and eventually all countries. During 2020 many countries reported increasing rates of infections and deaths within their populations. But the absence of international standard definitions of infection and related deaths, as well as accurate and widely accessible testing for the virus or serum antibodies that would prove an infection, initially made it difficult to understand the effects of the virus and its spread. Many of these inconsistencies could rapidly be corrected.

Understandably and predictably, many countries only slowly came to terms with the virus and its effects. On 11 March 2020, the WHO declared COVID-19 to be a pandemic, the first since that of swine flu in 2009. Numbers of confirmed cases and deaths showed a linear increase, although the case fatality rate fell (Table 13.13). In December 2023, the cumulative data were 773.82 million cases, 7.01 million deaths, and so a case fatality rate of 0.9%, indicating a plateau as the infection became endemic.

However, all three metrics varied greatly between countries (Table 13.14), reflecting differences in race, ethnicity, population demographics (age etc.), clinical features (co-morbidities such as obesity), general healthcare, and public health initiatives.

TABLE 13.13 Development of the global COVID-19 in three years of the pandemic

	Confirmed cases*	Deaths*	Case fatality rate
April 2020	0.89	0.05	5.6%
July 2020	10.47	0.55	5.2%
October 2020	34.65	1.09	3.1%
January 2021	83.61	1.92	2.3%
April 2021	128.70	2.94	2.3%
July 2021	182.11	3.96	2.2%
October 2021	234.10	4.80	2.0%
January 2022	289.11	5.46	1.9%
April 2022	487.56	6.16	1.3%
July 2022	545.45	6.33	1.2%
October 2022	615.44	6.55	1.1%
January 2023	734.84	6.72	0.9%
April 2023	762.32	6.90	0.9%

* Millions.

Data source: https://covid19.who.int/. Data refers to 1st of each month.

TABLE 13.14 Global data on COVID-19

	Population (10^6)	Cases (10^6)	Cases/population (%)	Deaths (10^3)	Deaths/10^3 population	Deaths/10^3 cases
World	8,031	765.2	9.5	6,921	0.86	9.0
Argentina	46.4	10.0	21.5	130.5	2.81	13.0
Australia	26.4	11.2	42.4	20.1	0.76	6.8
Brazil	216.9	37.4	17.2	701.5	3.23	18.7
China	1,455.2	99.2	6.8	121.0	0.08	1.22
Colombia	52.5	6.4	12.2	142.7	2.72	22.3
France	65.7	38.9	59.2	162.9	2.48	4.2
Germany	84.5	38.4	45.4	173.0	2.05	4.5
India	1,419.5	44.9	3.2	531.6	0.37	11.8
Indonesia	281.9	6.8	2.4	161.3	0.57	23.8
Iran	87.1	7.6	8.7	146.0	1.68	19.2
Italy	60.2	25.8	42.8	189.7	3.15	7.3
Japan	125.4	33.7	26.9	74.5	0.59	2.2
Mexico	132.9	7.6	5.7	333.9	2.51	44.0
Nigeria	221.6	0.27	0.1	3.1	0.01	11.8
Peru	34.3	4.5	13.1	220.1	6.41	48.9
Russia	146.1	22.8	15.6	398.4	2.73	17.4
South Africa	61.5	4.1	6.7	102.6	1.67	25.2
South Korea	51.4	31.2	60.7	34.5	0.67	11.1
Spain	46.8	13.8	29.5	120.7	2.58	8.7
UK	68.9	24.6	35.7	224.1	3.25	9.11
USA	336.6	103.3	30.7	1,124.1	3.34	10.9

Population data from https://www.worldometers.info/world-population/population-by-country/. COVID-19 data from https://covid19.who.int/, accessed 6 May 2023.

13.6.3 Public health

Early reports of the pathology of the infection from China during the spring of 2020 were rapidly confirmed in Europe: a standard flu-like illness with the potential to cause pulmonary disease, which in worse cases could progress to acute respiratory distress syndrome (ARDS, as in section 13.5.5) with potentially fatal pneumonia (section 13.5.3). It rapidly became clear that the virus, when compared to other respiratory viruses (principally influenza), was highly contagious and transmitted mostly by aerosols, prompting the use of face masks, hand/face washing and/or sanitizers.

As the pathogenicity of the virus became clear, governments advised and then mandated their populations to self-isolate (through lockdowns), and the most vulnerable to even more stringent shielding. This strategy was partly motivated by the fear that hospitals, and ICUs in particular, could become overwhelmed, which fortunately, in the initial phase, proved not to be the case. The clinical-scientific value of this process is difficult to quantitatively assess, but stringent public health measures taken by the Australian government may explain their low rates of infection, admission, and death (Table 13.14).

Formal recognition of the disease

Symptoms apart (see below), proof of infection relies on detecting the particular organism from a clinical sample, ideally a nasal and/or buccal swab, or a 'deep' lung swab. Once the SARS-CoV-2 genome became known, tests based on nucleic acids (polymerase chain reaction: PCR) and viral proteins (enzyme-linked immunosorbent assay: ELISA) were developed and rapidly disseminated to the scientific community for field testing. The latter provided the opportunity for self-testing (the lateral flow test), and so a form of diagnosis within 30 minutes. Although the USA's Food and Drug Administration expressed concerns about false negatives with certain kits, the UK's Medicines and Healthcare Regulatory Agency approved some tests for home use, and by June 2021 the UK's National Health Service recommended a rapid nasal/buccal test for those with symptoms, followed by PCR testing in those found to be positive. Later, blood tests were developed to enable the measurement of serum antibodies to SARS-CoV-2.

> **Cross reference**
>
> More information on the diagnostic technique PCR can be found in Chapter 16 of *Medical Microbiology*, another title in the Fundamentals of Biomedical Science series.

Key points

The COVID-19 pandemic is yet another global disease, but the first in a century to truly engage the world's scientific community, and the first to be the subject of such an intensive and coordinated investigation to effect a treatment.

The R statistic

The R statistic is a useful public health metric that refers to the theoretical reproductive rate of the virus in terms of its infectivity in a general population. R_0 describes no transmission to any other person (guaranteeing no further infections), R_1 to one other person, $R_{1.2}$ to 1.2 other people, and so on. Public health initiatives aim to bring the R statistics as close to zero as possible, so that the epidemic may be contained and then stopped; a rising R level calls for greater stringency as the epidemic is expanding. In the absence of preventative measures (i.e. vaccination) in the summer and autumn of 2020, the SARS-CoV-2 R value in the UK was estimated as being in the range 2.2–3.6. This compares with to 0.69 for MERS-CoV, consistent with the fact that this never became an epidemic. The 2009 H1N1 influenza pandemic was $R_{1.7}$, the 1918 influenza pandemic is estimated as R_2. Data from the UK indicated that lockdown is effective in reducing the R value of a population. In Melbourne, Australia, the mandatory wearing of face masks reduced the R value from 1.16–1.28 to 0.88–0.91.

SELF-CHECK 13.8

What is the pathophysiological link between COVID-19 and its clinical features?

13.6.4 Clinical aspects

Early stages of the infection

In common with many viral infections, there is a brief incubation period, widely reported as a median of 5 days, whilst symptoms usually develop after a median of 11.5 days. Early data pointed to 40–45% of those infected being asymptomatic, a figure subsequently reduced to 20%, whilst further data suggest a 40% chance of onward viral transmission prior to the onset of symptoms. This

TABLE 13.15 Leading clinical features of COVID-19

Severity (% present)	Clinical features
Mild (40%)	Mild fever, dry cough, sore throat, runny nose, nasal congestion, headache, sneezing, fatigue, muscle pain, tiredness, loss of smell, malaise, nausea, vomiting, diarrhoea, abdominal pain
Moderate (40%)	Fever (persistent or >37.8°C), dry cough, tachypnoea, some shortness of breath
Severe (15%)	Moderate fever, dyspnoea, tachypnoea, hypoxia, diarrhoea, vomiting, nausea, respiratory distress (respiratory rate >30/minute), finger oxygen saturation ≤93% resting, PaO_2/FiO_2 ≤300 mmHg, >50% lung involvement within 24 to 48 hours
Critical (5%)	Chest pain, severe shortness of breath with forced respiration possibly needing mechanical support, movement impairments, loss of speech, septic shock/sepsis, multiple organ dysfunction, ARDS, renal failure, acute cardiac injury, encephalopathy, rhabdomyolysis, arrhythmia

Modified from Salvamani S, Tan HZ, Thang WJ, et al. (2020) Understanding the dynamics of COVID-19; implications for therapeutic intervention, vaccine development and movement control, British Journal of Biomedical Science, 77: 4, 168–84, DOI: 10.1080/09674845.2020.1826136; NICE guideline NG191; COVID-19 rapid guideline: Managing COVID-19.

therefore (and worryingly) pointed to a high degree of undetected transmission. Table 13.15 summarizes clinical features classified by the severity of the disease. Most cases are mild, with a recovery time of 6 days.

For those sufficiently symptomatic as to warrant confirmation, the gold standard for diagnosis remained nucleic acid amplification testing by PCR, the genome of the virus having been determined. The most fruitful sample is likely to be lower respiratory tract material, including sputum, tracheal aspirate, and bronchial lavage, but for practical reasons an upper respiratory sample from the oropharynx and nasopharynx is used. Other methods for diagnosing a COVID-19 infection include CT and MRI, although one study reported a 25% false negative rate for CT, whilst X-ray is a useful but nonspecific tool for determining pneumonia (see 13.5.3).

Hospital admission and outcome

With a high probability of serious pulmonary disease, patients displaying sufficient symptoms are admitted to hospital for standard treatment for a viral infection. Data from China indicated a time to admission of around a week from the first onset of symptoms, with leading complications of dyspnoea, pneumonia, ARDS, and mechanical ventilation with ICU admission, each at sequential interval of one day, indicating an extraordinarily rapid disease progression. In a separate analysis, most of those who died did so an average of 18.5 days after admission, whilst survivors were discharged after a mean of 22 days.

An audit of over 80,000 UK hospital admissions with a mean age of 71, of whom 56% were male, showed that 81% had at least one co-morbidity (asthma, chronic cardiac, haematological, kidney, neurological and pulmonary disease, HIV/AIDS, history of malignancy, liver disease, obesity, rheumatological disorders, and smoking). Whilst in hospital, ~50% had developed at least one complication, the most common being renal (24%) and respiratory (18%), mostly developing in men. A problem with translating data of this nature is that only 68.5% of COVID-19-related deaths occurred in hospitals—25.6% occurred in care homes and 4.4% in private homes. This underlines the difficulty in comparisons with international data which may collect and report only on hospital deaths, thus loading the UK with an apparently worse death rate.

Should the infection escape the lung, other disease such as sepsis, pancreatitis, hepatitis, venous thrombosis, and arrhythmia may develop. Some of these may be secondary to a hyperimmune response, manifesting as a cytokine storm, whilst other blood markers (such as D-dimers) may predict disease severity and outcome, and so target management. If the general clinical picture deteriorates, transfer to intensive care is necessary (at a frequency of 1/16,000 of those testing positive), with the possibility of mechanical ventilator support for ARDS, which in the early stages of the pandemic was associated with a 37% mortality rate. Estimates of the mortality rate globally differ due to a lack of unified definition, but one source suggests a rate of 4% of those diagnosed, another gives a rate of 3.3% of those infected, and a third between 3.4% and 11%. The death rate for influenza is <0.1%. The full late consequences of COVID-19 infection (long covid, section 13.6.10) have yet to be determined.

SELF-CHECK 13.9

What are the signs and symptoms of an increasingly severe COVID-19 infection?

Clinical and demographic features

Unsurprisingly, increasing age is a major risk factor at all stages of the natural history of the disease: infection–hospital admission–intensive care–mechanical ventilation–death. Death rates in the first wave in those aged 80 and above were reported as around 21%, whilst in Italy in March 2020, 16.6% of deaths (the largest percentage) were in this age group, the rate in 70–79-year-olds being 9.6%. Others report that only 0.6–2.8% of deaths occur in those aged <65–figures that compare to 80% in the influenza pandemic of 2009, with a mean age of death of 37 years; and 95% in that of 1918, where the mean age of death was 27 years.

   Other risk factors for infection and admission are also unsurprising: obesity brings an odds ratio (95% confidence interval (CI)) of 2.91 (1.31–6.47) for infection (the reasons for which are pathogenetically unclear). Another risk factor is smoking, which may well be explained by the chronic and low levels of pulmonary disease compounded by the fact that it damages the endothelium, and so may lead it to be more susceptible to the virus. Other risk factors include hypertension (present in 17% of those who died), diabetes (8%), ischaemic heart disease (4%), cerebrovascular disease (2%), COPD (2%), chronic kidney disease (1%), and malignancy (1%). Additional data are summarized in Table 13.16

TABLE 13.16 Frequencies of those admitted with COVID-19 and those who died

Feature	Base sample (n ~ 6 million)	Admitted (n = 10,776)	Died (n = 4,384)
Male	49.9	55.3	57.4
Mean Age (year)	48	70	80
Body mass index ≤24.99	36.1	27.5	35.3
Body mass index 25–29.99	41.5	51.5	46.2
Body mass index ≥30	7.4	14.4	10.5
Smoker	16.9	7.4	5.7
Ex-smoker	21.2	34.5	39.6
Non-smoker	57.2	56.4	52.7
Normal renal function	96.1	75.7	66.8
Any chronic kidney disease	3.9	24.3	33.2
Respiratory cancer	0.21	1.21	1.39
Solid organ transplant	0.05	0.31	0.23
Type 1 diabetes	0.47	1.26	0.82

(Continued)

TABLE 13.16 Continued

Feature	Base sample (n ~ 6 million)	Admitted (n = 10,776)	Died (n = 4,384)
Type 2 diabetes	6.49	28.0	32.3
COPD	2.34	10.7	13.2
Asthma	13.6	16.2	13.3
Coronary heart disease	3.5	16.5	23.7
Atrial fibrillation	2.4	13.6	19.0
Congestive cardiac failure	1.2	9.3	13.1
Venous thromboembolism	1.7	7.0	8.7
Peripheral vascular disease	0.73	4.3	6.6
Stroke	2.1	12.4	18.4
Dementia	1.0	11.5	29.9
Parkinson's disease	0.2	2.0	3.1
Osteoporotic fracture	3.9	10.7	15.4
Rheumatoid arthritis or SLE	1.0	2.9	2.9
Severe mental illness	11.1	17.1	17.0
Epilepsy	1.3	3.2	3.6
White	64.5	63.0	67.2
Indian	2.89	3.93	2.99
Pakistani	1.89	2.30	1.57
Bangladeshi	1.44	1.61	1.57
Other Asian	1.82	2.30	1.30
Caribbean	1.14	3.64	3.47
Black African	2.47	4.23	2.78
Chinese	0.96	0.42	0.41
Other ethnic group	3.69	4.05	2.60
Not recorded	19.2	14.52	16.08

Data % (except age, mean (SD)). COPD = chronic obstructive pulmonary disease. Some data fail to sum to 100% as some 'data not recorded'.

Modified from Clift AK, Coupland CAC, Keogh RH, et al. Living risk prediction algorithm (QCOVID) for risk of hospital admission and mortality from coronavirus 19 in adults: national derivation and validation cohort study BMJ 2020: 371: m3731 doi:10.1136/bmj.m3731 (2020).

with a comprehensive and powered study showing selected demographic and medical characteristics of cases collected from 1,205 English General Practices during the spring and summer of 2020. Several of these (many of which are cardiovascular) show a clear trend from presence in the population, to admission to hospital, and then to death.

TABLE 13.17 Laboratory markers of adverse outcome

Increased	White blood cell count, neutrophil count, D-dimers, creatine kinase, creatinine, urea, aspartate aminotransferase, alanine aminotransferase, C-reactive protein, prothrombin time, activated partial thromboplastin time, serum calcium, IL-6, IL8, IL-10, ferritin, bilirubin, lactate dehydrogenase, procalcitonin, lactate dehydrogenase
Decreased	Lymphocyte count, platelet count, albumin, antithrombin

The laboratory

Early data from China, confirmed by those from Italy, pointed to increased levels of numerous laboratory analyses linked to more severe disease and death (Table 13.17). Of these, the literature favours D-dimers and a reduced lymphocyte count as major predictors of more severe disease and death. Notably, many factors are correlated, and reflect the acute phase response. A study from Spain of admissions to ICU reported that D-dimers ($p = 0.001$), the percentage of lymphocytes ($p = 0.009$), the platelet count ($p = 0.012$), and lactate dehydrogenase ($p = 0.013$) were the best laboratory predictors of 6-week mortality. A meta-analysis of over 29,000 patients found elevated troponins to bring an odds ratio of 3.17 for short-term mortality (Li et al. 2022).

COVID-19 and other disease

With cancer, cardiovascular disease, and respiratory disease being the leading causes of death in the pre-COVID-19 era (Table 13.18), it is perhaps unsurprising that patients with these co-morbidities and their treatments are likely to have an increased risk of infection, hospital admission, and death. Of 2,619 Italians (mean age 78, 32.6% female) who died of COVID-19 in early 2020, 28.7% had ischaemic heart disease, 22.3% had atrial fibrillation, 30.9% had diabetes, 20.4% had chronic renal failure, 15.9% had active cancer, 15.9% had dementia, 11% were obese, and 10.6% had suffered a stroke. However, the most common co-morbidity by far was hypertension, present in 68.5%. One co-morbidity was present in 14.9%, two were present in 21.3%, and three or more were present in 59.9% of those who died (Canevelli et al. 2020).

An early study from China (Yu et al. 2020) reported the incidence of a COVID-19 infection in those with cancer to be over twice that of patients free of cancer (0.79% and 0.37% respectively). This disparity extends to mortality, with a case fatality rate of 5.6%, over double that of the overall case fatality rate of 2.3% (Al-Shamsi et al. 2020). Elsewhere, the 1% incidence of COVID-19 in 1,590 cancer patients was over three times (0.29%) that of the overall local population, whilst 39% of cancer patients required intensive care or invasive ventilation, or died, compared to 8% of patients without cancer (Liang et al. 2020). Anti-tumour treatment within 14 days of a COVID-19 diagnosis brings a four-fold increased risk of developing severe disease (Zhang et al. 2020). A meta-analysis reported that COVID-19-positive patients with a haematological malignancy (e.g. leukaemia, lymphoma, myeloma) had an increased risk of all-cause- and COVID-19-related mortality compared to patients with a solid organ tumour (e.g., breast, lung, bladder cancer) (Hardy et al. 2023).

TABLE 13.18 Leading causes of death

	Cancer	Cardiovascular disease	Respiratory disease
Global	16.8%	32.2%	7.5%
England and Wales	28.4%	24.3%	13.7%
USA	28.3%	21.0%	5.5%

Data refer to the % of deaths in each disease group in terms of total deaths.

Cardiovascular disease (CVD, mostly of the coronary and cerebral arteries, and heart failure), was quickly identified as a major risk for COVID-19-related death. A meta-analysis showed that cardiovascular disease and cerebrovascular disease both brought an increased risk of mortality with pooled odds ratios of 3.59 and 3.11 respectively, figures that exceeded those of chronic renal disease, age, smoking, and chronic pulmonary disease (Reyna-Villasmil et al. 2022). These data reflect other meta-analyses, reporting such as that of Zuin et al. (2022), that pre-existing coronary artery disease brings an odds ratio of 2.61 for short-term mortality, and Zhang et al. (2022), who reported that ischaemic heart disease brings a pooled effect size of 1.27 (95% CI 1.17–1.38) for COVID-19 mortality. One of the larger meta-analyses, of over 3 million patients, reported that obesity brings an increased relative risk of severe disease and for mortality (Singh et al. 2022). Statins, taken by many with CVD, bring a reduced risk of death, with an odds ratio (95% CI) of 0.71 (0.55–0.92), but also of the need for ventilation (0.81 (0.69–0.95)), but not for ICU care (0.91 (0.55–1.51)).

Those carrying a transplant are required to take immunosuppressive therapy (steroids, tacrolimus, mycophenolate mofetil, etc.) to ensure the viability of their graft, which begs the question of whether they are more likely to suffer a COVID-19 infection, and whether they will be able to mount a sufficiently robust response to a vaccination. A multinational study by Webb et al. (2020) reported that among 151 adult liver transplant patients with a COVID-19 diagnosis requiring hospitalization, there were significantly fewer deaths than in a matched group not transplanted—19% versus 27% respectively. One explanation may be that the immunosuppression dampens any possible pathological cytokine storm. This does not, however, determine whether transplantation or immunosuppression are risk factors for infection, or for hospital admission. To some extent this is mirrored by a mortality rate of 21.8% in those hospitalized patients taking an IL-6 antagonist versus 25.6% in controls. Similarly, tacrolimus use was linked to a reduced mortality in liver transplant recipients (odds ratio 0.55 (95% CI 0.31–0.99)). However, a high degree of immunosuppression has been associated with an increased risk of severe disease.

HIV infection brings a significant (p < 0.01) adjusted hazard ratio (95% CI) of 1.69 (1.15–2.48) for mortality once admitted compared to those admitted but free of HIV. This is presumed to be the consequence of the patient's intrinsically impaired ability to mount an effective immunological defence.

Ethnicity

Other early data from the UK pointed to an increased incidence of COVID-19 in Black people, Asian people, and people from other ethnic minorities, compared to a White group. This disparity remains after adjusting for occupation, for reasons that are unclear, but there are clues. For example, compared to White people, Black people and Asian people carry a disproportionate burden of hypertension (increased 2.5-fold and 1.8-fold respectively) and diabetes (2.9-fold and 3.9-fold respectively) (Cappuccio et al. 1997). Section 12.4.2 discusses the role of obesity in South Asians, citing NICE CG189, whilst in 2020–1, overweight or obesity was present in 72%, 57%, 60.4%, and 37.5% of those of Black, Asian, White, and Chinese ethnicity, respectively. Furthermore, for patients admitted to hospital, obesity is a stronger risk factor for Black ethnic groups (Yates et al. 2021).

Other possible reasons are demographic and sociological, e.g. a greater likelihood of living in high-density housing in urban areas (Coleman et al. 2022). In August 2020, the ONS reported that 33.9% of people who were critically ill were from ethnic minorities. A breakdown of the excess risk of death in people from ethnic minority groups is shown in Table 13.19.

Obesity is more strongly associated with COVID-19 critical care admission, mechanical ventilation, and death for ethnic minorities (mostly South Asians and African Caribbeans) compared with White people. A study from the UK by Marwah et al. (2021) considered differences in laboratory parameters and outcomes between 229 White people and 216 people categorized as from Black, Asian, and Minority Ethnic (BAME) groups who were admitted for treatment of COVID-19 in 2020. After 28 weeks, there were 97 (42.4%) and 93 (43.1%) deaths, respectively. Laboratory factors predicting mortality are shown in Table 13.20. In multivariate analysis the neutrophil/lymphocyte ratio and urea/albumin ratio were independent predictors of death in the White group, whilst diabetes and urea/albumin ratio were independent predictors in people from BAME groups.

TABLE 13.19 Rate of death involving COVID-19 by ethnic group

Group	Males	Females
Black African	3.8 (3.5–4.2)	2.9 (2.5–3.4)
Bangladeshi	3.4 (2.9–4.0)	2.5 (2.0–3.2)
Black Caribbean	2.8 (2.6–3.1)	2.3 (1.9–2.5)
Pakistani	2.5 (2.3–2.8)	2.5 (2.2–2.9)
Other ethnic group	2.4 (2.2–2.6)	1.8 (1.6–2.1)
Indian	2.1 (1.9–2.3)	1.8 (1.6–2.0)

Data are age adjusted hazard ratio (HR) (95% CI) compared to White counterparts. All remain significant but with lower HRs after adjusting for geography, socio-economics, and health status. Fully adjusted death rates in Chinese men and women did not differ significantly from those of White men and women, and so are not shown.

© Office for National Statistics. Source: https://www.ons.gov.uk/. Licensed under the Open Government Licence 3.0.

TABLE 13.20 Laboratory factors predicting mortality in admitted patients

BAME group only	Increased white blood cell count, neutrophil count, HbA1c, potassium, creatinine, and troponin-I
White group only	Reduced eosinophil count, increased lactate dehydrogenase
Both groups	Increased mean cell volume, urea, sodium, reduced albumin, CRP, neutrophil to lymphocyte ratio, urea/albumin ratio. Reduced lymphocyte count
Neither group	Altered haemoglobin, monocyte count, platelet count, ferritin, alkaline phosphatase, alanine aminotransferase, bilirubin

Key points

Prediction of those likely to have more serious disease requires careful assessment of clinical features and laboratory markers.

13.6.5 Treatments

In the UK, the lead document, NICE's NG191, referred separately to management of an infection in the community and in hospital, a major initiative being emphasis on the former in order to prevent the latter. An example of this was the effective use of anti-SARS-CoV-2 monoclonal antibodies in those at high risk of admission. Box 13.3 describes the role of NICE in COVID-19 with other conditions.

Drugs

The anti-viral drugs remdesivir, lopinavir, favipiravir, interferon, and ritonavir have all been used with success in infections with HIV and hepatitis C virus, but have no effect on the course of an infection with SARS-CoV-2. However, they may have value in those with a severe viraemia, and possibly

BOX 13.3 Role of the National Institute for Health and Care Excellence (NICE)

This UK body published dozens of guidelines and other documents offering advice and recommendations regarding COVID-19, the leading report being NG191. Many others focused on specific conditions or situations, such as NG170 on COVID-19 and cystic fibrosis, NG187 on COVID-19 the use of vitamin D, and NG172 on COVID-19 in gastrointestinal and liver condition treated with drugs affecting the immune response.

those in intensive care. Early studies with chloroquine/hydroxychloroquine (CQ/HCQ, which has anti-inflammatory properties and is effective in rheumatoid arthritis) and colchicine (also used in certain inflammatory connective tissues diseases) suggested possible value in treating COVID-19. However, a meta-analysis of the use of HCQ/CQ in hospitalized patients showed no effect on mortality (10% in those on HCQ/CQ, 9% in the control group), whilst adverse events were more frequent in the HCQ/CQ group (0.39 v 0.29 per patient) (Di Stefano et al. 2022). Others found no effect of HCQ in non-hospitalized patients (Mitjà et al. 2023), and a meta-analysis found its use was even associated with an increased risk of mortality with a relative risk of 2.17 (95% CI 1.32–3.57, p = 0.002) (Singh et al. 2020). Similarly, a meta-analysis of over 16,000 subjects found no effect of colchicine on mortality, the need for ventilation, or the length of hospital stay (Lan et al. 2022).

The RECOVERY trial randomized patients hospitalized with clinically suspected or laboratory confirmed SARS-CoV-2 infection (of whom 56% had at least one co-morbidity) to usual care or usual care plus 6 mg oral or intravenous dexamethasone (an immunosuppressant and anti-inflammatory drug) for up to 10 days (RECOVERY Collaborative Group 2021a). After 28 days, 25.7% of those on usual care died, compared to 22.9% in those also taking dexamethasone—a significant difference (rate ratio 0.83, 95%CI 0.75–0.93, p < 0.001). This effect was more marked in those on mechanical ventilation (41.4% v 29.3%), and in those who were on oxygen but not ventilated (26.2% v 23.3%). The investigators also reported the benefit of the anti-IL-6 receptor antibody tocilizumab in hospitalized patients with hypoxia and severe inflammation (31% v 35% died, rate ratio 0.85, 95%CI 0.76–0.94, p = 0.0028), a possible mechanism being immunosuppression aimed at part of the cytokine storm (RECOVERY Collaborative Group 2021b). A meta-analysis by Piscoya et al. (2022) subsequently reported this drug to be linked to reduced all-cause mortality (relative risk 0.89, 95%CI 0.81–0.98, p = 0.03), the need for mechanical ventilation (0.80 95%CI 0.71–0.90, p = 0.001), and the length of hospital stay (1.92 fewer days, 95%CI 0.38–3.46, p = 0.01).

Janus kinase (JAK) is an enzyme that part-regulates cellular cytokine activation pathways—inhibitors such as baricitinib are used in inflammatory disease such as rheumatoid arthritis. The RECOVERY trial of baricitinib showed it to be linked to a shorter time in hospital (rate ratio 1.10 95%CI 1.04–1.15, p = 0.0002), reduced need for invasive mechanical ventilation (0.85, 0.73–0.99, p = 0.033), and a reduction in death (0.89, 0.80–1.00, p = 0.049), whilst a meta-analysis showed the use of baricitinib to be linked to a reduction in the risk of death (rate ratio 0.87 95%CI 0.77–0.99, p = 0.028), and the use of any JAK inhibitor to be linked to a 20% reduction in the risk of death (0.80, 0.72–0.89. p < 0.001) (RECOVERY Collaborative Group 2022a). Others suggest that the antibiotic azithromycin (Trial Collaborative Group 202, NICE NG191), anti-parasitic agent ivermectin (Popp et al. 2021), or vitamin D should not be used for routine care (NICE NG187). NICE NG191 and others (Perico et al. 2023) recommend the use of paracetamol or NSAIDs for treating the early symptoms of an infection.

Antibodies

Convalescent plasma from those who survived an infection and are presumed to have antibodies had been used with good effect in MERS and SARS infections, but has no benefit in those with COVID-19 (RECOVERY Collaborative Group 2021c). However, humanized monoclonal antibodies (mAbs), mostly directed towards the receptor-binding domain of the viral spike protein, are effective in various studies in patients with COVID-19 in the community, and some of them have been approved by NICE (NICE NG191, TA878). A meta-analysis of eight studies examining the combination of casirivimab with imdevimab in almost 85,000 subjects compared to over 300,000 on usual care found a positive effect on the likelihood of hospitalization (odds ratio 0.31 95%CI 0.20–0.48, p < 0.001) and of death (0.21 95%CI 0.06–0.68, p < 0.0001) (Gao et al. 2023). A single dose of sotrovimab given to high-risk patients with mild to moderate disease severity significantly reduced all-cause emergency department visits (adjusted relative risk 0.34 95%CI 0.19–0.63, p < 0.001), viral load (0.26, 0.12–0.59, p = 0.007), and progression to severe or critical disease (0.26, 0.12–0.59, p = 0.002) (Gupta et al. 2022). In ambulant subjects with mild or moderate COVID-19, the combination of bamlanivimab with etesevimab (each directed towards different regions of the SARS-CoV-2 spike protein) was linked to 2.1% of hospitalizations or deaths, compared to 7% in the placebo group (p < 0.001) (Dougan et al. 2021). A second pair of monoclonal antibodies, casirivimab and imdevimab, was trialled in patients admitted to hospital with COVID-19. Although there was no difference in deaths within 28 days of admission, subgroup analysis of those testing seronegative for SARS-CoV-2 found that the combination reduced the number of deaths (rate ratio 0.79 95%CI 0.69–0.91, p < 0.001) (RECOVERY Collaborative Group 2022b). A side-by-side comparison of bamlanivimab and etesevimab versus casirivimab and imdevimab in patients with mild to moderate COVID-19 found a primary outcome in 2.6% in the former and in 6.6% of the latter (odds ratio 2.67 95%CI 1.17–6.06, p = 0.01) (O'Horo et al. 2022).

A strength and weakness of monoclonal antibodies is their exquisite specificity. Those raised to spike proteins of the ancestral Wuhan variants of the virus are less likely to recognize the tertiary structure of mutated spike proteins of subsequent variants (section 13.6.9) and so may not necessarily be effective treatments.

Respiratory support

The principal clinical pathology being pneumonia, admitted patients are likely to be provided with oxygen-enriched air delivered to the nostrils. As finger pulse oximetry falls, the proportion of oxygen will be increased, and nasal delivery replaced with a mask. However, in view of the importance of this metric, a crucial caveat is that pulse oximetry in those with dark skin is less accurate at lower oxygen saturation, such as 80%, and may result in over-estimation, so that subjects may actually have a worse oxygen saturation than the metric delivers (Cabanas et al. 2022). As lung function deteriorates and ARDS develops (or is likely to), mechanical ventilation, requiring intubation, may be necessary. This intervention is, unsurprisingly, a risk factor for death.

13.6.6 Vaccines and vaccination

Vaccine development

Several pharmaceutical houses, some in partnership with academia, were quick to act, aiming at the receptor-binding domain of the spike protein. Several strategies were used:

- Inactivated viruses, the oldest technique first used for polio, sees an attenuated strain of the virus that is non-pathogenic but retains its antigenicity.

- A non-replicating viral vector that is not in itself pathogenic and cannot reproduce, but expresses at its surface viral antigenic proteins of choice or acts as a vector in delivering nucleic acid to its host cell.

- Recombinant DNA technology can generate only the relevant protein subunit parts of the virus that retain their antigenicity, as used for hepatitis B vaccines. This method can produce large amounts of protein from bacterial, yeast, mammalian, and insect expression systems cells that can be purified and formulated as a vaccine.

- A vector containing a section of viral messenger RNA (mRNA) that encodes the full-length spike protein. The mRNA is inserted and stabilized inside a vector of a lipid nanoparticle, composed of cholesterol, a synthetic amino-lipid, and other lipids. The lipid coat of the nanoparticles ensures fusion with a cell membrane, and so passage of the mRNA into the cell. Once there, it can interact with the host cell's ribosomal complex and so generate the spike protein.

The primary objective of most vaccine trials was to prevent deaths, and a secondary objective to prevent hospital admission. Clinical trials of most vaccines showed acceptable efficacy and side-effect profiles, permitting approval by regulatory authorities and accordingly a roll-out for mass population vaccination. In the process, data from millions of people were gathered, making sub-analyses possible, such as of the effects of co-morbidities. Systemic and local side effects of certain vaccines were lower in the community than in clinical trials. Most vaccines need two doses for maximum effect, but those with a history of a natural COVID-19 infection already have antiviral antibodies, and have some protection, such that only a single vaccination may be needed.

The vaccines—the first generation

Various vaccines have been licensed by authorities in a number of countries. Most provide protection up to 95% from end points that include hospital admission, intensive care, and death. Direct comparisons are difficult as trials failed to recruit subjects with comparable clinical and demographic profiles. For example, in the major trial of ChAdOx1, 39% of recruits were men, whereas the BNT162b2 trial recruited 51% men, and the Gamaleya/Sputnik trial 61% men. There were also differences in co-morbidities and in occupational risk factors. In the UK, two vaccines dominate:

- The Oxford/Astra-Zeneca vaccine, ChAdOx1, consists of a chimpanzee adenovirus that cannot reproduce, but which includes the DNA of the SARS-CoV-2 spike protein, which is expressed at the surface of the viral particle. Once inside the cell, the engineered DNA is translated and transcribed to produce a spike protein that provokes the immune system to generate antibodies and viral-specific T lymphocytes.

- The Pfizer-BioNTech vaccine, BNT162b2, is formed from the modified mRNA of the ACE2-binding protein spike, delivered in a lipid nanoparticle that merges easily with cell membranes. Two doses of this vaccine produces high titres of anti-SARS-CoV-2 neutralizing antibodies and robust antigen-specific CD8-positive and CD4-positive T cell responses.

Side-by-side comparisons of efficacy and side effects are rare, but some are available. A community study from Scotland reported a vaccine effect (in terms of preventing admission to hospital) of a single dose of ChAdOx1 to be 88% (95%CI 75–94) at 28–34 days post-vaccination, compared to 91% (85–94) in those receiving BNT162b2 (Vasileiou et al. 2021). A side-to-side comparison of the ChAdOx1 and BNT162b2 vaccines found no difference in the frequency of a severe COVID-19 outcome (hospital admission, death) (Agrawal et al. 2021).

The Moderna vaccine, mRNA-1273, delivered within a lipid nanoparticle, was approved by UK regulators on 8 January 2021, whilst Janssen/Johnson and Johnson's Ad26.CoV2 was approved in the UK on 28 March 2021. Other products included the Sinopharm, Sinovac-CoronaVac, Bharat Biotech BBV152 COVAXIN, Covovax, and the Nuvaxovid vaccines, all of which obtained emergency use listing by the WHO during 2021.

SELF-CHECK 13.10

What are the principal technological differences between the two major first-generation vaccines?

The vaccines—the second generation

The first generation of vaccines was based on the Wuhan strain of SARS-CoV-2. Mutations in this virus subsequently appeared, the principal variants being named Alpha, Delta, and Omicron, each of which had different infectiousness and pathology, leading to concerns of loss of effect of the vaccines. This was part-addressed by the development of a second generation of vaccines designed to give protection from these variants. Three such hybrid, or bivalent, vaccines, BNT162b2-Omicron, Spikevax Bivalent, and VidPrevtyn Beta, were approved in the UK during the latter part of 2022.

Side effects

No drug is free of side effects, and vaccines are no exception. These can be classified as general and specific. Local general side effects (to the site of the vaccination) include pain, swelling, redness, etc, whilst general systemic side effects include headache, fatigue, arthralgia, myalgia, chills, fever (often present in up to 10% of subjects), diarrhoea, nausea, vomiting, and loss of appetite, lymphadenopathy, dizziness, somnolence, insomnia, abdominal pain, pruritis, rash, sneezing, and tremor (<1%). However, there may also be rare side effects, developing in <0.1%, that are specific for a particular vaccine.

Within months of its release, cases of venous thromboembolism and thrombocytopenia were reported from those vaccinated with ChAdOx1, at a frequency of 0.00038%. This phenomenon—vaccine-induced thrombotic thrombocytopenia—is addressed in NICE document NG200. There were also reports of thrombocytopenia following BNT162b2 vaccination, but far more prevalent were cases of an allergic reaction, at a frequency of ~11 per million, that is, 0.0011%. These reactions were also present in the Moderna mRNA lipid nanoparticle, a candidate allergen being a component of the lipid coating of the nanoparticle, such as polyethylene glycol. Those suffering these side effects are recommended an alternative vaccine for additional dosing.

Mass vaccination in the UK

Responding to early epidemiological data of mortality, the UK government's response was to vaccinate people in order of risk, beginning in January 2021. The first priority were residents and staff in care homes for older adults, followed by those aged 80 or over as well as front-line health and social care workers. The next group was those aged 75 or more, then those aged 70 or more, and those deemed to be clinically extremely vulnerable. After those aged 65 and over, vaccinations of adults aged 16 to 65 in at-risk groups began, as they were deemed to be at increased vulnerability (Table 13.21).

Other risk groups include younger adults in long-stay nursing and residential care settings, adult household contacts of people with immunosuppression, those expecting to share living

TABLE 13.21 Groups at enhanced risk of a COVID-19 infection

Broad at-risk group	Examples of conditions
Chronic lung disease	Asthma, COPD, chronic bronchitis, emphysema, cystic fibrosis, interstitial lung fibrosis
Cardiovascular disease and its risk factors	Congenital heart disease, hypertension with cardiac complications, chronic heart failure, ischaemic heart disease, atrial fibrillation, peripheral vascular disease, a history of venous thromboembolism, diabetes, class III obesity (BMI >40)
Chronic kidney disease	Chronic kidney disease at stage 3, 4, or 5; chronic kidney failure; nephrotic syndrome; kidney transplantation
Chronic liver disease	Cirrhosis, biliary atresia, chronic hepatitis
Chronic neurological disease	Stroke, transient ischaemic attack, cerebral palsy, severe or profound learning disabilities, Down's syndrome, multiple sclerosis, epilepsy, dementia, Parkinson's disease, motor neurone disease
Immunosuppression	Chemotherapy or radiotherapy, solid organ or bone marrow/stem cell transplants, HIV infection, genetic disorders affecting the immune system (e.g. SCID), users of anti-TNF medications, alemtuzumab, ofatumumab, rituximab, steroids, protein kinase inhibitors, or PARP inhibitors, cyclophosphamide, mycophenolate mofetil, haematological malignancy (leukaemia, lymphoma, myeloma), and an inflammatory connective tissue disease (e.g. RA, SLE) requiring long-term immunosuppression
Severe mental illness	Individuals with schizophrenia or bipolar disorder, or any mental illness that causes severe functional impairment

accommodation on most days (and therefore for whom continuing close contact is unavoidable) with individuals who are immunosuppressed (defined as above), and those who are the sole or primary carer of an elderly or disabled person who is at increased risk of COVID-19 mortality and therefore clinically vulnerable. Vaccination of the remaining population continued with younger groups in 5 years bands down to the age of 18, followed by younger teenagers. Later initiatives in 2021 reduced age bands down through the teens, and lower. Key milestones are shown in Table 13.22. The third vaccination (often described as a 'booster') began in the autumn of 2021, and a fourth was offered to those clinically vulnerable (e.g. having had a transplant) in the spring of 2022 and extended to other groups in the autumn of that year. Further boosting of high-risk groups continued into the spring of 2023, in some cases with bivalent second-generation vaccines.

TABLE 13.22 **Progress in vaccinating the UK population**

1st of the month	First vaccination		Second vaccination		Third vaccination	
	%	Millions	%	Millions	%	Millions
2021						
January	4.0	2.2	0.7	0.4		
February	16.8	9.6	0.9	0.5		
March	35.6	20.5	1.5	0.8		
April	54.5	31.3	8.6	5.0		
May	60.0	34.5	26.7	15.3		
June	68.8	39.6	45.3	26.1		
July	78.3	45.0	58.1	33.2		
August	81.4	46.9	66.8	38.3		
September	83.7	48.1	74.8	43.0		
October	85.0	48.9	78.1	44.9	1.7	1.0
November	86.9	50.0	79.2	45.7	14.4	8.3
December	88.7	51.0	80.4	46.4	33.0	19.0
2022						
January	90.1	51.8	82.4	47.4	59.3	34.1
February	91.1	52.4	84.3	48.5	65.0	37.4
March	91.5	52.6	85.2	49.0	66.6	38.3
April	91.8	52.8	85.9	49.4	67.5	38.8
May	92.2	53.2	86.4	49.7	68.3	39.3
June	93.0	53.5	86.9	50.0	69.0	39.7
July	93.2	53.6	87.3	50.2	69.5	40.0
August	93.4	53.7	87.9	50.5	69.9	40.2
September	93.6	53.8	88.2	50.7	70.2	40.4

% is of those aged 12 and over, around 57.5 million people. UK data not published after September 2022, at which point 151 million doses had been administered. Data source: https://coronavirus.data.gov.uk/details/vaccinations. Licensed under the Open Government Licence 3.0.

Notably, despite considerable societal awareness, some 3.7, 6.8, and 17.1 million eligible people declined a first, second, or third vaccination. There are numerous potential reasons for this vaccine hesitancy or denial, some of which are based on scepticism, educational level, and conspiracy beliefs (Seddig et al. 2022). Further booster vaccination were subsequently offered to certain groups. By mid-December 2023, 69% of those aged over 65 invited to have a booster actually did so, some 7.8 million people.

Although all vaccines were licensed for sole use only for two doses only, logistics issues required some individuals to have their second dose with a different vaccine from their first, that is, a heterologous regime. The hypothesis that such a heterologous regime would have at least equal efficacy as the same vaccine used twice (i.e. a homologous regime) was tested in a formal trial, which found that a heterologous regime is well tolerated and improves immunogenicity compared to a homologous regime (Demirhindi et al. 2021).

Vaccination with co-morbidities

The demands of clinical trials often exclude certain groups (such as those transplanted), but widespread population vaccination programmes allow numerous sub-analyses for specific conditions. Numerous reports in the literature, and guidelines from learned societies, all advocate vaccinations for people with many pre-existing conditions, including the leading risk factors of cancer, cardiovascular disease, diabetes, and obesity, but also transplantation.

Studies with other viral vaccines (influenza; mumps, measles, and rubella; hepatitis viruses) in solid organ transplants point to variable responses, but on balance a reduced effect. Responses to a COVID-19 vaccine supported this view, where in one study only 54% of patients produced a detectable antibody response, assumed to be the case because of the required immunosuppression (Boyarsky et al. 2021). In other studies, 38% of heart recipients and 64% of lung recipients were non-responders after two doses of a vaccine (Hallett et al. 2021), studies that were confirmed in a meta-analysis that concluded that immunocompromised patients were 48% less likely to seroconvert (relative risk 0.52 95%CI 0.42–0.65, $p < 0.0001$) than immunocompetent patients (Nejad et al. 2022). These and other reports prompted practitioners to prescribe a third (booster) dose. In a placebo-controlled trial of a third dose of a vaccine in recipients of a solid organ transplant (35% thoracic, 65% abdominal), antibody and cellular responses all improved significantly ($p < 0.001$) compared to patients randomized to placebo (Table 13.23) (Hall et al. 2021). A third vaccination improved the proportion of spike-IgG antibodies in 42 heart transplant recipients from 33% to 57%, and achieved a 35-fold increase in the median antibody titre (Shaul et al. 2022). A similar result was reported in a cohort of 98 transplant patients, where a third dose increased median (inter-quartile range) anti-spike IgG antibodies 33-fold ($p < 0.0001$) (Saiag et al. 2022).

The same principle applies to those with such severe inflammatory disease—such as intractable rheumatoid arthritis (RA) or SLE—that immunosuppression is also called for. In one study the mean neutralizing antibody titre increased 7-fold after a booster in 48 patients with rheumatoid arthritis, 12-fold in 16 patients with SLE, and 3-fold in 154 patients with systemic sclerosis (all differences $p < 0.0001$) (Ferri et al. 2022). Similarly, median titres of anti-spike IgG antibodies increased 2.8-fold in 57 patients

TABLE 13.23 Response to a third vaccination in recipients of a solid organ transplant

Laboratory metric	Randomized to vaccination	Randomized to placebo	Difference (95% CI)
Antibody to the receptor binding domain (RBD) of the spike protein ≥100 u/ml	55%	18%	Adjusted relative risk 3.1 (1.7–5.8)
Median % virus neutralization	71%	13%	Inter-group difference 58% (11–76%)
Viral neutralization assay positive	60%	25%	Relative risk 2.4 (1.5–4.0)
Median SARS-CoV-2 specific T lymphocyte count per 10^6 CD4-positive cells	432	67	Difference 365 (46–986)

Data source: Hall et al. (2021).

with rheumatoid arthritis after their booster vaccination (Saiag et al. 2022). Whilst these data show that patients who are immunosuppressed are able to mount strong laboratory-defined anti-SARS-CoV-2 responses after a third dose, it remains to be seen whether this translates to improved clinical outcomes.

13.6.7 Genetics

A potential explanation for the spectrum of the severity of symptoms in Table 13.15 is the presence of risk factors such as age, obesity, and diabetes, but another is genetics. This may also part-explain the racial/ethnic differences in Tables 13.17 and 13.20.

HLA

The HLA locus is the major driver of responses to micropathogens, and links with certain alleles (HLA B*07:03, B*46:01, Cw*08:01, DRB1*03:01, and DRB1*12:02) and previous SARS infections were reported, whilst DRB1*11:01 and DQB1*02:02 were linked to susceptibility to MERS. Unsurprisingly, therefore, early (but small) studies also reported links, such as one between DRB1*15:01, DQB1*06:02, and B*27:07, and severe or extremely severe course of COVID-19. Another study linked HLA C*07:29 and B*15:27 with susceptibility to SARS-CoV-2 (Ovsyannikova et al. 2020). A study from Saudi Arabia found HLA-B*07, B*56, and B*57 to be protective of infection, and A*24 of infection and fatality, whilst A*01 was predictive of infection. HLA C*01 and DRB1*08 were linked to infection, and C*18 and DRB1*04 to protection. The strongest (p < 0.001) link with an infection was DRB1*08, which, compared to DRB*04, brought a relative risk (95% CI) of 2.32 (1.71–3.16), with frequencies of 4.8% and 0.7% respectively (Naemi et al. 2021). There may also be a racial/ethnic component in HLA class I alleles (Table 13.24).

> **Cross reference**
>
> Section 9.6.1 outlines the basics of immune response genes, and of HLA in particular.

A meta-analysis of 28 studies with over 13,000 SARS-CoV-2 positive patients found HLA-A*01, HLA-A*03, HLA-A*11, HLA- A*23, HLA-A*31, HLA-A*68, HLA-A*68:02, HLA-B*07:02, HLA-B*14, HLA-B*15.01, HLA-B*40:02, HLA-B*51:01, HLA-B*53, HLA-B*54, HLA-B*54:01, HLA-C*04, HLA- C*04:01, HLA-C*06, HLA-C*07:02, HLA-DRB1*11, HLA-DRB1*15, HLA-DQB1*03, and HLA-DQB1*06 to be linked to severity or mortality (Dobrijević et al. 2023). The reasons for these associations may be that certain HLA types bind SARS-CoV-2 peptides more avidly than other types, thus more efficiently presenting the antigen between leukocytes, so ensuring a good response.

Cytokines, other genes and loci

The cytokine storm, characterized by increases in IL-1β, IL-6, IL-10, IL-12B, TNFα, etc., is associated with more severe disease, especially of the lung (Fricke-Galindo and Falfán-Valencia 2021). A homozygous SNP in *IL6* at 7p15.3 brings an increased risk of severe COVID-19 with an odds ratio (95% CI) of 10.0 (2.02–33.02) (Ghazy 2023). Three SNPs in *IL-10* have each been linked to a daily death rate at p < 0.0001 (Abbood et al. 2023). Genome-wide analysis has revealed a relationship between the severe course of the disease and a cluster of six genes at 3p21.31. Of these, *SLC6A20* encodes the transport protein for ACE2, with certain SNPs linked to a more severe course

TABLE 13.24 Race/ethnicity and HLA class I alleles

Race/ethnicity	Protective alleles	Predisposing alleles
African and African American	A*12:03, B*15:03, C*12:03	
European	C*12:03	A*25:01
Hispanic and Mixed-race American	A*02:02, C*12:03	C*01:02
East Asian	C*12:03	A*25:01, B*46:01, C*01:02

Modified from Malkova A, Kudlay D, Kudryavtsev I, et al. Immunogenetic Predictors of Severe COVID-19. Vaccines 2021: 9; 211. https://doi.org/10.3390/vaccines9030211.

of COVID-19 (Ellinghaus et al. 2020). A meta-analysis pointed to SNPs in molecules involved in the interaction between SARS-CoV-2 and its target cell (ACE2 and TMPRSS2), and in interferon-induced transmembrane proteins, linked to susceptibility and severity of COVID-19 (Pecoraro et al. 2023). Those of blood group A and positive rhesus factor are likely to incur more severe disease, those of group O and negative rhesus factor enjoying a protection (Abuawwad et al. 2023)

13.6.8 Pathology of SARS-CoV-2 variants

A strength of the response to the pandemic was the collection and international sharing of data, including that of the genomes of various SARS-CoV-2 isolates, which allowed for tracking mutations. Data presented by the WHO, described as a 'dashboard', offered a summary of global figures on a number of metrics (https://covid19.who.int/). As the data were supplied by each country, the WHO was unable to verify their accuracy or comparability with those of other countries. Indeed, in May 2022 the WHO itself estimated 15 million global COVID-19 deaths, far exceeding the number given in their dashboard (6.27 million). Nevertheless, dashboard data show clear cyclical patterns of at least six peaks and troughs of varying intensity, described as waves.

The UK Government supplied its own daily and weekly data with several key metrics. These included the number of confirmed cases of COVID-19, admissions to hospital and intensive care, the number of patients requiring ventilation support, the number of deaths, and uptake of vaccinations. Selected data from this source also showed sequential peaks in these indices, punctuated by troughs, since described as waves, each characterized by its own variant of SARS-CoV-2.

The first wave

As the Wuhan strain spread around the UK, its effects were soon noted with infections, hospital admissions, and deaths. The first phase of around 31 weeks began in February 2020 peaked in early to mid-April with over 5,000 daily cases and over 1,000 daily deaths, but by July and August the daily rate of cases and deaths had fallen to 500–600 and less than 10 respectively. The absolute and relative (to cases) high daily death toll reflects the understandable unpreparedness of the health authorities (as in all countries) and quite possibly a failure to recognize the presence of the infection and accurately record data. Hospital mortality fell from 42% in the first few weeks to 16% towards the end of this period, mainly due to more rapid and intense treatment such as ventilation. It is recognized that the total number of deaths and infections were underestimated, perhaps 4- or 5-fold. Given the knowledge of variants of the influenza virus, and of the effects of mutation and natural selection, the appearance of a variant of the 'ancestral' Wuhan strain was, to many, unsurprising, and caused a second wave.

The second wave

The number of confirmed cases and deaths began to rise again in August and September 2020, the former peaking at 83,000 on 29 December 2020, the latter at 1,490 on 21 January 2021, a delay reflecting the natural history of a fatal infection. Notably, the 17-fold increased infection rate far exceeded that of the 30% increase in the case fatality rate, most likely reflecting improved clinical care and developing resistance of the population. Levels of daily cases and deaths fell to less than 2,000 and (once more) less than 10 respectively in May 2021, giving a wave duration of some 39 weeks. Although population vaccination had started towards the end of this period, numbers were as yet too small to have an effect.

The initial strains of the virus originating in Wuhan mutated to generate isotypes with structural changes to the protein coat, including the spike, often the objective of vaccine development. Key mutations to the receptor binding domain (Figure 13.10) included a deletion (H69/70) and certain SNPs (D614G, N501Y, and E484K), although there were many others. Three were of major public health interest, being described as 'variants of concern', and were named by a B number and a letter of the Greek alphabet. The B.1.1.7 variant (designated Alpha) had a 53% increased transmissability compared to the original Wuhan strain, with an R value of 1.75, more than adequately explaining the increased infection rate. Other variants were noted. The B.1.351 variant (Beta) was first identified in South Africa, with a related version emerging in Brazil, whilst P.1 (Gamma) emerged in Brazil, where it was a major public health issue. However, in the UK neither variants were major public health issues.

The third wave

The third wave began in June 2021, with an initial peak in daily cases of 62,000 in July, but unlike previous waves, case numbers failed to fall, but persisted at around 30,000–40,000 a day for around 26 weeks. Although the number of infections was still worryingly high, all other indices (hospital admissions, need for lung ventilation, death) were considerably lower, with a peak of 144 deaths on 17 September, a trough of 97 on 5 October, and a second peak of 192 on 30 October. These data most likely further reflect better community and hospital care, less pathogenic variants of the virus, and the developing effects of the mass vaccination programme (section 13.6.6).

A new SARS-CoV-2 variant, Delta or B.1.617.2, first identified in India, was the cause of this wave, and was defined by further changes to the genetic make-up, and so the structure, of the viral spike protein. Evidence points to an increased infectiveness relative to other variants, with an R value of 7, thus replacing the Alpha variant as the dominant species as early as July 2021. The Delta variant had an increased adjusted hazard ratio for hospital admission of 2.26 compared to the Alpha variant. But focusing only on data on admissions and deaths does not account for the issue of out-of-hospital morbidity.

The Omicron wave

Global genetic surveillance identified a further change in isolates from infected individuals. B.1.1.529, also known as Omicron, emerged in Botswana, Hong Kong, and South Africa in November 2021, and was of interest because it carried 32 spike protein mutations; most of these were SNPs, but there were also some deletions. There was concern that these mutations could have changed the transmissibility, infectivity, immune escape, and pathogenicity of this variant, as had previously been seen in other variants. Its estimated R value of around 10, which is around 3.2 times that of the Delta variant, explains why it spread rapidly, with over 15,000 cases within two weeks of its recognition. In the week commencing 15 December 2021, 12% of cases in intensive care units carried Omicron, rising to 100% in the week of 16 February 2022, thus completely displacing the Delta variant. This led to vastly increased rates of infection, with peaks at 272,000 on 31 December 2021 and 275,000 on 8 January 2022, and with a modest increase in the case fatality rate, peaking on 23 January 2022 at 306 deaths. However, as with other variants, these data do not provide information on how many people have been taken out of the workforce by an infection, a topic of particular relevance for front-line hospital staff, who were at high risk of infection and whose absence from work exacerbated the strain on the care of patients with any health issue. Despite this very high rate of infectivity, compared to the Delta variant, Omicron infection carried low adjusted hazard ratios of 0.38 for hospital admission, and 0.41 for the need for intensive care.

Given the developing knowledge of the virus and its ability to mutate, the appearance of a variant of the original Omicron (renamed as B.1.1.529/BA.1) in England in January 2022, named BA.2, was perhaps unsurprising. Although BA.2 had an increased growth rate compared to BA.1, preliminary analysis found no evidence of a greater risk of hospitalization. The BA.1 sub-wave waned during February 2022, with a trough of 25,000 cases on 28 February and 127 deaths on 7 March, following which both metrics increased, with a BA.2 sub-peak of 109,324 cases on 23 March, and 303 deaths on 10 April.

Additional Omicron variants BA.4 and BA.5 were designated variants of concern on 18 May 2022, both with growth rates exceeding that of BA.2, which they were expected to displace. As of 21 June 2022, 22.3% and 39.5% of new cases were estimated to be due to BA.4 and BA.5 respectively. They were likely causes of a new peak in cases, admissions, and deaths in England in July 2022 following a trough in these metrics in May 2022. By mid-August, daily new cases had fallen to less than 4,000 and deaths to less than 150, pointing to an end of this particular wave.

13.6.9 Infections, admissions, and deaths over the pandemic

Overview

Table 13.25 summarizes selected data on the variant waves, excluding the first (as data are likely to be of poor reliability), and tells several stories. Although the Delta variant caused a three-fold increased rate of infections compared with the Alpha variant, the fact that it caused about the same number of admissions and deaths indicates that it posed a much-reduced clinical danger to those infected.

TABLE 13.25 Data on COVID-19 in England and Wales

	Alpha wave	Delta wave	Omicron BA.1 wave	Omicron BA.2 wave
Dates (week of)	7 September 2020–31 May 2021	1 June 2021–12 December 2021	13 December 2021–28 February 2022	1 March 2022–12 June 2022
Duration (weeks)	39	27	12	15
Cases/week	77.5 (36.8–127.8)	243.1 (210.0–260.3) (< 0.001)	597.3 (286.4–834.8) (< 0.001)	79.5 (53.2–295.2) (< 0.001)
Admissions/week	6.2 (2.4–10.6)	5.6 (5.2–6.1) (0.473)	10.1 (8.0–13.3) (< 0.001)	7.7 (4.4–12.4) (0.272)
Deaths/week	1.2 (0.3–2.9)	0.8 (0.6–0.9) [0.135]	1.2 (0.9–1.7) [0.001]	1.0 (0.5–1.4) [0.164]
Admissions/case (%)	6.7 (6.1–8.5)	2.3 (2.1–2.7) (<0.001)	2.0 (1.1–3.1) (0.369)	8.9 (3.6–10.9) (<0.001)
Deaths/case (%)	1.39 (0.9–1.8)	0.31 (0.26–0.34) (<0.001)	0.28 (0.11–0.40) (0.637)	0.98 (0.35–1.19) (0.004)
Deaths/admission (%)	20.6 (13.7–23.7)	13.8 (10.9–15.4) (0.002)	12.5 (9.2–13.6) (0.494)	11.4 (9.0–14.8) (0.510)

Data per week are thousands, given as median and 95% CI, with Mann–Whitney p value compared to the previous variant.

© Office for National Statistics. Source: https://www.ons.gov.uk/. Licensed under the Open Government Licence 3.0.

Omicron BA.1 was more than twice as infectious as Delta, and with this increased number of infections brought an increase in the rate of admissions and deaths. However, crucially, the rates of admission per case, deaths per case, and deaths per admission were no different, implying that the virus itself was of equivalent pathogenicity once within its host. Conversely, its successor, BA.2 was markedly less infectious, equivalent to that of the Alpha variant, but had a higher rate of admissions and deaths than BA.1, although the number of deaths per admission was no different.

These data show that each variant has a unique pathological fingerprint in terms of both infectiousness and lethality. However, an important caveat is the effect of vaccinations, which became a factor during the middle of 2021, at which point the vaccination programme in the UK was estimated to have saved 60,000 lives. Nevertheless, an important analysis is the slow but steady fall in certain metrics, indicating better population handling of the virus, and, for death, much improved clinical care once in hospital, the latter evidenced by the fall in deaths per admission. The Spearman correlation coefficients between the wave sequence and the deaths per case was r = -0.31 (p = 0.002), and with deaths per admission the data were r = -0.43 (p < 0.001).

Further analyses: Waves compared

The mean ages of those who died in the first and second waves were 76 and 71 respectively, whilst those succumbing to an infection in the third wave and being admitted to hospital were younger than those of earlier waves, with a mean age of 68 years. One explanation for this trend is that the older, and so more susceptible, had died in the first two waves; another is that third-wave admissions had a

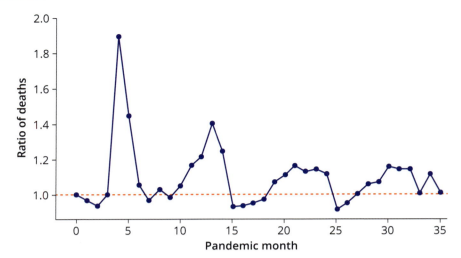

FIGURE 13.11
Deaths 2020–2 compared to 2015–19. Data are mean number of deaths per month in England and Wales during 2020–2 compared with the same month in 2015–19. Month 0 = January 2020, 10 = October 2020, 20 = August 2021, 30 = June 2022. Data points above line 1.0 describe an excess of deaths in 2020–2, points below the line as fewer deaths in 2020–2 compared to 2015–19.

Data © Office for National Statistics. Source: https://www.ons.gov.uk/. Licensed under the Open Government Licence 3.0.

high frequency of those not vaccinated: the vaccination process at the time had not reached all those aged under 30.

A weakness with the data in Table 13.25 (and elsewhere) is that they fail to adjust for the time of year, as more deaths are to be expected in the winter. Figure 13.11 shows the changes in the ratio of deaths in each month of the pandemic in relation to deaths in the same month in the 5 years preceding the pandemic. These data point to the expected excess deaths in April 2020 (Wuhan: Months 4–6) and January 2021 (Alpha: Months 11–14), and a prolonged and flattened peak in September–November (Delta), extending to December 2021 (Omicron BA.1: Months 19–24). However, the fewer deaths in January and February 2022 (months 25–26) are notable, a period when Omicron BA.1 was the most prevalent variant, implying reduced lethality. The ratio of deaths subsequently increased in the spring and summer of 2022 (from month 28), implying increased BA.2 mortality. However, it must be emphasized that there may be other causes of increased mortality. New variants of Omicron, such as XBB and BQ1, appeared in the Autumn of 2022, followed by variants CH.1.1 and XBB.15 towards the end of 2022. As of April 2023, the frequencies of Omicron variants in the UK were 44% XBB.1.5, 27% XBB, 8% CH.1.1, 4% BQ1, with the remaining 17% being variants that each represented <1%. The frequencies point to a developing landscape not dominated by a single variant (e.g. as was the case with Alpha and Delta), but a broad population of related variants whose precise individual infectiousness and lethality is difficult to quantify.

Key points

It is widely believed that the considerable reduction in the death rate of the third wave is due to vaccination. In May 2021, the National Institutes of Health estimated that the roll-out of vaccines for COVID-19 had prevented nearly 140,000 deaths in the United States alone. In the United Kingdom, the Health Security Agency estimated in September 2021 that vaccination had prevented around 123,000 deaths, over 230,000 hospitalizations, and around 24 million infections.

13.6.10 Further consequences of COVID-19

Effects on primary and secondary healthcare

An immediate consequence of the pandemic in the UK was a huge reduction in self-referral to accident and emergency, and to primary and secondary care for non-respiratory symptoms, as the public were urged to self-isolate. Other social avoidance methods also reduced the frequency of other conditions such as influenza, viral hepatitis, asthma, tonsillitis, bronchitis, laryngitis, conjunctivitis, chickenpox, and diarrhoeal diseases. However, a more worrying observation is that those who would normally have had treatment for a chronic condition (cancer, cardiovascular disease, and orthopaedic surgery) have been denied treatment as secondary care struggles with the logistical consequences of the virus.

These issues contributed to an extension of the waiting list for an NHS appointment, which in January 2022 stood at 6 million, with 312,000 people waiting for more than a year. It has been estimated that this backlog may take years to clear. Nevertheless, the greatly reduced pathogenicity of the Omicron variant contributed to a relaxation of public health measures (self-isolation, mandatory use of face masks, etc.), and their subsequent removal altogether in early 2022, around 2 years after the pandemic first gripped the country. Like many other respiratory viruses, SARS-CoV-2 entered its endemic phase.

Long COVID

If (as expected) the long-term effects of a severe infection are similar to those of other SARS/ARDS outcomes, in the medium term affected people are at risk from other health issues. These are likely to include pulmonary fibrosis and damage to the myocardium, and there are developing data on neurological impairment with cognitive decline. These are potential issues in those whose disease has been severe, possibly requiring intensive care. One of the difficulties for public health is the degree of relatively low-level morbidity in those with mild or moderate disease. In one meta-analysis, persistent symptoms were defined as those present for ≥60 days after diagnosis, symptom onset, or hospitalization, or ≥30 days after recovery from the acute illness or hospital discharge. Of almost 10,000 patients, 72.5% experienced at least one persisting symptom. Individual symptoms occurring most frequently included fatigue or exhaustion (40%), shortness of breath or dyspnoea (36%), sleep disorders or insomnia (29%), and cough (17%).

The design and quality of studies focusing on long COVID vary greatly, which has implications for their interpretation and often limits their direct comparability and combinability. There are major differences in how time zero—that is, the beginning of the follow-up interval—is defined, as well as in the follow-up lengths, the definitions of outcomes and illness severity, and the patient populations that participated in the studies. In the UK, NICE (NG188) and other bodies recommend the use of three terms for signs and symptoms that develop or are present after an infection and that cannot be ascribed to an alternative diagnosis, which are defined as follows:

- Acute COVID-19 for those signs and symptoms for up to 4 weeks

- Ongoing symptomatic COVID-19 where signs and symptoms are present from 4 to 12 weeks

- Post-COVID-19 syndrome, where signs and symptoms persist for 12 weeks or more after the acute infection.

Notwithstanding the above, the precise definition of long COVID and its long-term health consequences requires international consensus. New data continue to be presented. A study of over 1,000 post-COVID-19 subjects assessed 30 days after their diagnosis collected signs and symptoms into six groups: cognitive (the most common), pain-syndrome, pulmonary, cardiac, **anosmia-dysgeusia**, and headache.

13.6.11 COVID-19 Summary

On 5 May 2023, the WHO Emergency Committee on the COVID-19 pandemic reported that COVID-19 is now considered an established and ongoing health issue and therefore no longer constitutes a public health emergency of international concern, which effectively ended the pandemic. On 18 May 2023, NHS England stepped down the national NHS incident level to 3, whereby management shifts from a nationally coordinated to a regional level. SARS-CoV-1 and MERS-CoV were local infections that eventually dissipated, but information on these viruses was of little help in combatting SARS-CoV-2. The truly global pandemic spread as mathematical modelling predicted, impacting each country around the world. By this closure date, the WHO estimated that the virus was linked to almost 765 million confirmed cases and 6.9 million deaths globally, these being 9.5% and 0.085% respectively

TABLE 13.26 Deaths involving, or due to, influenza and pneumonia, and COVID-19, in England and Wales

Deaths involving . . .	January–June 2022	July–December 2022	January–June 2022
Influenza and pneumonia	44,314	42,196	51,506
COVID-19	20,659	10,533	12,982
COVID-19/Influenza and pneumonia	46.7%	25.0%	25.2%
Deaths due to . . .	**January–June 2022**	**July–December 2022**	**January–June 2022**
Influenza and pneumonia	9,475	10,533	14,732
COVID-19	13,741	8,067	7,935
COVID-19/Influenza and pneumonia	145.0%	76.6%	53.9%

Data © Office for National Statistics. Source: https://www.ons.gov.uk/. Licensed under the Open Government Licence 3.0.

of the world's population, a case fatality rate of 0.906%. In the UK, these figures were 24.6 million cases and 227,000 deaths, these being around 36.2% and 0.33% of the population respectively, with a case fatality rate of 0.922%. After 4 years of the pandemic/endemic, in December 2023 the WHO report 772 million confirmed cases and almost 7 million deaths, a case fatality rate of 0.907%.

As in any such infection, modelling predicted that the effects of COVID-19 would rise in a population with no herd immunity, and then fall, as was indeed the case. In the UK, all measurable metrics fell during the spring of 2023, and in England and Wales, deaths due to, or involving, influenza or pneumonia exceeded those of COVID-19 every week from January to November 2023 by an average factor of 4.9 and 2.0 respectively. Table 13.26 shows these data in the first and second parts of 2022, and the first half of 2023. These data point to a trend in the fall in deaths due to COVID-19 compared to pneumonia and influenza. However, these data do not consider the morbidity of these infections and the chronic effects of SARS-CoV-2—that is, long COVID—which are yet to be fully established. Nevertheless, the pandemic induced an unprecedented scientific, clinical, and public health revolution that should provide essential armament against future infections.

Chapter summary

- Diseases of the respiratory system are the fourth most common cause of death (61,679 deaths: 10.2% of all deaths in England and Wales in 2021), but as COVID-19 is effectively a respiratory disease, this made such diseases the third most common, with 131,605, and 21.6% respectively.

- Bronchitis, emphysema, and COPD together lead to the deaths of over 25,000 people annually.

- Although asthma is rarely fatal, it brings a major personal health burden, as does COPD.

- In most respiratory disease, the major clinical consequence is pneumonia, which may be fatal and may progress to acute respiratory distress syndrome.

- COVID-19, caused by SARS-CoV-2, was the most important public health issue for a century and requires mass vaccination becoming established as endemic in 2023.

 # Further reading

- Demirhindi H, Mete B, Tanir F, et al. Effect of heterologous vaccination strategy on humoral response against COVID-19 with CoronaVac plus BNT162b2: A prospective cohort study. Vaccines 2021: 10; 687.

- Dobrijevic Z, Gligorijevic N, Sunderic M, et al. The association of human leykocyte antigen (HLA) alleles with COVID-19 severity: A systematic review and meta-analysis. Rev Med Virol 2022: e2378.

- Salvamani S, Tan HZ, Thang WJ, et al. Understanding the dynamics of COVID-19; implications for therapeutic intervention, vaccine development and movement control. Br J Biomed Sci 2020: 77; 168–84.

- Clift AK, Coupland CA, Keogh RH, et al. Living risk prediction algorithm (QCOVID) for risk of hospital admission and mortality from coronavirus 19 in adults: national derivation and validation cohort study. Br Med J 2020: 371; m3731.

- Lu R, Zhao X, Li J, et al. Genomic characterisation and epidemiology of 2019 novel coronovirus: implications for virus origins and receptor binding. Lancet 2020: 395; 565–74.

- Cruz RJ, Currier AW, Sampson VB. Laboratory Testing Methods for Novel Severe Acute Respiratory Syndrome-Coronavirus-2 (SARS-CoV-2). Front Cell Dev Biol. 2020: 8; 468. Doi: 10.3389/fcell.2020.00468.

- Marwah M, Marwah S, Blann A, et al. Analysis of laboratory blood parameter results for patients diagnosed with COVID-19, from all ethnic group populations: A single centre study. Int J Lab Hematol. 2021: May; 3: 1243–51.

- Malkova A, Kudlay D, Kudryavtsev I, et al. Immunogenetic Predictors of Severe COVID-19. Vaccines 2021: Mar 3; 9(3): 211. doi: 10.3390/vaccines9030211.

- Vestbo J, Hurd SS, Agusti AG. Global strategy for the diagnosis, management and prevention of chronic obstructive pulmonary disease. Am J Respir Disease Crit Care Med 2013: 187; 347–65.

- Smith MC, Wrobel JP. Epidemiology and clinical impact of major co-morbidities in patients with COPD. Int J COPD 2014: 9; 871–88.

- Hosseini M, Almasi-Hashiani A, Sepidarkish M, et al. global prevlance of asthma-COPD overlap in the general population. Respir Res 2019: 20; 229.

- Dharmage SC, Perret JL, Custovic A. Epidemiology of asthma in children and adults. Frontiers Paeditr 2019: 7; 246.

- Kim KW, Ober C. Lessons learned from GWAS of asthma. Allergy Asthma Immunol Res 2019: 11; 170–87.

- Specjalski K, Jassem E. MicroRNAs: Potential biomarkers and targets of therapy in allergic diseases? Archiv Immunol Therap Exp 2019: 67; 213–23.

- Turner SW, Chang AB, Yang IA. Clinical utility of exhaled nitric oxide fraction in the management of asthma and COPD. Breathe 2019: 15; 306–16.

- Bakakos A, Loukides S, Bakakos P. Severe eosinophilic asthma. J Clin Med 2019: 8; 1375.

- Musher DM, Thorner AR. Community acquired pneumonia. New Engl J Med 2014: 371: 1619–28.

- Leone N, Bouadma L, Bouhemad B, et al. Hospital acquired pneumonia in ICU. Anaesth Crit Care Med 2017: 36; 83–98.

- Blann A, Heitmar R. A Guide to COVID-19 for Health Care Professionals. Cambridge Scholars 2022.

Useful websites

- National Institute for Health and Care Excellence: https://www.nice.org.uk/
- UK government statistics and guidance on COVID-19: https://www.england.nhs.uk/coronavirus/primary-care/about-COVID-19/
- World Health Organization data and guidance on COVID-19: https://www.who.int/emergencies/diseases/novel-coronavirus-2019

 Discussion questions

13.1 From a pathophysiological perspective, why it is unwise to classify cancer of the bronchus with cancer of the lung?

13.2 Compare and contrast asthma with chronic obstructive pulmonary disease.

13.3 How has the COVID-19 pandemic impacted on biomedical science and clinical practice?

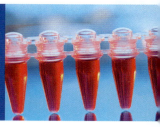

14

Disease of the genitourinary system

The urinary system consists of the kidneys, ureters, bladder, and urethra, together linked to 8,616 non-cancer deaths in England and Wales in 2021, the leading (53.4%) aetiology being urinary tract infection. The male reproductive system consists of the testes, vas deferens, prostate, seminal vesicles, Cowper's glands, and penis. The female reproductive system consists of the ovaries, fallopian tubes, uterus, cervix, vagina, vulva, and breasts. The two systems (excretion, reproduction) have very little direct physiological aspects in common, and where they do, this is the result of positional anatomy. The male urethra would function as efficiently by exiting between the scrotum and the anus, whilst there is no intrinsic reason why the female urethra does not empty into the vagina. In both cases, natural selection has determined it otherwise.

An estimated 220–500 people per million (around 0.036% of the population) may be described as intersex: they are not defined by the simple binary of male–female. Others do not identify with the sex they were assigned at birth and may be described as transgender. Which diseases of the genitourinary system someone is susceptible to therefore ultimately depends on their individual bodies, which can take many forms, depending for instance on whether or not a transgender man or woman chooses to have gender-confirmation surgery.

The urinary system is described in section 14.1, the male reproductive system in 14.2, whilst the female reproductive system is split into that not actively involved in pregnancy (gynaecology, section 14.3) and that which is (obstetrics, 14.4). The chapter concludes in section 14.5 with an examination of sexually transmitted disease.

Learning objectives

After studying this chapter, you should confidently be able to:

- explain the major disease of the urinary system
- understand the pathology of the male reproductive system
- describe major issues in gynaecology
- outline potential problems in pregnancy and the puerperium and their treatment
- recognize the principal sexually transmitted diseases, their consequences, and treatments

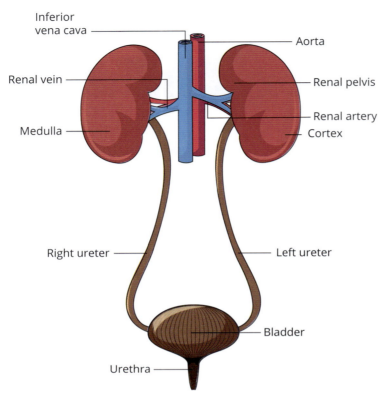

FIGURE 14.1
Anatomy of the urinary system.

14.1 The urinary system

The purpose of the urinary system tissues is to generate (the kidneys), pass (ureters), and store (the bladder) urine, and ensure its passage out of the body (via the urethra) (Figure 14.1). However, the kidneys also have numerous other functions.

14.1.1 The kidneys

Anatomy and physiology

These paired organs are located at the rear of the abdominal cavity, roughly at the point between the thoracic and lumbar vertebrae. They are supplied with oxygenated blood via the renal arteries, both branches of the aorta (Figure 14.1), whilst blood leaves via renal veins, which join the interior vena cava. The blood vessels connect with the kidney at the hilum, where the ureter leaves and where nerves and lymphatics enter. The gross internal architecture of the kidney consists of the renal pelvis (the intra-renal extension of the ureter), and parenchyma of around a million nephrons arranged in an inner medulla and an outer cortex. Above each kidney is an adrenal gland, discussed in section 15.3.

The nephron is a long tube where urine is formed. At one end is a Bowman's capsule, consisting of glomerulus into which a knot of arteriolar blood vessels form (Figure 14.2).

At the terminal end of the nephron, the tubule merges with others into a collecting duct, which in turn forms the ureters. The central portion of the tubules is in three parts: the proximal convoluted tubule, the loop of Henle, and the distal convoluted tubule, all in intimate contact with arterioles and venules. These and other structures conduct the major functions of the kidney:

- Regulation of blood electrolytes and osmolarity (Na^+, K^+, Ca^{2+}, Mg^{2+}, Cl^-, HPO_4^{2-}, other small molecules)

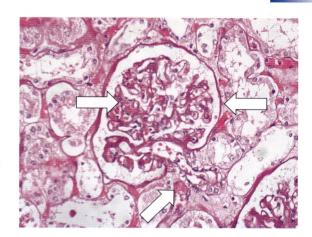

FIGURE 14.2
The glomerulus. The glomerulus is the circular body in the centre of the figure. The left arrow points to the knot of blood vessels, the right arrow to the circumferential Bowman's capsule, and the lower arrow to the entry of blood vessels into the glomerulus.

Credit: Ed Uthman. Retrieved from: https://www.flickr.com/photos/euthman/3423439397.

- Excretion of waste metabolites (urea, creatinine, ammonia/ammonium, bilirubin, uric acid)

- Maintenance of blood pH (H^+, HCO_3^-)

- Regulation of blood volume and blood pressure (conserving or eliminating water, release of renin and so regulation of the angiotensin–aldosterone pathway)

- Endocrine/metabolic function (calcitriol for vitamin D metabolism and erythropoietin for red blood cell production)

Epidemiology

Like many diseases, renal disease may be described as a silent killer: in 2021 the UK's Office for National Statistics (ONS) reported 8,224 non-malignant deaths due to genitourinary disease in England and Wales—slightly fewer than those 3,757 caused by renal cancer and 4,947 due to bladder cancer combined. In 2017, the Global Burden of Disease Study reported 1.23 million deaths due to chronic kidney disease (CKD: 2.2% of all deaths), the leading subgroup being CKD due to type 2 diabetes (28.4%) followed very closely by hypertensive CKD (28.2%), then glomerulonephritis (15.4%) and type 1 diabetes (6.3%), with other causes of renal disease causing 21.7% of deaths.

Pathology

The cell biology and aetiology of renal cancer are discussed in section 5.6, and so will not be duplicated, but much of clinical renal cancer has features in common with the clinical pathology of other aspects of renal disease. Of the non-cancer disease, the leading aetiologies are diabetic, hypertensive (both relatively easy to diagnose from blood tests and blood pressure respectively, and often to be expected—see section 7.4, Table 7.10, and Box 7.1), and immunological, both autoimmune and inflammation, which of the kidney is nephritis, and of the glomerulus is glomerulonephritis. However, primary kidney disease includes Alport syndrome (accounting for 1–2% of those requiring renal replacement therapy) and nephrotic syndrome (causing minimal change disease, membranous glomerulonephritis, and focal segmental glomerulosclerosis). The non-glomerular section of the nephron—the tubules—may also be subject to pathology, most as acute tubulointerstitial nephritis, which may be linked to drugs or to viral or bacterial infections.

The major atherothrombotic cause of renal disease is renal artery stenosis, which reduces blood flow into the kidneys. Metabolic disease leads to cholesterol emboli, calcification, and the formation of stones, which are analogous to gall stones. There may be other structural disease, such as cysts, one or more of which are present in 50% of those aged over 50. The principal genetic disease is autosomal dominant polycystic kidney disease (ADPKD), 85% of which is caused by mutations in *PKD1*,

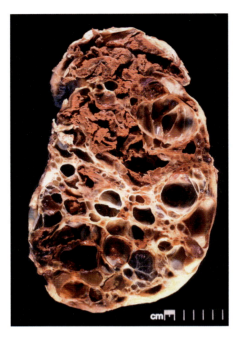

FIGURE 14.3
Polycystic renal disease. Markedly damaged kidney from a patient with autosomal dominant polycystic kidney disease.

From Ahmed (ed.) *Clinical Biochemistry*, second edition, Oxford University Press, 2016.

located at 16p13.3 and coding for polycystin-1, the other 15% being caused by mutations in *PKD2*, located at 4q22.1 and coding for polycystin-2. The most common inherited kidney disease, ADPKD has a prevalence of ~1,300/million and an incidence of ~180/million. Over half will require dialysis or transplantation, and account for ~6% of those needing dialysis as the disease destroys the renal parenchyma (Figure 14.3).

In addition to diabetes and hypertension, the kidney can also be the target of numerous other diseases, several of which we have explored elsewhere and will address in the text to come (Table 14.1). The pathology of renal disease can be classified in a pseudo-anatomical manner into pre-, true (or intrinsic), and post-renal disease (Figure 14.4). Pre-renal disease is due to insufficient blood entering the kidneys. Reasons for this include renal artery stenosis, thrombosis and/or aneurysm within the

TABLE 14.1   Diseases with a renal component

Disease	Section
Anti-phospholipid syndrome	11.6.3
Haemolytic uraemic syndrome	11.5.4
Myeloma and gammopathies	11.4.4
Polyarteritis nodosa (and other vasculitides)	10.3.5
Rheumatoid arthritis	10.3.2
Systemic lupus erythematosus	10.3.3
Systemic sclerosis	10.3.4
Thrombotic thrombocytopenia purpura	11.6.2
Pre-eclampsia	14.4.2
Diabetes	7.4

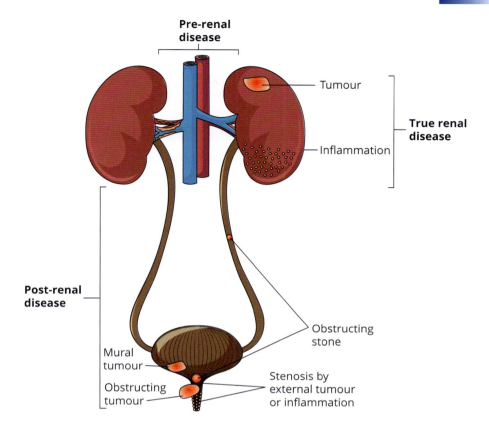

FIGURE 14.4

Types of renal disease. Renal disease can be classified as pre-, true, or post-renal disease, and reflects the aetiology of certain clinical aspects of its pathophysiology.

abdominal aorta, and low blood volume and/or pressure due to the poor cardiac output of heart failure. Not only will blood fail to be cleaned, but the kidney itself will become ischaemic.

True renal disease is that of the organ itself and includes cancer and inflammatory disease (nephritis). Post-renal disease follows from downstream issues. Should the urethra or ureters be stenosed or obstructed by a renal stone or tumour, or the bladder become full, urine will not be expelled, but will back up into the kidney, causing local damage.

The consequences of renal disease can be viewed in terms of the various functions of this organ:

- Altered levels of electrolytes and other molecules, and so osmolarity

- Increased levels of excretory products, principally urea, creatinine, and uric acid

- Loss of correct pH, resulting in acidosis (mostly) or alkalosis (less often)

- Increased or decreased water in the blood, leading to hypertension or hypotension, and so fluid overload (resulting in hypertension and oedema) or fluid insufficiency (resulting in dehydration)

- Loss of endocrine function, leading to vitamin D deficiency with potential bone disease (renal osteodystrophy), perhaps compounded by loss of calcium, and anaemia resulting from insufficient erythropoietin (section 11.1.1)

The endothelium of the glomerulus has a barrier function of a molecular sieve to retain molecules with a relative molecular mass of ~ 67 kDa, the approximate size of albumin. A trace amount of albumin may well be present in normal urine. However, should there be glomerulonephritis and endothelial damage, albumin will not be retained and pass into the urine, leading to falling levels of serum albumin, and so changes in the osmolarity of the blood, and proteinuria. Albumin can carry many ions and molecules (calcium, bilirubin, iron, and thyroxine), so that hypoalbuminaemia can have several metabolic consequences.

Key points

Renal disease can have many different causes, and has numerous consequences, such as loss of ion and water homeostasis, bone disease, and anaemia.

SELF-CHECK 14.1

Renal disease can be a component of many other diseases: how many can you name?

Presentation and diagnosis

Renal disease may present with any of a large number of non-specific symptoms that include back pain, nausea, vomiting, diarrhoea, itching, malaise, tiredness, etc., but it may also be asymptomatic. The leading sign is changes in frequency of urination and volume of urine released—an increase (polyuria), a gross reduction (oligouria: urine output <400–500 ml/day), or cessation (anuria)—and these are most likely to bring the patient to a practitioner and will help diagnosis. Proteinuria may be present should toilet water becoming frothy when urinating.

Dipstick urine analysis (sometimes called urinalysis) is a useful screening test in primary practice and in outpatient departments for a number of pathological features, some reflecting renal and/or bladder pathology (Table 14.2). With relatively poor sensitivity, a negative dipstick result does not exclude the particular pathology, but a positive result demands further investigation.

More sensitive dipsticks can determine a much lower level of protein (microalbuminuria), an early sign of glomerular damage. Other gross tests of urine include colour (e.g. red: potential haemolysis, orange: potential jaundice), clarity (cloudy: crystals, bacteria, leukocytes), and smell (the sweet or fruity smell due to ketones in diabetic ketoacidosis). The laboratory can measure many more analytes in urine, two of the most common being an exact determination of protein in the urine, often in all urine collected over 24 hours, and of the level of creatinine. The latter can be compared to the level of albumin in the urinary albumin:creatinine ratio (uACR), where a result of <3 mg/mmol is defined as normal to mildly increased (stage A1), 3–30 as moderately increased (stage A2), whilst ≥30 mg/mmol (stage A3) is taken to be significantly increased proteinuria. A result >220 mg/mmol defines the most severe form of renal disease—nephrotic syndrome. The laboratory can also perform microscopy for red and white blood cells, casts, and bacteria, which may also be cultured.

Cross reference

Chapter 6 in the volume on *Medical Microbiology* in this series provides a more detailed overview of urinary sampling and the processing of samples in the laboratory.

TABLE 14.2 Urine analysis

Analyte	Pathology
Bilirubin	Haemolytic anaemia, liver damage
Blood	Haemorrhage in the kidney and/or bladder
Glucose	Hyperglycaemia, often due to diabetes
Ketones	Potentially diabetic ketoacidosis
Leukocytes	Implies bacterial infection of the urinary system
Nitrite	Presence of coliform bacteria
pH	Acidity or alkalinity of urine, and so possibly of the blood
Protein	Increased glomerular permeability
Specific gravity	High levels imply dehydration

TABLE 14.3 CKD Stages

Stage	eGFR	Renal function
1	≥90	Normal or increased glomerular function
2	60–89	Mildly decreased
3A	45–59	Mildly to moderately decreased
3B	30–44	Moderately to severely decreased
4	15–29	Severely decreased
5	<15	End-stage renal failure

Units: ml/min/1.73m².

Diagnosis proceeds with blood tests, focusing on serum urea, sodium, potassium, and creatinine (i.e. U&Es). Increased urea implies AKI, whilst increased creatinine is the marker of choice in CKD. A more complex measure of renal function is the creatinine clearance test, which quantifies excretion of creatinine over a set period of time. However, the method most commonly used is the glomerular filtration rate (GFR), which is the rate of fluid filtration at the glomerulus and is estimated (hence eGFR) by an equation. NICE NG203 recommends the Chronic Kidney Disease Epidemiology Collaboration (CKD-EPI) equation, which calls for serum creatinine, age (as this is linked to falling renal function in health), sex, and race. Therefore, the eGFR is basically serum creatinine adjusted for age, sex, and race, with units of ml/min/1.73 m². Calculators are freely available online.

However, the place of race in the CKD-EPI and other equations has been questioned as it assigns, for the same level of serum creatinine, a higher (i.e. better) eGFR for Black people than it does for those who are not Black. A meta-analysis reported that removal of race adjustment improves bias, accuracy, and the precision of an eGFR result for Black people. Certain equations prefer to use Cystatin C in place of creatinine to assess renal function. (Umeukeje et al. 2022; Delanaye et al. 2022).

By convention, CKD is classified into stages according to the eGFR (Table 14.3), the final step described variously as end-stage kidney disease (ESKD), end-stage renal disease (ESRD), and established renal failure.

Diagnosis may be completed by a third set of investigations: imaging. X-rays can give information about calcification and stones, but ultrasound is far more informative, such as for the detection of cysts and, using colour Doppler, renal artery blood flow (peripheral artery disease is discussed in Chapter 8). Further information can be offered by CT and MRI, whilst contrast media can be introduced into the renal pelvis to determine dilations and possible obstructions. More complex investigations include renal arteriography and scintigraphy, the latter often with **technetium**. Once all pertinent information has been collected, the practitioners will determine a more precise diagnosis.

Having accounted for malignancy, diabetes, and hypertension, the greater part of renal disease that remains is inflammatory/immunological, mostly involving the glomerulus, therefore glomerulonephritis (Chapter 10), often characterized by a thickened basement membrane (Figure 14.5).

Many of the fine distinctions between the different forms of the disease (such as proliferative or segmental involvement) can only be made by ultrasound-guided biopsy. However, autoantibodies can be detected by indirect immunofluorescence on sections of renal tissues or cells fixed on to a glass slide (e.g. p-ANCA and c-ANCA—Figure 9.21). Many of these diseases are of a vasculitic nature, as described in section 10.3.5, where Table 10.12 and nearby text are relevant.

From a purely clinical perspective, disease can be classified as AKI or CKD, definitions of which stem from the rapid (or not) presentation or signs and symptoms, and levels of urea and creatinine. AKI implies a direct injurious factor, such as a toxin (ethylene glycol, contrast medium, NSAIDS, high-dose antibiotics); bacterial (*Staphylococcus*, *Pseudomonas*, and *Streptococcus* species), viral (Epstein–Barr virus, hepatitis B and C) or parasitic (malaria, toxoplasmosis, schistosomiasis) infection; or an obstruction within the previous week or 10 days. However, true renal infection is difficult to prove

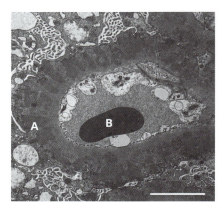

FIGURE 14.5

Membranous glomerulopathy. An electron micrograph showing a grossly thickened glomerular basement membrane (GBM) (A). The lumen contains a red blood cell (B), which would be expected to have a width of around 1μm, so that a crude estimate of the width of the GBM is also around 1μm, a figure around five times that of a normal GBM at 0.2 μm.

From Orchard & Nation, *Histopathology*, second edition, Oxford University Press, 2018.

unequivocally without a biopsy, although a urethral catheter can capture urine immediately leaving the kidney (discussed further in section 14.2.1).

The pathology of AKI is often acute tubular necrosis. In addition to changes in urination pattern (mostly reductions), unexpected signs and symptoms include **haematuria** and oedema, and potentially raised blood pressure. It is also possible that CKD may transform to AKI, as can be the case in lupus nephritis, and it may arise in a secondary manner from the infection of other organs (streptococcal throat infection, pneumonia, infective endocarditis). AKI is particularly prevalent in intensive care units and contributes to mortality. In COVID-19, AKI is the leading in-hospital complication.

The key laboratory result in an apparently uncomplicated AKI presentation is an increase in serum urea, which can be compared to serum creatinine. Serial measurements 24–48 hours apart will also help assess the extent of renal injury, and although serum creatinine may rise, the relative increase in urea will be greater. With the ease of measurement of U&Es, much CKD is discovered (or a suspicion confirmed) in the course of the investigations of other disease and is often linked to a raised creatinine with a normal urea. Additional routine investigation in both situations will be full blood count, serum albumin, and LFTs, and as the CKD stage increases, measuring calcium, phosphates, parathyroid hormone, and alkaline phosphatase is justified due to the increased risk of osteoporosis and osteodystrophy. Should there be evidence of bone disease or should it be suspected, dual-energy X-ray absorptiometry (DXA) scanning may be undertaken to determine the extent of the problem.

Key points

Renal disease can be classified as pre-, true, or post-, and these give clues to aetiology. Key investigations are urine analysis, blood tests, and imaging.

SELF-CHECK 14.2

Describe the value of urine analysis.

Medical treatment

Treatments of diabetic and hypertensive renal disease are discussed elsewhere, principally by the risk factors themselves. In the UK, practitioners are likely to refer to NICE documents NG148 on AKI, NG203 on CKD, and CG157 on hyperphosphataemia. These documents underline the importance of addressing risk factors where present, in addition to lifestyle and diet (i.e. minimal salt intake). CKD carries a weak increased risk (<2 fold) of venous thromboembolism (VTE) and nephrotic syndrome a moderate (2–9-fold) increased risk of VTE (regardless of whether being an inpatient), so anticoagulation will be considered. Those with a proven or strong likelihood of glomerulonephritis will be treated with immunosuppression (section 10.3.5).

As the CKD progresses, other functions of the kidney fail. NICE NG8 offers guidance on managing anaemia (potentially with erythropoietin, with a target haemoglobin of 110–20 g/l), whilst renal osteodystrophy (or risk thereof), can be addressed by calcium and/or vitamin D supplements (e.g. cholecalciferol 80–1,600 IU od, calcitriol 0.25 μg od), restriction of the intake of phosphates, and use of dietary phosphate binders (e.g. sevelamer 800 mg od). Low levels of calcium may lead to secondary hyperparathyroidism, which then leads to bone marrow fibrosis, osteoclast hyperactivity, and cyst formation. Raised serum alkaline phosphatase may reflect increased bone turnover. Increased urate retention (more so with hydration) may lead to gout, generally treated with 100 mg allopurinol od.

Surgical treatment

Should many of these approaches fail and the patient move to end-stage renal disease (ESRD: stage 5, eGFR <15 ml/min/1.73m^2), treatment is with renal replacement therapy, of which there are two types—dialysis and transplantation. There are two types of dialysis: continuous ambulatory peritoneal dialysis (CAPD) and haemodialysis. The former consists of infusing the patient's peritoneal cavity daily with a solution of electrolytes with glucose or dextrose to give an approximate osmolality of ~280 mOsm/l. The fluid is infused over 30–40 minutes via a permanent tap/drain in volumes of 1.5 to 2.5 litres for 4–6 hours. A major issue is cleanliness, and if this is imperfect, this may lead to bacterial peritonitis, most frequently with *Staphylococcus epidermidis*. This can be determined by features such as a fever, an acute phase response, and a neutrophil count >100 cells/ml, and treated with intraperitoneal antibiotics.

The haemodialysis machine effectively replaces the filtration aspect of the kidney. The patient is fitted with a permanent arteriovenous fistula, most often in the forearm, to allow large-bore needles to carry blood to and from the machine. High concentrations of undesirable ions and molecules diffuse across a semi-permeable membrane and into a dialysis fluid, which is discarded. Should dialysis be needed urgently, a large-bore double-lumen catheter is preferred, and inserted into a large vein, whilst in an ICU setting the jugular is a popular choice. A modification of haemodialysis by diffusion is haemofiltration, where waste products and water are removed, and the balance of fluid is made up by a suitable buffer. The pros and cons of peritoneal and haemodialysis are summarized in Table 14.4.

The ultimate renal replacement therapy is transplantation, as described in sections 9.8.5 (immunology and cell biology) and 10.4.3 (clinical aspects). The mortality rate on dialysis is 16.6 per 100 patient/years, whilst after transplantation this falls to 2.9 per 100 patient/years. The kidney is the most frequently transplanted solid organ, with two patients potentially treated by each donor (section 10.4.3), and is a valuable economical as well as clinical option as the cost of a single transplant is close to that of a year's haemodialysis.

Late effects of transplantation include infections resulting from the immunosuppression, and in view of this, patients will be prescribed prophylaxis with antibiotics and antivirals for at least the first

TABLE 14.4 Peritoneal dialysis and haemodialysis

	Pros	Cons
Peritoneal dialysis	CAPD can be done anywhere No needles Control over schedules No machine needed Less restricted diet May be done overnight	Needs to be daily Permanent catheter Risk of infection Storage and disposal Loaded abdomen
Haemodialysis	More efficient Less frequent	Tied to the machine for hours Indwelling arteriovenous fistula

3 months. The leading cause of death in renal transplant recipients is cardiovascular disease at ~30%, slightly more than the 24.3% in the general population (Table 1.2), whilst ~24% die from cancer, somewhat better than the 28.4% in the general population. A further risk is non-Hodgkin's lymphoma, and some 12% will develop post-transplant diabetes mellitus, which also contributes to mortality, whilst infections and their complications account for 13% of deaths.

Key points

Whilst the financial cost of a kidney transplant is easily justified, and in most provides a return to normal life, it comes at a long-term medical cost to the patient of risk of malignancy due to the required immunosuppression.

CASE STUDY 14.1 Nephrotic syndrome

A 65-year-old man mentions in passing to his wife that the water in the toilet is frothy after he has urinated. She passes this information to their daughter, a nurse, who insists on his immediate self-referral to his GP. On arrival, the practice nurse performs a urine dipstick analysis, which shows +++ proteinuria. The GP takes blood for FBC, U&Es, and LFTs, and sends it (with a sample of urine) to the local hospital, and phones the consultant nephrologist for an urgent appointment. Blood results are shown in Table 14.5.

A few days later the consultant performs a full examination and takes a medical history, the principal feature of the former being mildly swollen ankles and of the latter being coronary artery bypass grafting performed four months ago. The patient's blood pressure is 143/88 mm Hg, and pulse rate 65. Medications are aspirin 75 mg, atorvastatin 20 mg, enalapril 15 mg od, and allopurinol 100 mg od (for a history of gouty arthritis in the left knee). He also says he has lost at least half a stone (3.2 kg) in weight since his operation (his wife claims the weight loss is greater), with a current BMI of 22.4 kg/m², and adds that he has been quite tired, attributing this and the weight loss to late effects of the bypass.

The U&E part of the blood results (Table 14.5) is consistent with the dipstick and the sign of proteinuria, diagnosing CKD stage 4, but there is also marked hypercholesterolaemia (despite use of atorvastatin) and marked hypoalbuminaemia. Measurement of cystatin C is not thought to be worthwhile. The laboratory reports the ACR to be 250 mg/mmol, sufficient to change the diagnosis to nephrotic syndrome, with a presumed aetiology of membranous glomerulonephritis. The immunology laboratory reports normal levels of IgA (although this does not necessarily exclude IgG nephropathy).

He is sent for chest and abdominal X-rays and renal ultrasound (which all prove to be normal) and returns the following morning to a medical day-case ward for intensive treatment with a bolus of intravenous cyclophosphamide at 2.5 mg/kg body weight and 1 g methylprednisolone in a saline drip. He tolerates this reasonably well but is kept in for 24 hours.

Other medications are increased: atorvastatin to 80 mg od, and enalapril to 20 mg od, and to these are added bendroflumethiazide 5 mg od (aiming for blood pressures <135/85), azathioprine 50 mg od, and prednisolone 30 mg od. The allopurinol is temporarily stopped as it has an inhibitory effect on azathioprine. He is also prescribed high-protein diet supplements and a high-protein diet per se, aiming at 100g/day for a fortnight, until serum albumin and protein improve. Nephrotic syndrome brings an increased risk of VTE, so anticoagulation must be considered, but the patient is already on aspirin because of a risk of arterial thrombosis due to his cardiovascular disease.

After appointments with consultants in Haematology and Cardiology, and with fears of haemorrhage, he is prescribed a low dose of a direct-acting oral anticoagulant. Blood and urine tests are performed on alternate days, and by day 10 (see Table 14.5) those values previously abnormal have improved across the board (with blood pressure 138/83) and continue to do so. The steroids are reduced by 5 mg/day every three weeks and are maintained at 10 mg/day, and the atorvastatin is reduced to 40 mg od after 30 days. The azathioprine is reduced, and allopurinol restored. After 3 months most blood markers are broadly acceptable, but a residual proteinuria and raised creatinine remain, which prove to be a permanent legacy of the disease. Over the years that follow, his eGFR slowly falls, but stabilizes at around 40 ml/min/1.73m².

TABLE 14.5 Blood results in Case study 14.1

Analyte (unit)	Reference range	Result day 1	Result day 10	Result day 30	Result day 90
Haemoglobin (g/l)	133–67	128	131	142	143
Mean cell volume (fl)	77–98	92	90	91	90
White blood cells (× 10^9/l)	4.0–11.0	9.5	10.1	8.9	7.7
Platelets (× 10^9/l)	143–400	306	301	310	303
Red blood cells (× 10^{12}/l)	4.3–5.7	4.5	4.8	5.2	5.5
Haematocrit (l/l)	0.35–0.53	0.41	0.43	0.47	0.49
MCH (pg)	26–33	28.4	27.3	27.3	26.0
MCHC (g/l)	330–70	309	303	300	289
ESR (mm/Hr)	1–12	15	12	11	8
Sodium (mmol/l)	135–45	145	143	144	141
Potassium (mmol/l)	3.8–5.0	4.6	4.9	4.3	4.3
Urea (mmol/l)	3.3–6.7	11.4	9.5	7.3	6.3
Creatinine (µmol/l)	71--133	200	150	122	109
eGFR (ml/min/1.73m^2)	>90	29	41	53	61
CKD Stage	1–5	4	3B	3A	2
uACR (mg/mmol)	<3	250	100	45	10
ALT (IU/l)	5–42	40	37	38	35
AST (IU/l)	10–50	36	35	32	34
GGT (IU/l)	5–55	35	36	35	34
Alk Phos (IU/l)	94–320	120	121	116	117
Bilirubin (µmol/l)	<17	12	13	12	13
Urate (mmol/l)	0.1–0.42	0.96	0.66	0.44	0.35
Albumin (g/l)	35–50	18	22	29	36
Total protein (g/l)	63–84	48	53	58	60
Calcium (mmol/l)	2.2–2.6	2.0	2.1	2.3	2.4
Cholesterol (mmol/l)	<5.0	8.1	7.6	6.5	4.5
LDL (mmol/l)	<3.0	5.8	5.4	4.2	2.5
HDL (mmol/l)	>1.2	1.3	1.2	1.4	1.3
Triglycerides (mmol/l)	<1.7	2.1	1.9	2.0	1.5

MCH = Mean cell haemoglobin; MCHC = mean cell haemoglobin concentration. eGFR = estimated glomerular filtration rate. ESR = erythrocyte sedimentation rate. ALT = alanine aminotransferase; AST = aspartate aminotransferase. GGT = gamma glutamyl transferase, Alk phos = alkaline phosphatase. CKD = chronic kidney disease, LDL = low-density lipoprotein cholesterol; HDL = high-density lipoprotein cholesterol.

SELF-CHECK 14.3

SELF-CHECK 14.3

From the patient's perspective, what are the pros and cons of peritoneal dialysis versus haemodialysis?

14.1.2 The ureters, bladder, and urethra

Anatomy and physiology

These three organs have no vascular basis or homeostatic roles, being concerned with the transport and storage of the urine. Peristaltic contractions of the ureters (25–30 cm/10–12 inches long) help gravity move urine to the bladder, where a physiological vesicoureteric (vesical: pertaining to the bladder and nearby organs, vessels, and tissues) valve prevents the reflux of urine back up into the kidneys. The bladder itself resembles a large inverted pear, which can expand to ~800 ml, consisting of an inner mucosa of epithelia, a lamina propria, and outer sheets of detrusor muscles that form an inner sphincter muscle to regulate outflow to the urethra. The muscle wall has stretch receptors that form a spinal reflex arc to the inner sphincter to effect relaxation and so urination, although there is a second set of external sphincter muscles under voluntary control. The urethral lumen is lined with epithelia above a thin layer of smooth muscle cells. In the female it passes directly from the bladder to the vulva. In the male it passes through the prostate, where it is joined by ductal tubes bringing secretions of the prostate, the seminal vesicles, Cowper's glands, and testes (see section 14.2).

Pathology, presentation, and diagnosis

Cancer of these organs is described in Chapter 5. With very little atherothrombosis of these organs, the remaining pathology includes infections, obstructions, and an overactive ('irritable') bladder, which potentially leads to incontinence. As will be evident in sections that follow, presenting features (e.g. pain on urination and changes in pattern of urination) give firm clues to diagnosis, later supported by other investigations. Incontinence is its own diagnosis. Blood tests are generally unhelpful, unless in excluding renal disease and if any inflammation is so severe that its effects are systemic. Urine analysis may point to an infection, but a sample sent to the microbiology laboratory will be definitive, whilst a bladder biopsy may help, and will be processed by the histology laboratory. Imaging (X-ray, CT, MRI, and ultrasound) is essential in assessing the ease of passage, and the storage, of the urine. Diseases of these organs are best described according to aetiology: infection and obstruction.

Infections

Following respiratory tract infection, urinary tract infection is the second most frequent infection, with an incidence of 50,000/million/year and accounting for ~1.5% of primary care self-referrals. In England and Wales in 2021, UTIs were the leading cause of urinary system deaths, accounting for 56% and far exceeding the number of deaths attributed to the second most frequent cause, CKD, at 27%. UTIs present in around twice as many women as men (67%/33%) across all age groups, probably because of the much shorter urethra (4 cm/1.5 inches v 20 cm/8 inches) and its exit being close to the vagina and anus. An estimated third of women present with a UTI before the age of 24, 50% by age 35, and 70% of women will refer at some time in their lives, whilst 30% have recurrent UTIs, causing considerable anxiety, depression, and poor quality of life. An additional risk factor is the number of vaginal deliveries, as this increases the risk of pelvic floor damage. Some 40% of all hospital-acquired infections are UTIs, and most of these are linked to a catheter.

The leading causative organism of a UTI is *E. coli*, present in 70% of cases, especially those strains with adhesive flagellae/fimbrae that promote binding to epithelia. Other organisms include *Staphylococcus* species (~13%, principally *saprophyticus*), *Klebsiella* (13%), *Streptococcus* (11%), *Enterococcus* (5%), *Enterobacter* (5%), and *Proteus* (4%) species. A key point is whether the UTI is complicated by other disease/issues (diabetes, obesity, CKD, stones, sickle-cell disease), as this may have a bearing on management. Patients with long-term urinary catheters are at high risk of infection, mostly with *E. coli* and *Proteus mirabilis*.

Treatment with antibiotics will ideally be tailored to the organism(s) responsible for recurrent UTI. NICE offers guidance for numerous indications. For example, NG112 on the choice of antibiotics for recurrent UTIs suggests a single dose of an antibiotic when exposed to a trigger (such as sexual intercourse), or a smaller dose at nights: drugs include trimethoprim 200 mg/100 mg or nitrofurantoin (if eGFR ≤45) 100 mg/50 or 100 mg as first choice, and amoxicillin 500 mg/250 mg or cefalexin 500 mg/125 mg. Many of these antibiotics are recommended in several other guidelines and clinical situations (e.g. acute pyelonephritis, NG111, which also has other suggestions such as co-amoxiclav and gentamycin).

Dosing schedules differ in those aged <16 years, for men and women, in pregnancy (e.g. avoid nitrofurantoin), and in those with hepatic or renal impairment. Cranberry juice, D-mannose, lactobacillus, and chemicals can be used to increase the pH of the urine but are not universally favoured, although the use of topical oestrogen on the exit of the urethra in the vulva has most support (e.g. NICE NG112). Instillation of intravesicular hyaluronic acid and chondroitin sulphate (which may help restore damaged glycoaminoglycans) have been used with some success for recurrent UTIs, but also in interstitial cystitis and overactive bladder syndrome (see below).

Obstruction and stones

As with infections, there can be obstructions to any part of the urinary tract. Inflammation of any soft tissue results in swelling, so for a tube this will be an increase in the general diameter and a reduction in the lumen. So, in theory, urethritis will lead to a reduced urine flow, although in practice this is difficult to quantify. If present, treatment with antibiotics should be effective. Malignant growths, diverticulitis, and abdominal aneurysms in the lower peritoneal cavity may also obstruct urine flow down the ureters, whilst tumours within the bladder may obstruct urine entering and leaving—treatment is by surgery (Figure 14.4).

However, the clearest example of obstructions is by stones (urolithiasis: nephrolithiasis when in the kidney). These calculi, of which the general population has a 10% lifetime risk (although few of these are clinically important and many pass during urination, often to the surprise of the producer!), may be the product of abnormal diet, metabolism, and/or renal disease, notably in calcium, oxalate, and uric acid metabolism, caused in turn by hypercalcaemia, hyperoxaluria, and hyperuricaemia/hyperuricosuria, and there are mixed species (e.g. calcium tartrate tetrahydrate). Once suspected (dysuria, pain, etc., the latter treatable with NSAIDs), stones may be located by X-ray (which may also detect a dilated ureter (megaureter)), ultrasound, or MRI, but non-contrast MRI (although more expensive) is preferred (Figure 14.6). In some cases unexcreted urine may back up the ureters into the kidney, causing post-renal AKI.

Causes of stone development include dehydration, hypercalcaemia, obesity, infections, renal tubular acidosis, gout, and certain medications. Some (calcium phosphate (around 7% of all stones), struvite (magnesium ammonium phosphate, forming ~12% of stones)) form in alkaline urine, others (calcium oxalate (~75%), uric acid (~7%)) when it is acid. Hypocitraturia predisposes to stones as citric acid sequestrates calcium cations. The leading genetic cause is hypercysteinaemia (~1% of stones, resulting from mutations in *SLC3A1* (at 2p21) and *SLC7A9* (at 19q13.11)) and so cysteinuria, the stones generally being undetectable by imaging.

Stones may be expelled by medical treatment: alpha blockers and calcium channel blockers aid the passage of distal stones <10 mm in diameter. Surgical treatments for stones <20 mm include shockwave **lithotripsy** (the preferred option) and ureteroscopy, whilst renal stones >20 mm may need percutaneous nephrolithotomy. These treatments should be offered within 48 hours of presentation or admission if the pain is ongoing or not tolerated, and the stone is unlikely to pass out with urine. Once recovered, the stone should be analysed for chemical content and any issues addressed—such as diet, hydration, use of potassium citrate for calcium-based stones—so that

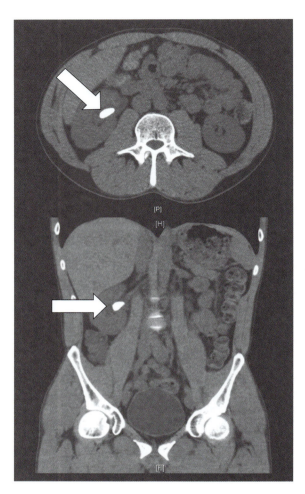

FIGURE 14.6
Kidney stone detected by CT. The single large stone is arrowed.

Kleinguetl C et al. Calcium Tartrate Tetrahydrate, Case Report of a Novel Human Kidney Stone. Journal of Endourology Case Reports, Dec 2017: 3(1). https://doi.org/10.1089/cren.2017.0118. ©2017, Mary Ann Liebert, Inc.

recurrence can be prevented. Thiazides may be required for recurrent calcium oxalate stones and hypercalciuria.

Lower urinary tract symptoms (LUTS)

Urinary system disease is challenging because of the numerous and diverse signs, symptoms, and syndromes. This is exemplified by an overactive bladder, bladder pain syndrome (BPS), and interstitial cystitis (IC), often collected under the umbrella of lower urinary tract symptoms (or syndrome; LUTS). These frequently overlap in clinical practice, in pathophysiology, and in the poor quality of life that they bring, such as urinary incontinence and sexual dysfunction (Figure 14.7). The key investigation is cystoscopy, with local anaesthetic to the urethra, the ultimate treatment for intractable disease being surgery to remove the bladder (cystectomy).

At presentation, a full medical history, blood tests, and urine analysis (and potentially imaging) are required to exclude other causes and provide a baseline. Treatments for an overactive bladder include anti-cholinergics, protamine sulphate, supervised physiotherapy-led exercises to strengthen the muscles of the pelvic floor, and minimizing caffeinated beverages. Pharmacological treatments include 2 mg bd tolterodine tartrate (which acts on muscarinic receptors), 5 mg od solifenacin

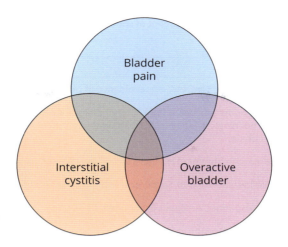

FIGURE 14.7

Relationships between overactive bladder, bladder pain, and interstitial cystitis. For illustration only: the true proportion of people in each category is unknown.

succinate (a competitive cholinergic receptor antagonist), and 50 mg mirabegron (an agonist of the β-3-adrenoreceptor, resulting in muscle relaxation). Vasopressin may be considered specifically to reduce nocturia in incontinence or with an overactive bladder.

Should these options fail and urodynamic investigation have shown detrusor muscle overactivity, there are surgical options, such as injection of 100 units of botulinum toxin type A into the bladder wall, percutaneous sacral nerve stimulation, and augmentation cystoplasty (enlarging the bladder with intestinal tissues). Despite the discussion of pathology brought about by infections of the entire urinary system (section 14.1.1), there may be a place for a bladder-specific microbiome, as treatment with antibiotics can be effective in improving the symptoms of an overactive bladder. Decreased *Lactobacillus* species is common in all three bladder diseases in this section, and is linked to increased pain, whilst *Proteus* is more often found in those with an overactive bladder.

BPS as an independent entity also lacks an international consensus and may be a diagnosis of exclusion. To add to the confusion, IC is increasingly being replaced by BPS, with the terms often used interchangeably. The prevalence of IC/BPS is estimated at 0.01% and 6.5%, with a wide geographical variation: 2.7–6.5% in American women, 0.6% in the UK, 0.3% in Finland, 0.26% in Korea, and 0.03% in Japan (Li et al. 2022). Early studies estimate that women are nine times more likely to have a diagnosis of IC/BPS than men, although in a managed care population this ratio falls to 5:1 (Davis et al. 2015).

As regards aetiology, the inflammation hypothesis is supported by reports of increased inflammatory cytokines IL-8, CXCL-1, and CXCL-8 in a patient's urine, such that measurement may enter clinical practice. However, it is possible that inflammation may develop from the effects of toxins on the urothelium in the absence of an infection. Hunner's ulcers/lesions (diagnosed by cystoscopy) are present in 5–10% of cases of IC and (if present) are likely to cause bladder pain as their removal reduces—or even abolishes—the pain. Other guidelines require the presence of haemorrhage (glomerulation) within the bladder upon cystoscopy, and symptoms for >9 months. In rare cases, biopsy may be needed to confirm the extent of detrusor damage.

Many of the treatments of BPS and/or IC have been described previously. One possibility is pentosan polysulphate 100 mg tds for BPS (NICE TA610), a further option being intravesical injection of botulinum toxin (as above); pentosan polysulphate (200 mg in 30 ml sterile saline for 30–60 minutes) and dimethyl sulphoxide (typically 50 ml for 30–60 minutes, once weekly for 6 weeks) may be used for IC. As in many diseases, management passes in an orderly manner through a series of stages that depend on severity of the condition and patient preferences (Table 14.6).

TABLE 14.6 Treatment of confirmed or uncomplicated bladder pain syndrome/interstitial cystitis

Stage	Treatments
1	Patient education, self-awareness, and care, etc., e.g. avoidance of caffeine, stress management
2	Structured physiotherapy for bladder and pelvic floor strengthening. Oral medications (e.g. amitriptyline, cimetidine, pentosan polysulphate, anticholinergics). Intravesical treatments (e.g. heparin, hyaluronic acid, dimethylsulphoxide, local anaesthetics, chondroitin sulphate)
3	Cystoscopy and hydrodistension. Treatment of Hunner's ulcers (resection or injected steroids such as triamcinolone 40 mg)
4	Botulinum toxin A for the bladder muscle wall. Sacral/posterior tibial nerve stimulation
5	Immunosuppression (e.g. cyclosporine). Analgesia as required
6	Substitution cystoplasty, uretheral diversion, cystectomy
All	Analgesics as required, support, and counselling

Data source: NICE, https://www.nice.org.uk/.

SELF-CHECK 14.4

What are the most common chemical components in renal stones?

Incontinence

Urinary incontinence is the uncontrolled release of urine. Some commentators distinguish stress incontinence (with a clear precipitating factor such as exercise or emotional stress) from urge incontinence (without a clear precipitating factor). In either case a full examination and investigations are required to determine the cause and any additional issues, such as overactive bladder. An immediate suggestion is to reduce caffeine intake, lose weight, and stop smoking if needed, and begin bladder training and pelvic floor exercises. The latter may be aided by short-term electrical stimulation. NICE NG123 offers guidance on assessing and managing urinary incontinence and pelvic organ prolapse in women aged 18 and over, and covers complications associated with mesh surgery. Medical treatments include agents such as mirabegron (a beta 3-adrenoceptor agonist, 50 mg od), whilst antimuscarinics fesoterodine, oxybutynin, solifenacin, tolterodine, and trospium are all effective. Surgical treatments for prolapse include mesh procedures, slings, colposuspension (lifting the neck of the bladder), urinary diversion (redirecting the urethra so it exits through the abdomen, requiring a bag to collect the urine), and artificial urinary sphincters.

Key points

There is considerable overlap in the signs, symptoms, and diagnoses of overactive bladder, bladder pain syndrome, interstitial cystitis, and incontinence. There is also much similarity in the management of these conditions, implying a commonality in pathophysiology.

14.1.3 Molecular genetics of urinary system disease

Leaving aside ADPKD, genome-wide association studies have identified over 50 germline gene loci with potential roles in this CKD. A leading candidate is *UMOD* at 16p12.3 coding for uromodulin (produced by cells of the loop of Henle), variants also being linked to eGFR as well as CKD. Several studies have independently reported variants of *CUBN* at 10p13 that code for a component of the intestinal receptor for the vitamin B_{12}/intrinsic factor complex. Other potential markers include *APOL1* (coding an apolipoprotein) and *MYH9* (non-muscle myosin heavy chain IIa).

Non-coding RNAs are linked to various aspects of renal disease. Several miRNAs are associated with different forms of glomerulonephritis, and although miRNAs -16, -25, -155, -201, and -638 are reduced in those with advanced renal failure, others, such as miR-233 and miR-146a-5p, are increased. miR-21, miR-17-92, and miR-200 may have roles in ADPKD, leading lncRNAs with renal involvement include MALT1 and HOTAIR in AKI, and blood and urine circular RNAs may be biomarkers of kidney disease. Other genetic markers include the autosomal dominant Alport syndrome (e.g. *COL4A3*, *COL4A4* at 2q36, and *COL4A5* at Xq22.3) whilst several genes (e.g. *APOL1* at 22q12.3, and *KAT2B* at 3p24.3) are linked to focal segmental glomerulosclerosis.

14.2 The male reproductive system

This system comprises the testes (the site of sperm production), epididymis (where sperm mature and are stored), the vas deferens (which conduct sperm to the urethra as it passes through the prostate), the seminal vesicles (contributing an alkaline secretion of fructose, coagulation components, and prostaglandins to the sperm, thus forming 60% of the semen), the prostate (making further contributions to the semen, such as citric acid and enzymes), and the Cowper's (or bulbourethral) glands. The latter secrete an alkaline mucus solution upon sexual arousal that neutralizes the acid urine in the urethra and lubricates the head (glans) of the penis to assist sexual intercourse. Sperm are conducted to the exterior through the urethra, which runs from the base of the bladder, through the prostate to the head of the penis. This cylindrical body is structurally composed of tubes of spongy tissues (two sets of corpora cavernosa and a corpus spongiosum) that, when engorged with blood, lead to an erect penis that can effect sexual intercourse (Figure 14.8).

Atherothrombosis is rarely present unless on a background of cancer or trauma—an unexplained deep-vein thrombosis may be the first sign of prostate cancer. The aetiology of the remaining diseases focuses on inflammation, infection, and hyperplasia, and there is overlap with the previous subsection as lower urinary tract symptoms (LUTS: problems with storage and urination) can be caused by prostate disease in addition to bladder, ureter, and urethral disease (NICE CG97). However, the major cause of LUTS is benign prostatic hyperplasia. In some cases, diagnosis is immediate (e.g. erectile dysfunction), in others it will require additional investigations (such as endocrinology), and accordingly, each major form of disease will be discussed separately.

Cross reference

Malignant diseases of this system are described in sections 4.5 (prostate) and 5.13.1 (other organs).

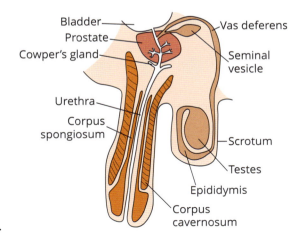

FIGURE 14.8
Male reproductive system.

14.2.1 Benign prostatic hyperplasia (BPH)

Aetiology and pathology

This is the principal non-malignant prostate disease, also described as benign prostatic enlargement. With a mass of some 7–16 grams (or ml) in young men, the prostate grows at around 2–3% a year due to a slow but steady increase in stromal smooth muscle and epithelial cells. The factors controlling the size of the enlargement (40 g in some, 200 g in others) have yet to be determined. The frequency of BPH in those in their 40s is <10%, thereafter rising by the 'rule of ten', with a frequency of 50% at age 50, 60% at age 60, 70% at 70, and 80% at 80 and above. A genetic component is suspected as first-degree relatives have a 4-fold to 6-fold increased risk, whilst twin concordance is in the range of 50–70%, which may be compared to that of prostate cancer in the range of 42–58%.

As the prostate is an exocrine gland, BPH is technically an adenoma, and it may be that certain prostate cancers develop from these hyperplastic cells and tissues as it is often found alongside overtly malignant tissues in excised tumours. BPH may be uni- or multinodular, or diffuse. The only unequivocal cause of BPH is obesity (BMI >35 brings a 3.5-fold increased risk), whilst lack of physical activity/exercise, testosterone, oestrogen, inflammation, and neurological factors may also be involved. Oestradiol induces an epithelial-to-mesenchymal transition in BPH epithelial cells *in vitro*, pointing to a potential mechanism.

Presentation and diagnosis

BPH in itself is asymptomatic, but it is the major risk factor for LUTS and may progress to prostate cancer. Indeed, many consider BPH to be incorrectly named as it is the leading cause of LUTS, the hyperplasia interfering with urine flow down the urethra, hence leading to bladder outflow obstruction (BOO). Leading genetic SNPs linked to LUTS include those in genes coding for the vitamin D receptor, angiotensin-converting enzyme, glutathione S-transferase, and telomerase reverse transcriptase. Some 50% of men aged over 50 and ~80% of men aged over 80 suffer from BPH-induced BOO and so LUTS, which itself is linked to erectile dysfunction. The frequency for moderate to severe LUTS caused by BPH is ~80/million men. However, if nodular, a BPH lesion may be distant from the urethra, and so not precipitate any LUTS symptoms. These can be any combination of the urinary system problems described in this section, and include post-urination 'dribble', that is, incontinence. Eventually these will be so severe that they will call for treatment.

Primary investigations will be urine analysis and digital rectal examination, but formal diagnosis is by ultrasound (which can estimate prostate volume (converted into grams: normally <20 g)). If this is unclear, an MRI will be required. Imaging will also provide details of the bladder and other organs and tissues that may impact into prostatic disease. Serum U&Es will be needed to exclude overt renal disease, and biochemical analysis of the urine for the albumin-to-creatinine ratio may be helpful. Urodynamics may also be employed—a normal flow rate is ~20 ml/s, a rate <10 ml/s is regarded as indicative of BOO, whilst a post-void residual volume of urine >100 ml is predictive of acute urinary retention.

Many men with BPH also have raised serum prostate specific antigen (PSA). This small enzyme (~34 kD), coded for by *KLK3* at 19q13.33, is present in serum, being likely to help liquify the coagulated semen once in the vagina, and possibly to liquify cervical mucus clearing a path for sperm to enter the uterus. Levels of PSA rise with age in parallel with the size of the prostate and are higher in those of African descent (as is frequency of prostate cancer, which is also linked to raised PSA).

Treatment

The international prostate symptom score (IPSS) is often used to guide management. It consists of seven questions on the frequency and quality of urination, answered on a scale of zero to five, where a score of 3–7 indicates mild, 8–19 moderate, and 20–35 severe narrowing of the urethra. It is available online (see Further Reading). Table 14.7 summarizes treatments of BPH/LUTS, the choice being made by the MDT and the patient, dependent on efficacy, invasiveness, and side effects.

TABLE 14.7 Treatment of BPH

Treatment	Method
5-α-reductase inhibitor	Blocks the conversion of testosterone to dihydroxytestosterone, thus reducing the effect of androgens
Alpha-blocker	Relaxes smooth muscle cells of the prostate and neck of the bladder by inhibiting sympathetic stimulation
TURP	Surgical removal of prostate tissue, passing specialized catheters up the urethra with a microscalpel
Laser therapy	Vaporization of prostate tissue by high-energy wavelengths
Implants	Mechanical implants placed in the prostatic urethra (stents), devices to keep the prostate away from the urethra (lifts), and creation of a new channel in the urethra
Prostatic artery embolisation	Prevents oxygen and nutrients reaching the tissues to prevent growth and promote apoptosis
Water vapour	Resection of prostate tissue using water vapour ablation
Botulinum toxin	Injected deep into the prostate. Inhibits exocytosis of neurotransmitters, so paralysis of the neuromuscular junction and reduction in smooth muscle cell activity
Prostatectomy	Removal of the prostate

Medical treatment focuses on three factors: the tone of the smooth muscle cells of the prostate and neck of the bladder, the enlarging adenoma causing obstruction, and the hypertrophy and irritability of the bladder muscle. The 5-α-reductase inhibitors finasteride and dutasteride are particularly effective, reducing prostate volume by 20–25% and clinical symptom progression by a third whilst improving sexual function. These drugs also improve the IPSS score by 30–45% and urinary flow by 15–50%. Alpha-blockers, which increase the maximal urine flow rate and improve symptoms, include prazosin, terazosin, doxazosin, tamsulosin, silodosin, and alfuzosin. However, these drugs may lead to hypotension and floppy iris syndrome, which frustrates cataract operations. Combinations of the two classes of drugs are particularly effective. Nocturia in BPH may be treated with vasopressin, possibly combined with an alpha-blocker.

There are several surgical options. TURP is linked with numerous complications (infection in ~12.5%, urethral stricture ~6%, bladder neck stenosis 4.6%, bleeding ~3%, retrograde ejaculation ~64%, and a recurrence rate of 15%) so that its use is falling as newer and safer techniques become available:

- Lasers: The pulsatile holomium system uses a CO_2/neodymium/YAF laser to simultaneously cut and cauterize tissue; the Greenlight laser uses potassium-titanyl-phosphate is a photo-selective vapourization system.

- The prostatic urethral lift places a mechanical implant in the prostate to keep the prostatic lobes under traction and away from the urethra. Stents (biodegradable or permanent) and other devices can be placed in the urethra and removed if necessary.

- Under ultrasound guidance, and under general anaesthesia, a stream of saline or water (to degenerate vapour or steam) can remove parenchymal tissue, a technique appropriate for those aged >50 with moderate to severe LUTS refractory to drug therapy, an IPSS >12, a maximum flow rate <12 ml/s and a prostate size of 25–80 ml.

- Prostate artery embolization, performed under local anaesthesia, consists of passing a catheter up a femoral artery, and so to prostatic arterioles, to deliver microparticles that completely occlude the vessels. The prostate will then undergo apoptosis and necrosis, and so shrink. However, failure rates are relatively high—around 19%.

14.2.2 Other reproductive system disease

An inflamed prostate (prostatitis) is likely to be enlarged and so may be diagnosed as BPH. The difficulty lies with obtaining prostate tissue in the absence of BPH or LUTS, and of prostate secretions uncontaminated by urine. Classification has been attempted with the descriptor of chronic prostatism/chronic pelvis pain syndrome (CP/CPPS), present in ~5% of men and characterized by frequent urination and interstitial cystitis, which account for 95% of all cases of prostatitis. CPPS encompasses pain elsewhere in the general pelvic/genital area, and so collects several diverse aetiologies. One treatment may be intra-prostatic injection of botulinum toxin A, whilst treatments cited in Table 14.7 may be valuable. The most commonly prescribed antibiotics are ceftriaxone, doxycycline, ciprofloxacin, and piperacillin/tazobactam.

> ## Key points
>
> **As with the complicated nature of bladder pain syndrome, overactive bladder, and interstitial cystitis, the practitioner is faced with difficulty in the precise definitions of BPH, LUTS, and prostatitis, and so the possible treatments.**

Pathology of the seminal vesicles, Cowper's glands, and vasa deferentia is rare. Endocrine disease of the reproductive system is described in section 15.4.1 of Chapter 15. The testes may be the subject of a hydrocoel (accumulation of fluid, possibly due to lymphatic blockage, leading to scrotal swelling), and the epididymis may be damaged by trauma or the subject of testicular torsion. A variant of a hydrocoel is a variocoel—swelling of a precise anatomical region, the pampiniform venous plexus. Inflammations of the testes (orchitis) and epididymis (epididymitis) are often found together, and many of the presenting (scrotal pain, and perhaps malaise and fever) and precipitating factors (microbial infections) are common. Ultrasound is the primary investigating tool. Epididymo-orchitis is reported as having an incidence of 2.45 per 1,000 men in the UK and may be secondary to other conditions such as Behcet's disease (see section 10.3.6). Both forms of scrotal disease can be caused by sexually transmitted diseases (STDs, section 14.5).

The leading aetiological feature of orchitis is viral, notably with coxsackievirus, varicella, echovirus, and cytomegalovirus. Mumps is the leading cause of adult orchitis, developing in ~25% of cases, and resolving within 10 days. In certain cases, there may be coinfection with the urinary system, often with *Escherichia coli*, *Klebsiella pneumoniae*, *Pseudomonas aeruginosa*, and with *Staphylococcus* and *Streptococcus* species. Following 10–14 days of treatment with antibiotics such as a fluoroquinolone (ciprofloxacin, ofloxacin, levofloxacin), trimethoprim-sulfamethoxazole, ceftriaxone, doxycycline, or azithromycin, the infection should resolve. Failure to address the orchitis will unsurprisingly lead to impaired fertility, testicular atrophy, a reactive hydrocoel, and epididymitis.

Most cases of epididymitis in men ages 20–39 are caused by an STD (section 14.5), others from vigorous exercise; from age 40 the leading cause is *Escherichia coli*. If bacterial, treatment is often as that for orchitis. Common features of epididymitis are high fever in 21%, **pyuria** (48%), and **bacturia** (49%).

14.2.3 Male infertility/subfertility

The frequently used diagnosis of infertility is often translated as complete inability to produce a viable infant, and so brings considerable personal and social implications for patients and their partners. In many cases this is true infertility (as in loss of testes to disease or trauma), but in many cases the couple can be assisted with a number of interventions in both sexes (not necessarily with *in vitro* fertilization etc.), proving the diagnosis to be incorrect. Many therefore prefer the diagnosis of subfertility, which gives hope (see Case Study 14.2 on page 587).

Physiology

The principal aspect of male physiology, the generation of sperm, requires a functional hypothalamic–pituitary–gonadal (HPG) axis and the positive effect of kisspeptin, which has roles in the hypothalamus, and extrahypothalamic roles in sexual and emotional behaviour. Under the direction of kisspeptin, gonadotropin-releasing hormone (GnRH) from the hypothalamus stimulates the pulsatile release of luteinizing hormone (LH) and follicle stimulating hormone (FSH) from the anterior pituitary (section 15.4). These act on the testes to prompt testosterone production and spermatogenesis. The second physiological feature, ejaculation of semen, requires a complex series of psychological and neurological stimuli that require an erect penis. Kisspeptin also has a role in libido, as it is effective in treating hypoactive sexual desire disorder, with increased penile tumescence and behavioural measures of sexual desire and arousal.

Pathology

The leading cause of sub/infertility is failure of the HPG. At ages 12–14, it will be notable and investigated as in the adult, an example being Kallman syndrome (present in ~21/million), which is due to GnRH deficiency and also brings loss of the sense of smell (anosmia). In the adult, there are several forms of testicular disease. Erectile dysfunction may follow psychological, metabolic, and/or anatomical issues. An example of the latter is damage to the nervous supply to the arteries feeding the corpus cavernosum: damaged nerves (perhaps by prostate surgery) will fail to instruct these arteries to dilate and so fill the penis with the blood required for an erection. The leading metabolic cause is diabetes, where the same nerves may be subject to neuropathy, and/or blood supply may be restricted by atheroma in the iliac and pudendal arteries.

Presentation

It is expected that regular sexual intercourse between a healthy man and woman will produce a pregnancy within a year, and in ~95% of cases, this is achieved. Failure to do so, if undesired, will lead to investigation in a specialized reproductive medicine unit, the couple having already made the diagnosis. A third of cases is attributed to the man, a further third to the woman, and a third to both, although these figures vary by location and with age.

Diagnosis

The initial investigation is of quality and quantity of sperm, requiring the production of two semen samples (a week apart), a physical examination (such as for gynaecomastia, itself linked to numerous causes such as hyperthyroidism, liver, and renal disease), medical history (e.g. mumps), and blood testing. The couple may also attend a clinic, having engaged in sexual intercourse in the previous 8–12 hours, for a post-coital test. The presence of sperm in the vagina is likely proof of erection and successful ejaculation, but this does not imply that the sperm are functional and at a sufficient concentration required for fertilization. Semen testing metrics include the volume of the ejaculate (on average, 3.7 ml, ranging from 1.5 to 5 ml), the number (15–200 million per ml, hence total sperm count per ejaculate >39 million) and motility of sperm (>50%), and the proportion of abnormal sperm (<30%, such as asthenozoospermia—abnormal flagellum, and globozoospermia—lack of an acrosomal cap), with <1 million white blood cells per ml. Blood tests focus on immunology and endocrinology.

In some cases, ultrasound of the scrotum and CT/MRI of the skull (for the pituitary) may be necessary. The former may find anatomical abnormalities in the testes, epididymis, vas deferens (obstruction, stricture, or absence) or prostate, or any excessive fluid, such as a hydrocele, or a malignancy.

SELF-CHECK 14.6

What are the two forms of inflammation of the contents of the scrotum?

Should the anatomy be normal and the semen analysis satisfactory, no other investigations are likely. However, should the semen analysis indicate a low (<15 million/ml: oligospermia) or absent (azoospermia) sperm count, and with the blood results, the practitioner can consider a number of options:

- Hypothalamic–pituitary disease is characterized by failure of the two organs to generate sufficient LH and FSH to induce spermatogenesis. LH acts on Leydig cells to induce the formation of testosterone; FSH acts on Sertoli cells to promote spermatogenesis, which also requires testosterone. Both interact via specific receptors. This syndrome is described as hypogonadotrophic hypogonadism.

- The testicular apparatus involved in spermatogenesis (Leydig and Sertoli cells) may be damaged, perhaps by external factors such as toxins and inflammation (see above), and as a consequence, testosterone levels will be low. As the hypothalamic–pituitary axis is uninvolved, LH and FSH will be normal, but are often increased as testosterone would normally regulate these hormones by negative feedback.

- Hyperprolactinaemia may point to a pituitary tumour, and prolactin itself has a negative influence on GnRH, produced by the hypothalamus to stimulate LH and FSH release by the anterior pituitary. Many antipsychotics (and other drugs) reduce levels of dopamine, a negative regulator of prolactin. Raised prolactin can also cause gynaecomastia.

- Sperm may be antigenic and, if exposed to blood (perhaps by trauma, surgery, or inflammation of the testes/blood barrier), can generate autoantibodies, detectable in the IgG-mixed anti-globulin reaction test, although IgA variants are available. A result of >50% for this test is present in ~3.5% of infertile men with a viable sperm count. Furthermore, antibody titres correlate with poor motility (but not with sperm DNA damage) and often associate with a negative post-coital test. High titres of antibodies are present in men with obstructive vasa deferentia and in chronic prostatitis.

Management

Replacement therapy with recombinant FSH (e.g. 75 IU on alternate days for 3 months) is effective in improving sperm concentrations and in achieving both natural and assisted pregnancy. Testosterone replacement therapy in hypogonadism is generally unsuccessful as the hormone feeds back to the HPG axis to suppress LH and FSH production, and ultimately leads to azoospermia in the majority of men. This is because the testosterone is needed in high concentrations (perhaps 1,500 nmol/l, some 75 times that in the blood) at the testicular site of spermatogenesis, intramuscular or subcutaneous injections, or transdermal patches failing to do so. Intra-testicular injections may still not get sufficient hormones to the correct site without damaging the delicate anatomy of the testes. However, testosterone replacement therapy may include injections of human chorionic gonadotropin (hCG), which binds to the LH receptor on Leydig cells, thus inducing them to produce testosterone.

Other medical options include GnRH, selective oestrogen receptor modulators (such as clomiphene and tamoxifen that increase LH and FSH by inhibiting the negative feedback of oestrogen on the hypothalamus and pituitary), and aromatase inhibitors (such as letrozole and anastrazole that inhibit the conversion of testosterone to oestradiol). Surgical options include harvesting sperm from within the epididymis for assisted reproductive techniques (*in vitro* fertilization, gamete intra fallopian transfer, intra-cytoplasmic sperm injection, etc.). NICE CG156 on 'Fertility problems: assessment and treatment' deals mostly with female infertility.

Molecular genetics

Chromosomal abnormalities are present in ~16% of men with azoospermia and ~2.5% in those with oligospermia (0.6% in the general population), and in ~5.5% of men with primary infertility and ~2.5% of men with secondary infertility. Klinefelter's syndrome (47,XXY) is present in ~3% of infertile men, trisomy 47,XYY in 0.1% of men presenting with infertility. The Y chromosome contains 60 genes in patterns that are markedly susceptible to deletions. The first intra-chromosomal lesion of relevance to infertility, azoospermia factor (AZF), is due to a number of separate point mutations, micro- and partial deletions of Yq, which are present in 1/2,000 to 1/3,000 males, and in 5% of men with azoospermia or severe oligospermia.

The sex-determining region, housing *SRY*, is found at Yp11.2, which codes for a transcription factor most active in the embryo to control development of structures such as the seminiferous tubules. Congenital bilateral absence of the vas deferens is present in ~1% of infertile men and is linked to mutations in the gene coding for the cystic fibrosis transmembrane conductance regulator. Loss-of-function mutation in genes coding for LH, FSH, and their receptors (such as *FSHβ* -211G>T, *FSHR* 2039A>G, *FSHR* -29G>A) in both sexes leads to reduced or absent levels/expression and so infertility. Mutations in the receptors are untreatable, but lack of the hormones can be treated by replacement therapy. *LHCGR* codes for a receptor that binds both LH and chorionic gonadotropin, mutations in which can have several consequences.

Key points

Investigation of male infertility focuses on physical examination, blood and semen analysis, and ultrasound. The primary cause is lesions in the hypothalamic–pituitary–gonadal axis, potentially treated by replacement therapy.

14.3 The female reproductive system 1: Gynaecology and the breast

This system consists of the breasts, ovaries, fallopian tubes, the uterus, the cervix, the vagina, and the vulva. The latter is composed of labia majora and minora, the clitoris and clitoral hood. Malignant disease of the breast (section 4.6), ovary (section 5.8), uterus (section 5.10), cervix, vagina, and vulva (section 5.13.1) together led to the deaths of 16,862 women in England and Wales in 2021, that being 24.4% of all female cancer deaths. By far the most important non-malignant disease of the ovary is polycystic ovary syndrome; that of the uterus is endometriosis.

14.3.1 The breast

The likelihood of autoimmune disease or atherothrombosis of this organ is diminishingly small. The leading non-malignant breast disease is inflammatory, and there are numerous other types of pathology.

Inflammatory breast disease

Mastitis (inflammation of the breast) is classified as 'in breastfeeding' (present in perhaps 10–33% of those doing so) or not. In both cases presentation is with any combination of redness, tenderness, warmth, swelling, itching and pain, and there may be systemic symptoms such as a fever. In breastfeeding, mastitis may be due to inefficient milk generation by lactiferous cells, ill-fitting underwear, blocked ducts, poor feeding technique with damage to the nipple (rarely, <1%), or an abscess (~5% of cases). In many instances breastfeeding councillors can rectify poor feeding technique, resolving the problem. An abscess may also underly non-lactating mastitis, and in both cases there may be a bacterial infection, often with a *Corynebacterium* or *Staphylococcus* species, treatable with antibiotics such as flucloxacillin, cicloxacillin, or cephalexin. Failure to treat it adequately may lead to an abscess or non-necrotizing granulomatous lobular mastitis, which would need to be treated with steroids and/or surgery.

Other breast disease

As the differential diagnosis of most of the symptoms and conditions that follow is breast cancer, they are likely to be investigated urgently, initially with ultrasound.

- An estimated 20% of women refer with breast pain (mastalgia) at some time in their lives. In the pre-menopausal woman, it may be part of her menstrual cycle, but most breast pain is present in those who are post-menopausal. In over 50% of cases no organic disease can be found, and in 75% the pain resolves spontaneously. No clear pathology has been described, but danazol and tamoxifen may bring relief.

- Non-lactating secretion of milk (galactorrhoea) is mostly caused by high levels of prolactin (that also cause amenorrhea), which may be due to certain medications (often antipsychotics and anti-hypertensives), pituitary adenoma, or hypothyroidism. When confirmed with a blood test (and possibly MRI), most cases are medically treatable by bromocriptine (a dopamine agonist: 2.5 to 15 mg od) or cabergoline (a dopamine receptor agonist: 0.25 to 1 mg twice weekly), and by surgery.

- Other symptoms of breast disease are hardening and stiffness, suggestive of underlying fibrosis; the aetiology of these is unclear. Lumps in the breast may not be malignant; they may be cysts or lipomas, investigated by ultrasound and treated by surgery if necessary—almost all are benign.

- Macromastia (large breasts) is technically not a disease, but the polar extreme of the natural distribution. Nevertheless, it brings pain to the chest, neck, shoulders and back, and (often) damage to the skin and subcutaneous tissues by the brassiere, in addition to psychological issues (self-social exclusion, depression, anxiety). Treatment is by weight management (if obese, often present) and surgery. Micromastia is generally not considered a medical issue.

SELF-CHECK 14.7

Describe the most common forms of breast disease.

14.3.2 Polycystic ovary syndrome (PCOS)

Polycystic ovary syndrome (PCOS) is the most common endocrine disorder in women of reproductive age with a frequency of 5–10%, depending on definition: some estimate the rate to be 25%. It is thought to be responsible for >25% of cases of female infertility.

Aetiology and pathophysiology

The aetiology lies in a triad of ovulatory dysfunction, hyperandrogenism, and polycystic ovaries (the Rotterdam criteria), although not all may be present: hyperandrogenism may be present in only 80% of PCOS cases. The greater proportion of the increase in circulating androgens arises from the ovary and a smaller fraction (perhaps a fifth) from the adrenals, although some sources suggest that each produces an equal amount of testosterone and androstenedione. Body mass index is an issue: 25% of PCOS cases in secondary care are overweight and 50% are obese, although this does not suggest overweight/obesity causes the PCOS—the reverse may be true. This is pertinent as testosterone and other androgens are present in adipocytes at markedly higher levels, and so may be a reservoir and/or ectopic producers of these molecules. PCOS may also arise after an infection.

The cysts themselves are not true cysts but are undeveloped ovarian follicles, although one or two cysts may be present with normal endocrinology. Nevertheless, PCOS accounts for >80% of oligomenorrhoea and >25% of amenorrhoea cases (excepting pregnancy). An additional issue is the possibility of the development of hyperinsulinaemia (present in ~50% of cases) and insulin resistance (~60%). Insulin can also act on the ovary to increase the release of androstenedione and testosterone and can also downregulate sex hormone-binding globulin (SHBG), which effectively increases the bioavailability of the androgens. Screening women with type 1/type 2 diabetes has revealed PCOS rates of ~30% and ~70% respectively.

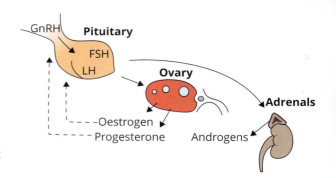

FIGURE 14.9

Role of LH and FSH in PCOS. GnRH: gonadotropin hormone releasing factor, FSH: follicle stimulating hormone, LH: luteinizing hormone. Solid lines: stimulation. Dotted lines: inhibition.

Much of the endocrinology of PCOS is understood (Figure 14.9). Arising in the hypothalamus, increased GnRH acts on cells of the anterior pituitary to effect the release of luteinizing hormone (LH) and follicle-stimulating hormone (FSH). GnRH itself is part-regulated by kisspeptin, a ~1.3 kDa single-chain product of *KISS1* at 1q32.1 and expressed in the hippocampus and other parts of the brain and the hypothalamus. It is the ligand for a ~43 kDa receptor coded for by *KISSR* at 19p13.3, and receptor–ligand binding initiates GnRH release, and so FSH and LH. *KISS1* and *KISSR* are expressed by the ovary, fallopian tube, and uterus at both mRNA and protein level. SNPs in *KISS1* have been linked to PCOS, and this may part-explain the PCOS characteristic of a raised LH with a low or normal FSH that together bring a raised LH:FSH ratio. These two hormones would normally regulate ovulation and the menstrual cycle, but high levels of LH will suppress ovulation, that being a feature of PCOS. A second aspect of the biochemistry of LH is the stimulation of the production of adrenal and ovarian androgens, principally dihydroxytestosterone, accounting for the hirsutism. The small hormone inhibin (section 15.4.1) feeds back to suppress FSH production in both sexes.

Presentation

PCOS may present with infertility (present in 40% of cases), miscarriage (62% of those attempting pregnancy), menstrual irregularities (75%: absent, sporadic, or heavy), hirsutism (hair on the face, back and chest, present in 70% of cases, the consequence of hyperandrogenism), and acne (~25%), whilst many women (>75%) are also overweight or obese. Unsurprisingly, anxiety, mood swings, and other psychological changes are likely. Faced with these signs and symptoms, the practitioner will request an abdominal and/or transvaginal ultrasound (Figure 14.10). More than 12 cysts generally define a polycystic ovary (>20 is not unusual), and as a result the volume of the entire ovary will be >10 cm^3.

The second source of crucial information are levels of FSH, LH, oestradiol, progesterone, kisspeptin, sex hormone-binding globulin, prolactin, and testosterone (reference ranges are given in Table 15.15). Given the link with diabetes, HbA1c and insulin may be requested, especially in the obese who are already at risk of this disease (Chapter 7, section 7.4). A lipid profile may also be requested, as may thyroid-stimulating hormone, thyroxine, and dehydroepiandrosterone.

Diagnosis

Before a final diagnosis, alternate diagnoses must be considered. These include late-onset congenital adrenal hyperplasia (11-hydroxylase and 21-hydroxylase deficiency), ovarian failure, hypothyroidism, hyperprolactinaemia, and Cushing's syndrome. Isolated androgenic hirsutism may be due to ovarian stromal hyperthecosis (hyperplasia of thecal cells), with raised testosterone, and often clitoromegaly and loss of head hair.

The principal laboratory results consistent with PCOS are raised LH and a raised LH/FSH ratio. Pituitary function in both sexes can be assessed by stimulation with exogenous GnRH, which should result in a measurable increase in LH and FSH. Increased serum PSA is a feature of PCOS, the source of which and any pathophysiology (if present) are unknown. Less puzzling is a potential role for anti-Mullerian hormone (AMH), expressed by ovarian granulosa cells, with serum levels being two to four

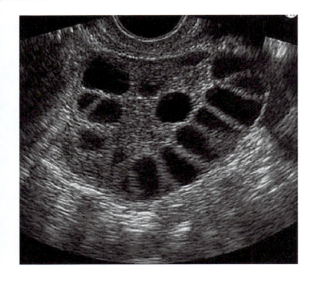

FIGURE 14.10

Ultrasound of PCOS. Grossly polycystic ovary with 15 cysts: a normal ovary may have up 10 follicles between 2 mm and 8 mm in diameter, the largest being that about to ovulate.

Hiremath PS & Tegnoor JR. Follicle Detection and Ovarian Classification in Digital Ultrasound Images of Ovaries: Advancements and Breakthroughs in Ultrasound Imaging, edited by GPP Gunarathne, IntechOpen, 2013. https://www.intechopen.com/chapters/45102. © 2013 The Author(s). Licensee IntechOpen.

times higher in women with PCOS than in healthy women. The link is that AMH inhibits the development of ovarian follicles by opposing the action of FSH. Details of AMH in males are to be found in section 15.4.1. Based on signs and symptoms, imaging, and laboratory findings, diagnosis of PCOS can be refined:

- Classical PCOS: clinical and/or biochemical evidence of hyperandrogenism, oligo- or anovulation, and ultrasound evidence of a polycystic ovary

- Endocrine PCOS: as classical, but lacking ultrasound evidence, hence only the biochemical and ovulatory issues arising from endocrine malfunction

- Ovulatory PCOS: ultrasound identifies a polycystic ovary; there are biochemical abnormalities, but ovulation occurs (although not necessarily with regularity)

- Non-endocrine PCOS: there is no evidence of hyperandrogenism, but there are ovulatory irregularities and ultrasound finds a polycystic ovary

Some practitioners have further refined diagnostic options for adolescents, citing a uterine bleeding pattern that is abnormal for the age or gynaecological age, with persistent symptoms for 1–2 years. Some use a lower LH/FSH ratio of >2 as part of the diagnostic criteria. A further definition is polycystic ovary morphology (PCOM), used where describing only the ovary phenotype.

Treatment

There is no specific cure for PCOS: the first-line treatment, where appropriate, involves weight loss, exercise, moderation of alcohol intake, and smoking cessation. Medical treatment focuses on the three major manifestations, some of which overlap.

Hirsutism can be addressed by regular shaving, plucking, bleaching, etc. Exogenous oestrogens (e.g. the contraceptive pill (OCP), preferably delivering a low dose of oestrogen) can help as they increase SHBG, thus suppressing circulating androgens, and also suppress androgen production by the ovary. Other antiandrogen treatments include finasteride (also a treatment for benign prostatic hyperplasia and androgenic alopecia in men), spironolactone (hypertension), and cyproterone acetate (prostate cancer, gynaecomastia). NICE ESUOM6 focuses on metformin in women not planning pregnancy, finding no difference in regimes containing metformin compared to co-cyprindiol (combining cyproterone and ethinylestradiol, an OCP) in controlling hirsutism or effects on acne. Flutamide (a treatment for prostate cancer) reduces acne and body hair and has lipid-lowering properties in addition to promoting ovulation, but has been linked to hepatotoxicity, requiring regular liver function tests.

In many cases the simplest (and often first-line) treatment for menstrual abnormalities is the OCP, which also helps with body hair and acne. Cyclicity can also be helped with metformin, which will also help alleviate possible hyperinsulinaemia.

Several of the drugs described previously can also help with fertility, but a specific treatment is direct ovarian stimulation with clomifene citrate 50 mg od, although this has numerous side effects. An alternative is the aromatase inhibitor letrozole (2.5 mg od), which has a better profile regarding pregnancy rates and live births than clomifene. However, care must be taken in order to avoid ovarian hyperstimulation syndrome.

The pregnant woman with PCOS is at increased risk of gestational diabetes, hypertension, and the premature delivery of a low-birthweight infant. Continuing management is further compounded by the risk of endometriosis and endometrial cancer, weight gain, dyslipidaemia (leading to non-alcoholic liver disease), metabolic syndrome, hyperinsulinaemia, and insulin resistance, the latter set developing into type 2 diabetes and then cardiovascular disease.

SELF-CHECK 14.8

What are the biochemical changes in PCOS?

Genetics

Family history is a risk factor for PCOS, with perhaps 20–25% of mothers and sisters of affected women testing positive, and a number of susceptibility loci has been identified at 2p16.3, 2p21, and 9q33.3. Twin studies estimate 72% of the risk of PCOS to be genetic. Mutations in genes coding for hepatic, adrenal, and ovarian enzymes involved in steroid metabolism in PCOS (Figure 14.12), such as CYP11a and CYP21, may be important. Lower levels of aromatase activity, required for the generation of oestrogen, in women with PCOS may be linked to mutations in *CYP19* on 15q21.2, which codes the relevant metabolic enzyme CYP19. Testosterone is converted into the more potent dihydrotestosterone by 3-oxo-5α-steroid 4-hydrogenases 1 and 2, coded for by *SRD5A1* at 5p15.31, and *SRD5A2* at 2p23.1 respectively: polymorphisms in both genes have been reported in PCOS. Other candidate genes are shown in Table 14.8.

There are numerous miRNAs linked to PCOS, most of which are overexpressed. Examples include increased expression of miR-9, miR-135a, and miR-18b in follicular fluid and granulosa cells, targeting *IL8*, *SYT1*, and *IRS2*, which may give clues as to cellular processes. miR-107 increases testosterone release when introduced into primary ovarian granulosa cells *in vitro*. There is ample evidence of potential roles for interactions between FSH and miRNAs in regulating granulosa cell proliferation and hormone secretion. Serum levels of miR-222, miR-146a, and miR-30c are increased in PCOS, suggesting possible roles as biomarkers.

TABLE 14.8 Genes with possible roles in PCOS

Gene	Location	Product/link
AR	Xq12	The androgen receptor
SHBG	17p13–p12	Sex hormone binding globulin
AMH	19p13.3	Anti-Müllerian hormone
FSHR	2p16.3	Follicle-stimulation hormone receptor
VNTR/INS	11p15.5	Polymorphism in *VNTR* associated with *INS*, coding for insulin
INSR	19p13.2	Insulin receptor
CAPN10	2q37.3	Calpain 10 (calcium-dependent cysteine protease), linked with insulin

There is similar evidence of the potential role of lncRNAs in PCOS. A micro-array found over 600 lncRNAs to be two-fold up- or downregulated in cumulus cells from PCOS cases compared to healthy cases obtained from *in vitro* fertilization ova. LncRNA RP11-151A6.4 may be important in insulin resistance, androgen excess, and adipose dysfunction in PCOS. RT-PCR of leukocyte nucleic acid found that lncRNA H19 is upregulated in PCOS and correlated with fasting glucose, but not with testosterone or insulin resistance. As with many other diseases, the coming decades will determine the importance of non-coding RNAs in PCOS.

Other ovarian disease

Premature ovarian failure/insufficiency is defined as a premature menopause before the age of 40. In many cases the cause can be predicted from medical history, such as Turner's syndrome, radiotherapy, chemotherapy, endometriosis, exposure to chemical toxins (cadmium, tricholoroethylene, etc.), peritonitis, and PCOS. Some 1% of women will experience this disease (0.1% aged under 30), and in 75% cases, no clear cause is found. Pathology identifies disease of the hypothalamus and/or pituitary (hence lack of GnRH, LH, and FSH) or ovaries (lack of oestrogens and progesterones), although a functioning adrenal will continue to generate these molecules. Symptoms are as of the natural menopause, e.g. amenorrhea, hot flushes, irritability, vaginal dryness, etc. Treatments include hormone replacement therapy (HRT), as in the menopause (bearing in mind the risk of venous thromboembolism), whilst vaginal oestradiol will help dryness of this organ. Loss of bone density may need to be addressed with oral calcium and vitamin D.

Ovarian torsion (equivalent to testicular torsion) is due to twisting, and may interact with nearby vessels, causing pain, nausea, and vomiting. Causes are unclear, but it is diagnosed by ultrasound, and the torsion is rectified by laparoscopic surgery. Many women feel the precise moment of ovulation, some with a slight and transient pain. However, should this pain be sharp, becoming dull, it may indicate an ovarian rupture, again requiring diagnosis by ultrasound and laparoscopic treatment.

14.3.3 Fallopian tubes

The principal disease of the fallopian tubes (Figure 14.11) is inflammatory (salpingitis), most commonly caused by an infection.

The most frequent organisms involved are those transmitted sexually (section 14.5), although infection may also follow a ruptured appendix and surgical interventions and so be caused by opportunists such as *Streptococcus pyogenes*. Acute salpingitis is accompanied by low abdominal pain, fever, and a vaginal discharge, and is often considered under the umbrella of pelvis inflammatory disease (PID). Fallopian tubes may be a conduit for retrograde menstruation causing endometriosis (section 14.3.4). Treatment is with antibiotics, whilst blockages may be formed from inflammatory leukocytes (pyosalpinx), a serous fluid (hydrosalpinx), or blood (haematosalpinx). All forms of fallopian tube pathology will be investigated by ultrasound, possibly MRI/CT, and potentially laparoscopy.

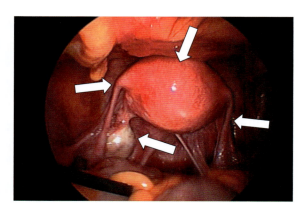

FIGURE 14.11

The fallopian tubes (left and right arrows), an ovary (lower arrow), and the uterus (upper arrow). Photograph taken at laparoscopy.

Science Photo Library.

14.3.4 Endometriosis and the uterus

Section 5.10 in Chapter 5 has details of uterine and endometrial cancer, the former including fibroids. The leading diseases of the uterus are heavy or prolonged menstrual bleeding (HMB, formerly menorrhagia) and endometriosis.

Aetiology and pathology

Endometriosis is growth of endometriotic glands and stroma outside the uterus: the ovaries, fallopian tubes, uterine ligaments, the abdominal wall, the urinary tract, and the external walls of the colon and rectum. Some estimate that it is present in 6–10% of women before the menopause and in 2–5% of post-menopausal women, where it is often linked to HRT or tamoxifen. Eighty per cent of cases in post-menopausal women involve the ovary. Perhaps a third (or more) of women with endometriosis are infertile, and conversely in a third of infertile women there is endometriosis, which is presumably causative. Aetiology has been linked to combinations of environmental, endocrine, obstetric, menstrual, and genetic factors. Where there is fever and an acute phase response, an immunological aetiology is likely (with inflammatory cytokines and leukocytes present in the peritoneal fluid), and as the healthy uterus cycles the loss of and replacement of the endometrium, aberrant angiogenesis may be suspected. The precise cell biology of the disease remains obscure, with several other competing but often overlapping theories. These are:

- Retrograde menstruation (outflow via fallopian tubes into the peritoneal cavity)

- Post-puberty development of primitive endometrial cells misdirected during organogenesis

- Inappropriate settlement and transformation of bone marrow-derived stem cells

- Migration of endometrial stem cells from the uterus

- Failure of apoptosis by endometrial cells

Some disease may be linked to increased aromatase activity (generating oestrogens) in cells from endometriosis tissues and from normal endometrial tissues, and the reduced incidence after the menopause also implies the involvement of oestrogens. Infection may be a cofactor in the development of endometriosis as there is often lower genital tract infection and pelvic inflammatory disease, with *Enterococci*, *E. coli*, *Gardnerella*, and *Streptococcus* species present.

Presentation and diagnosis

Primary symptoms include lower abdominal pain (present in ~75% of cases) and infertility (~33%); others include dysmenorrhea, dysuria, dyspareuria, and irregular uterine bleeding, whilst 20% of cases are asymptomatic. The numerous nonspecific signs and symptoms make the diagnosis of endometriosis a challenge, and it may take years to form correctly, especially in the post-menopausal. The gold standard diagnostic method is laparoscopy, ideally with histology, although developing transvaginal ultrasound and complex imaging techniques are becoming popular. Serum CA-125 is increased in endometriosis, but has poor sensitivity and specificity (a result <30 U/ml cannot exclude disease), but these can be improved by co-assessment of inflammatory cytokines IL-8 and TNF-α. Once diagnosis has been made, CA-19.9 may be useful as it correlates with disease severity.

Management

In the absence of a cure, treatment is of symptoms. Pain can be managed with NSAIDs, whilst medical management consists of GnRH agonists (such as elagolix (200 mg od) and leuprolide (intra-muscular 1 mg daily or depots of 7.5 mg monthly to 65 mg annually)), the OCP, aromatase inhibitors, and progestins, all of which are designed to suppress ovarian function and therefore reduce the likelihood of pregnancy. The weak steroid danazol (200–800 mg daily in divided doses) is often effective as it has anti-oestrogenic, anti-progestogenic, and anti-androgenic activity. Analogues of gonadorelin (a GnRH agonist) such as buserelin (300 μg intranasally tds), goserelin (3.6 mg every 6 months), and leuprorelin (3.75 mg subcutaneous or intramuscular) may be used. Surgical treatments include laparoscopic resection of lesions, although this may damage underlying tissues, such as the ovary. Assisted reproduction with intrauterine insemination, *in vitro* fertilization, and intra-cytoplasmic sperm injection can all be successful. For the most severe cases, hysterectomy in an option.

Genetics of endometriosis

Family and twin studies (concordance 0.52 in monozygotic twins, 0.19 in dizygotic twins) demonstrate a clear genetic component to endometriosis. Many reports describe the aberrant expression of proliferation-associated genes such as those of the homeobox and Wnt signalling pathways, and genome-wide association studies point to potential roles for *HOXA10* and *HOXA11*. There are also links with genes coding for the FSH receptor, the oestrogen receptor, and in sex steroid hormone pathways.

Downregulation of miR-200b has been described in endometriosis tissues and may be responsible for an epithelial-to-mesenchymal transition similar to that of carcinogenesis. Other miRNAs reputed to have a role include miR-34c-5p, miR-34b, and miR-9, all of which are downregulated. Upregulation of miR-210 may inhibit apoptosis by targeting signal transduction and activator of transcription-3 (STAT-3). Increased level of circular RNAs circ_0004712 and circ_0002198 may be potential novel biomarkers for the diagnosis of ovarian endometriosis, whilst LncRNAs CHL1-AS2, AC002454.1, LINC01279, and MALAT1 may all have some role in pathogenesis.

SELF-CHECK 14.9

What are the major anatomical and aetiological differences between PCOS and endometriosis?

Heavy menstrual bleeding

With a prevalence of 20–30% of women of reproductive age, HMB is one of the leading self-referrals to primary care, with peak referrals in adolescents and peri-menopausal women. However, some, taking a wider definition of bleeding, suggest the true rate may be a third of all adult women. The pathophysiology of abnormal vaginal bleeding is diverse and can be broadly categorized as idiopathic and objective. The former is where an obvious cause is unclear, but may include cancer, fibroids, PCOS, and endometriosis, as addressed in other sections. Objective HMB is essentially 'normal' cyclical menstrual bleeding that is markedly (>80 ml) in excess of or longer (>7 days) than normal bleeding (generally 10–40 ml, <7 days) in the absence of other pelvic disease, and so is reasonably easy to diagnose, even if the precise cause is obscure. The frequency of presentation in adolescents is high (35%) compared to adults and is quicker to resolve with fewer follow-up visits, most likely because natural endocrine cycling may take time to become established. The International Federation of Gynaecology and Obstetrics (FIGO) offers a classification system (Table 14.9).

An established cause is variants of von Willebrand disease, present in ~12%, another 8% having one of the other coagulation factor deficiencies (section 11.6.4). Overuse of anticoagulants has the same effect, as reported by over half of women using these agents. Where the bleeding is objective or due to fibroids, adenomyosis, or the use of intra-uterine devices, prostaglandin disorders may be causal. An important non-cycling cause is fibroids, present in 10% of women with HMB and 40% of those with severe HMB. The latter, if acute, may require hospitalization. Conversely, up to 30% of women with HMB may have associated fibroids, and accordingly, NICE NG88 recommends hysteroscopy or ultrasound as the first-line investigation. Wherever possible, any cause should be addressed, but treatments include the OCP (with >20 μg of ethinyl oestradiol, which can reduce menstrual loss by 50%) and the oral antifibrinolytic agent tranexamic acid (1–1.5 g two to three times a day). An alternative in young women is nasal desmopressin (300 μg on days 2 and 3 of the menstrual cycle), whilst some women respond to NSAIDs (such as mefenamic acid 500 mg tds), possibly as it inhibits the cyclooxygenase pathway. NSAIDs and antifibrinolytics may be used together but should be stopped if there is no improvement after 3 months.

TABLE 14.9 FIGO classification of heavy menstrual bleeding

Anatomical	Non-anatomical
P—Polyp	C—Coagulopathy
A—Adenomyosis	O—Ovulatory dysfunction
L—Leiomyoma	E—Endometrial
M—Malignancy and transformation	I—Iatrogenic
	N—Not otherwise classified

Munro MG, Critchley HO, Fraser IS , The two FIGO systems for normal and abnormal uterine bleeding symptoms and classification of causes of abnormal uterine bleeding in the reproductive years: 2018 revisions. Int J Gynecol Obstet, 143: 393–408. https://doi.org/10.1002/ijgo.12666. © 2018 International Federation of Gynecology and Obstetrics.

Surgical treatments include endometrial ablation, **dilatation and curettage**, and hysterectomy, although there may be complications such as uterine perforation, infection, and haemorrhage. Nevertheless, ablation is effective in controlling blood loss in ~90% of women. In some cases, surgery may be less effective than a uterine implant of levonorgestrel, which delivers an androgenic progesterone directly to the local endometrium, preventing proliferation, and can decrease menstrual loss by >90%. In women with fibroids, uterine artery embolization reduces blood loss by ~85%. Treatment is required as increased blood loss will inevitably lead to a normocytic and then, once iron stores have been depleted, a microcytic anaemia.

Other disease of the uterus

These include inflammatory endometritis, characterized by abdominal pain, fever, and vaginal discharge, from which a causative organism may be detected. It most often follows delivery of an infant and/or obstetric surgery and is generally treated with an antibiotic such as doxycycline, but may also be the consequence of a sexually transmitted disease.

Key points

PCOS and endometriosis are complex diseases, and the most common non-malignant gynaecological conditions. Each has an endocrine aetiology, so that the laboratory is an essential tool, although PCOS cannot be diagnosed without ultrasound.

14.3.5 The cervix, vagina, and vulva

Cancer of these organs is described in sections 5.13.1, the remaining aetiology being inflammation, which is in most cases the consequence of a sexually transmitted disease (STD, section 14.5). Those cases not directly attributable to an STD are likely the result of changes to the local flora and fauna (such as loss of *Lactobacillus* species and changes in pH), perhaps due to the effects of systemic antibiotics and/or the overgrowth by opportunistic organisms such as *Gardnerella vaginalis* and *Candida albicans*, the latter present in >90% of cases. Vulvovaginal candidiasis is believed to afflict ~75% of women at some time in their lives, whilst it recurs in ~8%. Presentation is with pain, itching, and vaginal discharge, which is often offensive and can be heavy. Treatment is with antifungals such as clotrimazole cream or pessary. Bacterial vaginosis may be treated with antibacterial creams and gels such as metronidazole (effective against certain gram-negative organisms such as *Bacteroides* species) and clindamycin.

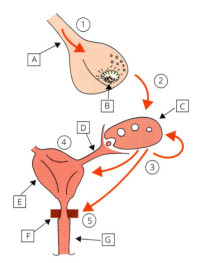

FIGURE 14.12
Fertility and infertility. Simplified diagram of steps required for conception and factors that can prevent it happening.

14.3.6 Infertility in the female

A successful conception is the product of a large and complex series of endocrine and physiological processes, a principal feature being the desire for sexual activity. The latter is part-controlled by kisspeptin, as treatment of hypoactive sexual desire with this hormone is effective in enhancing sexual brain processing (as it is also in men). Key biochemical steps are illustrated in Figure 14.12 (thick red arrows) and as follows:

1. Under the direction of kisspeptin, gonadotropin-releasing hormone (GnRH) from the hypothalamus acts on the anterior lobe of the pituitary.

2. LH and FSH travel from the anterior pituitary to the ovaries.

3. The hormones act to induce the release of oestrogen and progesterone to control the development of the follicle and the release of the ovum at ovulation.

4. In concert, oestrogen and progesterone regulate the formation of the endometrium and, should no blastocyst be implanted, its destruction.

5. Progesterone thins cervical mucus at mid-cycle to permit the entry of a sufficient number of sperm of sufficient quality into the uterus.

However, as discussed in previous sections, failure of any one of these steps will lead to infertility. The latter may be defined as the inability to produce a viable infant (as do 30–60% of pregnancies), and so includes implantation failure, spontaneous abortion/miscarriage, and stillbirth. Of those established pregnancies, pre-eclampsia and eclampsia develop in 10% of cases, and premature delivery affects a further 10%. Several of these issues will have been determined by the lack of signs and symptoms of sexual development in puberty, others in adult women. The thin black arrows in Figure 14.12 illustrate pathology as follows:

A. Failure of release of GnRH causes low or absent LH and FSH, and so hypogonadotropic hypogonadism. In some cases, the lesion derives from pathology of the hypothalamic neurons, and low oestradiol levels can suppress pulsatile GnRH release. Numerous mutations in various genes (such as *KISS1* (coding for kisspeptin) and *GNRH1* and their receptors *KISS1R* and *GNRNR*) cause GnRH deficiency. Fortunately, deficiency is amenable to replacement therapy.

B. Pituitary tumours are a known cause of reduced (and so hypogonadotropic hypogonadism) and increased levels of LH and FSH, but reduced LH and FSH may also be due to trauma or brain surgery. Fertility is likely to follow the hyperprolactinaemia result from a pituitary tumour as prolactin inhibits hypothalamic release of GnRH.

C. Even if present, LH and FSH may fail to stimulate the ovary should there be loss-of-function mutations in the genes for their respective receptors, leading to reduced oestrogen and progesterone.

There are reports of homozygous mutations in the oestrogen receptor in teenagers showing delayed puberty and absent breast development, but high levels of LH, FSH, and oestrogen. Many of these, and others, come together as ovarian failure (section 14.3.2), which in some cases is described as a premature menopause. PCOS may also cause infertility.

D. As fertilization occurs in the upper fallopian tubes, blockage of sections close to the uterus will lead to infertility (section 14.3.3).

E. The presence of fibroids, endometritis, and endometriosis are major causes of uterine infertility. As described in section 14.3.4, although the uterus is an immunologically privileged site, loss of tolerance to paternal antigens leads to miscarriage.

F. The viscosity of cervical mucus varies with the menstrual cycle. It may fail to become thin at midcycle, thus preventing sperm from entering the uterus. Such thick mucus is described as hostile.

G. Many forms of sexually transmitted disease (section 14.5) and other infections create a vaginal environment hostile to sperm. It may also cause pain during sexual intercourse (dyspareunia), thus preventing ejaculation. Vaginismus is spasmodic constriction of the muscle of the outer third of the vagina—so much so that penetration is difficult or impossible (H). Causes may be organic or psychological.

SELF-CHECK 14.10

What are the frequency, causes, and treatments of heavy menstrual bleeding?

The natural cause of infertility is the menopause (NICE NG23), and although not a disease, it can be associated with disease such as osteoporosis, which may be treatable with bisphosphonates, by raloxifene, and by teriparatide (a bioactive portion of parathyroid hormone). With the risk of a vertebral fracture being one in three and that of a hip fracture one in five, osteoporosis should be assessed. There is also a view that oestrogen helps protect women from cardiovascular disease, as the risk of this increases markedly after the menopause. Some women find relief from the symptoms of the menopause (hot flushes, loss of libido) in HRT, at a risk of venous thromboembolism and (in the long term) of breast and endometrial cancer.

CASE STUDY 14.2 *Challenges to fertility*

A couple in their 30s have been having regular unprotected intercourse for a number of years, but the woman has not become pregnant. Neither partner has any relevant medical or surgical history; the woman was on the OCP for around a decade, and both her sisters had untroubled and successful pregnancies. She reports a normal menstrual cycle, and an abdominal ultrasound reports normal ovaries and uterus. On referral to the local infertility clinic, extensive endocrine biochemistry tests in both are within the reference range, the woman cycling normally. A post-coital test is normal; andrology testing shows the man's semen volume, sperm count, and motility are all at the lower end of their reference ranges, but the proportion of abnormal forms is low. However, both his and a donor's sperm fail to penetrate a sample of the woman's mid-cycle cervical mucus. In further testing it is found to be highly viscous, and so is described as hostile. The couple are referred for *in vitro* fertilization.

At the specialist centre, all tests are confirmed, and the couple are called in to be prepared for their first *in vitro* fertilization cycle. However, the woman suffers a particularly bad congestive head and chest cold, so the couple do not report. She self-doses for several days with an expectorant cough mixture containing guaifenesin, which acts by increasing the volume of secretion of the trachea and bronchi, and by reducing their viscosity. This results in a relief of the chest symptoms, but contrary to the cough mixture insert, ten days later she reports an increase in the size of both breasts, which have become a cup size larger. A week later she fails to have her period and two weeks later a pregnancy test is positive. Although she miscarries, requiring aspiration, she becomes pregnant again the following month, and delivers a healthy boy, followed 2 and 4 years later by girls. The couple's medical team believe that the cough mixture also thinned the woman's hostile mucus, allowing the sperm through. The miscarriage probably permanently resolved the issue with her cervix.

14.4 The female reproductive system 2: Pregnancy

Pregnancy and childbirth may cause the death and/or disease of the mother and her child, examples of which, as classified by ICD-11, are shown in Table 14.10. In England and Wales, in 2021, there were 46 deaths due to pregnancy, childbirth, and the puerperium, and 1,783 neonatal deaths (56.1% boys) under 28 days (0.3% of all deaths). Globally, there were 1.7 million neonatal deaths (3.2% of all deaths). Leading causes were preterm birth (36.4%), encephalopathy due to birth asphyxia and trauma (29.9%), sepsis and other infections (11.4%), and haemolytic disease and other jaundice (2.8%). Other neonatal disorders were linked to 19.6% of deaths.

14.4.1 Miscarriage and stillbirth

It has been estimated that only 20–25% of unprotected conceptive copulations result in a live birth, early and late failures being due to factors such as embryo malformation and endometrial problems.

Miscarriage

Miscarriage (natural abortion) is the most common complication of pregnancy, with a reported rate of 12–24%, with recurrent miscarriages (≥3 losses) present in 1–2%. An estimated 12–15% of women wishing to become pregnant have a miscarriage at some point. However, the true rate is unknown as it may occur before the menstruation or may be thought of as an early menstruation. Accordingly, miscarriage may be defined as spontaneous loss of a conceptus after the cessation of menstruation and up to 20 weeks into the pregnancy, following which it is a stillbirth. In the UK the rate of miscarriage is ~125,000/year resulting in ~42,000 hospital admissions, whilst around 0.6% of pregnancies result in a stillbirth (around 4,000/year in the UK), compared to ~0.9% in the developing world.

Around 50–70% of early pregnancy losses are due to chromosomal abnormalities, of which 50–90% are aneuploidy, major forms being trisomy (52%, mostly trisomy 16), polyploidy (19%),

TABLE 14.10 International classification of diseases: Section 18: Pregnancy, childbirth, and the puerperium

ICD section (code)	Medical condition
Abortive outcome of pregnancy (JA00–0Z)	Abortion, ectopic pregnancy, molar pregnancy, missed abortion, blighted ovum, complications of pregnancy
Oedema, proteinuria, or hypertensive disorders (JA20–2Z)	In addition, pre-existing hypertension, pre-eclampsia superimposed on chronic hypertension, pre-eclampsia, eclampsia, gestational hypertension
Obstetric haemorrhage (JA40–4Z)	Whether antepartum, intrapartum, or postpartum
Certain specific disorders of pregnancy (JA60–6Z)	Excessive vomiting, venous complications, diabetes, malnutrition, infections of the genitourinary tract
Complications of labour or delivery (JB00–0Z)	Preterm labour or delivery, failed induction of labour, long labour, obstructed labour, umbilical cord complications, perineal laceration, retained placenta
Delivery (JB20–2Z)	Single spontaneous delivery, use of forceps or vacuum, caesarean section, multiple delivery
Certain obstetric conditions, not elsewhere classified (JB60–6Z)	Obstetric death; maternal infectious disease; sequelae of complication of pregnancy, childbirth, or the puerperium

Adapted from World Health Organization, ICD (international statistical classification of diseases and related health problems), 11th revision, 2022: https://icd.who.int/browse11/l-m/en#/http%3a%2f%2fid.who.int%2ficd%2fentity%2f714000734.

monosomy (15%), and structural abnormalities (6%) (Chapter 16). Three potential aetiologies dominate:

- Poor-quality embryos, in contrast to high-quality embryos, fail to secrete soluble factors promoting implantation.

- The endocrine theory suggests that miscarriage follows the inability of the endometrium to adequately host the blastocyst, which in turn may be due to poor endocrine signalling to the developing endometrium during the initial stage of the menstrual cycle.

- Immunological aspects of pregnancy and its related pathology, discussed in section 14.4.3.

Other causes include an anatomical abnormality such as uterine septa, adhesions and fibroids, or an incompetent cervix, and specific disease such as PCOS, endometritis, endometriosis, infections, uncontrolled diabetes, etc., some due to medications such as methotrexate and NSAIDs. Additional causes include thrombotic and autoimmune disease such as the anti-phospholipid syndrome, thyroiditis/hypothyroidism, inflammatory bowel disease, and SLE, the latter bringing a 17- to 20-fold increased risk of post-partum mortality. Nutrition is important, as shown by an increased risk of miscarriage in underweight, overweight, obesity, and bulimia nervosa/binge-eating disorders. Other risk factors include caffeine use (with dose–response), lifting a weight >20 kg daily, cigarette smoking (> 10/day), night work, and exposure to non-ionizing radiation (in healthcare workers).

Maternal infections with the potential to cause miscarriage include syphilis (14.9%, control group 2.3%), *brucella*, *gonococcus*, *listeria*, malaria, and toxoplasmosis. Age is a factor, with a J-shaped curve, lowest at age 27 (Figure 14.13).

Miscarriage occurs in ~12% of first pregnancies, in ~17% where there has been a previous stillbirth or neonatal death, and in 21% where there has been a previous miscarriage. The latter worsens with the number of miscarriages: 20% with a history of one miscarriage, 28% if two, and 42% if three or

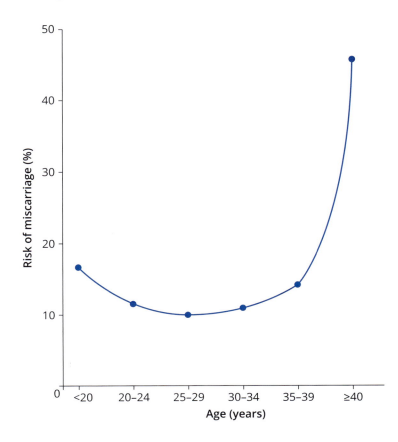

FIGURE 14.13
Relationship between women's age and the risk of miscarriage.

more (but see also section 14.4.3, where immunological aetiology is addressed). The age of the father, regardless of the age of the mother, is also important: the adjusted hazard ratio of spontaneous abortion where the father is aged ≥35 is 1.27 (95% CI 1.00–1.61) relative to that of a father aged <34, most likely due to accumulated genetic errors in spermatogenesis as the testes age (much as in oogenesis as the woman ages).

Some 25% of pregnancies are complicated by early bleeding, and of these ~30% miscarry. The concern is that marked bleeding may be placental abruption, so that great expertise is required in estimating the volume of blood lost. A serum progesterone level <6 ng/ml (19.1 nmol/l) reliably excludes a viable embryo, measurement of β-hCG can be informative, but a transvaginal ultrasound is essential and may detect other important features such as an ectopic pregnancy.

Around a quarter of miscarriages might be prevented by addressing modifiable risk factors, the most important being alcohol consumption. A potentially at-risk pregnancy (by virtue of vaginal bleeding) in women with a history of miscarriages may benefit from 400 mg micronized vaginal progesterone twice daily. An inviable conceptus can be managed with 200 mg oral mifepristone followed 24 hours later by 800 μg vaginal misoprostol or uterine aspiration, the preferred surgical treatment as it is associated with less pain and blood loss, and has a shorter duration.

Stillbirth

A leading cause of stillbirth is placental abruption: the detachment of the placenta from the uterine wall, estimated to occur in ~0.8% of pregnancies in the developed world, but in >3% in the developing world. Even if the abruption resolves, it is still linked to low birth weight, prematurity, and neonatal mortality. If it occurs late in the pregnancy, and with severe haemorrhage, an emergency caesarean section may be necessary and is a risk factor for disseminated intravascular coagulation and maternal mortality. Many of the causes of stillbirth are the same as those of a miscarriage (Table 14.11).

The parallel with miscarriage extends to an immunological hypothesis with late breakdown of tolerance of paternal antigens (section 14.4.3). The simplest example of this is the sensitization of the rhesus-negative mother to rhesus molecules of her infant at parturition with subsequent generation of anti-D antibodies. With potential to cross the placenta, these antibodies may cause haemolytic disease of the foetus and newborn. Despite all of the above, a precise factor for a stillbirth cannot be identified in ~40% of cases.

14.4.2 Oedema, proteinuria, and hypertension

With considerable clinical and pathophysiological overlap, the three conditions oedema, proteinuria, and hypertension together may have serious implications for mother and child in the form of pre-eclampsia and eclampsia. Of this triad, the principal is hypertension, which is present in 10% of all pregnant women and increases the risk of pre-eclampsia. In Asia and Africa, hypertensive disorders account for ~10% of maternal deaths, but in South America this figure is 25%. This underpins the guidance that all pregnant women should have their blood pressure checked regularly for gestational hypertension, potentially by themselves. Should gestational hypertension be found, monitoring is increased, other tests performed, and treatment initiated.

TABLE 14.11 Risk factors for stillbirth

Alcohol, smoking (both with an odds ratio (OR) of 1.36), increasing age (OR 1.75 for age >35), *Plasmodium falciparum* (OR 1.81: causing an estimated 20% of stillbirths in malaria-endemic subSaharan Africa), high blood pressure (4–9%), diabetes (3%), obesity, previous caesarean section (increasing the risk by 23%), hypothyroidism (0.83%), placental abruption (~12%), SLE (5%), previous multiple pregnancies, previous pregnancy loss, low socioeconomic status

Eclampsia

This serious and life-threatening condition is commonly characterized by seizures (similar to those in epilepsy) that last between 60 and 90 seconds and which have a frequency of <0.5% of pregnancies in the developed world and ~1% in the developing world. The seizures are often preceded by headache (in 80% of cases) and visual disturbance (45%), and less often by abdominal pain and ataxia/loss of balance control; they may occur ante-partum (~50%), intra-partum (~20%), and up to 6 weeks post-partum (~30%). The likely pathophysiology is a result of those of pre-eclampsia, notably hypertension, which may drive cerebral and neurological microvascular encephalopathology that causes the seizures.

Eclamptic seizures are a medical emergency: the woman is generally intubated and turned on the left side to reduce the risk of aspiration pneumonia, and she may lapse into a coma. Coming out of the seizure, some confusion is to be expected. A common immediate treatment is 4–6 g magnesium sulphate (which acts as an anticonvulsant) given intravenously over 15–20 minutes with maintenance doses of 2 g/hr, whilst there may often be use of sedatives and anxiolytics. Should SBP/DBP be >160/110 mm Hg, a first-line treatment may be 20 mg intravenous labetalol, 10 mg of nifedipine, and 5–10 mg intravenous hydralazine (mode of action of these drugs is described in section 7.2.6 of Chapter 7). Up-titration may be required if the target BP is not reached. Subsequent investigation may reveal pulmonary oedema on chest X-ray and ankle oedema, CKD with raised serum creatinine and proteinuria and liver disease.

SELF-CHECK 14.11

What risk factors for miscarriage can you recall?

Pre-eclampsia

This condition, present in 5–8% of pregnancies in their second and third trimester, rests on the triad of hypertension (SBP/DBP >140/90 mm Hg), oedema (Chest X-ray/examination), and proteinuria (albumin-to-creatinine ratio >8 mg/mmol).

The hypertension is generally defined as >140/90 mm Hg, but would be lower (e.g. 135/85) if there are risk factors such as hypertension in a previous pregnancy or a strong family history. Proteinuria may be detected on dipstick analysis, and if present, should be confirmed in the laboratory with a 24-hour urine collection >300 mg or (preferably) an albuminto-creatinine ratio, a result >8 mg/mmol being regarded as positive. Most authorities require all three to be present for a firm diagnosis. If undetected or untreated there is a risk that pre-eclampsia will develop into eclampsia. Like eclampsia, symptoms include headache and visual disturbance, but also severe pain just below the ribs, vomiting, and sudden swelling of the face, hands, or feet.

There are many risk factors for pre-eclampsia (Table 14.12); they are of varying penetrance and frequency. In first-time mothers, the risk is some 4%, and if present, this rises to ~15% in a second pregnancy and to 32% in third and subsequent pregnancies. From a public health perspective, the greatest population-attributable fractions reside (in order) with nulliparity, pre-pregnancy overweight, prior pre-eclampsia, pre-pregnancy obesity, and hypertension. Pre-eclampsia casts a long shadow, conferring a 2- to 4-fold risk of hypertension, a 2-fold risk of cardiovascular disease, and a 1.5-fold risk of stroke.

TABLE 14.12 Risk factors for pre-eclampsia

The antiphospholipid syndrome, autoimmune diseases (principally SLE), diabetes, use of assisted reproductive technology, pregnancy interval of >10 years, age >40, family history of pre-eclampsia, prior pre-eclampsia, prior stillbirth, CKD, PCOS, mental stress, smoking, infections, primiparity, multi-foetal pregnancy, obesity, a history of obstetric pathology

The dominant pathophysiology is abnormal vascular development of the placenta, with hypoxia, which in turn may result from disordered angiogenesis. A key laboratory finding is increased levels of the soluble form of FMS-like tyrosine kinase-1 (sFlt-1), which binds to the primary angiogenic mediator, vascular endothelial growth factor, and also placental growth factor. This results in generalized endothelial dysfunction, which leads directly to hypertension and so renal disease, accounting for the oedema and proteinuria. Further evidence of placental pathology is increased numbers of syncytiotrophoblast microparticles that may recruit and activate neutrophils and monocytes to further add to the pathology with the release of inflammatory cytokines.

Thus whilst the pathology linking the placenta with the triad is reasonably well understood and agreed upon, there are few concrete insights into to cause(s) of the placental issues. Numerous candidate genes have been proposed, including *PAI-1* (the most promising), *IL-10*, *VEGF*, *CTLA4*, *TGF-β1*, and genes of certain coagulation molecules. A genome-wide association study reported that a variant in the foetal genome near the Flt-1 locus is linked to pre-eclampsia, a finding that fits well with the value of measuring sFlt-1 in the serum of selected women. In at-risk women with hypertension, testing for placental growth factor (PlGF) may help exclude pre-eclampsia, whilst NICE DG23 offers guidance on testing, some of which includes measurement of sFlt-1. Levels of PlGF <12 pg/ml are regarded as highly abnormal, a result between 12 and 99 pg/ml abnormal, while a result ≥100 pg/l is regarded as normal.

Use of umbilical artery Doppler ultrasound can reduce the rate of unnecessary interventions such as caesarean section to rescue the infant. For the infant, pre-eclampsia brings a risk of intrauterine growth retardation, preterm birth, foetal distress, placental abruption, and stillbirth, and possibly hypertension and cardiovascular disease as an adult.

The only definitive treatment for pre-eclampsia is termination of the pregnancy or delivery of the foetus and placenta. Pre-eclampsia is responsible for ~25% of pre-term births, some of which will have been deliberately induced in cases of severe pre-eclampsia, an option extended to those pregnancies at term where there is gestational hypertension, or mild or severe pre-eclampsia. The principal medical treatment for high blood pressure is (in preferred order, and as in eclampsia) labetalol, nifedipine, or methyl-dopa to a target of 135/85 mm Hg. Expert advice will be needed for those already taking an ACE inhibitor, angiotensin-receptor blocker, or a thiazide, as these may cause congenital abnormalities. Should blood pressure remain high, planned early birth (induction) may be considered, and if so, a course of antenatal steroids and magnesium sulphate are options. Treatment with antihypertensives should be continued for the immediate post-partum period, and gradually downtitrated once normotension is achieved. Women at high risk or with one or more moderate risk factors should take 75–150 mg aspirin od from weeks 12–16 of their pregnancy, and foetal ultrasound will be regularly performed.

Key points

Pre-eclampsia is a major medical issue, but fortunately is reasonably easy to diagnose from blood pressure and proteinuria. If untreated, it leads to increased risk of pregnancy loss and eclampsia.

HELLP

This syndrome is defined by concurrent haemolysis, elevated liver enzymes (generally, transaminases), and a low platelet count (often cited as <100 × 10⁹/l), and is most prevalent in the third trimester with a frequency of 0.5–0.9% (see also 11.6.2). Leading symptoms include pain in the right upper quadrant of the epigastric area, nausea, and vomiting; the diagnosis is made in the laboratory. The liver disease may be linked to fibrin deposits and microthrombi, possibly contributing to the low platelet count. Some 30% of cases develop post-partum. The treatment is with corticosteroids, but blood transfusions may be necessary. One report points to magnesium sulphate infusions from admission for delivery to 24 or 48 hours post-partum (as with pre-eclampsia), whilst platelet transfusions may be required for severe thrombocytopenia (<20 × 10⁹/l).

14.4.3 The foetus as an allograft

The clear success of the presence of male genetic material in the uterus is due to immunological toler-ance. The uterus is an immunologically privileged site, para-uterine lymph nodes are enlarged, and the placenta produces 'industrial' doses of steroid sex hormones that have immunosuppressive quali-ties. Indeed, pregnancy brings relief from the symptoms of autoimmune diseases such as rheumatoid arthritis and thyroiditis, which unfortunately return post-partum. Other immunological defence is unaltered, such as vaccination against influenza SARS-Cov-2, and tetanus toxin.

The endometrium

NK cells, lymphocytes, macrophages, and dendritic cells are abundant within the endometrium, presumably with roles in endometrial loss and rebuilding. However, increased numbers in preg-nancy are thought to be recruited by the presence of paternal alloantigens, and/or to support spiral artery remodelling and trophoblast invasion. One reason why these cells fail to attack the tropho-blast is that the latter downregulates or even lacks certain of the HLA recognition molecules that the leukocyte requires in order to define non-self (section 9.6.1). However, the presence of antibodies to paternal HLA molecules in the blood of almost 50% of mothers proves that some sort of immune recognition is present. How (if at all) these antibodies impact on pregnancy is unknown, but they may mask paternal antigens, thus preventing them from being exposed to the woman's immune system.

Local leukocyte activity at the foetal/maternal interface is very likely to be moderated by the sustained expansion of CD4$^+$ CD25$^+$ FOXP$^+$ T$_{reg}$ cells, a process augmented by hCG produced by the syncytiotrophoblast. Macrophages and decidual NK cells can produce anti-inflammatory cytokines (IL-10, IFN-γ), whilst dendritic cells adopt a tolerogenic phenotype. Circulating T$_{reg}$ numbers fall in the luteal phase of the menstrual cycle, indicating a hormone effect that may be active during the pregnancy. Loss of control of T$_{reg}$ activity is frequently described in many differ-ent forms of reproductive pathophysiology, but also in autoimmune disease such as SLE and type 1 diabetes.

The male

Repeated exposure of the woman's genital tract to semen from the same man, which will include epithelial cells and cellular particles (sperm being HLA-negative) may condition her for tolerance of his antigens. The risk of pre-eclampsia correlates inversely with the duration of sexual cohabi-tation, the mean time to pregnancy being 7 months. In those with a history of prior spontaneous abortion, the incidence of pre-eclampsia is 50% lower in nulliparous women carrying a child of a long-standing sexual partner compared to women conceiving with a new sexual partner, and use of contraceptives (condom, cervical cap) preventing vaginal or uterine exposure to semen is also linked to pre-eclampsia.

Pre-eclampsia and miscarriage

A normal endometrium will reject poor-quality embryos, whilst that of women with recurrent preg-nancy loss accepts low-quality embryos, only to reject them at a later stage. This may explain why women who miscarry find it easier to become pregnant than women without a history of miscarriage. Tissue obtained from spontaneous abortions has fewer T$_{reg}$ cells than tissue from planned abortions, and those with a history of recurrent loss and low levels of T$_{regs}$ cells have more difficulty conceiving. The reported imbalance between circulating Th17 (more) and T$_{regs}$ cells (less) in unexplained recur-rent spontaneous abortion led to the unblinded subdermal immunization of 20 such women (with a history of 3–7 miscarriages) with donor or sexual partner lymphocytes. This intervention resulted in a reduction of Th17 cells, an increase in FOXP3 T$_{regs}$ cells, and a fall towards normal in the Th17:T$_{reg}$ ratio, and, crucially, resulted in an 83% continuing pregnancy rate.

Recurrent foetal loss is linked to reduced anti-inflammatory cytokines IL-35 (in serum) and galectin-1, which promote T cell apoptosis. Changes in other cytokines are presented in Table 14.13, and

TABLE 14.13 Cytokine changes in recurrent pregnancy loss

Pro-inflammatory/activation increased	Anti-inflammatory/activation decreased
IL-2, IL-7, IL-8, IL-12, IL-18, IFN-γ	IL-4, IL-10, IL-22, IL27, TGF-β, macrophage inhibitory factor

this aberrant cytokine network is likely to lead to an environment resulting in loss of tolerance. Furthermore, miscarriage is associated with an increased proportion of macrophages in the phagocyte (as opposed to regulatory) phenotype. Certain HLA-DR types are linked to recurrent pregnancy loss.

SELF-CHECK 14.12

What is the evidence that pregnancy brings a state of generalized immunosuppression?

Delivery

Traditional initiators of term delivery include stimulation of stretch receptors in the myometrium of the uterus, prostacyclins and endocrine signalling (cortisol, progesterone, the progesterone receptor, oxytocin, and possibly androgens), and signals from the foetus (such as lung surfactant and exosomes from the syncytiotrophoblast). There is now growing evidence of roles for immunological factors at term and pre-term delivery. Examples include pro-inflammatory cytokines in amniotic fluid and the infiltration of the myometrium and cervix by leukocytes. Pre-term birth (<37 weeks) is linked to polymorphism in inflammatory and coagulation genes, including those coding for TNF-α, TNF-β, IL-4, IL-6, IL-10RA, toll-like receptors, tissue plasminogen activator, and coagulation factors V and VII, the latter suggesting a thrombotic aetiology. In a normal pregnancy, pro-inflammatory cytokines and chemokines activate the NF-κB and AP-1 signal transduction pathways (amongst others) that increase the expression of genes promoting contractility, a process accelerated by increased levels of the cytokines.

Clinical and pathophysiological features of this section are summarized in Figure 14.14.

14.4.4 Other maternal disorders

Ectopic pregnancy

The frequency of an ectopic pregnancy (almost always within a fallopian tube) may be as high as 2%, and it is a major cause of morbidity and mortality, linked to 6% of maternal deaths. The foetus is inevitably compromised. The leading symptoms are abdominal pain (often sharp), nausea, and vomiting, sometimes arising before menstrual cessation, although most are diagnosed (by ultrasound and serum hCG) in the following 6 weeks. Risk factors include smoking, pelvic inflammatory disease, previous ectopic pregnancy, and previous tubal surgery. The danger lies with rupture, which requires urgent surgery. Treatment is one of three arms:

- Medical treatment of a mass ≤3.5 cm consists of 50 mg/m² of intramuscular methotrexate, the effect being monitored with falling hCG.

- Surgical treatment is considered should there be a high hCG and a large mass.

- Watchful waiting (or expectant management) of a small mass and low levels of hCG involves no active intervention. The hCG should fall to zero and the ectopic should resorb, but if neither happens, then surgery will be likely.

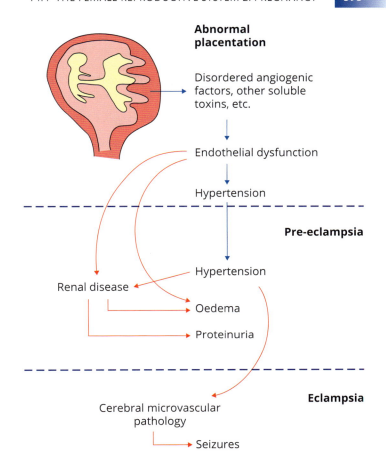

Abnormal placentation

FIGURE 14.14
Hypertension, pre-eclampsia, and eclampsia. A schematic of the link between abnormal placentation and the worsening clinical features of hypertension, pre-eclampsia, and eclampsia.

Thrombosis

Although venous thromboembolism (VTE) complicates <1% of pregnancies, this is around five times the risk of age-matched non-pregnant women, which rises to a 60-fold increased risk post-partum. Women with a history of VTE have a 3- to 4-fold increased risk of a further event in subsequent pregnancies, as most VTEs are due to mutations in genes coding for coagulation factors. Around 42% of pregnant women with a VTE are FVL positive and 17% are G20210A positive; more precise risks are shown in Table 14.14. The increased risk of thrombosis with lupus anticoagulant is 2- to 10-fold, whilst loss-of-function mutations in genes coding for natural anticoagulants are 4.7 for antithrombin deficiency, 4.8 for protein C deficiency, and 3.2 for protein S deficiency. For those

TABLE 14.14 Risk of VTE in pregnancy

	Homozygous	Heterozygous
FVL	34.4 (9.86–120.05)	8.32 (5.44–12.70)
Prothrombin G20210A	26.4 (1.24–559.3)	6.8 (2.46–18.77)

Data odds ratio (95% CI).

Data from Robertson L et al. Thrombosis: Risk and Economic Assessment of Thrombophilia Screening (TREATS) Study. Thrombophilia in pregnancy: a systematic review. Br J Haematol. 2006: 132; 171–96.

TABLE 14.15 Coagulation gene mutations and pregnancy complications

Complication	FVL	Prothrombin G20210A
Early pregnancy loss	Homozygous 2.71 (1.32–5.58) Heterozygous 1.68 (1.09–2.58)	Heterozygous 2.49 (1.24–5.00)
Late pregnancy loss	Heterozygous 2.06 (1.10–3.86)	Heterozygous 2.66 (1.28–5.53)
Pre-eclampsia	Homozygous 1.87 (0.44–7.88) Heterozygous 2.19 (1.46–3.27)	Heterozygous 2.54 (1.52–4.23)
Placental abruption	Homozygous 8.43 (0.41–171.2) Heterozygous 4.70 (1.13–19.59)	Heterozygous 7.71 (3.01–19.76)
Intra-uterine growth restriction	Homozygous 4.64 (0.19–115.68) Heterozygous 2.68 (0.59–12.13)	Heterozygous 2.92 (0.62–13.70)

Data are odds ratio (95% CI).

Data from Simcox LE, Ormesher L, Tower C, et al. Thrombophilia and Pregnancy Complications. Int. J. Mol. Sci. 2015: 16; 28418–28. https://doi.org/10.3390/ijms161226104.

women at high risk, anti-coagulation may be necessary, most often with a low-molecular-weight heparin.

Oral contraceptive pills based on ethinylestradiol (commonly 30–40 µg daily) bring an absolute risk of VTE between 1.4 and 2.8 events per 1,000 person/years, depending on the co-medication (risk in a 20–44-year-old woman not on the pill is 0.39). HRT with a combined oestrogen/progestogen pill or an oral oestrogen-only pill brings risks of 2.6 and 2.2 respectively (risk in a 45–54-year-old woman not on HRT is 1.0). Coagulation issues also impact into the pathology of the pregnancy. Elevated levels of D-dimers are associated with the severity of pre-eclampsia, which has been likened to a thrombotic microangiopathy (as in HELLP). Results of a meta-analysis of the relationships of FVL and prothrombin G2021A to certain pregnancy complications are shown in Table 14.15.

Diabetes

This major metabolic disease is a further important issue in pregnancy, as it is in other conditions (section 7.4), as perinatal and long-term infant outcomes following exposure to diabetes relate directly to glycaemic control during pregnancy. A leading potential pathophysiology is of insulin resistance, as pregnant women with type 1 diabetes need increasing daily doses of insulin to manage their hyperglycaemia: 35% more at weeks 23–8 and 70% more at weeks 33–6.

NICE NG3 is a key UK document for the management of diabetes in pregnancy. If present from conception, diabetes is linked to many complications such as miscarriage and pre-eclampsia, and to minimize risk of the latter, some recommend 75 mg aspirin od. For the foetus complications include neonatal hypoglycaemia and hyperbilirubinaemia, anencephaly, microcephaly, congenital heart disease, and stillbirth. Babies born to mothers with diabetes are larger than in health, with a high proportion of macrosomia (birth weight >4 kg), often leading to pre-term delivery and birth trauma for both, which may precipitate prolonged labour, caesarean section, and post-partum haemorrhage. All pregnant women take vitamin D supplements, not merely because it reduces fasting plasma glucose.

Women with diabetes planning to become pregnant should establish good glycaemic control (target HbA1c <6.5%/48 mmol/mol) before and during pregnancy, take 5 mg/day folic acid, and move to a body mass index (BMI) <27 kg/m². Women whose HbA1c is >10%/86 mmol/mol are advised not to become pregnant, whilst metformin and/or insulin should be used in the preconception period and after conception, and all other blood glucose-lowering drugs discontinued.

TABLE 14.16 Risk factors for gestational diabetes mellitus

Level of evidence	Increased risk
Convincing*	BMI 30–35 v normal, BMI >35 v normal, hypothyroidism
Highly suggestive*	BMI 27.5–30, snoring, sleep disorders breathing, PCOS, family history of diabetes
Suggestive**	BMI 25.0–27.5, low vitamin D and dietary total iron intake, subclinical hypothyroidism, extreme sleep duration, age, and menarche
Weak***	Serum ferritin, HLA-DQ2, DR13, and DR17, thyroid antibodies

* $p < 0.001$, ** $p < 0.01$, *** $p < 0.05$.

Modified from Giannakou K, Evangelou E, Yiallouros P, et al. (2019) Risk factors for gestational diabetes: An umbrella review of meta-analyses of observational studies. PLoS ONE 14(4): e0215372. https://doi.org/10.1371/journal.pone.0215372.

TABLE 14.17 Potential genetic links with gestational diabetes mellitus

Form		Genetic factor
Single nucleotide polymorphism		TCF7L2, ADIPOQ, MTNR1B, KCNQ1, KCNJ11, SLC30A8, CDKAL1, CAPN10, CDKN2A/2B
miRNA	Increased	miR-8a, miR-17-5p, miR-20a-5p, miR-21-3p, miR-19a/b-3p, let-7, miR-23b, miR-30b, miR-100
	No different	miR-16, miR-17, miR-19a, miR-19b, miR-29a, miR-146b, miR-210, miR-222, miR-518
	Reduced	miR-20a, miR-222, miR-29a, miR-132, miR222, miR-494

Note that some miR species appear in more than one box, potentially as they arise from different tissues and/or are derived from studies of low power.

Gestational diabetes mellitus (GDM)

Risk factors for gestational diabetes mellitus (GDM) include advanced maternal age, previous macrosomic infant >4.5 kg, and previous GDM; others are shown in Table 14.16.

Candidate protein biomarkers for GDM at around week 24 include retinol-binding protein 4, sex hormone-binding globulin, visfatin, fetuin-A, ficolin-3, CRP, IL-6, and leptin. Prospective biomarkers from weeks 24–8 include fibroblast growth factors 21 and 23, TNF-α, PAI-1, fetuin-B and follistatin. Some suggest roles for placental growth factor and for pregnancy-associated plasma protein-a, whilst adiponectin is associated with increased birth weight in GDM and in women with no diabetes (as are glucose and HbA1c, but also in diabetes). Potential genetic markers are shown in Table 14.17.

GDM develops in some 9–25% of women, generally in weeks 24–8 of the second trimester, and leads to the development of diabetes mellitus in 20–60% of mothers within 5–10 years and in her child in their adult life. However, like obesity, its frequency has been increasing over the last few decades.

A study from the USA found no diabetes-related deaths in infants weighing <2.5 kg, 5.8% of deaths in those weighing 2.5–3.99 kg, but in 27.8% of infants weighing 4 kg or more (i.e. macrosomia). Accordingly, all women are screened in antenatal care for diabetes with HbA1c and/or the oral glucose tolerance test and if negative, on serial occasions during pregnancy for GDM (section 7.4)

Women with diabetes are advised to keep their capillary plasma glucose at <5.3 mmol/l throughout the day, and at <7.8 mmol/l one hour after meals and <6.4 mmol/l two hours after meals. NICE recommends numerous interventions at defined points in the pregnancy, such as the following:

- 10 weeks: General education (diet, exercise, etc.), offering retinal and renal assessment, determination of HbA1c, and the means to self-determine blood glucose (those with previous history of GDM will have these blood tests and an OGTT)

- 20 weeks: Ultrasound scan for detecting foetal structural abnormalities

- 32 weeks: Ultrasound monitoring of foetal growth and amniotic fluid volume

- Weeks 37 and after: Consider the elective induction of labour or caesarean section and test for foetal wellbeing

Isophane insulin is the first-choice long-acting insulin; others include insulin detemir or insulin glargine (and analogues aspart and lispro), alongside advice on a healthy diet from a dietician. One scheme reported increasing doses of insulin proportional to the woman's weight during the pregnancy—for example, in weeks 14–26, the number of units/day would be 0.8 × her weight in kg, and in weeks 27–37 the dose would be 0.9 × her weight in kg. Other choices include metformin (850 mg bd or tds) and/or glibenclamide/glyburide (5–15 mg daily). Anti-hypertensive drugs targeting the angiotensin–aldosterone pathway should be stopped and replaced with alternatives (e.g. labetalol, nifedipine, methyldopa, diltiazem, hydralazine). This aspect is important as hypertension is a key aspect of pre-eclampsia, whilst statins should be stopped. Should a woman with diabetes present with hyperglycaemia or be unwell, she should be tested for diabetic ketoacidosis.

Women free of diabetes will be assessed regularly, and GDM will be diagnosed if they have a positive OGTT. Once diagnosed, practitioners will advise with interventions to reduce blood glucose, and pregnancy should not proceed longer than 40 weeks. Treatment of mild GDM with insulin (if required) and formal nutritional counselling and diet therapy result in less weight gain during pregnancy and reduced frequency of birth weight, infant fat mass, macrosomia, shoulder dystocia (obstructed labour), pre-eclampsia, and caesarean section.

Post-partum care of women with diabetes or GDM will be for up to 13 weeks with advice and blood tests. Those with diabetes should return to their previous medications as blood glucose levels dictate, and if breastfeeding, can resume metformin and glibenclamide. Women with GDM should stop their glucose-lowering therapy promptly after parturition but have an annual HbA1c test to check for new-onset diabetes.

The thyroid

It has long been established that overt thyroid disease is linked to adverse obstetric outcomes, and we now know that more subtle and sub-clinical hypothyroidism (present in ~2.5% of pregnant women) is associated with prematurity and foetal loss; the latter is also linked to maternal thyroid autoimmunity. Where iodine is plentiful, the thyroid increases ~10% in size during pregnancy, whilst in iodine-deficient areas this can be 20–40%. Women with anti-thyroid antibodies have a 2-fold increased risk of pregnancy loss (17% v 8.4%). Further details of thyroid pathology are presented in section 15.5.

Treatment of sub-clinical hypothyroidism reduces the risk of miscarriage to that of euthyroid women, arguing strongly for the screening for thyroid disease early in pregnancy. Furthermore, one guideline recommends that pregnant women take a 250 μg supplement of iodine daily and women planning pregnancy take 150 μg daily.

Thyrotoxicosis most commonly manifests as hyperthyroidism, and in turn the most common form of this is Graves' disease, present in 0.4–1.0% of women and in ~0.2% of pregnant women. However, transient gestational thyrotoxicosis is more common (1–3%) and characterized by raised thyroxine and low TSH. Uncontrolled hyperthyroidism is linked to miscarriage, low birth weight, stillbirth, and maternal congestive heart failure. Thyrotoxic women should be rendered euthyroid before becoming pregnant. However, certain anti-thyroid drugs are teratogenic, and so an alternative therapeutic and referral to an expert should be sought. In Graves' disease, the problem lies with suppressing the mother's thyrotoxicosis (and so anti-thyroid antibodies) with drugs that can, like the antibodies, cross the placenta and act on the foetus, and accordingly, the lowest possible drug doses should be used and the effects on thyroid markers monitored regularly. As in undiagnosed Graves' disease, suboptimal treatments may lead to neonatal thyrotoxicosis.

14.5 Sexually transmitted disease

Although over 30 bacteria, viruses, parasites, and fungi can be transmitted sexually, not all pose a serious health issue and many are not fatal, although they may bring considerable morbidity, notably to the liver (with hepatitis, cirrhosis, and hepatocellular cancer) and the urogenital system (with renal disease, chronic inflammation, pregnancy failure, and infertility). The World Health Organization (WHO) estimates that over one million curable sexually transmitted infections (STIs) occur each day. The Global Burden of Disease study reported almost 56 million STI-related deaths in 2017, of which slightly over a million (1.9%) were due to sexually transmitted diseases (STDs). Of these, 89.1% were due to HIV/AIDS, 10.6% to syphilis, and 0.3% to gonococcal infections. In the UK in the same year, there were 162 HIV deaths (0.03% of all deaths). Many of these can be prevented with barrier contraceptives, whilst circumcision protects men and their female sexual partners from herpes simplex virus type 2, human papilloma virus (HPV), and syphilis, and those female sexual partners from *Trichomonas vaginalis* and bacterial vaginosis (Bershteyn et al. 2022). A systematic review and meta-analysis found that male circumcision in sub-Saharan Africa is associated with a significantly reduced risk of HIV infection among men, with an adjusted relative risk of 0.42 (95% CI: 0.34–0.54%) (Weiss et al. 2000). The WHO offers excellent online guidelines on the STDs, each describing data on epidemiology and issues pertaining to public health.

14.5.1 Viral disease

HIV

The WHO places HIV in its own category, apart from the other STDs. In 2018 there were almost 38 million cases worldwide with 770,000 deaths (2%) and 1.7 million new infections. The risk of seroconversion after a heterosexual act is estimated to be twice as high in women as in men. In common with most other sexually transmitted pathogens, the cell biology and virology of HIV is understood. The virus targets T lymphocytes via CD4, CCR5, and CXCR4. Infection of the cell leads to its death and so an immunosuppression that, if untreated, can be lethal.

Treatment is with highly active anti-retroviral therapy, a cocktail of different drugs that can block entry of the virus and suppress replication by targeting its reverse-transcriptase and other enzymes. If the viral load is sufficiently eliminated or suppressed, some degree of immune function can be achieved which in many instances leads to an otherwise normal life with little other pathology. NICE NG60 describes testing for HIV in primary and secondary care and recommends testing those attending services for treating hepatitis B and C infection, lymphoma, and tuberculosis.

Cross reference
HIV is covered in detail in Chapter 12 of the volume on *Clinical Immunology* in this series.

SARS-CoV-2

The leading UK bodies, the Royal College of Obstetrics and Gynaecology and the Royal College of Midwives have published numerous excellent guidelines on SARS-CoV-2 in the context of pregnancy and women's health. Many of these reflect the care of those not pregnant, and all pregnant women are offered vaccination (including the booster) when their age group is called upon and should continue to breastfeed. There is no evidence that the virus affects fertility. Public Health Scotland reported no serious adverse effects of vaccination from over 19,000 pregnant women.

Other viruses

The WHO places hepatitis, herpes, and papilloma viruses in their own individual categories, and whilst many of the infection routes are sexual, a small but significant number are caused by non-sexual routes (as is the case for HIV). Isoforms 16 and 18 of the HPV are oncogenic and likely responsible for at least 75% of cases of cervical cancer; strains 6 and 11 cause 90% of genital warts. Globally, HPV infects 480 million women (Serrano et al. 2018)—some 12% of all women—and each year there are 600,000 new cases and 314,000 deaths from cancer of the cervix, most frequently in East Africa, South America, and West Africa. There is an effective vaccine, which in the USA over 6 years reduced the rate of all four strains by 64% in 14–19-year-old females and by 34% in those aged 20–24. There are two isoforms of herpes simplex virus (HSV). HSV-1, which in 2017 infected 3.7 billion people aged under 50, primarily infects the lips. The primary site of an HSV-2 infection (present in 417 million people) is the genitals, although each isoform can be transmitted to the other anatomical region by oral contact. HSV infection rarely has additional pathology outside these areas and may be treated with oral and/or topical antivirals.

Naturally there are co-infections: an estimated 5% of people with a hepatitis B virus infection also have a hepatitis C infection, whilst 60–90% of those infected with HIV are also infected with HSV-2 (probably consequent of their immunosuppression).

14.5.2 Bacteria

The WHO categorizes chlamydia, gonorrhoea, syphilis, and the parasite trichomonas into a separate group, considering them to be curable. They are estimated to be present in 127 million, 87 million, 6 million, and 156 million people respectively. Frequencies in 15–24-year-olds from the USA are shown in Table 14.18.

Common symptoms of bacterial infections include painful and purulent vaginal and urethral discharge, the latter also upon urination. However, there may be other non-specific symptoms such as a fever and a rash. These bacteria (and others) commonly cause pelvic inflammatory disease (PID), bacterial vaginosis, endometrial disease, and salpingitis.

TABLE 14.18 Frequencies of bacterial STIs in US 15–24-year-olds

Organism	Age	Females	Males
Chlamydia trachomatis	15–19	30,709	8,326
	20–4	37,790	15,586
Neisseria gonorrhoeae	15–19	4,821	2,808
	20–4	5,955	6,168
Treponema pallidum	15–19	33	89
	20–4	67	379

Data are prevalence per million.

Modified from Shannon CL and Klausner JD, The growing epidemic of sexually transmitted infections in adolescents: a neglected population. Current Opinion in Pediatrics: February 2018: 30(1); 137–43, doi: 10.1097/MOP.0000000000000578. © 2018 Wolters Kluwer Health, Inc.

These bacteria may drive out less pathogenic organisms normally part of the vaginal microbiome (such as *Lactobacillus* species), possibly providing a niche for other organisms (such as *Ureaplasma urealyticum* and *Gardnerella vaginalis*). Diagnosis is traditionally made by growth and isolation of the particular organism, but molecular genetic techniques are becoming the method of choice, with sensitivities and specificities close to 100%.

Chlamydia

Caused by *C. trachomatis*, this is the most frequent bacterial STI, infecting 4.2% of women and 2.7% of men, most of whom are asymptomatic. The proportions of PID, ectopic pregnancy, and tubal factor infertility (mostly salpingitis) that are attributable to chlamydia are ~20%, 4.9%, and 29% respectively in women aged 16–44. This translates to 171 cases of PID, 73 of salpingitis, and 2 of ectopic pregnancy per 1,000 infections in this age group. A major extra-genital complication is trachoma—an inflammatory infection of the conjunctiva, and an infection brings an increased risk of preterm labour (relative risk 1.35), low birth weight (1.52), and perinatal mortality (1.84). Infections can be treated with aziththromycin 1 g orally as a single dose or doxycycline 110 mg orally bd for a week. Other options include tetracyclin, erythromycin, and ofloxacin.

Syphilis

Infection with the spirochaete *Treponema pallidum*, causing syphilis, is of special interest as it can cause congenital defects at a rate of around 2/million of the population (globally ~4,730/million live births), with some 30% of pregnancies resulting in stillbirth or neonatal death. Infants born to syphilitic mothers are likely to be pre-term, underweight, and with signs that mimic neonatal sepsis. In Europe, the incidence in men is several times higher than in women (~55/million, ~15/million respectively). However, much of the increased rate in men is due to men who have sex with men (some 3,100/million in the USA, accounting for over half the number of cases) such that the rate in heterosexual men is around 29/million. A leading sign of infection is a single ulcer (chancre) or multiple lesions on the genitals 10–90 days after infection, which often resolve, followed weeks or months later by a fever, headaches, and a maculopapular rash. Diagnosis is made by recognition of the organism by microscopy, immunoassay, or molecular genetics. Treatments include intra-muscular doses of 2.4 million units of benzathine penicillin once or 1.2 million units of procaine penicillin for 10–14 days. Alternatives include oral doxycycline 100 mg bd for 14 days or intramuscular ceftriaxone 1g od for 10–14 days.

Gonorrhoea

The causative agent of gonorrhoea, *Neisseria gonorrhoeae*, may also infect the throat and the eye, causing conjunctivitis. In the UK it has a prevalence of some 810/million. Infection may be treated with dual therapy of ceftriaxone 250 mg intramuscular (IM) as a single dose plus azithromycin 1 g orally as a single dose, or with cefixime 400 mg orally as a single dose plus azithromycin 1 g orally as a single dose. Should there be local strain resistance, single therapy with a single dose of IM ceftriaxone 250 mg, oral cefixime 400 mg, or IM spectinomycin 2 g IM are options.

Tuberculosis

Although it is primarily a lung disease, over a quarter of those infected with tuberculosis have extra-pulmonary disease. Genital tuberculosis accounts for 7% and is an important factor in menstrual irregularity, dyspareunia, PID, and (in 3–16% of cases) infertility. Potential pathogenic mycobacteria include *Mycoplasma hominis* and *Mycobacterium genitalium*, causes of PID and bacterial vaginosis. *Mycobacterium tuberculosis* can affect the fallopian tubes (95–100% of cases), endometrium (50–60%), ovaries (20–30%), cervix (5–15%), myometrium (2.5%), and vagina/vulva (1%), and so is a differential diagnosis for many other diseases.

14.5.3 Parasites and fungi

The leading parasitic infection is with the flagellate protozoan *Trichomonas vaginalis*, with infections exceeding those of each of the major bacterial agents. Infection is characterized by the usual STD symptoms—vaginal or urethral discharge, general discomfort, and inflammation. Epidemiological data is flawed by the high (80%) asymptomatic rate, such that national frequencies vary from 1.5% to 26.2%, although most report <10%. Diagnosis has historically been by microscopy of a wet preparation (a sample from a swab of the vagina/vulva), but this has a high rate of false negatives. Superior methods include an antigen detection dipstick that can be used at point-of-care, part of a DNA hybridization probe that also tests for *Gardnerella vaginalis* and *C. albicans*, and a direct nucleic acid amplification test that has excellent sensitivity and specificity (both 95–100%). The first choice of treatment is with a drug of the nitromidazole family such as metronidazole. The leading fungal STD is infection with the yeast *C. albicans*, diagnosed by history, examination, and microbiology—molecular genetics are generally not required. It is part of the normal vaginal microbiome of many healthy women, and overgrowth may be secondary to, for example, use of antibiotics. Antifungal treatments include oral or topical clotrimazole, nystatin, and fluconazole.

Chapter summary

- Most non-malignant urinary system disease is of an infection and/or inflammation. Principal initial investigations are serum urea and electrolytes, and ultrasound; treatments are antibiotics, immunosuppression, renal replacement therapy, and transplantation.

- Leading diseases of the male reproductive system, after prostate cancer, are benign prostatic hyperplasia, prostatitis, and those leading to sub- or infertility.

- Major issues in the female reproductive system include those of the breast, polycystic ovary syndrome, endocrinology, and endometriosis, many of which also lead to sub- or infertility.

- Pregnancy may fail (a miscarriage) due to chromosomal/genetic issues with the foetus and a large number of risk factors in the woman. She herself may suffer a venous thrombosis, hypertension, pre-eclampsia, and eclampsia, and develop diabetes.

- Sexually transmitted diseases may be caused by viruses, bacteria, and parasites. All are treatable, and all are linked to sub- or infertility.

Further reading

- Moon A, Veeratterapillay, Garthwaite M, et al. Urinary tract infection management—do the guidelines agree? J Clin Urol 2018: 11; 81–7.

- Davis NF, Gnanappiragasam S, Thornhill JA. Interstitial cystitis/painful bladder syndrome: the influence of modern diagnostic criteria on epidemiology and on Internet search activity by the public. Transl Androl Urol 2015: Oct 4(5); 506–11.

- Malde S, Palmisani S, Al-Kaisy A, et al. Guideline of guidelines: bladder pain syndrome. Br J Urol Int 2018: 122; 729–43.

- Warner RM, Greenwell TJ. A comparison of the NICE and European Association of Urology guidelines for the assessment and management of incontinence in women. J Clin Urol 2018: 11; 88–100.

- Cox A. Management of interstitial cystitis/bladder pain syndrome. Can Urol Assoc J 2018: 12(6Suppl3); S157–60.

- D'Costa M, Pais VM, Rule AD. Leave no stone unturned: Defining recurrence in kidney stones formers. Curr Opin Nephrol Hypertnes 2019: 28; 148–53.

- Fortie G, de Geus HRH, Betjes MGH. The aftermath of acute kidney injury: a narrative review of long-term mortality and renal function. Critical Care 2019: 23(24); 1–11.

- Cañadas-Gare M, Anderson K, Cappa R, et al. Genetic susceptibility to chronic kidney disease. Fronts Genetics 2019: 10; 453: 1–16.

- Webster AC, Nagler EV, Morton RL, et al. Chronic kidney disease. Lancet 2017: 389; 1238–52.

- DeWitt-Foy, Nickel JC, Shoskes DA. Management of chronic prostatitis/chronic pelvis pain syndrome. Eur Urol Focus 2019: 5; 2–4.

- Srinivasan A, Wang R. An update on minimally invasive surgery for benign prostatic hyperplasia. World J Mens Health 2020: 28(4); 402–11, https://doi.org/10.5534/wjmh.190076.

- Krzastek SC, Smith RP, Kovac JR. Future diagnostics in male infertility: genetics, epidgenetics, metabolomics and proteomics. Translat Androl Urol 2020: 9(Suppl 2); S195–S205.

- Rosenfield RL, Ehrmann DA. The pathogenesis of polycystic ovary syndrome. Endocrine Revs 2016: 37; 467–520.

- Munro MG, Critchley HOD, Fraser IS, et al. The two FIGO systems for normal and abnormal uterine bleeding symptoms and classification of causes of abnormal uterine bleeding in the reproductive years. Int J Gynecol Obstet 2018: 143; 393–408.

- Thurston L, Abbara A, Dhillo WS. Investigation and management of subfertility. J Clin Path 2019: 72; 579–87.

- Alexander EK, Pearce EN, Brent GA, et al. 2017 Guidelines of the American Association for the Diagnosis and Management of Thyroid Disease During Pregnancy and the postpartum. Thyroid 2017: 3; 315–89.

- Luxi N, Giovanazzi A, Capuano A, et al. COVID-19 Vaccination in Pregnancy, Paediatrics, Immunocompromised Patients, and Persons with History of Allergy or Prior SARS-CoV-2 Infection. Drug Saf. 2021: 44; 1247–69.

- Serrano B, Brotons M, Bosch FX, et al. Epidemiology of HPV-related disease. Best Pract Res Clin Obststs Gynaecol 2018: 47;14–26.

Useful websites

- Guidelines by the European Association of Urology: https://uroweb.org/guidelines

- National Institute for Health and Care Excellence: https://www.nice.org.uk/

- British Association of Urological Surgeons: https://www.baus.org.uk/

- International prostate symptom score (IPPS): https://www.baus.org.uk/_userfiles/pages/files/Patients/leaflets/IPSS.pdf

- WHO International Classification of Diseases (ICD-11): https://icd.who.int/browse11/l-m/en

■   WHO recommendations for the prevention and treatment of pre-eclampsia and eclampsia: https://apps.who.int/iris/bitstream/handle/10665/44703/9789241548335_eng.pdf?sequence=1

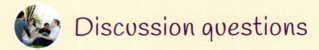 Discussion questions

14.1 How and why does treatment of bladder symptoms differ between men and women?

14.2 Defend the statement that failure to complete a pregnancy is often an immunological issue.

15

Endocrine disease

Almost all messages are transmitted from cell to cell in one of two ways—by connections via a third type of cell (the nervous system), and by molecules in the blood. Chapters 9 and 10 explain the importance of small chemical messengers (almost all being proteins, such as the inflammatory mediators cytokines) in the immune system. A second set of messengers (of both protein and non-protein composition: hormones) and the cells and organs that produce them make up the endocrine system. Inflammatory mediators such as interleukins (that also carry messages) are not part of this system as their role is in defence, whereas hormones are involved in homeostasis.

Compared to other major disease groups, the epidemiology of endocrine disease is poorly developed, partly because (with the exception of diabetes) it generally fails to impact directly into death. Neither the Global Burden of Disease Study nor the World Health Organization have published simple comparable data on these diseases, whilst the UK's Office for National Statistics (ONS) reported 7,662 deaths due to endocrine disease in England and Wales in 2021, this being 1.3% of all deaths. Of these, 95.8% were linked to diabetes, the second most frequent being (non-malignant) thyroid disease, with 190 deaths. But as always, these data fail to address the degree of morbidity that endocrine disease brings.

The endocrine system breaks down conveniently into its component cells and organs, and following an introduction (section 15.1), each will be discussed in turn. However, by far the greatest clinical and public health weight of endocrine pathology lies with diabetes, discussed in section 7.4. Many hormones are regulated by the hypothalamus and pituitary, as explained in section 15.2. Sections 15.3 to 15.8 will focus on the other endocrine organs and hormones. The role of the laboratory and reference ranges is described in section 15.9.

Learning objectives

After studying this chapter, you should confidently be able to:

- identify the organs and hormones of the endocrine system

- appreciate the importance of the pituitary and hypothalamus

- understand the roles of endocrinology in disease of the adrenals and gonads

- discuss disease of the thyroid and parathyroids

- recognize the consequences of high and low levels of growth hormone

- explain the biology and pathology of prolactin, melanocyte-stimulating hormone, and the pineal gland

- discuss the role of the laboratory in endocrine disease

15.1 Introduction to endocrinology

The endocrine system is composed of cells in defined organs and their low-molecular-weight products—hormones. The latter regulate the functions of other organs, many of which are involved in homeostasis and reproduction.

15.1.1 Organs and hormones

The location of the major organs of the endocrine system is summarized in Figure 15.1. These organs include the pituitary and hypothalamus (which are physically continuous, and so count as one unit), the thyroid, parathyroids, adrenals, kidneys, ovaries, testes, and pineal gland. Although not an organ, adipocytes also have endocrine function. These cells and organs secrete small, low-molecular-weight mediators that may be classified as protein and non-protein, as summarized in Table 15.1. Some non-protein hormones can be further classified as steroidal, others as amino-acid derivatives. This is relevant because steroid hormones, having lipophilic qualities, can pass more easily through a cell membrane, and accordingly their receptors are intra-cellular.

Upon binding their ligand, receptors pass into the nucleus where they interact directly with DNA, and accordingly are described as nuclear receptors. Receptors for other protein hormones are expressed at the cell membrane. The binding of a protein hormone ligand to its receptor sets off a series of messengers and pathways that result in changes within the cell, broadly as described in section 3.2, where Figure 3.1 is relevant. The effect of the hormone therefore depends not only on the amount being delivered by the blood, but also on the number and quality of receptors, defects in which will result in low hormone activity.

Some hormones, such as thyroid-stimulating hormone, act on only one tissue, whereas others, such as the thyroid hormones, are likely to act on every type of cell in the body. Others, principally growth hormone, are most active early in the life cycle and then settle down to a lower steady state activity in the adult, whilst others, such as the sex hormones, become more active in puberty under the direction of a biological clock, and in the female fluctuate regularly as part of the menstrual cycle.

Further classification of hormones involves those that act on adjacent or nearby cells of a different form (paracrine: such as stimulation of hydrochloric acid by a gastric parietal cell by histamine from a nearby mast cell), those that act on themselves (autocrine: such as insulin-like growth factors), and those that act distantly, being transported to their targets by the circulation.

15.1.2 Regulation of hormone release

In a stable physiological situation, levels of hormones are also stable, but levels can increase/decrease, and so are regulated by external stimuli/factors. Examples of this include stimulation of the production of parathyroid hormone by low levels of plasma calcium and the release of intestinal hormones

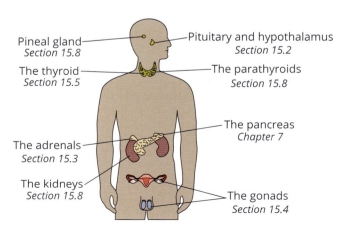

Pineal gland
Section 15.8

The thyroid
Section 15.5

The adrenals
Section 15.3

The kidneys
Section 15.8

Pituitary and hypothalamus
Section 15.2

The parathyroids
Section 15.8

The pancreas
Chapter 7

The gonads
Section 15.4

FIGURE 15.1

Organs of the endocrine system.

TABLE 15.1 Hormones

Protein hormones	
Origin	**Name**
Hypothalamus	All releasing and inhibiting hormones
Pituitary	Oxytocin, antidiuretic hormone, growth hormone, thyroid-stimulating hormone, adrenocorticotrophic hormone, follicle-stimulating hormone, luteinizing hormone, prolactin, melanocyte-stimulating hormone
Parathyroids	Parathyroid hormone
Pancreas	Insulin, glucagon, somatostatin
Kidneys	Erythropoietin, renin
Adipocytes	Leptin, adiponectin
Stomach and small intestines	Gastrin, secretin, somatostatin, cholecystokinin
The placenta	Human chorionic gonadotropin, placental lactogen, oestrogen, progesterone
Non-protein hormones	
Origin	**Name**
Adrenals	*Aldosterone, cortisol, androgens, oestrogens
Kidneys	*Calcitriol (vitamin D)
Testes	*Testosterone
Ovaries	*Oestrogen, progesterone
Thyroid	#Tri-iodothyronine, #tetra-iodothyronine
Adrenal medulla	#Adrenalin/epinephrine, noradrenalin/norepinephrine
Pineal gland	#Melatonin

* Steroids: therefore lipid soluble.

Amino acid derivatives.

by eating and the presence of food in the stomach. Responses such as these are described as reflex arcs, and are also present in the nervous system. However, the most common regulation is by negative feedback, where high levels of the downstream hormone inhibit its release by the parent organ and/or the hypothalamus/pituitary (Figure 15.2). There are many examples of this process in endocrinology, but it is also common in biochemistry in general, such as regulation of enzymes by substrates and products.

Key points

The principal components of endocrinology involve the production of a hormone that has effects on cells and organs, and whose levels are commonly regulated by negative feedback.

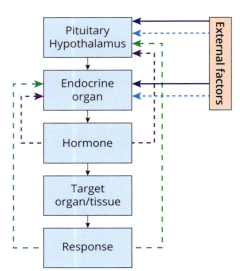

FIGURE 15.2

Regulation of pituitary hormone levels. Simple schematic in which levels of the hormone and/or its product feedback to those parent organs. Dashed lines—negative regulation. Solid lines—positive regulation.

15.2 The hypothalamus and pituitary

Although they have different functions, these two conjoined organs are generally treated as a single unit. Between them, they secrete hormones that regulate a large number of other organs and so, in turn, their own hormones. Accordingly, they have been described as the master controlling glands of the endocrine system. The pituitary is a bean-sized organ (12 mm × 8 mm) of two lobes (anterior and posterior) weighing some 500 mg, attached to the hypothalamus by a stalk (the infundibulum); it lies in a depression (fossa) in the sphenoid bone and is supplied with blood via the inferior hypophyseal artery.

15.2.1 The hypothalamus

Anatomy and physiology

As part of the limbic system, the almond-sized hypothalamus is connected to, and positioned inferior to, the thalamus by neural-rich tissue. It responds to neural signals from the thalamic area of the brain and levels of molecules in arterial capillaries from the superior hypophyseal artery by synthesizing and releasing hormones, and direct neural stimulation, that act on the pituitary.

Messages from the hypothalamus reach the pituitary in one of two ways. Some soluble releasing hormones are secreted into the capillary system that passes to the anterior lobe of the pituitary (also described as the adenohypophysis). These releasing hormones (RH) act specifically on pituitary cells to stimulate the release of their specific hormone (Table 15.2).

Other messages pass along modified nerves through the infundibulum to the posterior lobe (or neurohypophysis). Once stimulated by specific messages from the hypothalamus, the two lobes secrete different hormones that act on other organs. However, other factors can bypass the hypothalamus and act directly on the pituitary: low glucose, physical trauma, and high levels of IL-1 can also stimulate ACTH release.

The hypothalamus also produces inhibitors acting on the anterior pituitary. Somatostatin (a 1.6 kDa protein coded for by *SST* at 3q27.3) is produced by neuronal cells and passes to the anterior pituitary, where it inhibits the secretion of growth hormone. Somatostatin also has a role in glucose metabolism (section 7.4.1) and in the intestines (section 12.3.1, Table 12.5). Dopamine (an organic 153 Dalton derivative of phenethylamine and so product of a series of anabolic reactions) inhibits the release of prolactin, but as a neurotransmitter has important functions in other pathways, and its loss is linked to Parkinson's disease.

The hypothalamus has roles in non-endocrine physiology, namely in thermoregulation (sweating, shivering), blood pressure and heart rate control, satiety and feeding, and in higher cerebral functions of learning and memory.

TABLE 15.2 Releasing hormones and their targets

Releasing hormone	Target(s)
Corticotrophin-releasing hormone	Adrenocorticotrophic hormone (ACTH) Melanocyte-stimulating hormone (MSH)
Gonadotropin-releasing hormone	Follicle-stimulating hormone (FSH) Luteinizing hormone (LH)
Growth hormone-releasing hormone	Growth hormone (GH)
Prolactin-releasing hormone	Prolactin
Thyrotropin-releasing hormone	Thyroid-stimulating hormone (TSH)

SELF-CHECK 15.1

How many hormones can you name?

Pathology

The hypothalamus may be subject to tumours (primary or metastatic), autoantibodies, and toxins, and disease can follow head trauma and surgery. Loss of function of any or all of the microanatomical regions of the hypothalamus results in failure to produce releasing hormones and so in a lack of downstream stimulation of the pituitary and deficiency disease (discussed in the sections that follow). Both single and multiple deficiencies are known. A craniopharyngioma may involve the hypothalamus or pituitary.

15.2.2 The anterior pituitary

The divergent features of the two lobes of the pituitary amply justify separate discussions.

Physiology

Selected details of the hormones secreted by the anterior lobe of the pituitary are shown in Table 15.3, and further details of their biology are discussed in the sections that follow. FSH, LH, and TSH are heterodimers, as is the placental product human chorionic gonadotropin (hCG), and all share the same alpha subunit, coded for at 6q14–21.

TABLE 15.3 Hormones of the anterior pituitary

Hormone	Molecular mass (kDa)	Gene: location	Target
ACTH	4.5	*POMC*: 2p23.3	The adrenal glands
FSH	35.5	*CGA*: 6q14–21 and *FSHB*: 11p13	The gonads
GH	22.1	*GH1* and *GH2*: 17q22–24	Numerous cells and tissues
LH	30.0	*CGA*: 6q14–21 and *LHB*: 19q13.3	The gonads
MSH	2.0–3.0	*POMC*: 2p23.3	Melanocytes (brain?)
Prolactin	22	*PRL*: 6p22.3	Mammary glands
TSH	30	*CGA*: 6q14–21 and *TSHB*: 1p13	The thyroid

The biological activity of growth hormone (GH) is mediated by two almost identical isoforms coded for by a duplicated gene, whilst melanocyte-stimulating hormone (MSH) has three isoforms (α, β, and γ). These, with ACTH, are all derived from various post-translational cleavages of pro-opiomelanocortin, their precursor protein, which also produces β-lipotropin, γ-lipotropin, and β-endorphin.

The primary function of ACTH is to induce the synthesis and release of cortisol from the adrenal glands (section 15.3). FSH and LH both act on the gonads to induce their secretion of the sex hormones testosterone, oestrogen, and progesterone and so promote gametogenesis (section 15.4). TSH acts on the thyroid to direct the release of thyroid hormones tri-iodothyronine and tetra-iodothyronine (thyroxine) (section 15.5), whilst GH acts on many organs, tissues, and cells to direct the release of insulin-like growth factor (IGF), which itself acts on many other cells (section 15.6). Prolactin acts on the mammary glands to promote milk production, whilst MSH acts on the skin and hair to induce melanin production, and may have a role in sexual arousal and in the suppression of appetite (section 15.7). Relationships between the molecules of the anterior pituitary and their physiology are summarized in Figure 15.3.

Pathology 1: Overactivity synopsis

As with the hypothalamus, the leading known cause of pathology of both pituitary lobes is primary or metastatic cancer (90% of cases); other causes include head and brain trauma/surgery and (very rarely) infections and inflammatory disease (hypophysitis is estimated to be present in one in nine million).

Tumours resulting in overactivity (adenomas) and underactivity are both known to lead to exaggerated or absent hormone activity: 25% are non-secretors. Pituitary adenomas are surprisingly common:

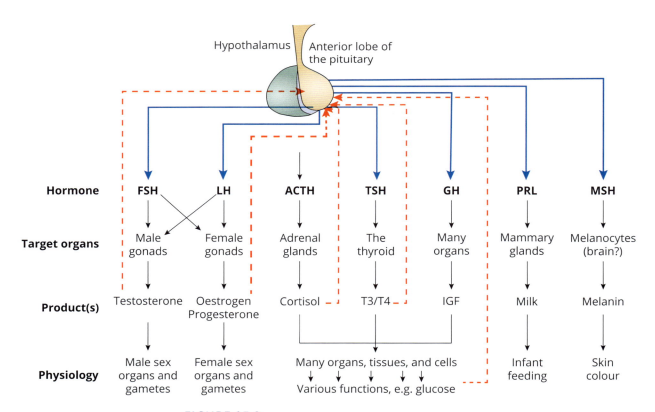

FIGURE 15.3

Actions of hormones of the anterior pituitary. Schematic of the links between anterior pituitary hormones, their target organs, products, and general physiology. Dashed lines represent regulatory feedback from products to the hypothalamus/pituitary. PRL = prolactin.

high-resolution MRI and post-mortem studies report a frequency of ~17% of pituitary tumours and accordingly may be 'incidenalomas'.

Large tumours (<1% of all pituitary tumours) can have other effects, such as visual field defects, as the pituitary is close to the optic chiasma, so that clinical investigation involves ophthalmology and visual perimetry. Tumours may also impact into hypothalamic function, and certain cranial nerves may also be affected. MRI is superior to CT as it can detect changes in pituitary mass, and, of course, the input of the laboratory is crucial.

The most common tumours are those leading to excess prolactin (hence prolactinoma, often leading to inappropriate milk secretion (galactorrhoea)), excess ACTH (leading to Cushing's disease—section 15.3.4), and excess growth hormone (leading to acromegaly—section 15.6). Tumours secreting FSH, LH, TSH, and MSH are very rare, and some secrete more than one hormone. Treatments for an overactive pituitary include:

- Surgery: through the bone at the roof of the mouth, or through the nasal cavity, although these may bring complications and hypopituitarism

- Medicines: Bromocriptine, quinagolide, cabergoline (dopamine agonists), octreotide, lanreotide (somatostatin analogues), and pegvisomant (GH receptor antagonist)

- External beam or proton beam radiotherapy, alone or as a post-surgical adjunct

Pathology 2: Underactivity synopsis

Single-hormone hypopituitarism may, as discussed, arise from hypothalamic disease, but also from disease of the pituitary, and has an incidence of 40/million/year. The most common are of FSH, LH and GH; multi-hormone deficiencies are infrequent whilst panhypopituitarism is very rare. Clinical disease reflects the nature of the deficiency but there are non-specific features such as tiredness, lethargy, and weight gain.

Mutations in several genes coding for transcription factors are responsible for multiple deficiency disease: PROP1 (at 5q35.3) is linked to deficiencies in LH, FSH, GH, prolactin, and TSH; POU1F1 (at 3p11.2) to deficiencies in GH, prolactin, and TSH; LHX3 (at 9q34.3) to deficiencies in FSH, LH, GH, prolactin, and TSH, and later in ACTH; whilst LHX4 (at 1q25.2) is linked to early deficiencies in GH and TSH, and later ones in ACTH. Incorrect activity of these genes, and so their respective transcription factor products, is likely to have roles in failure to produce adequate levels of the hormones, hence the clinical syndromes. Treatment is with replacement therapy with the particular hormone, generally of a recombinant form.

15.2.3 The posterior pituitary

The posterior lobe of the pituitary consists mainly of the axons and axon terminals of over 10,000 secretory neurons whose cell bodies are in the hypothalamus. The cell bodies synthezise two hormones that move down the axons to the axon terminals in the pituitary, where they are stored in vesicles and released into the circulation.

Oxytocin–physiology

This 1 kDa nonapeptide is coded for by OXT at 20p13. It is released by levels of oestrogen, activation of stretch receptors of the cervix, and contraction of the uterus before and after parturition to expel both the neonate and its placenta. Oxytocin is also released by stimulation of the nipple after childbirth, and accordingly it is closely linked with the actions of prolactin. There is also evidence that it has a role in social and sexual conditioning, but if so, this has yet to make a major impact into clinical practice. Its receptor, coded for by OXTR at 3p25, is found on certain parts of the central nervous system, mammary gland epithelial, and on the uterine myometrium and endometrium. Oxytocin acts on the latter tissues to prepare for milk production and to induce contractions of the uterus.

Oxytocin–clinical aspects

There is no common pathology related to this hormone. However, it does have an important role in obstetrics, where it can be used to induce prenatal contractions to determine the heart rate of the foetus and so its viability. This challenge can be determined by an intravenous bolus of oxytocin, or by manual stimulation of the nipples. It is also commonly used post-partum to accelerate the expulsion of the placenta.

Antidiuretic hormone (ADH)–physiology

Also known as vasopressin or arginine vasopressin (AVP), ADH is, like oxytocin, a 1 kDa nonapeptide coded for by *AVP* at 20p13. Indeed, the two hormones bear striking homology and are almost certainly an example of gene duplication (Figure 15.4)—so much so that oxytocin has a very mild anti-diuretic effect, also reducing the production of urine.

ADH acts on its target tissues via its receptor, of which there are three isoforms:

- V_{1A} (or V1) coded for by *AVPR1A* at 12q15–15, found mostly on vascular smooth muscle cells, platelets, uterine myometrium, and hepatocytes

- V_{1B} (or V3) coded for by *AVPR1B* at 1q32, present on cells of the anterior pituitary

- V2 coded for by *AVPR2* at Xq28, localized to renal distal convoluted tubules and collecting ducts, endothelial cells, and vascular smooth muscle cells

Binding of ADH to receptors on vascular smooth muscle cells induces their constriction, and so the part-regulation of blood pressure; its binding to receptors on platelets and endothelial cells (releasing von Willebrand factor) promotes thrombosis. However, its major role in physiology is in decreasing urine production. This is achieved by directing nephrons to increase their absorption of water by activating an isoform of the water pump aquaporin. This has an effect on blood volume (and so hypertension) and therefore osmolarity. High blood osmolarity is detected by hypothalamic osmoreceptors, resulting in synthesis of ADH, its passage down the axons, and release at the pituitary, whilst low osmotic pressure inhibits that process.

Other factors stimulating ADH production include physiological stress, pain, acetylcholine, nicotine, and certain drugs. Alcohol interferes with the generation of ADH, so that diuresis continues without regulation, leading to excess urination. ADH also acts on sweat glands, which has the potential to reduce water loss.

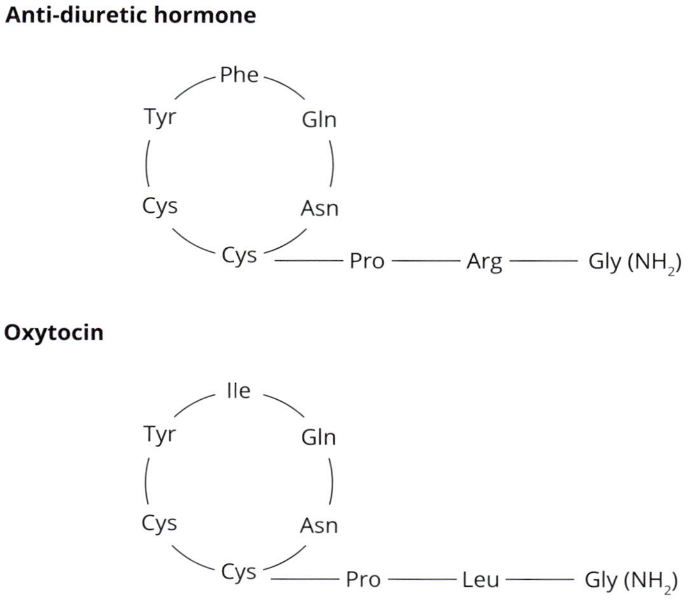

Anti-diuretic hormone

Oxytocin

FIGURE 15.4

Amino acid sequence of oxytocin and ADH.

Antidiuretic hormone (ADH)–pathology

The leading diseases of the posterior pituitary are diabetes insipidus and the syndrome of inappropriate diuresis (SIAD). The former is caused by either failure of the hypothalamus and pituitary to generate and release ADH (neurogenic diabetes insipidus), or failure of the kidney to respond to ADH (nephrogenic diabetes insipidus). The consequence of either is unregulated diuresis and so excessive urination (polyuria), with the physiological responses of polydipsia (as in diabetes mellitus). It may lead to potentially fatal dehydration if the water is not made up by drinking. A third form is excessive drinking of water, which may result from pathology of the hypothalamic thirst centre or a psychiatric disorder.

A hereditary form of the nephrogenic variant of diabetes insipidus is caused by loss of function of *AVPR2*, and so failure of renal cells to respond. An alternative mutation is in *ACQ2* at 12q13.12, coding for aquaporin 2, also causing the polyuria, which is associated with acute kidney injury and CKD. With a very short half-life, measurement of ADH is a problem, but its surrogate, with a long half-life, is copeptin, a 39 amino acid cleavage product of pre-pro-ADH, measurement of which can help diagnosis. Replacement therapy for the neurogenic form is desmopressin, a complex ADH mimetic also known as DDAVP, at a dose of 10 µg by nasal insufflation or 4 µg subcutaneously or intramuscularly.

SIAD is characterized by an absolute increase in body water, which can have a number of causes. The most frequent is high levels of ADH, often described as the syndrome of inappropriate ADH secretion (SIADHS). The consequence of increased ADH is reabsorption of water leading to low osmotic pressure and the dilution of the constituents of the blood, such as sodium, which may fall to 122 mmol/l (reference range 135–45). High levels of water in the blood can provoke a fatal cerebral oedema. Diagnosis is made in the laboratory with a low osmolality and hyponatraemia alongside high levels of sodium in a concentrated urine.

Causes of high ADH focus on the hypothalamus and pituitary with a gain-in-function malignant adenoma, certain drugs (including oxytocin), and infection. However, ADH may be produced from an ectopic source, most often a lung/bronchus carcinoma. Treatment is by restricting fluid intake, removal of the ectopic source and the antibiotic demeclocycline (initially 900–1,200 mg in divided doses, then a maintenance dose of 600–900 mg in divided doses). An additional agent is the selective vasopressin V2-receptor antagonist tolvaptan, starting at 15 mg od, rising as needed to 60 mg od.

Figure 15.5 summarizes the physiology of the posterior pituitary.

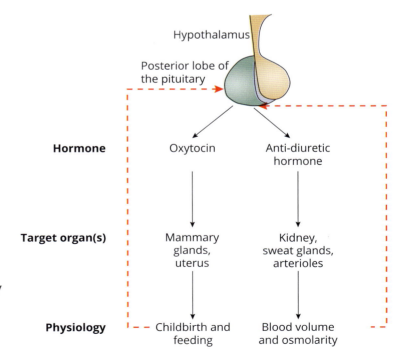

FIGURE 15.5
Actions of hormones of the posterior pituitary. Schematic of the links between posterior pituitary hormones, their target organs, products, and general physiology. Dashed lines represent regulatory feedback from products to the hypothalamus/pituitary.

Key points

The anterior and posterior pituitary release hormones that control the greater part of the endocrine system.

Explain, in simple terms, negative feedback.

15.3 The adrenals

The 4–6 gram paired adrenal glands are each loosely joined by connective tissue to the superior aspect of their kidney and have a flattened pyramidal shape. Each gland is composed of a central section (the medulla, making up some 10–20% of its mass) surrounded by an outer section (the cortex) and bounded by a capsule. With separate and precise functions, each has been described as an organ within an organ.

15.3.1 The adrenal cortex: physiology

This tissue is composed of three zones:

- an outer zona glomerulosa, secreting mineralocorticoids (making up 10% of the cortex);
- a central zona fasciculata, secreting mainly glucocorticoids (75–85%); and
- an inner zona reticularis, secreting masculinizing androgens (5–15%).

The zona glomerulosa

The principal **mineralocorticoid** product of this zone, aldosterone, part-regulates blood pressure and electrolytes by acting on receptors in the convoluted tubules to control sodium, potassium, and pH. When low blood pressure/blood volume is detected by renal juxtaglomerular cells, they secrete the enzyme renin, which converts liver-derived angiotensinogen (a ~55 kDa protein coded for by *AGT* at 1q42.2) in the plasma into the decapeptide angiotensin I. Angiotensin-converting enzyme (ACE), a component of the endothelial cell membrane, cleaves two amino acids from its substrate to produce angiotensin II. Upon arrival at the adrenals, angiotensin II binds its receptor (which is also present on blood vessels, renal tubules, pulmonary tissues, and parts of the central nervous system) to stimulate the complex metabolic processes resulting in the production of aldosterone. This hormone increases the reabsorption of sodium, and water follows by osmosis, resulting in an increase in blood volume. Aldosterone release is also stimulated by increased plasma potassium and inhibited by low levels of potassium. Angiotensin II also stimulates arteriolar smooth muscle cells, resulting in a vasoconstriction that may contribute to increasing the blood pressure, as shown in Figure 15.6. This figure also helps explain the mechanism of action of ACE inhibitors (ACEIs: enalapril, lisinopril, etc.) and angiotensin receptor blockers (ARBs: valsartan, olmesartan, etc.) in the control of hypertension (section 7.2).

The zona fasciculata

Following its release by hypothalamic CRH stimulation of pituitary corticotrophic cells, ACTH, also known as corticotrophin, is conducted by the blood to the adrenal cortex. Having arrived at the adrenals from the anterior pituitary, ACTH binds to its specific receptor (coded for by *ACTHR* at 18p11.21) where it acts to stimulate the production and release of cortisol (or corticosterone) from the adrenocortical zona fasciculata. The ACTHR is also present on some dermal fibroblasts and adipocytes, especially when the latter are differentiating, and is therefore likely to have a role in glucose metabolism.

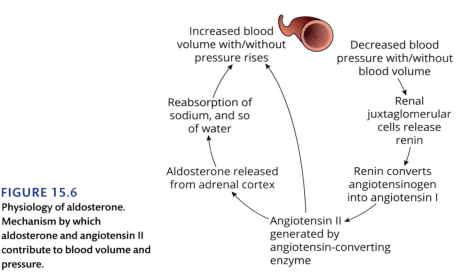

FIGURE 15.6

Physiology of aldosterone. Mechanism by which aldosterone and angiotensin II contribute to blood volume and pressure.

Increasing levels of cortisol feed back to the hypothalamus to suppress the release of CRH and so of ACTH from the pituitary. The secretion of CRH is not continuous: pulses leave the hypothalamus to stimulate ACTH levels, and these also have a diurnal variation, so that cortisol levels peak around 8–9 am and then fall to a trough around midnight.

Cortisol itself acts on various cells and tissues. In glucose metabolism it stimulates gluconeogenesis, glycogenolysis, proteolysis, and lipolysis, thus generally increasing plasma glucose. In the immune system it is anti-inflammatory in suppressing leukocyte function and so the release of cytokines such as IL-1 and IL-6. However, there is evidence that it also influences a host of other processes such as in connective tissue metabolism, renal function, and digestion. Plasma levels have a marked diurnal variation, being high in the morning and falling throughout the day, so that time-specific reference ranges are required.

The zona reticularis

In both sexes, the tissues of the zona reticularis synthesize and release relatively small amounts of androgens, the principal one being dehydroepiandrosterone (DHEA). In the adult male, the testes are the major site of production, whereas in the female, androgens are converted to oestrogens by other tissues. Post-menopausal levels of female sex hormones fall as the ovary becomes less active, but oestrogens continue to be produced by the adrenal cortex.

15.3.2 The adrenal medulla: physiology

The adrenal medulla is a modified ganglion of the sympathetic nervous system innervated by a branch of the splanchnic nerve, whose cells have lost their long axons so that only the cell bodies remain. These cells form clusters around blood vessels, enabling rapid release of their products upon stimulation from autonomic acetylcholine-releasing neurones. These hormone-producing chromaffin cells (so called because of their affinity for chromium salts) produce adrenalin (epinephrine) and noradrenalin (norepinephrine), which mediate the fight-or-flight response with increased heart rate, dilated pupils and bronchioles, higher blood pressure, etc. The metabolic pathway for these hormones begins with tyrosine and passes, via hydroxylation and decarboxylation, to dopamine (secreted in trace amounts), which is oxidized to noradrenaline (20%), and then adrenalin (80%). The latter two molecules may be broken down to metanephrine and normetanephrine and then, by monoamine oxidase, to vanillylmandelic acid before being excreted.

The physiology of the adrenal gland is summarized in Figure 15.7.

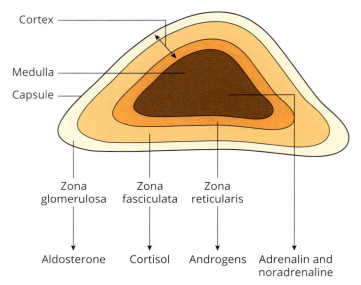

Cortex

Medulla

Capsule

Zona
glomerulosa

Zona
fasciculata

Zona
reticularis

Aldosterone   Cortisol   Androgens   Adrenalin and
noradrenaline

FIGURE 15.7
Key features of
adrenal physiology

Key points

The adrenal is composed of four distinct sections, each secreting specific hormones with
varied roles in several metabolic, homeostatic, and physiological processes.

15.3.3 Steroid biochemistry

It is appropriate at this point to briefly describe the biochemistry of the different steroid hormones, the metabolic pathways for which are both ordered and complex. The root molecule is cholesterol, which is synthesized primarily in the liver and serves as a substrate for enzymes (mostly of the cytochrome P450 family) that generate a number of molecules collectively described as progestogens in the zona glomerulosa. These 21-carbon atom molecules include pregnenolone, progesterone, 17α-hydroxypregnenolone, and 17α-hydroxyprogesterone. In the zona fasciculata, the enzyme 21-hydroxylase can convert the latter two progestogens to de-oxycorticosterone and 11-deoxycortisol respectively, followed by the action of 11-β-hydroxylase (in mitochondria) to yield corticosterone and cortisol respectively. Aldosterone is produced by the action of aldosterone synthase (coded for by *CYP11B2* at 8q24.3) on corticosterone, also in the mitochondrion.

The substrates for the 19-carbon atom androgens are 17α-hydroxypregnenolone and 17α-hydroxyprogesterone, converted to dehydroepiandosterone and androsterone respectively by 17,20 lyase. The latter two androgens are the substrates for 17β-hydroxysteroid dehydrogenase in the zona reticularis, generating androstenediol and testosterone respectively, the testosterone being converted to dihydrotestosterone by 5α-reductase. Implicit in several of these reactions is 3-β-hydroxysteroid dehydrogenase. The 18-carbon atom oestrogens are generated from androstenediol and testosterone, by aromatase (coded for by *CYP19A1* at 15q21.2), to osterone and oestradiol respectively. In the liver and placenta, both of these are converted to oestriol (Figure 15.8). As we shall see in section 15.5, knowledge of these pathways is required for an understanding of certain diseases.

Once in the plasma, a variable proportion of each steroid is generally carried non-specifically by liver-derived plasma proteins such as albumin, but in some cases most is transported by its specific protein. Examples include corticosteroid-binding globulin (CBG, or transcortin: coded for by *SERPINA6* at 14q32.13) for cortisol, and sex hormone-binding globulin (SHBG, coded for by SHBG at 17p13.1) for testosterone, oestriol, and other sex hormones. Differences in levels of these proteins can have regulatory and pathological implications: low SHGB in women is linked to hyperandrogenism and endometrial cancer so that investigation of the ratio of free to protein-bound sex hormone,

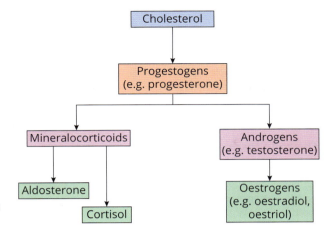

FIGURE 15.8
Simplified steroid metabolic pathways

especially in diagnosis of sub-fertility and PCOS (section 14.3), may be required. Non-steroid hormones thyroxine and insulin-like growth factor are also carried by their own specific binding proteins. In addition to carrying cortisol, CBG has a non-specific and potential role in insulin biology, it being downregulated by insulin and correlating with markers such as BMI and waist-to-hip ratio.

Defend the statement that the adrenal is actually four separate organs.

15.3.4 The adrenal cortex: pathology

Disease of the adrenal cortex is most easily discussed by taking an anatomical perspective of the three zones as shown in figure 15.7.

The zona glomerulosa and aldosterone

Overactivity: As aldosterone is the product of the renin–angiotensin pathway (as described above), high levels may be due to the hyperactivity of the pathway. This may arise from renal disease (that being the site of renin release), or idiopathic adrenal hyperplasia (perhaps the product of inflammation). The ratio of aldosterone to renin may be useful in diagnosis. Conn's syndrome is caused by an aldosterone-secreting adrenal adenoma and can be detected by CT/MRI. Most cases are caused by a mutation in *KCNJ5* (at 11q24.3, coding for a potassium channel), others being mutations in *ATP1A1*, *ATP2B3*, *CACNA1D* (also coding for membrane cation transporters) and *CTNNB1*, coding for β-catenin. miRNAs including hsa-miR-30e-5p, hsa-miR-30d-5p, and hsa-miR-7-5p have been linked to primary aldosteronism. Treatment of a Conn's adenoma is surgery. The consequences of hyperaldosteronism are sodium retention and potassium loss, leading to hypertension, often treated with the aldosterone antagonist spironolactone (section 7.2).

Underactivity: Low levels of aldosterone may be due to congenital adrenal hyperplasia, resulting from an autosomal recessive mutation in one of the genes coding for metabolic enzymes (section 15.3.3). The most common (>90%) form is due to a mutation in *CYP212A* (at 6p21.33, coding for 21-hydroxylase), present in ~750/million live births, and linked with hyperkalaemia. The second most frequent mutation is in *CYP11B1* (at 8q24.3, coding for 11β-hydroxylase), with a frequency of 10/million, and linked to hypokalaemia. Both mutations are characterized by low cortisol and aldosterone with increased androstenedione (see the section on the zona reticularis below). It may also arise from deficiency of aldosterone synthase (coded for by *CYB11B2* at 8q24.3), and if so, levels of its unmetabolized substrate, corticosterone, may give rise to other disease. Indeed, these (and other enzymes) are active in several pathways, and so congenital adrenal hyperplasia is likely to lead to other disease, such as in sexual development and function. 21-hydroxylase deficiency may also result from an autoantibody, present in 50/million, which also results in reduced levels of cortisol in addition to low aldosterone. Chronic renal failure may lead to aldosterone deficiency in its inability to produce renin.

Patients are likely to present with oedema, dehydration, and salt-craving, whilst the laboratory may report hyponatraemia and hyperkalaemia. Dependent on the precise enzyme lesion, replacement therapy may be possible. Fludrocortisone is a synthetic halogenated (with fluorine) pregnane corticosteroid that can be a substrate for other enzymes, so generating corticosterone and then aldosterone. Doses in the region of 50 to 100 µg/day are a common starting point, with up- or down-titration based on clinical and laboratory indices.

The zona fasciculata, ACTH, and cortisol

Overactivity: The leading pathology of the zona fasciculata is increased cortisol and takes two forms: Cushing's disease (present in ~12/million) and Cushing's syndrome. Depending on the cause of the increase, they may be classified as ACTH dependent or independent, and as endogenous or exogenous to the body.

- The leading endogenous cause of raised cortisol (~62% of cases, although some suggest up to 80%) is classical Cushing's disease, itself linked to excess secretion of ACTH from a pituitary adenoma, accounting for around 15% of all pituitary tumours that arise in around 2 per million per year and are prevalent in ~40/million, with women accounting for 75% of cases.

- Cushing's syndrome may have an aetiology independent of the pituitary. Increased ACTH may arise from an ectopic source, such as a bronchial, pancreatic, cervical, or lung carcinoma, or adrenal medullary hyperplasia (~5% of cases). Excess cortisol may be secreted independently of ACTH by an adrenal adenoma or carcinoma, or by adrenal nodular hyperplasia (~33% of cases).

- The primary cause of exogenous Cushing's syndrome is the use of steroids to treat a particular condition, such as those in inhalers for asthma (section 13.5.2).

Less than 5% of ACTH-secreting pituitary adenomas are familial, and include those linked to *MEN1* at 11q13.1, coding for menin, a tumour suppressor, and linked to the multiple endocrine neoplasia (MEN) syndrome (section 5.13.2). The remaining 95% of tumours involve mutations in any one of a large number of genes, many coding for a tumour suppressor or an oncogene. However, mutations in *USP8* at 15q21.2 and *USP48* at 1p36.12, coding for enzymes involved in ubiquitin metabolism, have been reported in ~38% and ~13% of adult corticotrophic tumours respectively. Of the cortisol-secreting adrenal adenomas and other neoplasia, mutations in *PRKACA* at 19p13.12 (coding a protein kinase), *PDE8B* at 5q13.3 (coding a phosphodiesterase), and *TP53* at 17p13.1 (coding a tumour suppressor active in many cancers) have been described. Both decreased (miR-493, miRNA-17-5p, miRNA-126-3p, and miRNA-126-5p) and increased (miRNA-150-5p and miRNA-223-3p) expression of miRNAs have been described in patients with endogenous hypercortisolism.

Cushing's disease/syndrome is characterized clinically by dozens of semi-specific signs, symptoms, and other features (Table 15.4), but these are dominated by obesity, hairiness, and abdominal striae (Figure 15.9).

However, there are numerous differential diagnoses that cause mild overactivity of the adrenals, often described by some as pseudo-Cushing's syndrome, and if suspected must be addressed by biochemical testing for ACTH and cortisol. These include any combination of obesity, alcoholism, depression, diabetes, and polycystic ovary syndrome (PCOS).

TABLE 15.4 Signs, symptoms, and clinical features in Cushing's disease/syndrome

Signs	Weight gain, thin and fragile skin, bruising, hairiness, abdominal striae, moon face, poor wound healing, Buffalo hump (fat pad between shoulder blades), acne, frontal hair loss (in women), pigmentation (in ACTH-dependent disease), plethora (red cheeks), abdominal obesity
Symptoms	Depression, anxiety, muscular weakness, erectile dysfunction, irregular or absent menstruation, fatigue, irritability, decreased libido, depression, insomnia, poor concentration, vertigo
Clinical features	Hypertension, osteoporosis, cardiomyopathy, glucose intolerance, diabetes

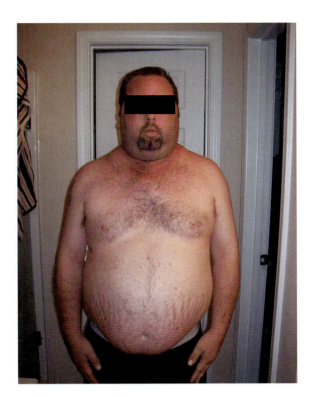

FIGURE 15.9

Cushing's disease/syndrome. Note the 'moon' face with double chin, central obesity, and striae.

© CSRF • Cushing's Support & Research Foundation.

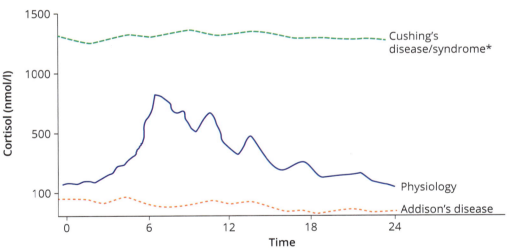

FIGURE 15.10

Physiological diurnal pattern in serum cortisol, and its absence in adrenal pathology. *High levels in endogenous disease: exogenous stimulation with an inhaler or oral steroids may produce a quasi-diurnal pattern.

A preliminary diagnosis made on clinical groups is confirmed in the laboratory with serum cortisol, which has a marked diurnal variation, so timing of the venepuncture is important in physiological investigations (Figure 15.10). However, very high levels of cortisol (and possibly ACTH) at any time of day point to a pituitary, adrenal, or other adenoma, whereas high levels of cortisol with low or even absent ACTH (due to extreme negative feedback) point to an overactive adrenal, probably an adenoma.

A 24-hour urinary cortisol >50–100 µg/1,000 nmol/l suggests Cushing's disease or syndrome. Very low levels at any time of day may reflect Addison's disease. A further test is that of dexamethasone: a 1 mg dose given at midnight will feedback to reduce ACTH levels and so abolish the physiological rise in cortisol the following morning. In Cushing's disease/syndrome, there is no or minimal reduction.

Once the root cause has been identified and located, perhaps with high-resolution imaging by MRI, treatment by surgery, generally via the transphenoidal route (through the nose) is the primary choice and evaluated by post-operative ACTH and cortisol. Remission rates can be as high as 90% for microadenomas, but long-term follow-up is necessary as 10-year recurrence rates can be >20%. Medical treatments include pasireotide (intramuscular 10 mg every 4 weeks), which targets the somatostatin receptor, and cabergoline, a dopamine agonist (2–3 mg od), whilst ketoconazole (400–1,200 mg in divided doses daily) inhibits cytochrome P450 enzymes (such as 11-β-hydroxylase) involved in steroidogenesis. Other drugs inhibiting steroidogenic enzymes include metrapone (500–6,000 mg two to four times a day) and mitotane (1.5–6 g tds); etomidate is useful in those with life-threatening hypercortisolaemia and is infused at 0.02–0.08 mg/kg per hour. Some of these, and radiotherapy, may also be used as adjuncts after surgery. Bilateral adrenalectomy may be required for refractory disease when other treatments have failed. Practitioners caring for those with medication-induced Cushing's syndrome (such as in severe inflammatory vasculitis) seek to reduce their patient's dependence, perhaps with alternative options, although for many this will be difficult.

Underactivity: The principal condition of failure of the production of cortisol is Addison's disease, with a frequency of ~5/million/year and present in ~120/million. The disease is generally chronic, insidious, and progressive, but its development can be abrupt—an Addisonian crisis—characterized by shock, severe dehydration, and hypotension. A crisis may be precipitated by an illness, stress, trauma, excessive adrenal suppression, such as with ketoconazole or the rapid withdrawal of, or insufficient use of, steroid medications. Crises develop at a rate of ~7/100 patient-years, and are fatal in 0.5/100 patient-years.

SELF-CHECK 15.4

What are the major causes of Cushing's disease?

As in excess cortisol, the key to understanding the aetiology of hypo- or acortisolaemia is the hypothalamus–pituitary axis. Rare failure of the hypothalamus and/or hypopituitarism or panhypopituitarism have been described previously, leaving adrenal pathology as the leading cause. As a paired organ, damage to or failure of one adrenal is not always a clinical issue, and some include Addison's disease as part of the primary adrenal insufficiency syndromes (which would therefore also include failure to produce other adrenal products) (Figure 15.12).

The most common cause of Addison's disease is autoimmunity, present in >50% of cases, often targeting 21-hydroxylase, and perhaps part of the autoimmune polyendocrine syndrome. Other causes include microbial infections (tuberculosis, *Streptococcus*, *Staphylococcus*, *Neisseria* species), congenital adrenal hyperplasia, insufficient cholesterol required to generate all steroids (Figure 15.8), and several genetic lesions. These include a defect in *ABCD1* at Xq28, coding for a peroxisomal membrane protein and causing deposits of very-long-chain fatty acids (X-linked adrenoleukodystrophy), mutations in *NR5A1* at 9q33.3, coding for steroidogenic factor 1, defects in which lead to adrenal failure, and *NR0B1*, at Xp21.2, coding for DAX1, a nuclear receptor, resulting in X-linked congenital adrenal **hypoplasia**. Primary adrenal insufficiency and autoimmune adrenal disease are linked to HLA haplotypes DR3-DQ2 and DR4-DQ8.

Patients are likely to present with general non-specific symptoms such as loss of appetite, salt craving, loss of consciousness, fatigue, muscle wasting, weight loss, sweating, diarrhoea, and vomiting. However, the leading sign is hyperpigmentation (Figure 15.11), arising from increased melanocyte-stimulating hormone, itself due to increased pituitary processing of the pro-opiomelanocortin, the ACTH precursor. Diagnosis is again made in the laboratory with measures of cortisol and (possibly) ACTH. However, as levels of cortisol show a marked diurnal variation, random levels are of no value, and therefore timing of the sample is crucial (Figure 15.10). Low levels of cortisol (e.g. <100 nmol/l at 8 am (reference range 200–800 nmol/l)) are strongly suggestive of adrenal insufficiency.

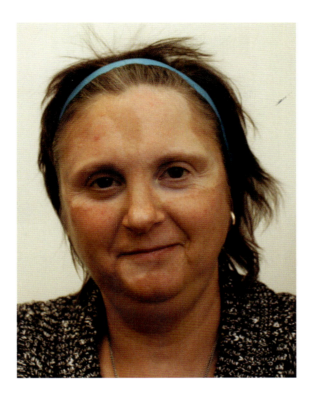

FIGURE 15.11
Hyperpigmentation in Addison's disease. Note the light and darker shades of the forehead.

Burton C, Cottrell E, Edwards J. Addison's disease: identification and management in primary care. Br J Gen Pract 2015: 65; 488–90. https://doi.org/10.3399/bjgp15X686713.© British Journal of General Practice 2015.

However, it is prudent to perform an important additional test to provoke the hypothalamic–pituitary–adrenal axis with corticotropin-releasing hormone, but more likely with a synthetic form of ACTH (synacthen, tetracosactide), and observe the result. Failure of synthetic ACTH (e.g. 250 μg intra-muscular) to provoke an increase in serum cortisol makes the diagnosis of Addison's disease. The laboratory will also report any electrolyte changes, principally hyponatraemia and/or hyperkalaemia, and hypoglycaemia.

Standard treatment is with lifelong replacement therapy, generally with oral hydrocortisone 0.4 to 0.8 mg/kg in children and 20 to 30 mg/day in adults, both in two or three divided doses, although there are other options, such as fludrocortisone 0.05–0.3 mg and 3 to 5 mg prednisolone daily. Increased doses may be advised in acute febrile illness, for major surgery, or in Addisonian crisis, where an intravenous infusion may be required. Use of dexamethasone is generally not advised. The exact dose of medication in long-term use should be titrated according to clinical effect and laboratory results. In view of the potential for additional autoimmune disease, patients are likely to be invited for annual testing for pernicious anaemia (vitamin B_{12}, intrinsic factor), coeliac disease (antibodies to tissue transglutaminase), autoimmune hepatitis (LFTs), type 1 diabetes (fasting glucose and HbA1c), and autoimmune thyroid disease (TSH, T3, thyroxine).

Figure 15.12 summarizes broad concepts in the physiology and pathology of the hypothalamus–pituitary–adrenal axis with respect to Cushing's and Addison's diseases.

The zona reticularis and androgens

In both sexes, the tissues of the zona reticularis secrete (as compared to the gonads) small amounts of weak androgens, mostly dehydroepiandrosterone (and its sulphate), but also androstenedione, possibly under the control of ACTH. Adrenal androgens contribute to the pre-puberty growth spurt, and then to axillary and pubic hair growth.

In the adult male, the production of adrenal androgens is insignificant compared to that from the testes, and in the adult female the androgens are converted into oestrogens by other tissues, and so contribute to general sexuality. After the menopause, when ovarian activity falls, all oestrogens arise from adrenal androgens.

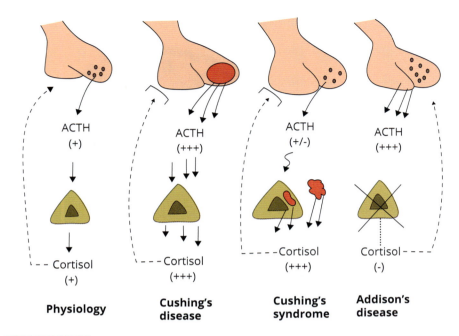

FIGURE 15.12

Physiology and pathology of Cushing's and Addison's diseases. Column 1: ACTH release by the pituitary acts on the adrenals to release cortisol, which itself feeds back to regulate levels of ACTH. Column 2: Over-production of ACTH by a pituitary adenoma results in over-stimulation of the adrenals, causing Cushing's disease. Feedback by high cortisol is ineffective. Column 3: An adrenal adenoma secretes excess cortisol, although this may arise from an ectopic (often malignant) source or medications, causing Cushing's syndrome. Feedback in this instance suppresses the release of ACTH. Column 4: An inactive adrenal cortex fails to produce cortisol, so the release of ACTH is unregulated and may be high in its attempt to stimulate the adrenals. Secondary hypoadrenalism due to hypopituitarism is also characterized by low levels of cortisol, but results from low levels of ACTH. Solid lines represent release and action of hormones, dashed lines feedback.

Overactivity: Most of the pathology of the zona reticularis has been previously described. Failure of the steroidogenic enzyme 21-hydroxylase to consume its progestogen substrates leads to their conversion to androgens by 17,20 lyase, which, in the female foetus, leads to ambiguous genital development (clitoromegaly, that may be mistaken for a penis). Children with partial defects in these enzymes, and so increased levels of sex hormones, may develop precocious puberty. Rarely, an adenoma may secrete excessive androgens that can further stimulate the reproductive system. As these tumours generally develop in those of later years, a reawakening of libido and secondary sexual characteristics may follow, possibly leading to utero-vaginal bleeding.

Underactivity: Almost all cases of adrenal hypoplasia and insufficiency are genetic and so develop shortly after birth, and generally extend to all parts of the adrenal cortex. Congenital adrenal hypoplasia is usually severe and associated with differences in sex development (absent genitalia) caused by gonadal failure, in addition to mineralocorticoid pathology. The hypothesis that ACTH is linked to secretion of adrenal androgens is supported by features of pituitary failure, wherein a lack of ACTH is associated with reduced androgen release (as does lack of ACTH and a form of Addison's disease).

15.3.5 The adrenal medulla: pathology

The leading pathology of the adrenal medulla is a phaeochromocytoma, essentially a tumour of nerve endings that over-secrete either or both of their catecholamine neurotransmitter products. It has an incidence of ~8/million/year. Tumours of the same cell biology found outside the adrenals are paragangliomas and can be found in diverse tissues (head, neck, thorax, bladder) at a frequency of ~3/million/year.

Most (~75%) of phaeochromocytomas arise spontaneously, and the remainder may be familial (and autosomal dominant). Both forms are linked to mutations in the von Hippel–Lindau tumour-suppressor gene (*VHL*) at 3p25.3, in *NF-1* at 17q11.2 causing neurofibromatosis-1, and in one of many mutations in genes (*SDH A* to *D*) coding for subunits of succinate dehydrogenase protein complex, and it may be part of the multiple endocrine neoplasia syndrome (see section 5.13.2).

Presentation and diagnosis

There are no specific established signs of a phaeochromocytoma, but the most common symptoms of headache, sweating, and palpitations, when combined, are suggestive. Other symptoms include pallor, weight loss, a high heart rate, flushing, tremor, anxiety, polyuria, chest pain, fever, and general malaise. Clinically, most patients have high blood pressure, often hypertension, and the laboratory may report hyperglycaemia and lactic acidosis. As many of these symptoms may be present in alternative diagnoses, such as the menopause, a carcinoid tumour, or thyroid cancer, a precise test is required.

Once more, the diagnosis is made in the laboratory with high levels of metanephrines and vanillylmandelic acid, the stable end-metabolites of adrenaline and noradrenalin, although certain drugs (notably tricyclic antidepressants and phenoxybenzamine; others include paracetamol, α-methyldopa, and sulphasalazine) increase catecholamine levels. As the hormones have a diurnal variation, a 24-hour collection is required, and can vary according to sitting or standing, and any recent stress or exercise. Given the psychological state most patients will be in during a practitioner consultation, there is a likelihood of false positive, and due to the rarity of the disease, analysis may be in a reference laboratory. An additional laboratory test is for chromogranin, a secretory protein coded for by *CHGA* at 14q32.12, although it may also arise from pancreatic beta islet cells. A parallel to the dexamethasone test to suppress cortisol is the trial of a dose of clonidine (an anti-hypertensive) which, in physiology, should reduce levels of catecholamines. Failure to suppress implies a phaeochromocytoma.

MRI (preferred) and CT can offer the precise location of the lesion (Figure 15.13), but positron emission tomography with F[18]deoxyglucose gives better sensitivity, especially as 10% of presentations are already associated with metastases, which bring a survival rate of 50% at 5 years.

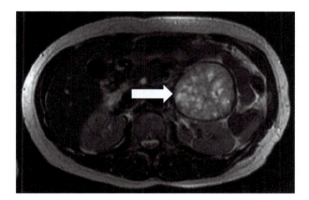

FIGURE 15.13

MRI showing left adrenal phaeochromocytoma (arrowed). The patient had a history of hypertension, and presented with hypertensive crisis and renal failure, prompting the MRI. Urinary normetanephrine was 19,043 µg/24 hours (reference range 75–325) and metanephrine was 74,510 µg/24 hours (24–96). The patient was admitted following hypertension-induced seizures with blood pressures of 205/138 mmHg. Whilst on ICU, she suffered a fatal right middle cerebral artery non-haemorrhagic infarction.

Vindenes T, Crump N, Casenas R, et al. Pheochromocytoma causing Cardiomyopathy, Ischemic Stroke and Acute Arterial Thrombosis: A Case Report and Review of the Literature Connecticut Medicine 2013: 77; 83–5. © 2022 Connecticut State Medical Society.

CASE STUDY 15.1 *Adrenal adenoma*

A 46-year-old man (non-smoker, body mass index 25.5 kg/m²) bought himself a blood pressure machine and was concerned to note that it gave SBP/DBP readings between 148–155 and 88–95 respectively. The GP confirmed these figures and found nothing else of note with no relevant medical or family history—urine analysis was normal. In accordance with NICE NG136, she advised generally on a healthy lifestyle, offered her patient 24-hour ambulatory blood pressure monitoring (ABPM), did a formal assessment of his cardiovascular risk (QRISK3 score 3.8%: 2.6% for a comparable healthy man), and took blood for tests of renal function (Table 15.5).

The only abnormality is the low serum potassium, which becomes 2.6 when testing is repeated, and the 24-hour ABPM confirms hypertension. The hypokalaemia with serum sodium at the top of the reference range is sufficient to consider Conn's syndrome, where high levels of aldosterone act on the renal tubules to reabsorb sodium but not potassium or hydrogen ions from the developing urinary filtrate, resulting in hypokalaemia. The movement of sodium is followed by that of water, and so increased blood volume and hypertension. Since aldosterone is produced by the zona glomerulosa of the adrenal cortex, this is the most likely location of the lesion, which may be general overactivity (a bilateral idiopathic hyperplasia), present in ~30% of cases; multiple adrenal nodules (10%); or an aldosterone-secreting tumour (60%). However, alternative pathology includes phaeochromocytoma (very rare: present in 1/1,000 cases of hypertension), renal artery stenosis and coarctation of the aorta, and all other causes of hypokalaemia considered. As the patient is not taking any drugs, any such effect (e.g. due to diuretics, beta-blockers, or methyldopa) can be eliminated. The appropriate test is for the ratio of aldosterone to renin (the renal enzyme converting angiotensinogen to angiotensin I), which must be performed in a specialist centre under controlled

TABLE 15.5 Renal blood tests in Case study 15.1

Test	Result	Reference range
Sodium	145 mmol/l	135–45
Potassium	2.5 mmol/l	3.8–5.0
Urea	4.2 mmol/l	3.3–6.7
Creatinine	88 µmol/l	71–133
Estimated glomerular filtration rate	86 ml/min/1.73m²	53–110

conditions: levels vary according to posture. Other tests include urine potassium (expected to be high in Conn's syndrome) and metanephrines (increased in phaeochromocytoma), whilst some measure ACTH as this has an impact of aldosterone, and others attempt to suppress a high aldosterone with exogenous sodium chloride or fludrocortisone.

The patient duly reports to clinic for the blood tests, which are taken in as stress-free an environment as possible. The analysing laboratory reports aldosterone in units of pmol/l, renin as nmol/l/hr (other laboratories may report as ng/dl), and results in the patient give an aldosterone-to-renin ratio of 40—well over the local reference threshold of 30. The baseline renin activity was 2.0 nmol/l/hr, well within reference levels, so that the high ratio is not due to very low renin with normal aldosterone.

The patient is quickly referred to imaging with ultrasound and CT/MRI, which detect a mass consistent with an adrenal adenoma. It is removed laparoscopically two weeks later, and a month later, his blood pressure is 132/77 mm Hg.

Key points

Study of the pathophysiology of the diseases named after Cushing and Addison leads the way to understanding the concepts of positive and negative feedback, and how they can be used in diagnosis and management.

Management

Treatment of choice is surgery, but expert care is required in case of pre- and peri-operation hypertension, with use of alpha- and beta-blockers and calcium channel-blockers. Patients will also be monitored post-operatively for rebound hypotension. Should surgery not be possible, palliative treatments include chemotherapy (with a triad of cyclophosphamide, vincristine, and decarbazine, and possibly tyrosine kinase inhibitors, cisplatin, and 5-fluorouracil), radiofrequency ablation, and radiotherapy.

What are the key features of Addison's disease?

15.4 Sex hormones

Failure to regulate, produce, and respond to primary (LH, FSH) and secondary (testosterone, oestrogen, progesterone) hormones has profound effects on sexual development and function in both sexes, as extensively described in sections 14.2, 14.3, and 14.4. Many of those diagnosed with differences in sex development are likely to have abnormalities in sex hormone regulation, and therefore potentially altered phenotypes and fertility problems. If a transgender patient receives gender-affirming hormone therapy with sex hormones, their healthcare team will need to take this into account when providing clinical care.

15.4.1 The male

Follicle-stimulating hormone (FSH)

FSH is required for sperm production in the testes by inducing Sertoli cells to produce androgen-binding protein, a 44 kDa protein coded for by *SHBG* 17p13.1 which has almost complete homology with sex hormone-binding globulin. The protein binds to and concentrates the testosterone and dihydrotestosterone generated by Leydig cells. High local levels of these androgens (far exceeding circulating levels) are necessary for spermatogenesis. Low levels of FSH may arise from hypothalamic disease, but more often from hypopituitarism, causes of which have been described in section 15.2.2.

Failure of FSH to support testosterone production leads to gonadal failure, itself linked to loss of facial and body hair growth, and to a reduction in red blood cell indices, often to those of women. In boys, this will lead to failure of the gonads to mature, with poor growth and delayed or absent puberty. A prolactinoma may cause sexual dysfunction via prolactin's inhibitory effect on GnRH and so low FSH (and LH), both of which (and sexual dysfunction) can be restored by treatment of such an adenoma with chemotherapy such as bromocriptine or cabergoline and/or surgery (see 15.7.1).

FSH is the Sertoli cell ligand for the FSH receptor, a 75 kDa transmembrane molecule coded for by *FSHR* at 2p16.3. As described in section 14.2.3, loss-of-function mutations in *FSHR* lead to infertility. Levels of FSH are also regulated by inhibin and activin, small (~13 kDa) proteins members of the TGF-beta superfamily consisting of various isoforms secreted by Sertoli and Leydig cells. Both are dimers, coded for by genes at 7p15–p13, 2q13 and 12q13, and are carried as complexes with plasma proteins α_2-macroglobulin and follistatin. Activin is also produced by the prostate, and serum levels are high in men with benign prostatic hyperplasia.

Kisspeptin

Details of this small hypothalamic hormone, with roles in physiology and sexual behaviour, are presented in Chapter 14, section 14.2.3. Mutations in the gene for kisspeptin and its receptor, and so failure of GnRH, have been linked to diverse phenotypes that range from partial sexual development to hypogonadism (micropenis and cryptorchidism) and but also to precocious puberty.

Luteinizing hormone (LH)

LH stimulates Leydig cells to produce androstenedione, dehydroepiandrosterone, and testosterone, and like FSH, levels feed back to the hypothalamus to suppress the releasing hormone. As with FSH, failure of the hypothalamus–pituitary–gonad axis leads to gonadal failure, such as hypopituitarism described above. However, this failure may be due to loss-of-function mutations in the receptor for LH caused by mutations in *LHCHR* (coding for a ~90 kDa glycoprotein at 2p16.3) where (once more) the parallel with loss of *FSHR* function applies.

Testosterone

Testosterone is the major product of a metabolic pathway that passes from cholesterol, via progestogens, androstenedione, and androstenediol, although the final product is a reduced form—dihydrotestosterone. This pathway is active in the adrenal cortex and testes, where in the latter testosterone is required for the maturation of spermatozoa in the epididymis. To facilitate this process, Sertoli cells are in close physical contact with developing spermatogenic cells.

Almost all testosterone leaving the testes is carried by sex hormone-binding globulin to various tissues and organs to stimulate secondary sexual characteristics, and feed back to the hypothalamus to suppress GNRH and so FSH levels. LH and FSH drive testosterone production *in utero* and for 6–12 months after birth, at which point levels fall until puberty, when secondary sexual characteristics develop and are thereafter maintained (Table 15.6). Levels of all three hormones remain high until middle age, whereupon those of testosterone fall (the andropause) with a concurrent rise in LH and FSH and (in some) loss of hair on the scalp.

Pathology

The leading male reproductive pathology is infertility/subfertility and recognizes two forms: primary hypogonadism (or hypergonadotropic hypogonadism), where the testes fail to produce testosterone and spermatozoa, resulting in high FSH and LH, and secondary hypogonadism, where the lesion lies with the hypothalamus and/or pituitary, and so low FSH and LH. The potential for these may be summarized in terms of Figure 15.14.

Whilst in theory fertility problems may arise from disease of the hypothalamus, this is rare. Markedly more frequent is pituitary disease with failure to generate LH and/or FSH, and other pituitary disease (principally prolactinoma). Failure of these hormones to correctly interact with their receptors is rare, but if present, leads to lack of testosterone and androgen-binding protein, and impaired spermatogenesis and secondary sexual characteristics. Table 15.15 shows typical reference ranges for blood tests in the investigation of male reproductive disease.

Anti-Müllerian hormone (AMH), a dimer with different isoforms, is coded for by *AMH* at 19p13.3, and secreted by Sertoli cells. In the male embryo, it prevents the Müllerian ducts from developing into a uterus, and serum levels fall consistently from birth to puberty. Both Leydig and Sertoli cells express a receptor for AMH, suggesting a paracrine effect, possibly in reducing steroidogenesis. AMH has a far greater role in female reproduction, notably in PCOS (section 14.3.2).

Reproductive pathology may also arise from infection, such as mumps, but also from sexually transmitted disease or trauma. Numerous other diseases, ranging from autoimmune disease and cirrhosis to influenza and smoking, can also impact on testicular function, and in some cases are reversible.

TABLE 15.6 **Effects of testosterone**

At puberty	Growth spurt; deepening of the voice; maturation of the penis, testes, and scrotum; development of facial, body, and pubic hair; increased muscularity and bone density; development of libido
As an adult	Maintenance of spermatogenesis, libido, muscularity, bone density, and body hair

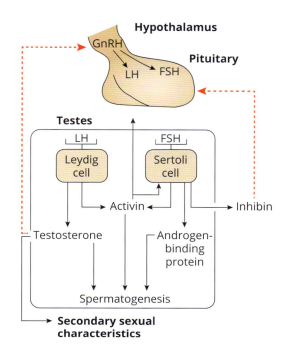

FIGURE 15.14
The biology of FSH, LH, and testosterone in the male. GnRH = gonadotropin-releasing hormone. Dashed lines represent negative feedback.

Management

Treatment will reflect the aetiology (i.e. primary or secondary hypogonadism). Pituitary disease is treated with chemotherapy or surgery. Treatment of low or absent FSH, LH, and testosterone is with replacement therapy, dependent on the particular aetiology. In rare instances, triple therapy may be indicated. However, in some treatment with testosterone may not stimulate spermatogenesis, as high levels are needed within the testes. Measurement of AMH can help diagnosis of male infants with ambiguous genitalia and may lead to a diagnosis of persistent Müllerian duct syndrome, often the result of mutations in *AMH* or its receptor, *AMHR2*, coded for at 12q13.13. Hypogonadism resulting from a *KISS1/KISSR* mutation can be successfully treated with exogenous GnRH. Spermatozoa may be harvested from the testes for methods in assisted reproduction.

Cross reference
Further details of the diagnosis and management of male infertility are described in section 14.2.3. Testicular cancer is discussed in more depth in section 5.13.1.

SELF-CHECK 15.6

Summarize the importance of the Sertoli and Leydig cells.

15.4.2 The female

Section 14.3 describes the pathology of female reproductive endocrine disease. Failure of sexual development at puberty and of fertility has been ascribed to all parts of the pathway, from kisspeptin through to oestrogen and progesterone. All major hormones are detectable at background levels in the child, and the integration of nutritional, growth, and metabolic signals in the adolescent (at the menarche, generally ages 12–14) lead first to the production of androgenic steroids by the adrenals and then to hypothalamic/pituitary puberty.

Rising levels of kisspeptin, GnRH, LH, and FSH drive the maturation of the ovaries and so the oestrogen production that in turn leads to the development of female sexual characteristics and their maintenance in the adult (Table 15.7).

Should these signs fail to appear by age 14, delayed puberty may be suspected and investigated. Conversely, precocious puberty is likely if these characteristics develop before the age of 8. In the adult, the cyclical (mean 28 days, ranging from 23 to 33 days) nature of reproductive endocrinology is well understood, being driven by higher brain centres, and may be broken down into four stages.

TABLE 15.7 Effects of oestrogen and progesterone

At puberty	Growth spurt; maturation of the ovaries, uterus, cervix, vagina, and vulva; initiation of the menstrual cycle; widening of the pelvis; development of pubic hair, subcutaneous fat, breasts, and libido
As an adult	Maintenance of secondary sexual characteristics and the menstrual cycle, oogenesis, libido

The menstrual phase

By consensus, the menstrual cycle starts with day 1 of menstruation, the breakdown of the endometrium resulting in a 50–150 ml loss of blood, endometrial cells, and mucus via the vagina, which lasts 4–5 days.

The follicular phase

Once shed, the endometrium is rebuilt over the next 10 days or so, with ovulation occurring in the middle of the cycle, generally around days 14–15. It is characterized by an early small increase in FSH, driven by hypothalamic GnRH, which acts on an ovary to initiate the maturation of the dominant (Graafian) follicle and the generation and release of oestrogen from granulosa cells. Activin stimulates granuloma cell development, follicle and oocyte growth and maturation, and stimulates the production of oestrone and oestradiol from androstenedione and testosterone respectively by aromatase. The endometrium proliferates to a thickness of 4–10 mm under the influence of follicular oestrogens, and towards the end of this phase, levels of oestrogen rise, peaking at or slightly before a peak in kisspeptin levels.

Ovulation

At around days 13–14, rising levels of LH peak, and some 9–18 hours later cause the rupture of the follicle and release of the egg, which passes into the fallopian tube. Many women experience a short, sharp pain associated with the ovulatory rupture. There is also a second small rise in FSH.

The luteal phase

Under the influence of LH, the remainder of the follicle is transformed into the corpus luteum, which secretes oestrogen and inhibin, and large amounts of progesterone. The next stage depends on whether the ovum is fertilized:

- Should an embryo implant into endometrium, the trophoblast secretes chorionic gonadotropin, which promotes the production of progesterone and oestrogen by the corpus luteum. These and other events lead to the promotion of the pregnancy.

- Lacking chorionic gonadotropin, the corpus luteum, with an intrinsic lifespan of 10–14 days, regresses into the corpus albicans. Having peaked at around day 21, declining levels of oestrogen, but primarily of progesterone from a regressing corpus luteum, result in the release of prostaglandins. These cause the spiral arterioles of the basal layer of the endometrium to constrict, so eventually causing the upper layers to become hypoxic and disintegrate, leading to menstruation.

These two options continue until the menopause at an average age of 51 years, and as a result of a reduction in oestrogen levels, those of FSH rise. Figure 15.15 illustrates changes in reproductive hormone in the course of the menstrual cycle (see Table 15.15).

Regulation of the menstrual cycle

In the follicular phase of the menstrual cycle, FSH promotes follicle development, with increasing numbers of granulosa cells expressing FSH receptors and the release of oestrogen. Initially, low levels of oestrogen feed back to inhibit pituitary FSH, but as oestrogen (without progesterone) from

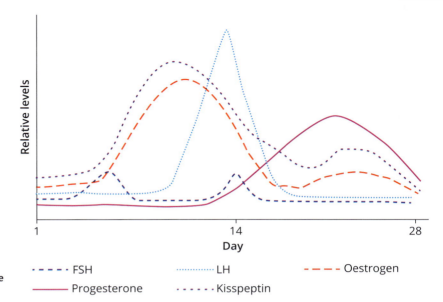

FIGURE 15.15
Reproductive hormones and the menstrual cycle. Changes in levels of reproductive hormones over the 28-day menstrual cycle. Note that FSH, oestrogen, and kisspeptin have biphasic patterns whilst LH and progesterone have only one peak.

the developing follicle rises, the feedback to the hypothalamus and pituitary becomes positive and induces the release of GnRH, FSH, and LH, the latter inducing ovulation.

Increased secretion of inhibin, oestrogen, and progesterone by the corpus luteum feeds back to the pituitary to suppress FSH and LH, and at the end of the cycle, low levels of progesterone and oestrogen induce the release of GnRH, FSH, and LH, thus starting a new menstrual cycle.

Pathology, diagnosis, and management

The leading pathological conditions of female reproduction are discussed in Chapter 14 and include those of the breast (section 14.3.1), PCOS (14.3.2), the fallopian tubes (14.3.3), the uterus and endometrium (14.3.4), the cervix, vagina, and vulva (14.3.5), and infertility (14.3.6 and parts of section 14.4). The menopause is not a disease, but can be associated with disease such as osteoporosis, as is discussed in section 14.3.6.

Key points

In both sexes, lesions in each part of the interactions between hormones, their receptors, and the associated organs and tissues can be linked to under- and overactivity of the reproductive system (Figure 15.16). This can occur in the embryo, with congenital disease, and may be evident at puberty and develop in the adult.

SELF-CHECK 15.7

What is the role of kisspeptin in male and female fertility?

15.5 TSH and the thyroid

Fertility problems may be present in up to 10% of couples, and since the leading aetiology is hormonal, the most common form of endocrine disease is in reproduction. The second most frequent is diabetes (~6%), the third being disease of the thyroid: at least 1–2% of the general population suffer from hyperthyroidism, and at least another 1–2% from hypothyroidism. Thus ~4% of the population

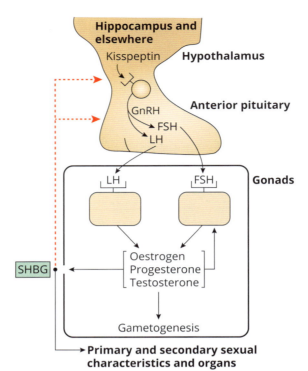

FIGURE 15.16
Interactions between hormones, their receptors, and the cells and organs of the reproductive system. GnRH = gonadotropin-releasing hormone, SHBG = sex hormone-binding globulin.

Cross reference
For a detailed discussion of thyroid cancer, see section 5.13.

are affected by thyroid disease, although it is likely that there are many cases of sub-clinical disease. The burden of disease falls mostly on women, where some form of thyroid disease has been estimated to be present in 3–5% of the population, higher in the elderly. In those locations where iodine is scarce, these figures are higher.

15.5.1 Anatomy and physiology

The thyroid is formed from right and left lateral lobes and a connecting section (the isthmus) from which a superior pyramidal lobe (pointing up the throat) may arise. Together, often described as butterfly-shaped, it is located anterior to the trachea. The greater part of the gland is formed of spherical bodies (follicles), the centre being composed of colloidal thyroglobulin, the outer sections by follicular cells placed on a basement membrane. The function of this gland is to produce and release the two thyroid hormones tri-iodo-thyronine (T3) and tetra-iodo-thyronine (T4, or thyroxine).

Blood levels of T3 and T4 are regulated at the pituitary and hypothalamus. At the latter, low T3/T4 (and also possibly a cold environment, hypoglycaemia, pregnancy, anaemia) prompts the release of thyrotropin-releasing hormone that passes to the anterior pituitary to stimulate the production and release of TSH (a ~24 kDa dimer coded for by *CGA* at 6q14–21 and *TSHB* at 1p13) into the circulation. When plasma levels of T3/T4 are sufficient, TSH levels are suppressed in classic negative feedback, whilst dopamine, somatostatin, and cortisol may also be inhibitory.

Under the influence of TSH, thyroid follicular cells synthesize and store a supply of ~100 days' worth of the thyroid hormones. Iodine is imported from the blood as sodium iodide by the sodium/iodine co-transporter, an 87 kDa transmembrane glycoprotein coded for by *SLC5A5* at 19p13.11. In the cell, tyrosine residues present in the thyroglobulin are sequentially iodinated by thyroperoxidase (TPO, a 105 kDa glycoprotein coded for by *TPO* at 2p24). T3 and T4 are exported by monocarboxylate transporter 8 (coded for by *SLC16A2* at Xq13.2) into the blood where, being lipid soluble, 70–90% are carried by thyroxin-binding globulin (of 660 kDa, coded for by *SERPINA7* at Xq22.3), the remainder by transthyretin (coded for by *TTR* at 18q12.1) and albumin (coded for by *ALB* at 4q11–22). However, some of both hormones are unbound, thus free T3 (fT3) and free T4 (fT4).

In the tissues, an iodine atom is removed from T4 by iodothyronine deiodinase, forming more T3, this being the more active isoform. At the cell membrane, T3 is conducted across the membrane by a transporter and binds a heterodimeric nuclear receptor (coded for by *THRA* at 17q11.2–12 and *THRB* at 3p24.1–22) in the cytoplasm and on the nuclear receptor, giving it genomic and non-genomic effects.

Most cells of the body (muscle, bone, fat, liver, neural, kidney, gonadal, adrenal, pituitary, hypothalamic, etc.) have thyroid hormone receptors, and as such they have a wide range of physiological functions:

- Control of the body's basic metabolic rate by stimulating the use of cellular oxygen to drive metabolism. This is linked to part-control of the body's temperature by promoting the production and use of ATP.

- Promoting metabolic pathways in protein synthesis, lipolysis, gluconeogenesis, and glycolysis.

- Supporting erythropoiesis and gut motility.

- Alongside growth hormone and insulin, participating in growth, particularly of connective tissues.

- By up-regulating β-receptors, enhancing the activity of adrenalin and noradrenalin, resulting in increased heart rate and force of left ventricular contraction (hence cardiac output) and in blood pressure.

Figure 15.17 summarizes the production, regulation, and cellular action of thyroid hormones.

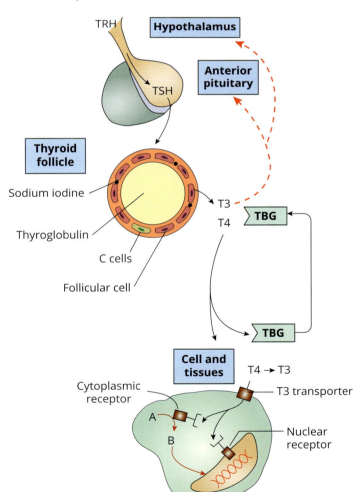

FIGURE 15.17

The production, regulation, and cellular action of thyroid hormones. TRH = thyrotropin-releasing hormone. TSH = thyroid-stimulating hormone. T3 = tri-iodothyronine. T4 = tetraiodothyronine. TBG = thyroxine-binding globulin. A, B =- second messengers, generally involving phosphoinositide-3-kinase. Dashed lines represent feedback.

Calcitonin

A small number of parafollicular (or C) cells are found between the thyroid follicular cells, and produce and secrete calcitonin. This small (~3.6 kDa) protein, coded for by *CALCA* at 11p15,2 part-regulates serum calcium (see section 15.8.1 on PTH).

15.5.2 An introduction to thyroid disease

Thyroid nodules are surprisingly common in the general population—~5% by palpation and >50% by ultrasound—and most are asymptomatic and benign. These nodules include lipomas, cysts, and those linked directly to thyroid disease, including cancers, so that further investigation is generally warranted. Endocrine thyroid disease in the adult is easily classified as underactivity (hypothyroidism) and overactivity (hyperthyroidism); NICE NG145: Thyroid disease: assessment and management, considers both. It recommends that TSH should be measured in those in whom non-pituitary disease is suspected, with subsequent measurement of fT4 if TSH is above the top of the reference range, and measure of fT3 and fT4 if TSH is below the bottom of the reference range. However, where a pituitary component is suspected, and in children and young people, measurement of both TSH and fT4 is recommended. In each form of thyroid disease, women outnumber men by factors of between 3:1 and 10:1. Autoantibodies to thyroid tissues are very common, and their presence in the absence of clear pathology is nonetheless a risk factor for the development of disease. IgG autoantibodies may cross the placenta to cause injury to the foetus, evident at birth, but clinical signs generally abate as the antibody becomes less effective in the neonatal period.

The two leading autoimmune conditions are Graves' disease in hyperthyroidism and Hashimoto's disease in hypothyroidism. Pathogenesis may begin with genetic and environmental factors with damage to thyroid cells and the release of auto-antigens not previously encountered by the immune system. Following the recognition of the auto-antigen by antigen-presenting cells, the gland may be infiltrated with auto-reactive T and B lymphocytes. Should Th1 cells predominate, cellular immunity develops and can lead to target cell destruction and apoptosis, and so Hashimoto's disease. Conversely, should Th2 cells dominate, the resulting humoral pathway sees the development of blocking or stimulating autoantibodies (such as to the TSH receptor) and so Graves' disease.

Thyroid disease may be present even when blood results are normal. Those with thyroid enlargement, but normal thyroid function, should be offered ultrasound imaging and fine-needle aspiration if a nodule is present, and cysts should be aspirated. Should a large growth threaten airways or other tissues, surgery or radioactive iodine ablation may be necessary.

Given the wide range of the effects of T3 and T4 on many different tissues and organs, pathology can develop in many different conditions. For example, sex hormone-binding globulin is increased in hyperthyroidism and decreased in hypothyroidism, potentially leading to changes in reproductive physiology, whilst extreme hyperthyroidism, a thyroid storm, can be life-threatening.

15.5.3 Hypothyroidism

This is the most common form of thyroid disease, with an incidence of 3,500/million/year. Its onset is often insidious with numerous minor signs and symptoms, striking 5–6 times as many women as men, generally in the 5th–7th decade.

Pathology

Hypothyroidism is the result of failure of the hypothalamus, pituitary, and/or thyroid gland itself to generate sufficient T3 and T4 (Figure 15.17). As thyroxine-binding globulin is a product of the liver, hypothyroidism may in theory arise from hepatic disease, but other carriers are generally able to cope. The causes of anterior hypopituitarism have been discussed in section 15.2.2, but are responsible for <10% of cases of hypothyroidism, the leading cause being issues regarding the thyroid itself.

Globally, the leading cause of hypothyroidism is lack of iodine, but where iodine is plentiful, the leading aetiology is inflammation, hence thyroiditis, with destruction of follicles. This may follow an infection with any of the common bacteria such as *Staphylococcus*, *Streptococcus* and

Klebsiella species, in which case the intrinsically healthy thyroid is attacked as part of the general inflammatory response. However, Hashimoto's thyroiditis is the leading inflammatory disease, the causative agents being autoantibodies to TPO (present in 95% of cases), thyroglobulin, or those which block the TSH receptor. These can lead to a Type II hypersensitivity (antibody-dependent cellular cytotoxicity: section 10.2.2) with complement-mediated cell destruction, also likely to lead to infiltration with leukocytes and so cell-mediated damage, possibly transforming into chronic lymphocytic thyroiditis. The local inflammation may cause fibrosis and the formation of granuloma and nodules. Hashimoto's thyroiditis can also lead to hyperthyroidism (see section 15.5.4). Autoantibodies to the sodium/iodine co-transporter have been described, but their significance is, at present, unclear.

Other causes of hypothyroidism include drugs (such as lithium, amiodarone (which has two atoms of iodine per molecule), interferon, dopamine and glucocorticoids), unexplained atrophy, and also the over-treatment of hyperthyroidism.

Presentation

Lack of iodine during pregnancy can lead to congenital iodine deficiency in the neonate, characterized by low birth weight, a protruding tongue, blunted reflexes, and muscle wastage. Later in childhood, there is poor growth and learning disability. In the adult, there may be a slowly developing enlargement of the thyroid—a goitre—due to hypertrophy resulting in an increase in the number and size of thyroid follicles. However, a goitre (which may be diffuse or nodular) may also develop in Hashimoto's thyroidism.

In the absence of a clear goitre, hypothyroidism may present with lethargy; dry, coarse, and pale skin; slow speech and mental function; low blood pressure; pallor; hoarse voice; constipation; feeling cold; weight gain; thinning of the hair; bradycardia; infertility; anaemia; and menstrual irregularities.

An established consequence of severe and/or long-term hypothyroidism is myxoedema, characterized by the deposition of subdermal mucopolysaccharides. Often on the face, this gives rise to fleshy skinfolds over the forehead, below the eyes, and on the cheeks and lips, and possibly a large tongue (macroglossa). Skinfolds may appear on the lower leg.

Diagnosis

Blood tests will show reduced levels of T3 and T4, and in almost all cases there will be an increase in TSH due to lack of negative feedback. However, should the aetiology follow hypopituitarism, there may also be low or absent TSH. Practitioners should consider measuring antibodies TPO in adults with a raised TSH. There is also a place for ultrasound, which can be very helpful, such that MRI and CT are not often required. Fine-needle aspiration or biopsy of a suspected nodule will help further a diagnosis, where an iodine-deficient goitre will show hyperplasia of thyroid follicle, in contrast to Hashimoto's thyroiditis where there may be infiltration with leukocytes, and the fixation of autoantibodies demonstrated by indirect immunofluorescence. The absence of serum autoantibodies does not exclude Hashimoto's disease as sero-negative cases are known. A further cause of thyroid nodularity is a lymphoma, although relatively rare. Measurement of serum iodine may be requested in those locations where dietary or drinking water iodine is low.

Management

A hypoiodoic goitre will regress when provided with iodine, and may well resume hormone production, but until such times, oral T4, as levothyroxine, is the standard treatment. In adults aged <65 and with no cardiovascular disease, a starting dose of 1.6 mg/kg is appropriate, but in those aged >65 or any adult with cardiovascular disease a starting dose of 25–50 mg od is common. In both cases, doses will be titrated depending on symptoms (if severe), but usually according to blood results, with rising T4 and falling TSH (if ever raised). The target is a level of TSH within its reference range, with levels checked every three months until achieved. However, symptoms persist in 5–10% of patients on levothyroxine, and these may benefit from liothyronine, a synthetic form of T3 (up to 60 μg od in two or three divided doses). In cases of myxoedema coma, the result of very low levels of T3/T4, drugs

CASE STUDY 15.2 Hypothyroidism

Despite increasing her daily walks and dieting, a 75-year-old woman continues to gain weight. As winter comes, she finds herself wearing scarves, gloves, and a hat in addition to a coat, whilst those around her do not feel the need to wrap up so thoroughly. As the winter progresses, she finds herself being increasingly cold, and is becoming more tired. She initially resists calls from her children to go to her GP, but does so as her hair begins to fall out.

The GP finds little concrete in a physical examination, with no major previous medical history, although his patient complains a little of back pain, for which she takes paracetamol. Other medications are combined calcium and vitamin D (Adcal, one tablet twice a day). Blood pressure is 138/76 mm Hg, body mass index (BMI) is 27.7 kg/m², and urine analysis is normal. The GP takes routine bloods and arranges for her to have a bone density scan.

Results of the blood tests are shown in Table 15.8. Renal and liver function are normal, but there are low levels of three of the four red cell indices that, allied to the symptoms, make a diagnosis of normocytic anaemia. ESR and anti-citrullinated antibodies are both a little raised, which is unsurprising, given the age of the patient. However, thyroid function is abnormal, with raised thyroid-stimulating hormone (TSH) and low levels of thyroxine, making a second diagnosis of hypothyroidism, which may be linked to the anaemia. Accordingly, the GP orders blood tests to assess his patient's iron status.

Treatment begins with 50 μg of levothyroxine od for 4 weeks, at which point TSH has come down to 4.5 mU/l. But the patient tells her GP that she feels no different, so he increases her tablets to 100 mg a day. Iron studies (total iron binding capacity, transferrin saturation) indicate that she could benefit from some iron, so the GP adds ferrous sulphate 200 mg twice a day for two weeks, then once a day. After three weeks on the higher dose of levothyroxine, the patient starts to feel better, although this may well be the effect of the spring weather. However, TSH is still too high at 3.8 mU/l, so the GP recommends 100 mg and 150 mg on alternate days. A month later, the haemoglobin level, free T4 and TSH are back in their reference ranges. The bone scan reported moderate osteoporosis.

Four months later, the patient's blood results are all in their reference ranges, and she is markedly more active and has lost 6 kg in weight (BMI now 25.4). However, she complains that her hair has not gone back to its previous thickness.

TABLE 15.8 Blood results for Case study 15.2

Analyte (unit)	Result	Reference range
Urea (mmol/l)	5.1	3.3–6.7
Creatinine (μmol/l)	78	71–133
Sodium (mmol/l)	140	135–45
Potassium (mmol/l)	4.5	3.8–5.0
Estimated glomerular filtration rate	66	36–74
Bilirubin (μmol/l)	12	<20
Alanine aminotransferase (IU/l)	29	5–42
Aspartate aminotransferase (IU/l)	32	10–50
Alkaline phosphatase (IU/l)	95	20–130
Gamma glutamyl transferase (IU/l)	65	0–100
Haemoglobin (g/l)	110	118–48
Red cell count (× 10¹²/l)	3.8	3.9–5.0
Mean cell volume (fl)	85	77–98
Haematocrit (l/l)	0.32	0.33–0.47
White blood cells (× 10⁹/l)	7.6	4.0–10.0
Neutrophils (× 10⁹/l)	5.2	2.0–7.0
Lymphocytes (× 10⁹/l)	1.6	1.0–3.0
Platelets (× 10⁹/l)	212	143–400
ESR (mm/Hr)	18	<10
Thyroid-stimulating hormone (mU/l)	5.5	0.2–3.5
Free T4 (pmol/l)	6	10–25
Anti-citrullinated protein antibody (U/ml)	25	<20

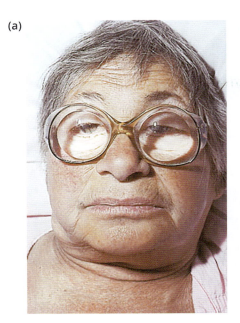

(a)   (b)

FIGURE 15.18
Myxoedema. Myxoedema (a) before and (b) after replacement therapy.

may be given via a naso-gastric tube, although the intravenous route is generally used. The results of replacement therapy in a case of myxoedema are shown in Figure 15.18.

For those with sub-clinical hypothyroidism and raised TSH levels, measurement of anti-TPO antibodies is recommended. Prescription of levothyroxine is recommended where the TSH is >10 mIU/l.

15.5.4 Hyperthyroidism

Although this variant (often described as thyrotoxicosis) has a markedly lower incidence than hypothyroidism, at around 800/million/year, it is still considerably more frequent than many other endocrine diseases, and nonetheless is linked to significant short- and long-term morbidity and mortality. Risk factors include certain drugs (many of which also cause hypothyroidism, such as amiodarone and steroids), alcohol, stress, and infections (such as with *Yersinia*), whilst one study reported an odds ratio of 3.3 for smoking. The most common manifestation of hyperthyroidism is Graves' disease, which accounts for 75% of cases. Much of the presentation, diagnosis, and management has common features with hypothyroidism, especially the laboratory and imaging.

Pathology

As with hypothyroidism, an overactive thyroid may arise from pathology of the hypothalamus/pituitary and the thyroid gland itself. As regards the former, the pituitary is generally the culprit lesion, with the excessive secretion of TSH, perhaps from an adenoma (Figure 15.19). Raised TSH will drive an otherwise innocent thyroid gland.

However, the majority of hyperthyroidism is due to pathology of the organ itself. Although there are several autoantibodies, the leading aetiology is an autoantibody that recognizes and so stimulates the TSH receptor, leading to high levels of T3 and T4, the clinical consequences of which are Graves' disease. Alternative causes of hyperthyroidism include forms of Hashimoto's thyroiditis, where instead of the autoantibody being destructive, it stimulates follicle cells independently of an autoantibody to the TSH receptor (as in Graves' disease), a tumour that secretes T3 and/or T4 and sub-acute thyroiditis, perhaps following an infection with a virus such as Coxsackie.

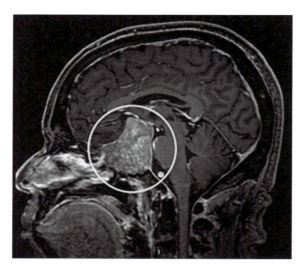

FIGURE 15.19
A large pituitary tumour. Skull MRI with contrast—pituitary tumour circled.

© 2022 New England Eye Center.

Presentation

The most common presenting signs and symptoms of hyperthyroidism include increased irritability and sweating, tremor, lethargy, breathlessness, muscle weakness, **tachycardia**, palpitations, arrhythmia, heat intolerance with warm extremities, and weight loss. As in hypothyroidism, there may be hyperplasia and so a goitre, and one or more nodules may be present. However, the classical clinical feature of advanced Graves' disease is exophthalmos (Figure 15.20), caused by the cross-reaction of the anti-thyroid autoantibody with muscles of the eyelids. This leads to the retraction of the eyelids, so exposing more of the white sclera, which also slows the blink reflex, resulting in more ocular damage. Other possible mechanisms include oedema and fat at the back of the eyeball, pushing it outwards.

Diagnosis

The primary diagnostic tool is the laboratory with measurement of T3, T4, and TSH which should point to the site of the lesion. Measurement of autoantibodies to the TSH receptor are required to confirm Graves' disease, and technetium scanning may be considered if the antibody result is negative. However, autoantibodies to TPO may also be present, and either antibody is present in ~75% of cases of Graves' disease. Ultrasound is also useful in investigating the thyroid, but MRI and CT are preferred

FIGURE 15.20
Exophthalmos. Retraction of the eyelids exposes more sclera.

Shutterstock.

for study of the pituitary (Figure 15.19). In some cases, isotope imaging can pick up micro-adenomas, and the precise nature of any nodules is likely to be defined by fine-needle aspiration or biopsy.

Management

Medical treatments for inflammatory hyperthyroidism include suppression with drugs such as carbimazole (initially 20–60 mg od in divided doses, then maintenance at 5–15 mg od) as a first-line therapy. Propylthiouracil (initially 300–600 mg od or in divided doses, then maintenance at 50–150 mg od) should be used in case of an adverse reaction to carbimazole, attempts to become pregnant, or a history of pancreatitis. In both cases, regular blood tests are required due to their ability to suppress the bone marrow, most often monitored by the neutrophil count, and this is reversible. Doses will be varied by their effect on T3 and T4. Radioactive iodine, typically I^{131}, can destroy follicular cells, thus reducing the ability of the gland to produce T3 and T4. This therapy may be a first-line option in those with multiple nodules, and if unsuccessful, lifelong antithyroid drugs are required. Radioactive therapy is difficult to control and reverse, and there is the danger of over-treatment and so a risk of hypothyroidism.

Should these approaches fail, surgery to remove a nodule, either lobe, or the entire organ is an option, as it is with malignant thyroid cancer. However, care must be taken to retain the parathyroid glands and avoid damage to the phrenic nerve, which runs close to the thyroid gland. NICE IPG499 offers guidance on minimally invasive video-assisted thyroidectomy.

Thyroid storm refers to a potentially lethal (in 10% of cases) acute exacerbation of hyperthyroidism where very high levels of T3 and/or T4 may be precipitated by any one of a dozen or more clinical situations such as a major cardiovascular event, anxiety, and psychological stress, burns, and severe infections bringing a fever. The Burch–Wartofsky point scale collects data on the temperature, atrial fibrillation, precipitating events, heart rate, symptoms of heart failure and central nervous system, or gastrointestinal or liver dysfunction. A high score points to the likelihood of a thyroid storm, and so urgent treatment (drugs as above plus beta-blockers and steroids).

15.5.5 The genetics of thyroid disease

Twin and family studies suggest a genetic aetiology for many forms of thyroid disease, with around 63% of the risk of Graves' disease being genetic (Lane et al. 2023) (Table 15.9). Having one affected sibling increases the risk of thyroid disease by 20–30%, having one parent affected by 10–25%, having a fraternal twin affected by 30–55%, and having a monozygotic twin affected by 10–40%.

Graves' disease (affecting around 3% of women and 0.5% of men) is linked to certain gene susceptibility loci in HLA and non-HLA molecules, including CD40, CTLA-4, protein tyrosine phoaphatase-22 (PTPN22), and for thyroglobulin and the TSH receptor. A GWAS study of 900 people with an autoimmune thyroid disease and 1,466 controls found *TSHR* (at 14q24–931, coding the TSH receptor) and *FCRL3* (at 1q23.1, coding for Fc receptor-like protein 3) to be linked to Graves' disease (Burton et al. 2007). One of the most commonly reported SNPs in *CTLA-4* brings a relative risk of around 1.8 for both Graves' and Hashimoto's diseases. Hyperthyroidism may also arise from a mutation in *THRB*, which can be familial or sporadic, leading to increased T3 and T4 (although TSH may be normal

TABLE 15.9 Twin concordance and thyroid disease

	Monozygotic twins	Dizygotic twins
Graves' disease	30–35%	2–5%
Hashimoto's thyroidism	50–55%	0–2%
Congenital hypothyroidism	1–20%	0–2%
Autoimmune thyroid disease	80–85%	30–40%
Thyroid nodules	50–60%	20–40%

or slightly increased), and is described as thyroid hormone resistance syndrome, which with other causes, has an incidence of ~22/million.

Unsurprisingly, the leading HLA locus is HLA-D; others include B35, B46, A2, and DPB1*0501 in Japanese; B46, DR9, DRB1*303, and DQB1*0303 in Hong Kong Chinese; DQA1*0501 in White Americans; and HLA-DRB3*202 in African Americans, although these may not apply to all populations (Ban and Tomer 2005). HLA-DR and B8 bring a relative risk of Graves' disease of 5.7 and 3.1 respectively, whilst HLA-DR3 has a frequency of 40–55% in Graves' disease patients, 15–30% in the general population, and so a relative risk of 3–4. In the UK, strong association of both DRB1*03 (relative risk 2.45) and DQA1*0501 (relative risk 2.26) has been reported, whilst the extended haplotype DRB1*03–DQB1*02–DQA1*0501 has been reported as bringing a relative risk for Graves' disease of around 2.6. Similarly, Hashimoto's thyroiditis has been linked to HLA-Dw3 (with a relative risk of 3.9), DRw3 (3.5), DR5 (3.2), haplotype DR3–DQw2 (2.2), DQA1*0402 (2.7), DRB4*0101 (4.5), haplotype DRB1*04–DQB1-0301 (4.0), and DRB1*03 (4.0).

miRNAs may help differential diagnosis of thyroid cancer, where levels of miR-197, miR-346, and miR-146b are upregulated and levels of miR-30d, miR-200, and mi26a are downregulated, whilst plasma miR-762 is increased (and correlates with fT3 and antibodies to the TSH receptor), and plasma miR-144-3p is decreased in Graves' disease. There may also be a place for miR-23a-3p in regulating Treg cells in Graves' disease, and for miR-210.3p in regulating FoxP3 mRNA in Hashimoto's thyroiditis. miR-451, miR-375, and miR-500a are associated with TSH levels, and miR-20a-3p is related to thyroglobulin antibodies in Hashimoto's disease. Whether these findings will have diagnostic or therapeutic use remains to be demonstrated.

SELF-CHECK 15.8

Figure 15.21 has features in common with Figure 15.12. How do these inform your understanding of clinical and laboratory endocrinology?

15.5.6 Summary of TSH and the thyroid

As summarized in Figure 15.21, thyroid pathology may be distilled to a number of processes. In column 1 (physiology), regulated TSH from the pituitary leads to T3/T4 release, high levels of which feed back (dashed lines) to suppress TSH levels. Column 2 represents hypopituitarism, with failure to produce TSH so that the thyroid in turn fails to produce T3/T4. In column 3, an autoantibody (Y shape) attacking the thyroid may stimulate excess levels of T3/T4, but in this illustration there is inhibition, suppression, or damage to the thyroid. Without negative feedback, levels of TSH are increased. Column 4 describes a pituitary adenoma secreting high levels of TSH, thus hyper-stimulating an otherwise innocent thyroid to secrete high levels of T3/T4. These high levels would normally feed back to suppress TSH, but in this case it fails. In column 5 a thyroid adenoma secretes high levels of T3 and T4 that feed back to the pituitary to suppress TSH.

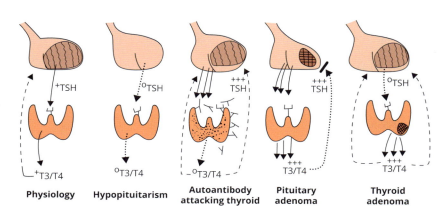

FIGURE 15.21

Key aspects of thyroid disease. Dashed lines: normal feedback. Dotted lines: failure of production or feedback. Shaded organs: normal function. Cross-hatched areas represent hormone-secreting adenomas.

CASE STUDY 15.3 Hyperthyroidism

A 52-year-old woman is taken to A&E by her son, who says that his mother has been losing weight over the past 6 months, with increasing shortness of breath and sweating, and in the past week she has developed headaches and nausea, and complains of feeling her heart pounding in her chest. The son has a home blood pressure device, which that morning recorded a blood pressure of 154/95 and a pulse rate of 134. The practitioner repeated this and found 165/102 and 157 respectively, ordering an immediate ECG. Blood analysis performed on the A&E near patient testing analyser found both CK and CK-MB to be at the top end of their reference range (the woman denied any chest pain), but normal full blood count and U&E. Bloods were sent to the laboratory for LFTs, calcium, magnesium, and thyroid function tests. The ECG found atrial fibrillation (AF), and the dangerously high haemodynamic parameters were immediately treated with three doses of 5 mg intravenous metropolol over 20 minutes, under blood pressure monitoring and ECG. These have the desired result of reducing the blood pressure and heart rate. Whilst the woman was in transit to the X-ray department for a chest X-ray, the laboratory rang back to say that the LFTs, calcium, and magnesium were all in their reference range, but that TSH was <0.1 mU/l (reference range 0.2–3.5), and fT4 was 60 pmol/l (9–23), giving a firm diagnosis of hyperthyroidism. On receipt of this result, she was noted to have mild exophthalmos and diffuse thyroid enlargement.

Hypertension is not a classical finding in hyperthyroidism, but AF is certainly a potential complication of both conditions (present in 15% of hyperthyroid cases), and hypertension is a major risk factor for AF. The theoretical question of causal aetiology of the triad therefore arises, as there is no relevant previous medical or family history. However, this is somewhat academic, as the immediate issues are the AF and hypertension, the latter being relatively easy to address, whilst fortunately (in uncomplicated AF) the heart rate would in any case be addressed with a beta-blocker. As amiodarone (as standard treatment for AF) may have thyrotoxic effects, the cardiologist advised against its use at this stage.

The patient was admitted to intensive care with a diagnosis of thyrotoxic AF under the care of a cardiologist and started on intravenous heparin (as AF brings a risk of venous thromboembolism), oral metropolol 50 mg bd, propylthiouracil 200 mg tds, and 10 drops of 2% potassium iodine tds. With this regime the patient reverted to sinus rhythm after 72 hours, and was transferred to a medical ward. With no further medical issues, she was discharged with a blood pressure/heart rate of 137/78/72 on metropolol, propylthiouracil, and oral anti-coagulation (warfarin, target INR 2–3), the latter stopped 6 weeks later. She was subsequently weaned off the beta-blocker and remained on thyroid-suppressive therapy with a TSH of 1.2 mU/l (reference range 0.2–3.5), and fT4 of 15 pmol/l (9–23). This case study is based on Parmar 2005.

Key points

Thyroid disease is so common, especially in the elderly (female), that thyroid function tests are often part of many routine early investigations, such as for type 1 diabetes or other auto-immune disease, or new-onset atrial fibrillation.

Cross reference

Further details of hypertension are described in Chapter 7, section 7.2, and of atrial fibrillation in Chapter 8, section 8.3.3.

15.6 Growth hormone and its targets

Height is of course a fascinating socio-scientific feature with a clear heritable component in that tall parents often have tall children, whilst genome-wide association studies have identified 697 SNPs at $p < 5 \times 10^{-8}$ that are related to growth. However, at a clinical-pathological level, growth factor is the only relevant factor.

15.6.1 Physiology

Also known as somatomedin, growth hormone, a 22 kDa glycoprotein coded for by *GH1* and *GH2* at 17q22–24, is secreted by somatotrophic cells, the most numerous in the anterior pituitary. Its release is regulated positively and negatively by two hypothalamic products. Growth hormone-releasing hormone (GHRH) is a 44-amino-acid peptide coded for by *GHRH* at 20p12, which, like ACTH, is secreted in a pulsatile manner, and so pituitary growth hormone release is also pulsatile. These pulses are markedly higher during sleep. Its action is opposed by growth hormone-inhibiting hormone, also known as somatostatin, a small peptide coded for by *SST* at 3q27.3. These two regulators are in turn controlled by a complex feedback mechanism involving growth hormone itself, insulin-like growth factor, sleep, exercise, and nutrition. A further regulator of growth hormone is gherlin, a product of gastric cells (section 12.2.1).

Growth hormone binds to its receptor (coded for by *GHR* at 5p13.1) on a number of target organs that include the bones, cartilage, neurons, and skeletal muscle, but primarily the liver. At the liver, the ligand–receptor interaction induces the production of insulin-like growth factors 1 and 2 (IGFs), coded for by *IGF1* (at 12p23.2, coding at 7.5 kDa product) and *IFG2* (at 11p15.5). IGF-2 is most active in the embryo and foetus, IFG-1 in the adult. Upon release from the hepatocyte, IGF-1 is carried by an insulin-like growth factor-binding protein (IGFBP), of which there are six isotypes of 24–45 kDa coded for by specific *IGFBR* variants, the most abundant being IGFBP3 coded by *IGFBP3* at 7p12.3, plasma levels being linked to (and possibly regulated by) growth hormone.

At its target cell, IFG1 binds to its receptor, coded for by *IFGR1* at 15q26.3, initiating numerous metabolic effects, the most notable being on cartilage (chondrocytes) and bone (osteoblasts), especially during puberty and shortly thereafter, but also in basic metabolic rate and increasing protein synthesis, partly by increasing amino acid influx into the cell. In the adult, IFG1 contributes to the mass of muscle and bone, and helps the repair of tissues following injury. Metabolic roles include promoting lipolysis in adipocytes, thus providing an energy source for the ATP required for growth. IFG1 also binds to the insulin receptor (coded for by *INSR* at 19p13.2), so participating in glucose homeostasis. Notably, low levels of glucose promote growth hormone release and inhibit somatostatin release, whereas high levels stimulate the hypothalamus to release somatostatin and suppress growth hormone release. Figure 15.22 summarizes major features of the physiology and cell biology of growth hormone and IGF1.

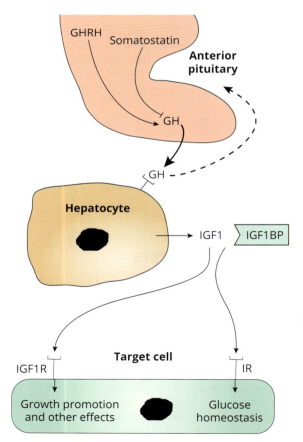

FIGURE 15.22

The physiology and cell biology of growth hormone and IGF1. GHRH: growth hormone-releasing hormone. GH: growth hormone. IGF: insulin-like growth factor. IGFBP: IGF binding protein. IGFR: IGF receptor. IR: Insulin receptor.

The polar extremes of growth hormone pathology are acromegaly/gigantism (acral: pertaining to the extremities of peripheral body parts), resulting from excessive growth hormone, and short stature/dwarfism, resulting from insufficient growth hormone.

15.6.2 Excess growth hormone

Disorders resulting from high levels of growth hormone have an incidence of 3–4/million/year and a prevalence of 65/million.

Pathology

The principal reason (>95% of cases) for excess growth hormone is a pituitary adenoma; others include lung, adrenal, or pancreatic tumours, and the multiple endocrine neoplasia syndrome (section 5.13.2). In the adult, this results in extended growth of bone, primarily of the skull, jaw, sternum, hands, and feet. Should the adenoma develop before the closure of the epiphyseal growth plates in the teenage years, there may also be extended long bones of all the limbs, leading to gigantism. X-linked acrogigantism is caused by a duplication in *GPR101* at Xq26.3.

As IGF1 has numerous metabolic roles, there may also be effects on glucose and lipid metabolism, and so type 2 diabetes and dyslipidaemia with associated risks of cardiovascular disease. The latter may be compounded by the development of biventricular cardiomegaly, which may predispose to heart failure. Should the adenoma be large, other pituitary functions may be adversely affected, the most common being mild to moderate hyperprolactinaemia, present in ~30% of patients.

Presentation

Children with excessive levels of growth hormone will be taller and also generally heavier than their peers, exceeding the respective 105th percentiles, or >3 standard deviations above the mean height for age, having controlled for the heights of the parents. Uncontrolled, this will lead to gigantism. In adults, the most specific presenting features of acromegaly include clothes no longer fitting and often an increase in weight. Patients often refer themselves when relatives point out changes in family photographs (Figure 15.23), such as an increasingly large jaw, hands, and feet (the need for ever-larger shoes), whilst practitioners may note prominent supra-orbital ridges (over the eyebrows). Other clues include the development of T2DM and hypertension, and in late disease there may be heart failure. There are also numerous non-specific signs and symptoms (Table 15.10).

Diagnosis

Major signs will call for blood tests. With its short half-life and pulsatile nature, measurement of growth hormone can be difficult, and accordingly, levels of IFG1 are preferred as levels are relatively stable with a half-life of some 15 hours. An alternative is to challenge the hypothalamus with a drink of a 75–100 g glucose solution, which should reduce growth hormone levels. A nadir serum growth hormone level <0.4 µg/l makes the diagnosis, although not all methods have this level of sensitivity, so that <1 µg/l in a random sample is often sufficient. CT and MRI are particularly useful in determining the size of an adenoma (Figure 15.19), and measurement of serum prolactin will be performed. Mean age at diagnosis is ~40 for men and ~45 for women.

TABLE 15.10 Non-specific signs and symptoms of acromegaly

Signs	Thick and/or oily skin; breathlessness; deepening of the voice; oedema; gaps between the teeth (as the jaw enlarges); acne; goitre; macroglossa; polyuria/polydipsia; enlarged lips, nose, and eyebrows
Symptoms	Headache, excessive sweating, tiredness and lethargy, menstrual irregularities, galactorrhea, sexual problems, visual disturbances, myalgia and arthralgia, obstructive sleep apnoea, carpel tunnel syndrome

1977

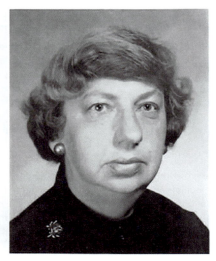

1981

1983

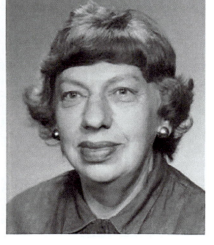

1988

FIGURE 15.23
Facial changes over time in acromegaly.

Reprinted from Molitch ME. Clinical manifestations of acromegaly. Endocrinol Metab Clin North Am. 1992: 21(3); 597–614. https://doi.org/10.1016/ S0889-8529(18)30204-4, with permission from Elsevier. ©Elsevier, 1992.

Management

The objective of treatment is to normalize growth hormone and IGF1 levels and address symptoms. As with other pituitary diseases, there are surgical (first line) and medical options (second line). The former takes the trans-sphenoidal route, but remission develops in 50–90% of cases, dependent on the size of the tumour. There is also a place for stereotactic or external beam radiotherapy (such as 25 fractions of 180 cGy each), but this targets poorly and may lead to hypopituitarism. There are three options for medical treatment:

• Somatostatin analogues such as long-acting octreotide (subcutaneous 20–40 mg at 4-week intervals) or lanreotide (60–20 mg every 28 days), which will suppress growth hormone production.

- Dopamine agonists such as cabergoline (0.5 mg od) or bromocriptine (2.5 mg at 2–3-day intervals rising to 5 mg 6-hourly). Both are also treatments for Parkinson's disease and hyperprolactinaemia (see section 15.7.1).

- Growth hormone receptor antagonism with subcutaneous pegvisomant at a loading dose of 80 mg and a maintenance dose of 10 mg od.

These generally control symptoms (tissue swelling, sweating, etc.) and arrest further bone growth, but will not address existing bone growth which is irreversible. Treatment of other disease (diabetes, hypertension) will proceed as discussed. Should obstructive sleep apnoea (in up to 70% of cases) be severe, surgery may be required to prevent airways constriction. With an increased colon length, polyps may arise, and accordingly 5-year colonoscopies are advised, as are 2-yearly cardiovascular system assessments.

15.6.3 Insufficient growth hormone

The two leading causes of short stature (dwarfism) are achondroplasia (70%) and growth hormone deficiency (GHD). In the neonate, the former is relatively easy to diagnose as the legs are often short (see section 16.2.3 of Chapter 16), but a small neonate may be the result of intra-uterine growth restriction or Turner syndrome. An alternative cause of a short stature is a mutation in, or only one copy of, the short-stature homeobox-containing gene, *SHOX*, coded for by duplication at both Xp22.33 and Yp11.32. True congenital GHD has a frequency of ~250/million live births, whereas new-onset adult GHD has a frequency of ~12/million.

Pathology

GHD has several potential aetiologies. There are numerous examples of loss-of-function mutations in *GH1*, *GH2* (both at 17q23.3, coding for growth hormone), and *GHRHR* (at 7p14.3, coding its receptor) that can lead to low or absent growth factor. As in all pituitary disease, disorders of pituitary development are linked to genes such as *POU1F1* (coding a transcription factor), and so hypopituitarism, which may be growth hormone-specific or combined pathology, which may be present in 25%. Half of those with Alström syndrome (a result of a mutation in *ALMS1* at 2p13.1) have GHD. There are also instances of growth hormone resistance (e.g. due to lesions in *GHR*), IFG-1 deficiency (*IGF-1*), and IGF1 insensitivity (*IGF1R*), all leading to short stature, and often other features such as mid-facial hypoplasia and **microcephaly**. A further cause is destruction of healthy growth hormone tissues by an adenoma of an alternative lineage.

Presentation

In the young, GHD will become increasingly suspected with growth rates persistently lower than those expected for the age. Adults are likely to complain of reduced exercise capacity and strength, increasing weight, and psychological issues such as depression and mood swings.

Diagnosis

A child's growth will be compared to that of standard height and weight charts. As with acromegaly, a blood test is the first investigation, where low or absent levels of growth hormone and/or IGF-1 µg/l are expected. A further test is the glucose challenge, which is negative in GHD, and the allied insulin challenge test that in health will also lead to an increase in growth hormone (typically >6.7 mg/ml), with stunted responses in partial GHD (3–6.7 ng/ml) and severe GHD (<3 ng/ml).

Management

Treatment with subcutaneous recombinant human growth hormone (rGH; somatotropin) at 23–39 µg/kg/day in children should start as soon as possible, in order to minimize the time needed to reach a target height. This treatment adds some 3 to 11 cm in height compared to untreated children.

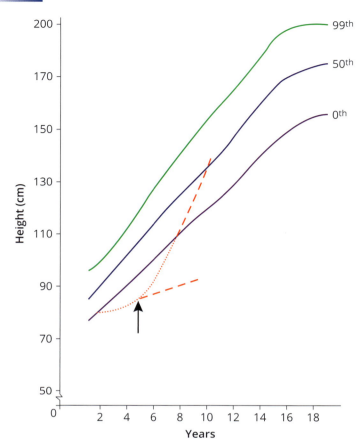

FIGURE 15.24

Male growth curves plotting height against age. The three lines are the expected growth patterns expressed as percentiles. The dotted red line represents a child with GHD, the arrow the point at which rGH was started. The dotted line thereafter reflects change in growth pattern and the dashed red lines extrapolation on treated and untreated growth patterns.

In transitioning to adulthood, a dose of 200–500 µg/day is appropriate, whilst those with adult-onset GHD may start on 150–300 µg/day, which is then adjusted according to patient response and IFG-1 levels. For those children in whom rGH is inappropriate, perhaps with a loss-of-function mutation in *GHR*, 40 µg/kg recombinant human IGF-1 may be injected subcutaneously twice a day, rising to a maximum of 120 µg/kg twice a day (Figure 15.24).

Although children born small for gestation age show catch-up within 2–3 years, those 10% that do not may be treated with rGH having passed their fourth birthday with 35 µg/kg/day. Similarly, girls with Turner syndrome may need rGH at 45–50 µg/kg/day, and those also given the steroid oxandrolone gain an additional 3 cm. Those whose growth has been compromised by chronic renal insufficiency would also be prescribed rGH at 45–50 µg/kg/day (NICE TA188: Human growth hormone (somatomedin) for the treatment of growth failure in children).

SELF-CHECK 15.9

What are the major presenting features of excessive levels of growth hormone in children and in adults?

Key points

The pathology of GH excess and deficiency are relatively easy to diagnose and treated with surgery/chemotherapy and replacement therapy respectively.

15.7 Prolactin and melanocyte-stimulating hormone

15.7.1 Prolactin

Physiology

Prolactin is coded for by *PLR* at 6p13, which generates three isoforms of 22 kDa, 48 kDa, and 150 kDa. Its receptor (*PRLR* at 5p13–p12) also binds to growth hormone and human placental lactogen, and is highly expressed by lactocytes, but also by several regions of the brain and reproductive tract, and by hair follicles. During pregnancy, receptor signalling drives the expansion of lactocyte clusters/alveoli and then milk production, as well as increased neurogenesis in the maternal brain. Prolactin is part of a collective that includes insulin, thyroxine, growth hormone, oestrogen, progesterone, and oxytocin that all act together to prepare the mammary glands for lactation, and to ensure milk flows when demanded by the infant. In the male, prolactin upregulates LH receptors on Leydig cells, thus promoting testosterone secretion and spermatogenesis.

The hypothalamus regulates prolactin by stimulatory (prolactin-stimulating hormone) and inhibitory (prolactin-inhibiting protein, also known as dopamine) signals. This inhibition wanes towards the end of the luteal phase of the menstrual cycle, so that the levels of prolactin rise, possibly explaining breast tenderness prior to menstruation. Rising oestrogen and progesterone levels in pregnancy stimulate prolactin production, and post-partum stimulation of the nipple inhibits the release of prolactin-inhibitory hormone. There is also evidence of interaction between prolactin and kisspeptin.

Pathology

A prolactin-secreting adenoma is the primary disease, leading to hyperprolactinaemia, which may have several consequences. They are the most common (35%) pituitary adenoma, with a prevalence of 100/million men and 300/million women. Many have mutations in *PTTG* at 5q33.3 (coding for pituitary tumour-transforming gene-1, with potential links to basic fibroblast growth factor expression) and *FGF4* at 11q13.3 (coding for fibroblast growth factor 4). In perhaps 10% of cases, there is co-secretion of growth hormone, whilst a further cause of hyperprolactinaemia is hypothyroidism—so this should be explored. Common iatrogenic causes include drugs that antagonize dopamine, such as metoclopramide and domperidone (both used to treat nausea and vomiting), and most antipsychotics.

Presentation

Hyperprolactinaemia is linked to several forms of reproductive disease, including infertility, menstrual and fertility disorders (by inhibiting the synthesis and release of LH and FSH, and so ovulation), erectile dysfunction, gynaecomastia, galactorrhoea, and loss of libido. In common with other adenomas, there may be weight gain, cranial nerve palsies, visual disturbance, and headaches. On further investigation there may be hydrocephalus, oligospermia (due to secondary hypogonadism), and osteoporosis. In children and teenagers there may be pubertal delay and growth arrest.

Diagnosis

Serum prolactin >5,000 mU/l is almost always due to prolactinoma, whilst a result of <1,000 mU/l is often associated with PCOS. Reference ranges are higher in women than in men (Table 15.15). MRI (with gadolinium) may be helpful in detecting adenomas, as it is in the investigation of all pituitary tumours. Given the complex nature of endocrine biology, many feel a comprehensive investigation of all hormones should be performed (especially T3/T4/TSH), in addition to visual field testing.

Management

In contrast to other pituitary adenomas, many practitioners consider a medical approach to be the first line of treatment, unless the tumour is compressing other tissues. Two dopamine agonists, cabergoline (as in the treatment of Cushing's disease/syndrome, acromegaly, and gigantism) and bromocriptine (also as in acromegaly and gigantism), may be used. Should medical treatment be insufficient, surgery and/or radiotherapy may be needed.

Hypoprolactinaemia

The primary cause of low serum prolactin is hypopituitarism; others include over-treatment of hyperprolactinaemia and anti-psychotics such as aripiprazole. Lack of prolactin leads directly to reproductive issues, principally failure to produce milk, but also erectile dysfunction, reduced ejaculate and premature ejaculation, and lack of maternal behaviour and of sexual arousal. A mutation in *IGSF1* at Xq26.1 leads to hypoprolactinaemia, but also to hypothyroidism and occasional GHD.

15.7.2 Melanocyte-stimulating hormone (MSH)

Physiology

The three (α, β, and γ) forms of MSH derive from a single protein product of *POMC* at 2p23. Although MSH has a major role in amphibian skin, its precise function in our species is unclear beyond a darkening of the skin, but the large amount of MSH receptors in the brain imply cerebral functions. There is evidence that MSH may have anti-inflammatory properties, with MSH receptors being present on leukocytes, and possible roles in eating disorders, being described as anorexigenic.

Pathology

In comparison to other endocrine diseases, the pathology of MSH is minor and brings little discomfort. Perhaps the only major aspect is that it causes the hyperpigmentation seen in Addison's disease. The mechanism involves increased pituitary processing of the ACTH precursor pro-opiomelanocortin following adrenal failure, leading to the excessive levels of MSH. Type 1 MSH receptors are overexpressed on melanoma cells, suggesting possible therapeutic targeting in this cancer, whilst technetium[99]-labelled MSH has been used to detect micrometastases. A consequence of erythropoietic protoporphyria (section 11.1.3) is skin damage following exposure to the sun (sunburn), which may be treated by subcutaneous synthetic MSH such as a 16 mg subcutaneous implant of afamelanotide.

SELF-CHECK 15.10

Describe the differences in the roles of prolactin in men and women.

15.8 Non-pituitary endocrinology

We conclude our exploration of endocrine pathophysiology by considering hormones whose production and activity do not directly involve the pituitary. Principal amongst these are the islets of Langerhans, producing insulin, glucagon, and somatostatin, described in section 7.4.1 of Chapter 7. The kidney has two major endocrine functions. In section 11.1.1 we discussed the importance of erythropoietin in red blood cell production, and how chronic kidney disease can lead to anaemia. The role of the kidney in part-regulating blood pressure via renin is described in sections 7.2 and 14.1.1, whilst intestinal hormones are described in section 12.3 and Table 12.6.

The standard pregnancy test relies on the production of human chorionic gonadotropin (hCG, a 37 kDa dimer, coded for by *CGA* at 6q14–21 and by *CGB* at 19q13.3) by the syncytiotrophoblast that promotes the corpus luteum and so the production of progesterone. The placenta also secretes progesterone, oestrogen, and placental lactogen, also known as chorionic somatomammotropin, a 22 kDa dimer coded for by *CSH1* and *CSH2* at 17q22–24 (as is growth hormone). It binds the prolactin receptor (hence mammo-) but also promotes foetal growth and nutrition. Other hormones that remain can be classified with relative ease.

15.8.1 The parathyroids

The parathyroids are of crucial importance for the integrity of bone, in which 99% of our calcium (equivalent to 1 kg), 85% of our phosphates, and 55% of our magnesium are found. These tissues participate in the homeostasis of these minerals by secreting parathyroid hormone (PTH). The major clinical feature of parathyroid disease is abnormal levels of blood calcium.

Anatomy and physiology

The four (very rarely, five) parathyroid glands (each with a mass of ~40–70 mg) are located in the posterior superior (upper) and inferior (lower) aspect of the lobes of the thyroid gland. The major ('chief') cell within each gland generates and secretes PTH, an 84 amino acid polypeptide of ~9.5 kD coded for by *PTH* at 11p15.3. It acts on its target cells via a specific receptor of which there are two types: *PTHR1* coded for at 3p21.31 and expressed primarily on bone and renal cells, and *PTHR2* coded for at 2q34, present on testicular, placental, pancreatic, and brain cells.

PTH has a number of functions: it acts as the major regulator principally of blood calcium (as Ca^{2+}), but also of phosphates (a mixture of HPO_4^{2-} and $H_2PO_4^-$) and magnesium (Mg^{2+}). This it achieves in several independent but linked processes:

- In the bone, PTH increases the number and activity of bone-degrading osteoblasts, thus increasing local and blood calcium and phosphates.

- In the kidney it slows the rate at which the three ions pass into the urine. As more phosphates are lost in the urine than are gained by bone resorption, this effectively reduces blood phosphates and increases the calcium and magnesium.

- PTH decreases the renal reabsorption of bicarbonate, leading to a more acidic environment that raised blood calcium.

- An additional role is in vitamin D metabolism: it promotes the formation of the active form of the vitamin, 1,25-dihydroxycalciferol (also known as calcitriol) by the kidneys. Calcitriol promotes the absorption of calcium (and phosphates) in the intestines and from the renal tubules, promotes bone resorption by osteoblasts, and increases levels in the blood (see Box 15.1).

BOX 15.1 Vitamin D

This vitamin is unique in its class as it is not linked solely to diet, and is synthesized by a complex pathway involving several organs:

1. 7-dehydrocholesterol from the liver and ergosterol from the diet are converted to cholecalciferol (vitamin D_3) and ergocalciferol (vitamin D_2) respectively by UV light in the skin

2. These two metabolites, supplemented by minor intakes from the diet, are converted to 25-hydroxycholecalciferol in the liver.

3. In the kidney, and under the direction of PTH, there is further hydroxylation by 1-α-hydroxylase, resulting in the formation of 1,25-dihydroxycalciferol (calcitriol).

Section 12.4 discusses nutritional and pathological aspects of vitamin D, where the failure of this pathway leads to hypocalcaemia and potential bone disease. NICE PH56 discusses supplementation of vitamin D in specific population groups.

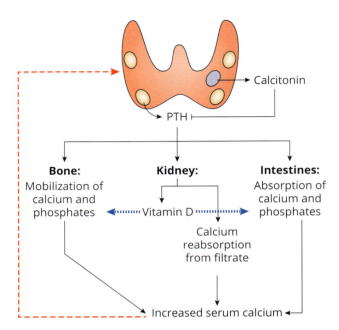

FIGURE 15.25
Physiology of PTH. Dashed lines: feedback. Dotted lines: additional influences of vitamin D.

Levels of blood calcium are regulated principally by negative feedback at the parathyroid gland by calcium-sensing receptors (coded for by *CaSR* at 3q13.3–q21.1) that, via second messengers, instruct the exocytosis of PTH-containing vesicles. Hypercalcaemia suppresses the secretion of PTH, whereas hypocalcaemia increases its release. A second level of regulation is the secretion of calcitonin from the parafollicular C cells in the thyroid, which effectively counters the effects of PTH, such as in reducing calcium absorption by the intestines and suppressing osteoclast activity. However, low and high levels of magnesium may also (respectively) stimulate or inhibit PTH release. Figure 15.25 summarizes the physiology of PTH.

Pathology

Hyperparathyroidism, characterized by raised PTH, can be primary, secondary, or tertiary—these three together have an incidence of ~450/million with a female:male ratio of around 3:1. As so often, the leading cause of a primary hyperparathyroidism (in 80% of cases) is a single adenoma, often linked to second-messenger kinase *PIK3CA* at 3q26.32 (mutations in which are also present in <33% of breast cancers) and in *CCND1* at 11q13.3, coding for cyclin D1. Parathyroid hyperplasia is present in 15%, multiple adenomas in 4% and a carcinoma in 1%. Parathyroid tumours may also be part of the multiple endocrine neoplasia syndrome.

These tumours secrete excess levels of PTH, with effects on calcium and other ions that may be life-threatening. Loss-of-function mutations in *CaSR* are known, and so is failure to sense and therefore regulate calcium. When heterozygous, these are asymptomatic and generally do not impact on PTH or plasma calcium, but homozygotes suffer from type I familial hypocalciuric hypercalcaemia (FHH) due to unregulated PTH release, underlining the role of feedback. Type II FFH is caused by a mutation in *GNA11* at 19p13.3, coding for a guanine nucleotide-binding factor. These patients also have low magnesium. Very rare mutations in *PTH* result in decreased bioactive PTH, hypocalcaemia and hyperphosphataemia.

The leading cause of secondary hyperparathyroidism is renal disease, the second vitamin D deficiency, and these are often combined. Failure to resorb sufficient calcium from the tubular filtrate will lead to hypocalcaemia, the response of the parathyroids being to secrete excessive PTH to rectify the deficiency, with one consequence being calcium leaching out of bone. Increased PTH may also follow bariatric surgery and gastrointestinal disease. If untreated, the prolonged hypocalcaemia can lead to tertiary hyperparathyroidism, characterized by autonomous PTH production that is independent of

Cross reference

The diagnosis, assessment, and management of hyperparathyroidism is discussed in NICE NG132.

BOX 15.2 Parathyroid hormone–related protein (PTHrP)

Although coded for on a different chromosome (12p11.22), the 20 kDa 170 amino-acid peptide PTHrP has striking homology with PTH, especially in the region that binds *PTH1R*, and so also promotes hypercalcaemia. It was so named following its discovery in certain cancers, where it was linked to increased serum calcium, but where the parathyroids were unaffected. We now know that it has many physiological roles, such as in embryogenesis, endochondral bone growth, epithelial–mesenchymal interactions, keratinocyte differentiation, beta cell proliferation and insulin production, and mammary gland development. Furthermore, many of its fragments have biological activity (such as in cardiomyocyte contractability and in promoting osteogenesis) and arise from malignancies. A microdeletion of 12p, which includes PTHrP, causes **brachydactyly**, and is characterized by short stature and learning difficulties.

calcium, with raised PTH, calcium, and phosphates. The PTH analogue, parathyroid hormone-related protein (PTHrP), is discussed in Box 15.2.

Hypoparathyroidism, characterized by a low PTH, has a prevalence of ~300/million, the most common (~57%) aetiology being neck surgery, others being damage caused by infections and inflammatory/autoimmune disease. Milk-alkali syndrome is characterized by the ingestion of large amounts of calcium, often with an absorbable alkaline (hence milk) that leads to hypercalcaemia and so low PTH. Pseudohypoparathyroidism is a rare (prevalence 7/million) condition caused by a mutation in *GNAS1* located at 20q13.32 and coding for a G-protein second messenger, such that the parathyroids fail to respond to plasma calcium.

Presentation

The leading clinical features of hyperparathyroidism and hypoparathyroidism are generally those of hypercalcaemia and hypocalcaemia (Table 15.11). For example, lacking the PTH to recover calcium from urinary filtrate, patients with hypoparathyroidism may present with the product of hypercalciuria, that is, renal calcium-based stones, present in ~22% of cases.

Purpura and petachiae are often the result of hypocoagulopathy as calcium is a required cofactor in the coagulation cascade (section 11.2.3), whilst certain cases of cardiac arrhythmias such as a prolonged Q–T interval can be life-threatening and possibly call for calcium infusions (section 8.3.4). Trousseau's and Chvostek's signs are accounted for by muscular hyperexcitability, whilst bone changes may be attributable to loss of calcium and so structural weakness that may bring spontaneous fractures. However, there are numerous other causes of abnormal calcium, as indicated in Table 15.12, which therefore may be differential diagnoses for parathyroid disease. Vitamin D deficiency may lead to the bone diseases rickets and osteomalacia, both linked directly to hypocalcaemia and so a secondary hyperparathyroidism.

TABLE 15.11 Signs and symptoms of calcium pathology

	Hypercalcaemia	Hypocalcaemia
Signs	Nausea, vomiting, anorexia, renal stones, muscle atrophy, twitching and/or tremor, fractures, kyphosis, coma	Cataracts, petechiae, purpura, convulsions, cardiac arrhythmias, tetany, bronchial and laryngeal spasm, Trousseau's and Chvostek's signs
Symptoms	Abdominal pain, constipation, polyuria, polydipsia, muscle weakness, bone pain, apathy, depression, irritability, memory loss	Parasthesia (especially around the mouth), depression, memory loss, hallucinations, anxiety

Modified from Ahmed N (ed.), *Clinical Biochemistry*, second edition, Oxford University Press, 2016.

TABLE 15.12 Causes of abnormal blood calcium unrelated to PTH

Hypercalcaemia	Hypocalcaemia
Excessive vitamin D uptake, acromegaly, malignancy (such as myeloma and bone metastases), Paget's disease, hyperthyroidism, lymphoma, Addison's disease, thyroxicosis, long-term immobility	Vitamin D deficiency, chronic kidney disease, acute pancreatitis, rhabdomyolysis, neck surgery, malabsorption (e.g. inflammatory bowel diseases), hyperphosphataemia, hypomagnesaemia, hypoalbuminaemia
Drugs: thiazide diuretics, lithium vitamin A	Drugs: calcitonin, bisphosphonates

Modified from Ahmed N. (ed), *Clinical Biochemistry*, second edition, Oxford University Press, 2016.

Diagnosis

The key investigations are the laboratory and imaging, following which fine-needle aspiration under ultrasound direction may be performed. Microadenomas may be detected by MRI, and for further sensitivity, technetium[99]-methylisobutylisonitrile (sestamibi) is an option. Alongside PTH, measurement of calcium is very important. Perhaps 50% of calcium is carried by albumin, and accordingly the calcium result may need to be corrected should levels be low, such as <40 g/l, or high, such as >45 g/l. Alternatively, free ionized calcium can be determined by an ion-specific electrode. Of further value is alkaline phosphatase, elevated levels of which indicate severe bone disease, whilst serum angiotensin-converting enzyme may be helpful in the diagnosis of sarcoid, in which increased calcitriol drives hypercalcaemia and therefore low PTH (section 15.8.1). Pseudohypoparathyroidism is characterized by high PTH and phosphates alongside low calcitriol and calcium, whilst in some circumstances assessment of calcium in the urine can be informative.

Management

Parathyroidectomy is the treatment of choice for an adenoma, which in most cases is curative, the remaining parathyroid tissues adjusting their responses. However, in many instances there is a rebound hypocalcaemia, hypomagnesaemia, and hypophosphataemia due to rapid and extensive bone rebuilding—described as the hungry bone syndrome—such that calcium supplements may be necessary. The hypercalcaemia of malignancy will be a matter for physicians and surgical oncologists, and hypercalcaemia >3.5 mmol/l should be treated promptly with vigorous rehydration (4–6 l immediately followed by 34 l/day for several days), intravenous bisphosphonates (such as pamidronate 60–90 mg in saline), prednisolone (30–60 mg od), intravenous calcitonin (100 units tds/qds), and oral phosphates.

One of the problems with treating the hypocalcaemia of hypoparathyroidism is that calcium may be rapidly lost in urine. Thiazide diuretics have been shown to decrease urinary calcium, and so increase blood levels, thus reducing the need for calcium and vitamin D supplements (often a synthetic analogue such as alfacalcidol), and sodium restriction may help. Many patients and their practitioners find replacement therapy with (typically) 100 μg of recombinant PTH(1–84) delivered subcutaneously on alternate days to be acceptable (despite e.g. musculoskeletal and gastrointestinal side effects), not only in biochemical terms, but also in bone mineral density of the lumbar spine. Acute, symptomatic hypocalcaemia (e.g. <1 mmol/l) must be treated urgently with 10 ml intravenous 10% calcium gluconate in dextrose, with additional boluses as recovering blood levels demand.

Figure 15.26 summarizes key features of the pathology of the parathyroids.

Phosphates

The total body store of phosphates is around 500–800 g and circulates in the plasma as free anions (~80%), bound to proteins such as albumin (~15%) and in a complex with calcium and magnesium. Blood levels depend on a balance between absorption, movement in and out of bone, and losses in the faeces (10%) and in the urine (90%), where it is reabsorbed by a sodium/phosphate co-transporter.

Levels of this transporter are upregulated by low phosphate and by vitamin D, thereby increasing phosphate reabsorption, but reabsorption is inhibited by PTH, hypercalcaemia, and

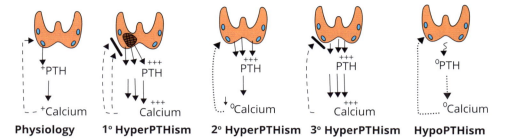

FIGURE 15.26

A summary of the pathology of the parathyroids. The first column from the left shows the physiology of PTH and calcium. Column 2 shows an adenoma over-secreting PTH and so driving hypercalcaemia. Negative feedback is ineffective. In column 3, low levels of calcium, perhaps due to renal disease or vitamin D deficiency, fail to feed back to the parathyroid so that levels of PTH rise. In column 4, the parathyroid is refractive to hypercalcaemia, and secretes PTH autonomously. In column 5, the gland fails to secrete PTH, leading to hypocalcaemia that in turn leads to a failure to feed back.

hyperphosphataemia. It is also part-controlled by fibroblast growth factor 23 (FGF23), at ~29 KDa peptide coded for by *FGF23* at 12p13.32 and secreted by osteoblasts, osteocytes, and bone marrow erythroid precursors. In renal tubules, FGF23 binding its receptors leads to downregulation of sodium/phosphate transporters and so phosphaturia, and it also suppresses vitamin D processing, all leading to a reduction in serum phosphates.

The most common cause of hyperphosphataemia is renal failure, where it is an issue in end-stage renal disease, but it can also be caused by drugs such as thiazides, furosemide, steroids, and penicillin; by a high ingestion of phosphates (typically, laxatives); and by increased vitamin D. Underlining the complexity of endocrinology, increased serum phosphates may be present in acromegaly, hypoparathyroidism, and thyrotoxicosis, and result in a hypocalcaemia. Intercellular levels of phosphates are relatively high (as is potassium) so that widespread cell lysis (of tumour cells by chemotherapy and of red blood cells in haemolysis) also leads to raised plasma levels. Long-term hyperphosphataemia can result in calcification of blood vessels and heart valves, being a product of calcium and phosphate, and renal osteodystrophy. Mutations in FGF23 and/or its receptor lead to a form of hyperphosphataemic familial tumoural calcinosis, which manifests as skin and subcutaneous nodules of hydroxyapatite and/or calcium carbonate, **hyperostosis**, inflammation, and ocular disease. Treatment of hyperphosphataemia is with dialysis and intestinal phosphate binders such as sevelamer (one or two 800 mg tablets, bd/tds, depending on plasma levels).

Hypophosphataemia has three leading aetiologies: insufficient phosphate in the diet and/or malabsorption, increased urine phosphate (possibly driven by raised PTH), and movement of phosphates from plasma into cells (as in hungry bone syndrome and re-feeding for malnutrition). Leading risk factors include diabetic ketoacidosis, poor nutrition, alcoholism and sepsis, and the excessive use of phosphate binders and antacids. Patients with profoundly low phosphates are likely to present with neurological symptoms (ranging from muscle weakness and paraesthesia to seizures) and an altered mental state, and prolonged deficiency leads to bone disease such as osteopenia and osteomalacia. Treatment is with oral replacement therapy, such as 30–80 mmol of phosphate od, titrated against serum levels.

Key points

The biochemistry of calcium, phosphates, magnesium, PTH, vitamin D, and calcitonin is complex and overlapping, and all involve the kidneys and bone. For example, low levels of phosphates and high levels of calcium are linked to a fall in magnesium levels.

Magnesium

This electrolyte is an essential cofactor for up to 300 enzymes, is a component of bone, and is required for nerve and muscle function, but also in vitamin D metabolism. Most plasma magnesium (~55%) is the free ionized form, around a third is carried by albumin, the remainder being complexed with citrate and phosphate. It is not directly regulated by a particular hormone, but by reabsorption from renal tubules, as are phosphates. Causes of low levels include dietary deficiency, an important issue as it has been estimated that 60% of adults fail to achieve their recommended daily allowance (men 420 mg/day, women 320 mg/day (360–400 mg/day if pregnant)), and absorption is reduced in vitamin D deficiency. Medications linked to reduced serum magnesium include proton pump inhibitors and H_2 blockers (leading to a more alkaline intestine), antibiotics, and diuretics (causing increased loss in urine). Reabsorption of magnesium from renal tubules is inhibited by alcohol and in both forms of diabetes.

Like hypocalcaemia and hypokalaemia, hypomagnesaemia is often present in ventricular arrhythmias and tachycardias such as **torsades de pointes** with prolongation of the Q–T interval. This is reflected in other clinical features, such as of the central nervous system (seizures, depression, migraine, agitation, and psychosis), neuromuscular system (the signs of Chvostek and Trousseau, muscle weakness, tetany, tremor, and dysphagia) and intestines (nausea and vomiting). Low serum magnesium also reduces levels of PTH.

Hypermagnesaemia may be caused by excessive intake of over-the-counter products, lithium carbonate, antidepressants such as sertraline and amitriptyline, and potassium-sparing diuretics such as amiloride and spironolactone. It may also be present in hypothyroidism and Addison's disease. Signs and symptoms include muscle weakness, cardiac arrhythmias, dizziness and falls due to hypotension, lethargy, and confusion. Treatments should address the cause and may include intravenous fluids with a diuretic and/or calcium gluconate or calcium chloride. Table 15.13 summarizes features of phosphates and magnesium.

15.8.2 The pineal gland

This small body lies to the rear of the brain, just above the cerebellum on the opposite side of the brain stem to the pituitary, as part of the epithalamus (Figure 15.27). Its function is to secrete melatonin, an indoleamine derivative of serotonin, which plays a part in setting the body's biological clock, and the sleep-wake cycle, more being secreted in darkness and when asleep. Accordingly, it has been viewed by some as a treatment for sleep disorders. One study showed that subjects fell asleep 7 minutes earlier and stayed asleep 8 minutes longer whilst others reported 27 minutes and 20 minutes respectively in children with attention deficit hyperactivity disorder (NICE ESUOM2).

TABLE 15.13 Features of phosphates and magnesium

	Phosphates	Magnesium
Mass in bone (mmol)	17,000	750
Causes of high levels	Increased oral intake and vitamin D, reduced renal loss, hypoparathyroidism, acromegaly, increased cellular release	Increased intake, use of laxatives and antacids, reduced renal loss, hypothyroidism
Clinical features of high levels	Tetany, calcification, renal osteodystrophy	Respiratory difficulties, hypotension, arrhythmia, nausea, vomiting
Causes of low levels	Decreased intake, vitamin D deficiency, increased renal loss, primary and secondary hyperparathyroidism, increased cellular intake, alcoholism	Decreased intake, malabsorption, increased renal loss, hypercalcaemia, hypoparathyroidism, increased gastrointestinal loss
Clinical features of low levels	Parasthaemia, ataxia, muscle weakness, osteomalacia	Ataxia, depression

Modified from Ahmed N. (ed.) *Clinical Biochemistry*, second edition, Oxford University Press, 2016.

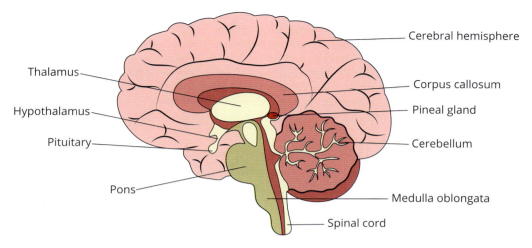

FIGURE 15.27
Cross section of the brain, showing location of the pituitary, hypothalamus, and pineal glands.

Others point to possible use in shift-workers and in post-traumatic brain disorders: doses of 0.1–10 mg are reported. A dose of 3 mg daily for 6 days is reported for jet lag and 2 mg od for primary insomnia in the over-55s, whilst NICE NG71 suggests considering melatonin in patients with Parkinson's disease who have rapid eye movement sleep behaviour disorder.

Hypofunction of the pineal gland is exceptionally rare, being linked to *PAX6* at 11p13, coding a protein involved in brain formation. Pineal tumours comprise 0.5% of all intracranial tumours, with symptoms of headaches, nausea, vomiting, ataxia, and visual disturbances; treatment is via surgery. MRI studies have shown small pineal glands in those with anorexia nervosa, depression, attention deficit disorder, psychosis, and chronic migraine.

15.8.3 Adipocytes

Adipocytes are not merely fat-storage bodies. These cells also secrete leptin (which increases satiety, free fatty acid oxidation, and insulin sensitivity) and adiponectin (which also increases free fatty acid oxidation and insulin sensitivity, but reduces hepatic glucose production). There is a link with fertility, overweight, and underweight as these cells can synthesize oestrogens from androgens. Whilst section 7.4 refers to the two hormones in type 2 diabetes, section 12.4.2 describes their roles in digestion and adiposity.

Other adipocyte hormones include resistin (as 12 kDa peptide coded for by *RETN* at 19p13.2, also secreted by monocyte/macrophages), with a role in insulin sensitivity, and retinol-binding protein-4 (a 21 kDa peptide coded for by *RBP4* at 10q23.33), which is also produced by the liver and promotes adipose tissue inflammation, so contributing to insulin resistance. Other hormones with minor links to adipocytes and obesity include visfatin, omentin/intelectin-1 (which both increase insulin sensitivity and are expressed by many other cells), and chemerin (which negatively regulates FSH-induced follicular steroidogenesis and has roles in adipocyte differentiation). The adipocyte is also of interest to immunologists as it secretes TNF-α, IL-6, IL-8, IL-17, and TGF-β.

15.9 The laboratory in endocrinology

The laboratory is an essential tool in the diagnosis and management of all endocrine diseases. In some cases whole-blood glucose can be measured, but in general the preferred sample is serum: the local laboratory will advise. Most protein hormones are measured by immunoassay, whilst mass spectrometry is often used to measure steroid hormones. In many cases national quality-control organizations have standardized results, but there is variability according to particular methods, so that local reference ranges must be consulted. Reference ranges for hormones and related molecules are shown in Tables 15.14 (not sex variable) and 15.15 (sex variable).

Reference ranges may vary with analytical technique, time, and geographical location. As with all reference ranges, the practitioner must work to their local guidelines. Some vary with body mass index, ethnicity, and race. Which reference range is most appropriate for transgender men and women depends on whether they receive gender-affirming care that affects hormone levels; if they receive hormone therapy, this generally targets the reference ranges of their affirmed gender.

TABLE 15.14 Reference ranges for sex-neutral hormone levels

Analyte		Reference range
ACTH		5–20 pmol/l
ADH		1–5 pmol/l
Adrenaline		<140 pg/ml
Cortisol	At 09.00 hours	140–690 nmol/l
	At 24.00 hours	80–350 nmol/l
Dehydroepiandrosterone		1–5 nmol/l
Dopamine		<30 pg/ml
Gastrin		<40 pmol/l
Glucagon (fasting)		<50 pmol/l
GH		<10 mU/l
Insulin-like growth factor (IGF-1)		90–292 ng/ml
IGF-1-binding protein		2.2–7.9 µg/ml
Insulin (fasting)		2–10 mU/l
Noradrenaline		70–1,700 pg/ml
Metanephrine		<99 pg/ml
Pancreatic peptide		<100 pmol/l
Parathyroid hormone		1–6 pmol/l
Somatostatin		<150 pmol/l
Tetra-iodothyronine (T4)	Free	10–25 pmol/l
	Total	60–160 nmol/l
Tri-iodothyronine (T3)	Free	4.0–6.5 pmol/l
	Total	1.2–2.3 nmol/l
TSH		0.2–3.5 mU/l
T4 binding globulin		1.2–3.0 mg/dl
Vasoactive intestinal peptide		<30 pmol/l
Vitamin D/calcitriol (as 25(OH)D)		>75 nmol/l

TABLE 15.15 Reference ranges for sex-specific hormone levels

Androstenedione	Female	2–8 nmol/l
	Male	2–12 nmol/l
Calcitonin	Female	<4.5 ng/l
	Male	<11.5 ng/l
Dehydroepiandrosterone sulphate	Female	3–8 μmol/l
	Male	2–10 μmol/l
FSH	Follicular phase	4–13 IU/l
	Mid-cycle	5–22 IUl
	Luteal phase	2–8 IU/l
	Post-menopausal	26–135 IU/l
	Male	1–11 IU/l
LH	Follicular phase	3–13 IU/l
	Mid-cycle	14–96 IU/l
	Luteal phase	1–11 IU/l
	Post-menopausal	8–59 IL/l
	Male	1–8 IU/l
Oestradiol	Female pre-menopausal	100–1,500 pmol/l
	Female post-menopausal	50–150 pmol/l
	Male	50–200 pmol/l
Progesterone	Follicular phase	<5 ng/ml
	Mid-cycle	5–20 ng/ml
	Luteal phase	10–50 ng/ml
	Post-menopausal	<4 ng/ml
	Male	<1 ng/ml
Prolactin	Female	103–497 mU/l
	Male	86–324 mU/l
Sex hormone-binding globulin	Female	40–120 nmol/l
	Male	20–60 nmol/l
Testosterone	Female	1–3 mmol/l
	Male	10–30 nmol/l

SELF-CHECK 15.11

The spellings of melanin and melatonin are similar. Does this reflect their pathophysiology?

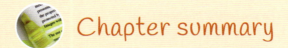

Chapter summary

- The endocrine system is composed of a complex collection of organs and molecules that influence many metabolic, homeostatic, and reproductive processes.

- Levels of many hormones are controlled by negative feedback of the hypothalamus and pituitary, the latter having an anterior and posterior lobe. A leading cause of increased levels is an adenoma, which may be treated surgically.

- The anterior lobe of the pituitary secretes FSH, LH, ACTH, TSH, GH, prolactin, and MSH.

- The posterior lobe of the pituitary secretes oxytocin and ADH.

- The adrenal gland consists of three zones in the cortex (secreting aldosterone, cortisol, and androgens) and a medulla (secreting adrenaline and noradrenaline).

- Excess production of cortisol leads to Cushing's disease and syndrome, whilst failure of its production leads to Addison's disease.

- Failure of the pituitary to secrete FSH and LH leads to hypogonadism, failure to produce gametes, and so infertility.

- Insufficient TSH at the thyroid, or failure of the entire gland, leads to hypothyroidism (the leading clinical syndrome being myxoedema), whilst high TSH levels and inflammation/autoimmune disease of the gland leads to hyperthyroidism (often accompanied by exophthalmos).

- GH stimulates levels of insulin-like growth factor, which in high levels leads to gigantism in the young and in acromegaly in the adult. Low levels in the young lead to short stature.

- Prolactin promotes mammary gland development and stimulates milk production, but increased production leads to fertility problems in both sexes. Increased levels of MSH result in a darkening of the skin.

- PTH is an important regulator of blood calcium, acting on bone, the intestines, and the kidneys. Increased levels drive hypercalcaemia, although this may also arise from other pathology, whilst low levels lead to low blood calcium.

Further reading

- Higham CE, Johannsson G, Shalet SM. Hypopituitarism. Lancet 2016: 388; 2403–15.

- Tomkins M, Lawless S, Martin-Grace J, et al. Diagnosis and Management of Central Diabetes Insipidus in Adults. J Clin Endocrinol Metab 2022: 107; 2701–15.

- Vindenes T, Crump N, Casenas R, et al. Pheochromocytoma causing Cardiomyopathy, Ischemic Stroke and Acute Arterial Thrombosis: A Case Report and Review of the Literature. Connect Med 2013: 77; 83–5.

- Dagnistani MAH, Dagnistani MAZ, Dagnistini MAM, et al. Relevance of *KISS1* Gene Polymorphisms in Susceptibility to Polycystic Ovary Syndrome and Its Associated Endocrine and Metabolic Disturbances. Br J Biomed Sci 2020: 77; 185–90.

- Barthel A, Benker G, Berens K, et al. An update on Addison's disease. Exp Clin Endocrinol Diabetes 2019: 127; 165–70.

- Nishioka H, Yamade S. Cushing's disease. J Clin Med 2019: 8; 1951; doi:10.3390/jcm8111951.

- Dong YH, Fu DG. Autoimmune thyroid disease: mechanism, genetics and current knowledge. Eur Review Med Pharm Sci 2014: 18; 3611–18.

- Bacuzzi A, Dionigi G, Guzzetti L, et al. Predictive features associated with thyrotoxic storm and management. Gland Surg 2017: 6; 546–51.

- Chaker L, Blanco AC, Jonklaas J, et al. Hypothyroidism. Lancet 2017: 390; 1550–62.

- Adelman DT, Liebert KJP, Nachtigall LB, et al. Acromegaly: the disease, its impact on patients, and managing the burden of long-term treatment. Int J Gen Med 2013: 6; 31–8.

- Bilezikian JP, Bandeira L, Khan A, et al. Hyperthyroidism. Lancet 2018: 391; 168–78.

Useful websites

- International Diabetes Federation: https://idf.org/

- National Institute for Health and Care Excellence: https://www.nice.org.uk/

- Addison's Disease Self-Help Group: https://www.addisonsdisease.org.uk/

- Endocrine Society: https://www.endocrine.org/ (Many excellent guidelines.)

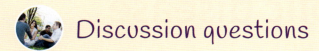 Discussion questions

15.1 What are the major causes and consequences of diseases of the adrenal gland?

15.2 Why should we take an interest in the Müllerian ducts?

16

Chromosomal, genetic, and metabolic disease

As introduced in Chapters 1 and 2, disease may be caused by factors external to the body (such as tobacco smoke, a diet lacking certain vitamins, and microbial pathogens), or by factors within the body, caused by abnormal genes. Many of these diseases have been discussed in previous chapters, such as haemophilia in Chapter 11 and acromegaly in Chapter 15. The purpose of this chapter is to describe those non-neoplastic, non-immunological diseases and syndromes that arise from pathology of the chromosomes and the genes within them. An estimated 7 million babies (around 10% of all global births) are born each year with either a congenital abnormality or a genetic disease, errors that can be classified by the size of the molecular lesion. On a large scale, these are changes in the numbers of entire chromosomes; on a modest scale, changes of sections of DNA within and between chromosomes, and small changes in the sequences of nucleotides in a section of DNA. There may also be changes to the external aspects of DNA, described as epigenetics.

This chapter will focus first (section 16.1) on links between different types of chromosome abnormalities and disease (such as Down's syndrome). The following four sections focus on specific gene defects: those of a single gene causing disease in a single organ or organ system (16.2); a single gene causing disease in several organs/organ systems (16.3); multiple genes causing disease in a single organ or organ system (16.4); and multiple gene defects causing disease in several organs or organ systems (16.5). Metabolic disease caused by gene defects will be discussed in section 16.6, and the chapter will conclude in section 16.7 by examining the most common global form of genetic disease—haemoglobinopathy.

Learning objectives

After studying this chapter, you should confidently be able to:

- explain the structure and classification of chromosomes, and instances where errors in chromosome numbers or sections lead to disease

- understand how one or more genetic abnormalities can lead to disease in a single or in multiple organs

- provide examples of how genetics influences metabolism

- describe the pathophysiology of haemoglobinopathy as a genetic disease

16.1 Chromosomes

The UK's Office for National Statistics (ONS) reports deaths due to congenital malformations, deformations, and chromosomal abnormalities that together were linked to 1,309 deaths in England and Wales during 2021. Of these, the biggest groups were in those aged under 1 year (160 deaths) and in those aged 60–64 (164 deaths). But, as elsewhere, these data ignore the considerable morbidity these diseases bring. However, before we look at chromosomal disease in detail, it is prudent to revisit key features of these macromolecules.

16.1.1 Chromosome numbers and structure

Chromosomes are complex macromolecules formed of chromatin, a complex of DNA and proteins called histones. The latter are globular proteins, of which there are five families, H1–H5. DNA coils around an octamer of pairs of H2A, H2B, H3, and H4, whilst a single H1 molecule provides support, and as a result the 2 metres of DNA are packaged more efficiently. The combination of a section of 146 base pairs of DNA and the eight histones it wraps around is a nucleosome (see Figure 3.11).

Enclosed tightly within the nucleus, chromosomes cannot easily be visualized. However, individual chromosomes can be viewed when they are 'free' in the cytoplasm when the cell is in the metaphase stage of the cycle, where each chromosome has been replicated. Here, replicated chromosomes appear to be an 'X' shape (n.b. this has nothing to do with the X chromosome that defines sex), comprising two sister chromatids linked at the middle by the centromere. A common misconception is that this 'X' is the true representation of a chromosome that is resting, or is involved in protein synthesis, whereas in reality the chromosome is a single long molecule.

Chromosomes are numbered by size, and may be classified sequentially in groups of two to seven with a letter of the alphabet, giving, for example B4, B5, and B6. The correlation between the number of base pairs in each chromosome and the number of genes it contains is unsurprisingly very good, with a correlation coefficient of 0.84. Individual chromosomes can also be identified by the location and size of certain bands when stained by dyes. The most common staining is with the dye Giemsa (hence G-bands); another is with quinacrine (hence Q-bands). The dyes are taken up with greater or lesser affinity in those regions whose local nucleotides differ in that highly condensed heterochromatin tends to be rich in adenine and thymine, and stains darkly with Giemsa, whereas loosely packed euchromatin, dominated by guanine and cytosine, stains poorly, giving bands of light and heavy staining. These bands allow a numbering system that in turn provides for further identification of sites of a particular sequence. When collected together, the 46 chromosomes, arranged in their 23 pairs, are represented as a karyogram (Figure 16.1), whilst a karyotype is a shorthand of the chromosomal make-up, e.g. 46XY.

It has long been established that much of the DNA in the chromosomes fails to code for messenger RNA, and more recently non-coding RNA has been discovered. Sections of coding DNA are described as exons, sandwiched between intervening non-coding sections, hence introns. Other non-coding sections are the telomeres (at the ends of each chromosome) and centromeres (sections involved in mitosis). The telomere is characterized by 300–800 repeats of the sequence TTAGGG, which form a loop, and is effectively a tie-off of the ends of the chromosome. These repeats are shortened by 50–200 base pairs per cell cycle, a process that creates a barrier to malignancy as it limits the lifespan of transformed cells.

Contrary to its name, the centromere may not always be in the middle of the chromosome, as there is variation in its position. In some chromosomes, it lies in the centre of the entire macromolecule (metacentric), but in others it is asymmetrical, being slightly offset (submetacentric) or markedly offset (acrocentric). It follows that there may be 'long' and 'short' sections of the chromosome either side of the centromere. This difference helps to give information about the location of a particular section of DNA, where the short arm is designated p (from the French for small—petit) and the long arm designated q (as q follows p in the alphabet).

The individual G-bands themselves, and their particular region, can be numbered, starting from the centromere, and moving towards the telomere. Bringing these systems together provides the means to localize a gene to a given position on the chromosomes. For example, the location of the section of genes coding for certain haematological products can be mapped to locations within chromosome 1 (Figure 16.2).

Cross reference

Read more about the correlation coefficient and how to interpret it in section 2.4.3.

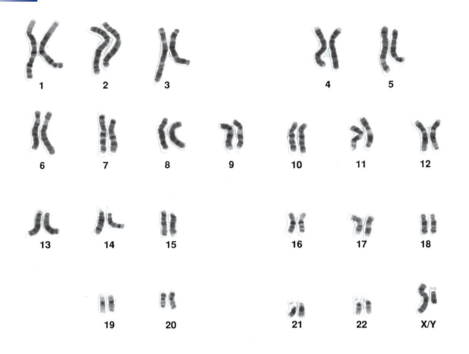

FIGURE 16.1

Karyogram of the human chromosome complement. The numbering system goes from left to right, top to bottom. On the bottom right are the unpaired X and Y chromosomes; all others are autosomes.

Courtesy of National Human Genome Research Institute, Public domain, via Wikimedia Commons.

Mitochondria contain their own DNA (mtDNA) not as a single strand bounded by telomeres, but as a circle. Of the 37 genes, 13 code for subunits of ATP synthase, cytochrome C oxidase, and NADH dehydrogenase; 22 code for a transfer RNA; two code for ribosomal RNA subunits; and one codes for a humanin, a 21-amino-acid peptide believed to have a role in protecting from Alzheimer's disease and having anti-apoptotic activity. Table 16.1 summarizes these aspects of chromosomes.

We will address the pathology of chromosomes and genes according to a spectrum, first considering changes in whole numbers of chromosomes, and then changes with and between chromosomes, concluding the spectrum with microchanges to the nucleotide sequences of particular genes. The latter may be as few as two or three nucleotides, but the most frequent are changes in a single nucleotide, hence single-nucleotide polymorphism, or SNP, of which there are estimated to be 11 million instances in the genome, one per 300 base pairs.

In some cases, a SNP will not result in a change in protein structure (e.g. the codons AAA and AAG both code for lysine), but in others, the SNP (e.g. leading to codon AAT) will result in a different amino acid (asparagine), and so probably a different protein product. The average number of SNPs per chromosome is around 59,000, ranging from 130,000 in chromosome 1 to 4,200 in the Y chromosome.

Mutations are not only found in the DNA nucleotide sequence: epigenetics considers other changes to chromosomes that are on the outer surface of the macromolecule. In many cases these are changes to the exposed section of DNA, often a methylation, and in others there are changes to the histones of the nucleosome.

Cross reference

For a closer look at epigenetics, see section 3.8.4.

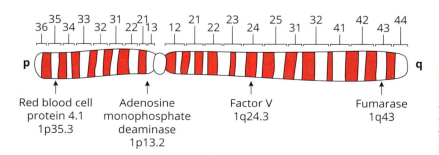

Red blood cell protein 4.1
1p35.3

Adenosine monophosphate deaminase
1p13.2

Factor V
1q24.3

Fumarase
1q43

FIGURE 16.2

Location of certain genes on chromosome 1. A representation of the structure of chromosome 1 showing the location of four genes. The size and position of the bands is for illustrative purposes only.

TABLE 16.1 Selected features of chromosomes

Chromosome and group		Centromere position	Chromosome size (million base pairs)	Number of genes	Example of gene/disease link
1	A	Metacentric	243.5	3,000	Glaucoma
2	A	Submetacentric	241.4	2,500	Colon cancer
3	A	Metacentric	199.2	1,900	Lung cancer
4	B	Submetacentric	190.6	1,600	Achondroplasia
5	B	Submetacentric	180.4	1,700	Asthma
6	C	Submetacentric	170.4	1,900	Epilepsy
7	C	Submetacentric	154.4	1,800	Cystic fibrosis
8	C	Submetacentric	143.1	1,400	Lymphoma
9	C	Submetacentric	135.2	1,400	Tangier disease
10	C	Submetacentric	132.7	1,400	Refsum disease
11	C	Submetacentric	134.5	2,000	Diabetes
12	C	Submetacentric	131.6	1,600	Phenylketonuria
13	D	Acrocentric	112.1	1,200	Wilson's disease
14	D	Acrocentric	103.1	1,200	Alzheimer's disease
15	D	Acrocentric	100.1	1,200	Marfan syndrome
16	E	Metacentric	89.4	1,300	Crohn's disease
17	E	Submetacentric	79.3	1,600	Breast cancer
18	E	Submetacentric	73.1	600	Pancreatic cancer
19	F	Metacentric	61.9	1,700	Haemochromatosis
20	F	Metacentric	61.7	900	ADA deficiency
21	G	Acrocentric	43.5	400	Down's syndrome
22	G	Acrocentric	45.0	800	Neurofibromatosis
X	Sex	Submetacentric	152.5	1,400	Haemophilia
Y	Sex	Acrocentric	53.8	200	Infertility
Mitochondrial		–	0.016	37	Myopathy

ADA = Adenosine deaminase.

Data sources: Genes and Disease, National Center for Biotechnology Information (US), https://www.ncbi.nlm.nih.gov/books/NBK22266/, and Jackson M, Marks L, May GHW, et al. The genetic basis of disease. Essays Biochem. 2018: 62; 643–723.

SELF-CHECK 16.1

Explain the meaning of 15p25.5.

16.1.2 Errors of chromosome number

Despite the checks for errors during the cell cycle, chromosomes can be incorrectly duplicated, leading to a deviation from the correct number of chromosome (46: euploidy) with too many (47 or more) or not enough (up to 45) chromosomes. Both of these deviations are described as aneuploidy. The principal cause of this abnormality is failure of the chromosomes to separate correctly at metaphase/anaphase, leading to non-disjunction (Figure 16.3).

At the metaphase plate, the pair of sister chromatids separates at the centromere, and each is drawn to one of the poles of the replicating cell. Should these sisters fail to separate, both may be drawn to one pole, and none to the other pole, leading to one cell with 47 chromosomes and another with 45. Common causes of aneuploidy include agents known to damage DNA and cells (i.e. that are cytotoxic) including X-rays, tobacco smoke, pesticides, and other industrial chemicals.

Non-disjunction generally happens in meiosis or very soon after fertilization and is present in all cells of the body. However, it may also occur at a later stage on embryogenesis, when it results in only some of the body cells being aneuploidic and others being euploidic, leading to mosaicism. Individuals with these abnormalities of aneuploidy generally reach adulthood but may be affected by an assortment of physical and mental problems. It has been speculated that there may in fact be many cases of aneuploidy that fail to implant and so are not reported. Trisomy 16, for example, is lethal and the affected foetus aborts spontaneously.

Turner syndrome

When a chromosome is missing, the lesion is a monosomy, as in cell D in Figure 16.3, and so is a hypoaneuploidy. The best known and only major example of this is Turner syndrome, where only one copy of the X chromosome is present in women, so that their sex chromosomes can be represented as XO, giving an overall genotype as 45XO (Figure 16.4). With a frequency of around 290/million live female births, practitioners in the delivery suite may well recognize certain features in the neonate, and several others develop over childhood to prompt the blood test for karyotyping. Signs and symptoms include a short stature, a thick neck with skin folds to the shoulders, a small jaw, elbow deformity, widely spaced nipples, a low hairline at the back of the head, short fingers, low-set ears, and lymphoedema (Figure 16.5).

Many abnormal skeletal features can be attributed to a mutated short-stature homeobox gene (*SHOX*) at Xp22.33. In adolescence there may be delayed and/or low-grade puberty with hypogonadism and no or minimal menstruation. In adulthood there may be cardiological issues with conduction and repolarization abnormalities, atrial and ventricular septal defects, a bicuspid aortic valve (it should

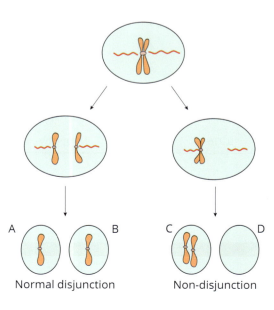

A   B   C   D

Normal disjunction   Non-disjunction

FIGURE 16.3

Non-disjunction. The top represents the alignment of a pair of sister chromosomes (a chromatid) at the metaphase plate. Those on the left separate and move towards each pole of the cell in anaphase, resulting in one chromosome in each of the daughter cells A and B. On the right, non-disjunction has occurred, and both chromosomes are drawn towards one pole, resulting in both being in daughter cell C, and none in daughter cell D.

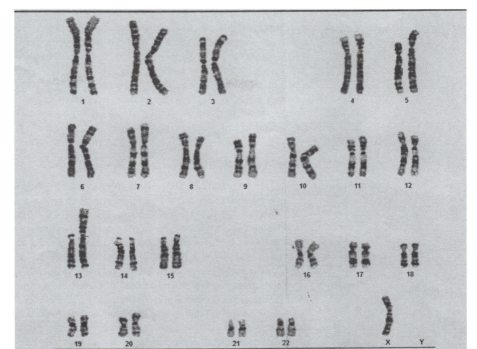

FIGURE 16.4

Karyogram of Turner syndrome. Note that there is only one X chromosome.

Mousavi S, Amiri B, Beigi S, et al. The value of a simple method to decrease diagnostic errors in Turner syndrome: a case report. J Med Case Reports 15: 79; (2021). https://doi.org/10.1186/s13256-021-02673-0. Licensed under Creative Commons Attribution 4.0 International License, http://creativecommons.org/licenses/by/4.0/.

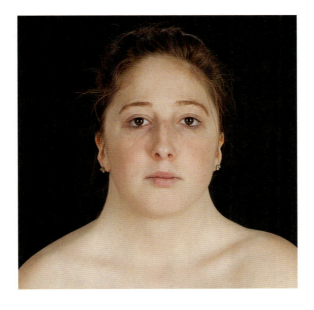

FIGURE 16.5

Turner syndrome. Note the neck webbing to the shoulders and low-set ears.

Science Photo Library.

be tricuspid), and aortic **coarctation**. Together, these contribute to nearly 50% of the excess morbidity in this disease, requiring regular monitoring, whilst there is an increased frequency of autoimmune disease (such as Crohn's disease) and **scoliosis**. Ultrasound may report a single horseshoe kidney. The short stature can be treated with growth hormone, such as 0.35–0.55 μg/kg/day, whilst puberty can be induced with oestrogen/progesterone replacement therapy, possibly with transdermal patches of 17-β oestradiol at around age 11–12, starting at ~5 μg/day, rising over 2–3 years to 100 μg/day, or in an oral form. Neck webbing is amenable to surgery.

Although 45XO monosomy is the most common and severe karyotype of Turner syndrome (~50%), less severe phenotypes based on a mosaic (45,X/46,XX, ~25% of cases) and other structural abnormalities are recognized, notably isochromosome X, being two copies of the q arm, hence iXq. An important differential diagnosis is Noonan syndrome, which has a frequency of 500/million, and which in both sexes also brings a mild variable phenotype of short height, widely spaced eyes, low-set ears, a small jaw, and a short neck. There may also be congenital heart disease, renal anomalies, lymphatic malformation, and coagulopathy. The genetic aetiology is pathogenic variants of *PTPN11* (at 12q24.1, coding for a protein tyrosine phosphatase, and the most common at 50% of cases), others being in *SOS1* (2p21, whose product phosphorylates guanidine diphosphate, 10%), *RAF1* (2p25, coding a serine/threonine protein kinase, 10%) and *KRAS* (12p12.1, coding an enzyme that dephosphorylates guanidine triphosphate, <5%).

Down's syndrome

This trisomic abnormality and probably best-known chromosomal disease results from the formation of an extra copy of chromosome 21 (hence it is also known as G21 trisomy) leading to the genotype 47XX and 47XY (Figure 16.6). With a global frequency of ~1,100/million—in some populations it may reach 1,800/million—the incidence rises steadily with the age of the mother: 700/million at age 20, 1,040/million at 30, jumping markedly to 12,000/million at age 40 and 23,000/million at age 50. It has been estimated that 50–75% of foetuses with Down's syndrome are lost before term. The most frequent age of death for people with Down's syndrome is 60–64 years.

The neonate may exhibit **hypotonia**, upslanted fissures of the eyelid, folds at the back of the neck, a single palmar flexion crease, and clinodactyly (curvature) of the fifth finger. Later in life the phenotype develops into a short stature with characteristic facial features (often with slanted eyes, a large tongue and flat bridge of the nose, so that the eyes seem far apart (Figure 16.7)), overweight/obesity, congenital heart disease, obstructive sleep apnoea, and endocrine disease (often of the thyroid).

Later in life there is an increased (2–5%) risk of leukaemia (particularly acute myeloid, due to mutations in *GATA1*). The learning disability is likely due to a lack of grey and white matter within the central nervous system, which is in turn linked to four overexpressed genes, one of which, *APP*, at 21q21.3, codes for amyloid precursor protein, the major protein responsible for Alzheimer's disease. The frequency of dementia rises with age to >50% after age 40. With good care, life expectancy can be at least 70 years.

An in-utero sign, detectable by ultrasound, is nuchal translucency (a collection of fluid under the skin of the foetal neck), whilst maternal biochemical markers include low oestradiol, high human chorionic gonadotropin (hCG), high inhibin A, low maternal alpha feto-protein, and low pregnancy-associated plasma protein-A (PAPP-A). Together, these markers (often described as a Bart's, or triple, test) give a score which may be interpreted to estimate the likelihood of a foetus having Down's syndrome. For full confirmation, amniocentesis and chorionic villus sampling are required, whilst

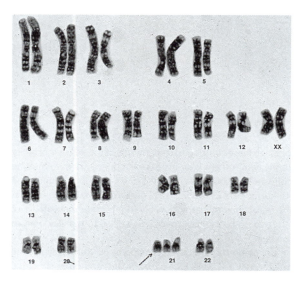

FIGURE 16.6

Karyogram of Down's syndrome. Note that there are three copies of chromosome 21.

Credit: Down's syndrome karyotype 47,XX,+21. Wessex Reg. Genetics Centre. Attribution 4.0 International (CC BY 4.0).

FIGURE 16.7

Down's syndrome.

Shutterstock.

estimation of foetal cell-free DNA in the maternal circulation is a developing field that offers a weighted pooled detection rate of >99% and a false positive rate of <0.1% for trisomy 21.

SELF-CHECK 16.2

Explain non-disjunction.

Klinefelter syndrome

A further trisomy is an extra X chromosome in men (hence XXY, Figure 16.8), described as Klinefelter syndrome, which has a relatively high frequency of 1,000/million. The phenotype is effectively normal and so this figure is likely to be an underestimate: some suggest that the true rate is 1,500/million, although there is considerable variation by geography. Some 86% of those with the syndrome exhibit 47XXY and mosaics comprise 6%, with the remainder having 48XXXY, 48XXYY, and 49XXXXY with diminishing frequency but increasingly severe abnormalities. There may also be 47,iXq,Y, a structurally abnormal X isochromosome.

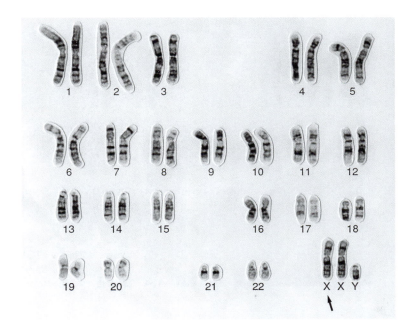

FIGURE 16.8

Karyogram of Klinefelter syndrome. Note that there are two X chromosomes and a Y chromosome.

Credit: Klinefelter's syndrome karyotype 47,XXY. Wessex Reg. Genetics Centre. Licence: Attribution 4.0 International (CC BY 4.0).

Klinefelter syndrome may only be suspected at puberty and adolescence as most young men are tall, but a major sign is small testes. Others include poor growth of body hair, low muscle mass, gynecomastia in the absence of overweight/obesity (possibly driven by oestradiol formed from aromatase), **hypertonia**, elbow dysplasia, and speech or language difficulties. There may also be hypogonadism with low testosterone (accounting for low libido and erectile dysfunction), raised luteinizing hormone and follicle-stimulating hormone, and non-obstructive **azoospermia**. Accordingly, adults may be diagnosed as part of sub-fertility investigations, such that long-term testosterone-replacement therapy may be effective.

Key points

The most common abnormalities of chromosome numbers are Turner (45XO), Down's (G21 trisomy), and Klinefelter syndromes (47XXY), all caused by non-disjunction.

Other trisomies

Jacobs syndrome is defined by 47XYY (Figure 16.9) and is present at the same frequency as Klinefelter syndrome at 1,000/million live male births. Most (72%) are true trisomics, whilst 9% are mosaics. With a median age of presentation around age 17, variable phenotypic features include large testes, a tall stature, an abnormally large head, and an increased distance between the eyes. Patients are also at risk of asthma, autism, seizures, and infertility, although those who are fertile can father normal children.

The karyotype of trisomy Triple X syndrome is 47,XXX, and it also presents with a frequency of ~1,000/million, 53% being true trisomics, 29% being mosaics. The phenotype may be only of a tall female, but many also have any combination of a risk of learning disabilities; delayed development of speech, language, and motor skills; weak muscle tone; microcephaly; eyes being far apart (hypertelorism); clinodactyly (bent fingers); behavioural and emotional difficulties; seizures; and kidney abnormalities.

Trisomy of chromosome 18 defines Edwards' syndrome, and is associated with low maternal AFP, low oestradiol, low hCG, normal inhibin A, and low PAPP-A. With a frequency of 200/million, key features include neurological pathology, growth restriction, and numerous organ malformations that bring a median survival of 10 days. A differential diagnosis of Edwards' syndrome is Patau syndrome, a trisomy of chromosome 13 which also has a frequency of 200/million live births. Survival is also very poor, with a median of 7–10 days, but with aggressive medical and surgical treatment, this can be extended to two years.

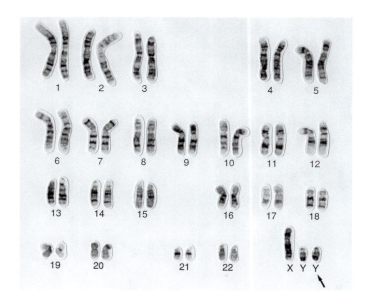

FIGURE 16.9

Karyogram of Jacobs syndrome. Note that there are two copies of the Y chromosome.

Credit: XYY syndrome karyotype 46,XYY. Wessex Reg. Genetics Centre. Licence: Attribution 4.0 International (CC BY 4.0).

16.1.3 Errors of sections of chromosomes

Section 3.5 describes changes to the structure of chromosomes (translocations, inversions, duplications, and deletions) that lead to various forms of solid organ tumours, whilst section 11.4 describes these changes in blood cancers such as leukaemia and lymphoma. Several types of abnormalities are known (Figure 16.10).

Translocations

Three forms of translocation are relevant: those that swap sections between two different chromosomes, and so are reciprocal; a form where two homologous chromosomes fuse; and a third where two different chromosomes fuse (Figure 16.10). The latter two cases are described as Robertsonian translocations, generally restricted to acrocentric chromosomes, and where the centromere is often lost. This translocation causes ~7.5% of cases of Down's syndrome, such as when 21q is translocated onto chromosome 14, giving t(14q21q) and mosaics. De la Chapelle syndrome is characterized by the translocation of part of the Y chromosome responsible for determining male sex (*SRY*) onto an X chromosome, resulting in a 46,XX male. With a frequency of ~45/million, the phenotype is essentially male but there is impaired sexual development at puberty.

Inversions, duplications, and insertions

Generally, inversions are relatively rare, perhaps the most common being that which causes the haemorrhagic disease haemophilia (section 16.2.1); another is an inversion in *CCM2* at 7p13 that causes cerebral cavernous malformations (as do deletions), generally of blood vessels, which may cause a stroke when damaged or ruptured.

Numerous cancers display duplications of a variety of genes, a common abnormality being of *MYC* in breast, cervical, ovarian, and oesophageal cancers. Duplications in 16p11.2 have been described in schizophrenia. They are also relatively common in Duchenne muscular dystrophy (~10% of cases, generally in exons 2–10) (see below), and are the cause of myotonic dystrophy, with mutations in *DMPK* at 19q13.32 or in *CNBP* at 3q21.3. However, there are numerous instances of the physiological duplication of genes, an excellent example being those of haemoglobin, which undoubtedly arose from a single ancestral gene (section 16.7.1).

A small number of nucleotides may be added (i.e. inserted) into a sequence following incorrect repair by polymerases, leading to changes such as a frameshift mutant (e.g. where a SNP insertion of

FIGURE 16.10

Abnormalities within and between chromosomes. Translocation (a) is a standard reciprocal, (b) is homologous Robertsonian, and (c) is heterologous Robertsonian. Deletion (a) is of an internal section, (b) is of a terminal arm. Insertion (a) is of a large section of DNA, as may occur in a translocation, but in pathology may be insertion of alien DNA such as from a virus, and (b) is a small number of nucleotides incorrectly inserted by repair or other mechanisms.

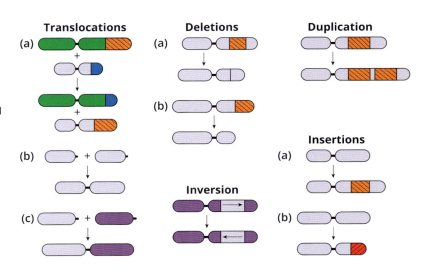

an adenosine (A) residue into ATC/GGC/ATA/GGA gives ATC/GGA/CAT/AGG, and so an alternative amino acid sequence) or a stop codon (such as TAG, TAA, or TGA). Several diseases are characterized by the insertion of a variable number of sets of three nucleotides (such as CAG, CTG, CGG, and GAA) that range from 39 to several thousand copies and are described as repeat expansion disorders. The major pathological insertion of nucleic acids by a retrovirus such as HIV is a further example, whilst genetic engineers/editors may attempt to deliberately insert semi-artificial DNA by site-directed mutagenesis and CRISPR/Cas-9 technology into, for example, a stem cell in order to correct a genetic disease.

Deletions

There are hundreds, perhaps thousands, of rare deletions resulting in a variety of diseases. These include cri du chat syndrome (French for 'cat-cry', as neonates have a malformed larynx), with a frequency of 20/million live births and an aetiology of deletions around 5p15.2–15.3. The 1p36 deletion syndrome (generally from 1p36.13–36.33 to the telomere), with a frequency of ~130/million, brings numerous physical and mental abnormalities. Some cases of Turner syndrome result from a partial deletion in the X chromosome (e.g. Xp and Xq), and with cri du chat syndrome, are classified by some as monosomies.

Other deletions include those in 7q11.23 that lead to Williams syndrome, present at a frequency of ~70/million; Weiss–Kruszka syndrome, resulting from a deleted *ZNF462* (coding for a zinc finger protein) in 9p31.2; Malan syndrome, caused by loss of *NFIX* (encoding a nuclear factor) and *CACNA1A* (coding a membrane calcium channel) from 19p13.2–13.12; and schizophrenia, with deletions in 15q11.2 and 22q11.2.

Other diseases with a chromosomal component include cystic fibrosis (section 16.3.1), whilst abnormalities in *PMP22* at 17p12 that codes for peripheral myelin protein 22 are unusual because a deletion causes hereditary neuropathy with liability to pressure palsy, and a duplication causes an alternative neuropathy: Charcot–Marie–Tooth type 1A disease. Deletions cause also two other notable diseases: muscular dystrophy and DiGeorge syndrome.

SELF-CHECK 16.3

How would you classify the chromosome abnormalities in Turner, Down's, and Klinefelter syndromes?

Muscular dystrophy

Muscular dystrophy is an umbrella term for a small number of diseases that present with a frequency of ~250/million and with a broadly similar clinical and pathophysiological background, that is, poor muscle function in all major groups, as damaged myocytes are replaced by fibrosis and adipocytes. This leads to immobility and eventually to dilated cardiomyopathy and heart failure, the most common cause of death. The root issue is with dystrophin, a 427 kDa protein whose main function is to maintain the stability, strength, and flexibility of muscle fibres. The 427 kDa dystrophin protein is coded for by *DMD* at X-21.2–21.1, and of length 2.2 Mbp with 79 exons, making it the largest known human gene (~0.1% of the genome). Consequently, it is vulnerable to mutation, mostly (75%) of the deletion of one or more exons, the remainder being SNPs, the nature of which defines the form of the disease.

Loss of dystrophin results in myocyte damage with necrosis and apoptosis, and so increased serum markers creatinine, imidazole acetic acid, isohomovanillic acid, creatine phosphokinase, aspartate aminotransferase, alanine aminotransferase, and lactate dehydrogenase, with reduced creatine and guanidinoacetic acid. These may have a role in discriminating different forms of the disease. Increased expression of certain non-coding RNAs and increased serum miRNAs may provide tools for diagnosis, management, and treatment.

Duchenne muscular dystrophy (DMD, ~200/million male births) is the most frequent form of the disease, generally presenting before the age of 5 with abnormal gait, delayed speech, and increasing loss of muscle function and scoliosis culminating with the need for 24-hour ventilation and so a life expectancy of 30–40 years. Most disease (75% of cases) is caused by a deletion spanning one or more exons, causing premature stop codons and so a truncated product. Disease severity depends on the particular lesion: those flanking exon 44 cause a slightly milder phenotype, whereas loss of the first 10 exons leads to a severe phenotype. SNPs may also introduce a premature stop codon or nonsense mutations, and although the genetic lesion can be heritable, in a third of cases it is due to a *de novo* mutation.

Becker muscular dystrophy (BMD, ~50/million) has a much slower rate of disease progression than DMD, and spans a spectrum ranging from the need for a wheelchair by the third decade to remaining ambulant at 60. The leading (~80%) genetic lesion is again a deletion, the remainder being SNPs and other changes, and these generally maintain the reading frame so that the deficiency in dystrophin (and the disease in general) is less severe. Exon deletions generally bring a relatively mild phenotype, which may be asymptomatic.

Treatments of muscular dystrophy aim to retain muscle strength (perhaps with steroids such as prednisolone up to 40 mg od, or deflazacort, up to 36 mg od) and to reduce the development of cardiomyopathy (often with angiotensin-converting enzyme inhibitors). The sheer size of the gene is a barrier to effective viral gene replacement therapy.

DiGeorge syndrome

DiGeorge syndrome is caused by deletions in 22q11–13, ranging from 1.5 mB (~4% of cases) to 3 Mb (~90%), a key absent gene being *TBX1* at 22q11.21, which codes for a transcription factor active in embryonic development, particularly of the pharyngeal pouches. Also described as velocardiofacial syndrome, it presents in 250/million, although prenatal screening suggests this may be as high as 1,000/million pregnancies, implying a high rejection rate *in utero*. The disease is phenotypically variable. Major manifestations are malformations of the bones and structures of the chest (cardiac abnormalities such as of the aortic arch, and Fallot's tetralogy in 75%) and neck (that cause difficulty with speech and swallowing, thyroid and parathyroid disease with consequent hypocalcaemia, present in 35%). Thymic aplasia, present in ~60%, can lead to the absence of T lymphocytes and so a form of immunodeficiency with recurrent infections (section 10.1.1).

In common with other chromosomal disease (such as Down's syndrome), there is developmental delay, learning difficulties, and characteristic changes to the face and mouth. The latter include a cleft palate, pharyngeal incompetence, a bulbous nose, hooded eyelids, a small mouth and jaw, ear anomalies, and a prominent nasal bridge (Figure 16.11). Renal abnormalities are present in 14% of cases, whilst the deletion also removes a number of miRNA and lncRNAs species, the full significance of which is as yet unclear.

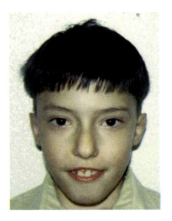

FIGURE 16.11
DiGeorge syndrome. This phenotype has a bulbous nose and a small jaw.

Digilio M, Marino B, Capolino R, et al. Clinical manifestations of Deletion 22q11.2 syndrome (DiGeorge/Velo-Cardio-Facial syndrome). Images Paediatr Cardiol. 2005 Apr: 7(2); 23–34. PMID: 22368650; PMCID: PMC3232571. © Images in Paediatric Cardiology.

> ## Key points
>
> **The major large intra- and inter-chromosomal changes are translocations and deletions, causing disease such as muscular dystrophy and DiGeorge syndrome.**

16.2 Single-gene effects on a single organ or organ system

We now move from macro-changes to entire chromosomes and large section of chromosomes, to micro-changes to sequences of DNA, which range from one (i.e. a SNP) to a handful of nucleotides. All changes can be described as sporadic (i.e. with no clear indication such as family history or exposure to a toxin) or inherited, which may be dominant or recessive, autosomal or sex-linked, or mitochondrial. A full spectrum of the number of mutated genes and the extent of the resultant pathology can be described. Some mutations produce changes that affect only a single organ, or several organs and perhaps organ systems. Conversely, certain well-characterized conditions of a single organ or organ system may be brought about by several genes, whilst other diseases may involve multiple organs and be influenced by more than one gene (Figure 16.12). We will work through this model as follows: Section 16.2 will focus on single genes working on single organs/systems, section 16.3 on single genes working on several organs/systems, section 16.4 on several genes working on a single organ/system, and section 16.5 on several genes working on several organs/systems. Metabolic disease is considered in section 16.6.

16.2.1 Haemorrhagic disease

Loss of haemostasis leads to thrombosis or to haemorrhage, perhaps the most well-known example of the latter being haemophilia. Caused by mutations in *F8*, a very large gene of 186 Kbps and 26 exons at Xq28 (very close to the telomere), this disease has a frequency of 200/million males, although, due to X-chromosome **lyonization**, it may also appear in females. With over 150 mutations defined, many are SNPs, small deletions, or inversions, but over half are inversions in intron 22.

Considerably more frequent is von Willebrand's disease (vWD), which is present in 10,000/million and caused by mutations in *VWF* at 12p13.31, a gene of ~175 K base pairs. This generates a very large monomeric molecule (~234 kDa) that polymerizes to give a plasma protein that may have a mass of 1,000 kDa. Such a large gene is naturally the target for many different mutations (mostly SNPs and small–moderate deletions) that (unlike haemophilia) lead to considerable heterogeneity in presentation and management. Despite the high prevalence, most of those with vWD have relatively mild disease, the most severe variant (whose phenotype resembles haemophilia) being present in only 100/million. The gene coding for Factor IX (*F9*) at Xq27.1 may house SNPs and small microsatellite inserts, both causing a defective mature molecule and so haemorrhage. It presents at a frequency of ~20/million. Clinical and laboratory aspects of these coagulopathies are described in section 11.6.4.

16.2.2 Huntington's disease

This late-acting condition of the nervous system, generally asymptomatic before the age of 35, is caused by a mutation (two thirds being one or two SNPs) in *HTT* at 4p16.3, normally coding for 350 kDa protein huntingtin. The exact function of this molecule is unknown, but it certainly has a role in neurons, where it is highly expressed. Most mutations in *HTT* that are associated with the disease have a large number of extra CAG nucleotide repeats, giving rise to an abnormal protein that accumulates inside the cell, causing disruption.

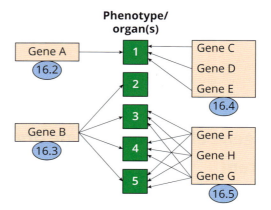

FIGURE 16.12

Forms of genetic disease. Phenotypes/organs(s) include the coagulation system, the nervous system (including the brain; with cognitive and/or motor/sensory disease), the eyes, the immune system, the skeleton, the muscles, the endocrine system (such as the adrenals), the reproductive system, and metabolic disease. Numbers 1–5 refer to different potential types of disease (blood, skeletal, endocrine, etc.).

With a prevalence of ~100/million, clinical features include problems with motor skills (such as dyskinesis/ataxia, hence chorea), cognitive decline, and psychiatric changes (depression, anti-social behaviour, risk of suicide) that are progressive and which lead to death a median of 16.5 years after diagnosis. Treatments are aimed at symptoms. Dyskinesis may be addressed with neuroleptics halo-peridol and olanzapine, or benzodiazapines, tetrabenazine, and valproic acid. In England and Wales, 129 women and 132 men died from this disease in 2021.

16.2.3 Achondroplasia

The various forms of fibroblast growth factor (FGF) are macrophage-derived cytokines acting on their target cells (principally fibroblasts, chondrocytes, and osteoblasts) via a tyrosine-kinase recep-tor (FGFR), of which there are also several forms. One such receptor, FGFR3, also known as CD333, is coded for by *FGFR3* at 4p16.3. A gain-in-function mutation in *FGFR3* increases the expression of the receptor, the consequences of which include inhibition of chondrocyte proliferation resulting in abnormal growth plate activity and non-linear bone growth. The most common manifestation of this, achondroplasia, results in shortening of the limbs and so a short stature, typically around 4 feet, and changes to the shape of the skull. Intellectual development is generally unimpaired, and its estimated prevalence is 36–60/million.

16.3 Single-gene effects on several organs or organ systems

This section considers mutations in a single gene that has widespread effects on the body, of which one dominates.

16.3.1 Cystic fibrosis

Epidemiology

Estimates of the frequency of cystic fibrosis, mainly an inherited autosome recessive disease, vary from ~160/million to 330/million, giving carrier rates of 1/28 and 1/40 respectively. However, there is marked variance in different populations: 740/million in Ireland, 250/million in the USA, 40/million in Finland, 20/million in India, and 3/million in Japan. Rates may fall with general public awareness and carrier screening.

Genetics

The aetiology resides in the mutation of *CFTR*, located at 7q31.2, which codes for the 1,480 amino acid cystic fibrosis transmembrane conductance regulator (CFTR), an ATP-binding cassette (ABC) trans-porter that regulates the movement of chloride anions across the cell membrane. The ABC transporter super-family includes hundreds of molecules of a similar structure that transport a variety of ions, vita-mins, drugs, metabolites, and other molecules into and out of cells. *CFTR* is ~189 K base pairs long with 27 exons, and some 2,000 mutations have been reported. The most common (~70%) is homozygosity for δF508, a deletion of three nucleotides (CCT) coding for a phenylalanine (F being its international abbreviation) residue at position 508 and resulting in a loss of function of the product. With δF508 on one chromosome, the second harbours any of a group of (mostly) SNP mutations responsible for changes to amino acids at certain positions in the mature CFTR molecule, these being G542X (glycine at position 542 to any amino acid, in around 6% of cases), N1303K (asparagine at position 1303 to lysine, 4%), R1162X (arginine at position 1162 to any amino acid, 4%), and G551D (glycine at position 551 to aspartic acid ~3%).

Diagnosis and clinical course

The result of all of these mutations is decreased secretion of chloride anions, and so a resultant increased resorption of sodium. This in turn leads to increased reabsorption of water, and so a thickening of secretions, which becomes critical where the secretions are of mucus. The median age at diagnosis is 7 months, and diagnosis is generally based upon prolonged neonatal jaundice, lung infections, prolonged meconium ileus (sticky black faeces) following delivery, failure to thrive, poor weight gain, immunoreactive trypsinogen, and increased sweat chloride (typically 70–100 mEq/l, reference range <60), but the genetic test is definitive. By the teenage years, the disease will have involved many organs and organ systems:

- In the lungs, thickened mucus in bronchial airways produces a phenotype that resembles chronic obstructive pulmonary disease (COPD), involving a purulent sputum with persistent bacterial infections with organisms such as *Pseudomonas aeruginosa*, *Haemophilus influenzae*, and *Klebsiella pneumoniae*, and so a classical neutrophil-based chronic inflammation with chronic bronchitis.

- Secretions from the sinuses lead to infections and so to chronic sinusitis and rhinitis.

- In the pancreas (with fibrosis and cysts, hence the name), the thickened mucus blocks digestive secretions and leads to a failure to neutralize acidic gastric chyme, resulting in abdominal pain, malabsorption, and greasy faeces. Pancreatitis is common, and there may be diabetes resulting from autodigestion.

- Obstruction of the biliary tree has the potential to cause obstructive cirrhosis with increased bilirubin and increased portal hypertension leading to oesophageal varices and splenomegaly.

- Increased secretion of sodium in sweat gives the skin its characteristic salty nature and a degree of dermatitis.

Other manifestations include osteopenia/osteoporosis, arthropathy, iron-deficient anaemia, kidney stones, and subfertility.

Management

Treatments are aimed at the lungs, with physiotherapy (to clear the mucus), antibiotics, bronchodilators, and mucolytics, whilst the deterioration in pulmonary function can be monitored with forced vital capacity (section 13.5.1). Pancreatic enzymes and vitamin supplements are often given. Drug treatments aimed at the CFTR protein can be useful in certain subtypes of the disease. These include 150 mg ivacaftor bd, which may be effective in those with the G551D or other variants, the combination of up to 800 mg oral lumacaftor and 500 mg ivacaftor daily in homozygous disease, and 100 mg tezacaftor daily. These drugs, which support the CRTF protein and so chloride transport, are taken in different combinations. However, as pulmonary failure is the most common cause of death, lung transplantation may be necessary, and stem cell therapy is a distant hope. With advances in medical care, survival rates are continually improving, presently around the age of 47, with some aged over 60—these rates differ by genetic variant.

Cross reference

NICE NG78 provides comprehensive guidance on the diagnosis and management of cystic fibrosis.

> ## Key points
>
> Cystic fibrosis is perhaps the most common and clinically the most wide-ranging condition consequent to a single-gene defect. Successful treatment requires the attention of a broad school of specialists.

SELF-CHECK 16.4

What is the major biochemical problem with cystic fibrosis?

16.3.2 Neurofibromatosis type 1

Mutations in the large (350 Kb, 62 exons) *NF1* at 17q11.2 code for an abnormal variant of the 320 kDa neurofibromin 1 that normally part-regulates certain intracellular second messengers, notably those of the Ras–MAP kinase pathway (see section 3.2 and Table 3.1). Failure to do so leads to the development of numerous dermal neurofibromas that are characteristic of the disease. Of the >1,300 mutations, most are SNPs and deletions, resulting in a disease with a frequency of ~350/million.

In many cases, the presenting feature in infancy is café-au-lait spots, comprising collections of melanocytes, but soon after disease appears in various forms. A further diagnostic sign is axillary or **inguinal** freckles. Musculoskeletal abnormalities include deformities of the skull, vertebrae, and long bones (principally the tibia, which may cause unequal limb length), with muscle weakness, the latter due to motor nerve dysfunction. A further aspect of neurological disease is formation of sporadic multiple benign tumours (neurofibromata) often arising from Schwann cells (hence schwannoma). Ocular problems may arise from changes to the eye socket, nodules in the iris, and **gliomas** along the optic nerve. There is a high degree of autism and attention deficient hyperactivity disorder (ADHD), with risk of other changes to the central nervous system including epilepsy, learning difficulties, and depression.

The only pharmaceutical treatment is with selumetinib (generally 20 mg/m^2), a drug that effectively blocks the second messenger pathway that would normally be the role of neurofibromin 1. Mutations in a related gene, *NF2* at 22q12.2, lead to multiple tumours of the central nervous system.

16.3.3 Marfan syndrome

This disease, with a frequency of ~150/million, is caused in 95% of cases by a defect in *FBN1*, a large gene of 65 exons at 15q21.1, which in health codes for profibrillin. This molecule is cleaved by furin convertase to give the 340 kDa structural protein fibrillin-1 and asprosin (a hormone released by adipocytes that part-regulates blood glucose). SNPs, premature terminations, and deletions in *FBN1* have all been described and variously lead to truncated or otherwise malformed fibrillin-1 that fails to form functional elastic and non-elastic microfibrils. There may also be a place for mutations in *FBN2* at 5q23–31, which codes for fibrillin-2, a likely participant in elastogenesis and cofactor with fibrillin-1.

The consequences of dysfunctional fibrillin-1, as with many such diseases, runs a spectrum with abnormal connective tissue in different organ systems. In the heart, there may be aortic regurgitation (possibly arising from a malformed aortic root), atrial fibrillation, mitral valve regurgitation, and dilated cardiomyopathy (section 8.3.2), whilst in the peripheral circulation there may be thoracic and abdominal aortic aneurysm, which may dissect or rupture. Skeletal features have a highly variable age of onset and include joint laxity, long bone growth (with elongated arm span), and arachnodactyly (long fingers), as well as anterior chest wall and vertebral column deformities. Ocular disease includes dislocation of the lens, corneal flatness, **myopia** and stretching of the retina (which may cause detachment), and **strabismus**. Non-specific facial features include a long head (dolichocephaly), exophthalmos, and an arched palate with tooth crowding resulting from a small jaw.

16.3.4 Fragile X syndrome

With a frequency of 260/million males and 145/million females, and a carrier status of ~1/190 women and ~1/525 men, this disease is the most common monogenic cause of autism spectrum disorder and the prevalent inherited cause of mild to severe learning disability, second to Down's syndrome in terms of general mental impairment.

Leading phenotypic features in the neonate include a prominent jaw, a broad **philtrum**, enlarged testes, a long and narrow face, large and low-set ears (Figure 16.13), and flexible fingers. As the child ages, delayed speech, autism, cognitive deficits, otitis media, strabismus, seizures, and hyperactivity may be noted. In the adult there maybe tremor, ataxia, mitral valve prolapse, deteriorating mental functions, premature ovarian failure, and PCOS. Differential diagnoses include Sotos syndrome, Prader–Willi syndrome, and Klinefelter syndrome.

FIGURE 16.13
Fragile X syndrome. The phenotype has large, low-set ears, and a wide philtrum.

Peter Saxon, CC BY-SA 4.0 <https://creativecommons.org/licenses/by-sa/4.0>, via Wikimedia Commons.

16.3.5 Ataxia–telangiectasia

This autosomal recessive disease focuses on two aspects, the ataxia (impaired movement coordination), and the telangiectasia (dilated capillaries appearing as red, spider-like dermal lesions, most noticeably on the sclera) although many other signs and symptoms may be present. The earliest sign is often poor gait in childhood, which deteriorates until a wheelchair is needed; later developments include poor growth, delayed puberty, cerebellar atrophy, and diabetes. There may also be tremor, **dystonia**, and involuntary muscle spasms (myoclonus). The disease is the second most common autosomal recessive childhood ataxia (after Friedreich's ataxia) and the most common genetic ataxia developing in the first decade.

The genetic lesion, present in ~14/million, resides with *ATM* at 11q22.3, which codes for a serine/threonine kinase that is a crucial part of a DNA repair system. As a consequence, loss-of-function mutations in *ATM* lead to a high (25–30%) risk of malignancies, particularly of lymphocytes. Allied to the latter is hypogammaglobulinaemia, with resultant infections of the mucus membranes. Diagnosis is based on the clinical picture, often supported by increased serum alpha-fetoprotein, but confirmed by molecular genetics. Treatment is palliative, the median survival following diagnosis being 22 years.

16.3.6 X-linked adrenoleukodystrophy

This disease, with a strikingly variable clinical spectrum, is characterized by over 600 mutations in *ABCD1* at Xq28, which normally codes for an intracellular transport protein: the leuko- refers to white neural matter, not white blood cells. The protein carries coenzyme-A-activated long-chain fatty acid (≥22 carbon atoms) esters to peroxisomes for degradation by β-oxidation. The mutated variant results in (often inflammatory) lipid accumulation in virtually all tissues: in one post-mortem case the mass of C26 fatty acids in cerebral white matter was 39-fold greater than that in normal brains.

In almost all males (~50/million) X-linked adrenoleukodystrophy results in childhood adrenocortical insufficiency that resembles Addison's disease and progresses in the adult to gait difficulties, bowel and bladder problems, myopathy, and peripheral neuropathy. In a subset, fatal cerebral demyelination and cerebral adrenoleukodystrophy may develop, but others remain pre-symptomatic for up to five decades. It follows that some cases may be misdiagnosed as Addison's disease. Around 50% of female carriers (thus 10/million) can also develop myelopathy and peripheral neuropathy, but at a later age, and <5% have adrenocortical insufficiency.

Diagnosis is based on presence of high levels of long-chain fatty acids in a biopsy. But as other rare conditions cause this form of lipid disease, and whilst gross changes to cerebral white matter may be detectable by MRI, the ultimate diagnosis is by molecular genetics. In some cases, the disease process may be decelerated by dietary modification, but physiology continues to generate the fatty acids.

16.3.7 Friedreich's ataxia

This multi-system disease most commonly affects 'individuals of European, North African, Middle Eastern, or Indian origin and is not described in those of sub-Saharan African, Amerindian, or East Asian origin' (Parkinson et al. 2013). In Western European populations the average frequency of 30/million and a carrier rate of ~1/85 mask considerable geographical variation—4/million in north and east Europe to 50/million in the south-west. Almost all cases (~98%) are caused by increased numbers (600–900) of a homozygous GAA triplet insertion in *FXN* at 9q21.11, normally coding for frataxin, a protein that migrates to the mitochondrion. Its function is unclear but may relate to iron/sulphur transport and/or storage.

The phenotype is variable, but focuses on the central and peripheral nervous systems, the common features being gait and limb ataxia, dysarthria (inability to communicate coherently with speech), and loss of lower limb reflexes leading to the need for a wheelchair. The disease generally arises at age 10–16 (although there is a variant of late (> age 25) and very late (>40) development), and other symptoms include increasing difficulty with day-to-day activities such as **dysphagia** and loss of both bladder and bowel control, with hyperactivity. There may also be higher cerebral dysfunction manifesting as poor memory with difficulty in reading, and anxiety and depression. Almost all of these can be ascribed to cerebellar (with loss of grey and white matter) and spinal cord atrophy, as demonstrated by MRI. Ocular and hearing impairment is common, as is scoliosis, which adds to movement difficulties.

The major non-neurological issue is cardiomyopathy with left ventricular hypertrophy and a reduced ejection fraction. Later, there may be atrial fibrillation contributing to worsening systolic function and eventually heart failure, which accounts for over half of the deaths. This may be compounded by diabetes, present in ~15% of patients. A large retrospective study reported the mean age of death at 36.5 years, but with considerable variation (age range 12–87).

Management is with a range of supporting therapists (physio-, occupational, speech and language) and specialists (orthopaedic, urology, psychiatric). The patient will need regular monitoring for neurological, musculoskeletal, cardiac, and other potential pathology. Pharmacotherapy includes

CASE STUDY 16.1 *Marfan syndrome*

A tall (over 2 m), underweight young male asylum seeker with limited English language skills is examined by a health-care professional upon arrival at a processing centre. She finds that he has a deformity of the bones of the chest (which appears flat) and very long fingers, and although blood pressure is good (129/77), his pulse rate is quite high (75 beats per minute). She refers him on for further investigations.

A few weeks later the man is examined in more detail, with additional findings of a slight curvature of the spine and nipples spaced far apart. Although an ECG is normal, it confirms the tachycardia, and the practitioner finds abnormal heart sounds with her stethoscope. She also notes that just above the left clavicle, the skin is pulsating, and indeed finds a pulse, and that the man has widely spaced and protruding teeth. At this stage it is possible that many of these findings are due to injuries and childhood malnutrition. Nevertheless, she refers the patient (as he has now become) upwards to consultant level.

At hospital, an interpreter is waiting, and the patient is X-rayed and bloods are taken. The former confirms skeletal and jaw abnormalities; the latter are normal. But the X-ray also points to abnormalities in the ascending and descending aorta, and long, bony toes. Ultrasound and MRI imaging find aortic and left carotid/subclavian aneurysms, the latter accounting for the pulse. The consultant forms a preliminary diagnosis of Marfan syndrome, based also on dolichocephaly (the head length-to-width ratio exceeds the reference range), and this is confirmed in the molecular genetics laboratory with the finding of a mutation in *FBN1* that predicts a malformed fibrillin-1 molecule, the molecular basis of the pathology.

Further investigations find no ophthalmic abnormalities and no cranial or abdominal aortic aneurysms, but cardiac echocardiography reports mitral valve prolapse. The cardiac surgeon does not consider the thoracic aneurysm to be worthy of immediate action, but the patient is anticoagulated in view of a potential stroke and reviewed regularly. As the aneurysm is likely to grow, surgery will eventually be inevitable.

memantine (5 mg od, then uptitrated (also used in Alzheimer's disease)), acetazolamide (250–1,000 mg od in divided doses, as for epilepsy), or clonazepam (up to 1 mg/day, also used for epilepsy) for muscular spasms etc., with anti-muscarinics such as oxybutynin (5 mg bd or tds), tolterodine (4 mg od), and solifenacin (5 mg od) for urinary incontinence. Heart disease may call for anticoagulants and ACE inhibitors.

SELF-CHECK 16.5

Which three conditions described in section 16.3 are present at the highest frequency, and which of these three brings the most diverse pathology?

16.4 Multiple-gene effects on a single organ or organ system

16.4.1 Retinitis pigmentosa

The phenotype of retinitis pigmentosa, present in 225/million, is of a progressive loss of night and peripheral vision that in a small number eventually leads to blindness. It is the most common inherited retinal disease. One form of the disease is X-linked, so overall, slightly more men are affected. The aetiology may combine with other disease (i.e. it is syndromic—25% of cases, often part of the Usher syndrome, itself linked to 10 genes), but most (i.e. non-syndromic) follow mutation in genes coding for parts of rod and cone photoreceptors, of which there are many types. One commentator reports links with over 90 genes (Bhardwaj et al. 2022), and another suggests the most prevalent mutations are in *RHO* (30.7% of cases, located at 3q22.1, and coding for rhodopsin), *PRPF31* (8.9%, located at 19q13.42, coding for splicing factor hPRP31), and *RPGR* (8.1%, located at Xp11.4, coding for a GTPase regulator) (Daiger et al. 2014).

There is no specific treatment, but most advise vitamin A supplements, and the avoidance of bright lights (with use of sunglasses) is recommended.

16.4.2 Alzheimer's disease

Epidemiology

Alzheimer's disease, a leading cause of dementia, is a disease of the nervous system that can lead to death. In 2021 it was linked directly to 7,078 male deaths (2.4% of all male deaths) and 14,412 female deaths (5.0% of all female deaths) in England and Wales. The size of the population of people living with Alzheimer's disease overshadows all other diseases, and it is the most common neurodegenerative disease, being present in the UK at a rate of ~9,000/million. Related strongly to age, these figures translate to around 1 in 15 of those aged 65 and over; a small (<5%) but significant proportion aged <65 present with early-onset disease.

Data on differences in the frequency of Alzheimer's disease in the UK according to ethnicity are sparse. However, data from the USA on the incidence of the condition in those aged 65–74 show it to be 1.7% per person-year for African Americans and 0.4% in White Americans, rising to 4.4% and 2.6% respectively in those aged 75–84. Ethnicity within Black populations may also be important: a study found the age-adjusted incidence of the disease in Black Nigerians to be 1.4% versus 6.2% in African Americans. However, it is unclear whether this difference between the two groups is attributable to differences in ancestry or environmental factors. The incidence in rural India has been reported as 4,700/million, which is around half that in the UK (Weiner 2008). It is unclear what these data mean for the multi-ethnic population of the UK. But given the high incidence (especially in the elderly) and marked differences in mortality linked to sex in the UK and to race and ethnicity in the US, there is a clear public health issue.

Pathology

The pathological aetiology is of the deposition of extracellular plaques composed primarily of β-amyloid, produced by the cleavage of the amyloid precursor protein (coded for by *APP* at 21q21.3). The cleaving enzyme is a complex of β-secretase (coded for by *BACE1* at 11q23.3) and γ-secretase, formed of subunits presenilin-1 (coded from by *PSEN1* at 14q24.2), presenilin enhancer 2 (*PSENEN* at 19q13.12), nicastrin (*NCSTN* at 1q23.2), and anterior pharynx-defective 1 (*APH1A* at 1q21.2). Certain miRNAs (miR-29c, miR-107, miR-124, and miR195) may have a role in BACE1 levels. Abnormal deposits of amyloid (hence amyloidosis, of which there are over 30 forms, often causing widely varying organ damage) in the brain are postulated to be a major cause of the higher cerebral function in this disease.

Molecular genetics has found a strong link between the gene coding for apolipoprotein-E (apoE at 19q13.32) and Alzheimer's disease. The apoE molecule, which plays a role in lipid metabolism (see section 7.3.1), has three isoforms: apo ε2, ε3, and ε4. There is marked variability of the impact of the homozygous apoε4 genotype (i.e. ε4/ε4) on Alzheimer's disease in different populations. Among Japanese people the genotype brings an odds ratio of over 40; the odds ratio is around 12 in White people, 5 in African American people, and not significant in Hispanic people. Those heterozygous for apo ε4 (such as ε4/ε2) have an intermediate risk of the disease (Belloy et al. 2019). The precise pathophysiological reason for this relationship is unclear, although it may relate to lipid transport in the brain.

Presentation and diagnosis

The disease is characterized by a progressive deterioration in memory, spatial orientation, language, understanding, and cognitive functioning, and it is the leading (in 70% of cases) cause of dementia—others being vascular dementia, dementia with **Lewy bodies** (see section 16.4.3), and the dementia of Parkinson's disease. Diagnosis is based on a series of signs, symptoms, and assessment by a trained healthcare professional who will consider a number of features, all of which have poor sensitivity and specificity. The key difficulty is differentiating the early stage of the disease from normal ageing (Table 16.2). As the diagnostic value of symptoms is of modest effectiveness, robust biomarkers are actively sought. These may include cerebrospinal fluid (CSF) or serum β-amyloid, total tau, phosphorylated tau, neurofilament light, and the ratio of CSF to blood albumin.

Management

There are no clear and successful treatments, although maintaining active cognitive function (through puzzles, crosswords, card games, etc.) and psycho-social interactions are often cited in suppressing stage progression. Similarly, at present there are no clear medications to prevent, suppress, or reverse the disease, although there may be a place for microRNAs as potential therapeutic targets. However, in a trial of a monoclonal antibody directed towards amyloid-β in patients with early symptomatic Alzheimer's disease and amyloid/tau pathology, those randomized to active treatment had a significantly slower clinical progression of their illness (Sims et al. 2023).

TABLE 16.2 Stages of Alzheimer's disease

Stage	Features
Normal ageing	Lapses in memory, failure to recall fine details, slow in searching for words in conversation. All generally described as normal aged forgetfulness
Early disease	Deterioration in the above, plus misplacing or losing a valuable object, difficulty organizing or planning, problems in social or work settings in performing new tasks
Moderate disease	Deterioration in the above, plus changes in sleep pattern, difficulty in choosing season-appropriate clothing, increased tendency to wander off and so become lost, failure to recognize distant family and friends
Marked disease	Deterioration in the above, plus increased need for full-time care, loss of bladder and bowel control, difficulty in communications, failure to recognize close family and friends

Key points

Alzheimer's disease is one of the most prominent genetic conditions, not merely as a cause of death, but also due to its major personal and social effects. A successful and targeted treatment remains elusive, possibly because of the numerous gene defects that bring about the phenotype.

16.4.3 Parkinson's disease

Epidemiology

Parkinson's disease, the second most common neurodegenerative disease, has a frequency of ~1,500/million that rises with age: it is very rare in those aged <40, and appears in 1–2% of those aged >60 and 4–5% of those aged >85. The disease is seen in 50% more men than women, and was named on the death certificates of 4,390 men and 2,876 women in England and Wales in 2021.

Pathology

The first gene to be linked to this disease (*SNCA* at 4q22.1) codes for α-synuclein, the abnormal aggregated structural component of Lewy bodies, followed by perhaps a dozen more. These include *PRKN* (or *PARK2*) at 6q26, coding for Parkin (a ubiquitin ligase); *PARK5* at 4p13, coding for ubiquitin C-terminal hydrolase; *PINK1* at 1p36.12, coding for PTEN-induced kinase 1; *LRRK2* at 12q21 (the most prevalent), coding for leucine-rich repeat kinase 2; and *GBA1* at 1q22, coding for glucocerebrosidase. Certain of these genes may be part-regulated by miRNAs and by lncRNAs, such as HOTAIR and *LRRK2*. There is also evidence of a link with HLA: alleles such as B*40:01, C*03:04, DRB1*04:04, and DQA1*03:01 are protective, whereas B*07:02, C*07:02, DRB1*15:01, and DQA1*03:01 bring risk of the disease.

Despite these associations, only ~10% of cases have a monogenic origin, and several may interact alongside other disease processes (such as inflammation) and environmental factors (where smoking and caffeine-users are protected), reminiscent of a multi-hit hypothesis. Notably, some genes (such as MAPT) are also of interest in Alzheimer's disease.

Presentation and diagnosis

The disease is characterized by the progressive loss of motor control over muscles, manifesting as slow movement (bradykinesia, most noticeable in walking), rigidity, tremor, and an unstable posture. Differential diagnoses include multiple system atrophy, dementia with Lewy bodies, and progressive paralysis. The pathology is principally of the degenerative loss of dopaminergic neurons in parts of the brain that results in a fall in dopamine levels, and accordingly symptoms often respond to levodopa. This therapy is so successful that failure to respond implies a secondary cause such as vascular (with ischaemic cerebrovascular disease), drug-induced (anti-psychotics), toxin-induced (heavy metals manganese and iron), and malignancy. Diagnosis is based on symptoms, whilst ultrasound may report increased echogenicity of the **substantia nigra**, and tumours and vascular infarcts may be excluded by MRI.

Management

As dopamine does not cross the blood–brain barrier, standard treatment is with levodopa, often together with a carbidopa, a dopa decarboxylase. The dose is generally up-titrated from a single tablet combining 100 mg of levodopa with 25 mg of dopa decarboxylase tds, although other tablets with different combinations are available. Other treatments are shown in Table 16.3.

Prognosis depends on aetiology, early use of therapy, and age at onset of symptoms (late-onset having a faster progression and earlier cognitive decline). Potential markers to detect cognitive decline include α-synuclein, total tau, and amyloid-β-40. As the disease progresses, dementia, depression, hallucinations, sleep disorders, and psychosis may develop. Mortality is generally a decade after diagnosis.

TABLE 16.3 Treatments for Parkinson's disease

Pharmaceutical group	Mode of action	Examples/dose
Dopamine agonists	Stimulate dopamine receptors	Pramipexole 0.125 mg tds Ropinirole 0.25 mg tds Bromocriptine 10–30 mg od
Catechol-O-methyltransferase (COMT) inhibitors	As COMT degrades dopamine, inhibition increases dopamine levels	*Entacapone 200 mg up to 8 times a day, tolcapone 100 mg tds
Monoamine oxidase (MAO) inhibitors	As MAOs decrease dopamine metabolism, inhibition increases dopamine levels	Selegiline 5 mg od Rasagiline 0.5–1 mg od
Anti-cholinergics	Block acetylcholine receptors	Trihexyphenidyl 0.5–1 mg bd, Benztropine 0.5–2 mg bd Amantadine 100 mg bd/tds

* Used alongside levodopa. As always, practitioners must consult their local pharmacopeia.

SELF-CHECK 16.6

Briefly describe the pathophysiology of Parkinson's disease.

16.4.4 Red blood cell morphology

Section 11.5.4 describes genetic conditions that lead to a haemolytic anaemia, several being changes in red cell morphology. Those due to the haemoglobinopathies are described in section 16.7.

Hereditary spherocytosis, present in ~200/million of those of European ancestry, results from mutations in any gene that codes for elements of the cytoskeleton of the red blood cell. The most common lesions are in *ANK1* at 8p11.21, coding for Ankyrin 1; *SPTA* at 1q23.1 and *SPTB* at 14q23.3, coding for the alpha and beta chain spectrins respectively; and *EPB42* at 15q15.2, coding for band 4.2.

In hereditary elliptocytosis, elliptoid red cells may be caused by mutations in *SPTA1* and *SPTB*, but also by mutations in *EPB41* at 1p35.3, coding for protein 4.1, or in *SCL4A1* at 17q21.31, coding for band 3. Penetrance is variable, as other factors are involved, and so rates of presentation differ around the world: 4/million in the USA but up to 10,000/million in West Africa, and up to 10% of some populations in East and Southeast Asia. Most cases are autosomal dominant.

Most cases of hereditary xerocytosis, present at a frequency of ~20/million (possibly an underestimate), are due to missense mutations in *PIEZO1* (at 16q24.3, coding the Piezo channel) or in *KCNN4* (19q13.31, coding the Gardos channel), both channels being cation pumps. Heterozygous autosomal dominant mutations in either gene confer gain-of-function status and lead to dehydrated red cells that are deficient in potassium, chloride, and water such that the cells appear as stomatocytes.

16.4.5 Iron overload

As we have no mechanism for the active excretion of excess iron, it is stored in various organs that are eventually damaged by these deposits (section 11.7).

Hereditary haemochromatosis (HH)

The most common (~90%) genetic cause of iron overload, ~80% of cases of autosomal recessive hereditary haemochromatosis (HH) are caused by a G>A (hence cysteine>tyrosine) SNP mutation (C282Y) in *HFE* at 6p22.2, coding a membrane component (HFE (H = high, Fe = iron)) that normally associates with β-2 microglobulin. The location of *HFE* within the MHC (major histocompatibility complex (section 9.6.1)) is notable and indeed is linked to haplotypes HLA A*03, B*35 and A*01, B*08 (and other alleles) in certain populations.

The mutation impairs the ability of HFE to bind to β-2 microglobulin, leading to failure to activate *HAMP* at 19q13.12, which codes for hepcidin. Low levels of hepcidin therefore fail to regulate (in this case, inhibit) iron passage from the enterocyte into the plasma, resulting in the hyperferraemia, the key laboratory feature, and ultimately iron overload and HH.

Heterozygous C282Y is present in 6% of Europeans (60,000/million), and although they have no clinically evident disease, with a penetrance of 3%, iron levels may be raised in ~25%. In its homozygous form HH is present in 0.4% (4,000/million) of Europeans and has a penetrance of some 14%. The large disparity between the presence of the mutation and clinical disease implies modifier genes and/or environmental factors (Box 16.1). Other SNPs causing dysfunctional HFE include those of C>G (histidine to aspartic acid) SNP H63D (heterozygous in ~20%, homozygous in ~1%) and S65C (A>T) (heterozygous in ~2.5%).

A juvenile form of HH is recognized, caused by mutations (a) in *HJV* (also known as *HFE2*) at 1q21.1, coding for haemojuvelin, part of a second pathway regulating hepcidin, and (b) in *HAMP*, which codes for hepcidin.

Other reasons for iron overload

While hereditary haemochromatosis accounts for the largest proportion of genetically based cases of iron overload, there are also some other genetic reasons for it. These include:

- Mutations in *FPN1* (*SLC40A1*), coding for ferroportin, which (a) impair the iron-exporting ability of ferroportin, particularly in macrophages, and (b) render ferroportin unresponsive to the inhibitory effects of hepcidin so that it exports all absorbed iron into the plasma.

- Acaeruloplasminaemia, a rare recessive disorder caused by a loss-of-function mutation in *CP* at 3q24–q25.1 and resulting in impaired ferroxidase activity, leading to excessive iron deposits, low transferrin saturation, and mild microcytic anaemia.

- Atransferrinaemia, a recessive disorder caused by transferrin deficiency due to mutations in *TF* at 3q22.1, coding for the 76 kDa glycoprotein transferrin. Characteristics include very low to undetectable plasma transferrin, leading to impaired erythropoiesis and a microcytic hypochromic anaemia.

16.4.6 Spinocerebellar ataxia (SCA)

Also described as cerebellarspinal ataxia, spinocerebellar ataxia (SCA) is characterized by the slow loss of motor function; there is loss of control of walking (68% of cases), of the hands, of speech and eye muscles, and dysarthria (together present in 50%). Over 47 types have been described that together bring a global frequency of ~27/million, caused by 35 genes with autosomal dominance, the most frequent of which are summarized in Table 16.4. The common feature is the increased number of

BOX 16.1 Balanced polymorphisms

It has been argued that the high frequency of a potentially damaging gene must bring a benefit in natural selection. In the case of *HFE* this may be the ability of heterozygotes to maintain higher iron levels (or other trace metals) that hundreds of thousands of years ago may have been an advantage, possibly more so for women (who need a higher intake of iron) than for men. The same argument has been made for the high frequency of Factor V Leiden, which may minimizes blood loss at childbirth in heterozygous women (see Box 11.2). The advantageous role of heterozygosity in various genes involved in red blood cell protection from malaria is established (section 16.7).

TABLE 16.4 Genes causing spinocerebellar ataxia

Type	Gene	Location	Product function
SCA1	*ATXN1*	6p22.3	DNA-binding protein
SCA2	*ATXN2*	12q24.12	mRNA translation regulator
SCA3	*ATXN3*	14q32.12	Deubiquinating enzyme
SCA6	*CACNA1A*	19p13.13	Voltage-dependent calcium channel
SCA7	*ATXN7*	3p21.1–12	Transcription regulator
SCA17	*TBP*	6q27	Transcription factor
DRPLA	*ATN1*	12p13.31	Transcription regulator

trinucleotide CAG repeats (as in Huntington's disease and spinal bulbar muscular atrophy), which correlate inversely with the age of onset, generally between the ages of 20 and 40. The rate of progression of each type of SCA is highly variable, may develop at any age, and is due to deterioration of the motor coordination centres in the cerebellum.

Some types of the disease (e.g. SCA11) are relatively benign with no reduction in lifespan, whilst others (e.g. SCA2 and SCA3) have an average duration of 10 years, and many have semi-specific clinical features. For example, dystonia is common in SCAs 3, 14, 17, 20, and 35; tremor in SCAs 12, 15, and 27; and psychiatric symptoms in SCAs 2 and 17. Many cases may be collected together according to broad phenotypes. There are no specific treatments and management is according to the severity of symptoms, although some patients respond to physiotherapy and high-intensity coordinative training. Anti-sense nucleotides and small interfering RNAs are a potential treatment.

16.4.7 Albinism

With two major features—the skin and the eyes—oculocutaneous albinism (OCA) is the result of autosomal recessive mutations in any one of 19 genes responsible for seven phenotypes. The overall global frequency of ~50/million varies considerably between populations, such as 28/million in the United States and 6,300/million in the Indigenous peoples of Panama and Columbia. Four subtypes of OCA are known:

- OCA1, globally the most common, has a prevalence of 2,500/million in most populations but is rare in African Americans. It is linked to mutations in *TYR* at 11q14.3, which codes for tyrosinase, the key enzyme in the production of melanin.

- OCA2 is caused by mutations in *OCA2* at 15q11.2–q12, coding for melanocyte-specific transporter protein. Although the prevalence in the United States is 28/million, this rises to 100/million in African Americans, and in southern parts of Africa it can reach 256/million.

- OCA3 is the product of mutations in *TYRP1* at 9p23, coding for tyrosinase-related protein; it is rare in White Europeans and Asians, but is present in 118/million Africans.

- OCA4 is also rare in white Europeans but is present in 12/million Japanese. It is caused by a mutation in *MATP* at 5p13.3 that codes for membrane-associated transporter protein (Grønskov et al. 2007).

The common feature of the different forms of albinism is lack of pigmentation of the skin, brought about by loss-of-function mutations in genes needed to produce melanin that darkens the skin. Deprived of the dark skin and ability to tan, OCA patients are not protected from the effects of ultraviolet light that at least cause severe sunburn and at worst oxidative damage to DNA, and so skin cancer, notably basal cell carcinoma (~75%) and squamous cell carcinoma (~25%). OCA can also bring considerable psychosocial problems, and the use of sunglasses, sunscreen (≥SPF 30), and other photoprotective measures is crucial.

SELF-CHECK 16.7

What causes albinism and why is it a problem?

16.5 Multiple-gene effects on several organs or organ systems

16.5.1 Atherosclerosis

The multi-faceted nature of this disease is apparent and dominated by risk factors. The major pathophysiological focus of type 2 diabetes mellitus is years, perhaps decades, of hyperglycaemia resulting in insulin resistance, and links with obesity. However, at least 10 genes may contribute to the initiation and/or severity of the disease. These include *PPARG* (with roles in glucose and lipid metabolism); *TCF7L2* (insulin secretion); *HNF1A* and *HNF1B* (development of the liver and pancreatic islets); *KCNJ11* and *KCNQ1* (glucose-stimulated insulin secretion); *FTO* (fat mass and obesity: see below); and *SCL30A8*, *CDKN2A/2B*, and *CDKAL1*, whose precise roles in diabetes are not as clear-cut.

Section 7.2.2 describes the role of mutations in genes that may contribute to hypertension and its treatment, and it is possible that several, alongside non-genetic factors, act in concert. However, certain conditions can be traced to specific gene defects, such as those coding for four subunits of the epithelial sodium channel that cause the high blood pressure in Liddle's syndrome.

Section 7.3.3 describes familial dyslipidaemias, the principal one being familial hypercholesterolaemia (FH). Mutations in the 18-exon *LDLR* at 19p13.2 code for a defective receptor that ensure levels of serum LDL-cholesterol >12 mmol/l in those 1/million with homozygous disease and levels generally between 7–10 mmol/l in those 2,000/million (i.e. 1/500) with heterozygous disease.

16.5.2 Ehlers–Danlos syndromes

This heterologous group of conditions, which have a frequency of some 130/million, is characterized by any combination of abnormalities in the joints, ligaments, and skin. The leading signs are overextendable skin, atypical scar formation, and hyperflexibility, the latter often bringing spontaneous dislocations. With a frequency of 200/million, the dominant genetic lesions are those that code for collagen and other components of connective tissues. However, the heterogeneity of the clinical presentation and associated genetic links has required a complex classification into 13 subgroups (Table 16.5).

Minor diagnostic signs of the classical form include easy bruising; soft, fragile, and doughy skin; hernia; and a family history of these features, whilst 90% of cases harbour abnormalities in *COL5A1* or *COL1A1*. In contrast, no clear genetic link has been reported in the hypermobile variant, so that diagnosis relies on clinical features and personal and family history, the major criterion being generalized hypermobility, such as ≥5 joints from 9. These two forms comprise 90% of cases of Ehlers–Danlos syndromes; the third most common is the vascular form, at ~7% of cases. Major features of this variant include family history, arterial rupture at an early age, thin and translucent skin with bruising, sigmoid colon perforations, and third-trimester uterine rupture.

Given the diversity of signs and symptoms, differential diagnoses of Marfan syndrome and Loeys–Dietz syndrome may be considered. The latter is often linked to widely spaced eyes (hypertelorism), an abnormal uvlua, cleft palate, strabismus, club foot, and numerous other features, all potentially linked to mutations in *SMAD3* and genes coding TGF-β receptors. Treatment of Ehlers–Danlos syndromes (and, indeed, Loeys–Dietz syndrome) varies with subtype, and many include surgery, whilst education is effective in making patients aware of potential dangers.

TABLE 16.5 Classification of Ehlers–Danlos syndromes

Subtype		Inheritance	Genetic basis	Location	Protein
I	Classical	AD	*COL5A1, COL1A1*	9q34.3, 17q21.33	Collagens types I and V
II	Classical-like	AR	*TNXB*	6p21.33–32	Tenascin XB
III	Cardiac-valvular	AR	*COL1A2*	7q21.3	Collagen type 1
IV	Vascular	AD	*COL3A1, COL1A1*	2q32.2, 17q21.33	Collage types I and III
V	Hypermobile	AD	*LZTS1*	8p22–21.1	Leucine zipper putative tumour suppressor
VI	Arthrochalasic	AD	*COL1A1, COL1A2*	17q21.33, 7q21.3	Collagen type I
VII	Dermatosparaxic	AR	*ADAMTS2*	5q35.3	ADAMTS2
VIII	Kyphoscoliotic	AR	*PLOD1, FKBP14*	1p36.3–2, 7p14.3	LH1, FKBP22
IX	Brittle cornea	AR	*ZNF469, PRDM5*	16q24.2, 4q25–26	ZNF469, PRDM5
X	Spondylodysplastic	AR	*B4GALT6 and 7, SLC39A13*	5q35.3, 5q35.3, 11p11.2	β4GalT7, β3GalT6, ZIP13
XI	Musculocontractural	AR	*CHST14, DSE*	15q15.1, 6q22.1	D4ST1, DSE
XII	Myopathic	AD or AR	*COL12A1*	6q13–14.1	Collagen type XII
XIII	Periodontal	AD	*C1R, C1S*	12p13.31, 12p13.31	C1r, C1s

AD: autosomal dominant, AR: autosomal recessive.

Modified from Malfait F et al. The 2017 international classification of the Ehlers–Danlos syndromes. Am J Med Genet Part C Semin Med Genet 175C: 8– 26. © 2017 Wiley Periodicals, Inc.

16.5.3 Overweight and obesity

It has long been assumed (and often remains so) that a large component of overweight (body mass index 25–29.9 kg/m^2) and obesity (≥30 kg/m^2) is simply the result of an excessive calorie intake unbalanced by calories used in exercise. Whilst this is certainly the case for the great majority, there are several predisposing genes, and genome-wide association studies indicate at least 75 obese-susceptibility loci.

Prader–Willi syndrome

The multi-system disease Prader–Willi syndrome has three potential aetiologies: a 5–6 Mb deletion in paternal 15q11–13 (~70% of cases), maternal uniparental disomy 15 (with no paternal chromosome 15: ~25%), and imprinting defects in either gamete, often with hypermethylated histones and DNA (~2%), with rare SNPs making up the balance. With a prevalence of ~60/million, it is the leading cause of life-threatening obesity, and may be observed in the neonate with severe hypotonia and poor feeding, followed in infancy with **hyperphagia**, short stature (consequent to growth hormone deficiency),

(a)
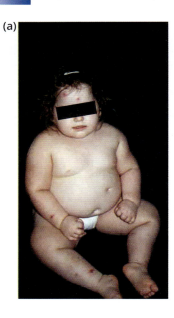

(b)
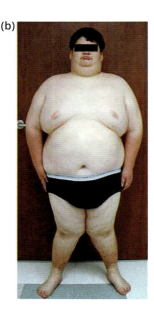

FIGURE 16.14
Prader–Willi syndrome.

Cassidy S, Driscoll D. Prader–Willi syndrome. Eur J Hum Genet, 2009: 17; 3–13. https://doi.org/10.1038/ejhg.2008.165 (Figure 2 a & b). © and reprinted by permission, Springer Nature.

learning difficulties and behavioural challenges, and often self-harm. In childhood, people with this disease will often develop a thin upper lip and down-turned mouth, almond-shaped eyes (perhaps with strabismus), have small genitals, and will be constantly hungry, and so develop hyperphagia. This can be expected to increase into the teenage years, where there may be delayed or incomplete puberty; adulthood obesity causes type 2 diabetes in >25% of cases (Figure 16.14).

Mapping of the leading locus (15q11–13) has identified several candidate genes/loci These include *SNURF* or *SNRPN* at 15q11.2 (which includes the imprinting centre) coding for a nuclear protein, genes for several small nucleolar RNAs (*SNORDs*, particularly *SNORD116*) that may target cellular RNAs for methylation or alternative splicing, and *MAGEL2*, as SNPs can cause class c Prader–Willi symptoms. There may also be a role for long non-coding RNA *IPW*, whilst the methylation of *SNRPN* may be a starting point for genetic diagnosis. Notably, *UBE3A* is also present in 15q11–13, so that Prader–Willi syndrome may be a differential diagnosis for Angelman syndrome, which is found at a frequency of ~65/million and is linked to SNPs in, or absence of, this gene.

One mechanism to part-explain the hyperphagia is high levels of the gastric and pancreatic hormone ghrelin (coded for by *GHRL* at 3p25.3) that stimulate the appetite centres in the brain. The hormone may act by reducing the sensitivity of the vagus so that it is less likely to react to gastric distensibility. There may also be roles for increased cholecystokinin, adiponectin, and leptin. The latter may be implicated as leptin-deficient individuals develop a complex phenotype with hyperphagia and so obesity, and treatment with leptin reverses these features.

The short stature can be treated with recombinant growth hormone, and the desire to eat can be ameliorated by specialist diets that are bulked up with indigestible material to give the impression of satiety. Medical treatments include orlistat (a reversible inhibitor of gastric and pancreatic lipase resulting in reduced dietary fat absorption), octreotide (a long-acting somatostatin analogue to decrease gherlin), and liraglutide or exenatide (glucagon-like peptide 1 receptor agonists that increase insulin secretion). Surgical treatments (e.g. sleeve gastrectomy) are successful in weight reduction, with reduction in ghrelin and increased glucagon-like peptide-1. Major causes of death are respiratory failure with right ventricular hypertrophy and other cardiovascular disease.

Fat mass and obesity-associated gene (FTO)

The fat mass and obesity-associated gene, also known as FTO, is located at 16q12.2, which codes for α-ketoglutarate dependent dioxygenase, and functions in the demethylation of nucleic acids and accordingly may regulate other genes by this process. Although homozygous loss-of-function mutations cause severe growth retardation and developmental delay, there are no clear phenotypic changes

CASE STUDY 16.2 *Ehlers–Danlos syndrome*

In the early 1970s, a 21-year-old woman ('A' in Figure 16.15) returns from hospital with her 3-year-old child ('B'), who was treated for a major skin abrasion requiring stitches, having fallen off his bicycle. Upon learning of this, the woman's own mother remarked how she often took the woman herself to the family doctor with knocks and bruises calling for bandaging, but that she grew out of it. The woman's legs and hips bear testament to this damage with several scars.

As B grows up, his parents note that their son also seems to bruise easily and has skin damage, and—when falling out of a tree—he dislocates a shoulder, at around which time he is joined by a sister ('C' in Figure 16.15) and later a brother. In his teenage years in the mid-1980s, B continues to suffer injuries (such as at school sports) that seem to be more traumatic than would have been expected. When he complains that both feet hurt after a cross-country run, A takes her son to a podiatrist, who diagnoses flat feet. The mother is also questioned and eventually examined, and flat feet are also found. At this point she recalls numerous problems with finding comfortable shoes and that she was never able to wear high heels comfortably. Noting the scars on both of his patients' legs, the practitioner casually asks if there is a genetic problem in the family. A few days later, the GP agrees, and refers the family to a geneticist at a nearby university teaching hospital.

The geneticist meets the entire family, takes a thorough medical history, and an examination reveals a number of common features between certain family members. Each of the probands A, B, and C shares thin, almost translucent skin that is fragile, easy to break, and slow to repair. Their skin is unlike that of the other two family members (A's mother has died). Furthermore, each proband demonstrates previously unexplored hypermobility of finger, elbow, and hip joints far in excess of other family members and nearby hospital staff, although the skin itself is not particularly hyper-extendable. Consulting with rheumatologist colleagues, the geneticist makes a preliminary diagnosis of a form of Ehlers–Danlos syndrome, and in further investigation invites a surgeon to take a skin sample very carefully from each family member. A histopathologist tests the samples for the presence of collagen, found to be poorly developed in the probands, but normal in other family members. This finding supports the preliminary diagnosis of one of the (at that time) seven recognized forms of the syndrome.

In the decades that follow each proband learns to live with the syndrome, whilst slow and steady advances in biomedical science make progress in defining the various sub-syndromes. On the family front, proband C (who now also has flat feet and thin skin) has a daughter who appears to be free of the syndrome. Proband B first has a daughter who is apparently not affected by the syndrome, and then a second one who displays flat feet, hypermobility, and soft and translucent skin that cuts and bruises easily. A's grandson has a normal phenotype.

By 2010, molecular genetics have identified many genes linked to certain of the sub-syndromes, which now number 13 (Table 16.5). However, although an advance, molecular genetics have not fully clarified the classification system. For example, the clinical phenotype of marked joint hypermobility is present in types I (classical, with mutations in *COL5A1*), III (hypermobility, *TNXB*), and VIIa (**arthrochalasia** multiplex congenita, *COL1A1*). Similarly, thin translucent skin and marked bruising are part of types IV (vascular, *COL3A1*) and VIIc (**dermatosparaxis**, *ADAMTS2*). Therefore, it could be argued that this family may fall into any of several subtypes. However, there are other links. As a neonate, the youngest proband, C, required surgery for pulmonary valve stenosis, a pathology linked to mutations in *COL1A2*, whilst her grandmother ('A', now aged 70) suffers a dislocation of the hip with resultant arthritis (this being a characteristic of type VIIa) and a perforated bowel (rupture of the bowel being a characteristic of type IV). Fortunately, the family seems to have avoided types VIII and XIII (characterized by periodontal loss) and type IX (brittle cornea). It is therefore clear that there is considerable scope for the further understanding of this syndrome, but whatever that entails for this case study, the fact that the family phenotype is autosomal recessive implies the genetic lesion(s) lie(s) with types II, III, VII, XI, or XII, although some of these may be excluded by other testing.

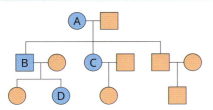

FIGURE 16.15

Genetic family tree of Case study 2. Probands A–D show signs of Ehlers–Danlos syndrome (circles: females; squares: males).

to metabolism. However, heterozygosity is found in both the lean and the obese. Nevertheless, SNPs that cluster in the first intron of *FTO* have been very strongly (p < 10^{-50} and p < 10^{-60} in European-ancestry populations) linked with BMI, although there is a degree of racial and ethnic variation. Although the frequency of candidate genes is relatively large, their penetrance into the variance in BMI is overall very small.

16.6 Metabolic disease

Much of this chapter so far has focused on the effects of a gene on an organ, an organ system, or the whole body. The present section will discuss examples of cases where a mutation has a precise outcome in terms of a metabolic pathway, and so a precise molecule or molecules. Many are regarded as inborn errors of metabolism and are based on biochemistry, and the leading—and often also most common—diseases are named after those who first described them, whilst others are classified by the substrate, its enzyme, the product, and/or clinical features. As a group, they were linked to 1,480 deaths in England and Wales in 2021.

16.6.1 Phenylketonuria

Phenylalanine hydroxylase (PAH), coded for by *PAH* at 12q22–24.1, catalyses the formation of tyrosine from phenylalanine. Deficiency in the enzyme, perhaps due to loss of function in one of the >950 variants in *PAH* (~60% being missense) therefore leads to insufficient tyrosine (required for dopamine, melanin, and noradrenaline) and excess levels of phenylalanine, which is converted to phenylpyruvate. The latter is converted to a ketone *in utero*, and excreted via the placenta, but appears in neonatal urine, hence the autosomal recessive disease phenylketonuria (PKU). However, increased phenylalanine (and so PKU) may also be due to a deficiency of tetrahydrobiopterin, a cofactor to PAH.

Unaddressed, PKU leads to widespread and irreversible pathology of the central nervous system with microcephaly, motor deficits, eczema, autism, seizures, development problems, and a reduced intelligence quotient. PKU appears in the European population at a frequency of ~100/million, with wide geographical variation (10/million in Finland, 250/million in Turkey), and is relatively easy to test for, leading to neonatal screening. This may be by the Guthrie test; other methods include immunoassay and tandem mass-spectrometry. One guideline suggests a cut-off point of serum phenylalanine >600 µmol/l should be treated, the leading method being dietary avoidance of phenylalanine for life, with potential for tyrosine supplementation.

16.6.2 Lysosome storage disorders

As implied, this large group of more than 20 disorders is characterized by defects in genes coding for enzymes that generally have roles in lysosomal catabolism. Consequences include failure to generate a product and failure to degrade a substrate, which subsequently builds up within the lysosome, possibly resulting in the rupture of the organelle, release of its enzymes, and so damage to the cell. All (except those X-linked) are autosomal recessive.

Lysosomal acid lipase deficiency

This autosomal recessive disease, sometimes described as Wolman disease and present at a frequency of between 6 and 25/million, is caused by mutations in *LIPA* at 10q23.2–23.3, the enzyme product normally catabolizing lipids such as cholesterol esters and triglycerides within hepatic and macrophage lysosomes. Loss-of-function mutations in *LIPA* lead, in its most severe form—Wolman's syndrome—to build-up of these substrates in intestinal tissues with consequent malabsorption, failure to thrive, and fatty faeces in infants. However, milder phenotypes may present later. The most common genetic lesion is a splice-junction mutation, present in ~50% of cases, the remainder comprising over 100 other variants.

Developing pathology includes dyslipidaemia, lipid deposits in the spleen (with splenomegaly) and pancreas, with fatty liver, jaundice, and so hepatic failure. Liver failure is the most common cause of death, typically before the age of 40, although in some there may be accelerated atherosclerosis secondary to raised serum LDL-cholesterol, requiring statins. However, the latter will not ameliorate the fatty liver, the only treatment for this being transplantation. Intravenous enzyme replacement therapy with sebelipase 1 mg/kg once every other week is generally well tolerated and effective in reducing serum liver function tests and lipids, and liver volume and fat content.

Gaucher disease

Cerebrosides are a group of glycosphingolipids with important functions in several cell processes, such as the cell membrane. β-glucocerebrosidase, coded for by *GBA1* at 1q22 and found in the membrane of the lysosome, hydrolyses its substrate glucocerebroside. Any of 400 mutations in *GBA1* that lead to a reduction or absent β-glucocerebroside lead to the rare autosomal recessive Gaucher disease, one of the several lysosomal storage diseases. It has a frequency of around 20/million (higher in culturally inbred populations) and a carrier status of 1 in 20, and is the most common lysosomal storage disease.

Impaired enzyme activity leads to build-up of the glucocerebroside in macrophages and neurons and is likely to account for the major clinical features of splenomegaly, hepatomegaly, bone lesions (e.g. pain, thrombocytopenia), and neurological impairment, although these vary in each of the three subtypes. Type 1 (adult) disease is most common (~90%), and is characterized mostly by visceral disease, with little or no neurological damage, whereas neurological impairment is the leading feature in types 2 (neonatal: severe) and 3 (juvenile: variable) (~5% each). Those with certain heterozygous or homozygous SNP mutations in *GBA1* are at risk of Parkinson's disease as lack of β-glucocerebrosidase leads to the build-up of α-synuclein oligomers and fibrils in neurons and so the formation of Lewy bodies.

Diagnosis is based on the presence of leukocyte enzyme activity although flow cytometry of blood monocytes is more accurate. Molecular genetics are rarely called upon. Other biomarkers include chitotriosidase, chemokine ligand 18, glucosylsphingosine, and ferritin. Treatment focuses on intravenous recombinant enzyme replacement with imiglucerase (e.g. 60 units/kg every two weeks) and velaglucerase (60 units every other week), and on substrate reduction with a glucocerebroside synthase inhibitor such as miglustat (e.g. 100 mg tds). Bone marrow transplantation is no longer offered, and gene transfer into haemopoietic stem cells has been disappointing but may improve. Life expectation is mildly decreased in type 1 disease, considerably in type 2 disease (age ~3–5) and markedly in type 3 disease (age 25–35).

Rare lysosome storage disorders

With a frequency of <5/million, these include Tay–Sachs disease, characterized by any one of more than 100 mutations in *HEXA* at 15q23, coding for the α subunit of β-hexosaminidase A, an enzyme that cleaves gangliosides; Fabry disease, caused by a deficiency in α-Galactosidase, coded for by *GLA* at Xq22.1, which degrades globotriaosylceramide; and Niemann–Pick disease, resulting from mutations in *SMPD1* at 11p15.4, which codes for sphingomyelin phosphodiesterase 1. Lack of function of the particular enzymes leads to deposits in the substrates that can be fatal.

Mucopolysaccharidoses (MPS)

The lesion for this group of diseases is mutation in genes whose enzymes digest mucopolysaccharides/glycoaminoglycans such as chondroitin sulphate, keratan sulphate, and hyaluronic acid, found in connective tissues throughout the body and in many organs. As a group, their frequency is ~40/million, but this varies in subtypes, type IV at 4–13/million and type VII at <3/million, and most are autosomal recessive (for a more detailed overview of the subtypes, see Sun 2018). As with all lysosome storage diseases, the excess of undigested substrate leads to clinical features, which vary by severity, rate on onset, and organ/system involvement. These include seizures, disease of the ears/eyes, dementia, depression, ataxia, and other neuromuscular disease (e.g. tremor) and skeletal abnormalities, often of the spine and skull. The best-known MPS, type I, also known as Hurler syndrome, is caused by any one of some 200 mutations in *IDUA* at 4p16.3 that codes for α-L-iduronidase, and has a prevalence of around 10/million. A loss of function in this gene leads to reduced or absent enzyme activity, and so build-up of its substrate within lysosomes that eventually burst, liberating other functional lysosomes into the cytoplasm and the associate pathological consequences. Management includes replacement therapy with laronidase (a genetically engineered form of iduronidase) given intravenously at a dose of 100 U/kg weekly. The second most frequent MPS, type II, or Hunter syndrome, has a prevalence of 8/million and is caused by a loss-of-function mutation in *IDS*, located on the X chromosome and coding for iduronate 2-sulphatase. It can also be treated by replacement therapy with the recombinant enzyme idursulphase.

Other lysosome storage diseases

Many other gene-directed enzyme deficiencies are known to cause a variety of disease and syndromes. These include α- and β-mannosidosis, characterized by failure to digest mannose-rich oligosaccharides and caused by mutations in *MAN2B1* and *MAN2B2* respectively, resulting in insufficient α- and β-mannosidase, respectively; and fucosidosis, resulting from *FICA1* mutations and so lack of fucosidase, which degrades complex carbohydrates rich in fucose.

16.6.3 Glycogen and glucose diseases

The metabolism of these two molecules is complex and therefore subject to many inborn errors of metabolism. Several of these are described as glycogen storage diseases, of which there are numerous forms that relate to the formation of glycogen or its metabolism and degradation back to glucose, and together they have an overall frequency of around 38/million. Therefore, the leading issue is with the availability of glucose, and so hypoglycaemia with its attendant problems (exercise intolerance, low muscle tone, etc.), particularly in the liver (with hepatomegaly) and muscles, but also the kidney and intestines, whilst in infants there may be failure to thrive. Many of these enzymes operate in the cytoplasm and some in lysosomes, which may technically be classified as lysosome storage disease.

Pompe disease

Pompe disease has the highest frequency (varying globally from 25 to 227/million) of the glycogen storage diseases and is characterized by deficiency of acid α-glucosidase, coded for by *GAA* at 17q25.3. Glycogen build-up in lysosomes leads (in common with other disease of this group) to damage to skeletal and cardiac muscle (often hypertrophic cardiomyopathy), hepatocytes, and neurons with allied signs and symptoms of peripheral myopathy, hypotonia, hepatomegaly, and hypoglycaemia with raised creatine kinase and lactate dehydrogenase. Infants will fail to thrive, feed with difficulty, and have delayed motor development with enzyme levels <1% of normal. In those with later-onset disease, where levels are 1–30% that of normal, cardiac involvement is less common but motor impairment is frequent. Enzyme replacement therapy is possible with intravenous recombinant α-glucosidase (myozyme) at a dose of 20 mg/kg once every two weeks.

Other glycogen storage diseases

Key features of the other major glycogen storage diseases are shown in Table 16.6, listing the most prominent and well-described diseases. Some (e.g. type X, phosphoglycerate mutase, and type XIII, enolase) are part of the glycolytic pathway. Clinical features, progression, and prognosis vary markedly: type IV is generally fatal in infancy, whereas type VI management consists of addressing the hypoglycaemia with supplements, dyslipidaemia with statins, hyperuricaemia with allopurinol. Where the disease is severe and/or has progressed to malignancy, transplantation is an option.

TABLE 16.6 Glycogen storage diseases

Type	Syndrome	Gene	Location	Enzyme
0	–	GYS1 (cardiac and skel-etal muscle), GYS2 (liver)	19q13.3, 12p12.2	Glycogen synthase
I	Von Glerke	G6PC	17q21.131	Glucose-6-phosphatase
II	Pompe	GAA	17q25.3	Acid α-glucosidase
III*	Corl/Forbes	AGL	1p21.2	4-α-glucanotransferase
IV**	Anderson	GBE1	3p12.2	1-4-α-glucan branching enzyme
V	McArdle	PYGM	11q13	Myophosphorylase
VI	Hers	PYGL	14q22.1	Glycogen phosphorylase
VII	Tarui	PFKM	12q13.11	Phosphofructose kinase

* Glycogen debranching enzyme deficiency.

** Glycogen branching enzyme deficiency.

Galactosaemia

Galactose, the monosaccharide product of lactose, can be metabolized to glucose by a short pathway involving three key enzymes, these being galactose-1-phosphate uridyl transferase (coded for by *GALT* at 9p13), galactokinase (*GALK* at 17q24), and UDP-galactose epimerase (*GALE* at 1p35–36). Loss-of-function variants of *GALT* fail to complete this metabolism, leading to increased levels of galactose, and so galactosaemia, a condition with a frequency of around 32/million in European populations, but ~2/million in East and Southeast Asia. Breast milk is rich in lactose, and 40% of the infant's calories derive from its hydrolysis into glucose and galactose, so that *GALT* mutations result in a serious calorific shortfall and hypoglycaemia. These in turn lead to poor weight gain and feeding, vomiting and diarrhoea, hypotonia, lethargy, and hepatocellular damage. Subsequent pathology may include cataracts, neutropenia, *E. coli* sepsis, cerebral oedema, and renal failure, whilst >80% of affected women suffer primary ovarian insufficiency. Diagnosis using red cells is relatively straightforward with standard biochemical enzymology, and treatment focuses on avoidance of lactose and galactose. Figure 16.16 summarizes key aspects of glucose/glycogen metabolism of this section, showing the points where the different types of disease operate.

SELF-CHECK 16.8

What is the major cellular difference between the lysosomal diseases and the glucose/glycogen diseases?

16.6.4 Other metabolic disease

Menkes disease

With a frequency of ~10/million, Menkes disease results from mutations in *ATP7A* at Xq21.1 that codes for an ATPase enzyme that is part of the complex importing copper into the cell, and transporting it within the cell. In most cases, a deletion or insertion in part of *ATP7A* produces a non-function truncated or missense product such that copper cannot enter the cell. Thus, copper is over-deposited in some tissues (intestines, kidney) but is insufficient in others, mostly the central nervous system.

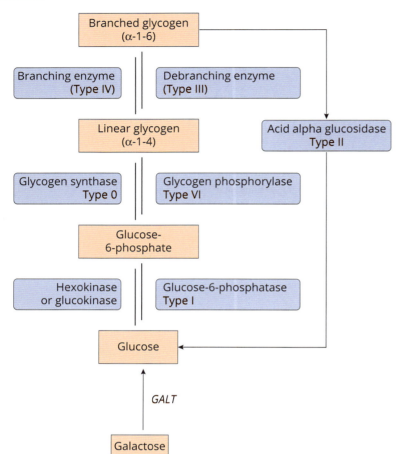

FIGURE 16.16

Summary of the biochemistry of glucose/ glycogen disease. The simplified metabolic pathway shows points at which particular types of glycogen storage diseases (types 0–VI) exert their effect and the position of *GALT* in causing galactosaemia.

Enzymes and other molecules requiring, or with roles in, copper metabolism include cytochrome c oxidase, superoxide dismutase, caeruloplasmin, and β-hydroxylase. The phenotype of Menkes disease is variable: the most severe form is apparent in infancy with hypotonia and failure to thrive, with progression to poor physical and mental development, and seizures, and leads to death within 3 years. The semi-specific sign is characteristic brittle, colourless, tangled, kinky, and fragile hair (Figure 16.17). At the polar extreme is occipital horn syndrome (protuberances on the occipital bone at the lower rear of the skull), wherein some *APT7A* mRNA is functional, and the signs and symptoms are less severe and delayed in comparison to the severe phenotype.

Wilson disease

This second form of copper disease is caused by mutations in *ATP7B* at 13q14.3 and is therefore chromosomally unrelated to *ATB7A* and Menkes disease (at Xq21.1). However, the product of *ATP7B* is 57% homologous to that of *ATP7A* and is also important in copper metabolism. Any of the 700 reported SNPs (such as H1069Q, present in ~65% of patients) or deletions in the gene for this copper-transporting protein lead to failure to export the metal. Thus, copper accumulates within the cell, particularly those of the liver (which has the highest expression of *ATP7A*) and brain (where copper levels may be increased 10-fold), causing liver failure and psychiatric, neurological, and related symptoms respectively, generally developing between the ages of 5 and 35 at a frequency of ~30/million.

Early changes in liver function tests, primarily bilirubin, point to a diagnosis, supported by copper biochemistry (including caerulopasmin and 24-hour urinary copper), imaging, and ultimately

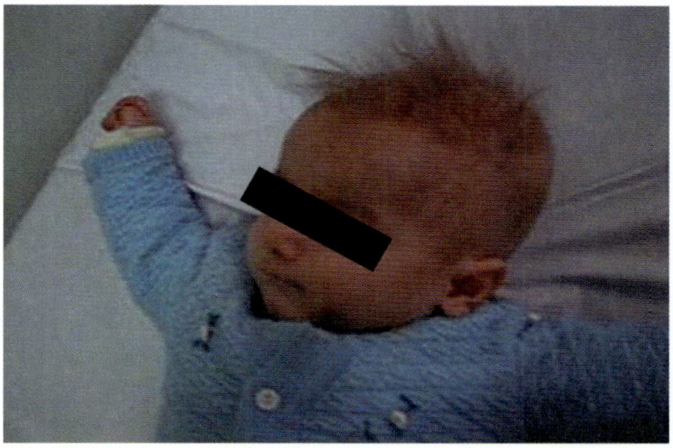

FIGURE 16.17
Neonatal hair in Menkes disease. The serum copper level of this patient was 15 µg/dl (reference range 70–150) and serum caeruloplasmin 58 mg/l (reference range 187–322). The neurological development was impaired, and he died of a respiratory infection aged 14 months.

Barzegar M, Fayyazie A, et al. Menkes Disease: Report of Two Cases. Iranian Journal of Pediatrics 2007: 17; 388–92.

molecular genetics. Ophthalmological signs include Kayser–Fleischer rings and sunflower cataracts. If untreated, this disease is fatal, but fortunately copper chelators such as penicillamine (1.5–2 g in divided doses daily) and trientine (450–950 mg in divided doses daily) are often highly effective in alleviating symptoms and reversing pathology; they require monitoring for effect. An additional treatment is to avoid foods rich in copper, such as chocolate, mushrooms, nuts, and shellfish, whilst antidepressants, mood-stabilizers, and anti-psychotics may be called for. Ultimately, liver transplantation may be required.

Gilbert's syndrome

Uridine diphosphate-glucuronosyltransfer-1, coded for by *UGT1A1* at 2q37.1, transforms lipophilic substrates into hydrophilic glucuronides, thereby facilitating their elimination in bile, urine, and faeces. Mutations in this gene, present in perhaps 60,000/million of the general population and generally autosomal recessive, lead to modestly increased levels of total (~10%) and unconjugated (~20%) bilirubin, a benign condition described as Gilbert's syndrome. Given this high frequency, it has been argued that the syndrome simply represents the polar extreme of the normal range. Intense physical exercise and sports, dysmenorrhea, pregnancy, surgery under general anaesthetic, and systemic infections may all induce a mild jaundice, but all other liver function tests are normal. The *in vitro* antioxidant properties of bilirubin are reputed to account for a mortality rate lower than that of the general population.

Ornithine transcarbamylase deficiency

Ornithine transcarbamylase, coded for by *OTC* at Xp11.4, is a key mitochondrial enzyme in the urea cycle, where it catalyses the transfer of an amino group (itself derived from ammonia) to ornithine to form citrulline (Figure 16.18).

Loss-of-function mutations (present in 15/million, mostly deletions) in *OTC* lead to failure to generate citrulline and a build-up of ornithine and ammonia. Should the salvage pathway transferring the ammonia to orotic acid be insufficient, hyperammonaemia may develop. In the neonate, this may arise 24–48 hours after birth, marked by lethargy and poor feeding, and may be acutely fatal. Late-onset forms have variable and non-specific signs and symptoms (atypical behaviour, headache, nausea, seizures, vomiting) depending on the extent of the hyperammonaemia.

Most cases display unequivocal biochemistry, but molecular genetics may be required. Treatments include avoidance of nitrogen in the diet and use of oral sodium benzoate (such as 500 mg tds) whilst acute episodes benefit from intravenous sodium benzoate (neonate and child; 250 mg/kg over 90 minutes). The disease is curable with liver transplantation.

A differential diagnosis for ornithine transcarbamylase deficiency is the HHH syndrome (hyperornithinaemia, hyperammonaemia, homocitrullinaemia) caused by lack-of-function mutations in *SCL25A15* at 13q14, coding for ornithine translocate, an enzyme required for the passage of ornithine into the mitochondrion (Figure 16.18).

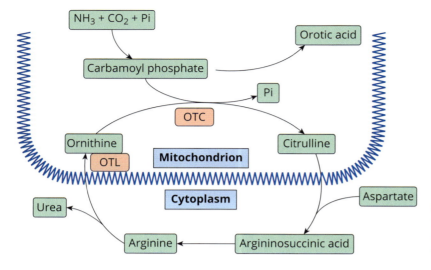

FIGURE 16.18

The urea cycle. OTC = ornithine transcarbamylase, OTL = ornithine translocase.

Homocysteine, cysteine, and methionine

The close relationship of these three amino acids warrants some attention. The biochemistry is summarized in Figure 16.19. The key products are methionine, which feeds into protein synthesis, and cysteine, a semi-essential amino acid that is also required in protein synthesis, but has other roles such as in the generation of the antioxidant glutathione and as a major component of metallothionines that transport certain cations. Four enzymes are of interest. Methylene tetrahydrofolate reductase (MTHFR) coded for by *MTHFR* at 1p36.22 controls the generation of methyl-THF from methylene-THF, a proton being donated by NADP. The methyl-THF donates its methyl group to homocysteine, forming methionine and THF, a reaction regulated by methionine synthase (which requires vitamin B_{12}), coded for by *MTR* at 1q43.

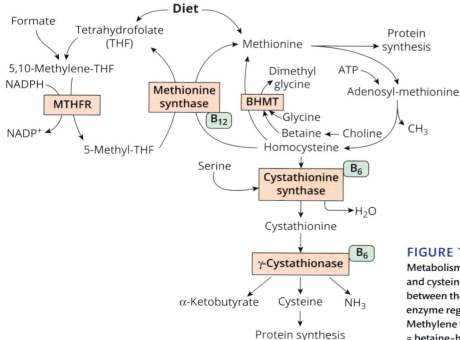

FIGURE 16.19

Metabolism of homocysteine, methionine, and cysteine. Summary of the relationship between the three amino acids and key enzyme regulators (boxed in red). MTHFR = Methylene tetrahydrofolate reductase. BHMT = betaine–homocysteine methyltransferase.

Homocysteine can be regenerated by the demethylation of methionine; the methyl group is transferred to an acceptor such as phosphatidyl ethanolamine and with the formation of the intermediate s-adenosylmethionine. Homocysteine is merged with serine to form cystathionine and water under the control of cystathionine synthase, coded for by *CBSL* at 21p12, which requires vitamin B_6. The cystathionine is metabolized to ammonia, α-ketobutyrate and cysteine under the control of γ-cystathionase, coded for by *CTH* at 1p31.1.

Mutations in any of these genes have metabolic consequences and so generally clinical features. The most common are in *MTHFR* and lead to increased levels of homocysteine in the blood (hyperhomocysteinaemia) and urine (hyperhomocysteinuria), both demonstrated with ease by the laboratory. The most severe (and very rare) manifestations of hyperhomocysteinaemia, brought on by very low or absent MTHFR enzyme, are often evident in infants, including accelerated atherosclerosis. The most common mutation, the SNP C677T, sees the conversion of a cytosine (C) to thymine (T) at position 677 in the gene. This results in an alanine to valine substitution, and so, in homozygotes, a 50–60% reduction in the activity of MTHFR. This substitution has a marked global variation, the frequency of homozygosity being 7.8–13.2% in seven European countries (18% in Italy), 10.7% in Australia, 2.2% in Yemen, and 0% in sub-Saharan Africa. In White and Black Brazilians frequencies are 10.3% and 1.6% respectively, whilst in the USA, frequencies for White and Black Americans are 11.9% and 1.2%. These data translate to frequencies of heterozygous carriers of C677T of around 35% in White populations, and 6–14% in Black populations, making it one of the most common hereditable metabolic mutations (Botto and Yang 2000).

A meta-analysis also reported marked geographical variation in the prevalence of hyperhomocysteinaemia: 73.1% in Iran, 62.3% in West Africa, 37.2% in China, 19.1% in Canada, and 6.9% in the USA (Zeng Y et al. 2021). However, the low level of hyperhomocysteinaemia in the latter is countered by the report of a prevalence of 26.1%, which varied by ethnicity: 24.6% in Latin Americans, 31.2% in White Americans (significantly higher), 22.5% in Black Americans, and 19.1% in Asian Americans. This study also found levels of the amino acid to be significantly higher (by 15%) in men compared to women (Carmel et al. 1999). The Other SNPs in *MTHFR* include A1298C and T1317C, both minimally increasing serum homocysteine.

In comparison, the factor V Leiden (FVL) mutation has a frequency of 5%, but the fact that neither C677T nor FVL bring a large risk of disease is due to their low penetrance into their particular pathological phenotype. However, both genes may bring disease if there are additional risk factors, which in the case of hyperhomocysteinaemia include other errors in the same pathways and low levels of the required B vitamin co-factors. Indeed, hyperhomocysteinaemia may arise despite a fully functioning *MTHFR* due to deficiency in these vitamins that will in turn lead to hypoactive methionine synthase, cystathionine synthase, and γ-cystathionase. Increased levels may also arise from the A2756G SNP in *MTR*, coding for methionine synthase, or from the T833C SNP in *CBSL*, coding for cystathionine synthase.

Whatever the aetiology, there is little firm evidence that hyperhomocysteinaemia is even a minor contributing factor to cardiovascular disease or venous thromboembolism, which themselves have far stronger environmental and genetic risk factors. Despite considerable effort, any outstanding link between C677T and any major pathology has yet to be reported. Treatments include ensuring adequate dietary folate and B vitamins, whilst some advocate betaine (trimethylglycine), a choline derivative which, via betaine-homocysteine methyltransferase, can generate methionine independently of methionine synthase, and also dimethylglycine (Figure 16.19).

Dyslipidaemia

Lipid biochemistry is described in section 7.3, increased LDL-cholesterol being a major risk factor for atherosclerosis. Serum cholesterol is the sum of that taken in the diet and that synthesized by the liver, which is under genetic control of many genes. In the general population, this leads to the bell-shaped curve of a normal distribution, high levels being described as polygenic hypercholesterolaemia. Mutations in a small number of genes also produce hypercholesterolaemia. As described in section 6.6.1, the leading condition is familial hypercholesterolaemia (FH), caused by mutations in any one of three genes:

- *LDLR* at 19p13.1–13.3, coding for the LDL receptor; by far the most common (~80-90% of cases) and present in the UK as a heterozygous form in 2,000–4,000/million, homozygous at around 1/million (NICE CG71), although the frequency is significantly higher in other populations, sometimes as a result what has been termed the 'founder effect' (Box 16.2)

BOX 16.2 The founder effect

By the late 18th century, north-east Canada was home to around 70,000 French settlers, centring on the city of Quebec, and with subsequent immigration, around 7 million now claim this heritage. The frequency of heterozygous familial hypercholesterolaemia in this population exceeds that of the population of France by over 3-fold (1/154 versus 1/500). This founder effect may be explained by the likelihood that the original settlers had a high frequency of mutations in *LDLR*, which expanded within the close-knit community to form the present-day increased incidence of hypercholesterolaemia. Founder effects have also been described in increased rates of breast cancer in Mexican women with European heritage compared to women with an Indigenous ancestry, whilst mutation 943ins10 in *BRCA1* in African American women originated in West Africa.

Cross reference

For a closer look at the studies on the founder effect that Box 16.2 is based on, see Mszar et al. 2020, Couture et al. 1999, and Hobbs et al. 1987 for hypercholesterolaemia in Canada, and Fejerman et al. 2010 and Pal et al. 2004 for breast cancer in Mexican and African American women respectively.

- *apoB* at 2p24–23, coding for the apoB molecule such that it does not interact with, and so activate, the LDL receptor; accounts for ~5% of FH cases

- *PCSK9*, coding for PCSK9, an essential co-factor for the LDL receptor; linked to <1% of cases

ApoE, coded for by *APOE* at 19q13.32, has three major isotypes, ε2, ε3, and ε4, the latter being a risk factor of hyperlipidaemia and atherosclerosis, whilst the ε2 variant is linked to type III hyperlipoproteinaemia.

CASE STUDY 16.3 Gilbert's syndrome

A 25-year-old male is taken to surgery under general anaesthetic for multiple fractures in the ribs, clavicle, and femur resulting from a road traffic accident. With the potential of a punctured lung, he is ventilated and held in a drug-induced coma in intensive care for 48 hours, towards the end of which he is noted to have mild sclerotic jaundice. A check on admission LFTs did indeed reveal a raised serum bilirubin that was presumed to have been the product of the accident, and the high level explained the jaundice (Table 16.7). This sign disappeared shortly afterwards.

Recovery went well and all LFTs moved towards admission levels after 96 hours, but blood taken at an outpatient follow-up revealed the bilirubin level to be high. Subsequent molecular genetic studies revealed an appropriate mutation in *UGT1A1* sufficient for a diagnosis of Gilbert's syndrome.

TABLE 16.7   Serial liver function tests for Case study 16.3

	Reference range	Admission	+24 hours	+48 hours	+96 hours	+8 weeks
ALT	5–42 IU/l	30	35	40	35	30
ALP	20–130 IU/l	45	60	52	42	43
AST	10–50 IU/l	23	30	29	25	22
Bilirubin	<17 μmol/l	20	31	46	38	27
GGT	5–55 IU/l	21	25	27	24	22

SELF-CHECK 16.9

Mutations in genes leading to hyperhomocystinaemia, Gilbert's syndrome, and familial hypercho-lesterolaemia are present in the population at very high levels. Why don't they all cause a great deal of disease?

16.7 Haemoglobinopathy

This disease provides the opportunity to look at the diverse effects of mutations in closely related genes that often have similar clinical features and are very likely to be the most common specific forms of chromosomal disease in our species. With six genes generating several alternative clinical pheno-types, this disease would otherwise be discussed in section 16.5, but its sheer global scale justifies its own section. This section will focus on genetic aspects: clinical and laboratory features are described in section 11.5.5.

16.7.1 Introduction

Each year, between 300,000 and 400,000 babies are born—half of them in the Democratic Republic of Congo, India, and Nigeria—with serious haemoglobin disorders, the most common of these being homozygous sickle-cell disease (220,000), haemoglobin SC disease (55,000), and β-thalassaemia major (20,000) (Williams and Weatherall 2012; Chakravorty and Dick 2019). Some 90% of these infants survive into adulthood, with 7% of the world's population (over 560 million) being carriers (Kohne 2011), a figure that overshadows all other genetic disease.

The basic genetics of the various types of Hb molecules are described in section 11.1.3. Four genes and products are of interest: *HBA1* and *HBA2* at 16p13.3 code for α-globin, *HBB* at 11p15.4 codes for β-globin, *HBD* at 11p15.4 codes for δ-globin, and *HBG1* and *HBG2* at 11p15.4 code for γ-globin. Two molecules of α-globin and two of β-globin form HbA ($\alpha_2\beta_2$), two molecules of α-globin and two of δ-globin form HbA$_2$ ($\alpha_2\delta_2$), whilst α-globin and γ-globin form foetal haemoglobin (HbF) ($\alpha_2\gamma_2$) and are a clear example of gene duplication.

The leading chromosomal and genetic disease of this molecule is haemoglobinopathy, which causes a haemolytic anaemia (section 11.5.5), and is due principally to mutations in *HBA1*, *HBA2*, and/or *HBB*. These can occur either on both copies of the relevant chromosome or on only one, giving homozygous and heterozygous disease respectively, and most can be classified in two groups:

- Quantitative defects, principally deletions in those sections of chromosomes 16 and 13 that code for α-globin and β-globin, resulting in a reduced mass of globin protein produced: dozens (and perhaps hundreds) of amino acids are missing. These defects lead to the disease of thalassaemia.

- Qualitative abnormalities, generally of a SNP where the variant codon generates an alternative amino acid and so a structurally different globin molecule. There is no shortage of globin protein itself; the abnormality lies with the individual component amino acids. The primary consequence is sickle-cell disease, although there are other manifestations.

Key points

The haemoglobinopathies are the most common global genetic disease and bring a considerable degree of morbidity. Drug treatments are improving, especially for symptoms, but for many, blood transfusion is the only effective remedy for the chronic haemolytic anaemia.

16.7.2 Thalassaemia

This disease can be classified as α-thalassaemia, in which α-globin synthesis is reduced or absent, and β-thalassaemia, with reduced or absent β-globin synthesis. There are four variants of α-thalassaemia, as we have two copies of *HBA* on each of two chromosomes:

- A one-gene mutation, represented by $-/\alpha/\alpha/\alpha$, in which α-globin output is 75% of normal, is generally clinically and haematologically silent.

- Mutations in two genes, either on the same ($-/-/\alpha\alpha$) or different ($-/\alpha/-/\alpha$) chromosomes is alpha-thalassaemia minor, also known as α-thalassaemia carrier, or trait. In both cases, α-globin production is 50% of normal and there is a mild anaemia.

- Deletion or functional inactivity of three genes, that is, $-/-/-/\alpha$, is HbH disease (HbH being a β-globin tetramer). It presents clinically as a type of thalassaemia intermedia and is characterized by levels of α-globin that are 25% of normal.

- Complete loss of all four α-globin genes ($-/-/-/-$, thus no α-globin) results in haemoglobin Bart's (four γ-globin molecules), also called hydrops foetalis syndrome. It is incompatible with life, and death usually occurs *in utero* or soon after birth.

β-thalassaemia is characterized by deficient or absent β-globin, caused by SNPs or deletions in either copy of *HBB*, of which over 200 have been described. Those where β-globin is absent are described as β^o, those where some β-globin is present as β^+, leading to a number of possibilities. The genotypes of heterozygous disease include $\beta^o\beta$, $\beta^+\beta$ or perhaps $\beta^o\beta^+$, so that the overall cell phenotypes become combinations such as $\alpha\alpha\beta^o\beta$ and $\alpha\alpha\beta^+\beta$. The homozygous mutation in both genes may therefore include $\alpha\alpha\beta^o\beta^o$, $\alpha\alpha\beta^+\beta^+$, and the composite $\alpha\alpha\beta^o\beta^+$.

The clinical expression of these variants depends very much on the extent of the respective deletion or SNP, and so on how much mature and functioning β-globin is present in the cell. Whilst protein electrophoresis may be helpful, ultimate diagnosis is with molecular genetics. Nevertheless, each thalassaemia is managed individually according to red blood cell indices and the severity of the disease.

16.7.3 Sickle-cell disease

The disease is caused by a SNP in *HBB* resulting in an abnormal form of the β-globin molecule, referred to as β^s. Since there are two copies of *HBB*, if both are mutated, and so homozygous, the genotype of the entire Hb molecule is HbSS, hence $\alpha_2\beta^s\beta^s$, or $\alpha_2\beta_2^s$, but if only one is a mutation, and so heterozygous, the genotype is HbAS, or $\alpha_2\beta^s\beta$ (Figure 16.20).

The product of the mutated *HBB* is a different form of the β-chain protein, where the normal amino acid glutamic acid (DNA triplet GAG) is replaced by a different amino acid, valine (GTG). The mutated β-globin retains the ability to combine with the second normal β-globin molecule and the two α-globins, the entire haemoglobin molecule now being named haemoglobin S. However, this

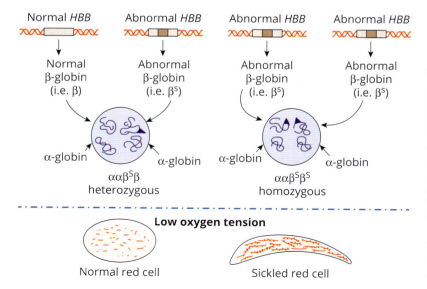

FIGURE 16.20

Sickle-cell disease. On the left, a normal *HBB* produces a normal β protein that combines with the product of an abnormal *HBB*, that is, a β^s-globin. These combine with two α-globin genes to give the heterozygous HbAS. On the right, two mutated *HBB* genes each produce a mutated β^s-globin, which combine with normal α-globins to give homozygous sickle haemoglobin, that is, HbSS. Under most conditions of marginally low oxygen tension, the cell with HbAS retains its morphology, whereas the $\alpha_2\beta^s\beta^s$ haemoglobin forms insoluble polymers that cause the characteristic shape of the red cell.

mutated haemoglobin is unable to carry oxygen with the same efficiency as does haemoglobin A, causing an anaemia.

Furthermore, under certain circumstances, the abnormal haemoglobin S polymerizes to such an extent that the red blood cell adopts a curved shape, giving the condition its name—sickle-cell disease. A curious feature of this pathology is that it manifests with great variety in people with exactly the same mutation: in some it can be relatively mild whilst in others it can be debilitating with a wide range of clinical problems, such as venous ulcers and bone pain.

This mutation is of further scientific interest because those individuals with a sickle mutation in only one of their beta globin genes (that is, $\beta\beta_s$) are relatively protected from malaria, and therefore have an evolutionary advantage over those with both their beta genes mutated ($\beta_s\beta_s$) or with normal beta genes ($\beta\beta$). This heterozygote advantage has been driven by the high frequency of sickle-cell disease in those areas of Africa and Asia where malaria is endemic, such that in some areas 15% of the population are heterozygous. In sub-Saharan Africa, children with HbAS are 90% less likely to develop severe and complicated *Plasmodium falciparum* malaria than HbAA children ($p = 1.6 \times 10^{-225}$), at a cost of a 50–90% under-5 mortality rate in those with sickle-cell anaemia. The same principle is very likely to explain the high frequency of other genes leading to changes in the cytoskeleton of the red cell (section 16.4.4).

SELF-CHECK 16.10

What is the major genetic difference between the two major forms of haemoglobinopathy?

16.7.4 Other haemoglobinopathy

Although HbS is the dominant β-globin disease, there are others. An alternative mutation in *HBB* leads to DNA triplet AAG and so lysine, denoted HbC and β^C. HbE is also caused by a mutation leading to a lysine substitution, leading to β^e, but at a different location, and can reach a frequency of 35% in Southeast Asia. Combinations of genes, such as HbS and HbC (i.e. HbSC (the genotype being $\beta^s\beta^c$)), are relatively common, whilst heterozygotes in other conditions, such as HbE with both α- and β-thalassaemia, are known.

There may also be fixed gene mutations, an example being the unequal meiotic cross-over between homologous δ- and β-globin genes (*HBD* and *HBB*), resulting in the formation of a fused $\delta\beta$-globin gene. This 'Lepore' gene when heterozygous clinically resembles β-thalassaemia trait with a mild anaemia, but when homozygous presents as does homozygous β-thalassaemia with markedly more severe disease.

 Chapter summary

- Chromosomal abnormalities are common: around half of all spontaneous abortions carry abnormalities, and have a frequency of around 5,000/million live births.

- Major chromosomal disease includes Turner (XO), Down's (G21 trisomy), Klinefelter (XXY), and Jacobs (XYY) syndromes.

- Muscular dystrophy and DiGeorge syndrome are some of the major diseases caused by deletions.

- Single-gene defects lead to von Willebrand's disease, neurofibromatosis type 1, fragile X syndrome, cystic fibrosis, and haemophilia.

- Several genes often contribute to a single clinical phenotype, the primary examples being Alzheimer's disease and diseases of iron overload, whilst atherosclerosis typifies a multi-system disease caused by several genes.

- The most common genetic metabolic diseases include Gilbert's syndrome, hyperhomocysteinaemia, and familial hypercholesterolaemia, although only in the latter two cases are homozygotes likely to suffer serious adverse clinical symptoms.

- Haemoglobinopathy, in its various manifestations, is the most common global genetic disease and is a likely product of selection pressure by malaria.

 # Further reading

- Morrow BE, McDonald-McGuinn D, Emanuel BS, et al. Molecular genetics of 22q11.2 deletion syndrome. Am J Med Genet A 2018: 176; 2070–81.

- Shankar RK, Backeljauw PF. Current best practice in the management of Turner syndrome. Ther Adv Endocrinol Metab 2018: 9; 33–40.

- Bonomi M, Rochira V, Pasquali D, et al. Klinefelter syndrome (KS): genetics, clinical phenotype and hypogonadism. J Endocrinol Invest 2017: 40; 123–34.

- Bull MJ. Down syndrome. New Engl J Med 2020: 382; 2344–52.

- Gao Q, McNally EM. The dystrophic complex: structure, function, and implications for therapy. Compr Physiol 2015: 5; 1123–239.

- Zeng Y, Li FF, Yuan SQ, et al. Prevalence of Hyperhomocysteinemia in China: An Updated Meta-Analysis. Biology (Basel). 2021 Sep 26: 10(10); 959. doi: 10.3390/biology10100959.

- Meesrer JAN, Verstraeten A, Schepers D, et al. Differences in manifestations of Marfan syndrome, Ehlers–Danlos syndrome, and Loeys–Dietz syndrome. Annals Cardiothorac Surg 2017: 6; 582–94.

- Neuner SM, Tcw J, Goate AM. Genetic architecture of Alzheimer's disease. Neurobiol Dis 2020: 143; 104976. Doi: 10.1016/j.nbd.2020.104976.

- Polissidis A, Petropoulou-Vathi L, Nakos-Bimpos V, et al. The future of targeted gene-based treatment strategies and biomarkers in Parkinson's disease. Biomolecules 2020: 10; 912. Doi:10.3390/biom10060912.

- Hollerer I, Bachmann A, Muckenthaler MU. Pathophysiological consequences and benefits of HFE mutations. Haematologica 2017: 102; 809–17.

- Lood RJF, Yeo GSH. The bigger picture of FTO—the first GWAS-identified obesity gene. Nat Rev Endocrinol 2014: 10.51–61.

- Sun A. Lysosomal storage disease overview. Annals Transl Med 2018:6; 476. Doi: 10.21037/atm.2018.11.39.

- Sullivan R, Yau WY, O'Connor E, et al. Spinocerebrellar ataxia: an update. J Neurol 2019: 266; 533–44.

- Cook A, Giunti P. Friedreich's ataxia: clinical features, pathogenesis, and management. Br Med Bull 2017: 124; 19–30.

- Williams TN, Weatherall DJ. World distribution, population genetics, and health burden of the hemoglobinopathies. Cold Spring Harb Perspect Med 2012 Sep 1: 2(9); a011692. doi: 10.1101/cshperspect.a011692.

- Achour A, Koopmann TT, Baas F, et al. The Evolving Role of Next-Generation Sequencing in Screening and Diagnosis of Hemoglobinopathies. Front Physiol. 2021 Jul 27; 12: 686689. Doi: 10.3389/fphys.2021.686689.

● **Malfait F, Francomano C, Byers P, et al. The 2017 international classification of the Ehlers–Danlos syndromes. Amer J Med Genet Part C 2017: 175C; 8–26.**

Useful websites

- Information on Alzheimer's disease by the Alzheimer's Society: **https://www.alzheimers.org.uk/about-dementia/types-dementia/alzheimers-disease**

- Foundation for Prader-Willi Research: **https://www.fpwr.org/**

- Down's Syndrome Association: **www.downs-syndrome.org.uk**

- Sickle Cell Society: **http://www.sicklecellsociety.org/**

- United Kingdom Thalassaemia Society: **https://ukts.org/**

- Umbrella site for numerous genetic diseases: **https://geneticalliance.org.uk/our-members/**

 Discussion questions

16.1 Ehlers–Danlos syndrome has 13 subtypes. Devise a system with fewer subtypes by merging those with similar features.

16.1 Which genetic diseases can be most easily identified with biochemical testing, and why?

Glossary

Acanthosis nigricans Blue or black patches of skin discolouration.

Adenocarcinoma A tumour that arises from epithelial glandular tissues.

Adipocyte A cell that stores fat.

Adipocytokines Cytokines such as leptin and adiponectin released from adipocytes with possible metabolic activity.

Adiponectin A hormone product of an adipocyte with roles in glucose and lipid metabolism.

Alkylating agent A form of cancer chemotherapy that alkylates (adds an alkyl group to) DNA, thereby interfering with its function.

Allergens Substances that provoke an allergic response.

Alpha-feto protein (AFP) A protein produced by the foetus, such that high levels in the adult imply an abnormality such as liver cancer.

Alzheimer's disease A slowly progressing brain disease, with a variable course. Characterized initially by features such as loss of memory and disorientation, later by dementia and psychosis.

Ampulla of Vater The anatomical point where the bile duct and pancreatic duct merge, and empty their products into the duodenum.

Amylase An enzyme that digests carbohydrates.

Amyloidosis Build-up of abnormal proteins (amyloids) in tissues and organs.

Anaerobic In the absence of oxygen.

Anaphylactic Pertaining to an acute allergic response, with any number of symptoms, which may be fatal.

Anastomosis The joining together of two tubes.

Anderson–Darling test A statistical test to help determine the distribution of a data set.

Androgen A male sex hormone, such as testosterone.

Angina Chest pain of a cardiac origin.

Angioedema Swelling of the skin, often of the face and often induced by an inflammatory process (in contrast to 'standard' oedema).

Angiogenesis The process of the generation of new blood vessels.

Angiography A procedure for imaging the integrity of arteries.

Angioplasty A procedure for reducing an arterial stenosis.

Anosmia–dysgeusia Loss of smell combined with a taste disorder in which all foods are perceived to taste sour, sweet, bitter, or metallic.

Aplasia Failure of, or reduction in, growth.

Apoptosis A programmed process by which a cell initiates its own death, often described as cellular suicide.

Aromatase An enzyme involved in the synthesis of oestrogen.

Arrhythmia An irregular heartbeat.

Arthralgia Pain in the joints.

Arthritis Inflammation of the joints.

Arthrochalsia Severe joint hypermobility.

Ascites Excess fluid within the abdominal cavity.

Asthma A disease of immunological hypersensitivity to foreign particles such as pollen, causing respiratory symptoms.

Asymptomatic Being free of any clear symptoms of disease.

Ataxia-telangiectasia A rare genetic condition characterized by ataxia (poor movement coordination) and telangiectasia (localized patches on the skin formed from dilated capillaries).

Atherogenesis The formation of an atheroma, or of atherosclerosis.

Atheroma The mass of cholesterol-rich material within the arterial wall that is the basis of atherosclerosis.

Atherosclerosis Formation of cholesterol-rich atheroma within arteries, leading to conditions such as coronary artery disease.

Atrio-ventricular node A focus of cells that passes electrical impulses from the sinoatrial node to the bundle of His.

Atrophic Referring to organs that have wasted away due to underuse or a direct pathological process.

Atrophy The wasting away of a group of tissues or an organ, often by a disease or by inactivity.

Auerbach's plexus A branching network of nerves supplying various intestinal muscles.

Autoantibody An antibody directed towards an antigen on the host's own cells, molecules, or tissues.

Autoimmune disease A diverse set of conditions in which the body's immune system attacks its own cells, tissues, and organs.

Autoimmunity A diverse collection of conditions characterized by the attack of the body's immune system on its own organs and tissues.

Autophagy The self-digestion of certain organelles and proteins.

Autosome A non-sex chromosome.

Axillary Pertaining to the axilla (armpit).

Azoospermia The absence of sperm.

Bacturia Bacteria in the urine.

Benign prostatic hyperplasia Excessive growth of the prostate that has no major pathological consequence.

Bisphosphonate A class of drug that improves bone density by inhibiting the osteoclasts that resorb bone.

Brachydactly Short fingers.

Brachytherapy The placement of sealed sources of radioactive elements near a tumour: therefore a form of radiotherapy.

Bradford Hill criteria A set of nine principles to help determine whether a particular feature (such as a virus) causes a disease, first developed by Sir Austin Bradford Hill.

B-type natriuretic peptide Secreted by ventricular cardiomyocytes, increased levels imply a disease process such as heart failure.

Caesarean section The surgical procedure through which a foetus is delivered from the uterus via the anterior abdomen.

Cancer A progressive disease of the abnormal growth of a group of cells that, unless treated, is invariably fatal (see also neoplasm).

Carcinoembryonic antigen (CEA) A protein produced by foetal intestinal cells, such that high levels in the adult imply an abnormality such as colorectal cancer.

Carcinogenesis The process of the development of a cancer.

Carcinoma A cancer that arises from epithelial cells.

Cardiomyocyte A muscle cell of the heart.

Chagas disease Condition caused by the protozoal parasite *Trypanosoma cruzi*.

Chi-square test A statistical test to determine if a difference between two sets of categorical data is meaningful.

Cholestasis Blockage of the bile duct, perhaps by a gall stone or inflammation.

Chordae tendineae Strands of fibrous tissues connecting the papillary muscles to the mitral and tricuspid valves, thereby preventing their inversion.

Chylomicrons Micro-globules composed of varying concentrations of cholesterol subtypes and triglycerides.

Chyme The semi-digested food material that leaves the stomach and is further digested in the intestines.

Cirrhosis Fibrotic scar tissue in the liver, formed by the result of a disease process such as alcoholism.

Cis- and trans-isomers Different forms of the same molecule that define alternative shapes: in this case the cis- form leads to an angled molecule whereas the trans- isomer leads to a linear molecule.

Coarctation A stricture or narrowing, often of the aorta close to the heart.

Colitis Inflammation of the colon.

Colostrum The first type of milk produced after childbirth, particularly enriched for bioactive components (as opposed to later nutritional milk).

Columnar epithelial cells A type of epithelial cell that is taller than it is wide, hence 'column'. It can be found in the intestines and elsewhere.

Complement A collection of proteins that come together to form pores in cell membranes and that have other immunological activities.

Computer-assisted tomography (CAT, or CT) A technique that provides images of the organs and tissues of the body from complex two-dimensional X-ray analyses.

Confidence A statistical term defining how confident we are of the accuracy of a result.

Corpus cavernosum Paired masses of spongy tissue in the shaft of the penis that, when engorged with blood, enable an erection.

Correlation A precise statistical test and term to determine any significant link between two sets of continuously variable data sets.

Corticosteroids A group of steroid hormones synthesized in the adrenal cortex; the most well-known is cortisol.

Cranial nerve A group of 12 nerves that arise directly from the brain

Craniofacial Referring to the skull and face.

Creatinine An amino acid marker of renal function.

Crohn's disease An inflammatory disease of the intestines whose aetiology is unknown; the full width of the section of the intestine is often involved.

Cyanosis A bluish-purple tinge to the skin implying hypoxia or ischaemia.

Cystectomy Removal of the bladder.

Cystic fibrosis A genetic disease caused by the excessive production of thick mucus in organs such as the lungs, pancreas, and intestines.

Cystitis Inflammation of the urethra.

Cytokine Small inter-cell signalling molecules with metabolic and immunological activity.

D-dimer A breakdown-product of a clot, such that high levels imply a high degree of clot removal.

Debridement Removal of dead or dying tissues.

Deep-vein thrombosis A condition where clots form in the veins of the calf, thigh, and groin.

Deletion The process of removal of a section of a chromosome.

Dermatosparaxis Skin elasticity and fragility.

Dexamethasone A commonly used glucose-based steroid that functions primarily as an immunosuppressant.

Diabetic ketoacidosis A serious metabolic condition of diabetes characterized by high levels of ketones in the blood, and by acidosis.

Diagnosis The process by which the presence of a particular disease is determined.

Digital gangrene Destruction of the fingers resulting from damage and obstruction of blood vessels and so profound hypoxia.

Dilatation and curettage Surgical removal of the endometrium by scooping/scraping.

Diuresis Production of urine.

Diuretic A drug that induces the formation of water-enriched urine.

Diverticulum Small pockets, in-tuckings, or bulges in the intestines, bladder, or urethra.

Doppler A physical effect of changes in the frequency of a waveform relative to the motion of bodies (in this case, blood cells).

Down's syndrome A genetic condition caused by the presence of an extra chromosome and linked to growth restriction and learning disability.

Duplication The process where a section of a chromosome is duplicated within that chromosome.

Dyslipidaemia Abnormal levels of lipids (such as cholesterol) in the blood.

Dysphagia Difficulty in swallowing.

Dysplastic/dysplasia Referring to abnormal growth of cells: may transform into a cancer.

Dyspnoea Literally, abnormality in breathing, but taken to be shortness of breath.

Dystonia Uncoordinated muscle activity.

Echocardiography A non-invasive technique for assessing the structure, function, and status of the heart muscle, chambers, and valves.

eGFR An important metric by which renal function is determined, based on creatinine.

Electrocardiography A non-invasive technique to determine abnormal features of the beating of the heart.

Electrolytes In this context, sodium and potassium.

Embolic Causing an embolus, that is, a circulating mass of material.

Emphysema Replacement of functioning alveoli by air-filled spaces, leading to a reduction in gas exchange.

Endobronchial Within the bronchus.

Endocarditis Inflammation of the inner layer of the heart.

Endocrine The system of ductless glands that secrete hormones directly into the blood.

Endometriosis A condition in which endometrial cells of the uterus move to the fallopian tubes, ovaries, and (rarely) the abdominal cavity.

Enteric Pertaining to the intestines.

Enterochromaffin Cells found in the intestines that part-regulate their function by secretion of serotonin.

Enterocyte An intestinal epithelial cell.

Enthesitis Inflammation of the entheses, the point where tendons and ligaments meet with bone.

Eosinophilia An eosinophil count above the top of the reference range.

Epidemiology The study of the process(es) by which disease appears in a population.

Epigenetics Alterations to histones and the DNA that do not involve the nucleotide sequence, but which do influence gene activity.

Erythema Patchy redness of the skin.

Erythroblast An intermediate in the production of red blood cells. The suffix 'blast' defines the cell as immature and is important in blood cancer.

Erythropoietin A renal hormone promoting red blood cell production.

Eustachian tube A tube linking the middle ear to the nasopharynx.

Exocrine A gland that secretes its product (such as an enzyme or mucus) via a duct.

False negative Failure to find a difference when one is truly present.

False positive The finding of a difference where one should never occur.

Fibrates A group of drugs that decrease triglycerides and increase HDL.

Fistula An abnormal connection between two hollow spaces, such as the rectum and vagina.

Gammopathy A form of pathology in which there are abnormal gamma-globulins.

Gastrula A stage in the differentiation of the embryo.

Genome-wide association study A complex analysis in molecular genetics that compares sections of DNA of those subjects with and without a certain condition or disease.

Germ cell A cell giving rise to ova or sperm.

Gestational Pertaining to pregnancy.

Glioma A tumour of neuron-supporting glial cells.

Glomerulonephritis Inflammation of the glomerulus.

Gout A destructive arthritis caused by the effects of sodium urate crystals within the joint and surrounding tissues.

Haemangioma A mass formed by an abnormal focus of blood vessels.

Haematinics The collective noun for haematological micro-nutrients, notably iron and certain vitamins.

Haematuria Blood in the urine.

Haemochromatosis Pathological deposits of iron in various organs.

Haemoglobinopathy Disease resulting from one or more genetic defect(s) in genes coding for haemoglobin.

Haemolytic anaemia A disease of the destruction of red blood cells.

Haemophilia An inherited disease where the ability to form a blood clot is impaired, leading to bleeding (haemorrhage).

Haemopoiesis The generation of blood cells in the bone marrow.

Haemoptysis Coughing up blood.

Haemostasis The physiology of blood coagulation—a balance between thrombosis and haemorrhage.

Haptocorrin A protein that carries vitamin B_{12} through the upper digestive tract, where it is given up to gastric intrinsic factor.

Hazard ratio Likelihood of a survival-related outcome (the hazard, such as death) being present in one group compared to another.

Heart failure Failure of the heart to deliver to the body the blood it needs for its physiological functions.

Helminth In this context, parasitic worms such as liver fluke and *Schistosoma* species.

Hepatocyte A liver cell.

Hepatomegaly An enlarged liver.

Hilum The point at which vessels and nerves enter/leave an organ.

human chorionic gonadotropin (hCG) A hormone normally produced by the trophoblast and so raised in pregnancy, but also increased in some reproductive malignancies.

Huntington's disease An inherited and ultimately fatal disease of the progressive loss of brain function, characterized by variable signs and symptoms.

Hydronephrosis Distension of part of the kidney by build-up of urine whose exit has been obstructed.

Hyperbilirubinaemia Levels of bilirubin in the blood above the top of the reference range.

Hypergammaglobulinaemia Levels of gamma-globulin above the top of the reference range.

Hyperglycaemic/hyperglycaemia High levels of glucose in the blood.

Hyperinsulinaemia High levels of insulin in the blood.

Hypernatraemia Serum sodium levels above the top of the reference range.

Hyperostosis Increased production/growth of bone.

Hyperparathyroidism An overactive parathyroid gland secreting excess levels of parathyroid hormone.

Hyperphagia Consumption of food in excess of that required to maintain healthy physiology.

Hyperplasia Excessive growth of a group of cells within a tissue or an organ.

Hypersensitivity Reactions brought about by an overactive immune system.

Hyperthyroidism An overactive thyroid gland.

Hypertonia High/increased muscle tone.

Hypertrophic Describes an organ that has enlarged.

Hypokalaemia Serum potassium levels below the bottom of the reference range.

Hypoplasia Failure of an organ or tissues to grow or develop.

Hypotonia Low/weak muscle tone.

Hypoxic Relating to hypoxia, a state of low levels of oxygen.

Immunocytochemical Referring to a laboratory method in which tissues are highlighted by an enzyme linked to an antibody: the coloured product of the enzyme's substrate defines the presence of the antigen being sought by the antibody.

Immunodeficiency An underactive immune system.

Inguinal Of the groin.

Insulinoma A tumour that secretes insulin.

Interferon A family of small molecules with immune-related activity.

IL-6 Interleukin-6, an inflammatory cytokine.

INR International normalized ratio, a measure of the prolongation of the prothrombin time.

Inter-quartile range A measure of the variation of a set of data whose distribution is non-normal.

Intestinal polyp An abnormal growth of intestinal tissues projecting into the lumen, often a precursor to a malignancy.

Intravesical Within the bladder.

Inversion The process where a section of a chromosome is removed and replaced in the opposite direction.

Ischaemic Of tissues and organs in a state of low oxygen (i.e. hypoxia).

Jaundice A yellow colouring of the skin caused by the deposition of high levels of bilirubin.

Ketogenesis The formation of ketones.

Knudson hypothesis The hypothesis that cancers are the product of two independent processes that happen in sequence.

Koch's postulates A set of four criteria to help determine whether a particular feature (such as a virus) causes a disease.

Lactocytes Cells that produce milk.

Leiomyomatosis A condition in which there is a (generally benign) tumour of smooth muscle cells.

Leukocyte A white blood cell.

Leukocytosis A white blood cell count above the top of the reference range.

Leukopenia A white blood cell count below the bottom of the reference range.

Leukopoiesis The generation of white blood cells.

Lewy body Aggregation of abnormal proteins within a neurone.

Ligand A small molecule that binds to a receptor, and so initiates a particular cellular process.

Lipase An enzyme that digests lipids.

Lithotripsy Fragmentation of calculi (stones) by extra-corporeal acoustic shock waves or invasively by a laser.

Lupus anticoagulant First described in a patient with SLE: an autoantibody that is anticoagulant *in vitro* but procoagulant *in vivo* with a prolonged APTT.

Lupus nephritis Inflammation of the kidney in systemic lupus erythematosus (SLE).

Lymphadenectomy Removal of enlarged lymph nodes.

Lymphadenopathy Literally, pathology (disease) of lymph nodes, but taken to be swollen or enlarged lymph nodes.

Lymphatics A system of vessels linking lymph nodes which delivers tissue fluid (lymph) to the circulation.

Lymphoma A tumour of B lymphocytes that develops within a lymph node.

Lymphopenia A lymphocyte count below the bottom of the reference range.

Lyonization Process by which one of the X chromosomes in the female is inactivated.

Magnetic resonance imaging (MRI) A technique that uses magnetic fields and radio waves to produce images of the body's soft internal organs.

Mann–Whitney U test A statistical test to determine if a difference between two sets of continuously variable data, one or both of which have a non-normal distribution, is meaningful.

Meninges Membranes that surround the brain and spinal cord.

Mesenchymal Of the mesenchyme, that is, embryonic connective tissues giving rise to cells of the skin, blood vessels, and bone.

Mesothelioma A tumour of the mesothelium, an extended membrane that surrounds organs such as the lung, intestines, and heart.

Meta-analysis The pooling of several small studies to provide increased power that may bring about (or refute) a significant result.

Metabolic syndrome A condition defined by any three of central adiposity, high blood pressure, low high-density lipoprotein, high triglycerides, and high blood glucose.

Metamyelocyte An immature cell in the penultimate stage of the neutrophil maturation process.

Metastases Tumours that arise in organs distant from the primary site of the cancer, and so are secondary growths.

Metastasis The process by which the 'parent' cancer sends out, or seeds, 'child' tumour cells, which form secondary growths in distant organs.

Micelle In this context, a small globule of fats and lipoproteins that transform into chylomicrons.

Microbiome The sum of all the microbes in a defined setting, for example in the intestines or the lungs.

Microcephaly A smaller-than-expected head—and thus brain.

Mineralocorticoid A group of corticosteroid hormones (the principal one being aldosterone) that part-regulate blood water and electrolyte levels.

Mucocutaneous candidiasis Infection of the mucus membranes and skin with the fungal yeast *Candida*.

Muscular dystrophy A group of rare progressive genetic diseases of skeletal muscle proteins and fibres.

Myalgia Muscle pain.

Mycobacterial Pertaining to organisms such as those causing tuberculosis and leprosy.

Myeloma A malignancy of B lymphocytes in the bone marrow.

Myocardial infarction Damage to a portion of the myocardium due to failure of the coronary circulation to provide oxygenated blood.

Myopia Near- or short-sightedness.

Neoadjuvant Primary treatment in advance of further treatment.

Neoplasm Translates as 'new growth', but in this case the growth is abnormal and may (if malignant) cause disease (e.g. cancer).

Nephrectomy Removal of the kidney.

Nephropathy Disease of the kidneys.

Neuroglia Non-neuronal cells in the brain.

Nodal Pertaining to lymph nodes.

Non-coding RNA Forms of RNA that do not code for proteins, but have other regulatory activity.

Non-normal distribution A data set in which most of the individual data points lie to the left or the right of the set as a whole.

Normal distribution A data set in which most of the individual data points lie in the middle of the set as a whole.

Normocytic anaemia An anaemia in which the size of the red blood cell is within the reference range.

Nulliparity The position of ever having had children.

Obstructive cholestasis Blockage or narrowing of the bile duct.

Odds ratio A comparison of the odds (likelihood) of a particular event happening in one group compared to the odds of it happening in another group.

Oedema Congestive swelling of an organ or tissues caused by excess tissue fluid.

Omenectomy Removal of the omentum, a mass of abdominal fatty tissue.

Oncogene A gene that causes cancer.

Oncovirus A virus that causes cancer.

Orthopaedics A surgical specialty focusing on correcting abnormalities in bone, joints, and other connective tissues.

Orthopnoea Shortness of breath when lying down.

Osmolality An overall measure of the molecular (e.g. glucose) and ionic (e.g. electrolytes) composition of plasma.

Osteoarthritis A localized inflammatory disease of (mostly) the hips and knees, often linked to overweight and obesity.

Osteopenia Lack of bone, often a precursor to osteoporosis.

Osteoporosis A disease of the loss of bone density, sometimes linked to spontaneous fractures.

Oxidative phosphorylation Generation of ATP in the mitochondrion by the electron transport chain.

Palliative Referring to terminal care, typically in a hospice.

Pancreatitis Inflammation of the pancreas.

Papillary Of a papilla—a small outgrowth resembling a nipple.

Parenchyma The bulk of cells within an organ.

Parkinson's disease A brain disease, with a variable course, characterized by progressive loss of motor function, for example tremor.

Paroxysmal nocturnal dyspnoea Intermittent shortness of breath at night.

Pathognomic The defining pathophysiological feature of a disease.

Pathophysiology The impact that a disease state (pathology) has on the workings of the body (physiology).

Percutaneous Through the skin.

Pericarditis Inflammation of the membranes surrounding the heart.

Peritoneal cavity A cavity that lies between membranes surrounding the intestines and membranes enclosing the abdominal organs.

Peritonitis Inflammation of the peritoneum.

Phaeochromocytoma A neuroendocrine tumour of the adrenal medulla.

Phagolysosome An intracellular organelle where a foreign body brought in by phagocytosis is fused with a lysosome, ensuring its digestion.

Philtrum A narrow groove from the upper lip to the base of the nose.

Phosphorylation A common signalling process of adding an atom of phosphorus to a molecule, which is often then activated.

Plasmapheresis A procedure for removing a substance (such as cryoglobulins and LDL-cholesterol) from the plasma.

Pleurisy Inflammation of the pleurae (membranes that line the lung).

Pleuritis Inflammation of the membranes surrounding the lungs.

Pneumonia Inflammation of the alveoli of the lung.

Polycistron A series of genes within a small location.

Polycythaemia An increase in the number of blood cells in the blood (most often red blood cells).

Positron emission tomography (PET) A technique that uses radioactive tracers to detect abnormal metabolic processes within the body.

PQRST A representation of five key aspects of the heartbeat.

Priapism Pathological state of an erect penis in the absence of sexual stimulation.

Proctitis Inflammation of the terminal section of the rectum and/or anus.

Prostatitis Inflammation of the prostate.

Proteasome A high-molecular-weight intra-cellular protein complex whose functions include degradation of damaged or unnecessary proteins.

Proto-oncogene A gene that is not an oncogene in itself, but may become so by being altered or fused with another gene.

Pruritis Itching.

Pulmonary embolism A clot in the lungs that obstructs blood flow and is often fatal.

Pyuria White blood cells in the urine.

Reactive oxygen species (ROS) Toxic molecular variants of oxygen that can cause damage to proteins and nucleic acids—and so disease.

Receptor A membrane-bound (with links to the cytoplasm) or cytoplasmic molecule that binds a ligand and in doing so has a role in metabolic regulation.

Relative risk The extent to which one group is more or less at risk of an outcome compared to a reference group.

Reticulocytosis A reticulocyte count above the top of the reference range.

Retinoblastoma A tumour of the retina.

Rhabdomyolysis Damage to skeletal muscle cells.

Rheology The study of the physical properties of blood and plasma.

Rheumatoid arthritis A common autoimmune disease characterized by chronic inflammation, leading to the destruction of bone and joints.

Sarcoidosis A disease of unknown aetiology characterized by the formation of granulomas in various organs.

Sarcomere A unit of a skeletal or cardiac muscle.

Scoliosis An abnormal curvature of the spine, with deviation to the left or right.

Second messenger A small molecule or cation that passes messages from the receptor to other molecules within the cytoplasm or to the nucleus.

Secretagogue A substance that induces the secretion of another substance.

Signal transduction The entire process of transmitting an external signal that has a measured effect on the cell.

Single nucleotide polymorphism (SNP) A change in a single nucleotide of a sequence of DNA (often described as a mutation) that has downstream repercussions for the cell.

Sinoatrial node A focus of cells on the heart that initiate the cardiac cycle through electrical impulses.

Sphincter muscle A circular muscle regulating the passage of material, such as from the oesophagus to the stomach, and from the bladder to the urethra.

Spirometry A method for assessing lung function, involving the measurement of forced exhalation.

Splenomegaly An enlarged spleen.

Squamous epithelia A type of epithelial cells that are flattened, such as those lining the alveoli.

Standard deviation A measure of the variation of a set of data whose distribution is normal.

STEMI S-T wave elevation myocardial infarction.

Stent A plastic or metal tube inserted into a blood vessel or other tube to ensure the latter is kept open (often into the coronary arteries).

Stoma A constructed pore enabling the exit of faeces and urine directly from the intestines or bladder, respectively.

Strabismus Failure of the eye to align correctly.

Stromal Cells that can differentiate (develop) into connective tissue cells.

Student's t test A statistical test to determine if a difference between two sets of continuously variable data with a normal distribution is meaningful.

Substantia nigra A part of the brain that coordinates movement.

Substrate-level phosphorylation Generation of ATP or GTP in the cytoplasm and mitochondrion by the direct transfer of a phosphate group.

Sympathetic nervous system The part of the autonomic (unconscious) nervous system whose actions mediate the fight-or-flight responses.

Syncopy Fainting or other loss of consciousness.

Systemic lupus erythematosus (SLE) An autoimmune disease variously affecting the skin, joints, liver, bone marrow, and kidneys.

Tachycardia Increased heart rate.

Technetium A heavy metal whose gamma-emitting isotope 99 is commonly used in nuclear medicine imaging.

Thoracentesis An invasive procedure for removing air or fluid from the pleural space.

Thrombocytopenia A number of platelets below the bottom of the reference range.

Thrombocytosis A platelet count above the top of the reference range.

Thrombolysis Breakdown of clots formed in blood vessels using medication.

Thymectomy Surgical removal of the thymus.

Thyrotoxicosis A clinical syndrome resulting from excessive production of thyroid hormones.

TNF Tumour necrosis factor, a cytokine with pro-inflammatory and other activities.

Toll-like receptor A membrane-bound pattern-recognition structure often present on sentinel cells that focuses on features of pathogenic microbes.

Topoisomerase An enzyme that releases the torsion in DNA strands during replication and related processes.

Torsades de pointes A rare form of ventricular tachycardia.

Transcription factor A small molecule that initiates transcription.

Translocation The process where a section of one chromosome is transferred to another (and vice versa, as it is often reciprocal).

Tumour suppressor A gene or protein whose action is to prevent a cell from becoming a tumour.

Ulcerative colitis An inflammatory disease of the mucosa of the colon and frequently the rectum and bile duct, often characterized by ulcers.

Ultrasound A non-invasive technique for assessing the structure, function, and status of internal tissues and organs.

Vasculitis Inflammation of the blood vessels.

Vasoconstrictor A factor that constricts (narrows) blood vessels.

Vasodilator A factor that dilates (opens up) blood vessels.

Vasopressin Also known as anti-diuretic hormone; it regulates water movement into urine.

VEGF Vascular endothelial growth factor.

Vitiligo An autoimmune disease characterized by loss of pigmentation of the skin.

References

Chapter 1

Eurostat. (2023). Causes of Death Statistics.

Xu JQ et al. (2022). Mortality in the United States, 2021, NCHS Data Brief No. 456. Hyattsville, MD: National Center for Health Statistics. https://dx.doi.org/10.15620/cdc:122516.

Chapter 2

Wilson-Barnes SL, Lanham-New SA, and Lambert H. (2022). Modifiable risk factors for bone health and fragility fractures. Best Practice & Research Clinical Rheumatology 2022: 36; 101758.

Chapter 4

De Feo MS et al. (2022). Breast-specific gamma imaging: an added value in the diagnosis of breast cancer, a systematic review. Cancers (Basel) 2022 Sep 23: 14(19); 4619. doi: 10.3390/cancers14194619.

Li J et al. (2019). Comparative efficacy of first-line ceritinib and crizotinib in advanced or metastatic anaplastic lymphoma kinase-positive non-small cell lung cancer: an adjusted indirect comparison with external controls. Curr Med Res Opin 2019 Jan: 35(1); 105–11. doi: 10.1080/03007995.2018.1541443.

Mattiuzzi C, Sanchis-Gomar F, and Lippi G. (2019). Concise update on colorectal cancer epidemiology. Ann Transl Med 2019 Nov: 7(21); 609. doi: 10.21037/atm.2019.07.91.

Sharif A et al. (2017). The development of prostate adenocarcinoma in a transgender male to female patient: could estrogen therapy have played a role? Prostate 2017 Jun: 77(8); 824–8. doi: 10.1002/pros.23322.

Soria JC et al. (2017). First-line ceritinib versus platinum-based chemotherapy in advanced ALK-rearranged non-small-cell lung cancer (ASCEND-4): a randomised, open-label, phase 3 study. Lancet 2017 Mar 4: 389(10072); 917–29. doi: 10.1016/S0140-6736(17)30123-X.

Chapter 5

Halaseh SA et al. (2022). A review of the etiology and epidemiology of bladder cancer: all you need to know. Cureus 2022 Jul 27: 14(7); e27330. doi: 10.7759/cureus.27330.

Rebbeck TR et al. (2002). Prophylactic oophorectomy in carriers of BRCA1 or BRCA2 mutations. N Engl J Med 2002 May 23: 346(21); 1616–22. doi: 10.1056/NEJMoa012158.

Chapter 6

Evangelou E et al. (2018). Genetic analysis of over 1 million people identifies 535 new loci associated with blood pressure traits. Nat Genet 2018 Oct: 50(10); 1412–25. doi: 10.1038/s41588-018-0205-x.

Tragante V et al. (2014). Gene-centric meta-analysis in 87,736 individuals of European ancestry identifies multiple blood-pressure-related loci. Am J Hum Genet 2014 Mar 6: 94(3); 349–60. doi: 10.1016/j.ajhg.2013.12.016.

Van de Vegte YJ et al. (2019). Genetics and the heart rate response to exercise. Cell Mol Life Sci 2019: 76(12); 2391–409. doi: 10.1007/s00018-019-03079-4.

Zilbermint M, Hannah-Shmouni F, and Stratakis CA. (2019). Genetics of hypertension in African Americans and others of African descent. Int J Mol Sci 2019 Mar 2: 20(5); 1081. doi: 10.3390/ijms20051081.

Chapter 7

Gopal DP, Okoli GN, and Rao M. (2022). Re-thinking the inclusion of race in British hypertension guidance. J Hum Hypertens 2022: 36(3); 333–35. doi: 10.1038/s41371-021-00601-9.

Helmer A, Slater N, and Smithgall S. (2018). A review of ACE inhibitors and ARBs in Black patients with hypertension. Ann Pharmacother 2018 Nov: 52(11); 1143–1151. doi: 10.1177/1060028018779082.

Lane D, Beevers DG, and Lip GYH. (2002). Ethnic differences in blood pressure and the prevalence of hypertension in England. J Hum Hypertens 2002 Apr: 16(4); 267–73. doi: 10.1038/sj.jhh.1001371.

Law MR, Wald NJ, and Rudnicka AR. (2003). Quantifying effect of statins on low density lipoprotein cholesterol, ischaemic heart disease, and stroke: systematic review and meta-analysis. BMJ 2003 Jun 28: 326(7404); 1423. doi: 10.1136/bmj.326.7404.1423.

Messerli FH. (1989). Management of essential hypertension in the black patient: profiling as the initial approach to treatment. J Natl Med Assoc 1989 Jan: 81(1); 17–23.

Ogrotis I, Koufakis T, and Kotsa K. (2023). Changes in the global epidemiology of type 1 diabetes in an evolving landscape of environmental factors: causes, challenges, and opportunities. Medicina 2023: 59(4); 668. doi: 10.3390/medicina59040668.

Oldroyd J, Banerjee M, Head A, et al. (2005). Diabetes and ethnic minorities. Postgrad Med J 2005 Aug: 81(958); 486–90. doi: 10.1136/pgmj.2004.029124.

Unger T et al. (2020). 2020 International Society of Hypertension Global Hypertension Practice Guidelines. Hypertension 2020 Jun: 75(6); 1334–57. doi: 10.1161/HYPERTENSIONAHA.120.15026.

Whelton et al. (2018). 2017 ACC/AHA/AAPA/ABC/ACPM/AGS/APhA/ASH/ASPC/NMA/PCNA Guideline for the Prevention,

Detection, Evaluation, and Management of High Blood Pressure in Adults: executive summary: a report of the American College of Cardiology/American Heart Association Task Force on Clinical Practice Guidelines. Circulation 2018 Oct 23: 138(17); e426–e483. doi: 10.1161/CIR.0000000000000597.

Wu P, Wilson K, Dimoulas P, et al. (2006). Effectiveness of smoking cessation therapies: a systematic review and meta-analysis. BMC Public Health. 2006 Dec 11: 6; 300. doi: 10.1186/1471-2458-6-300.

Chapter 9

Blann AD and Heitmar R. (2022). A Guide to COVID-19 for Health Care Professionals. Cambridge Scholars Publishing, 2022.

Dobrijević Z et al. (2023). The association of human leucocyte antigen (HLA) alleles with COVID-19 severity: a systematic review and meta-analysis. Rev Med Virol 2023 Jan: 33(1); e2378. doi: 10.1002/rmv.2378.

Chapter 10

Baughman RP et al. (2016). Sarcoidosis in America: analysis based on health care use. Ann Am Thorac Soc 2016 Aug: 13(8); 1244–52. doi: 10.1513/AnnalsATS.201511-760OC.

Bjornevik K et al. (2022). Longitudinal analysis reveals high prevalence of Epstein–Barr virus associated with multiple sclerosis. Science 2022 Jan 21: 375(6578); 296–301. doi: 10.1126/science.abj8222.

Brito-Zerón P et al. (2020). Epidemiological profile and north-south gradient driving baseline systemic involvement of primary Sjögren's syndrome. Rheumatology (Oxford) 2020 Sep 1: 59(9); 2350–9. doi: 10.1093/rheumatology/kez578.

Burns, JE and Graf EH. (2018). The brief case: disseminated *Neisseria gonorrhoeae* in an 18-year-old female. J Clin Microbiol 2018 Apr: 56(4); e00932-17. doi: 10.1128/JCM.00932-17.

Jamalyaria F et al. (2017). Ethnicity and disease severity in ankylosing spondylitis a cross-sectional analysis of three ethnic groups. Clin Rheumatol 2017 Oct: 36(10); 2359–64. doi: 10.1007/s10067-017-3767-6.

Piazza MJ and Gonzales-Zamora JA. (2020). Acute septic elbow monoarthritis with associated Neisseria gonorrhoeae bacteraemia: an uncommon presentation of an old disease. Infez Med 2020 Jun 1: 28(2); 253–7.

Singh JA et al. (2016). 2015 American College of Rheumatology Guideline for the Treatment of Rheumatoid Arthritis. Arthritis Rheumatol 2016 Jan: 68(1); 1–26. doi: 10.1002/art.39480.

Sjöholm AG et al. (2006). Complement deficiency and disease: an update. Mol Immunol 2006 Jan: 43(1–2); 78–85. doi: 10.1016/j.molimm.2005.06.025.

Smolen JS et al. (2020). EULAR recommendations for the management of rheumatoid arthritis with synthetic and biological disease-modifying antirheumatic drugs: 2019 update. Ann Rheum Dis 2020 Jun: 79(6); 685–99. doi: 10.1136/annrheumdis-2019-216655.

Wang CH and Lu CW. (2019). Images of the month 2: disseminated gonococcal infection presenting as the arthritis-dermatitis syndrome. Clin Med (Lond) 2019 Jul: 19(4); 340–1. doi: 10.7861/clinmedicine.19-4-340.

Ward MM et al. (2019). 2019 Update of the American College of Rheumatology/Spondylitis Association of America/Spondyloarthritis Research and Treatment Network Recommendations for the Treatment of Ankylosing Spondylitis and Nonradiographic Axial Spondyloarthritis. Arthritis Rheumatol 2019 Oct: 71(10); 1599–613. doi: 10.1002/art.41042.

Chapter 11

Barton JC, Edwards CQ, and Acton RT (2023). HFE hemochromatosis in African Americans: Prevalence estimates of iron overload and iron overload-related disease. Am J Med Sci 2023 Jan: 365(1); 31–6. doi: 10.1016/j.amjms.2022.08.015.

Guralnik J et al. (2022). Unexplained anemia of aging: etiology, health consequences, and diagnostic criteria. J Am Geriatr Soc 2022 Mar: 70(3); 891–9. doi: 10.1111/jgs.17565.

Chapter 13

Abbood SJA, Anvari E, and Fateh A. (2023). Association between interleukin-10 gene polymorphisms (rs1800871, rs1800872, and rs1800896) and severity of infection in different SARS-CoV-2 variants. Hum Genomics 2023 Mar 7: 17(1); 19. doi: 10.1186/s40246-023-00468-6.

Abuawwad MT et al. (2023). Effects of ABO blood groups and RH-factor on COVID-19 transmission, course and outcome: a review. Front Med (Lausanne) 2023 Jan 12: 9; 1045060. doi: 10.3389/fmed.2022.1045060.

ACTIV-3/TICO Study Group. (2022). Efficacy and safety of two neutralising monoclonal antibody therapies, sotrovimab and BRII-196 plus BRII-198, for adults hospitalised with COVID-19 (TICO): a randomised controlled trial. Lancet Infect Dis 2022 May: 22(5); 622–35. doi: 10.1016/S1473-3099(21)00751-9.

Al-Shamsi HO et al. (2020). A practical approach to the management of cancer patients during the novel coronavirus disease 2019 (COVID-19) pandemic: an international collaborative group. Oncologist 2020 Jun: 25(6); e936–e945. doi: 10.1634/theoncologist.2020-0213.

Agrawal U et al. (2021). COVID-19 hospital admissions and deaths after BNT162b2 and ChAdOx1 nCoV-19 vaccinations in 2·57 million people in Scotland (EAVE II): a prospective cohort study. Lancet Respir Med 2021 Dec: 9(12); 1439–49. doi: 10.1016/S2213-2600(21)00380-5.

Boyarsky BJ et al. (2021). Antibody response to 2-dose SARS-CoV-2 mRNA vaccine series in solid organ transplant recipients. JAMA 2021 Jun 1: 325(21); 2204–6. doi: 10.1001/jama.2021.7489.

Cabanas AM et al. (2022). Skin pigmentation influence on pulse oximetry accuracy: a systematic review and bibliometric analysis. Sensors (Basel) 2022 Apr 29: 22(9); 3402. doi: 10.3390/s22093402.

Canevelli M et al. (2020). COVID-19 mortality among migrants living in Italy. Ann Ist Super Sanita 2020 Jul–Sep: 56(3); 373–7. doi: 10.4415/ANN_20_03_16.

Cappuccio FP, Cook DG, Atkinson RW, et al. (1997). Prevalence, detection, and management of cardiovascular risk factors in different ethnic groups in south London. Heart 1997 Dec: 78(6); 555–63. doi: 10.1136/hrt.78.6.555.

Coleman P et al. (2022). COVID-19 outcomes in minority ethnic groups: do obesity and metabolic risk play a role? Curr Obes Rep 2022: 11(3); 107–15. doi: 10.1007/s13679-021-00459-5.

Dobrijević Z et al. (2023). The association of human leucocyte antigen (HLA) alleles with COVID-19 severity: a systematic review and meta-analysis. Rev Med Virol 2023 Jan: 33(1); e2378. doi: 10.1002/rmv.2378.

Dougan M et al. (2021). Bamlanivimab plus etesevimab in mild or moderate Covid-19. N Engl J Med 2021 Oct 7: 385(15); 1382–92. doi: 10.1056/NEJMoa2102685.

Ellinghaus D et al. (2020). Genomewide Association Study of Severe Covid-19 with Respiratory Failure. N Engl J Med 2020 Oct 15: 383(16); 1522–34. doi: 10.1056/NEJMoa2020283.

Ferri C et al. (2022). Absent or suboptimal response to booster dose of COVID-19 vaccine in patients with autoimmune systemic diseases. J Autoimmun 2022 Jul: 131; 102866. doi: 10.1016/j.jaut.2022.102866.

Fricke-Galindo I and Falfán-Valencia R. (2021). Genetics insight for COVID-19 susceptibility and severity: a review. Front Immunol 2021 Apr 1: 12; 622176. doi: 10.3389/fimmu.2021.622176.

Ghazy AA. (2023). Influence of IL-6 rs1800795 and IL-8 rs2227306 polymorphisms on COVID-19 outcome. J Infect Dev Ctries 2023 Mar 31: 17(3); 327–34. doi: 10.3855/jidc.17717.

Gupta A et al. (2022). Effect of sotrovimab on hospitalization or death among high-risk patients with mild to moderate COVID-19: a randomized clinical trial. JAMA 2022 Apr 5: 327(13); 1236–46. doi: 10.1001/jama.2022.2832.

Hall, VG et al. (2021). Randomized trial of a third dose of mRNA-1273 vaccine in transplant recipients. N Engl J Med 2021 Sep 23: 385(13); 1244–6. doi: 10.1056/NEJMc2111462.

Hallett AM et al. (2021). SARS-CoV-2 messenger RNA vaccine antibody response and reactogenicity in heart and lung transplant recipients. J Heart Lung Transplant 2021 Dec: 40(12); 1579–88. doi: 10.1016/j.healun.2021.07.026.

Hardy N et al. (2023). Mortality of COVID-19 in patients with hematological malignancies versus solid tumors: a systematic literature review and meta-analysis. Clin Exp Med. 2023 Feb 16: 1–15. doi: 10.1007/s10238-023-01004-5.

Hillus D et al. (2021). Safety, reactogenicity, and immunogenicity of homologous and heterologous prime-boost immunisation with ChAdOx1 nCoV-19 and BNT162b2: a prospective cohort study. Lancet Respir Med 2021 Nov: 9(11); 1255–65. doi: 10.1016/S2213-2600(21)00357-X.

Li Y et al. (2022). Myocardial injury predicts risk of short-term all-cause mortality in patients with COVID-19: a dose-response meta-analysis. Front Cardiovasc Med 2022: 9; 850447. doi: 10.3389/fcvm.2022.850447.

Liang W et al. (2020). Cancer patients in SARS-CoV-2 infection: a nationwide analysis in China. Lancet Oncol 2020 Mar: 21(3); 335–7. doi: 10.1016/S1470-2045(20)30096-6.

Marwah M et al. (2021). Analysis of laboratory blood parameter results for patients diagnosed with COVID-19, from all ethnic group populations: a single centre study. Int J Lab Hematol 2021 Oct: 43(5); 1243–51.

Modi AR and Kovacs CS. (2020). Hospital-acquired and ventilator-associated pneumonia: Diagnosis, management, and prevention. Cleve Clin J Med 2020 Oct 1: 87(10); 633–9. doi: 10.3949/ccjm.87a.19117.

Naemi FMA, Al-Adwani S, Al-Khatabi H, et al. (2021). Frequency of HLA alleles among COVID-19 infected patients: preliminary data from Saudi Arabia. Virology 2021 Aug: 560; 1–7. doi: 10.1016/j.virol.2021.04.011.

Nazerian Y et al. (2022). Role of SARS-CoV-2-induced cytokine storm in multi-organ failure: molecular pathways and potential therapeutic options. Int Immunopharmacol 2022 Dec: 113(Pt B); 109428. doi: 10.1016/j.intimp.2022.109428.

Nejad MM et al. (2022). Immunogenicity of COVID-19 mRNA vaccines in immunocompromised patients: a systematic review and meta-analysis. Eur J Med Res 2022 Feb 12: 27(1); 23. doi: 10.1186/s40001-022-00648-5.

Ovsyannikova IG et al. (2020). The role of host genetics in the immune response to SARS-CoV-2 and COVID-19 susceptibility and severity. Immunol Rev 2020 Jul: 296(1); 205–19. doi: 10.1111/imr.12897.

Piscoya A et al. (2022). Efficacy and harms of tocilizumab for the treatment of COVID-19 patients: A systematic review and meta-analysis. PLoS One 2022: 17(6); e0269368. doi: 10.1371/journal.pone.0269368.

Reyna-Villasmil E et al. (2022). Association of patients' epidemiological characteristics and co-morbidities with severity and related mortality risk of SARS-CoV-2 infection: results of an umbrella systematic review and meta-analysis. Biomedicines. 2022 Sep 29: 10(10); 2437. doi: 10.3390/biomedicines10102437.

Rider AC and Frazee BW. (2018). Community-acquired pneumonia. Emerg Med Clin North Am 2018 Nov: 36(4); 665–83. doi: 10.1016/j.emc.2018.07.001.

Saiag, E et al. (2022). The effect of a third-dose BNT162b2 vaccine on anti-SARS-CoV-2 antibody levels in immunosuppressed patients. Clin Microbiol Infect 2022 May: 28(5); 735.e5–735. e8. doi: 10.1016/j.cmi.2022.02.002.

Seddig D et al. (2022). Correlates of COVID-19 vaccination intentions: attitudes, institutional trust, fear, conspiracy beliefs, and vaccine skepticism. Soc Sci Med 2022 Jun: 302; 114981. doi: 10.1016/j.socscimed.2022.114981.

Shaul AA et al. (2022). Improved immunogenicity following the third dose of BNT162b2 mRNA vaccine in heart transplant recipients. Eur J Cardiothorac Surg 2022 Sep 2: 62(4); ezac145. doi: 10.1093/ejcts/ezac145.

Vasileiou E et al. (2021). Interim findings from first-dose mass COVID-19 vaccination roll-out and COVID-19 hospital admissions in Scotland: a national prospective cohort study. Lancet 2021 May 1: 397(10285); 1646–57. doi: 10.1016/ S0140-6736(21)00677-2.

Webb GJ et al. (2020). Outcomes following SARS-CoV-2 infection in liver transplant recipients: an international registry study. Lancet Gastroenterol Hepatol 2020 Nov: 5(11); 1008–16. doi: 10.1016/S2468-1253(20)30271-5.

Yates, T et al. (2021). Obesity, ethnicity, and risk of critical care, mechanical ventilation, and mortality in patients admitted to hospital with COVID-19: analysis of the ISARIC CCP-UK cohort. Obesity (Silver Spring) 2021 Jul: 29(7); 1223–30. doi: 10.1002/oby.23178.

Zhang L et al. (2020). Clinical characteristics of COVID-19-infected cancer patients: a retrospective case study in three hospitals within Wuhan, China. Ann Oncol 2020 Jul: 31(7); 894–901. doi: 10.1016/j.annonc.2020.03.296.

Chapter 14

Bershteyn, A et al. (2022). Understanding the evolving role of voluntary medical male circumcision as a public health strategy in Eastern and Southern Africa: opportunities and challenges. Curr HIV/AIDS Rep 2022 Dec: 19(6); 526–36. doi: 10.1007/s11904-022-00639-5.

Delanaye P et al. (2022). The new, race-free, Chronic Kidney Disease Epidemiology Consortium (CKD-EPI) equation to estimate glomerular filtration rate: is it applicable in Europe? A position statement by the European Federation of Clinical Chemistry and Laboratory Medicine (EFLM). Clin Chem Lab Med 2022 Oct 24: 61(1); 44–7. doi: 10.1515/ cclm-2022-0928.

Li J, Yi Z, Ai J. (2022). Broaden horizons: the advancement of interstitial cystitis/bladder pain syndrome. Int J Mol Sci 2022 Dec: 23(23); 14594. doi: 10.3390/ijms232314594.

Serrano B, Brotons M, Bosch FX, et al. (2018). Epidemiology and burden of HPV-related disease. Best Pract Res Clin Obstet Gynaecol 2018 Feb: 47; 14–26. doi: 10.1016/j. bpobgyn.2017.08.006.

Umeukeje EM et al. (2022). Systematic review of international studies evaluating MDRD and CKD-EPI estimated glomerular filtration rate (eGFR) equations in Black adults. PLoS One 2022: 17(10); e0276252. doi: 10.1371/journal.pone.0276252.

Chapter 15

Burton PR et al. (2007). Association scan of 14,500 nonsynonymous SNPs in four diseases identifies autoimmunity variants. Nat Genet 2007 Nov: 39(11); 1329–37. doi: 10.1038/ ng.2007.17.

Lane LC, Wood CL, and Cheetham T. (2023). Graves' disease: moving forwards. Arch Dis Child 2023 Apr: 108(4); 276–81. doi: 10.1136/archdischild-2022-323905.

Parmar MS. (2005). Thyrotoxic atrial fibrillation. MedGenMed 2005 Jan 4: 7(1); 74.

Chapter 16

Belloy, ME, Napolioni V, and Greicius MD. (2019). A quarter century of APOE and Alzheimer's disease: progress to date and the path forward. Neuron 2019 Mar 6: 101(5); 820–38. doi: 10.1016/j.neuron.2019.01.056.

Bhardwaj A, Yadav A, Yadav M, et al. (2022). Genetic dissection of non-syndromic retinitis pigmentosa. Indian J Ophthalmol 2022 Jul: 70(7); 2355–85. doi: 10.4103/ijo.IJO_46_22.

Botto LD and Yang Q. (2000). 5,10-Methylenetetrahydrofolate reductase gene variants and congenital anomalies: a HuGE review. Am J Epidemiol 2000 May 1: 151(9); 862–77. doi: 10.1093/oxfordjournals.aje.a010290.

Carmel R et al. (1999). Serum cobalamin, homocysteine, and methylmalonic acid concentrations in a multiethnic elderly population: ethnic and sex differences in cobalamin and metabolite abnormalities. Am J Clin Nutr 1999 Nov: 70(5); 904–10. doi: 10.1093/ajcn/70.5.904.

Chakravorty S and Dick MC. (2019). Antenatal screening for haemoglobinopathies: current status, barriers and ethics. Br J Haematol 2019 Nov: 187(4); 431–40. doi: 10.1111/bjh.16188.

Couture P et al. (1999). Fine mapping of low-density lipoprotein receptor gene by genetic linkage on chromosome 19p13.1–p13.3 and study of the founder effect of four French Canadian low-density lipoprotein receptor gene mutations. Atherosclerosis 1999 Mar: 143(1); 145–51. doi: 10.1016/ s0021-9150(98)00267-6.

Daiger SP, Bowne SJ, and Sullivan LS. (2014). Genes and mutations causing autosomal dominant retinitis pigmentosa. Cold Spring Harb Perspect Med 2014 Oct 10: 5(10); a017129. doi: 10.1101/cshperspect.a017129.

Fejerman L et al. (2010). European ancestry is positively associated with breast cancer risk in Mexican women. Cancer Epidemiol Biomarkers Prev 2010 Apr: 19(4); 1074–82. doi: 10.1158/1055-9965.EPI-09-1193.

Grønskov K, Ek J, and Brondum-Nielsen K. (2007). Oculocutaneous albinism. Orphanet J Rare Dis 2007 Nov 2: 2; 43. doi: 10.1186/1750-1172-2-43.

Hobbs HH et al. (1987). Deletion in the gene for the low-density-lipoprotein receptor in a majority of French Canadians with familial hypercholesterolemia. N Engl J Med 1987 Sep 17: 317(12); 734–7. doi: 10.1056/NEJM198709173171204.

Kohne E. (2011). Hemoglobinopathies: clinical manifestations, diagnosis, and treatment. Dtsch Arztebl Int 2011 Aug: 108(31–2); 532–40. doi: 10.3238/arztebl.2011.0532.

Mszar R et al. (2020). Familial hypercholesterolemia and the founder effect among Franco-Americans: a brief history and call to action. CJC Open 2020 Jan 25: 2(3); 161–7. doi: 10.1016/j.cjco.2020.01.003.

Pal T, Permuth-Wey J, Holtje T, et al. (2004). BRCA1 and BRCA2 mutations in a study of African American breast cancer patients. Cancer Epidemiol Biomarkers Prev 2004; 1794–9.

Parkinson MH et al. (2013). Clinical features of Friedreich's ataxia: classical and atypical phenotypes. J Neurochem 2013 Aug: 126 Suppl 1; 103–17. doi: 10.1111/jnc.12317.

Sims JR et al. (2023). Donanemab in early symptomatic Alzheimer disease: the TRAILBLAZER-ALZ 2 randomized clinical trial. JAMA 2023 Aug 8: 330(6); 512–27. doi: 10.1001/jama.2023.13239.

Sun A. (2018). Lysosomal storage disease overview. Ann Transl Med 2018 Dec: 6(24); 476. doi: 10.21037/atm.2018.11.39.

Williams TN and Weatherall DJ. (2012). World distribution, population genetics, and health burden of the hemoglobinopathies. Cold Spring Harb Perspect Med 2012 Sep 1: 2(9); a011692. doi: 10.1101/cshperspect.a011692.

Index

b refers to boxes; c refers to case studies; f refers to figures; t refers to tables